GAS FLOW AND CHEMICAL LASERS

GAS FLOW AND CHEMICAL LASERS

Edited by

Michele Onorato
Polytechnic Institute of Turin
Turin, Italy

PLENUM PRESS • NEW YORK AND LONDON

Library of Congress Cataloging in Publication Data

International Symposium on Gas Flow and Chemical Lasers (4th: 1982: Stresa, Italy)

Gas flow and chemical lasers.

"Proceedings of the Fourth International Symposium on Gas Flow and Chemical Lasers, held September 13–17, 1982, in Stresa, Italy"—Verso of t.p.
Includes bibliographical references and index.
1. Gasdynamic lasers—Congresses. 2. Chemical lasers—Congresses. 3. Laser beams—Congresses. 4. Optics—Congresses. I. Onorato, Michele. II. Title.
TA1695.I54 1982 621.36′6 83-27016

ISBN 978-1-4615-7069-1 ISBN 978-1-4615-7067-7 (eBook)
DOI 10.1007/978-1-4615-7067-7

Proceedings of the Fourth International Symposium on Gas Flow and Chemical Lasers, held September 13–17, 1982, in Stresa, Italy

Softcover reprint of the hardcover 1st edition 1984
A Division of Plenum Publishing Corporation
233 Spring Street, New York, N.Y. 10013

SCIENTIFIC ADVISORY COMMITTEE

SYMPOSIUM SPONSORS

Consiglio Nazionale delle Ricerche (CNR); European Office of Aerospace Research and Development (EOARD); European Research Office (ERO); Office of Naval Research - London Branch (ONRL); Politecnico di Torino

EXECUTIVE COMMITTEE

SESSION CHAIRMEN

C.M.Bowden, USA; J.D.Daugherty, USA; B.Forestier, France; R.W.Gross, USA; A.Hertzberg, USA; H.E.Hügel, West Germany; A.V.La Rocca, Italy; O.A.Volkovitsky, USSR; J.F.Wendt, Belgium

SYMPOSIUM CHAIRMAN

M.Onorato, Italy

PREFACE

This volume contains eight invited papers and seventy-three contributed papers presented at the Fourth International Symposium on Gas Flow and Chemical Lasers, which was held in Stresa, Italy, from September 13 to 17, 1984.

The purpose of the Symposium was to record and discuss current research developments and applications related to high power lasers. Papers were solicited from the entire spectrum of activities, including basic physics, aerothermodynamics, new laser media, laser design, diagnostic techniques, laser propagation, interaction phenomena, applications.

The Symposium was an opportunity for scientists and engineers representing all these disciplines to come together to report their recent work, to exchange ideas and to provide an up-to-date account of international progress in these fields.

The contributed papers were reviewed by the members of the Scientific Advisory Committee, who also took responsibility for formulating the program of invited lectures.

As editor I wish to express my appreciation and my gratitude to people and organizations that made this Symposium a success: the Members of the Scientific Advisory Committee, the Members of the Local Organizing Committee, the Sponsors and the Symposium Secretary. A grateful acknowledgement is expressed to Dr. M. Sandra Oggiano for assistance in the organization of the Symposium and publication of these proceedings.

Michele Onorato

CONTENTS

INVITED LECTURES

CHEMICAL LASERS

GASDYNAMIC LASERS

HIGH AVERAGED POWER SYSTEMS

HIGH ENERGY OPTICS. ADAPTIVE OPTICS

MATERIAL PROCESSING AND APPLICATIONS

LASER PROPAGATION AND LASER MATTER INTERACTION

CHEMICALLY PUMPED ELECTRONIC TRANSITION LASERS

Petras V. Avizonis

Air Force Weapons Laboratory

Kirtland Air Force Base, New Mexico

For a number of years we have been examining the ability to partition the chemical reaction energy directly into electronic transitions. The obvious advantage with this approach is that energy, which comes basically from chemistry, is partitioned directly into electronically excited states rather than going through an electrical power generation step with its associated inefficiencies. The disadvantage of such pumping schemes is that the excitation energy is limited to that of the chemical bond, 3 to 4 ev. Thus, unless multiple collisions are invoked, only electronic states below 4 ev could be pumped by such energy partitioning.

A successful demonstration of such an approach has been the oxygen/iodine transfer laser. Oxygen molecules are produced from a chemical reaction directly in the a'Δ state with 100% quantum efficiency. This is a highly forbidden state (60 minutes radiative lifetime), and the only way that lasing can occur from such electronic energy reservoir is to transfer such energy to a radiating state. This occurs in the resonant transfer of energy to iodine atoms (the $^2P_{1/2}$ state) which has a weak magnetic moment and radiates 1.315μ photons (to the $^2P_{3/2}$ state).

The success of the O_2/I laser has an indicated "roadmap" for direct partitioning of chemical energy into electronic states. What has become evident both experimentally and from quantum mechanics is that for such partitioning to be successful, the desired electronic state in the product molecule has to be forbidden and the reaction pathway to such a product molecule/ state has to be spin allowed. Under these conditions, and only these, will energy partition uniquely to a given/desired electronic state effectively. It also means that although we can store electronic energy effectively, we cannot lase it. We thus need to transfer such energy by collisions to a lasing atom or molecule whose optical transition is more allowed. Consequently we are talking, in most cases, of mixing lasers.

Oxygen/oxygen-like molecules (such as NF, NH, etc.) appear to be good electronic energy storage reservoirs, but poor lasing species because of forbidden transitions between low lying electronic and ground states and also because of unfavorable Frank Condon factors. Thus energy transfer needs to be accomplished to a "lasing species." Such molecules may well be halogens or interhalides such as IF. In order to be able to predict how all this may happen we need to understand the interaction between O_2 and Iodine. Such energy transfer process is expected to be similar to the other species that are isoelectronic to O_2 for energy shortage and isoelectronic to I_2 for lasing. The fact that this energy transfer also dissociates I_2 into iodine atoms in the O_2/I laser is incidental to the energy transfer process and unique to I_2 species.

The mechanism of energy transfer between O_2 (a'Δ) and I_2 (X'Σ) is not completely defined at this point. It is clear that the process proposed by Derwent and Thrush[1]

$$O_2\ (b'\Sigma) + I_2\ (X'\Sigma) \rightarrow O_2\ (X^3\Sigma) + 2I\ (^2P_{3/2}) \quad (1)$$

is not important, at least for chemically produced O_2 (a'Δ), O_2(b'Σ). This process was effectively disproved by Muller and Houston[2] where they pumped selectively the O_2 (b'Σ v = 0) state by a die laser and observed an I_2 dissociation rate which was found to be much too low to account for the chemically produced O_2/I_2 dissociation measurements.

The first indication of the process involved was by S. Davis, et al[3] where he observed that the iodine dissociation rate in a system where O_2(a'Δ) is produced chemically does not depend upon O_2(b'Σ). This second electronic O_2 level was quenched by H_2O, yet the I_2 dissociation was not significantly impacted. In fact, it was determined that the dissociation rate increased with increasing I_2 concentration (Figure 1) requiring between 16 and 2 O_2(a'Δ) molecules to accomplish such dissociation as I_2 concentration was raised. This suggested an autocatalytic process which was elegantly analyzed by Heidner, et al[4] and also by Lilenfeld[5]. It is summarized below

$$O_2\ (a'\Delta) + I_2\ (X'\Sigma) \rightarrow O_2\ (X^3\Sigma) + I_2^* \quad (2)$$

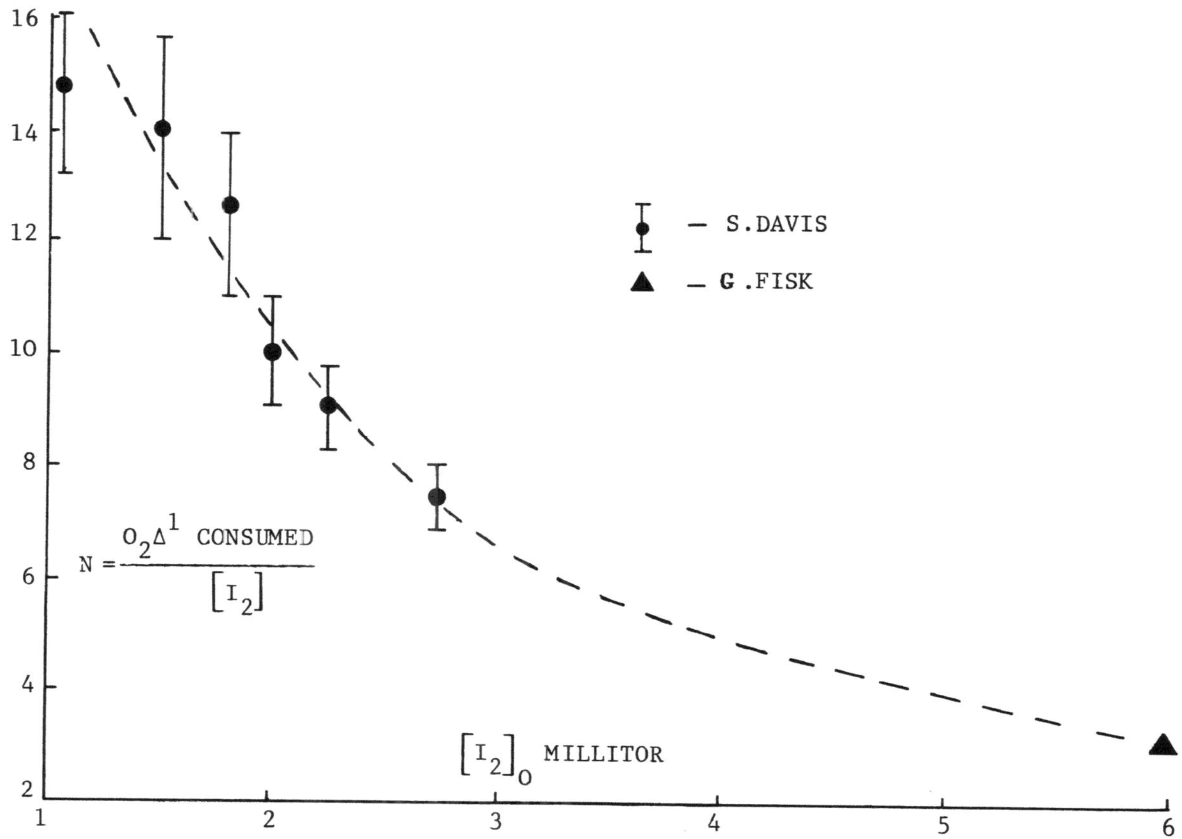

Fig.1 Dissociation of I_2 by $O_2(^1\Delta)$ (Experimental)

$$O_2\ (a'\Delta) + I_2^* \rightarrow O_2(X^3\Sigma) + 2I\ (^2P_{3/2})\quad \text{(slow)} \tag{3}$$

$$I(^2P_{3/2}) + O_2\ (a'\Delta) \rightarrow O_2\ (X^3\Sigma) + I\ (^2P_{1/2}) \tag{4}$$

$$I(^2P_{1/2}) + I_2^* \rightarrow 3I\ (^2P_{3/2}) \tag{5}$$

Process 5 starts the chain. The I_2^* intermediate can either be a high vibrational level of the I_2 ground state or one of the electronic A states. We do not believe that it is a high ground vibrational level because addition of SF_6 does not impact the I_2 dissociation rate. SF_6 is expected to thermalize vibrational non-Boltzman distributions because its vibrational modes are close to high vibrational state spacing of I_2. (I_2 $v = 20$ is 180 cm^{-1}). H_2O, on the other hand, should have no impact on high vibrational I_2 levels because its lowest vibrational mode is at 1595 cm^{-1}. It has been determined that H_2O does indeed reduce the rate of dissociation, yet SF_6 has no impact.

It is thus suggested that the intermediate I_2 state is either the $A'^3\Pi_2$ or the $A^3\Pi_1$. Strong emission at 10709 cm^{-1} (from $A^3\Pi_1$ state) has been observed. There is a problem of accessing such states by $O_2('\Delta)$. It would require $O_2('\Delta)$ in $v = 1$ for $A'^3\Pi_2$ or $v = 2$ for $A^3\Pi_1$. Such vibrational nonthermal populations have been shown to result from O_2 ($b'\Sigma$) deactivated by H_2O, resulting in $O_2(a'\Delta)$ $v > 0$ populations, and by rapid v-v exchanges between excited and ground state O_2.

Examples of such v-v exchange processes are indicated below as observed by Muller and Houston[6]

$$O_2\ (b'\Sigma v = 1, 2) + O_2\ (X^3\ \Sigma,\ v = 0) \rightarrow O_2\ (b'\Sigma v = 0) + O_2\ (X^3\Sigma v = 1, 2) \tag{6}$$

$$O_2\ (X^3\Sigma v = 1) + O_2\ (a'\Delta v = 0) \rightarrow O_2\ (X^3\Sigma v = 0) + O_2\ (a'\Delta v = 1) \tag{7}$$

We thus propose that O_2 ($a'\Delta$) molecules would pump an intermediate state (the A electronic state) of the lasing molecules which then would either dissociate (as in I_2 because of the predominance of repulsive states above A or achieve B electronic state population as is possible in some of the interhalides.

Another developed electronic energy reservoir which has been demonstrated is the NF molecule in its $a'\Delta$ state. It is created chemically from the reaction of $H+NF_2$ with very high efficiency. NF ($a'\Delta$) state is highly forbidden in terms of transitions to the $X^3\Sigma$ ground state, thus providing an excellent energy reservoir at 1.5 ev.

The classical reaction to produce NF in the excited state

$$N_2F_4 + \text{heat} \rightarrow 2NF_2 \quad (8)$$

$$NF_2(^2B) + H(^2S) \rightarrow NF\ (a'\Delta) + HF\ (X'\Sigma) \quad (9)$$

has been studied by Herbelin and Cohen[7], and Cheah and Clyne[8]. Reaction 9 has sufficient exothermicity to access the first electronic level for $v = 0$ at 11939 cm^{-1} and has a high rate constant, $K_9 \cong 2.5 \times 10^{-11}$ cm^3/mol-sec. Clyne[8] has confirmed that it branches to the NF (a'Δ) state almost with unity ratio through the subsequent reactions with hydrogen atoms.

$$H(^2S) + NF\ (a'\Delta) \rightarrow N(^2D) + HF\ (X'\Sigma) \quad (10)$$

$$H(^2S) + NF\ (X^3\Sigma) \rightarrow N(^4S) + HF\ (X'\Sigma) \quad (11)$$

The nitrogen produced in Reactions 10 and 11 were measured by resonance fluorescence and no products of reaction[11] (ground state N) were detected. Thus it was concluded that only the NF (a'Δ) state was formed.

There are two problems with production of NF by Reaction 8. First, excess H will deplete NF (a'Δ) state by Reaction 10, and secondly it is hard to produce atomic hydrogen.

Pritt and Coombe[9] examined the azide reaction to produce NF. First, HN_3 is produced by reaction of H_2SO_4 with NaN_3. Thus

$$F + HN_3 \rightarrow HCl + N_3;\ K_{12} = 1.5 \times 10^{-12}\ cm^3/\text{molec-sec} \quad (12)$$

$$F + N_3 \rightarrow NF\ (a'\Delta) + N_2;\ K_{13} = 2 \times 10^{-12}\ cm^3/\text{molec-sec} \quad (13)$$

they were able to generate excited NF without having to invoke hydrogen atom reactions. Reaction 13 is exothermic by about 55K cal/mol, thus providing more than enough energy to access the a state of NF (30 Kcal/mol). As may be seen from Figure 2, NF (a'Δ) would be almost impossible to lase because there is no noticeable potential curve displacement, and it has a radiative lifetime of about 5.6 seconds. Normally no "a" state vibrational distribution may be seen, as equilibration is very rapid (reaction times of the order of 100 microseconds begin to show vibrational structure).

Reaction 12 could be accomplished with Cl rather than F in order to minimize the accumulation of excess N_3. Excess N_3 leads to reactions of the type

$$N_3 + N_3 \rightarrow \text{products};\ k \sim 10^{-11}$$

and depletes the yield of NF (a'Δ).

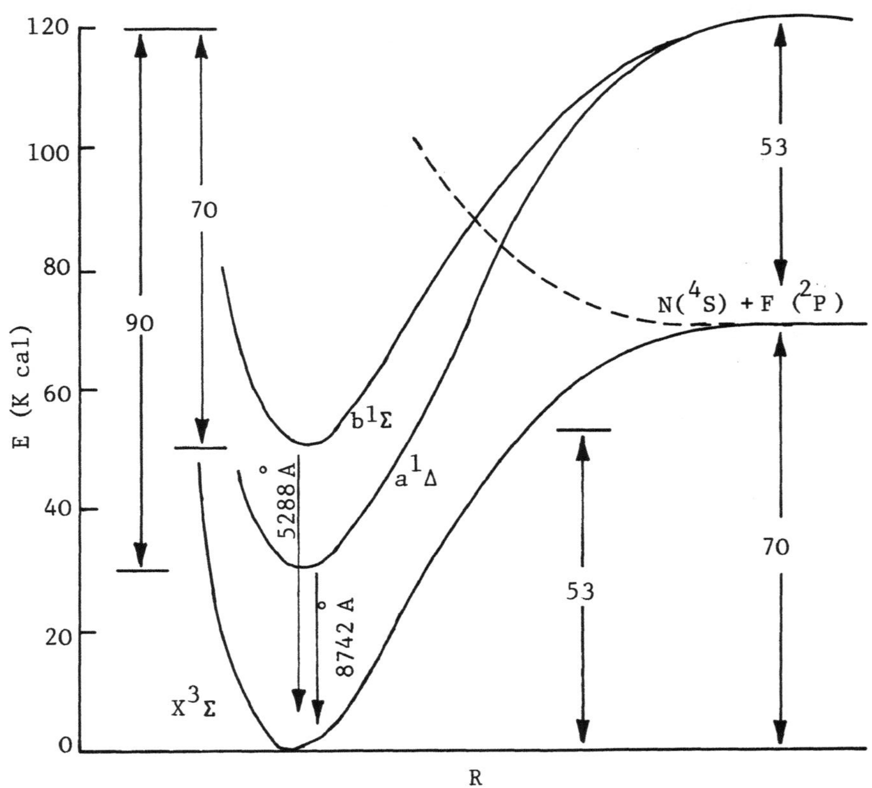

Fig.2 Potential energy diagram of NF

The NF (a'Δ) state can be pumped to the NF (b'Σ) state by collisions with HF (v>2) or with $I(^2P_{1/2})$. The most effective process was that of resonant transfer from excited iodine where the rate constant was essentially estimated to be gas kinetic[10].

$$I(^2P_{\frac{1}{2}}) + NF\ (a'\Delta) \rightarrow I(^2P_{3/2}) + NF\ (b'\Sigma)$$

$$K_{14} > 2 \times 10^{-10}\ cm^3/mol\text{-}sec \qquad (14)$$

The energy mismatch is only 110 cm^{-1} in the favorable direction, and excited iodine can be obviously made from $O_2(a'\Delta)$. The direct transfer of energy from oxygen is much slower ($K \sim 10^{-13}$/mol-sec). The "b" state radiative lifetime is substantially smaller (23 m-sec as estimated by Clyne[8]), but as seen from Figure 2, the Frank Condon factor is still negligible. Thus the "b" state of NF can be considered as another electronic energy reservoir, or a possible two-level lasing state (b→X).

As one looks beyond O_2/I laser, another interesting candidate is the IF molecule as a radiator in the green to red band for a $B^3\Pi$ (0+) to X'Σ transition. The wavelength spread comes from various vibrational levels of the B state (Fig. 3). Also, the radiative lifetime of the B state of 8×16^{-6} is favorable for lasing. IF (B→X) has already been made to lase through optical pumping[11]. Davis has also pumped it by multiple collisions with $O_2(a'\Delta)$[12], and it appears that it may be possible to pump it with NF* which has 50% more energy than $O_2(a'\Delta)$. Similar concepts would be also applicable to energy transfer for other interhalogen molecules, in that their B state potential energy surfaces are well displaced from that of the ground state, and the optical transitions of B→X are only weakly forbidden. Unfortunately, the location of the intermediate A' state is not known in most cases.

Initial experiments to produce IF from florine/iodine reactions were of the $F_2 + I_2 \rightarrow$ products type. The reaction rate for the overall process was determined by Whitefield and Davis [13] to be very slow, $K = 1.9 \times 10^{-15}$ cm^3/molecules-sec. Weak emission of the A→X and B→X states has been observed, but in general, the yield of excited states was very low.[14]. Coggiola, et al[15] performed a definitive molecular beam experiment where the reaction mechanism was concluded to be as follows:

$$I_2 + F_2 \rightarrow I_2F + F \qquad \Delta H < +4\ Kcal/mol \qquad (15)$$

$$I_2F + F \rightarrow 2IF \qquad \Delta H = -64 Kcal/mol \qquad (16)$$

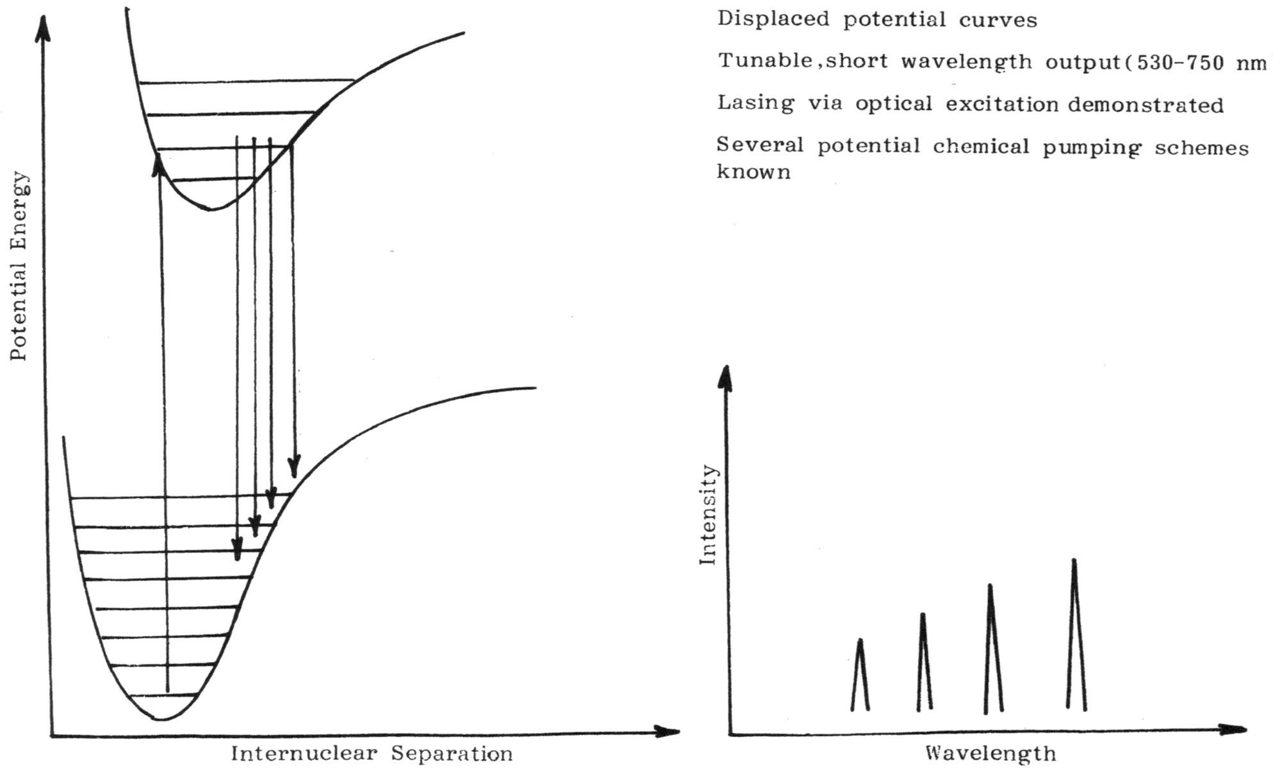

Fig.3 Iodine monofluoride as a visible chemical laser.

It is presumed that Reaction 15 is very slow, giving the low overall rate constant. Coombe and Horne[16] studied the reaction

$$I + O_2F \rightarrow IF^* + O_2 \quad \Delta H = -52Kcal/mol \qquad (17)$$

which has enough energy to access the IF ($A^3\Pi$,) state. Both A→X and B→X emission was observed, but photon yields were not determined. The reaction of

$$F + I_2 \rightarrow IF + I \qquad \Delta H = 30.2Kcal/mol \qquad (18)$$

was determined by Trickl and Wanner[14] to yield a bimodal distribution of vibrational population in the IF ($X'\Sigma$) state. They postulated that a long-lived I_2F complex decayed into $v = 0$, 1 levels, where the rapid dissociation (second mode) gave non-thermal distribution of levels $v = 10$ to 18 each at about the 1% level ($v = 19$ is the exoergicity limit of the reaction). Davis, et al[12], studied the reaction

$$2F + I_2 + O_2\ (a'\Delta) \rightarrow 2IF\ (B^3\Pi_o) + O_2\ (X^3\Sigma) \qquad (19)$$

Photon yields of 0.3 - 0.4% were determined. There are two possible mechanisms for this process. The first mechanism involves the reaction

$$F(^2P_{3/2}) + I_2\ (X'\Sigma) \rightarrow IF(X'\Sigma,v) + I^2P_{3/2}) \qquad (20)$$

with a vibrational IF population out to $v = 18$[14]. Then

$$IF\ (X'\Sigma,v) + O_2\ (a'\Delta) \rightarrow IF\ (A^3\Pi,\ A'^3\Pi_2) + O_2 \qquad (21)$$

Both states could be accesed by O_2 ('Δ) from vibrationally hot IF ground states. The $A'^3\Pi_2$ state has not been definitely assigned but has been tentatively identified as the terminal state for an electrically excited D'→ A' IF laser[17]. It is presumed to lie between 12000 and 16000 cm^{-1} and just below the A state (Fig. 4). The last pumping step would be accomplished by a second O_2 ('Δ).

$$IF\ (A^3\Pi_1;\ A'^3\Pi_2) + O_2\ (a'\Delta) \rightarrow IF\ (B^3\Pi_o) + O_2\ (X^3\Sigma) \qquad (22)$$

This mechanism is supported in that when I_2 + F was injected upstream of the O_2 ('Δ), the yield of IF (B) was significantly reduced. The explanation would be that the nonthermal ground vibrational population produced by Reaction 20 would be thermalized. The overall low yield could be explained in that it is limited to a maximum of several percent because only a few percent of IF from Reaction 20 is in sufficiently high vibrational states for O_2 ('Δ) to access the IF (A) states. Furthermore, for IF originating in the $A^3\Pi_1$ state, collisions with O_2 (a'Δ) in Reaction 22 would drive

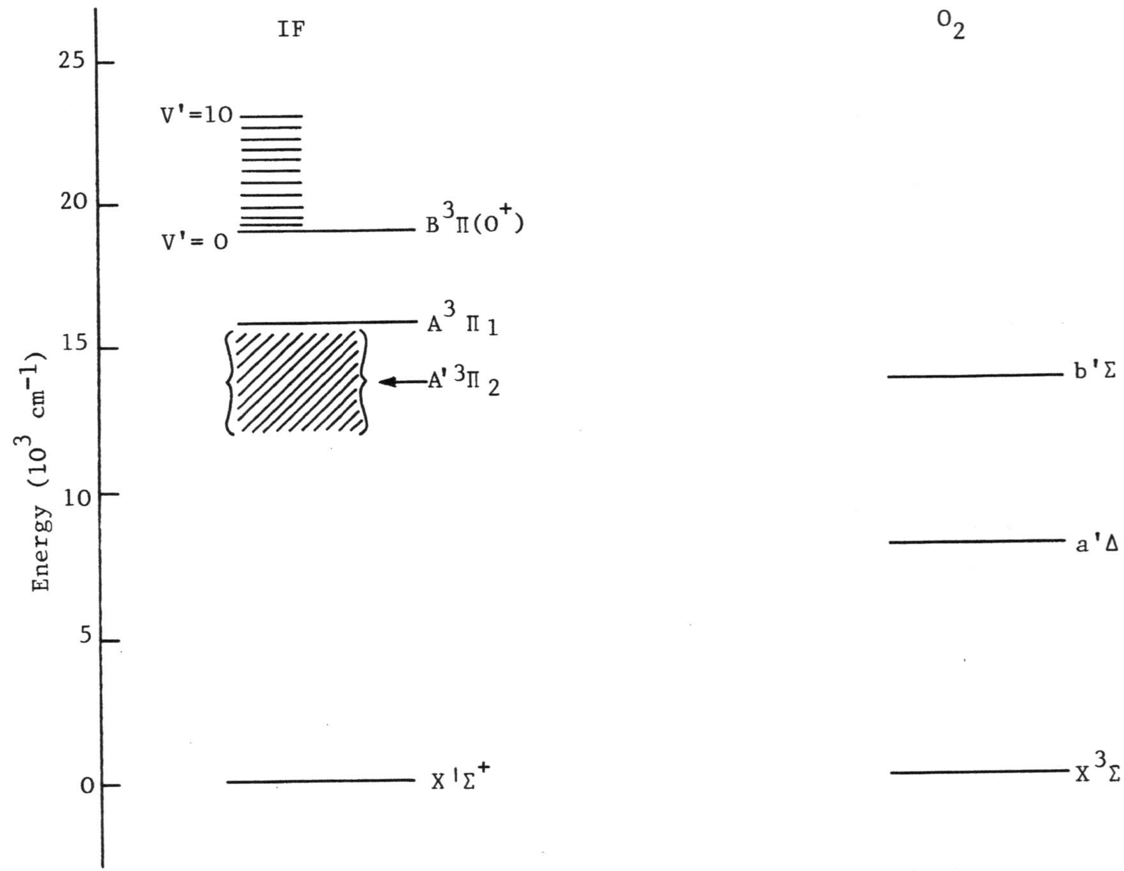

Fig.4 IF/O_2 energy levels

IF to 23580 cm^{-1}, which is above the $B^3\Pi_0$ state predissociation level of 22,300 cm^{-1}. This would further reduce the yield of IF in the B state. Also, it is known that O_2 in ground state is an efficient quencher of IF (B). Thus, a few tenths of a percent yield is not unexpected.

The second possible mechanism is one where

$$O_2\ (a'\Delta) + I\ (^2P_{3/2}) \rightarrow O_2\ (X^3\Sigma) + I\ (^2P_{1/2}) \qquad (23)$$

excited I is produced first, which then proceeds to react with F_2

$$I(^2P_{1/2}) + F_2\ (X'\Sigma) \rightarrow IF\ (A^3\Pi_1/A'^3\Pi_2) + F\ (^2P_{3/2}) \qquad (24)$$

This reaction is exothermic by 17900 cm^{-1}, thus both A states can ben accessed. Secondary pumping by O_2 (a'Δ) could then follow to produce the IF (B) state. The low yields can again be explained by pumping into the B state above the predissociation level and rapid quenching by ground O_2.

The first mechanism might be expected to dominate at stoichiometric conditions because Reaction 9 is extremely rapid ($K_{19} = 4.5 \times 10^{-11}$ cm^3/mol-sec). Reaction 23 is also gas kinetic but a bit slower than 19, yet the 2nd mechanism might become dominant under conditions of excess O_2 (a'Δ) and F_2, although the rate of Reaction 24 is not known.

An interesting possibility for pumping IF has been suggested by Pritt and Wodarczyk[18] where energy transfer from electronically excited NF is invoked. What has been observed is that in a flame of I_2 + F + HN_3, a strong and dominant emission of IF (B→X) occurs. A possible pumping mechanism could be:

$$NF\ (b'\Sigma) + IF\ (X'\Sigma) \rightarrow NF\ (X^3\Sigma) + IF\ (B^3\Pi_0) \qquad (25)$$

The energy mismatch is less than a KT. A variation of this process is to produce N_3 by reacting Cl + HN_3 whose rate, as discussed before, would match much better the rate of F+N_3, thus preventing the accumulation of N_3.

The alternative to 25 would be:

$$2NF\ (a'\Delta) + IF\ (X'\Sigma) \rightarrow NF\ (X^3\Sigma) + IF\ (B^3\Pi_0) \qquad (26)$$

This would not be as desirable since this energy transfer process would pump IF to 24000 cm^{-1}, somewhat above the predissociation limits.

As it was mentioned before, NCl and NBr in the "a" and "b" electronic states has been made chemically. Coombe and al[19], identified and assigned the a'Δ states for NCl and NBr to have T_e of 2660 cm^{-1} and 9226 cm^{-1} respectively, where as the $b^1\Sigma$ states are known to be at 14984 cm^{-1} and 14787 cm^{-1}, respectively. Chemical production of NCl or NBr has been studied extensively[9] and the followling set of reactions appears to be most suitable:

$$F + HN_3 \rightarrow HF + N_3 \quad K_{27} = 1.5 \times 10^{-11}\ cm^3/molec\text{-}sec \tag{27}$$

$$N_3 + Cl \rightarrow NCl\ (a'\Delta) \quad K_{28} = 1.3 \times 10^{-11}\ cm^3/molec\text{-}sec \tag{28}$$

$$NCl\ (b'\Sigma) \quad K_{28_b} = 2.2 \times 10^{-12}\ cm^3/molec\text{-}sec$$

$$N_3 + Br \rightarrow NBr\ (a'\Delta) \quad K_{29} = 7.2 \times 10^{-12}\ cm^3/molec\text{-}sec \tag{29}$$

NBr (b'Σ) - not known

Unlike NF, the "b"state of these molecules is a product of the chemical reaction rather than collisional pumping, and in both cases their yields were 16%. The "a" state had a yield of 80-85% as defined by $\{NX\}/\{HN_3\}$.

Since both these states are produced with high chemical efficiency (85/15 mixture) their spectroscopy suggests that the b'Σ state of NBr/NCl could be used to pump the NF (A or A') state, whereas the a'Δ state of NBr/NCl then could pump the NF (A/A') state to NF ($B^3\Pi_0$).

Recently, NH has aroused our interest as an electronic reservoir. This was due primarily to the work by Hack and Horie[20]. They observed strong NF ($b'\Sigma \rightarrow X^3\Sigma$) emission in a reacting mixture of $F + NH_3 + O_2$ ('Δ). The reagents for this reaction, $F + NH_3$, are very nice from the point of view of handling and the first step of the reaction is well known to yield NH_2 in ground state,

$$F + NH_3 \rightarrow NH_2\ (^2B) + HF(v) \tag{30}$$

and vibrationally excited HF. It was further noted that with no O_2 ('Δ) there was no NF (b→X) emission, implying that the product at the end of $F + NH_3$ reaction sequence has to be NF in the a'Δ state. One may postulate then the reaction sequence as follows:

$$F(^2P_{3/2}) + NH_2\ (^2B) \rightarrow NH\ (a'\Delta) + HF\ (X'\Sigma)$$

or

$$\rightarrow NH\ (X^3\Sigma) + HF\ (X'\Sigma) \tag{31}$$

$$NH\ (X^3\Sigma \text{ or } a'\Delta) + F_2\ (X'\Sigma) \rightarrow NF\ (a'\Delta) + HF\ (X'\Sigma) \tag{32}$$

The sequence could be stopped at Reaction 31 to produce NH. Both products in either ground state or a'Δ are spin allowed. The fact that Reaction 32 has to generate NF (a'Δ), spin conservation arguments would dictate that NH for Reaction 32 and consequently from Reaction 31 has to produce NH in the a'Δ state. The analagous reaction

$$H(^2S) + NF_2(^2B) \rightarrow NF\ (a'\Delta) + HF$$

produces NF with 100% yield in the excited state. Thus, it is hoped but not proven yet, that NH ('Δ) could be another chemically created electronic energy reservoir with energy storage at 12,566 cm^{-1}.

In summary, we have reviewed both electronic energy reservoirs and possible molecular lasing species. The energy reservoirs are summarized in Table I, and molecular lasing species in Table II. As one examines both tables, what is immediately clear is that a good pumping/lasing match would be NCl or NBr for IF, as discussed before. NF in the "b" state would pump by single collision both IF and BrF and Cl_2 very adequately, but so far chemical production of "b" state has not occurred.

Finally, one should mention atoms as recipients of energy transfer and consequent laser species. Iodine is an obvious example. Another might be Bi atoms for the $6^4S_{3/2} \rightarrow 6^2D_{3/2}$ transition with resonant energy transfer from NF (a'Δ) state.

Table I. Electronic Energy Reservoirs

Species	$a'\Delta$ T_e cm^{-1}	$b'\Sigma$ T_e cm^{-1}	τrad $\tau a/\tau b$	Comment
O_2	7918	13195	60 min	Chem prod of a - 100% yield Collisional "b" state
NF	12003	18877	5.6 sec/23 msec	Chem prod of a - 100% yield Collisional "b" state
NCl	9240	14984	16 msec/.630 msec	Chem prod of a - 80-85% yield Chem prod of b - 16% yield
NBr	9246	14787	?/.1msec	Chem prod of a - 80-85% yield Chem prod of b - 16% yield
NH	12566	21202	?/18ms	Possible chem production of "a" state

Table II. Lasing Molecules

Species	A' cm^{-1}	A cm^{-1}	B cm^{-1}	$\tau a/\tau b$	Predissociation of B cm^{-1}
IF	14,000 ?	15,706	19,052	?/8 μsec	22,330
BrF	?	17,385	18,272	?/60 μsec	21,000
BrCl	?	?	16,879	?/40 μsec	18,035
Br_2	?	13,905	15,902	300/10 μsec	16,500
Cl_2	17,160	17,440	17,809	?/300 μsec	20,000

REFERENCES

1. R. G. Derwent and B. A. Thrush, Trans. Faraday Soc., 53; 162 (1972) and references within

2. D. F. Muller, R. H. Young, P. L. Houston, and J. R. Wiesenfeld, Applied Physics Letters, 38;404 (1981) and references within

3. P. Alsing, G. Simmons and S. Davis (Air Force Weapons Laboratory), in publication (1982)

4. R. Heidner, C. E. Gardner, G. I. Segal and T. M. El-Sayed, submitted to J. Phys Chem, (1982)

5. H. Lilenfeld, McDonald Douglas Research Labs, unpublished work (1982)

6. D. F. Muller and P. L. Houston, J. Phys Chem, 85;3563 (1981)

7. J. M. Herbelin and N. Cohen, Chem Phys Letters, 20;605 (1973)

8. C. T. Cheah and M. A. A. Clyne, Journal of Photochemistry, 15;21 (1981)

9. A. T. Pritt, Jr. and R. D. Coombe (Air Force Weapons Lab), TR-81-2111 (1981) and Int J.Chem Kinetics, 12;741 (1980)

10. J. M. Herbelin, M. A. Kwok and D. J. Spenser, J. Appl Phys, 49;3750 (1978)

11. S. J. Davis and L. Hanko, Appl Phys Lett, 37;692 (1980)

12. P. D. Whitefield, R. F. Shea and S. J. Davis, submitted to J. Chem Phys (1982)

13. P. D. Whitefield and S. J. Davis, Chem Phys Lett, 83;44 (1981)

14. T. Trickl and J. Wanner, J. Chem Phys, 74;11 (1981)

15. J. Valentini, M. J. Coggiola and Y. T. Lee, Int J. Chem Kinetics, 8;605 (1978) and references within

16. R. D. Coombe and R. K. Horne, J. Phys Chem, 83;2470 (1979)

17. M. Diegelman, K. Tbhla, F. Rebentrost and K. L. Kompa, J. Chem Phys, 73;1233 (1982)

18. A. T. Pritt and F. J. Wodarczyk, Rockwell International Science Center, private communication (1982)

19. A. T. Pritt, D. Patel and R. D. Coombe, J. Mol Spec, 87;401 (1981)

20. W. Hack and O. Horie, Chem Phys Letters, 82;327 (1981)

OPTICAL RESONANCE TRANSFER LASERS FOR MODE CONTROL*

Philip K. Baily, John H. S. Wang, Jack Finzi,
and Richard C. Smith

Electro-Optical Engineering Division
Hughes Aircraft Company
El Segundo, California, USA

Optical Resonance Transfer Laser[1] (ORTL) devices have been under development at Hughes Aircraft Company for a variety of potential applications for the past five years. One of the most promising applications is that of the production of high quality mid-IR laser radiation capable of diffraction limited far field performance. While chemical lasers are a very efficient source of photons, it is difficult to efficiently produce diffraction limited output beams. With the ORTL approach, standard chemical laser technology is utilized to efficiently produce HF or DF radiation with no concern about mode control, and the ORTL is utilized to convert this radiation to a high quality laser beam. The ORTL output is expected to exhibit diffraction limited performance because none of the traditional causes of medium degradation in lasers (e.g., electric discharges, chemical reactions, supersonic flow) are present in the ORTL system.

Of course, in order for this approach to be viable, high efficiency conversion of chemical laser radiation to ORTL radiation is required. For this reason, intermolecular systems such as DF/HBr[2], HF/HCN[3], and HF/DF[3] are less favorable than the intramolecular HF[4] system where rotational equilibration rather than collisional energy transfer creates the population inversion. Previously reported work[4-7] has all utilized a chemical laser output beam to pump a separate ORTL system. This paper reports

* Work performed in part under the sponsorship of Defense Advanced Research Projects Agency and monitored by the Naval Research Laboratory.

intracavity HF ORTL experiments where the ORTL medium is internal to the chemical laser resonator. The energy outcoupling from the chemical laser is provided by the ORTL load and ORTL power is extracted along a separate optical axis.

The intracavity experimental arrangement is shown schematically in Figure 1. The chemical laser had a gain medium 50 cm long and 0.3 cm high and typically operated with an output power of about 500 watts. In order to efficiently couple chemical laser radiation into the ORTL and to prevent ORTL oscillation in the chemical laser resonator, spectral control was required. The chemical laser resonator mirrors were coated with wavelength dependent coatings which provided high reflectivity at wavelengths below 2.90 microns and very low reflectivity at longer wavelengths, limiting chemical laser oscillation to short wavelength transitions. The ORTL medium was heated prior to entering the chemical laser irradiation region in order to enhance the absorption on the transitions with wavelengths larger than the chemical laser resonator cutoff value. A preheated fixed bed heat exchanger was utilized to heat the HF/He or HF/SF_6 ORTL medium to temperatures up to 550° K. The ORTL gain length was 10 cm and the absorption length 3 cm.

Typical data is shown in Figure 2 where ORTL output power is shown as a function of HF partial pressure for a 295°K inlet temperature HF/SF_6 ORTL configuration with the outcouplings shown. Typical ORTL spectra are shown in Figure 3. ORTL performance for mode control applications is usefully characterized in terms of the overall efficiency which is defined as the ORTL output power divided by the power extracted from the chemical laser (power absorbed by the ORTL cell and by the chemical laser optical elements). This efficiency may be expressed as the product of input efficiency and conversion efficiency. These two are primarily determined by the experimental geometry and the ORTL kinetics respectively. The input efficiency is defined as the power absorbed by the ORTL cell (measured as the sum of ORTL output power and residual medium heating) divided by the extracted chemical laser power. The conversion efficiency is defined as the ORTL output power divided by the power absorbed by the cell. While values of 90% and greater have been measured individually for both the input and conversion efficiency in different configurations, it is difficult in a small scale configuration to arrange conditions such that both are simultaneously large. Figure 4 plots the efficiencies associated with the data shown above. As HF partial pressure increases, the absorption of chemical laser radiation and the input efficiency increase. At the same time, the energy required to reach threshold increases and conversion efficiency decreases.

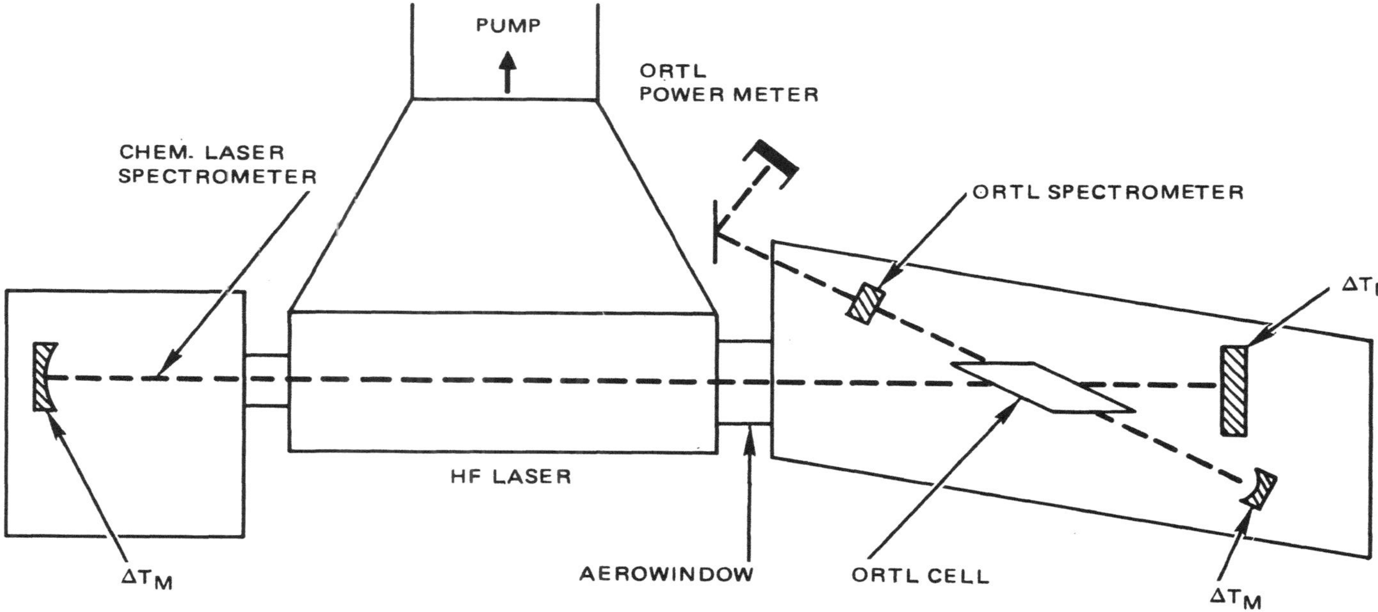

Figure 1 Intracavity HF ORTL experiment

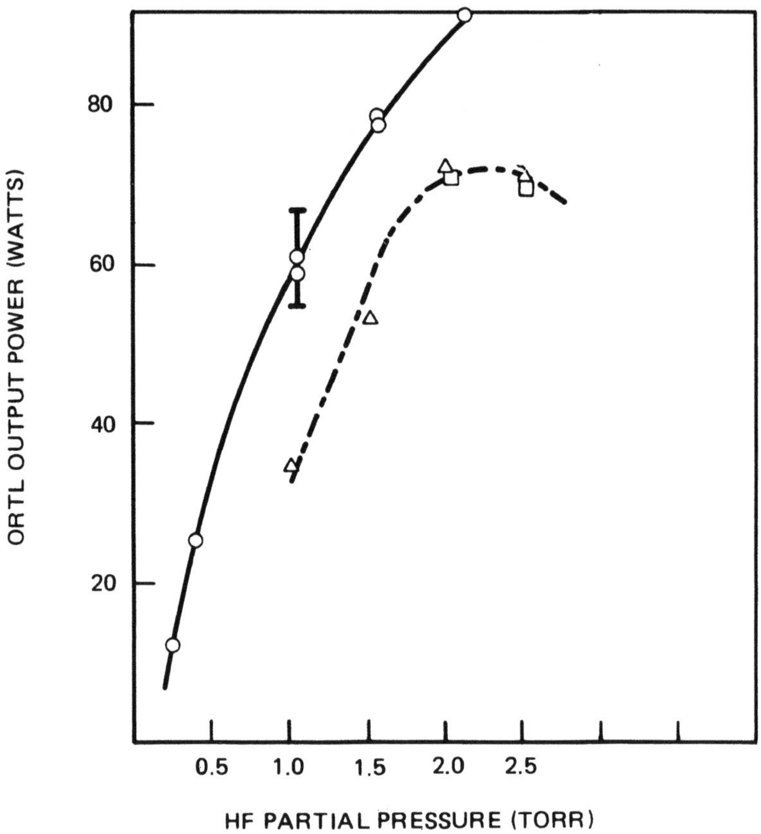

Figure 2 ORTL power as a function of HF pressure at 5 and 3% outcoupling.

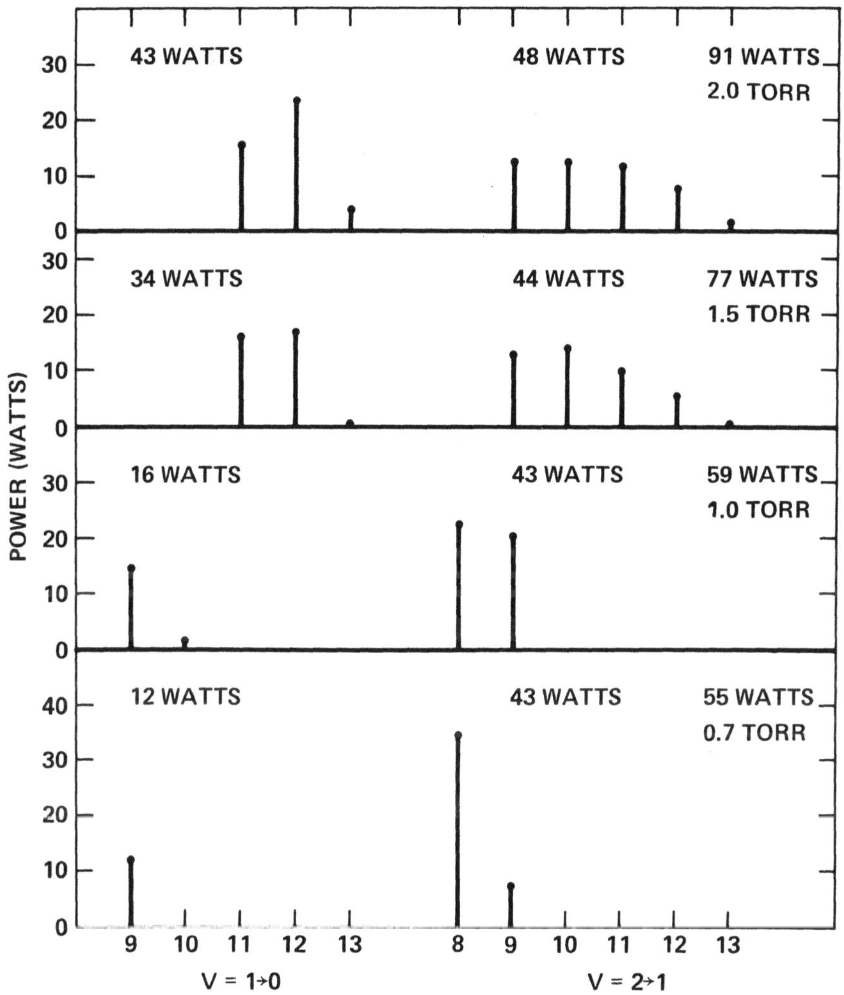

Figure 3 ORTL spectra at different HF partial pressures.

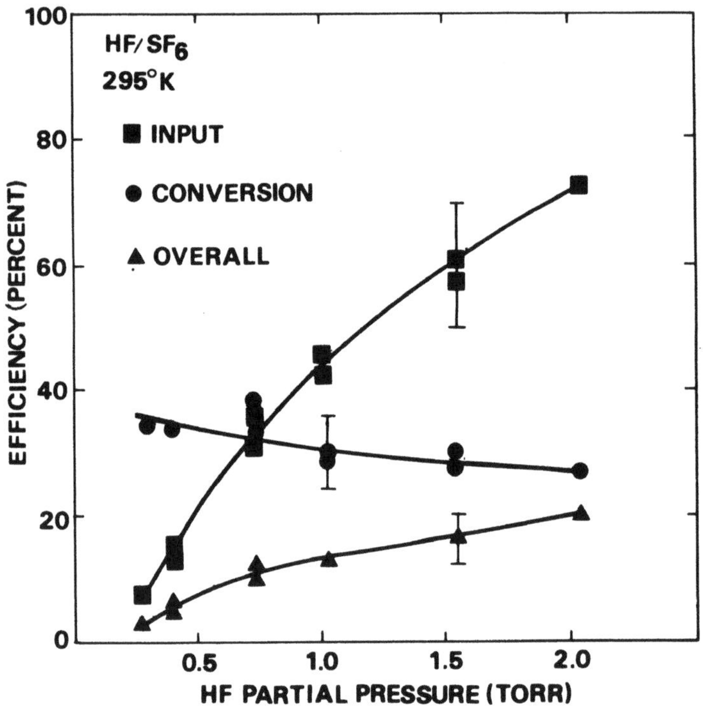

Figure 4 ORTL efficiency versus HF pressure.

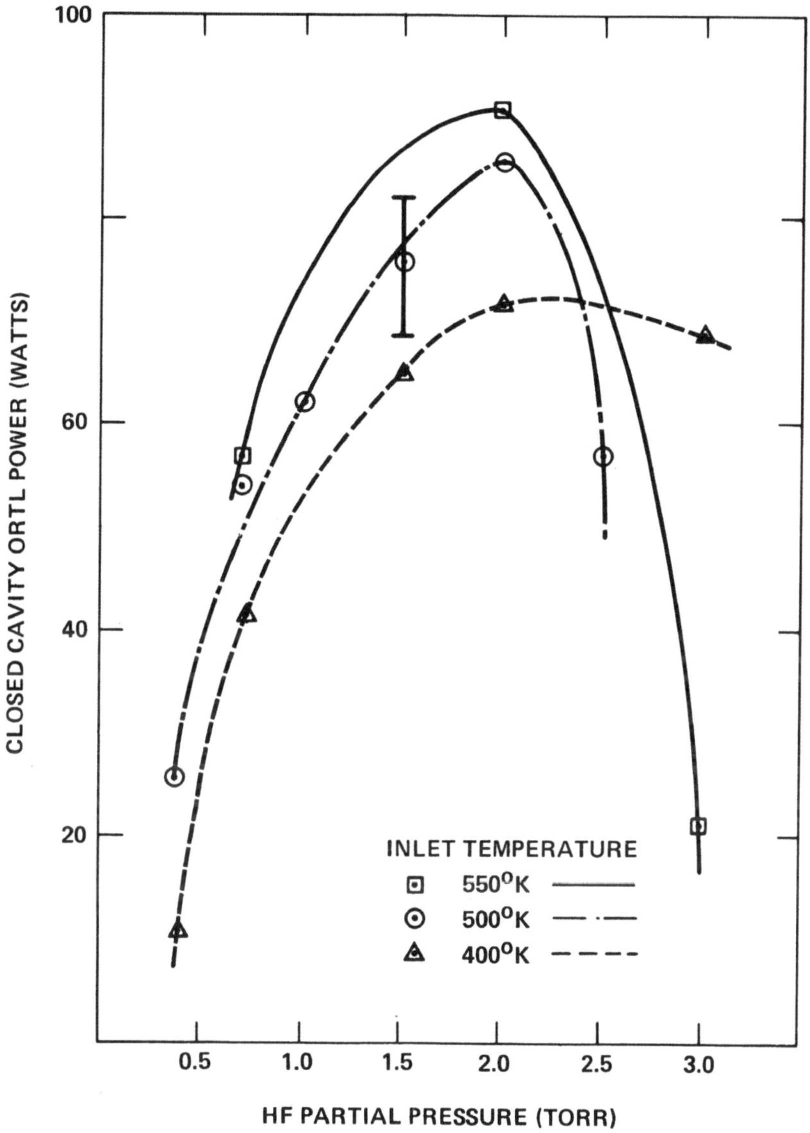

Figure 5 ORTL output variation with HF pressure and temperature (400°K - 550°K inlet).

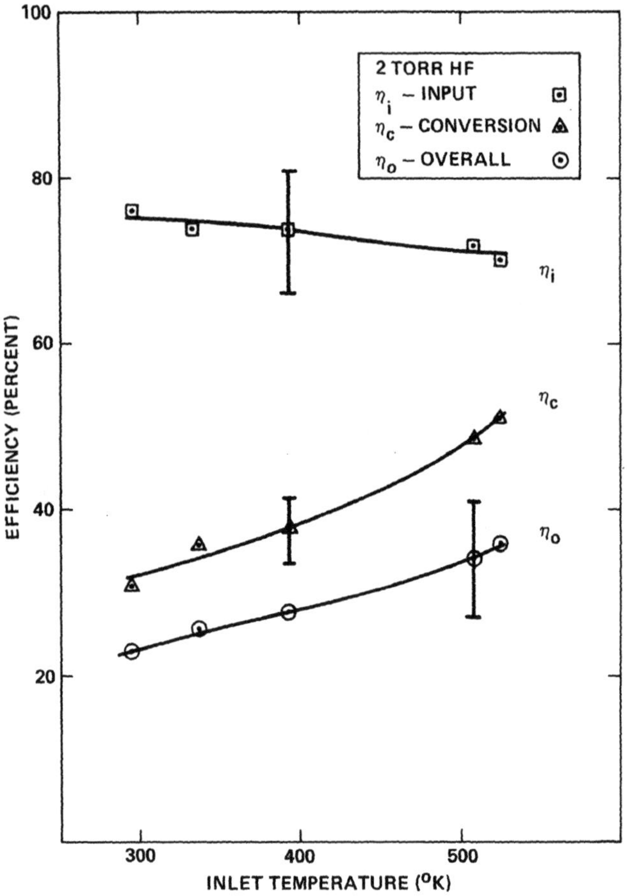

Figure 6 ORTL efficiency as a function of inlet temperature.

ORTL performance for the HF/SF_6 system was investigated over a range of inlet temperatures. Typical data is shown in Figure 5 and a summary of the observed efficiencies at 2 Torr HF partial pressure is shown in Figure 6. Apparently, at lower inlet temperatures, a significant amount of the absorbed energy is being used to heat the ORTL medium to a temperature at which significant gain is present on the lasing transitions. As the inlet temperature increases, this energy usage is less and the conversion efficiency increases.

Data with helium diluent at an HF partial pressure of 1.5 Torr and an inlet temperature of 295°K was obtained as a function of ORTL flow velocity. This data, along with an HF/SF_6 point for reference is shown in Figure 7. Equivalent performance was obtained although the associated ORTL temperature rise was 50% higher.

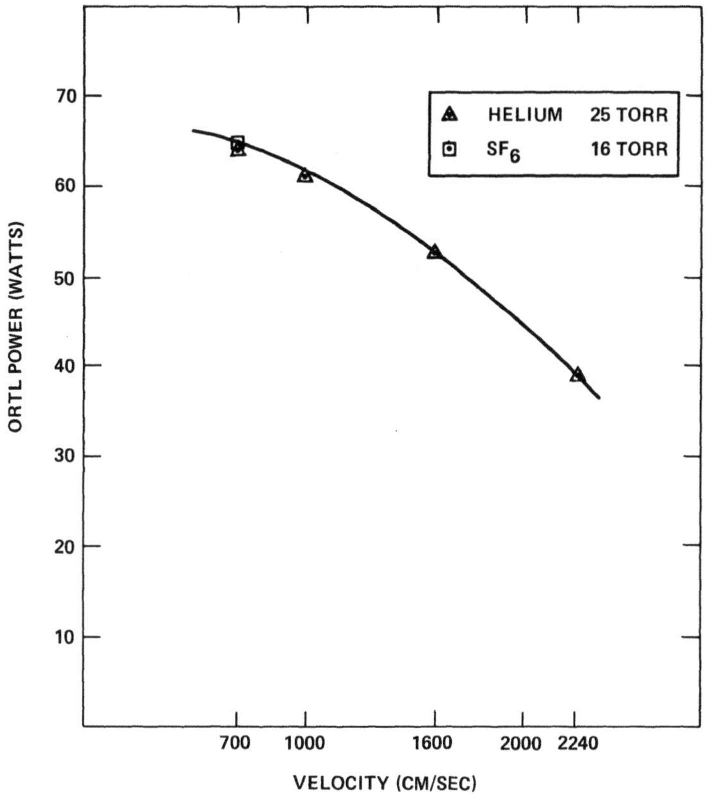

Figure 7 ORTL performance with He diluent.

Small signal gain measurements were performed over a limited set of ORTL conditions. The experimental arrangement and typical data are shown in Figures 8 and 9. The maximum observed small signal gain was approximately 2.8%/cm on the $P_2(9)$ transition.

A comprehensive model of the physical processes in the ORTL cell has been developed. The model includes spatially inhomogeneous optical pumping and stimulated emission and the important kinetic processes: V-V anharmonic pumping, V-T deactivation, and R-R transfer. Geometric optics and a Fabry-Perot resonator are assumed. This model has been combined with a semi-empirical chemical laser simulation for analysis of the intracavity HF/SF_6 ORTL data. Examples of a comparison between model and experiments are shown in Figures 10 and 11. The ORTL power is indicative of overall efficiency. The discrepancy between measured and predicted ORTL power at low pressure appears as a discrepancy in conversion efficiency. Because of the inherent experimental difficulties, the kinetic rates[8] used in the ORTL model have fairly large error bars. Perhaps adjustment of the rates would produce an improvement in the agreement between experiment and

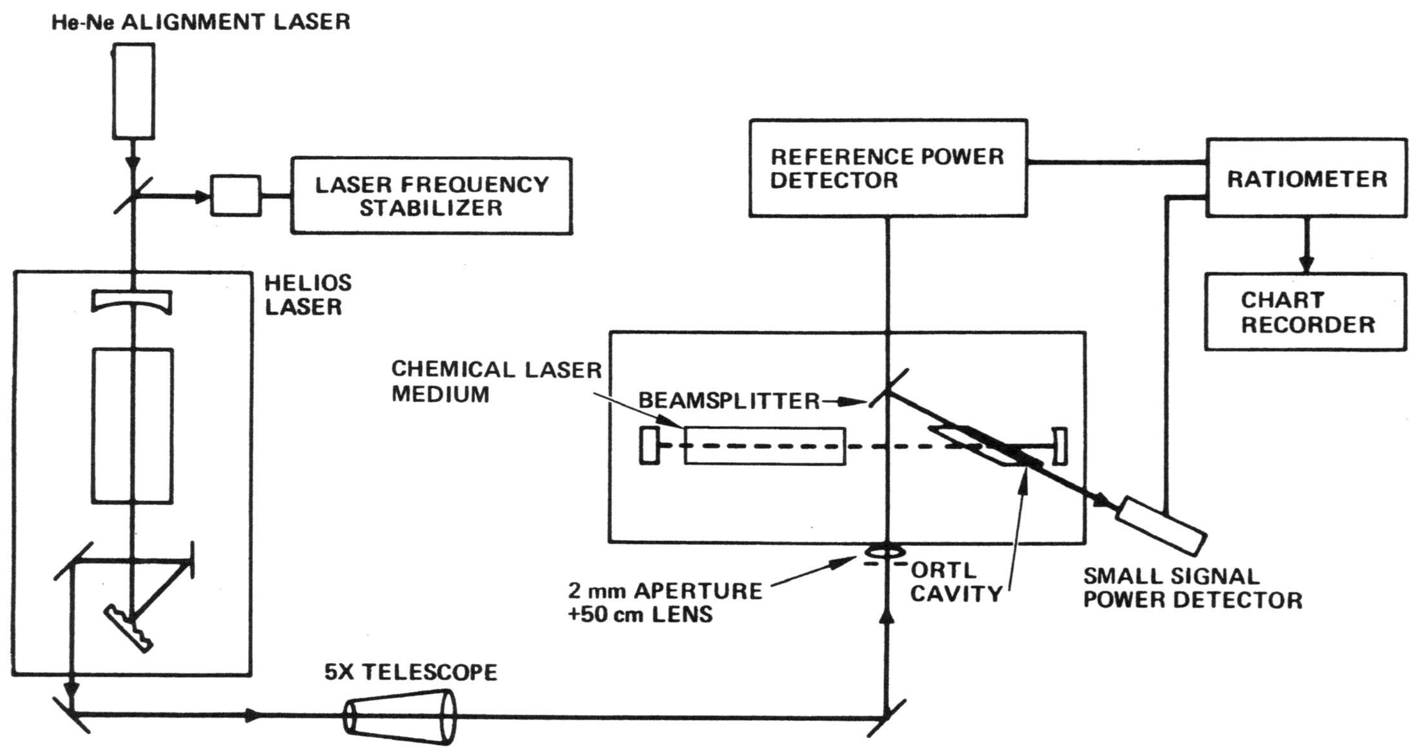

Figure 8 Small signal gain measurement apparatus.

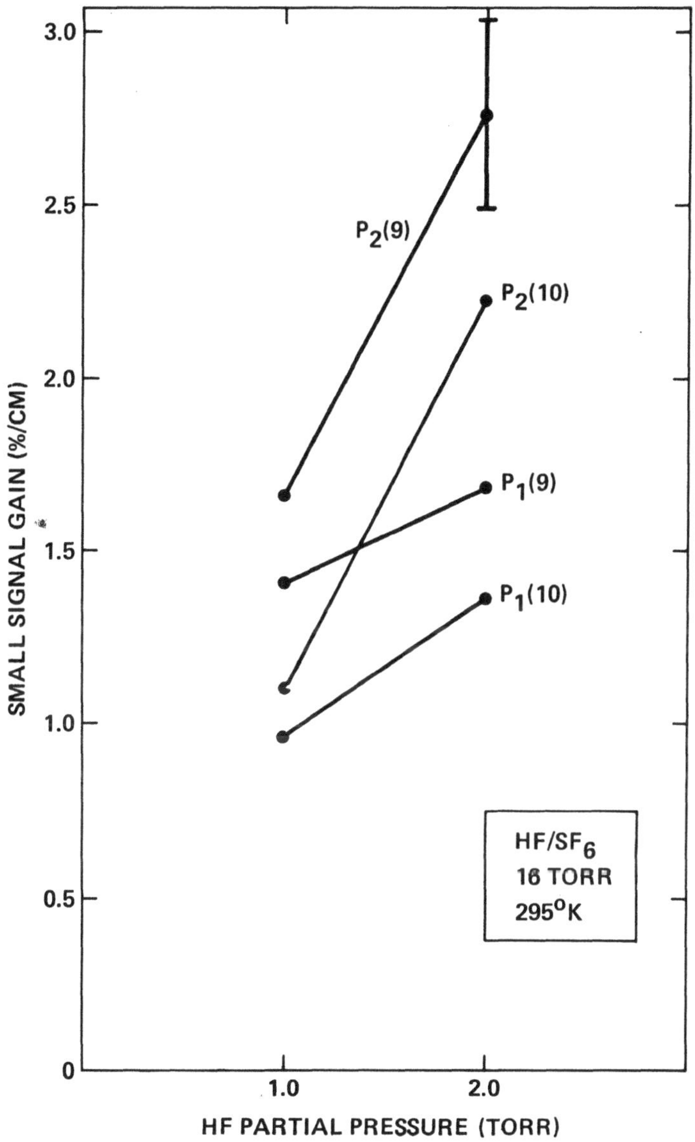

Figure 9 ORTL small signal gain.

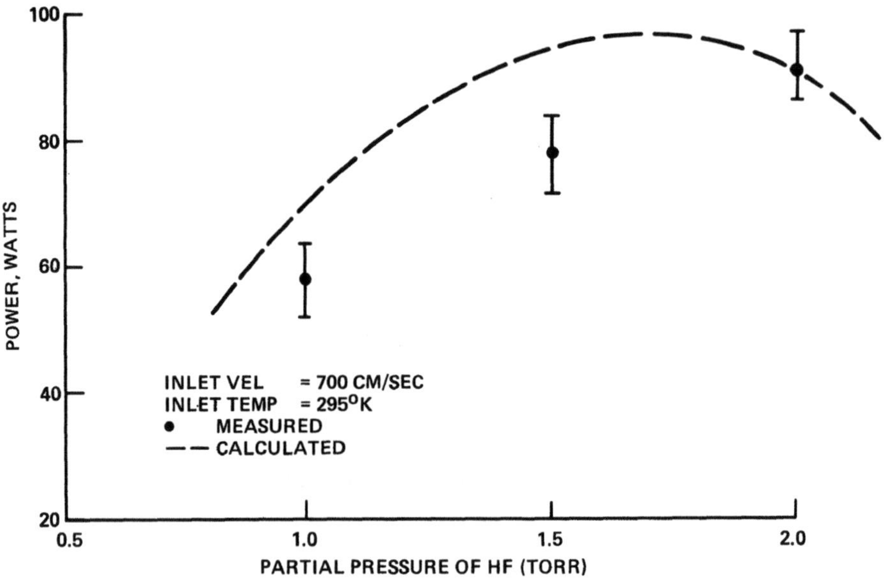

Figure 10 ORTL power versus HF pressure.

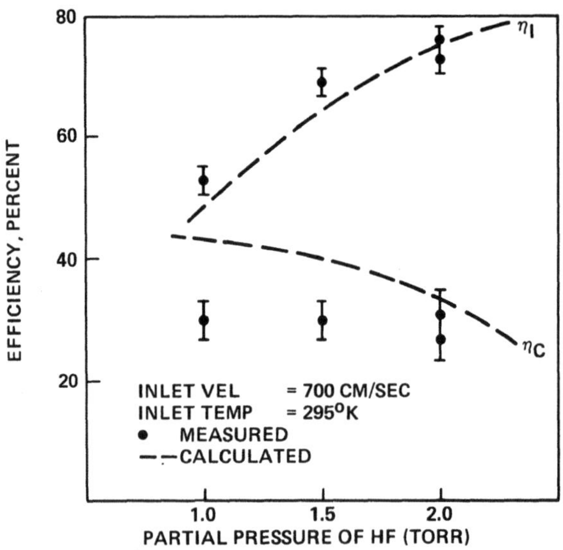

Figure 11 ORTL inlet and conversion efficiencies.

theory. A comparison of observed and predicted HF/SF_6 ORTL performance at 2 Torr HF partial pressure is shown in Figure 12. The agreement here is excellent. Finally, a comparison between predicted and measured gain and absorption is shown in Figure 13. The agreement on the $V=2 \rightarrow 1$ transition is quite good, but there seems to be some discrepancy in the $V=1 \rightarrow 0$ case.

In summary, the intracavity physical processes are now qualitatively well understood and in most cases there is also quantitative agreement between model predictions and observed data. Temporally stable spectrally controlled intracavity ORTL operation has been demonstrated in the laboratory. Overall efficiencies up to 38% have been achieved with a maximum intracavity HF ORTL output to date of 152 watts. While ORTL output beam quality has not yet been adequately assessed experimentally, it does appear that the ORTL approach to achieving good chemical laser beam quality is feasible.

This work described here was possible because of the foundation laid by previous efforts at Hughes Aircraft Company. In this

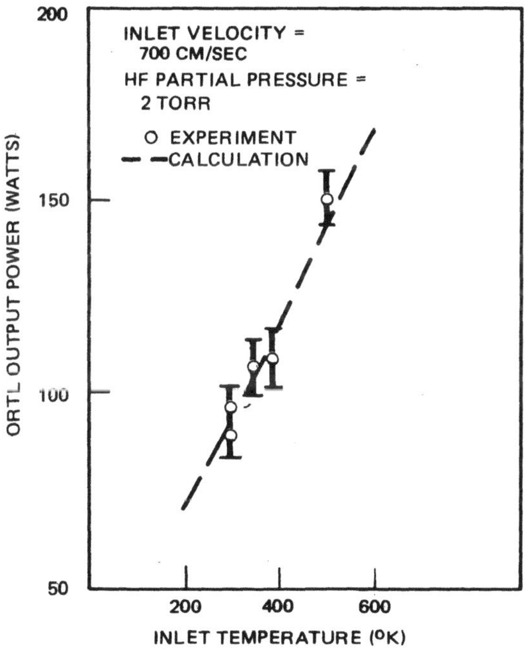

Figure 12 ORTL output power versus inlet temperature.

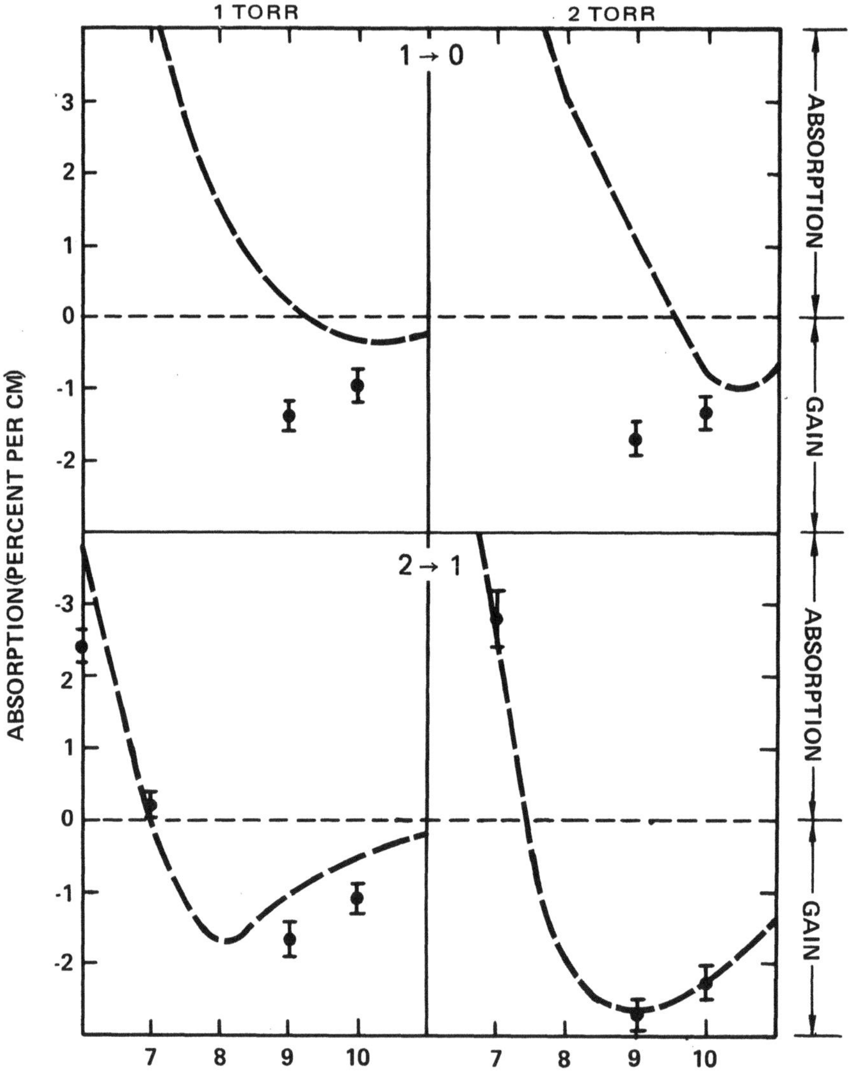

Figure 13 ORTL small signal gain.

regard, the authors would like to acknowledge the contributions of F. N. Mastrup, G. W. Holleman, K. K. Hui, C. M. Lovejoy, and H. Injeyan, as well as the assistance of D. Bruns, L. Williams, J. Paranto, and J. Jacobsen in performing the experiments. We are also grateful for the insight gained in technical discussions with W. Whitney, W. Watt, R. Sepucha, M. M. Mann, and E. R. Peressini.

REFERENCES

1. J. H. S. Wang, J. Finzi, and F. N. Mastrup, Appl. Phys. Lett 31, 351 (1977).
2. K. K. Hui, P. K. Baily, J. Finzi, J. H. S. Wang, and F. N. Mastrup, Appl. Optics 19, 831 (1980).
3. J. Finzi, J. H. S. Wang, K. K. Hui, P. K. Baily, G. W. Holleman, and F. N. Mastrup, IEEE J. Quantum Electron. QE-16, 912 (1980).
4. J. H. S. Wang, J. Finzi, P. K. Baily, G. W. Holleman, K. K. Hui, and F. N. Mastrup, Appl. Phys. Lett 36, 24 (1980).
5. J. H. S. Wang, J. Finzi, P. K. Baily, G. W. Holleman, and K. K. Hui, Journal de Physique 41, C9-463 (1980).
6. C. M. Lovejoy, J. Finzi, J. H. S. Wang, and P. K. Baily, Conf. on Lasers and Electro-Optics (CLEO '81), Washington, D.C. 1981.
7. J. Finzi, J. H. S. Wang, C. M. Lovejoy, and P. K. Baily, 4th International Conf. on Lasers and Applications (Lasers '81), New Orleans, La. 1981.
8. J. J. Hinchen and R. H. Hobbs, J. Chem. Phys. 65, 2732 (1976).

OPTICAL PHASE CONJUGATION IN LASER RESONATORS

C. R. Giuliano, R. C. Lind, T. R. O'Meara, and G. C. Valley

Hughes Research Laboratories
3011 Malibu Canyon Raod
Malibu, California 90265

ABSTRACT

The use of phase conjugating "mirrors" in oscillating optical structures has been a subject of recent interest, both experimentally and theoretically. One of the attractive features of a phase conjugate resonator, PCR, is its ability to provide an output wave whose wavefront depends only on the quality of the output coupler and is essentially independent of intracavity aberrations. Several concepts for PCR's will be presented in which either degenerate four-wave mixing or stimulated Brillouin scattering is employed as the phase conjugate mirror; advantages and drawbacks of each will be discussed.

INTRODUCTION

Optical phase conjugation is an area that has received a considerable amount of attention in the past several years. The earliest reports of optical phase conjugation by workers in the Soviet Union [1,2] were followed by several years of relatively little activity until reports suggesting potential applications began to appear in the open literature. In particular, the possible application to adaptive optics[3] was one area which generated a great deal of interest because of the potential for correcting for atmospheric turbulence for high energy lasers and for compensating optical train distortions and pointing errors in laser fusion systems. A number of other potential applications have been pointed out as well[4,5] but will not be discussed in this paper.

The interesting property of conjugate wave generation (also referred to as wavefront time reversal or phase reversal) that we will discuss here can be summarized in the following statement. When a wave enters a conjugator, a new wave is created traveling in the opposite direction with the sign of the transverse phase opposite to that of the incident wave. Hence, a diverging wave striking a conjugator at an angle θ will leave as a converging wave retracing the same path as the incident wave. Thus a conjugator can be thought of as a peculiar kind of mirror that combines reflection with phase reversal.

When one thinks of a conjugator as a type of reflector, it becomes logical to ask what would happen if such a reflector were incorporated as an element in an optical resonator - the subject of this paper. In the following pages we will briefly describe the properties of phase conjugate resonators, PCR's, and point out why they are of interest. We will briefly describe two of the most serious candidates for conjugators, the nonlinear optical phenomena degenerate four wave mixing, DFWM, and stimulated Brillouin scattering, SBS. This will be followed by a description of different types of PCR's along with a discussion of some practical issues in their construction.

BACKGROUND

Phase conjugate resonators have interesting characteristics which differ significantly from those of conventional resonators (CR's).[6-10] For example, and most important for this paper, the PCR can compensate for intracavity distortions. One can show that, when light is extracted from the "ordinary mirror" end of the PCR, the transverse phase of the wave depends only on the output mirror's figure (its detailed surface shape); the phase will not depend on any other sources of distortion within the body of the resonator.[9] This feature has been demonstrated qualitatively in laoboratory experiments.[11] Another contrasting feature of PCR's compared with CR's is that a PCR is always stable even if the ordinary mirror is convex.[6] This property gives PCR's the ability to extract energy effectively in situations where conventional resonators require an unstable resonator design (i.e. large Fresnel numbers).

Another unique property is that a PCR with an "ideal" conjugate mirror will not have longitudinal modes that depend on cavity length. In a PCR the phase that is accumulated as the wave propagates from the ordinary mirror to the conjugate mirror is subtracted by the same amount on the way back to the ordinary mirror; in one round trip, the net accumulated phase is always zero. Consequently, a PCR of length L can support any wavelength consistent with the bandwidth of the gain medium and the conjugate mirror itself.

Hence, a PCR oscillating at a given wavelength will continue to oscillate at that wavelength, independent of variations of the cavity length. This in contrast to an ordinary resonator, whose spectral output will exhibit "mode hopping" and frequency drift as the cavity length changes.

If the phase conjugator mirror, PCM, is a four wave mixer FWM (see below) there can be transverse modes but, in contrast to a conventional resonator whose modes are spaced by c/2L in frequency, the FWM PCR has modes that are spaced by c/4L and occur in pairs symmetrically spaced around the central frequency.[6,7,11] The central frequency always corresponds to the pump frequency of the PCM and is independent of the cavity length.

The types of PCR's that will be discussed in this paper utilize PCM's that are derived from either degenerate four wave mixing, DFWM, or stimulated Brillouin scattering, SBS. It is therefore appropriate to discuss briefly the physics of these nonlinear phenomena to the extent that they affect the properties of PCR's.

Four wave mixing is a nonlinear process in which three input waves mix to yield a fourth (output) wave. The three input waves consist of two planar counter propagating pump waves, labeled E_f and E_b (f for forward and b for backward), and a probe wave, E_p, entering at an arbitrary angle to the pump waves. All three couple through the nonlinear susceptibility, $\chi^{(3)}$, to yield a fourth wave, E_s, which is proportional to the spatial complex conjugate of E_p. The third order polarization - which yields the conjugate wave, $P_{nl} = \chi^{(3)} E_f E_b E_p{}^*$ - is proportional to the product of the amplitudes of the three input waves.

Stimulated Brillouin scattering involves the generation of a coherent acoustic wave when an intense optical wave interacts with a nonlinear medium. The mechanism generally acts through electrostriction - that is, the tendency of the medium's density to increase in proportion to the electric field intensity. Above some threshold intensity a light wave allowed to propagate in a nonlinear medium produces an intense back-scattered optical wave whose frequency is down-shifted by an amount equal to the acoustic frequency.

Analysis and experimental observation indicate that under appropriate conditions the scattered wave is the conjugate of the incident wave. Consequently an aberrated input wave yields an equally aberrated but conjugate scattered wave. Thus when the reflected wave retraces the incident path, the medium removes from it whatever phase errors were introduced in the first pass.

These two conjugation phenomena may be compared in terms of their impact as phase conjugate mirrors, PCM's.

- o Conjugation by SBS has an intensity threshold; DFWM does not.
- o DFWM requires separate pumps; SBS does not - hence an inherent simplicity for SBS.
- o SBS is accompanied by a frequency shift with each scattering step; DFWM produces a conjugate wave at the same frequency as the incident wave. This difference implies certain limitations for SBS PCR performance (see below).
- o DFWM can occur with conjugate reflectivities (E_s/E_p) exceeding unity - SBS reflectivities can only approach unity.

These comparative properties will be discussed in the light of how they affect the practical performance of phase conjugate resonators.

FOUR-WAVE MIXING PCR's

The first category of resonators to be described is that in which phase conjugate mirror (PCM) is generated by four-wave mixing (FWM). In this class there are three generic types, starting with the simplest (1) external laser pumped; (2) self-pumped and (3) self-starting and self-pumped.

External Laser Pumped PCR's

The basic configuration of this type is shown in Figure 1. This configuration is similar to conventional resonators where now a PCM pumped by an external laser replaces the normal mirror at the right hand end of the resonator. The threshold condition for such a resonator is the same as for a conventional resonator where the reflection coefficient of the PCM is the phase conjugate reflectivity determined by the four-wave mixing process. If it is assumed that the four-wave conjugate signal is produced by a 2-level saturable absorption process the output power of such a resonator can be calculated simply as[11]

$$P_{OUT} = AI_{sc}(1-R)(R\,R_{PCM}G_o^2-1)/RG_o \text{ Watts}$$

where I_{sc} is the conjugator saturation intensity (assumed to be much smaller than the saturation intensity of the gain medium) $G_o = e^{g_o L}$, R is the reflectivity of the output coupler, R_{PCM} is the reflectivity of the PCM and A is the area of a resonator mode.

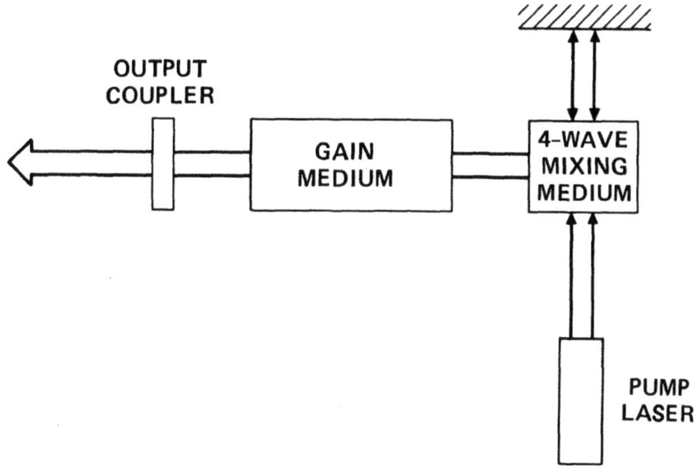

Fig. 1. Schematic diagram of an external laser-pumped FWM PCR.

This type of resonator is inherently inefficient when one considers the requirement for external pumps whose powers are typically greater than the PCR's output power. However, because of its relative simplicity the external laser-pumped configuration has been used for a number of significant experimental demonstrations of the unique properties of PCR's. A brief summary follows.

The first demonstration of a pulsed PCR was made in which stable oscillation was sustained even with a convex mirror.[6] Lind and Steel[11] were the first to demonstrate a cw PCR with an intracavity gain medium. They also showed that when an aberrator was placed inside the PCR, the output through the ordinary mirror was diffracted limited, while the output through the PCM was severely aberrated. The output frequency of the PCR was shown to be locked to that of the pump laser and independent of cavity length. Finally, they observed for the first time the paired half-axial modes separated by c/4L from the central frequency.

Another oscillating structure can be obtained in which the intracavity gain medium is absent. This occurs when the PCM reflectivity is substantially in excess of unity. This configuration has been used to demonstrate properties of PCR's[12] but, because it is an empty resonator structure, its principal appeal appears to be scientific rather than practical, especially for high power applications.

Self-Pumped PCR's

In this category the pumps for the four-wave mixer are not obtained by a separate laser but are derived from a portion of the useful output of the PCR itself. In its simplest form such a configuration is shown in Figure 2. It is assumed that an external laser is used in some fashion to start the oscillation by providing the pumps for the FWM and then switching this laser out of the system. (Schemes for self starting will be described later).

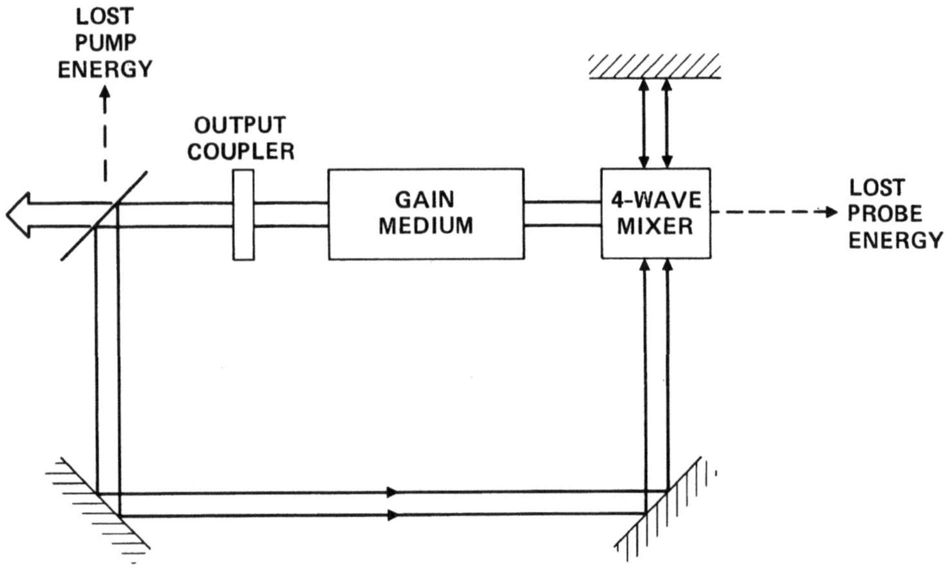

Fig. 2. Schematic diagram of a self-pumped FWM PCR indicating lost pump energy and lost probe energy.

Two key issues related to such a design that come to mind immediately are the efficiency of such a system and the beam quality of the pumps. Clearly the reason for incorporating a PCM in a resonator is to improve beam quality and thus increase the far-field irradiance. But since, in a self pumped configuration, some of the energy of the PCR is required to pump the PCM the question of whether one gains in energy delivery is a key factor in PCR design. In addition, the pumps must be plane waves (or conjugates of each other) to realize the conjugate properties of such PCR's.

Other factors must be considered for such PCR designs: (1) frequency control devices - gratings, etalons, etc. - must be included in the pump path if precise frequency control is needed to obtain efficient operation of the PCM. This would be required if a resonant nonlinear process such as saturable absorption were used to generate the PCM as in Na vapor;[11] (2) Pump depletion - that is in general, DFWM theory considers the probe (the intracavity beam in a PCR) intensity to be less (say a factor of 10 typically) than the pump intensity. It is clear that a PCR will not operate in this mode and in fact all beams will be of the same magnitude so that the effect of this "pump depleted" mode on the efficiency of PCR's must be carefully evaluated; (3) pump recovery - examining the diagram in Figure 2 we see that the counterpropagating pump of the FWM is lost to the PCR lowering the efficiency of the system. However, this energy can be recovered by the use of alternate geometries. An example of such a configuration is given in Figure 3. Here the pumps are obtained by the action of the tuned Fabry-Perot that surrounds the FWM. In this case the Fabry-Perot is tuned such that very little energy is reflected back from the resonant structure. (4) Probe recovery - again referring to Figure 2 we see that probe energy is lost to the system by transmission through the PCM. Recovering this energy represents a challenge to PCR design and is a current area of research at Hughes. In principle it is easy to return this energy to the PCR cavity but it will not be the conjugate signal. Thus the resonator produced will have an output that has a mixture of non-conjugated and conjugated beams.

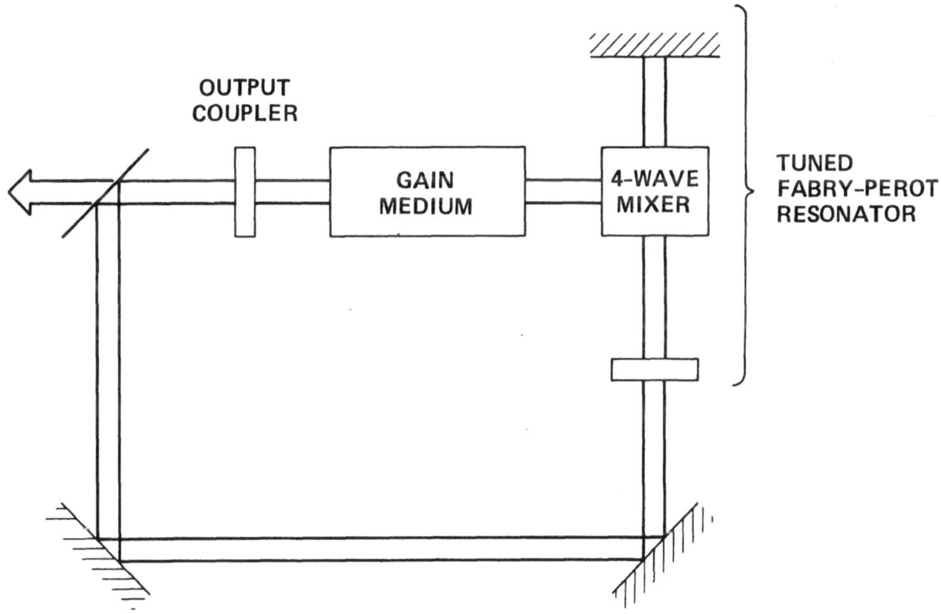

Fig. 3. Self-pumped FWM PCR using tuned Fabry-Perot resonator to minimize pump losses.

There are many other possible geometries and configurations which are variants on the theme shown but will not be described here. However, a somewhat different implementation is the synchronously pumped geometry described by Vanherzeele, Van Eck and Siegman.[13] This scheme uses pulses in a cavity timed to generate the various pump and probe pulses. It is a very clever and complex scheme which shows that pulse compression can be achieved in such a configuration but offers no promise for intracavity aberration correction since it extracts the useful output from the conjugator end.

Self-Starting and Self-Pumped PCR's

In this category all the features of the self-pumped configurations described previously are retained but in addition the resonator also has the ability to reach threshold itself - that is self-starting. Again in its simplest form such a configuration is shown in Figure 4. Here we make use of the "lost" transmitted probe by inserting an E/O switch and total reflector in that path. Oscillation is then initiated between the output coupler and the total reflector as in a conventional resonator. The FWM is in general a low loss device so that threshold is easily reached in the cavity formed by the output coupler and the total reflector. Initially no pump energy is available however at some point in time sufficient output will have been obtained such that the pumps will be of a magnitude that the FWM will be generating a strong conjugate signal so that the PCR can oscillate. At that time the E/O switch is operated removing the total reflector from the optical path. This type of resonator configuration is currently under investigation at Hughes.

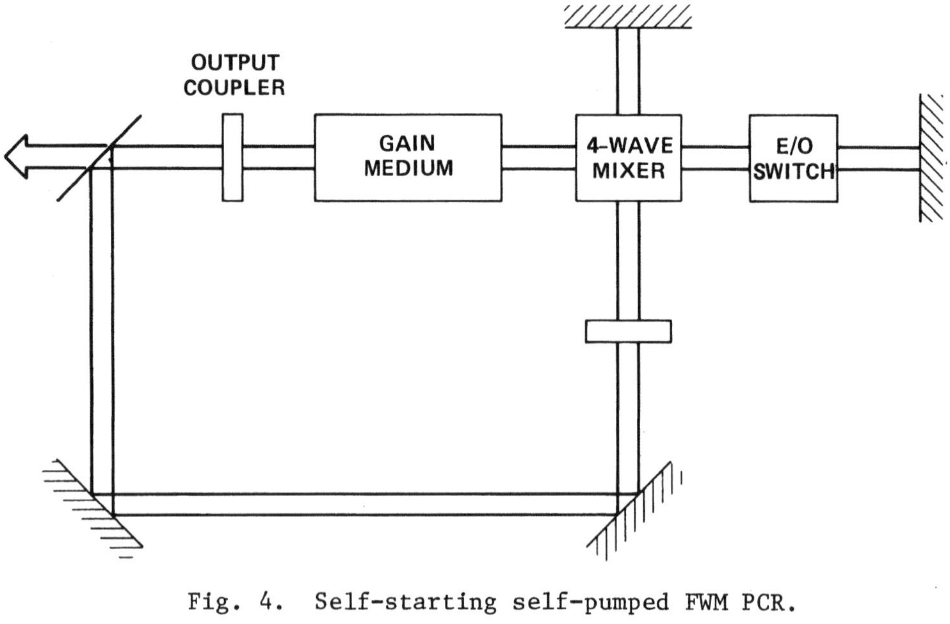

Fig. 4. Self-starting self-pumped FWM PCR.

PCR's USING STIMULATED BRILLOUIN SCATTERING

An alternate approach to PCR's is one in which the PCM is derived from SBS. This approach is conceptually simpler than the one that uses DFWM in that it does not require the use of auxiliary pumps. However, in spite of this apparent simplicity, it has certain challenges that must be met before it can be implemented in a viable way.

First, there is the problem of initiation. In contrast to DFWM, SBS will occur only after a certain threshold intensity is reached. Thus, a situation must be set up so that the intracavity flux can build up to a point at which SBS takes over. This can be realized conceptually by considering Figure 5 in which the SBS cell is situated inside an ordinary optical resonator. The idea here is that once the oscillation builds up in the ordinary resonator, the SBS mirror takes over and oscillation continues between it and the output coupler. Thus, the system starts out as an ordinary optical resonator and converts to a PCR once SBS threshold is reached inside the cavity.

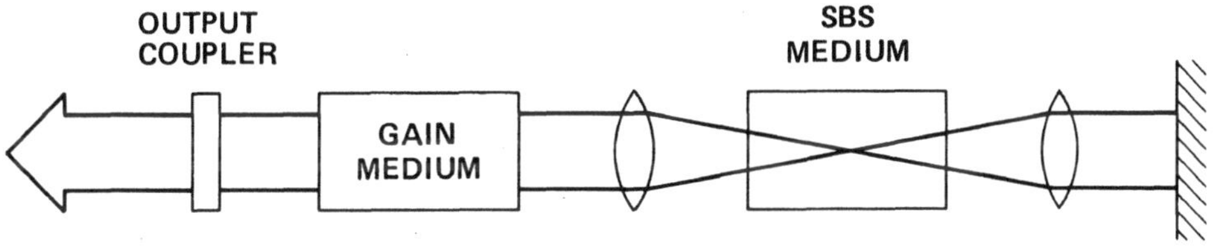

Fig. 5. Schematic diagram of SBS PCR.

More important than the initiation process is the fact that SBS occurs with a frequency down-shift equal to the frequency of the coherent acoustic wave generated in the process. This shift, equal to 2 v_s/c, where v_s is the speed of sound in the nonlinear medium, amounts to about one part in 10^5 and gives rise to a spectral output that sweeps downward in frequency with each successive pass through the resonator. Thus, a PCR with an SBS conjugator, once initiated, will eventually die out as the successive frequency shifts cause the circulating wave to walk off the gain curve of the active medium. Operating with this constraint, the SBS PCR is best suited for pulsed operation. The duration of the pulse is determined by the width of the gain curve, the degree of output coupling, and the cavity length. Moreover, one must accept operation which gives an inverse "chirp" for an output which may or may not be tolerable depending on the application.

Note that for several types of gain media, e.g. most gas lasers, the size of the SBS shift is larger than the width of the gain line. Hence, the "oscillation" will essentially die out after one pass through the resonator and the PCR as such is not truly realizable. For lasers that have relatively broad gain profiles (e.g. Nd: glass, ruby, most dyes, and excimers) many hundreds of SBS shifts can be sustained before oscillation ceases due to the frequency walk-off.

There are several approaches to the frequency walk-off problem, some more complex than others. The simplest conceptually is to move the SBS medium towards the incident optical wave at a speed large enough to cancel the frequency shift, i.e. the speed of sound. For gaseous media Mach 1 flows are within the state-of-the-art but pose a challenge for gas pressures that work best for efficient SBS (1-10 atm). For liquids and solids the challenge is greater and may require mechanically sophisticated implementations such as high speed rotating discs or synchronized reciprocating designs. If these challenges can be met, the promise of cw or quasi-cw PCR's using SBS conjugators may be realized.

Another approach to the frequency walk-off problem is the double SBS ring resonator configuration shown in Figure 6. This approach is more complex in concept and in implementation but may have promise for some applications. It was inspired by the work of Basov and Zubarev[14] and further elucidated by the work of Abrams, Giuliano and Lam.[15] Here we show the SBS medium pumped by a separate pump laser. The pump laser is chosen so that its intensity exceeds SBS threshold in the nonlinear medium. The circulating fields in the ring resonator are not large enough to reach SBS threshold but nevertheless undergo phase conjugation as long as the pump waves are present. As described[14,15] the nonlinear process that takes place is a kind of resonant-enhanced four-wave mixing in which the pump waves are the incident and SBS-scattered pumps externally imposed. Here, the counter-clockwise wave has frequency ω and the clockwise wave, $\omega-\delta$. Instead of the successive down-shift that takes place in the linear SBS resonator, here we have alternating up-and-down shifts and the walk-off problem is avoided. We also note as shown by Basov and Zubarev[14] that high quality pumps are not required for this process to work. This is because, as in the general case of four-wave mixing, the counterpropagating pumps are conjugates of each other and, for this reason, the signal and probe waves are also conjugates.

PCR's USING SHIFTED GRATING CONJUGATORS

A third category of PCR's is that which makes use of the

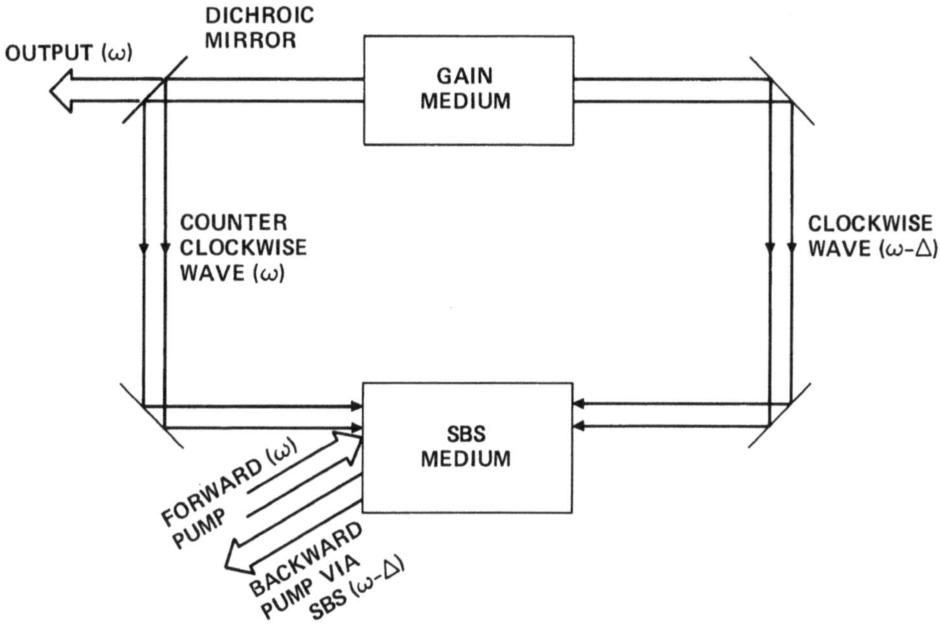

Fig. 6. Pumped SBS PCR that eliminates frequency walk-off; note down-shifted clockwise wave and up-shifted counter-clockwise wave.

shifted grating effect which occurs in certain nonlinear materials. Here, the refractive index grating which is produced in these materials is shifted laterally from the intensity grating by a phase shift determined by the properties of the material. Under some conditions, this shift can be exactly one-half of the grating period. Such a shift has very interesting consequences for two-wave interference patterns because the shift allows preferential energy transfer from one wave to the other. Shifted grating materials used for four-wave mixing have allowed construction of novel devices[16] and, in fact, a very significant demonstration of using a shifted-grating conjugator in a PCR that does not use external pumps[17] has been constructed. Figure 7 shows a schematic diagram of the PCR, which works as follows. Lasing is initially induced between mirror M_1 and beam splitter BS. Light transmitted through the beam splitter causes oscillation in the PCR consisting of the crystal and mirrors M_3 and M_4. This oscillation is due to the shifted grated process described above. M_2 is used to assist in the buildup of oscillation. With oscillation established between M_3 and M_4, the beam splitter and M_2 are removed.

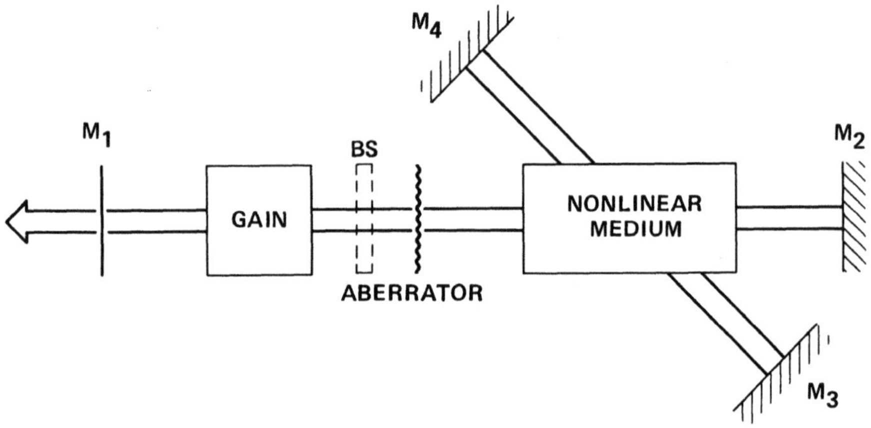

Fig. 7. Schematic diagram of a PCR using a shifted grating nonlinear medium.

At the present time, only photorefractive materials have operated as shifted grating PCR's. Because of the slow time response of such materials, these PCR's have only slowly varying aberration correction capabilities. Efficiency considerations of such PCR's have not been examined to date.

PRACTICAL ISSUES IN THE CONSTRUCTION OF PCR's

The key attraction of a PCR is its ability to produce a nearly diffraction limited output beam with extremely large intra-cavity aberrations while providing efficient mode filling for large gain volumes. The size of the aberrations that can be corrected is limited by the finite transverse size of the laser resonator and conjugator and by nonreciprocal propagation effects such as may occur in a saturated gain medium. Aberration correction has been analyzed analytically and computationally.[9,18-20] Numerical results from a three dimensional, saturated gain calculation are shown in Figure 8. This figure shows Strehl ratio and fractional outcoupled power as a function of the standard deviation of the phase produced by an aberration screen placed at the conventional plane mirror side of a PCR. We see that for the particular resonator chosen, the fractional outcoupled power degrades by only about 25% for $\sigma_\phi = 6$ radians. The curve for the CR corresponds to a resonator that has two plane parallel mirrors having the same separation as the PCR. At zero phase error the values of P_{out}/P_{stored} for the two resonators are different because their transverse modes are not the same. The key comparison between the two curves is that the

CR curve degrades substantially more rapidly than that for the PCR. In fact the dashed portion of the CR curve corresponds to a region in which there was doubt whether a converged solution has been obtained in the computer simulation.

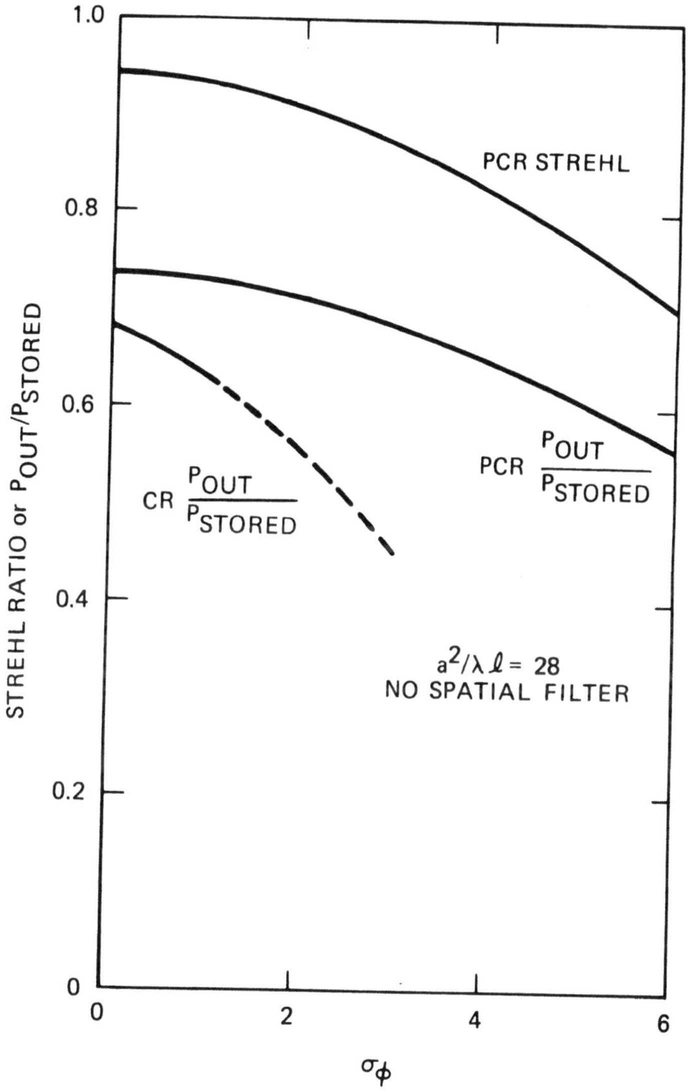

Fig. 8. Results of computer simulations comparing P_{out}/P_{stored} of a PCR compared to a CR and PCR strehl ratio as a function of rms phase error.

The Strehl ratio, defined as outcoupled far-field irradiance with abberrations divided by far-field irradiance of a uniform amplitude and phase beam, also degrades by about 25% at $\sigma_\phi = 6$. Thus, one concludes that even in the face of very bad aberrations, a PCR can provide good performance. It is important to note that as the Fresnel number of the resonator (square of radius divided by wavelength times resonator length) decreases, the degradation in PCR performance increases. Computer simulations show that this degradation becomes substantial at Fresnel numbers less than about 5.

Figure 9 shows results for computer-simulated PCR's with and without the use of intracavity spatial filters. The original motivation for the use of a spatial mode selector in a PCR was based on the notion that a unique transverse mode might not be established without one. In fact it can be shown that for a PCR having infinite lateral extent any arbitrary wave is a stable solution. A PCR of finite extent however does possess unique transverse modes. This

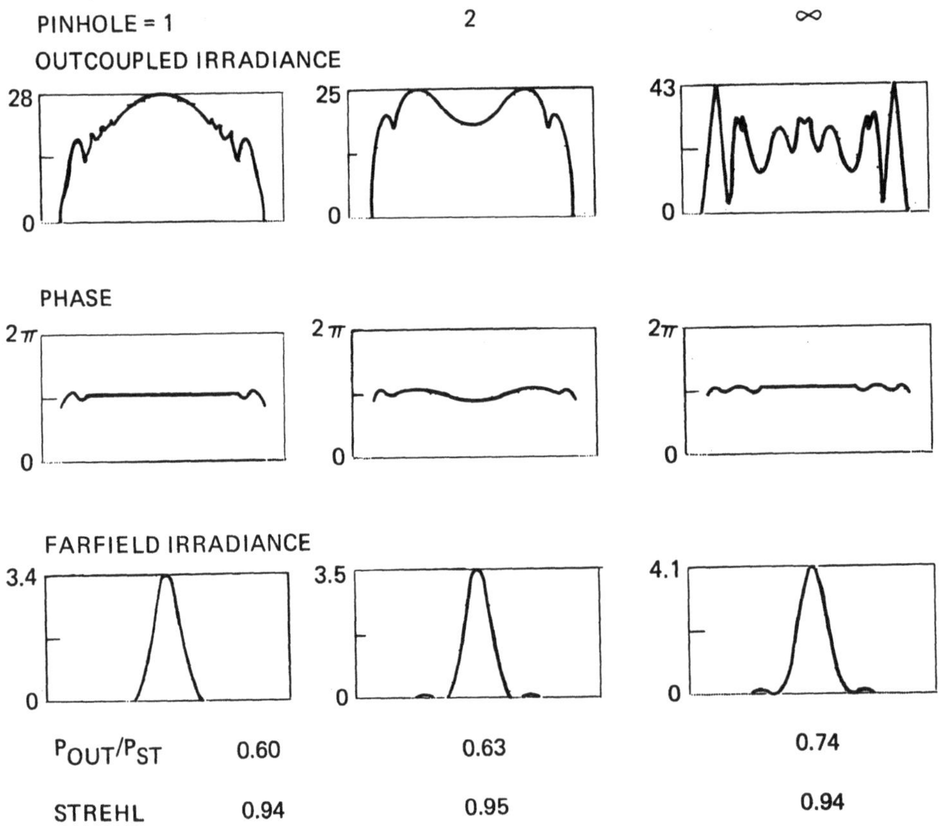

Fig. 9. Computer simulation results showing outcoupled irradiance, output phase, and far-field irradiance for different intracavity spatial filters.

occurs because of the diffraction losses at the edges of the mirrors which, because of their finite extent, act as spatial filters.

Note the following features in the simulation results shown in Figure 9 for a PCR with a saturable gain medium. (1) The phase of the wave at the output coupler is well behaved. (2) The outcoupled irradiance is spatially smoother with the spatial filters than without. (3) The far-field Strehl ratios are comparable for all three cases. (4) The extracted power, P_{out}/P_{stored}, is appreciably increased for the case without the spatial filter.

In spite of the aberration correction capability PCR's must still provide an acceptable level of extraction from the available gain medium. The two major differences in extraction between PCR's and CR's are the transverse mode shapes or filling factors and mirror losses. PCR's may offer the opportunity to fit a transverse mode to a given gain medium but must use a phase conjugate mirror as one reflecting element, and current phase conjugate mirrors are limited in efficiency. For SBS the principal efficiency issue is a reflectivity less than unity which results in some fraction of the resonator output lost through the PCM. For FWM PCR's the need to provide pumps operating in a strongly depleted mode combined with the loss through the PCM limits the overall efficiency.

Phase conjugate mirrors using SBS have been demonstrated with reflectivities between 50 and 70%. Efficient operation of a resonator with a mirror with such a reflectivity requires both a high gain laser medium and a very low value of feedback coupling, that is the resonator must have a very low Q. The gain requirements are quantified in Figure 10 for a 70% reflectivity mirror. The PCR system efficiency in these curves is defined as the ratio of the net output power to the power stored in the inverted population. It is a generalized extraction efficiency. The output coupling mirror reflectivity has been optimized for maximum extraction efficiency at every gain value on this curve. The predicted performance ignores transverse mode filling factor losses and spatial hole burning effects.

At first thought it might appear that four-wave mixer conjugators could minimize the lossy mirror problem since reflectivities up to and exceeding unity have been measured and reported in the literature. Unfortunately, there are major problems with the use of such conjugators in PCR's. The high reflectivities occur as a consequence of nonlinear diffraction coupling of the pump waves into the conjugated signal direction, and thus the magnitude of the required pump powers is given by the required conjugate signal power divided by the diffraction efficiency of the four-wave mixer. In typical high reflectivity experiments the pump powers have exceeded the signal powers by at least an order of magnitude. Clearly the pumps most operate at power levels which can easily exceed the PCR output power - an untenable situation.

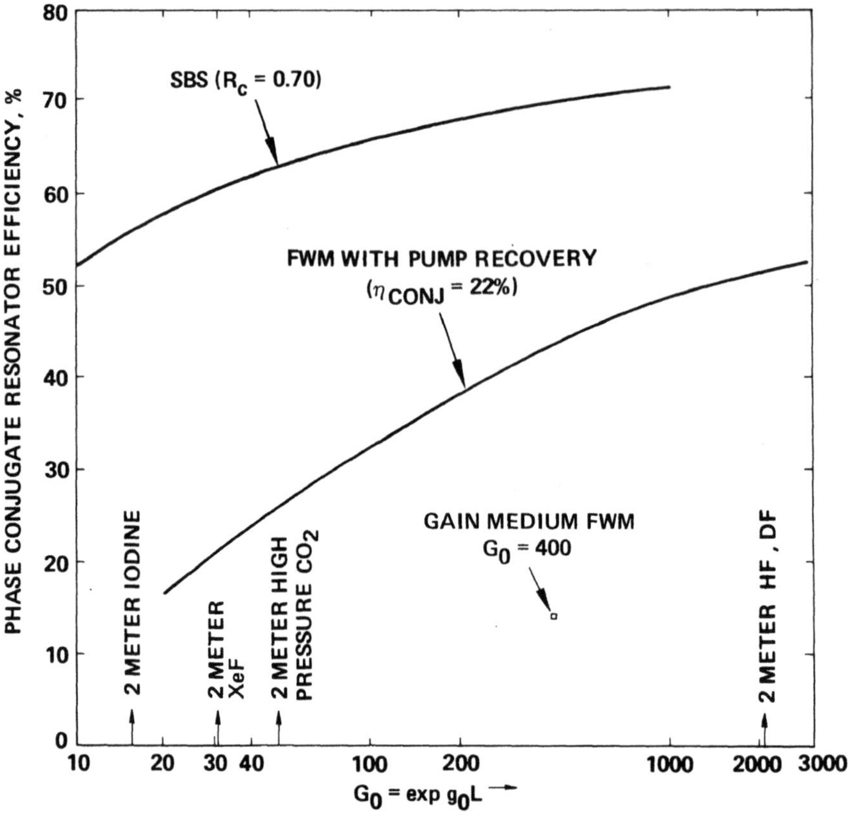

Fig. 10. System efficiency as a function of gain for SBS and FWM PCR's.

To summarize then, the key parameters to be optimized for a PCR with a FWM PCM is the operating efficiency, not the reflectivity. This efficiency is defined as the conjugated signal output power divided by the sum of the input powers required to product it. The FWM efficiency issues addressed in this paper are based on a 2-level saturable absorber model in which the nonlinearity is established by maintaining a substantial excited state population. The two factors which limit the efficiency are (1) the power required to maintain the excited state population and (2) the power transmitted through the PCM end of the resonator.

In order to achieve high efficiency in a two-level saturable absorber FWM one must operate far off resonance with a probe wave sufficiently strong that the pumps operate in a strongly depleted mode. There are few experiments and no published analysis which describes operation under these conditions. However, W.P. Brown of our laboratories[21] has recently analyzed and programmed computer solutions to mixers operating under such conditions. Using this code, the pump and probe levels have been perturbed to maximize the operating efficiency as defined above yielding a maximum efficiency of 18% without pump recovery and 22% with full pump recovery.

Figure 10 shows the results for the FWM PCR optimized as described above. Note that very much higher gains are required to realize the same system efficiency for FWM PCR's than resonators with SBS mirrors. Use of an active gain medium for a four-wave mixer avoids the external power requirement for the pumps since the pump fields are generated via a separate laser extraction of the inverted population of the mixer medium. However, one requires a very large energy investment in this inverted population, an investment that would otherwise be available to the basic laser itself. Similar calculations have been made for saturable gain FWM's (i.e. an inverted 2-level system) and one such result is shown in Figure 10. Thus, no improvement in system efficiency is achieved by using an active gain medium four-wave mixer. Because of the simplifying assumptions, these efficiencies are found to improve monotonically with increasing g_oL. In actual fact, unsaturated absorption coefficients or parasitic oscillations will provide an upper limit on the G_oL which can effectively be employed.

While PCR's can provide substantial improvement in the beam quality of a laser as shown this must be done with reasonable efficiency since little is accomplished by turning an efficient resonator with poor beam quality into a very inefficient PCR with good beam quality. A wave of poor beam quality can always be converted into a wave of good beam quality and lower power by using a spatial filter. Thus a reasonable figure of merit for comparing PCR's to conventional resonators is far-field power in a diffraction limited bucket. Given these problems a different way of comparing the performance is to use a so-called break-even beam quality factor.[22]

Take two resonators with equal stored power, one a conventional resonator, CR, and the other a PCR. The power delivered to the far field in a diffraction-limited spot depends upon two factors - the beam quality β and the resonator extraction efficiency, η. It is the tradeoff between these two quantities that we will use to illustrate under what conditions one would favor a PCR over a CR. We define break-even to occur when the peak far-field irradiance of the two resonators is equal. Hence $\beta^2_{Breakeven} = \eta_{CR}/\eta_{PCR}$ where we take $\beta = 1$ for the PCR (i.e. diffraction limited beam quality).

The curves in Figure 11 show $\beta^{CR}_{Breakeven}$ as a function of G_o for both DFWM and SBS PCR's. Conventional resonators whose β's lie above the curves are unfavorable compared to the PCR's. For example, suppose we had a conventional resonators with $G_o = 100$ and $\beta = 1.4$. The curves tell us that the SBS PCR would give a higher far-field irradiance than the conventional resonator, whereas the FWM PCR would not.

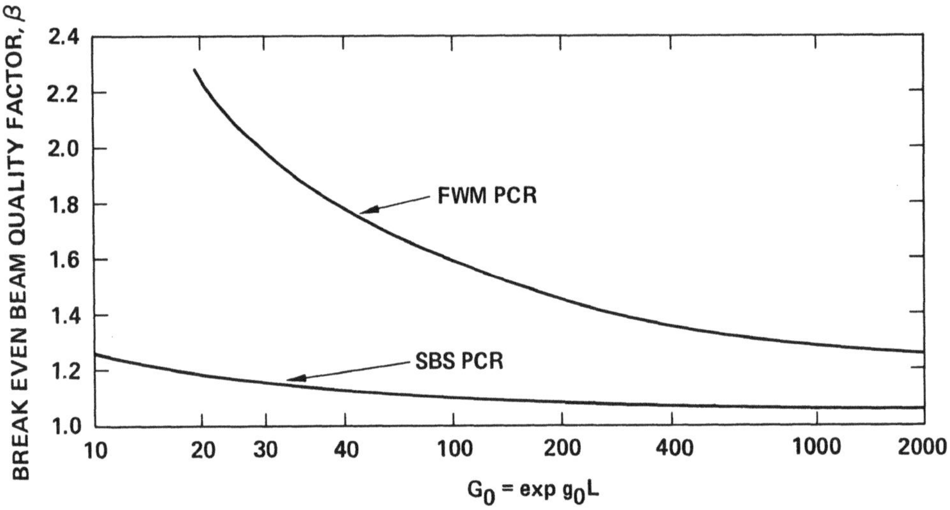

Fig. 11. Breakeven beam quality factor, β, as a function of gain for FWM and SBS PCR's.

CONCLUSION

The future of phase conjugate resonators as practical devices, especially for high power applications, lies in the issues of improved efficiency. Several questions arise in this regard that only further research can answer. First, to what extent can SBS conjugators be made to operate at higher reflectivities than have been demonstrated to date (i.e. 50 to 70%)? We know that SBS as a nonlinear process has been observed with reflectivities approaching unity. This has not occurred in the geometries used in conjugation experiments in which the input beams are aberrated as well. The reasons for the differences are not known. The challenge for SBS is to understand what are the fundamental limits on conjugate reflectivity; it is clear that the attractiveness of an SBS PCR hinges strongly on this issue, especially for lasers having low to moderate gains.

A similar question can be asked for FWM PCR's or, more generally, PCR's that employ real-time holography to do the conjugation. What are the fundamental factors that determine the efficiencies of these conjugators, especially under conditions of pump depletion for FWM? Remember that the FWM model used in the results presented here were for a two-level saturable absorber FWM, the only one that has been analyzed thoroughly. Other nonlinear mechanisms such as a purely dispersive interaction or the photorefractive effect may offer more promise for higher efficiency PCM's.

Finally, we wish to emphasize that all experimental demonstrations of PCR's to date have been carried out using low power laboratory lasers. Many of the practical issues surrounding high power operation of PCR's (e.g. thermal loading in the PCM, nonlinear propagation effects) and their impact on efficiency and conjugation fidelity have yet to be addressed. We believe that these issues could be best resolved through experimental research using high power lasers.

REFERENCES

1. B. I. Stepanov, E. V. Ivakin, A. S. Rubanov, Sov. Phys. Doklady 16, 46 (1971).
2. B. Ya.Zel'dovich, V. I. Popovichev, V. V. Ragul'skii, F. S. Faizullov, Sov. Phys. JETP 15, 109 (1972); O. Yu Nosach. V. I. Popovichev, V. V. Ragul'skii, F. S. Faizullov, Sov. Phys. JETP 16, 435 (1972).
3. V. Wang, C. R. Giuliano, Opt. Lett. 2, 4 (1978).
4. C. R. Giuliano, Physics Today 34, 27 (1981).
5. Optical Engineering 21, March-April (1982), Special Issue on Nonlinear Optical Phase Conjugation.
6. J. AuYeung, D. Fakete, D. M. Pepper, A. Yariv, IEEE J. Quantum Electron. QE-15, 1180 (1979).
7. P. A. Belanger, A. Hardy, A. E. Siegman, Appl. Opt. 19, 602 (1980).
8. J. M. Bel'dyugin, M. G.Galushkin, E. M. Zemskov, Sov. J. Quantum Electron. 9, 20 (1979).
9. J. F. Lam, W. P. Brown, Opt. Lett. 5, 61 (1980).
10. P. A. Belanger, Opt. Eng. 21, (1982), and references therein.
11. R. C. Lind, D. G. Steel, Opt. Lett. 6, 519 (1980).
12. J. Feinberg, R. W. Hellwarth, Opt. Lett. 5, 519 (1980); 6, 257 (1981).
13. H. Vanherzeele, J. L. VanEck, A. E. Siegman, Opt. Lett. 6, 467 (1981); Appl. Opt. 20, 524 (1981).
14. N. Basov, I. G. Zubarev, Appl. Phys. 20, 261 (1979).
15. R. L. Abrams, C. R. Giuliano, J. F. Lam, Opt. Lett. 6, 131 (1981).

16. J. O. White, M. Cronin-Golomb, B. Fischer, A. Yariv, Appl. Phys. Lett. 40, 450 (1982).
17. M. Cronin-Golomb, B. Fischer, J. Nilsen, J. O. White, A. Yariv, Appl. Phys. Lett. 41, 219 (1982).
18. I. M. Bel'dyugin, A. P. Pogibel'skii, Soviet J. Quantum Electron. 10, 1271 (1980).
19. A. Hardy, IEEE J. Quantum Electron. QE-17, 1581 (1981).
20. G. C. Valley, D. Fink, to be published in J. Opt. Soc. Am., April (1983).
21. W. P. Brown (Private communication).
22. T. J. Karr, H. J. Hoffman, Proceedings of SPIE, Volume 293 (August, 1981).

LASER MATERIAL PROCESSING

Gerd Herziger

Institut fuer Angewandte Physik
Technische Hochschule Darmstadt
6100 Darmstadt FRG

INTRODUCTION

Laser material processing investigated since 1963 [1,2] in numerous papers and reviewed by several authors [3,7] is still a strongly developing field. A typical arrangement for laser material processing is shown in Fig. 1. The parameters characterizing the processing properties of the laserbeam are wavelength λ, and intensity $I(r,t)$, the latter including mode structure, polarization and temporal modulation of the beam.

Compared to other techniques laser processing shows a special particularity: By placing a reflecting target in the focal plane of the system optical feedback is introduced resulting in stochastic fluctuations of the temporal and spatial behavior of the laserbeam and its mode properties, respectively [8]. Therefore, optical feedback must be carefully avoided to achieve proper processing results.

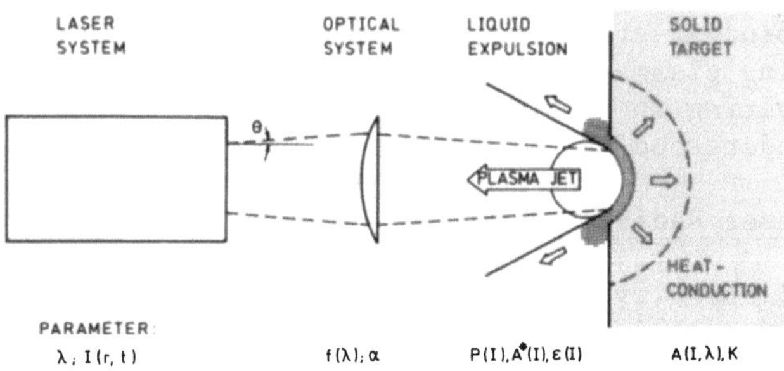

Fig.1 Scheme of laser material processing

Elimination of optical feedback is possible by inserting Faraday rotators between laser and target. In special cases a combination of linear and circular polarizers may be useful.

Laser radiation of power P focused by an optical system to the focal radius r_F yields an intensity I on the target surface: $I=P/\pi r_F^2$. The absorbed laser energy $W_A = \iint A_\lambda(I) I_\lambda(t) dFdt$ is used for machining purposes in different ways. For a survey of material processing it is useful to distinguish between two intensity regimes: Low-intensity processing ($I < 10^6$ W/cm^2) is preferrably the domain of heat treatment, such as annealing, alloying, transformation hardening, melt quenching, and welding. Analytic treatment is performed by the three-dimensional heat equation with processing parameters depending on temperature and materials. Compared to conventional heat treatment laser processing shows higher cooling and heating rates in combination with a smaller depth of the heat affected zone. As a result data of conventional heat treatment are in general not applicable to laser processing without modification.

An illustrating example which may stand for the various types of heat treatment is the transformation hardening by laser: The hardening temperature is well above that of induction hardening. There is also a characteristic difference in the dependence of hardening temperature on the carbon content of the probe. The probability of nucleation and the rapidity of diffusion of the carbon atoms are exceeding the conventional data. The result is that laser hardening differs from conventional hardening in some important details. This accordingly applies to the other types of laser heat treatment. The majority of open problems is in the field of material science, concerning the measurement of the optimum processing parameters. As an example for laser heat treatment Fig. 2 shows the hardness in the cut through a sliding surface. Theoretically predicted values of transformation hardening are compared with experimentally realized data[9].

High intensity laser radiation $|10^6 < I(W/cm^2) < 10^9|$ is preferably used for material removing processing, such as drilling and cutting, and - to a certain extend - for cladding, glazing and shock hardening. In this intensity region the material is mainly removed in gaseous form. Characteristic phenomena associated with the material transport are the appearance of a partially ionized plasma and an intensity-dependent absorption. The interaction of the laser beam with the expanding plasma as well as the resulting feedback to the motion of the melting in the working area are of high complexity and not yet fully understood.

Absorption of Laser Radiation

In the low intensity range ($I \ll I_c$) the absorptivity A_λ for a given laser wavelength λ is an optical constant of the targets material: $A = \text{const}$ ($I \ll I_c$). The fraction of the laser power absorbed by the target and the resulting machining efficiency are both proportional to A_λ. Unfortunately, the absorptivity of many of the materials that are interesting for laser machining is rather poor in the visible and infrared part of the spectrum. This especially refers to the highly reflecting metals, and most of the transparent hard materials such as sapphire and diamond, respectively.

Most of the materials show an increase of absorption for wavelengths $\lambda < 250$ nm due to interband-transitions or plasmon excitation respectively. Hence, short wavelength lasers - e.g. excimer lasers - can work materials of rather poor absorptivity in the visible and infrared at a convenient efficiency.

Insulators show increasing absorption in the far infrared due to photon-phonon interaction. The reflectivity R of conductors is proportional to their conductance σ: $R \alpha \sigma_{\sim}$. Therefore, all metals show a high reflectivity and because of $A \cong 1-R$ a correspondingly low absorption A, yielding a poor efficiency of laser machining from the visible to the far infrared region of laser radiation. Frequently layers of high absorptivity are deposited on the target surface in order to increase the absorption. Usually, these layers have a rather poor electrical and thermal conductivity. Most of these layers have an amorphous structure with low density and heat capacity, respectively.

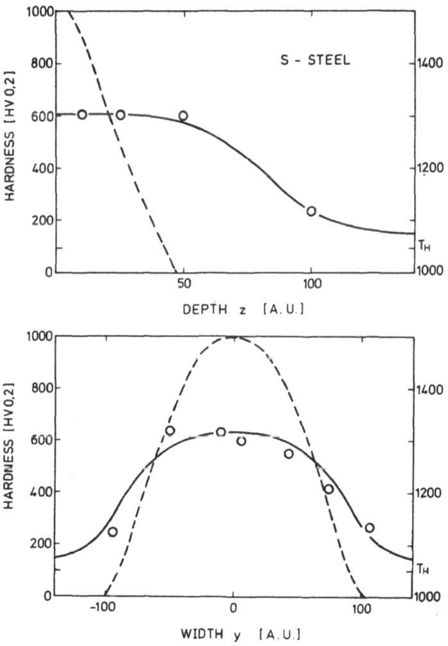

Fig.2 Temperature gradient (dashed) and hardening profile of laser-hardened sample (carbon steel 1%) by 3-dim. computer code 9 .

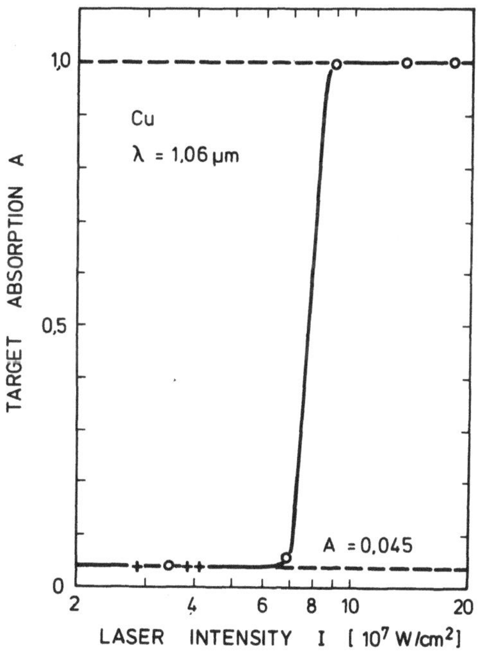

Fig.3 Absorption A versus incident laser intensity I for Cu as obtained from drilling experiments and corresponding reflection measurements

Consequently, the laser energy absorbed in the layer is transferred to the target with poor efficiency. Absorption layers are of practical importance mainly for transformation hardening with CO_2 lasers. Another method to increase the target's absorption can be expected from the excitation of polaritons [10]. Quantitative results of energy transfer are not yet available.

Intensity-dependent Absorption

As the laser intensity on the target's surface approaches a critical value I_c the absorption instantaneously rises to a higher level. The characteristic shape of A(I) as shown in Fig. 3 has been verified for a large number of solid targets independently of laser wavelength and target material. The critical intensity is in the range $10^7 < I_c(W/cm^2) < 10^9$. As a thumbs rule for most of the materials a typical value of $I_c \simeq 10^8$ W/cm^2 may be assumed. Thus, the intensity-dependent absorption is an interesting method to increase the absorptivity of highly reflecting metals and transparent dielectrics as well, thus increasing the efficiency of laser machining. By irradiating a target at critical intensity I_c during a time interval t_c a plasma of high luminance can be observed in a volume $\delta V = \pi r_F^2 L_c$ of the target surface. Simultaneously the reflection of metallic targets as well as transmission of transparent media drop to a minimum value. As the time of irradiation increases ($t > t_c$) the plasma expands towards the incident laser beam.

In experiments characteristical combinations of L_c and t_c are observed, which may be combined to a critical velocity $v_c = L_c/t_c$:

$$L_c \simeq 5.10^{-4}\text{cm}; \quad t_c \simeq 2{,}5\cdot 10^{-7}\text{s}; \quad v_c \simeq 2.10^3\text{cm/s}$$

Combining I_c with these data, a critical energy density ε_c for onset of anomalous absorption may be introduced:

$$\varepsilon_c = I_c t_c / L_c = I_c / v_c \, . \tag{1}$$

With $I_c = 10^8$ W/cm^2 there is $\varepsilon_c = 5\text{x}10^4$ /cm^3, which is close to the energy ε_v, necessary for evaporation of a target with density ρ, heat capacitance c, and melting- and vaporization heat E_m, E_v, respectively:

$$\varepsilon_v = \rho\{c(T-T_o) + E_M + E_v\} \quad \varepsilon_c \tag{2}$$

There is no model existing to explain the anomalous absorption. Most of the experimentally observed phenomena may be explained by assuming a laserinduced plasma in the target surface. The laser energy is transferred into the target by inverse bremsstrahlung of the plasma.

Extremely pure surfaces, as performed by cracking in uhv or thin layers evaporated in uhv, have critical intensities about $10^{10} < I_c(W/cm^2) < 10^{13}$. The breakdown mechanism is assumed to be the same as on normal surfaces: by generation of a laser-induced plasma [12]. The considerably higher values of the critical intensity are explained by different generation mechanisms of the first free electrons. Mechanically machined surfaces in normal atmosphere - even in the case of isolators - typically have a small density of quasi-free electrons that can start the avalanche. On the contrary, at an ultra clean undisturbed surface quasi-free electrons do not exist. They have to be generated by multiquantum effects or field emission, respectively [13].

Maximum Intensity for Laser Machining

The maximum laser intensity is limited by gas breakdown in front of the target. Gas breakdown induced by laser radiation first observed in 1963 [14] has since been investigated in numerous papers reviewed by several authors [15,20]. The gas breakdown appears either in the surrounding gas atmosphere or in the evaporated target material. Characteristic feature of the breakdown is the development of a plasma at the breakdown intensity I_B. First free electrons of density n_i are accelerated by inverse bremsstrahlung until their kinetic energy reaches the ionization limit. The number of carriers is increased by avalanche effect, yielding a plasma of critical electron density n_B at the time t_B:

$$n_B = \varepsilon_o kT_e/e^2 r_F^2 \tag{3}$$

ε_o : dielectric constant of the vacuum

e : electronic charge

kT_e : thermal energy of the electrons

n_B is the value of electron number density, where the Debye-length Λ_D equals the focal radius r_F.

$$t_B = (v_i - v_L)^{-1} \ln(n_B/n_i) \tag{4}$$

v_i, v_L: production and loss rate of electrons, respectively.
The absorptivity of the plasma at normal incidence is given by [22]:

$$\alpha = \frac{2\omega}{c_o}\left[\frac{1}{2}\left\{\sqrt{\left[1-\left(\frac{\omega_p}{\omega}\right)^2\frac{1}{1+(\nu_c/\omega)^2}\right]^2+\left[\frac{\nu_c}{\omega}\left(\frac{\omega_p}{\omega}\right)^2\frac{1}{1+(\nu_c/\omega)^2}\right]^2}\right.\right. - \left.\left.\left[1-\left(\frac{\omega_p}{\omega}\right)^2\frac{1}{1+(\nu_c/\omega)^2}\right]\right\}\right]^{1/2} \tag{5}$$

ω_p : plasma frequency

v_c : collision frequency

ω : laser frequency

c : speed of light (vacuum)

Starting from rate equations for electron number density, its momentum and energy, the breakdown intensity I_B [23] yields the classical rf-approximation of the inverse bremsstrahlung

$$I_B = \frac{\varepsilon_o m_e c_o}{e^2} \left[\left[\frac{2m_e}{M} E_o + \beta E_a \right] v_c^2 + \left[\frac{E_i + E_o}{t_B} \ln \left(\frac{n_B}{n_i} \right) \right] v_c^1 + \right.$$

$$+ \frac{kT_e}{m_e \Lambda^2} (E_i + E_o) + \left[\frac{2m_e}{M} E_o + \beta E_a \right] \omega^2 + \tag{6}$$

$$\left. + \left[\frac{E_i + E_o}{t_B} \omega^2 \ln \left(\frac{n_B}{n_i} \right) \right] v_c^{-1} + \frac{kT_e}{m_e \Lambda^2} (E_i + E_o)\, \omega^2 v_c^{-2} \right]$$

ε_o : dielectric constant of the vacuum

m, M : electron- and ion-mass

e : electronic charge

E_o : average kinetic energy of the electrons

$\beta = v_c'/v_c$: fraction of inelastic collisions

E_a : average excitation energy

v_c : collision frequency

Λ : diffusion length

$I_B(p)$ is shown in Fig. 4 for three characteristic laser wavelengths. The theoretical curves are in correspondence to the experimental values. Remarkable is the low value of I_B for the wavelength of the CO_2 laser. Hence, abrasive machining with CO_2 lasers is generally accompanied by a gas breakdown in front of the target. The values plotted in Fig. 4 are valid for single mode lasers, far away from bordering surfaces. Close to the surface of reflecting targets the breakdown intensity decreases to about 1/4 I_B due to superimposing of the field strengths.

A further decrease of I_B has been observed for multi-mode lasers. This is due to short time fluctuations of the intensity, yielding peak intensities of up to orders of magnitude above the average value. Onset of plasma generation has been observed at about 10^5.. 10^6 W/cm^2, when multimode CO_2 lasers were used for welding or transformation hardening.

Considering the breakdown phenomena observed at the surface of a solid target it appears to be useful to transfer the avalanche breakdown model to the bulk material. The main differences are the higher densities and collision rates, respectively. Hence, losses by diffusion and inelastic scattering ($\beta \to 0$) can be neglected. Taking into account that laser- and collision frequencies are of the same order of magnitude ($\omega/\pi \overset{\sim}{=} v_c$) eq.6 reduces to:

$$I_B \cong \frac{\varepsilon_0 m}{e^2} \left\{ \frac{2me}{M} E_0 \omega^2 + \frac{E_i + E_0}{t_B} \pi\omega \ln(n_B/n_i) \right\} \to I_c \tag{7}$$

The evaluation of typical machining materials and laser wavelengths respectively, yields breakdown intensities of $10^7 < I_B(\mathrm{W/cm^2}) < 10^9$ corresponding to the measured data within the order of magnitude.

Velocity of Material Removing Processing

Focusing a laserbeam of sufficiently high intensity $I \sim I_c$ on the surface of a bulk material the boundary between the liquid and the solid phase is moving with speed v into the material. v depending generally on the intensity I of the laserbeam, is introduced as the velocity of material removing processing. Its obvious meaning is that of a drilling velocity, or by introducing

$$v_s = v_p (r_F/d) \tag{8}$$

that of a cutting velocity in a material of thickness d. Fig. 5 shows the characteristic slope of the processing velocity v(I) as it is observed for the materials investigated. Conspicuous is a sharp step of v(I) at the critical intensity I_c. To simplify the discussion of v(I) it is useful to divide Fig. 5 into 4 regimes.

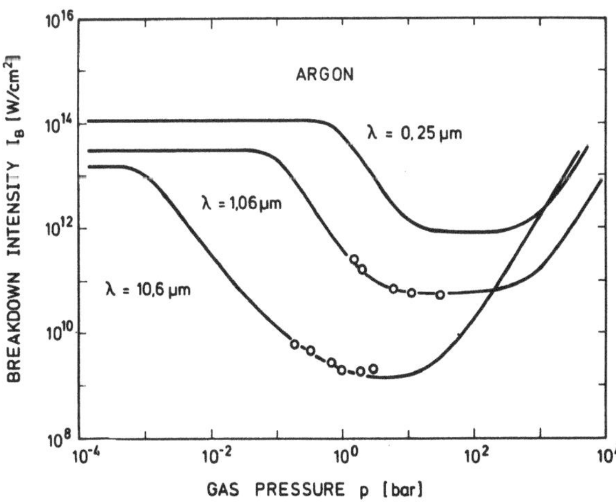

Fig.4 Breakdown intensity I versus gas pressure in Argon. Comparison of theory (eq.9) and experiment (Nd-YAG laser[23] CO_2 laser[24], both single mode)

I. Threshold regime: Material processing starts at a material-specific value of the intensity. At threshold the absorbed laser power is compensated by the losses of heat conduction. For different materials the threshold is shifted to higher intensities as heat conduction and evaporation heat are increasing. The threshold regime is distinguished by evaporation from the surface and a processing velocity v increasing proportional to the intensity: $v \propto AI$ where the slope of v is dominated by heat conduction.

II. Liquid expulsion regime: Due to the increasing energy density in the molten zone of depth d_m, the material is removed explosively in a mixture of vapour and liquid particles. As the liquid particles need no evaporation energy the processing speed is further increasing.

III. Regime of abnormal absorption: This regime is distinguished by a stepwise increase of the absorption, an accordingly increasing power input and evaporation rate, respectively. The depth of the melt takes its minimum value, the liquid expulsion vanishes and the processing velocity reaches its upper theoretical limit:

$$\hat{v}_P = c_m I \quad ; \quad c_m = 1/\varepsilon_v \tag{9}$$

Machining experiments performed with excimer- and YAG-lasers have realized this value, whereas with CO_2 lasers only 50% of the maximum speed could be achieved. The reason for that is the beginning of target shielding by the partially ionized plasma cloud in front of the surface.

IV. Shielding regime: Due to the increasing intensity the absorption of the laserbeam by the plasma in front of the target is dominating. The target is heated indirectly by radiation and heat transfer from the plasma [28]. The geometry of the working area is not further controlled by the diameter of the laserbeam but by the expansion of the plasma, which may exceed the focal spot considerably.

With CO_2 lasers presently available the critical intensity is not achievable in cw operation. Hence, to increase the power density

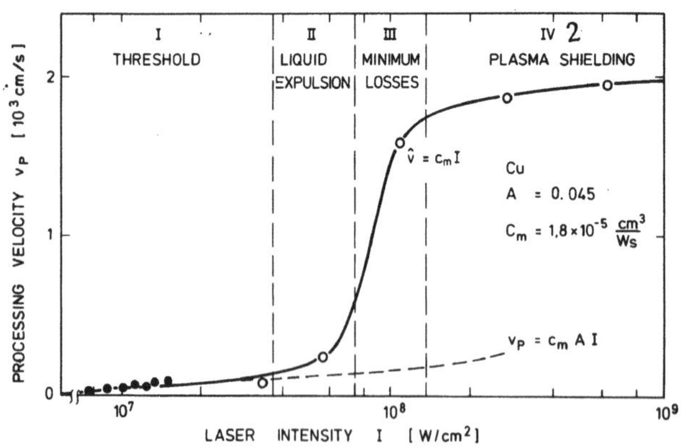

Fig.5 Processing velocity v_P in Cu-targets versus incident laser intensity I. Data from Nd-YAG laser drilling experiments[26,27]

at the target surface the addition of a working gas is commonly used. The power density is increased by the resulting reaction heat and the kinetic energy of the flowing medium which enhances the liquid expulsion out of the molten zone. The increase of power density depends on the material, the working gas, the resulting chemical reaction and the hydrodynamical parameters of the flowing working gas. The cutting speed of stainless steel worked by CO_2 laser can be enhanced by a factor of 6 by using O_2 as a working gas. The exothermal reactions

$$Fe+1/2O_2 \rightarrow FeO-3.43kJ/g; \quad 3FeO+1/2O_2 \rightarrow Fe_3O_4-1,29kJ/g$$

and the enforced liquid expulsion by the pressure of the working gas contribute in the same order of magnitude [29]. Using a working gas that supports laser cutting, the depth δ of the roughness of the machined surface increases with the thickness of the material. The empirically measured relation $\delta/d \simeq 10^{-2}$ is obviously due to a hydrodynamical instability.

The polarization of the laserbeam generally affects the laser processing, in particular the laser cutting of metals. Due to the dynamics of the cutting process the cutting edge generally is not perpendicular to the surface but exhibits an angle increasing with the depth of the cut. With increasing deviation from the normal incidence the difference between the reflectivity of the parallel and normal polarization is enhanced. Hence, cutting speed and -width are different for different modes of polarization [30]. The effect of the polarization on the cutting process is intensified by the optical feedback.

Material Transport

The laser induced plasma in the target surface expands towards the incident laserbeam, thus acting as a transport vehicle for the

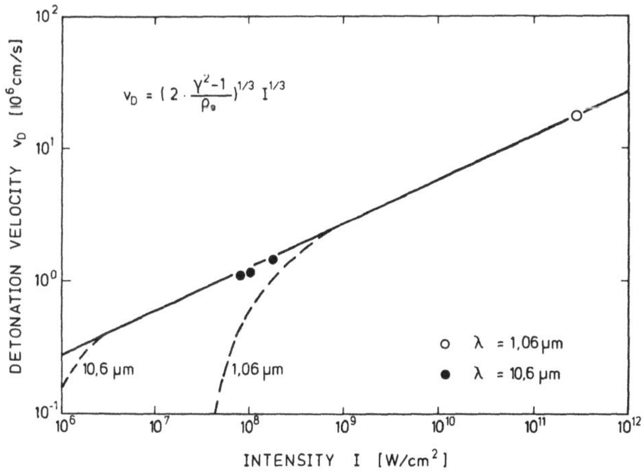

Fig.6 Velocity v_D of laser induced detonation wave versus intensity I, as obtained from eq.11, compared to experimental data: ○ Nd-YAG laser[23], ● CO_2 laser[24]. Broken lines are multi-mode experiments.

material to be removed. CO_2 laser processing at I_c supplies such a high energy density to the surface region, that the plasma expansion can be approximated by a laser supported detonation wave. Using the ideal gas approximation and neglecting losses by ionization, heat conduction and friction, the velocity of an one-dimensional shockfront reads [31]:

$$v_D = \{2(\gamma^2-1)I_c/\rho\}^{1/3} \tag{10}$$

Eq.10 is corresponding to the experimentally measured values of v_D |Fig. 6|. In experiments with YAG lasers the intensity has to be increased up to 10^{10} W/cm^2 until the measured values are in agreement with eq.10 indicating the wavelength dependence of the laser plasma interaction. At a larger distance the one-dimensional LSD wave is changing over into a spherical expansion travelling with sound speed.

Multimode lasers show an irregularly fluctuating intensity in time and space and a corresponding fluctuation in the plasma heating and shielding. The buildup of plasma bowls in front of the target is a frequently observed phenomenon. Fig. 7 shows schematically the development of a plasma structure which is typical for CO_2 lasers with unintentional selflocking: Pulse 1 is just reaching the breakdown intensity I_B inducing a plasma with energy density ε_B. An LSD-wave travels at speed v_D in the undisturbed gas producing a trace of ionized molecules. Succeeding pulses of lower intensity are generating successive new plasma discs at the front of the LSD-wave. Pulse 4 is below the LSD-wave threshold and no plasma breakdown is observed. The recoil of the expanding plasma results in a pressure p on the target[31]

$$p = \{p_o/(1+\gamma)\}^{1/2} I^{2/3} \tag{11}$$

Typical values for CO_2 lasers at critical intensity are $p \simeq 10^2$ bar and can be increased to values of as much as 10^6 bar by increasing the laser intensity. The pressure load by short wavelength laser at I_c is sufficiently below

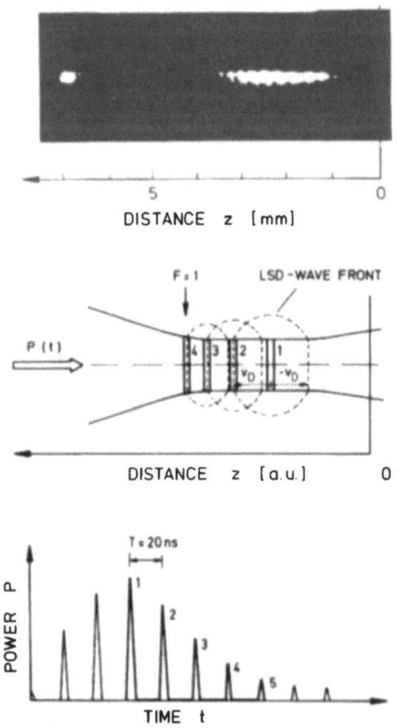

Fig.7 Plasma generation in front of a solid target located at z=0
Top: time-integrated side-on photograph of the laser-induced plasma [24]
Center: schematic graph of plasma discs generated by a solf-locked pulse train.
Bottom: laser pulse train.

the CO_2 values due to the wavelength dependence of laser plasma interaction.

The pressure acts on the melt at the surface, accelerating the liquid particles in radial direction so that they are leaving the melt in form of a cone (Fig. 1) the aperture of which becomes smaller with increasing penetration depth. The expelled liquid particles are deposited in the surrounding of the working area. By proper shaping of melt depth, intensity, and pressure these unwanted depositions can be largely suppressed. Multimode operation exhibits an irregular pressure distribution, fluctuating in time, which often results in chaotic depositions at the edges of the working zone.

Efficiency and Quality of Machining

The efficiency of material removing processing is defined as the ratio of the energy E_v, which is necessary to evaporate the mass ρdV, to the incident laser energy E_L:

$$\eta = \varepsilon_v \delta V/(\pi r_F^2 \, I\delta t) = E_v/E_L \tag{12}$$

$\eta(I)$ is shown in Fig. 8. The maximum efficiency is nearly independent from material and wavelength close to the critical intensity. With excimer- and Nd-lasers values of $\eta = 1$ are approximately reached. Maximum efficiency for CO_2 laser machining is about $\eta \simeq 0.5$ due to the shielding of the target by evaporated material. For $I < I_c$ the efficiency drops due to reflexion losses and heat con-

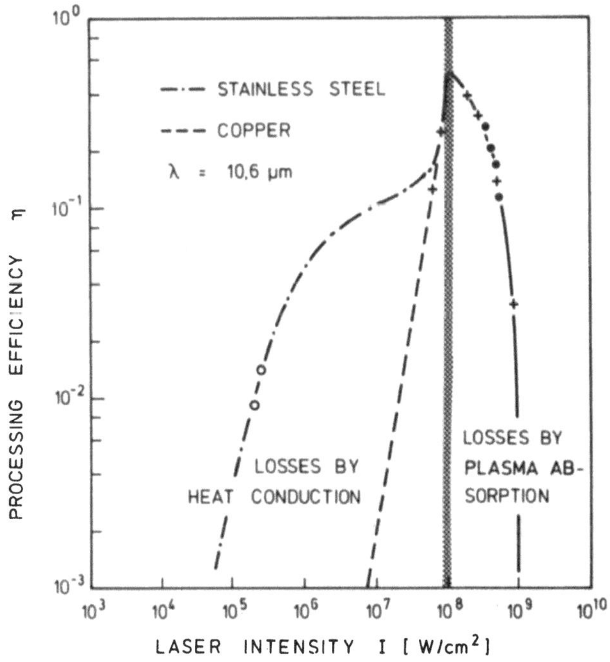

Fig.8 Processing efficiency η versus intensity I. Data from drilling experiments with CO_2 lasers[24]

duction. The specific material properties make significant contributions to the shape of the curve. The decrease for $I>I_c$ is due to increased shielding of the target by the plasma. The energy transfer from laser radiation to the target goes via radiation and heat contact to the plasma [28]. Important quality marks, such as the deviation from the ideal geometry or structural changes by liquid depositions in the working area, reach optimum values for $I = I_c$. Depositions of liquid particles at the inner walls or at the edges of the working area largely diminish at $I = I_c$, as the material is almost completely expelled in the plasma phase, so that the kinetic energy of the particles is large enough to prevent attachment and resolidification at the inner walls of the working zone.

In addition to the shielding of the plasma we have to consider its lensing effect. Assuming a spherical plasma cloud of refractive index n with a radius expanding according to $R(t)\ v_D t$, its focal length may be approached by:

$$f(t) = n(t)R(t)/\{n(t)-n_o\};\ n = n_o\{1-(w_p/w)^2\}^{1/2} \tag{13}$$

The measured focal lengths are in the order of a few mm or a fraction of it. Hence, plasma lensing mainly disturbs microdrilling.

Conclusion

The physical background of laser processing in the low intensity range, e.g. the various types of heat treatment, hardening, and welding is fairly well understood today. Starting point of any analysis is the three-dimensional heat equation with temperature dependent material parameters and material dependent transformation temperatures. Today's efforts are focused to find the optimum processing parameters for the various kinds of materials and processing methods, respectively.

In the high intensity range ($I > 10^6$ W/cm^2) the physical background of laser processing is worked out only partially. In particular, the interaction between laserbeam and laserinduced plasma turns out to be of considerable influence on quality and efficiency of the machining. The efficiency of laser processing reaches its maximum, the quality its optimum close to the critical intensity I_c. Laser processing at intensities close to I_c turns out to be an acceptable compromise between the typical defects of low intensity processing ($I \ll I_c$) such as liquid depositions, heat loading around the working area, low efficiency due to poor absorption, and losses by heat conduction, respectively, and the characteristical disturbances of high intensity operation ($I \lesssim I_c$), such as plasma lensing, pressure load of the working area, and a decreasing efficiency due to shielding of the target by the expanding plasma. It should be noted that the experimental results and the assignment to corresponding intensities are valid only in the case of single mode lasers. In multimode operation the threshold of gas breakdown as well as the threshold of surface breakdown is typically lowered by orders of magnitude. Laser processing by multimode lasers is adequate to processing by single mode lasers if efficiency, reproducibility and quality of the processing are not effected by the local and temporal intensity fluctuations. Hence, multimode operation is often permissible for heat treatment and welding - with the exception of deep welding - as long as no gas breakdown is occurring.

REFERENCES

1 J.F. Ready, Appl. Phys. Lett. 3, 11 (1963)
2 W.I. Linlor, Appl. Phys. Lett. 3, 210 (1963)
3 C. De Michelis IEEE QE-6, 630 (1970)
4 J.F. Ready, Effects of High Power Laser Radiation, AP (1971)
J.F. Ready, Industrial Application of Lasers, AP (1978)
5 F.T. Arrechi, E.O. Schulz-Dubois Laser Handbook, North Holland Publ. (1972)
6 W.W. Duley, CO_2 Lasers, Effects and Applications, AP (1976)
7 M. Newstein, N. Solimene, IEEE QE-17, 2085 (1981)
8 H. Eichler, G. Herziger, Z. Angew. Physik 23, 297 (1967)
9 K. Wissenbach, Verhandl. DPG (VI) 17, K 29 (1982)
10 F. Keilmann, Y.H. Bai, Verhandl. DPG (VI) 17, K 32 (1982)
11 W. Peschko, Thesis, TH Darmstadt (1981)
12 N. Bloembergen, IEEE J-QE-10, 375 (1974)
13 Tozer, B.A.; Phys. Rev. 137, A 1665 (1965)
Keldysh, L.V.; Sov. Phys. JETP 20, 1307 (1965)
14 R.G. Meyerand, A.F. Haught Phys. Rev. Lett. 11, 401 (1963)
15 C. De Michelis, IEEE QE-5, 188 (1969)
16 I.P. Shkarofsky, RCA Review 35, 48 (1974)
17 C.G. Morgan, Rep. Prog. Phys. 38, 621 (1975)
18 G. Bekefi, Principles of Laser Plasmas J. Wiley and Sons (1976)
19 J.M. Meek, J. D. Craggs, Electrical Breakdown of Gases, Wiley (1978)
20 L. Friedland, Phys. Rev. A 12, 2024 (1975)
21 C.H. Chan et al., J. Appl. Phys. 44, 1179 (1973)
22 P. Mulser et al., Physics Reports 6C,3, 189 (1973)
23 G. Brumme, Thesis, TH Darmstadt (1977)
24 H.P. Schäfer, Diplom Thesis, TH Darmstadt (1980)
25 M.C. Gover, Optics Comm. 36, 43 (1981)
26 V. Allmen, Thesis, University of Berne (1975)
27 P. Reinhardt, Diplom Thesis, TH Darmstadt (1980)
28 A.N. Pirri et al. A IAA J. 16, 1296 (1977)
29 K. Wissenbach, Verhandl. DPG (VI) 16, KP 19 (1981)
30 F.O. Olsen, DVS Berichte 63, 197, (1980)
31 Yu.P. Raizer, JETP 21, 1009 (1965)

THE OXYGEN-IODINE CHEMICAL LASER

William E. McDermott

Advanced Radiation Technology Office
Air Force Weapons Laboratory
Kirtland Air Force Base, New Mexico 87117

I will discuss four general areas of the oxygen-iodine chemical laser. First, the basic physics and chemistry of the device; second, kinetic measurements of rate constants in both the free iodine atom regime, and third, the region in which I_2 dissociation is occurring. These kinetic experiments will then be compared with real laser systems. Finally, I will describe the diagnostics being used in the oxygen-iodine system.

The oxygen-iodine chemical laser operates due to the fast, nearly resonant transfer between the first excited state of oxygen and the spin orbit components of the iodine ground state. This reaction, which comes to the equilibrium in the device, is slightly downhill which accounts for its greater than unity equilibrium constant. Because of this equilibrium constant and the degeneracies in the iodine atom, if an excited oxygen to total oxygen of greater than 1/6 is produced, then an inversion exists on the iodine atom.

The key questions which must be answered to build an efficient oxygen-iodine chemical laser are how to produce large fractions of excited oxygen, and how to produce iodine atoms.

The production of $^1\Delta$ oxygen occurs through the heterogeneous gas-liquid reaction between chlorine and basic hydrogen peroxide. This reaction has a unit quantum efficiency and the long lifetime and stable character of excited oxygen allows separation from the liquid phase. Yields of greater than 82% excited oxygen have been observed.

The details of the chemistry of excited oxygen production are not well known. However, it is suspected to involve the direct reaction of chlorine in solution with the hydroperoxy anion.

I will now describe the kinetics in the free iodine atom regime; i.e., that regime in which all the I_2 that has been put into the flow has been dissociated.

In the absence of contaminants from the generator, which are largely chlorine and water, and when the I_2 has been totally dissociated, the performance of the laser depends only on four reactions.

$$O_2\ (^1\Delta) + I \xrightarrow{k_f} O_2\ (^3\Sigma) + I^*$$

$$\xrightarrow{\beta_f\, k_f} O_2\ (^3\Sigma) + I$$

$$I^* + O_2\ (^3\Sigma) \xrightarrow{k_r} I + O_2\ (^1\Delta)$$

$$\xrightarrow{\beta_r\, k_r} I + O_2\ (^3\Sigma)$$

$$O_2\ (^1\Delta) + I^* \xrightarrow{k_3} O_2\ (^1\Sigma) + I$$

$$O_2\ (^1\Delta) + O_2\ (^1\Delta) \xrightarrow{k_2} O_2\ (^1\Sigma) + O_2\ (^3\Sigma)$$

The first two reactions are the forward and reverse transfer rates. Their nonresident rates are also indicated here. The third reaction is the feedback reaction between $^1\Delta$ and the I* which produces a $^1\Sigma$, and finally the pooling reaction between two $^1\Delta$'s. All of these reactions have been redetermined and found to be independent of temperature and to be near the previously accepted values.

For example, the energy pooling rate between two $^1\Delta$'s has been measured at several places, including the McDonnell Douglas Research Laboratory and The Aerospace Corporation, at a number of temperatures. The rate is essentially independent of temperature and equal to 2×10^{17} cm^3/mo/sec.

It was originally proposed by Derwent and Thrush that the I atoms were produced in the $^1\Delta$ oxygen flows by the reaction of I_2 with $^1\Sigma$ oxygen which had been produced by pooling processes. However, Houston, et al, have recently remeasured this rate and it is much slower than thought by Derwent and Thrush and is therefore too slow to account for the observed rate of dissociation of I_2 in excited oxygen flows. Furthermore, both in our own laboratory and at The Aerospace Corporation, we found that I_2 dissociation occurs in the presence of significant amounts of water which is a strong deactivant of O_2 $^1\Sigma$.

It should be noted that originally Arnold, et al[1], proposed that the I_2 was dissociated by the $^1\Delta$ oxygen. The $^1\Delta$ oxygen producecd an excited state of I_2 and Arnold suggested that this excited state of I_2 could be the A' $^3\Pi$ 2u state.

Recent observations at Aerospace, however, show the following results:

(1) "Initiator" Effects Important

(2) Dissociation Profiles Exhibit a Substantial Induction Time

(3) Small Amounts of H_2O Strongly Affect Rate
Large Amounts have a Proportionally Smaller Effect

(4) Induction Time $$(t_{\frac{1}{2}})^{-1} = \frac{k\ \{^1\Delta\}^2\ \ \{I_2\}^{0.7}}{\{^3\Sigma\}}$$

Implies a "Chain" Mechanism Involving I* as the Chain Carrier (Lilenfeld)

It should be noted that Result 4 implies a chain mechanism in which the I* is the chain carrier. Thus the dissociation is autocatalytic in iodine.

The dissociation process is quite complex and a number of the potential processes which may occur in the I_2 dissociation region are:

$$O_2\ (^1\Delta) + O_2\ (^1\Delta) \xrightarrow{k_2} O_2\ (^1\Sigma) + O_2\ (^3\Sigma)$$

$$O_2\ (^1\Delta) + I^* \xrightarrow{k_3} O_2\ (^1\Sigma) + I$$

$$O_2\ (^1\Delta) + I \underset{k_{-4}}{\overset{k_4}{\rightleftharpoons}} O_2\ (^3\Sigma) + I^*$$

$$O_2\ (^1\Delta) + I_2(X) \underset{k_{-5}}{\overset{k_5}{\rightleftharpoons}} I_2^\dagger + O_2\ (^3\Sigma)$$

$$O_2\ (^1\Delta) + I_2^\dagger \xrightarrow{k_6} 2I + O_2\ (^3\Sigma)$$

$$I^* + I_2\ (X) \underset{k_{-7}}{\overset{k_7}{\rightleftharpoons}} I_2^\dagger + I$$

$$I_2^\dagger + I^* \xrightarrow{k_8} 3I$$

$$I_2^\dagger + M \xrightarrow{R_{I_2^\dagger}} I_2 \quad (X) \quad + M$$

There are three possibilities for the intermediate state. First, an electronic state of I_2, a second, vibrationally-excited ground state I_2, and third, a complex between oxygen and iodine. There are arguments against each one of these potential intermediates.

To summarize the recent work on the I_2 dissociation process, the rate has been found to be proportional to $(^1\Delta)^2$ and also proportional to the amount of I_2 which is being dissociated. The process is quite efficient, consuming between 3 and 5 $^1\Delta$'s/I_2 molecule dissociated. The dissociation rate is faster than we have previously esimated.

I will now discuss how these kinetic measurements actually relate to real systems. At the Air Force Weapons Laboratory, the line broadening of excited iodine by ground state oxygen has been measured and found to be 7.3 $\pm$ 2.2 MH_2/Torr. Despite the resonant interaction between I* and O_2 ground state, the broadening is exactly what one would predict from a collision theory analysis.

The saturation behavior of the laser is as predicted. The efficiency of about 40% obtained in the experiments at TRW matches the theoretical curve.

An extensive amount of diagnostics has been developed for this system. These diagnostics are summarized as follows.

O_2 $(^1\Sigma)$	Emission 762 nm
H_2O	Diode Laser Absorption 2600 nm
	White Light Absorption 1400 nm
	Indirect $\sim \frac{\{^1\Sigma\}}{\{^1\Delta\}^2}$ 2600 nm
Cl_2	Laser Absorption (He-Cd)
	Line Absorption (Hg-Xe ARC) 335nm

As you see, every species of importance in the laser device can be measured in at least one way, and in some cases, in several different ways.

In conclusion, the O_2/I laser technology is approaching maturity (with the exception of the I_2 dissociation mechanism) where scaling and engineering of large devices may be contemplated.

[1]Arnold, Finlayson & Ogvyzlo, J. Chemical Physics, 44;2529 (1966)

AIR FORCE LASER RESEARCH

Michael Salkind

Director of Aerospace Sciences
Air Force Office of Scientific Research
Washington, D.C. 20332 USA

INTRODUCTION

The Air Force has played a major role in the development of laser technology from its inception with the invention of the maser (Ref 1). Figure 1 is a diagram of the principal historic events in the evolution of the laser. The first operational laser was a low power solid ruby laser. This led to glass, semiconductor, and dye lasers which find applications in metrology, communications, diagnostics, and guidance. High power lasers evolved from the gas laser, the most important of which are the electric discharge laser (EDL) and the gas dynamic laser (GDL). The invention of the chemically pumped laser (CL) led to a new branch of high power laser. More recent developments which offer high power potential are the free electron laser (FEL) and the excimer laser. This paper summarizes these developments, along with pertinent basic research contributions and indicates current and future directions of Air Force laser research.

HISTORICAL DEVELOPMENT

Gordon, Zeiger, and Townes (Ref 1) employed a nonuniform electrostatic field (Fig 2) to focus upper state ammonia molecules into a resonant cavity and defocus lower state molecules away from the cavity thus creating the basic population inversion necessary for stimulated emission. Independently, Basov and Prokhorov (Ref 1a, b) proposed a similar concept. The extension of this concept to infrared and optical wavelengths was proposed by Schawlow and Townes (Ref 2). They proposed the use of parallel plates to create a resonant cavity and optically pump solid state devices. Prokhorov

(Ref 3) also suggested the use of parallel plates to create a cavity for optical wavelengths. Although Gould did not publish his research, the patent courts have recently recognized (Ref 4) his contribution to the invention of the laser.

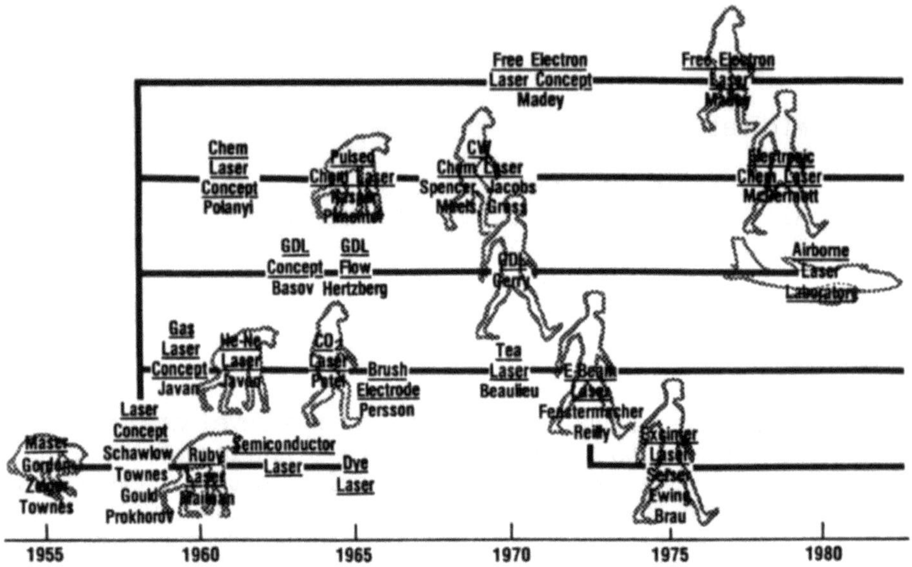

Figure 1

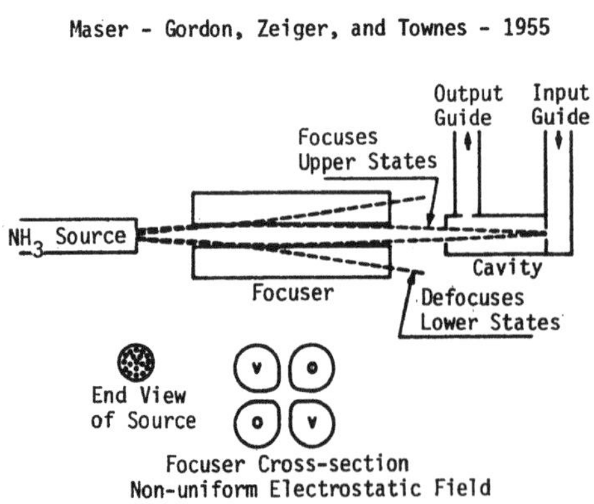

Figure 2

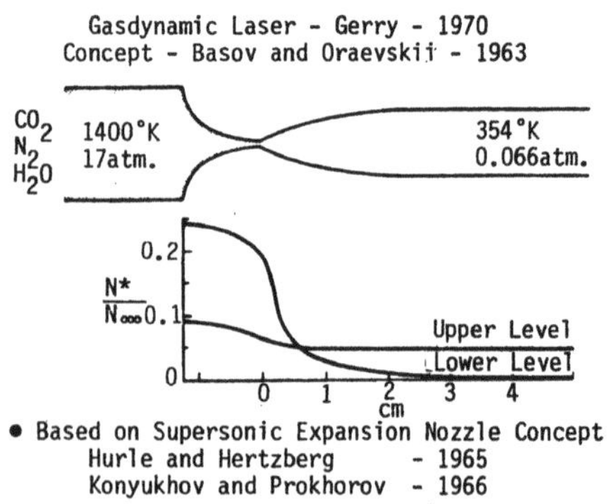

Figure 3

The first demonstration of laser action is attributed to Maiman (Ref 5) who pumped a ruby rod with a flash lamp and generated coherent radiation at 6943 angstroms. Javan (Ref 6) proposed the creation of a population inversion in the gas phase and demonstrated the concept with a He-Ne laser (Ref 7). A significant contribution was made by Patel (Ref 8,9) who identified the transfer of vibrational energy from N_2 to CO_2 and resulting infrared lasing at 10.6 microns using an electric discharge for pumping. This transition was a critical step in the evolution of high power lasers. Itwas exploited with the development of the gasdynamic laser (Fig 3) by Gerry (Ref 13). A critical contribution which led to the development of the gas dynamic laser was made by Basov and Oraevskii (Ref 10) who suggested that differential relaxation rates of species in a heated gas mixture could result in a population inversion. Hurle and Hertzberg (Ref 11) and Konyukhov and Prokhorov (Ref 12) suggested that this could be accomplished using a supersonic expansion nozzle. A gasdynamic laser system is currently being evaluated in a flying test-bed called the Airborne Laser Laboratory (Fig 4). This system will provide insight into the problems of integrating a relatively high power laser with a pointing and tracking device and maintaining a beam on a selected aimpoint. The gasdynamic laser stimulated many basic research challenges in fluid mechanics including understanding the shock structures, mixing and boundary layer behavior, and nozzles and diffuser configurations (Ref 15, 41). A significant contribution was the aerodynamic window (Ref 14) seen in Fig 5. This method of separating the lower pressure laser cavity from the outside precludes the absorption and distortion problems of solid windows.

Figure 4

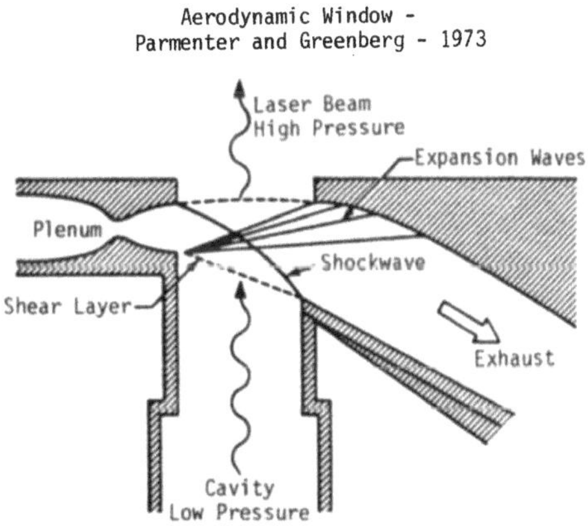

Figure 5

The possibility of chemical pumping was first proposed by Polanyi (Ref 16) for HCl. This was followed by the first demonstration of an HCl pulsed chemical laser (Fig 6) by Kaspar and Pimentel (Ref 17). They used a flash lamp to initiate the action by photodissociation of Chlorine. The amplification, due to chemical pumping, is seen in Fig 6. The first CW chemical laser (Fig 7) was based on the continuous production of fluorine atoms by dissociation of an SF_6, N_2 mixture in an arc heater (Ref 18, 19). CW chemical lasing was also obtained by reacting supersonic streams of HCl with F_2 in a shock tunnel (Ref 20). Critical basic research issues supporting the development of the chemical laser included an understanding of the distribution (branching ratios) of the excited species (Ref 21) and minimizing losses in the collision processes (Ref 22). In addition, fluid mechanics issues similar to those found in gasdynamic lasers are applicable to chemical lasers.

Pulsed Chemical Laser - Kasper and Pimentel - 1965

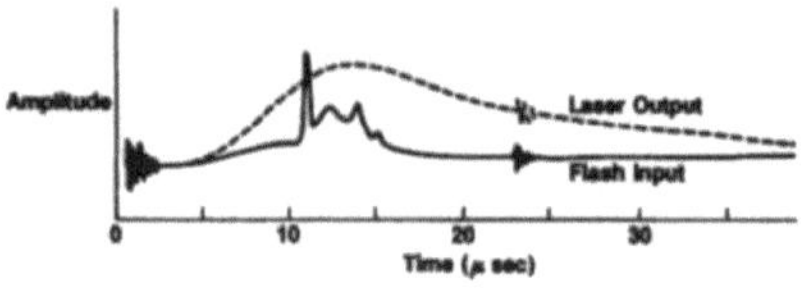

$H + Cl_2 \rightarrow HCl^* + Cl$

Figure 6

CW Chemical Laser - Spencer, Jacobs, Mirels, Gross - 1969

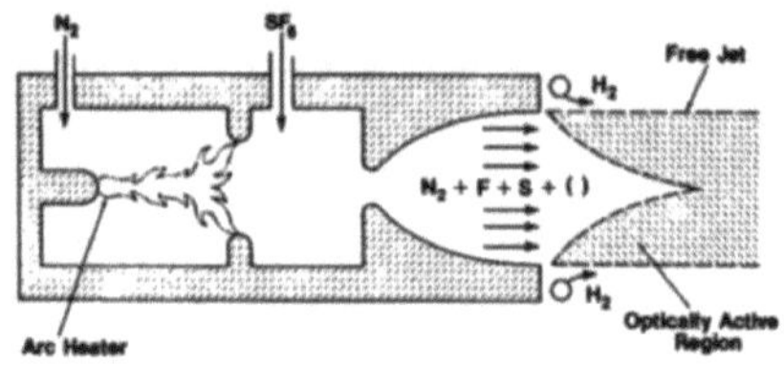

$H_2 + F \rightarrow HF^* + H$

Figure 7

An important recent development is the electronic transition CW chemical laser (Ref 23). As seen in Fig 8, the oxygen singlet delta state created chemically is used to pump iodine atoms to create the population inversion. The transition of iodine to the lower state emits at 1.3 microns. The oxygen singlet delta is also used to generate higher energy oxygen singlet sigma which dissociates the iodine molecule.

During the period when gasdynamic and chemical lasers were undergoing accelerated development, the electric discharge laser development was slowed by the difficulty of obtaining and sustaining a uniform well behaved plasma throughout a substantial volume. Key research contributions were made by Persson (Ref 24) who demonstrated a uniform, high electron density, low electron temperature plasma over a substantial volume by using a brush electrode, and Beaulieu (Ref 25) who demonstrated the transversely excited CO_2 laser which could be operated at pressures up to atmospheric. The attainment of high power EDL's required the production of a low electron temperature, uniform large volume plasma by having a separate pre-ionization discharge. This was accomplished by the electron beam controlled discharge (Ref 26, 27) and the pulser-sustainer discharge (Ref 28, 29).

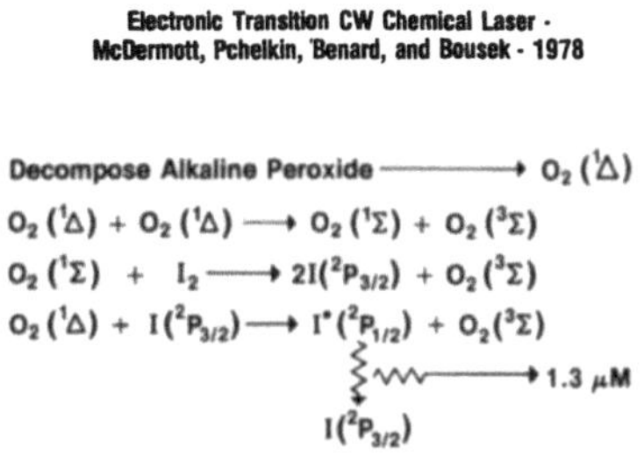

Figure 8

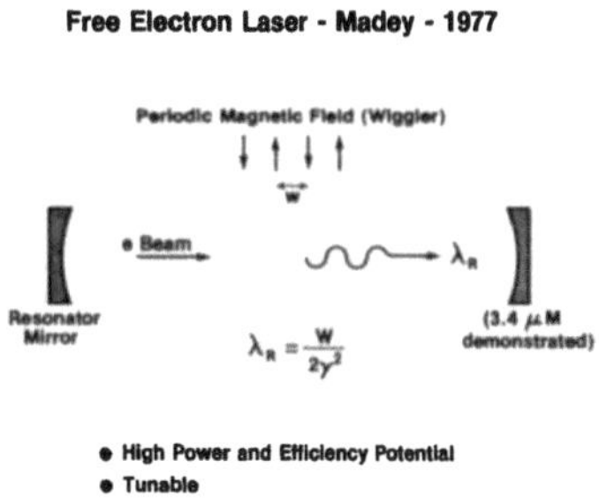

Figure 9

Excimer lasers rely on the creation of diatomic molecules which are strongly bound in the upper state but are weakly bound or preferably repulsive in the lower state. The relatively short life of the lower state compared with that of the upper state produces the population inversion. Setser and co-workers (Ref 30, 31) generated excimers of noble gas monohalides by electrical discharge and suggested the possibility of lasers based on this concept. Ewing and Brau (Ref 32, 33) demonstrated lasing for KrF,XeCl, and XeF using an electron beam discharge and projected reasonable efficiencies.

A recent development is the free electron laser developed by Madey (Ref 34, 35, 36). As seen in Fig 9, electrons accelerated to relativistic velocities are injected into a resonant cavity and interact with a periodic magnetic (wiggler) field. The stimulated emission is coherent at a wavelength dependent upon the wavelength of the magnetic field and the electron velocity. This allows the device to be tunable to the desired wavelength over a wide range. Madey demonstrated radiation at 3.4 microns (Ref 36) but suggested the potential for emission at wavelengths as low as one micron.

CURRENT EMPHASIS

During the past two and one-half decades, the Air Force Research program has supported a wide variety of activities in support of laser technology. As seen in Fig 10, these cover the range of scientific disciplines including physics, chemistry, materials, fluid mechanics, reactive flow phenomenon, control laws, and optics. The emphasis has been on the development of gasdynamic and chemical lasers (Ref 37-41) much of which has been reported at prior meetings of this group.

High Energy Laser
Major Basic Research Areas

Thermochemical Data, Aerodynamics of Fine Scale Interacting Nozzles, Efficient Pressure Recovery, Prediction of Flow Field Disturbances, Structure of Radial Flows, Chemical Laser Concept, Supersonic Diffusion Laser Development, HF/DF Chemical Kinetics, Nozzle Development, Performance Models, Reactant System Selection, Rotational Nonequilibrium, Efficient Pressure Recovery, Anomalous Dispersion, Iodine Laser Development, Pulsed Chemical Laser Development, Visible Chemical Laser Development, Aerodynamic Window, Electric Discharge Laser Phenomenology, Free Electron Laser Development, Beam Path Aerodynamics, Control Algorithm Development, Atmospheric Absorption & Cooling, Electro-optical Sensor Development, Adaptive Control Techniques (Kalman Filter), Flow Field Diagnostics, Laser Supported Detonation/Combustion, Repetitively Pulsed Damage Phenomena, Phase Conjugation Techniques

Figure 10

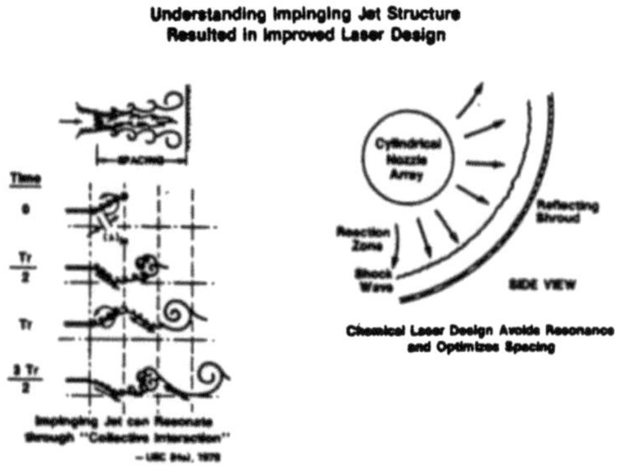

Figure 11

A recent example of fundamental fluid mechanics research supporting laser technology is seen in Fig 11. Ho and Nosseir (Ref 42, 43) studied the impingement of jets on a plate such as occurs during interaction of turbine engine exhaust with flap surfaces on short take-off and landing (STOL) aircraft. The jet contains a periodic vortex structure which produces an acoustic field which feeds back from the reflected surface and can cause a resonance. The resonances and associated pressure peaks occur periodically at specific distances from the jet exit. By considering this phenomenon in the design of a cylindrical laser, the distance of the reflecting shroud from the nozzle array can be selected to avoid destructive acoustic-fluid interactions.

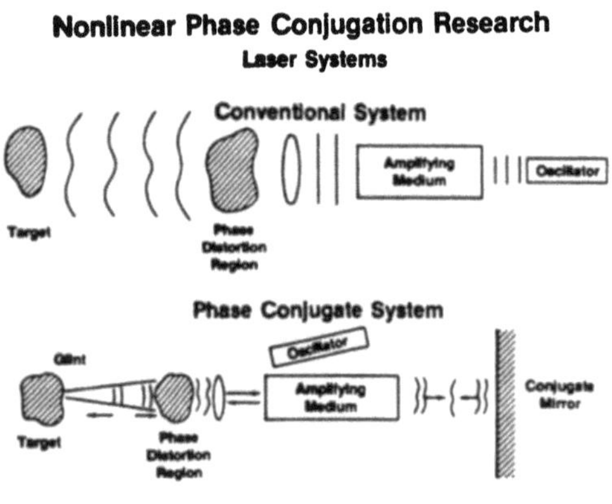

Figure 12

Air Force Laser Research
Current Thrusts

- Shorter Wavelength Chemical Lasers
- Free Electron Laser
- Enhanced Mixing
- Aerodynamic Windows
- Diagnostics
- Non-linear Optics

Figure 13

Another current area of research is in nonlinear optics as seen in Fig 12. By using adaptive optical techniques, the distortion between the laser and the target can be measured and the conjugate distortion introduced into the laser beam. Early contributions in this area were made by Stepanov, Zeldovich, Nosach, and their co-workers (Ref 44, 45, 46). More recently, Hellworth (Ref 47), Giuliano (Ref 48) and Siegman and co-workers (Ref 49) have investigated phase conjugate optics.

Current thrusts in Air Force laser research are summarized in Fig 13. We are seeking chemical reactions that provide shorter wavelength lasers. We are examining the capabilities and limits of the free electron laser. Fluid mechanics research is focused on mixing phenomena and aerodynamic windows. We are also emphasizing diagnostics and adaptive optics.

ACKNOWLEDGEMENT

The author would like to thank Hayes Bryan, Howard Schlossberg, Don Ball, Leroy Wilson, Pete Avizonas, Abe Hertzberg, and Hans Liepmann for their help in preparing this paper.

REFERENCES

1. J P Gordon, H J Zeiger, and C H Townes, "The Maser - New Type of Micro-wave Amplifier, Frequency Standard, and Spectrometer," Phys. Rev.., 99: 1264-74, (1955).

1a. N Basov and A Prokhorov, JETP, 27: 431 (1954).

1b. N Basov and A Prokhorov, Proc. Acad. Sci. (USSR), 101: 47 (1954).

2. A L Schawlow and C H Townes, "Infrared and Optical Masers," Phys. Rev. 112: 1940-49 (1958).

3. A M Prokhorov, JETP, 34: 1658 (1958).

4. R G Gould, Reported in Science, 198: 379-381 (28 Oct 1977).

5. T H Maiman, "Stimulated Optical Radiation in Ruby," Nature, 187: 493-4. (6 Aug 1960).

6. A Javan, "Possibility of Production of Negative Temperature in Gas Discharges," Phys. Rev. Letter, 3 (No 2): 87-9 (July 1959).

7. A Javan, W R Bennett, Jr, and D R Herriott, "Population Inversion and Continuous Optical Maser Oscillation in a Gas Discharge Containing a He-Ne Mixture," Phys. Rev. Letters, 6(3): 106 (1 Feb 1961).

8. C K N Patel, "Continuous-Wave Laser Action on Vibrational-Rotational Transitions of CO_2," Phys. Rev., 136: A1187 (1964).

9. C K N Patel, "Selective Excitation Through Vibrational Energy Transfer and Optical Maser Action in N_2-CO_2," Phys Rev Letters, 13: 617-9 (Nov 1964).

10. N G Basov and A M Oraevskii, "Attainment of Negative Temperatures by Heating and Cooling of a System," Soviet Physics JETP, 17: 1171-2 1963).

11. I R Hurle and A Hertzberg, "Electronic Population Inversions by Fluid-Mechanical Techniques," Phys. of Fluids. 8: 1601-7 (1965).

12. V K Konyukhov and A M Prokhorov, "Population Inversion in Adiabatic Expansion of a Gas Mixture," Soviet JETP Letters. 3: 286-8 (1966).

13. E T Gerry, "Gasdynamic Lasers," IEEE Spectrum, 7: 51-8 (Nov 1970).

14. E M Parmentier and R A Greenburg, "Supersonic Flow Aerodynamic Windows for High-Power Lasers," AIAA Journal, 11: 943-9 (July 1973).

15. W H Christiansen, D A Russell, and A Hertzberg, "Flow Lasers," Annual Rev of Fluid Mech., 7: 115-139 (1975).

16. J C Polanyi, "Proposal for an Infrared Maser Dependent on Vibrational Excitation," J Chem Phys, 34: 347-8 (1961).

17. J V V Kaspar and G C Pimentel, "HCl Chemical Laser," Phys. Rev. Letters 14: 352-5 (8 Mar 1965).

18. D J Spencer, T A Jacobs, H Mirels, and R W F Gross, "Continuous-Wave Chemical Laser," Int. J. Chem. Kinetics, 1: 493-4 (1969).

19. D J Spencer, H Mirels, T A Jacobs, and R W Gross, "Preliminary Performance of a C W Chemical Laser," App. Phys. Letters, 16: 235-7 (15 Mar 1970)

20. J R Airey and S F McKay, "Supersonic Mixing Chemical Laser," App. Phys. Letters, 15: 401-3 (15 Dec 1969).

21. N Jonathan, C M Melliar-Smith, and D H Slater, "Initial Vibrational Energy Level Populations Resulting from the Reaction $H+F_2$ as studied by Infrared Chemiluminescence," J. Chem. Phys. 53: 4396-7 (1970).

22. R M Osgood, Jr., P B Sackett, A Javan, "Measurement of V-V Transfer Rate from HFv = 3 Using Simultaneous Optical Pumping on the HFv = 2⟶1 and v = 1⟶0 Bands," Appl. Phys. Lett., 22: 254-6 (1973).

23. W E McDermott, N R Pchelkin, D J Benard, and R R Bousek, "An Electronic Transition Chemical Laser," App. Phys. Lett., 32: 469-70 (1978).

24. K B Persson, "Brush Cathode Plasma - A Well-Behaved Plasma," J. Appl.Phys., 36: 3086-94 (1965).

25. A J Beaulieu, "Transversely Excited Atmospheric Pressure CO_2 Lasers," App. Phys. Lett., 16: 504-5 (1970).

26. C A Fenstermacher, M J Nutter, W T Leland, and K Boyer, "Electron-Beam-Controlled Electrical Discharge as a Method of Pumping Large Volumes of CO_2 Laser Media at High Pressure," App. Phys. Lett. 20: 56-60 (1972).

27. E R Pugh, J Wallace, J H Jacob, D B Northam, and J D Daugherty, "Optical Quality of Pulsed Electron-Beam Sustained Lasers," Appl Optics, 13: 2512-7 (1974).

28. J P Reilly, "Pulser/Sustainer Electric-Discharge Laser," J. Appl. Phys., 43: 3411-6 (1972).

29. J P Reilly, "High-Power Electric Discharge Laser (EDL's), Astronautics and Aeronautics, 13: 52-63 (Mar 1975).

30. J E Velazco and D W Setser, "Bound-Free Emission Spectra of Diatomic Xenon Halides," J. Chem. Phys., 62: 1990-1 (1975).

31. J E Velazco, J H Kolts, and D W Setser, "Quenching Rate Constants for Metastable Ar, Kr, and Xe atoms by Fluorine containing Molecules and Branching Ratios for XeF^* and KrF^* Formation," J. Chem. Phys. 65:

32. J J Ewing and C A Brau, "Laser Action on the $^2\Sigma^+_{1/2} \rightarrow {}^2\Sigma^+_{1/2}$ Bands of KrF and XeCl," App. Phys. Lett.. 27: 350-2 (1975).

33. C A Brau and J J Ewing, "354-nm Laser Action on XeF," App. Phys. Lett. 27: 435-7 (1975).

34. J M J Madey, "Stimulated Emission of Bremsstrahlung in a Periodic Magnetic Field," J. App. Phys., 42: 1906-13 (1971).

35. L R Elias, W M Fairbank, J M J Madey, H A Schwettman, and T I Smith, "Observation of Stimulated Emmision of Radiation by

Relativistic Electrons in a Spatially Periodic Magnetic Field," Phys. Rev. Lett. 36: 717-20 (1976).

36. D A G Deacon, L R Elias, J M J Madey, G J Ramian, H A Schwettman, and I Smith, "First Operation of a Free-Electron Laser," Phys. Rev. Lett. 38: 892-4 (1977).

37. A Hertzberg, "High-Power Gas Lasers: Applications and Future Developments," J of Energy, 1: 331-346 (1977).

38. D Russell, "Gasdynamic Lasers," Astronautics and Aeronautics, 13: 50-55 (June 1975).

39 W Warren, Jr, "Chemical Lasers," Astronautics and Aeronautics, 13: 36-49(April 1975).

40. J E Broadwell, "Effect of Mixing Rate on HF Chemical Laser Performance," Appl. Optics, 13: 962-7 (1974).

41. L E Wilson, "Deuterium Fluoride CW Chemical Lasers," J De Physique, 41: C9-1 through C9-8 (1980).

42. C-M Ho and N S Nosseir, "Dynamics of an Impinging Jet. Part 1. The Feedback Phenomenon," J. Fluid Mech., 105: 119-142 (1981).

43. N S Nosseir and C-M Ho, "Dynamics of an Impinging Jet. Part 2. The Noise Generation," J. Fluid Mech., 116: 379-391 (1982).

44. B I Stepanov, E V Ivakin, A S Rubanov, Sov. Phys. Doklady 16: 46 (1971).

45. B Ya Zel'dovich, V I Popovichev, V V Ragul'skii, F S Faizullov, Sov. Phys.JETP, 15: 109 (1972).

46. O Yu Nosach, V I Popovichev, V V Ragul'skii, F S Faizullov, Sov. Phys. JETP, 16: 435 (1972).

47. R W Hellworth, J. Opt. Soc. Amer., 67: 1 (1977).

48. C R Giuliano, "Applications of Optical Phase Conjugation," Phys. Today 27-35 (Apr 1981).

49. A E Siegman, P A Balanger, A Hardy, "Optical Resonators Using Phase-Conjugate Mirrors," in Optical Phase Conjugation, R A Fisher, ed., Academic Press, to be published.

FLOW LASER DIAGNOSTIC TECHNIQUES

R.I. Soloukhin

Heat and Mass Transfer Institute
BSSR Academy of Sciences
220728 Minsk, USSR

INTRODUCTION

The great diversity of thermo-and gasdynamic states and substantial departures from thermodynamic equilibrium typical of molecular gas flow lasers specify the methods of the optical diagnostics of such systems. The most important characteristic of the inverse gas laser medium is the unsaturated gain coefficient, α, of unit length of the active medium, which is connected with the integral gain coefficient of the entire length of the active medium, $\kappa = P_{out}/P_{in}$, by the relation $\kappa = \exp[(\alpha-\delta)L_0]$ where δ is the coefficient of distributed losses (scattering) and L_0 is the active medium length. Except for the so-called method of calibrated losses, the gain coefficient is usually measured by passing through the active medium a monochromatic radiation with the same quantum transitions as those in the induced radiation of the medium. The remainder parameters of the diagnostic laser medium and of that tested (i.e. density, temperature, composition) may in this case differ substantially. The analysis of the conditions typical of gasdynamic and electric-discharge CO_2 lasers shows that the effective widths of the tested and diagnostic line contours may differ several-fold. This leads to a partial overlapping of lines and, consequently, to certain errors in the measurements of gain. Also, an important characteristic of the inverse molecular laser medium is the temperature of a particular vibrational state of the molecules. That the vibrational and rotational temperatures could be determined, the distributions of the gain (absorption) coefficient over a wide range of separate IR-spectrum lines are measured. The paper presents an analysis of some reasons for the errors occurring in determination of the gain coefficient for molecular gas laser media and describes the methods allowing simultaneous measurements of the vibrational and rotational temperatures in a CO_2 laser medium by means of a tuned diagnostic laser with high time resolutions.

GAIN COEFFICIENT

The unsaturated gain coefficient at the center of a separate line of the vibrational-rotational transition depends, as exemplified by the CO_2 molecule, on the number of the rotational sublevel, j, the variables being the population of levels, the Einstein coefficients, A_{21}, and the collision half-width, $\Delta\nu_c$:

$$\alpha = \frac{\lambda^2 A_{21}}{4\pi^2 \Delta\nu_c c} (N \frac{g'}{g} - N').$$

It is convenient to express A_{21} in terms of the product of the vibrational and rotational components (see, e.g. [1-3]):

$$A_{21} = \frac{64\pi^4 \nu_{21}^3}{3hc^3 g_1} S(j)(R_v)^2 F(m)$$

where S(j) is the Hunley-London factor which, to a fair approximation, is equal to j for the P-branch and to j+1 for the R-branch, g_1 is the statistical weight of the lower level, $(R_v)^2$, the squared matrix element of the dipole moment of transition (independent of j) and F(m) is the factor to account for the V-R interaction (usually having magnitudes near unity). The relationships for the collision half-widths are of the form:

$$\Delta\nu_c = (0.1254-0.98.10^{-3}j)(\frac{300}{T})^n p \qquad \text{(P-branch)},$$

$$\Delta\nu_c = (0.1301-1.13 \times 10^{-3}j)(\frac{300}{T})^n p \qquad \text{(R-branch)}.$$

Here, p is in atm; $\Delta\nu_c$, in cm^{-1}, and the exponent n for CO_2 is taken to be unity according to [4,5].

Thus, it is seen that with relation to the diagnostics of the laser active medium when a monochromatic radiation produced as a result of quantum transitions of the same or nearly the same frequencies propagates through the test volume, the relationship between the line widths and the overlapping of the line profiles of the excited medium radiation and of the probing laser beam are of prime importance. The measured value of the gain (or absorption) coefficient depends on the extent of the cross-coupling of these lines, as is evidenced by the typical versions shown graphically in Fig. 1 and analyzed in detail in [6]. In the first two versions, the single-mode laser illuminates the active medium of the other single-mode laser. The optimum version is that having the half-width of the amplification- or absorption-line contour, $\Delta\nu_0$, much in excess of the half-width of the generation-line contour, $\Delta\nu_0'$. The estimates made in [6] have shown that here the effect of "detuning", $\nu_0-\nu_0' = \delta\nu$, can be easily taken into account. The resulting error, ε , has the form:

$$\varepsilon = |[1 + 4(\frac{\delta\nu}{\Delta\nu_0})^2]^{-1}-1|.$$

Specifically, this version suggests itself when analyzing the results of measurements of the gain coefficient for the nitrous oxide laser illuminated by the probing CO_2-laser [7]. For the case of two Doppler broadened contours with the coincident centers of lines the error is evaluated to be:

$$\varepsilon = |\sqrt{\{\frac{1}{2}[1 + (\frac{\Delta\nu_0'}{\Delta\nu_0})^2]\}} -1|.$$

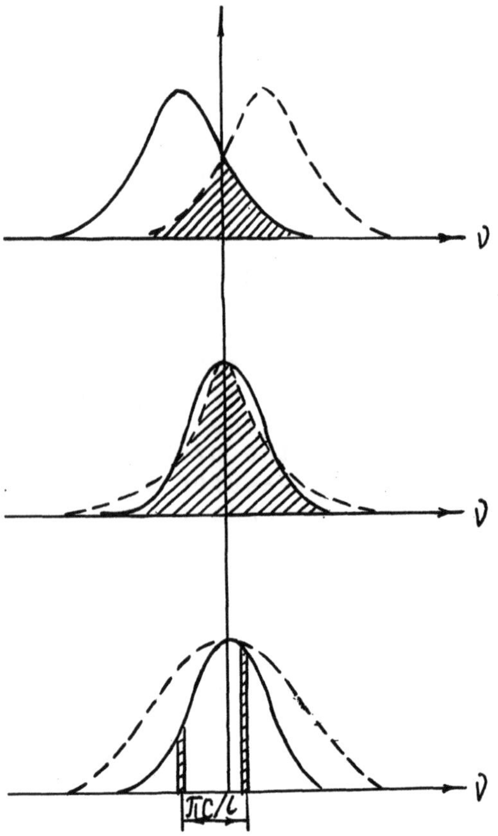

Fig. 1. Typical spectrum line overlapping regimes for probing laser beam and laser medium under study. First and second cases correspond to shifted and coincident positions of the line centre in a single-mode regime, the third case illustrates the multimode regime.

A great number of other versions are possible. The major difficulty in the analysis is posed by the account for the multimode generation structure of laser radiation. Thus, for the CO_2-laser active medium 1 m long the interval between the adjacent modes is about 150 Mc, while the spacing between the adjacent rotational lines is about 600 Mc. For the most favourable version, $\Delta\nu_0 \gg \Delta\nu_0'$, to be realized, it is necessary that the pressure of the diagnostic laser medium be below that of the medium studied. Therefore, for the flow lasers having the pressures of about 100 torr and upwards, a small error in the measurements of the gain coefficient is provided by the discharge lasers having the pressures of 20-30 torr and the Doppler broadened contour, $\Delta\nu_0'$, of about 50 Mc.

Finally, as regards the operational characteristics and state of the medium of flow lasers, it should be noted that the influence of the Doppler effect is small there, $\nu_0 u/c$ ~ 100 Kc, while the influence of the temperature and pressure gradients that "deform" the velocity distribution function of molecules and, consequently, the line contour is estimated to be of about 100 c [8].

MEASUREMENT OF ROTATIONAL AND VIBRATIONAL TEMPERATURES

The expression for the gain coefficient can be conveniently restated in the following form [9]:

$$\log \left\{ \frac{\alpha}{F_2(j)} \frac{A_{21}(j_*)}{A_{21}(j)} \left[\frac{\lambda(j_*)}{\lambda(j)} \right]^2 \frac{\Delta\nu_c(j)}{\Delta\nu_c(j_*)} \right\}^2 =$$

$$= \log \left[\frac{A_{21}(j_*)\lambda^2(j_*)}{T^{2-n}\,\Delta\nu_c(j_*)} C_1 \right] - \frac{j(j+1)B}{T}$$

where

$$C_1 = \xi_{CO_2} \frac{Bp^2}{2\pi^2 ckQ_v} \left[\exp\left(-\frac{\theta_3}{T_3}\right) - \exp\left(-\frac{\theta_1}{T_{12}} - \frac{F_1(j)B}{T}\right) \right]$$

j_* is the arbitrary but fixed quantum number; Q_v the partition function of the vibrational state; B the rotational constant; F_1=2j and 2(j+1), while F_2=2j-1 and 2j+3 for the P- and R-branch, respectively. The estimates show that C_1 is weakly dependent on j, therefore, the left-hand side of the above relation (α is measured as a function of j(j+1)) is a straight line, the inclination tangent of which determines the rotational (virtually translational) temperature of the medium. In the case of $T_{12} \simeq T$, the expression for C_1 offers the possibility to determine the vibrational temperature of the upper laser level, T_3 [9]. In a more general case, $T_{12} \neq T$, the vibrational temperatures are determined by measuring the gain coefficient for, at least, one of the lines of the transition 02°0 → 00°1 (9.6 μm), while the unknown temperatures T_3 and T_{12} are determined from the relationship [10,11]:

$$C_1' = \xi_{CO_2} \frac{Bp^2}{2\pi^2 ckQ_v}[\exp(-\frac{\theta_3}{T_3}) - \exp(-\frac{2\theta_2}{T_{12}} - \frac{F_1 B}{T})]$$

which, together with the expression for C_1, comprises a system of two transcendental equations. As an example, Fig. 2 presents the experimental data on the dependence of $K=\log[\alpha(\frac{\lambda_*}{\lambda_j})^2 \frac{A_{21}^*}{A_{21}} \frac{\Delta\nu_c}{\Delta\nu_c^*}]$ on $j(j+1)$ obtained in a gas-dynamic laser with a selective thermal excitation (description of the device is given in [12,13]) for the R--branch (10.6 μm). The same system was used to measure simultaneously the values of T_{12} on the P20 line in the 9.6 μm region [10,11]. The data show that at T_0=1800 K, p_0 = 7 atm and nozzle area ratio A/A_* = 30 the measured (translational) gas temperature was 320 K, while the vibrational temperatures were T_3=1510 K and T_{12}=415 K, respectively, which is in a good agreement with the analytical prediction.

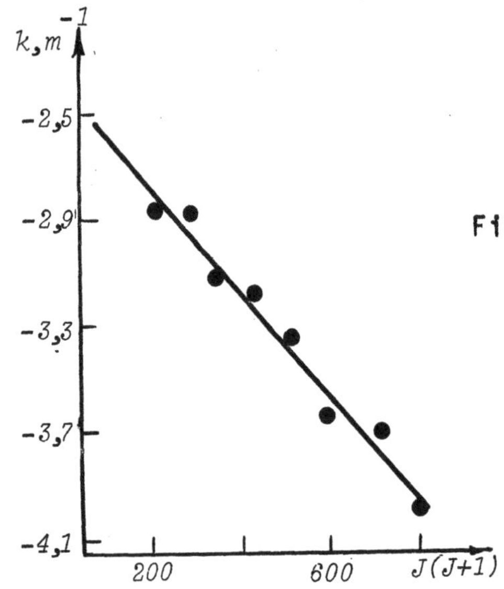

Fig. 2. Typical experimental data to determine vibrational and translational temperatures in the mixed flow GDL [11].

It should be noted that the above methods used to determine the vibrational temperatures of CO_2 molecules are based on the assumption of the Boltzmann population distribution on the rotational transition levels. This distribution can be violated when separate absorption lines of the fundamental transition overlap with the lines of transitions from the upper levels (the so-called "hot" transitions). The estimates show that these deviations for the CO_2 molecules are most appreciable at $p \sim 2$ atm and $T \sim 700$ K [10].

The vibrational temperature, T_3, of the axisymmetric CO_2 mode can be determined by measuring the absolute value of the IR-radiation intensity of the laser medium within the 4.3 μm region and presuming that the symmetric and bending modes of CO_2 are in equilibrium with translational degrees of freedom, $T_{12} \simeq T$ [14-17]. Simultaneous measurement of the CO_2 spontaneous radiation intensity in the 4.3 μm region and of the gain coefficient of the 001 → 100 transition has made it possible to determine the populations and, consequently, the vibrational temperatures of the both operational laser transition levels in the absence of equilibrium between T_{12} and T. That the method could be realized, it is necessary to determine the spectral absorption and, hence, the spectral intensity of radiation in the selected band as a function of the vibrational temperatures in the respective equilibrium or non-equilibrium conditions. The calibration of the optical system is usually carried out by measuring the equilibrium radiation of the medium heated in the incident shock wave under the conditions that are close to those in the GDL nozzle. Of the results obtained especially noteworthy is the fact that the combined bending and symmetric stretch mode (1,2) is substantially out of equilibrium in the absence of, or small, concentrations of the "catalyst" (He, H_2): $T_{12} \simeq 1.5\text{-}2T$ at $T_0 = 800\text{-}3000$ K and $p_0 = 5\text{-}25$ atm [15]. An appreciable population of the combined mode was also observed on expansion of the reaction products of $CO:N_2O$ mixtures diluted with helium and nitrogen [16].

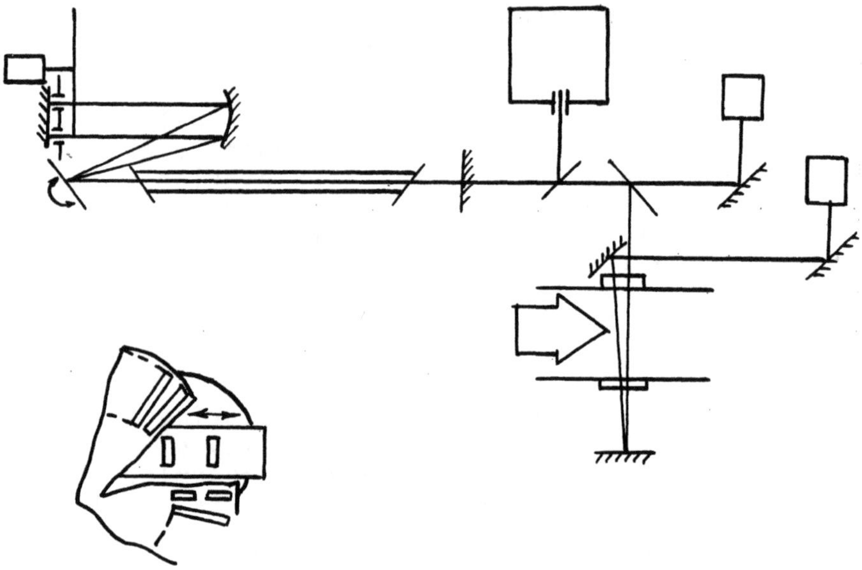

Fig. 3. Schematic diagram of the fast tuned CO_2-laser developed for the shock tube studies [20]. No turn of the grating is provided with the use of an intercavity rotating disc device selecting successively single rotational band lines. A schematic of "mask" used for selection of two separate lines is shown below.

In order to measure the vibrational and rotational temperatures in the GDL with a high time resolution, the authors of [18-20] have suggested a system involving a diagnostic CO_2 laser, in which a change in the generation frequency does not require the tuning of the optical cavity. The system allows a five-to-six orders of magnitude increase in the speed of spectral tuning to separate single rotational lines and enables one to make this speed sufficient for multiple measurements of temperatures for the time of a quasi-stationary flow in the shock driven GDL nozzle flow. The lines are separated by spatial, rather than angular, selection (Fig. 3). Leaving the echelette grating, a "fan" of the monochromatic rays arrives at a concave mirror, which in the plane forms a line of focuses that correspond to an assembly of lines comprising the laser medium amplification spectrum. Close to the focal plane the system has a plane mirror which reflects the radiation backwards. By using the system of mobile and fixed masks, it is possible to vary the frequency according to any specified program. This system allows one, for example, to achieve a successive generation of two lines in each of the slits with the disc rotating at a frequency of up to 12000 min^{-1}.

A typical oscilloscope record of the GDL process is shown in Fig. 4. Both traces show the alternating pulses of laser radiation from two vibrational-rotational transitions bounded by fixed slits. The lower trace corresponds to the control set of spikes, while the upper one to that which has passed through the amplifying medium. There is still no discharge of the hot nitrogen at the start of the trace and so there is no gain. The duration of the developed flow is 250-350 μsec, depending on the operating conditions. A pair of far-apart lines P(18) and P(30) on the transition $00^01 \rightarrow 1^000$ has been used in measurements. With the temperature of the active medium being about 300 K, the temperature for the selected lines was measured accurate to 5-10% by the above method. It can be noted that the measurement of

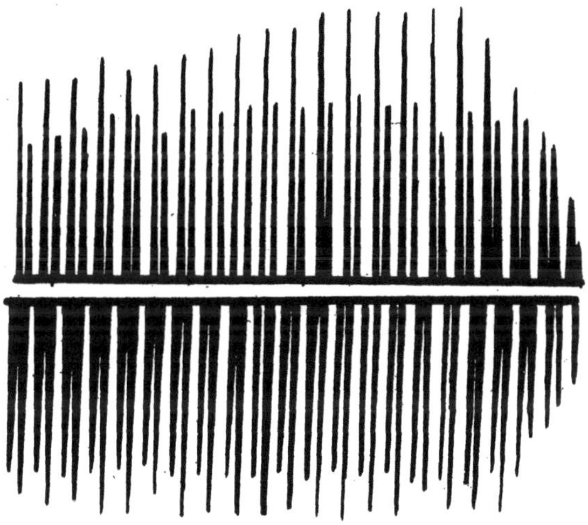

Fig. 4. Typical oscilloscope record of alternative two-line gain measurements signals obtained in the shock tube driven GDL flow [20]. The whole sweep duration is 2 ms. The upper trace shows the amplified beam signals whereas the lower one is the incident beam control recording. Each of two successive spikes gives the gain coefficient for P(18) and P(30) spectrum lines.

the gain coefficients for the intermediate lines does not markedly affect the accuracy of temperature determination with the use of a plot as shown in Fig. 2 but only unjustifiably increases the bulk of intermediate calculations and worsens the time resolusion of the method.

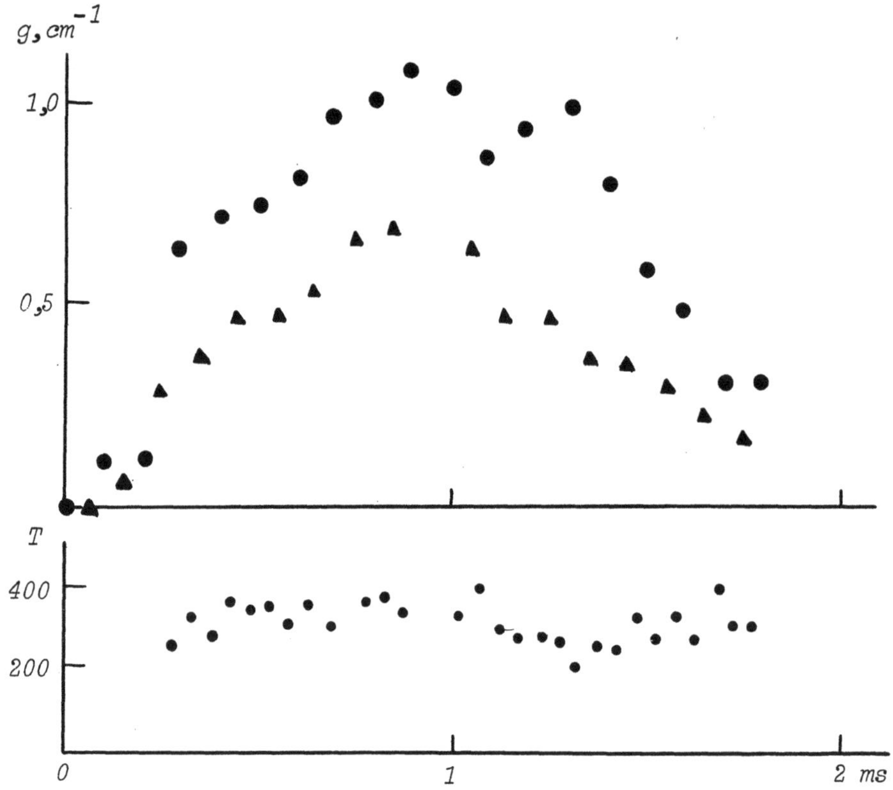

Fig. 5. Experimental data on the shock tube GDL flow development (correspond to the record shown in Fig. 4). In the first graph, the upper points correspond to the P(18) line, the lower one, to P(30). The results of the gain data adoption are shown on the lower graph, thus giving the rotational (i.e. translational in this regime) temperature as a function of time. No gain is recorded during first 200 μs.

As an example, Fig. 5 shows a change in time of the measured gain coefficient on the line P(18) (the upper set of points) and P(30) line (the lower set) - for one of the experiments. A separate graph below shows the corresponding data on a change in the rotational temperature of the medium. As seen, it is approximately constant within the entire period of the quasi-steady GDL nozzle flow in a shock tube.

Finally, it is necessary to mention the system developed for the "panoramic" measurement of the gain coefficient and vibrational temperatures [21]. It is based on the use of a thermovision device and was employed for the two-dimensional diagnostics of a lengthy area of active media in a DC electric discharge CO_2 flow laser.

CONCLUDING REMARKS

It should be noted in conclusion that the possibility for the direct measurement of temperature and, hence, of the frozen energy concentrations for the vibrational excitation of different states of gas laser molecules is of great importance for the analysis of the basic operational processes in flow lasers since this enables one to determine the efficiency of energy conversion in such systems with account for the relaxational losses in the process of laser medium preparation and cavity operation in the lasing regime.

ACKNOWLEDGEMENT

The author is indebted to Yu.A. Yacobi, N.A. Fomin, O.V. Achasov and S.S. Novikov for valuable discussions and contributions.

REFERENCES

1. G. Herzberg, "Molecular spectra and molecular structure," parts I (2nd ed.) and II, D. Van Nostrand Co., Inc., Princeton (1950;1945).
2. S.S. Penner, "Quantitative molecular spectroscopy and gas emissivities", Addison-Wesley Publ. Co., Inc., Reading, Massachusetts (1949).
3. M.A. Elyashevich, "Atomic and molecular spectroscopy", Fizmatgiz, Moscow (1962).
4. A.R. Strilchuk and A.A. Offenberger, High temperature absorption in CO_2 at 10.6 μ, Appl. Opt. 13:2643 (1974).

5. R.I. Soloukhin and N.A. Fomin, Resonance 10.6 μ CO_2 absorption in a shock wave, Zh. Prikl. Mekh. Tekhn. Fiz. No. 1, 42 (1977).
6. R.I. Soloukhin and Yu.A. Yacobi, Comments on optical gain measurements, Zh. Prikl. Mekh. Tekhn. Fiz. No. 3, 3 (1974).
7. O.V. Achasov and D.S. Ragosin, Study of the CO_2-laser beam absorption in N_2O, in: "Thermophysical processes in power designs," S.A. Zhdanok and Yu.V. Porodsinsky, ed., p. 12, Heat and Mass Transfer Institute, Minsk (1982).
8. V.L. Kolpashchikov, R.I. Soloukhin and T.A. Syroyezhko, Spectrum characteristics of thermally nonuniform resonance-amplifying media, Zh. Prikl. Spektr. 29: 1006 (1978).
9. P.V. Avizonis, R. Dean and R. Grotbeck, Determination of vibrational and translational temperatures in gasdynamic lasers, Appl. Phys. Lett. 23: 375 (1973).
10. O.V. Achasov, S.A. Labuda, R.I. Soloukhin and N.A. Fomin, Rotational and vibrational temperature measurements using tunable CO_2-laser, Fiz. Goreniya i Vzryva No. 6, 57 (1979).
11. N.A. Fomin and R.I. Soloukhin, "Laser diagnostics of molecular states in vibrationally excited gas flow systems," in: Proc. Colloque International Bertelot-Vieille-Mallard-Le Chatelier, publ. by Sect. Franc. Comb. Inst., p. 539, Bordeaux (1981).
12. V.N. Croshko, N.A. Fomin and R.I. Soloukhin, Population inversion and gain distribution in mixed flow systems, Acta Astronautica 2: 929 (1975).
13. A.V. Krauklis, V.N. Croshko, R.I. Soloukhin and N.A. Fomin, Generation regimes in GDL with thermal excitation and supersonic flow mixing, Fiz. Goreniya i Vzryva No. 5, 792 (1976).
14. N.N. Kudriavtsev, S.S. Novikov and I.B. Svetlichnyi, Experimental determination of carbon dioxide (001) level temperature in a nonequilibrium flow of CO_2+N_2+H_2O (He) mixtures, Fiz. Goreniya i Vzryva No. 2, 205 (1977).
15. N.N. Kudriavtsev, S.S. Novikov and I.B. Svetlichnyi, Method of vibrational temperature measurements in gasdynamic CO_2-lasers, Quant. Electr. 6: 690 (1979).
16. N.N. Kudriavtsev, S.S. Novikov and I.B. Svetlichnyi, CO_2 vibrational temperature measurements in CO+N_2O gasdynamic laser, Acta Astronautica 6: 391 (1979).

17. N.N. Kudriavtsev and S.S. Novikov, A study of infrared radiation of vibrationally excited CO in the 4.7 μm band and CO_2 in the 4.3 μm and 2.7 μm bands, Revue Phys. Appl. 16: 49 (1981).
18. R.I. Soloukhin, Yu.A. Yacobi and E.I. Viazovich, Fast tunable CO_2-laser design, Soviet Patent Bull. No. 11, Patent No. 594842 (1979).
19. Yu.A. Yacobi, Scanning of laser generation spectra by the intercavity space filtration, Quant. Electr. 8: 555 (1981).
20. P.V. Grigoriev, S.A. Murikov, R.I. Soloukhin and Yu.A. Yacobi, Rotational temperature measurements in GDL, Chem. Phys. (Russian) No. 2, 216 (1982).
21. S.P. Vagin, S.S. Vorontsov, R.I. Soloukhin, A.A. Shepelenko and Yu.A. Yacobi, Panoramic spectroscopy of lengthy lasing media, Zh. Prikl. Spektr. 29: 224 (1978).

PULSED-LASER/MATERIAL INTERACTION

J. Woodroffe

Avco-Everett Research Laboratory, Inc.
Everett, MA 02149

INTRODUCTION

The general field of pulsed-laser/material interaction is vast, even excluding conditions pertaining to inertial-confinement fusion. This paper is not intended as a complete review of the field, but rather as a summary of two specific examples chosen to be representative of the work in the field with emphasis on testing carried out on lasers at Avco-Everett.

The field can be broken out as shown in Table I: parameters of the pulsed laser, material type, and interaction type. In this paper we will discuss mechanical and thermal interactions of pulsed lasers at 10.6μ, 3.8μ, 0.35μ and 0.25μ with aluminum, and the thermal interaction of pulsed lasers at 10.6μ with glass-fiber reinforced plastic composites ("fiberglass").

The phenomenology of rep-pulsed laser damage to aluminum and to fiberglass will be discussed first, followed by the interaction data. The results are summarized in the last section.

PHENOMENOLOGY

One of the dominant features of pulsed lasers, especially in the infrared, is that they produce an air plasma. The plasma is generated by vaporization of a small amount of aluminum from defects on the surface.

The aluminum vapor is then ionized by the laser light and then, in turn, the air is ionized. The air plasma achieves very high temperature and very high pressures, and it is a very efficient radiator. This plasma can lead to enhanced thermal coupling to metals in the infrared (above what would be expected from ordinary absorptivity at the laser wavelength) by absorbing the infrared laser light and re-radiating in the ultraviolet where metals are good absorbers. Because the plasma is at high pressure for a significant length of time, a significant impulse can also be generated.

The phenomenology of aluminum rep-pulsed laser damage is shown in Figure 1, with the dominant path shown in heavy lines. Starting at the top of the figure, the rep-pulse laser radiation is shown (which may or may not generate a plasma depending on the flux level). The plasma or direct absorption causes energy to be absorbed on the surface; thermal conduction carries this energy into the bulk of the material. Under conditions of intense irradiation,not all of the absorbed energy can be carried into the bulk material; then prompt mass removal by ablation and melt splash can occur. It there is prompt mass removal, it in turn can generate an impulse. This impulse can also be generated by the plasma. The impulse can lead to mechanical damage. The combination of the residual energy left in the bulk after the pulse is over and mechanical damage leads to thermo-mechanical damage through heat-up to the point of mechanical failure.

The phenomenology of fiberglass rep pulsed laser damage is rather different (Figure 2). A plasma can again be present or absent, and impulse and mechanical damage can be generated as for aluminum. However, since in this case the material is thermally insulating, conduction is not a dominant mechanism for depositing residual energy. Rather, it is the fact that these materials are not surface absorbers, but have a finite in-depth absorption length of the order of 10μ, which leads to residual energy deposition. In turn, cumulative mass removal occurs through interpulse pyrolysis, combustion, melt removal, and delamination. For fiberglass, delamination is the dominant mechanism that leads to penetration.

Table I
Pulsed-Laser/Material Interaction

PULSED LASER	PULSE-LENGTH	~μSEC'S (1/2 - 20 SEC)
	FLUX	~10^5 - 10^8 W/CM^2
	FLUENCE	~1 - 100 J/CM^2
	WAVELENGTH	10.6μ(CO_2), 3.8μ(DF), 0.35μ(XeF), 0.25μ(KrF)
MATERIAL	METALS (ALUMINUM, STEEL, TITANIUM)	
	NON-METALS	
	– CERAMICS	
	– COMPOSITES	
	–OPTICAL MATERIALS	
INTERACTION	THERMAL (DIRECT ABSORPTION, PLASMA COUPLING)	
	MECHANICAL (VAPOR RECOIL IMPULSE, PLASMA IMPULSE)	

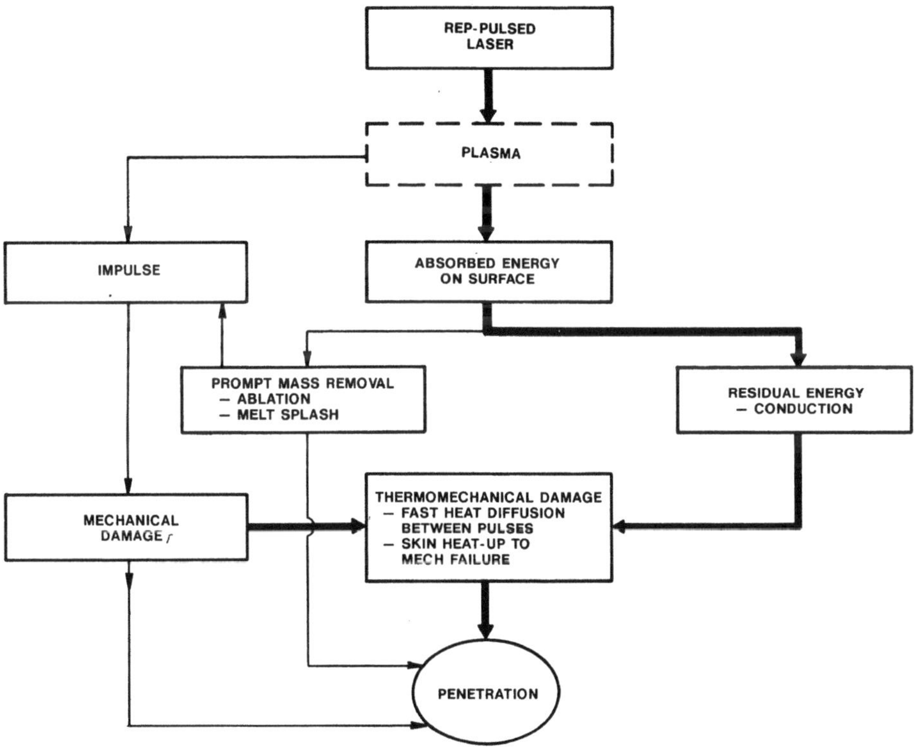

Figure 1 PHENOMENOLOGY OF ALUMINUM REP-PULSED LASER DAMAGE

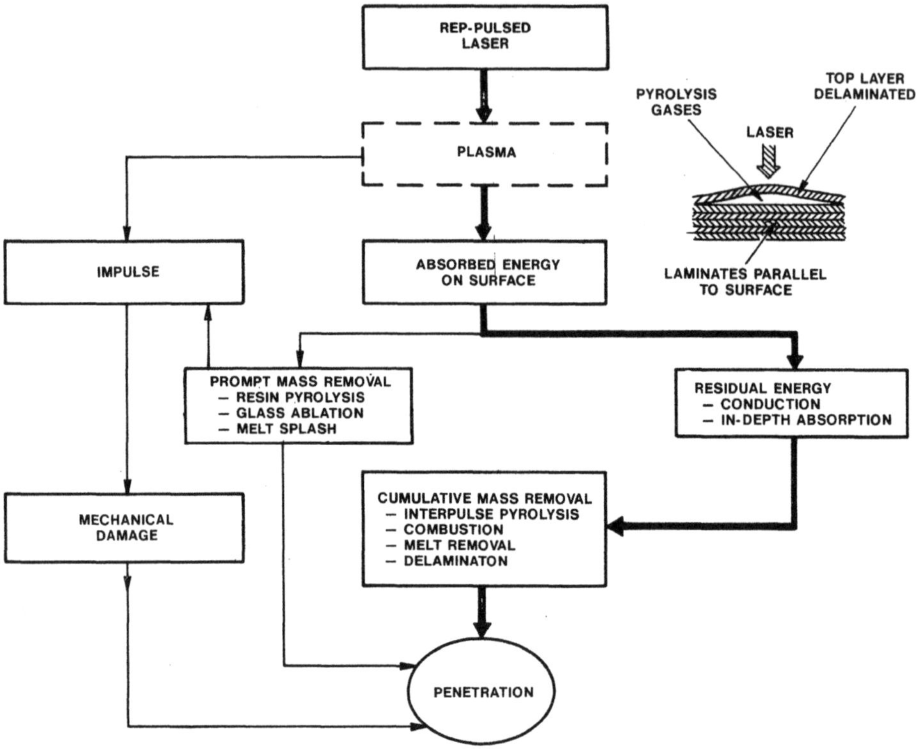

Figure 2 PHENOMENOLOGY OF FIBERGLASS REP-PULSED LASER DAMAGE

ALUMINUM

Figure 3 shows that plasma confinement time is the dominant parameter in plasma/surface interaction. The pulsed laser beam generates a laser supported combustion wave, which, in turn, leaves a dense plasma behind it. The plasma is "inertially confined" for a time that corresponds to that for a rarefaction wave to travel from the edge of the spot to the center of the target. This time is the spot radius divided by the plasma acoustic speed. The parameter $\hat{\tau}$ is the ratio of the pulse length to the plasma confinement time; it impacts the efficiency of both mechanical (impulse) and thermal coupling, as will be discussed below.

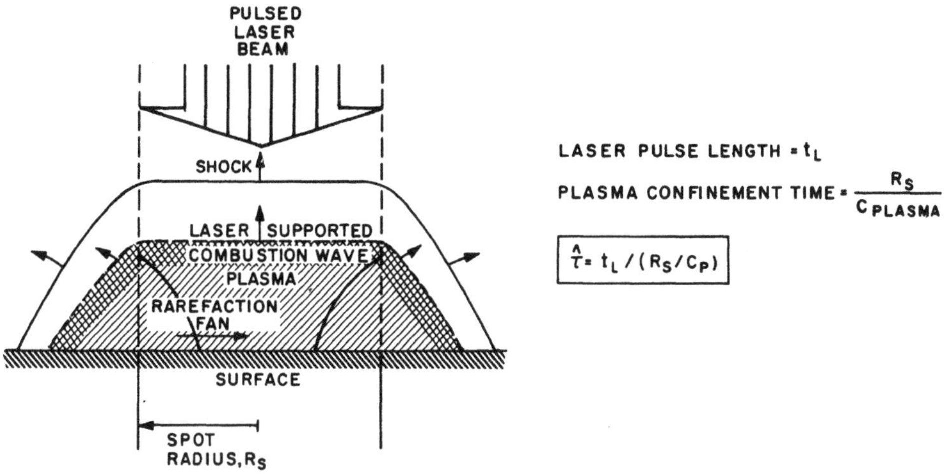

Figure 3 RATIO OF LASER PULSE LENGTH TO PLASMA. CONFINEMENT TIME IS A DOMINANT PARAMETER

Figure 4 shows the importance of $\hat{\tau}$ to air-plasma impulse delivery. We have plotted the ratio of laser-produced impulse to the product of the initial surface pressure and pulse length.(1) It can be seen that, when plotted versus the $\hat{\tau}$ parameter, this ratio can range from very high values (or order 10-100) to values below one.

Figure 5 shows measured values of surface pressure under plasma conditions versus flux for a variety of materials at both CO_2 (2) and DF(3) wavelengths. Generally, the data scatters between the laser supported detonation wave predictions and the laser supported combustion wave predictions.(4) One would expect that at very high fluxes, the pressure measurements would lie along the LSD line, as observed.

In Figure 6 we show that vaporization can also generate impulse, but much less effectively. The ratio of impulse to fluence plotted versus flux shows that with an air plasma values of order 15-30 can be achieved, whereas with vaporization at either 10μ(5) or 0.35μ(6) the values are about 2-4.

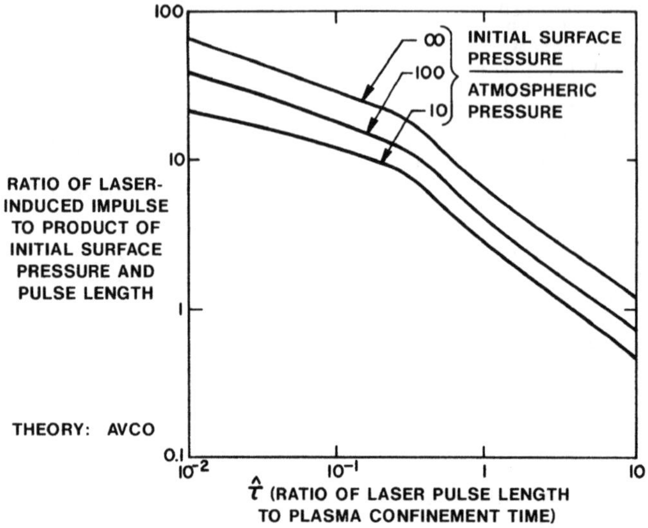

Figure 4 AIR PLASMA IMPULSE DELIVERY BY A PULSED LASER SHOWS $\hat{\tau}$ IMPORTANCE

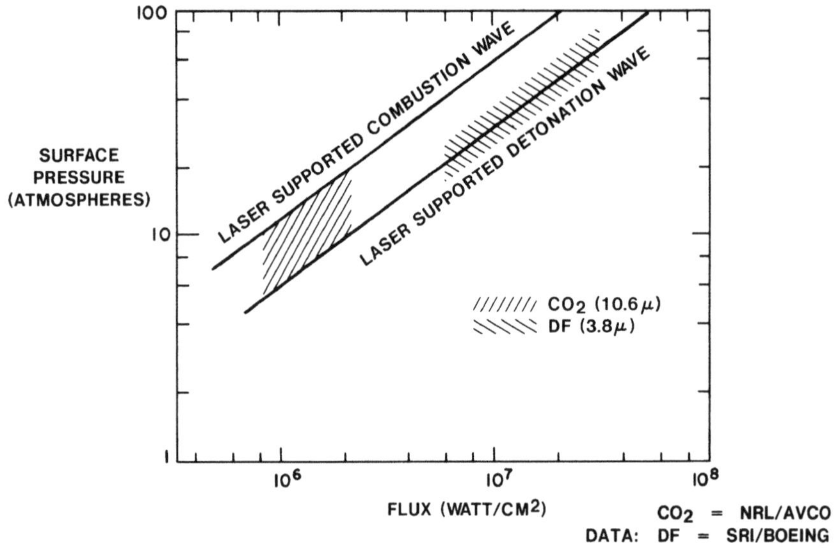

Figure 5 MEASURED PLASMA SURFACE PRESSURE LIE BETWEEN LSC AND LSD PREDICTIONS

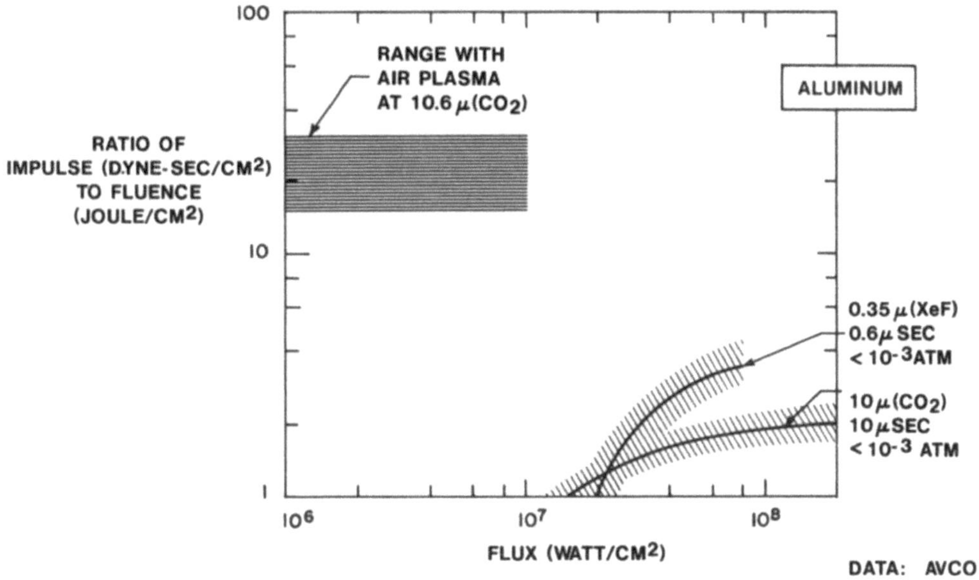

Figure 6 IMPULSE CAN BE GENERATED BY VAPORIZATION, BUT LESS EFFECTIVELY

Figure 7 shows that the absorptivity of aluminum is a strong function of wavelength. One would expect, then, based on the bare metal absorptivity, to see a significant difference in thermal coupling across the range of wavelength from 10 μ down to a fraction of a micron.

Figure 8 shows that both the laser wavelength and plasma formation dominate thermal coupling. The net thermal coupling is plotted versus flux for conditions corresponding to an ambient pressure of 1 atm, for several wavelengths. At 10 microns, at low fluxes the coupling is about what would be expected on the basis of Figure 7. However, once the plasma threshold is reached, very much higher couplings are observed.(2) There is a gradual decrease in coupling as the flux increases; this is purely due to plasma effects (4) and it is not connected with changes in $\hat{\tau}$. If $\hat{\tau}$ were large, one would also expect lower coupling. At 3.8 μ (DF wavelength), the same kind of behavior is observed.(7) Here the plasma threshold lies at higher fluxes and

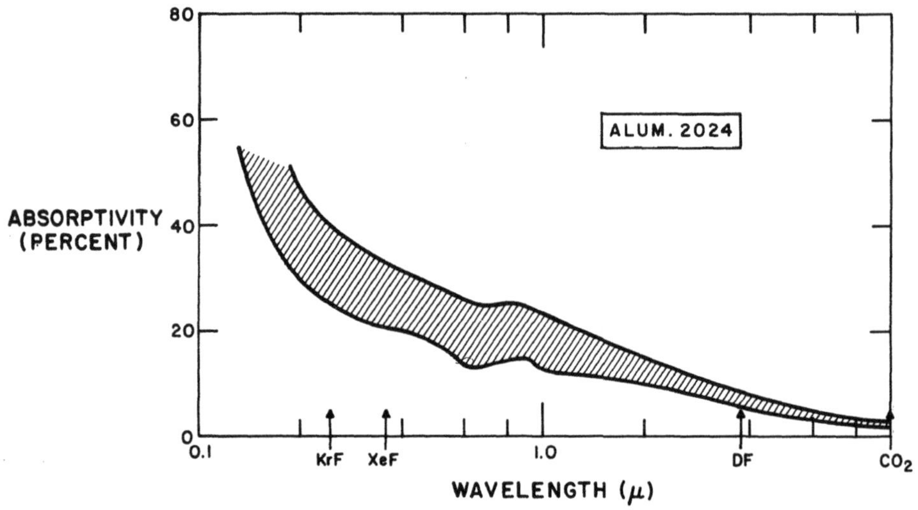

Figure 7 ABSORPTIVITY OF ALUMINUM IS A STRONG FUNCTION OF WAVELENGTH

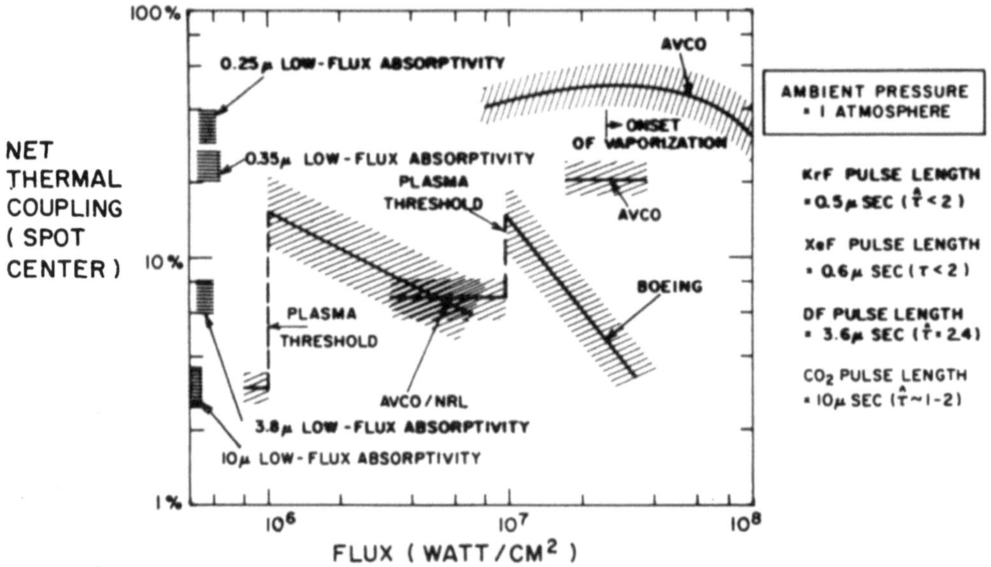

Figure 8 LASER WAVELENGTH AND PLASMA FORMATION DOMINATE THERMAL COUPLING

there is a sharp decline in coupling caused, in this case, by vaporization of the aluminum under the intense plasma heating. For shorter wavelengths (6,8) the coupling appears to be dominated by the bare metal thermal coupling. In these cases, vaporization occurs before plasma formation. In the case of KrF in particular vaporization leads to a decline in the net energy deposited, since the vapor carries away a large amount of energy.

The vacuum thermal coupling is strongly dependent on the formation of a metal vapor plasma in the infrared, as shown in Figure 9. For CO_2 (9), vapor plasma begins to appear at a higher flux than would be required to form an air plasma under ambient conditions; beyond that point the coupling is high and remains high up to very high fluxes. For XeF (6), the onset of plasma threshold does not appear to have an effect on the coupling. That may be due in this case to the fact that the experimental spots were small, so that there were counter-balancing effects between plasma enhancement and $\hat{\tau}$.

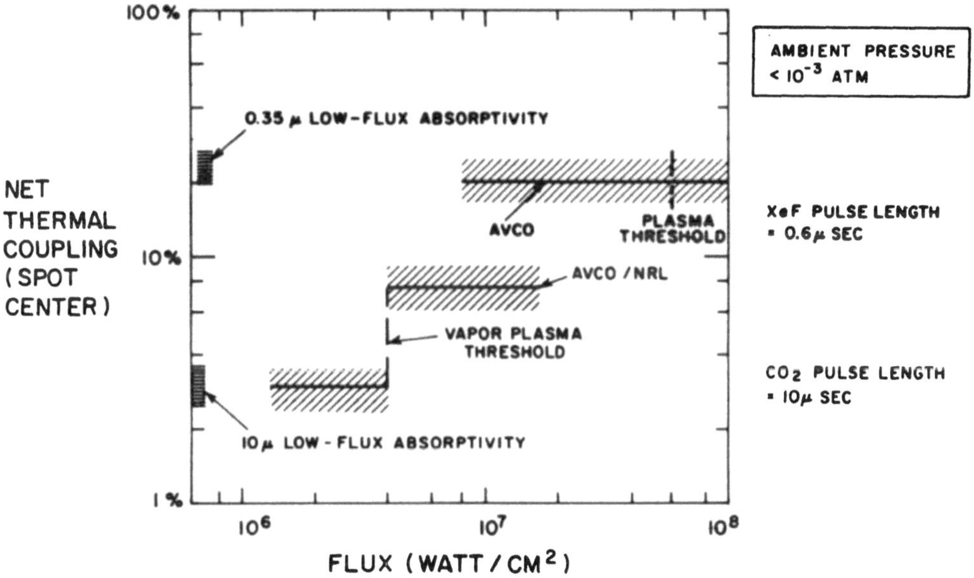

Figure 9 ALUMINUM VACUUM THERMAL COUPLING STRONGLY DEPENDENT ON FORMATION OF METAL VAPOR PLASMA, IN INFRARED

Table II
Pulsed-Laser/Aluminum Interaction

- **PLASMA FORMATION DOMINANT IN IR**
  - **– LARGE IMPULSE AND THERMAL COUPLING**
- **RATIO OF PULSE LENGTH TO PLASMA CONFINEMENT TIME < 1-2 REQUIRED**
- **NET THERMAL COUPLING INCREASES AT SHORTER WAVELENGTHS**
- **VACUUM IMPULSE LOWER THAN FOR AIR PLASMA**
- **VACUUM THERMAL COUPLING HIGH FOR CO_2 IF FORM METAL VAPOR PLASMA**

Table II summarizes the results on the interaction between a pulsed laser and aluminum. Plasma formation is dominant in the infrared, leading to large impulse and thermal coupling; the ratio of pulse length to plasma confinement time needs to be small for good coupling; thermal coupling increases at shorter wavelength; vacuum impulse is lower than for air plasma; and vacuum thermal coupling is high for CO_2 if a metal vapor plasma is formed.

FIBERGLASS

As was discussed earlier in connection with Figure 2, the dominant interaction of a pulsed CO_2 with fiberglass is thermal delamination which occurs when the rear of the layer reaches a critical temperature. Figure 10 shows the number of pulses to the onset of delamination (for a particular layer)versus the fluence per pulse. As one can see delamination asymptotes as the in-depth absorption limit is reached, at about 7-10 J/cm^2/pulse. This limit is due to the onset of vaporization of the glass; beyond that point, the vapor car-

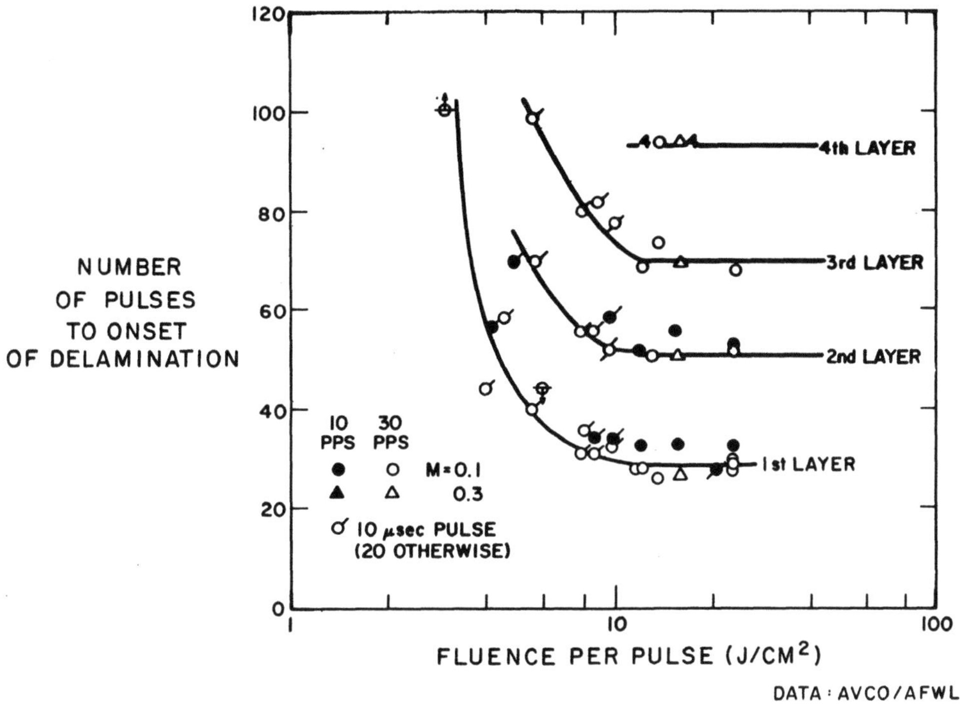

Figure 10 DELAMINATION ASYMPTOTES AS IN-DEPTH ABSORPTION LIMIT IS REACH (ONSET OF VAPORIZATION)

ries away the energy delivered by the laser. Figure 11 presents theoretical calculations (10) showing that delamination is much more efficient than ablation, but is limited by heat transfer: the time to remove one layer decreases with average flux until a limiting heat transfer time is reached. At that point, even on a steady state basis, the surface is at the vaporization temperature and therefore the rear of the layer cannot achieve the delamination temperature until a finite time has passed.

Figure 12 shows a comparison of the model with a set of data[10] showing the model predicts the thickness effect properly. The model has been extrapolated to high average flux in Figure 13, showing that fast penetration can be achieved by delamination. The time to remove one layer can be as short as a fraction of a second. The band is due to the fact that this extrapolation is based on data at low average fluxes and low rep rates, containing a certain amount of uncertainty.

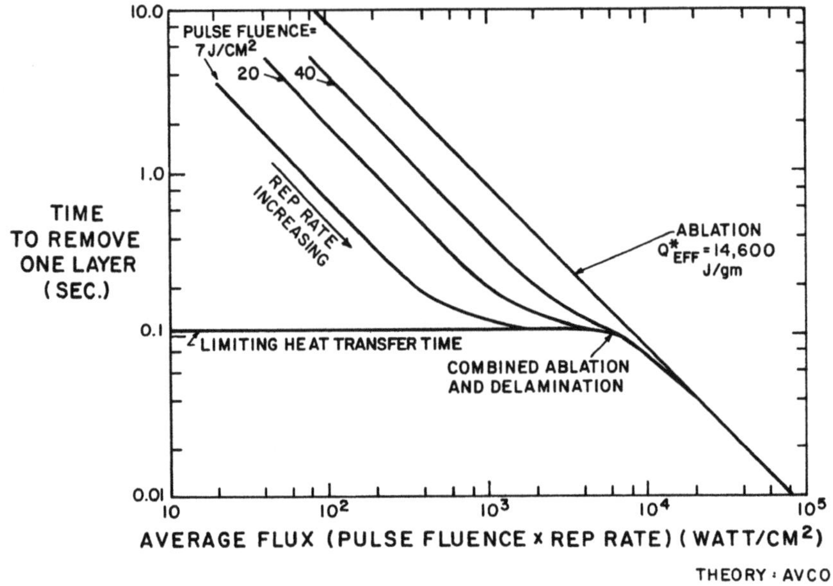

Figure 11 DELAMINATION IS MORE EFFICIENT THAN ABLATION, BUT LIMITED BY HEAT TRANSFER

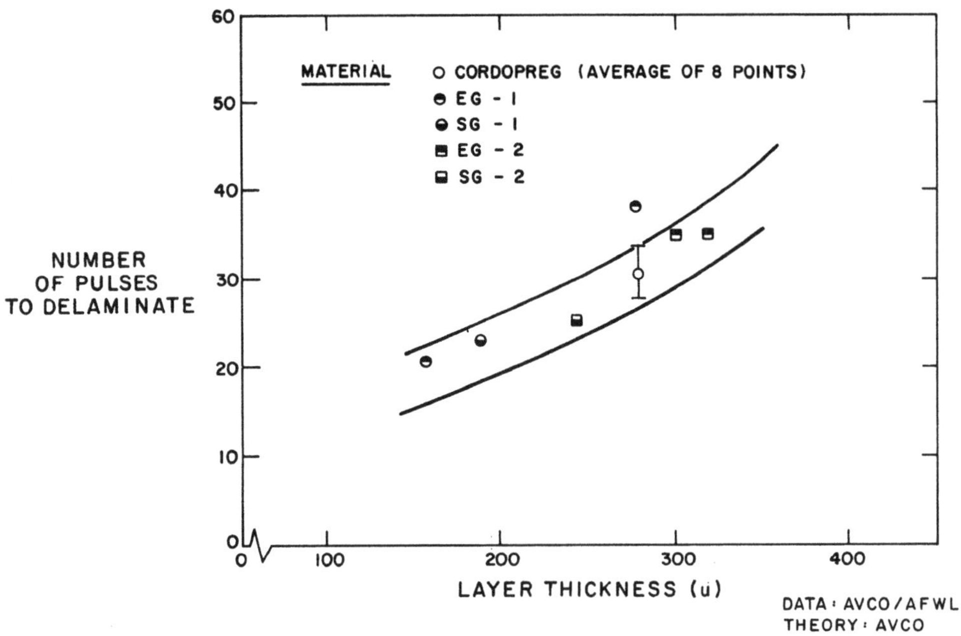

Figure 12 MODEL PREDICTS THICKNESS EFFECT

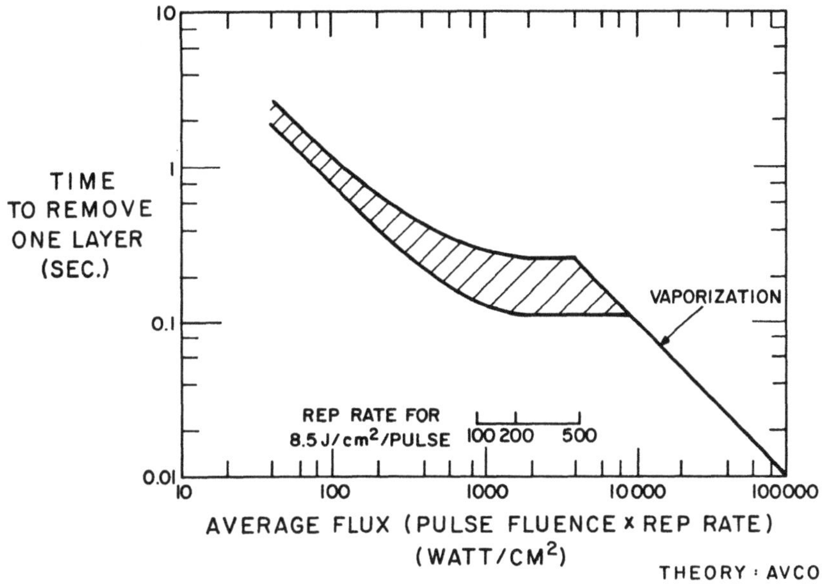

Figure 13 MODEL EXTRAPOLATED TO HIGH AVERAGE FLUX PREDICTS FAST PENETRATION BY DELAMINATION

To summarize the observations in fiberglass, indepth absorption dominates thermal deposition, slow heat transfer leads to multiple pulse cumulative effects, and delamination is the dominant damage mechanism.

SUMMARY

Table III summarizes the results presented in this paper for the interaction of rep-pulsed lasers with aluminum and fiberglass.

Table III
Pulsed-Laser/Material Interaction

- ALUMINUM
  - PLASMA FORMATION DOMINANT IN IR (LARGE IMPULSE & THERMAL COUPLING)
  - RATIO OF PULSE LENGTH TO PLASMA CONFINEMENT TIME < 1-2 REQUIRED
  - NET THERMAL COUPLING INCREASES AT SHORTER WAVELENGTHS
  - VACUUM IMPULSE LOWER THAN FOR AIR PLASMA
  - VACUUM THERMAL COUPLING HIGH FOR CO_2 IF FORM METAL VAPOR PLASMA
- FIBERGLASS
  - IN-DEPTH ABSORPTION DOMINATES FLUENCE DEPOSITION
  - SLOW HEAT-TRANSFER LEADS TO MULTIPLE-PULSE CUMULATIVE EFFECTS
  - DELAMINATION IS THE DOMINANT DAMAGE MECHANISM

REFERENCES

1. Reilly, J.P., Ballantyne, A., and Woodroffe, J. AIAA J. 17, 1098 (1978).
2. McKay, J., Schriempf, J., Cronburg, T., Eninger, J., and Woodroffe, J., Appl. Phys. Lett. 36, 125 (1980).
3. Holmes, B., Maher, W. and Hall, R., J. Appl. Phys. 51, 5699 (1980).
4. Pirri, A.N., Root, R.C., and Wu, P.K.S., AIAA J. 16, 1296 (1978).
5. Shui, V., Young, L., and Reilly, J., AIAA J. 16, 649 (1978).
6. Duzy, C., Woodroffe, J., Hsia, J., and Ballantyne, A., Appl. Phys. Lett. 37, 542 (1980).
7. Maher, W., and Hall, R., AIAA 14th Fluid and Plasma Dynamics Conference, Paper AIAA-81-1211, June 23-25, 1981, Palo Alto, CA.
8. Woodroffe, J., Hsia, J., and Ballantyne, A., Appl. Phys. Lett. 36, 14 (1980).
9. McKay, J., Schriempf, J., Woodroffe, J., and Cronburg, T., AIAA 16th Thermophysics Conference, Paper AIAA-81-1152, June 23-25, 1981, Palo Alto, CA.
10. Ballantyne, A., Woodroffe, J., Cronburg, T., Rudder, R., and Bina, M., AIAA J., 19, 731 (1981).

NUMERICAL SOLUTIONS OF COMPLETE AND REDUCED NAVIER-STOKES EQUATIONS FOR SUPERSONIC CHEMICAL LASER FLOW MODELING

V.K. Baev and V.I. Golovichev

Institute of Theoretical and Applied Mechanics
Academy of Sciences of the USSR
Novosibirsk 630090

INTRODUCTION

The use of nonequilibrium, high-speed, convective flows for producing a continuous laser action has recently led to the creation of various CW diffusion-type laser systems. What is required is a widely applicable computer model for analyzing the ducted flow configurations in the mixing zone of two chemically reacting coaxial streams with strong lateral pressure gradients. Computational efficiency is also a matter of paramount importance.

The analysis is limited to the numerical modeling of mixing and chemical reactions between gas components of the chemical diffusion-type HCl-laser in the base region produced by nozzle trailing edges of planar or axisymmetric geometry. One nozzle (A) injects vibrationally excited hydrogen, while the other (B) contains products of diluted molecular chlorine thermal dissociation. It is possible in principle to realize vibrational energy transfer from excited $H_2(v)$ molecules to $HCl(v)$ molecules produced in chemical reactions. We also examine the influence of additional fuel feeding by injecting a cold sonic jet of hydrogen through the nozzle (C).

In the general case the jet static pressures are not matched and the pressure gradients induced by compression and expansion waves as well as by chemical heat release greatly influence the energy conversion process. The latter phenomenon is a matter of importance for HCl chemical laser operation since vibrationally excited $HCl(v)$ molecules can be produced only in the "hot" branch of a chemical mechanism and the degree of Cl_2 dissociation cannot be arbitrarily increased. A "hot" branch is characterized by essential heat release that leads at a low concentration of diluent to "thermal blockage" of the flow. The use of a nozzle bank with a base relief region has proved[1] to be an effective means for preventing this phenomenon. The presence of the base relief region favors reducing of the pressure level due to the expansion fan interaction with a reacting flow immediately downstream of the base. Such a coupling of hydrodynamics and chemical kinetics has been examined using both time-dependent and steady-state numerical models.

COMPUTER MODEL

A system of complete Navier-Stokes equations supplemented with the mass and energy conservation laws in the form of the relevant partial differential equations was used to model the flow of a multicomponent mixture with finite-rate physical and chemical processes. These equations are well known and are presented here for reference:

$$\partial\rho/\partial t = -\nabla\rho\underline{v} \tag{1}$$

$$\partial(\rho\underline{v})/\partial t = -\nabla\rho\underline{vv} - \underline{\nabla}p + \nabla\eta\underline{\nabla v} \tag{2}$$

$$\partial\rho_i/\partial t = -\nabla\rho_i\underline{v} + \nabla D_i\underline{\nabla}\rho_i + \{\partial\rho_i/\partial t|_{chem}\} \tag{3}$$

$$\partial E/\partial t = -\nabla E\underline{v} - \nabla p\underline{v} + \nabla\lambda\underline{\nabla}T + \{\partial E/\partial t|_{chem}\} \tag{4}$$

where ρ, $\rho\underline{v}$, E, and p are the total mass, momentum, and energy densities, and pressure, respectively. The ρ_i and D_i are the mass density and the "effective" diffusion coefficient of the individual species. The quantities η and λ represent viscosity and thermal conductivity of the gas mixture. The effective diffusion coefficients D_i for individual species, depending in a complicated manner on the mixture composition and the binary diffusion coefficients for all the pairs of species, were approximated by simplified relations presented in Ref. 2. A stricter formulation using Stephan-Maxwell relations seems to be impractically complex, though it gives us a possibility of overcoming the difficulties associated with a correct description of species diffusion with highly different diffusion coefficients D_i. A mathematical statement of the problem suggests the specification of kinetic equations in the form of a phenomenological mass action law in accordance with a mechanism of kinetic processes or chemical reactions, reaction rate coefficients, and thermal properties of mixture components. Details of the physical model are presented in Ref. 3.

The finite-difference analogues of the general conservation laws in terms of the complete Navier-Stokes equations are solved by using the partially implicit (RICE) method[4] that employs a concept of operator splitting with respect to the physical processes.[5] The solution procedure for this predictor-corrector scheme is as follows:

a) guess a value of the pressure, e.g., the pressure at the last integration station;
b) integrate the two momentum equations (2) subject to the imposed boundary conditions by a symmetric ADI scheme to get the predicted values of the velocity components;
c) find the pressure corrections p' from the Poisson equation;
d) correct the velocity components by using the gradients of the scalar function associated with pressure corrections to get the new values of u, v, p, ρ;
e) continue this process until a convergence is obtained.

This procedure eliminates the Courant sound speed criterion on the magnitude of the time step, a restriction inherent in explicit compressible flow computer codes.

The mixture energy equation is then solved explicitly using the updated values of the density and the velocity components and neglecting the heat release from chemical reactions. The convective, diffusive, and chemical reaction contributions of the species transport equations are taken into account consecutively, and the final values of the species densities, mixture energy, and pressure are obtained.

The finite-difference calculations of the steady-state "parabolized" flow problems are naturally performed by a time-dependent method, and a special approach[6] is developed for reducing the amount of the computer time required. However, the governing conservation laws in the model of parabolic approximation of Navier-Stokes equations can be formally obtained when employing the steady-state ($\partial/\partial t = 0$) analogue of the system (1)-(4) by omitting the second derivatives of the unknown functions with respect to a marching spatial variable coinciding with a main flow direction. This approach is rather simple for numerical operating (parabolic equations can be derived), sufficiently economical (a marching scheme of numerical integration can be constructed), and incorporates, when applied to the analysis of two-dimensional turbulent flows, the basic concepts of semi-empirical turbulence theories.

These steady-state equations were solved by a finite-difference, "iterative space-marching" method which can be considered as a combination of the Patankar-Spalding[7] procedure for predicting the two-dimensional "parabolic" flows with the so-called "partially-parabolic" method[8] of handling "elliptic" flow without recirculation. In this numerical procedure, which represents also the predictor-corrector method, the governing equations are solved identically at each mesh point irrespective of the local flow condition (subsonic or supersonic). However, the elliptic and hyperbolic behaviors of the subsonic and supersonic regions are expressed through a "convective velocity" procedure suggested in Ref. 8. According to this procedure, the ρu momentum component is calculated as follows:

$$
\begin{aligned}
(\rho u)_e &= \rho_E u_e, \quad \text{for } M_e < 1 \\
(\rho u)_e &= \rho_p u_W, \quad \text{for } M_e \geqslant 1
\end{aligned} \tag{5}
$$

where the points e, E, p, W are typical locations of the unknown functions on the finite-difference mesh.

The Poisson equation for the pressure corrections is reduced in this case to a parabolic one:

$$
(A_N + A_S + A_E + A_W)P'_p = A_N P'_N + A_S P'_S + \varepsilon_p \tag{6}
$$

where A_N, A_S, A_E, A_W are the mesh functions defined at the typical mesh points (see Ref. 8), and ε_p represents the so-called kinematic compatability error in the region surrounding the mesh point.

These numerical procedures have proved to be successful in treating the two-dimensional chemical laser flows in a supersonic mixing-controlled device. The calculations for optical gain coefficients on selective transitions and output parameters of the laser system can be performed in the usual way.

NUMERICAL RESULTS

An example of calculations deals with the numerical modeling of mixing and chemical reactions between the components of the chemical diffusion-type HCl-laser in the base region produced by the nozzle trailing edges of planar and axisymmetric geometry. Figure 1 shows the geometric configuration of the nozzles under study. The products of chlorine thermal dissociation diluted by inert heat-absorbing gas (N_2) enter the lasing cavity at Mach number 4.3, a static temperature of 650 K, and a pressure of 7.5 torr, producing in the case of planar geometry the near wake behind the backward step with a slit in its bottom through which cold (300 K) partially excited hydrogen is injected at sonic conditions and at the elevated pressure of 250 torr. The Reynolds number defined by the flow conditions on the axis of symmetry and the obstacle size is equal to 3000. These conditions are specific for chemical HF-laser operation. Such a mixing design was examined in Refs. 3 and 4 and suffers from the shortcoming that the process of reaction product formation proceeds partially in the Cl/Cl_2 stream inside the oxidizer nozzle where the recirculation zone has been pushed owing to the excess pressure on the fuel side. However, at these parameters the flow conserves the supersonic

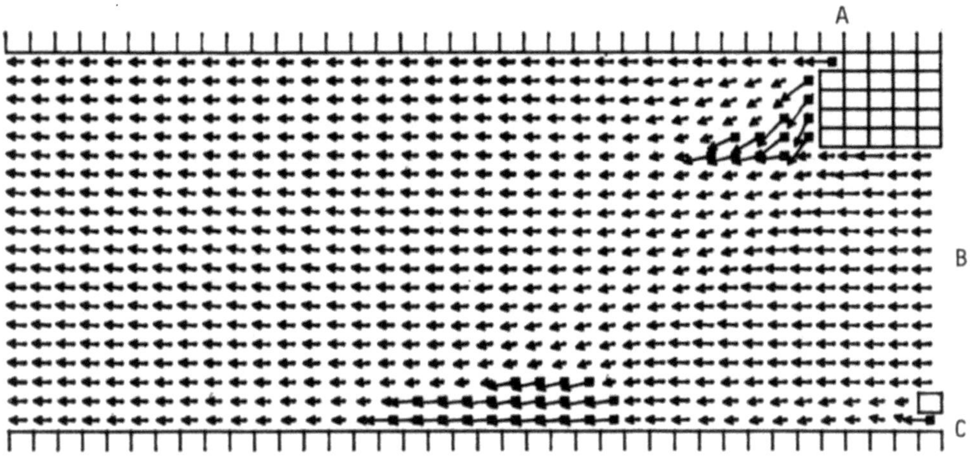

Fig. 1. Schematic and velocity vector plot of a steady plane flow.

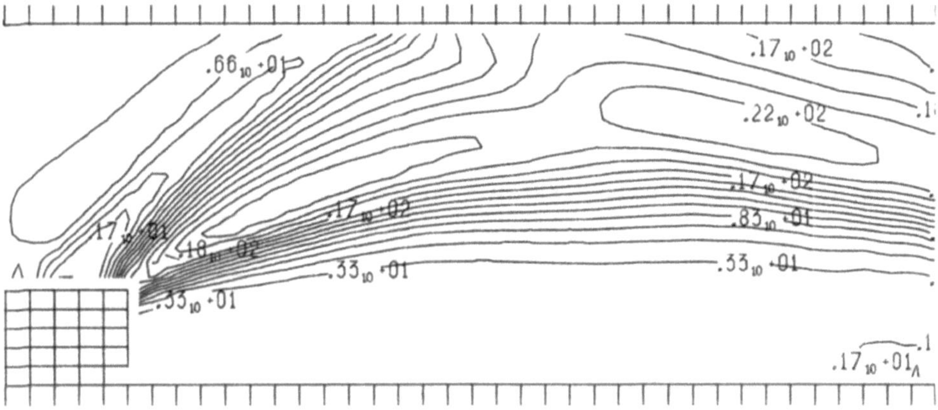

Fig. 2. Contour plot of temperature (T/100, steady plane flow).

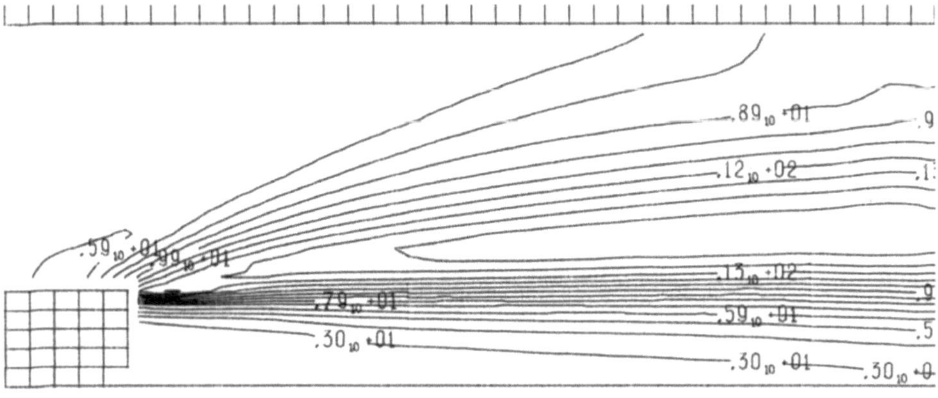

Fig. 3. Contour plot of temperature (steady axisymmetric flow).

jet-like nature, and calculations carried out with the help of a modified RICE computational program[4] are not expensive. A fuel jet in a duct expands like a supersonic stream passing through the corner expansion waves. The shock produced by the nozzle trailing edge reflects from the upper wall of the duct and together with a weak recompression shock "separates" the base region so that the base static pressure decreases. The steady-state picture of a plane flow (in the absence of a nozzle C) is shown in Fig. 2 by contours of equal temperature values. The secondary compression of the flow takes place mainly in the wake circumference in the oblique shock that becomes regular after its reflection from the plane of symmetry. The flow remains supersonic within the whole computational region ($z/H = 5$, z is the longitudinal distance, H is the height of the obstacle). Chemical reactions proceed in the regions of compressed flow where a sufficient degree of the mixing process completion is achieved, and relatively weak intensity of chemical processes is characterized by temperature distributions. Similar temperature distributions for the axisymmetric flow are represented in Fig. 3. In this case the flow is more homogeneous with respect to the mixture composition and pressure; the flame front is almost parallel to the axis of symmetry of the flow.

In Figs. 4 and 5 the results of numerical modeling of a flow originating under extra fuel feeding by injecting a cold sonic underexpanded (jet pressure ratio = 5) hydrogen jet are shown by contours of equal fuel concentration and pressure values. The pressure distributions at successive moments of time are indicative of the oblique shock formation during the initial stage of flow development, but the shock regular reflection from the upper wall of the duct does not occur since the steam in this flow region turns out to be "thermally blocked" due to the additional heat release.

The effective deceleration of the supersonic flow by heat release is accompanied by a noticeable increase in pressure in the flow region that adjoins the wake plane of symmetry (or a lower wall of the duct). The elevated pressure, however, cannot be "transferred" to the base region since the realized flow structure does not contain the subsonic region that couples the base region and the elevated pressure zone. The relative locations of subsonic regions are presented in Fig. 1. The velocity vectors in these regions are marked by squares.

Similar results for the axisymmetric flow are shown in Figs. 6 and 7. In this case the stream is "thermally blocked" in the vicinity of the axis of symmetry, but subsonic regions remain "isolated" and the pressure and temperature recovery slightly influence the averaged optical property of a medium.

A detailed chemical mechanism was included into consideration on the basis of the "parabolized" Navier-Stokes equations. The

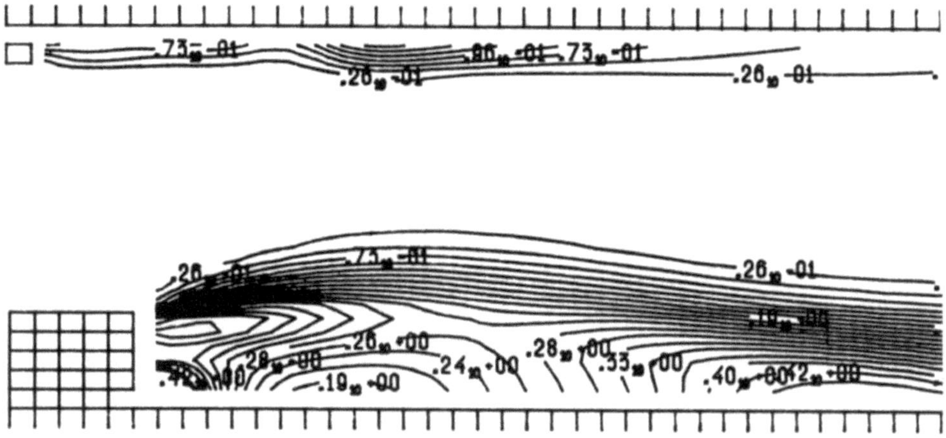

Fig. 4. Contour plot of fuel density (steady plane flow).

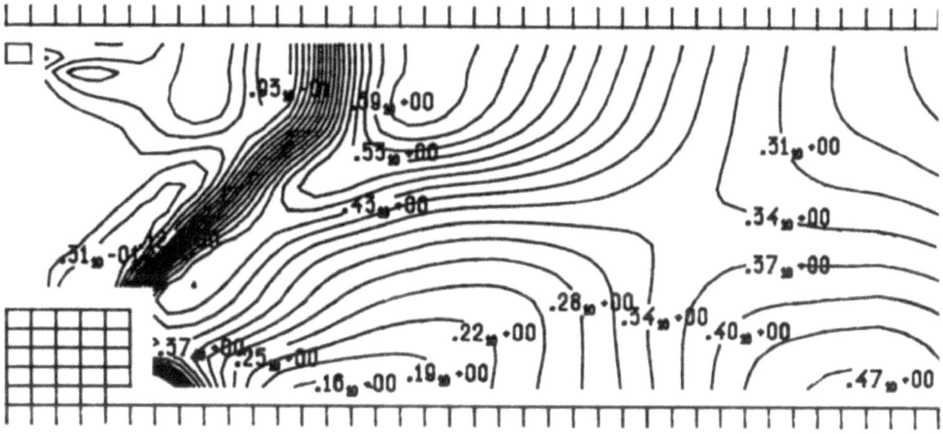

Fig. 5. Contour plot of pressure (steady plane flow).

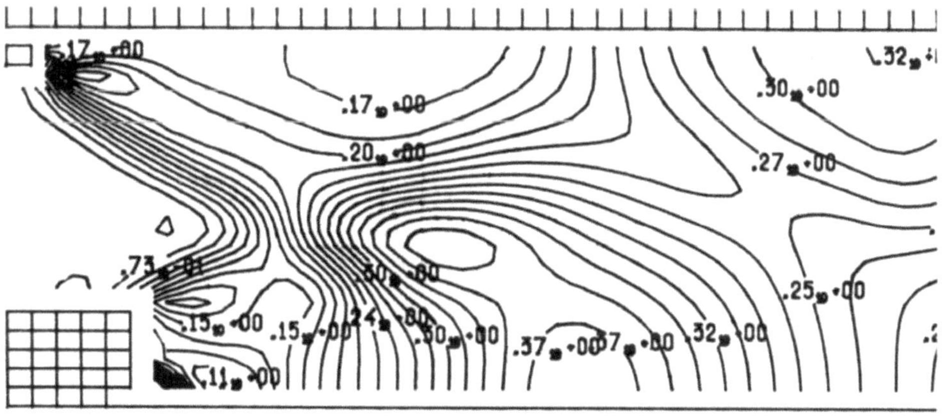

Fig. 6. Contour plot of pressure (steady axisymmetric flow).

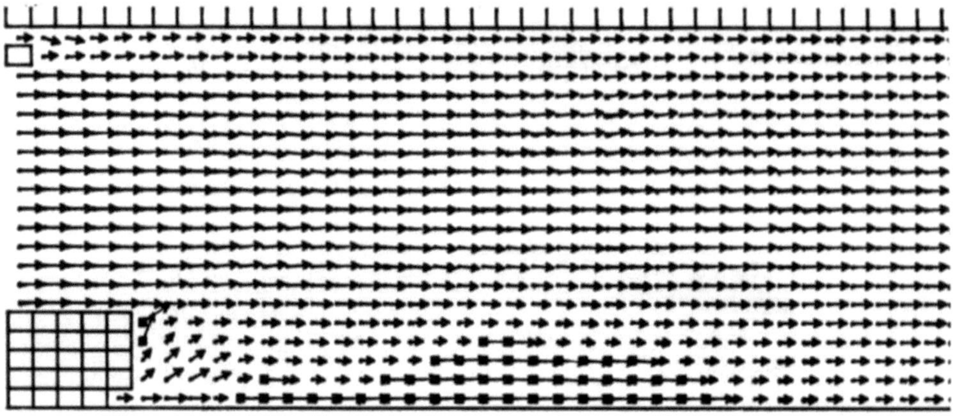

Fig. 7. Velocity vector plot of a steady axisymmetric flow.

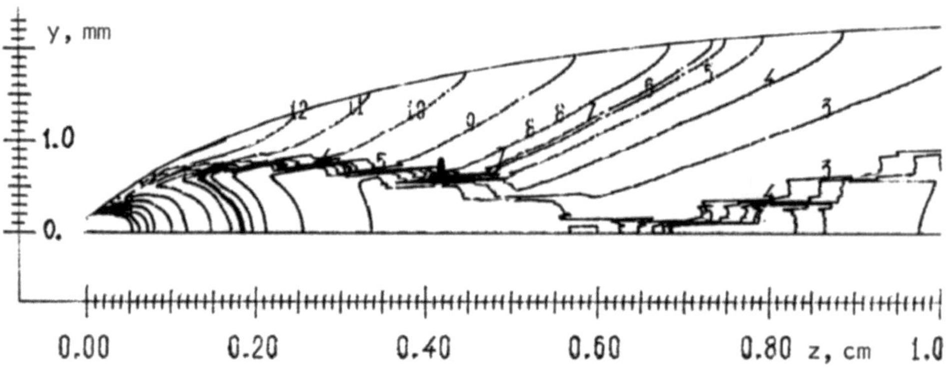

Fig. 8. Contour plot of normalized pressure for parabolized flow; 3 - .712, 12 - 6.34

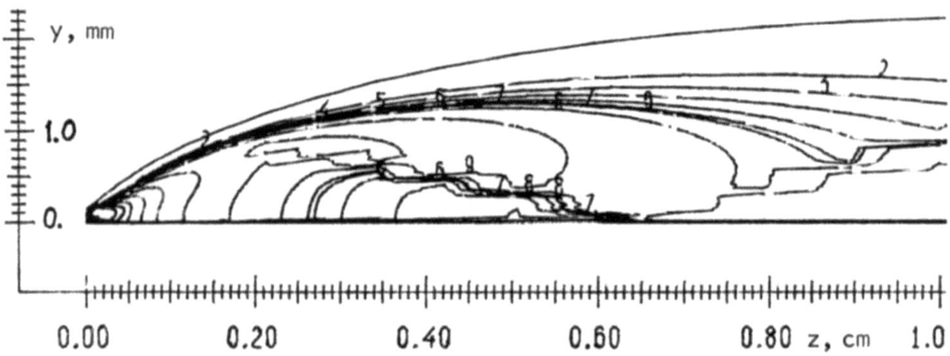

Fig. 9. Contour plot of effective viscosity coefficient; $2 - .213_{10} - 3$, $9 - .97_{10} - 3$.

advanced k-ε model of turbulence was adopted to account for the turbulent mixing effects. This model enables one to describe satisfactorily an appreciable transverse flow motion induced by the oblique shock that forms a specific diamond-shaped structure downstream of the flow. The gas-dynamic nature of the flow is illustrated in Figs. 8 and 9, which present values characterizing the averaged (pressure) and fluctuating (effective viscosity, $\mu = c_\mu \rho k^2/\varepsilon$) flow parameter of two reacting jets mixing.

A considerable rise in the turbulent transport takes place in the vicinity of the oblique shock as a result of lateral and longitudinal pressure gradients. This phenomenon leads to enhancement of the mixing process to a noticeable extent.

CONCLUSIONS

The numerical illustrations present informative results on the gas dynamics of reacting flows typical of chemical diffusion-type lasers.

The authors are indebted to Professor N.N. Yanenko for support of this work.

REFERENCES

1. Grohs and G. Emanuel, Gas dynamics of supersonic mixing lasers, in: "Handbook of Chemical Lasers," R.W.F. Gross and J.F. Bott, eds., Wiley, New York (1976).
2. E.S. Campbell and R.M. Fristrom, Reaction kinetics, thermodynamics and transport in the hydrogen-bromine system, Chemical Reviews, Vol. 58, No. 2 (1958).
3. V.K. Baev, V.I. Golovichev, et al., Numerical modeling of a chemically driven H_2-HCl transfer laser. Paper presented at the VIIIth ICOGER, Heat and Mass Transfer Institute, Minsk, USSR (9181).
4. W.C. Rivard, O.A. Farmer, and T.D. Butler, RICE: a computer program for multi-component chemically reactive flows at all speeds, LASL Report LA-5812 (1975).
5. V.M. Kovenya and N.N. Yanenko, The splitting-up method in gas dynamics problems, Nauka, Novosibirsk (1981). (In Russian)
6. O.A. Farmer and J.D. Ramshaw, A method for rapid attainment of the steady state in two-dimensional time-marching supersonic flow calculations, Comp. Phys., Vol. 24, No. 1 (1977).
7. V.K. Baev, V.I. Golovichev, and V.A. Yasakov, Two-dimensional turbulent flows of reacting gases, Nauka, Novosibirsk (1976). (In Russian)
8. A.K. Singhal and D.B. Spalding, A two-dimensional, partially parabolic procedure for axial-flow turbomachinery cascades, HTS/76/20 (1976).

A PULSED, MULTI-KILOHERTZ-REPETITION-RATE, SUPERSONIC HF CHEMICAL LASER

D. Chuchem

Laser Unit, Nuclear Research Centre - Negev
Beer-Sheva, Israel

S. Rosenwaks

Department of Physics, Ben-Gurion University
Beer-Sheva, Israel

ABSTRACT

Operation of a pulsed, multi-kilohertz-repetition-rate, supersonic HF chemical laser is described. Continuously flowing reactants are premixed and then expanded through a supersonic nozzle to the laser cavity where a pulsed electrical discharge initiates the chemical reaction producing the lasing species. Average power of 12 watts has been obtained from a hydrogen/sulfur-hexafluoride/helium mixture. It is suggested that this apparatus can be used to overcome some of the inherent problems in achieving quasi-continuous laser operation based on chain reaction in fluorine/hydrogen mixtures.

INTRODUCTION

The currently existing HF chemical lasers may be divided into two main classes:[1]

The first class is based on in situ mixing of an oxidizer (F atom) and a fuel (H donor) gas, producing a continuous output. The oxidizer containing gas is atomized by means of electrical discharge (microwave,[2,3] RF,[4] or DC[5]), by thermal decomposition (arc heaters[6] or combusters[7]), or by chemical reactions,[8] to produce reaction initiating F atoms. The rate of the chemical reaction is diffusion dominated and thus this class of lasers must be operated at comparatively low pressures (of the order of 1-20 Torr), so that the rate of production of vibrationally excited HF molecules will exceed the rate of their collisional deactivation.

The second class is based on the use of a premixed fuel and oxidizer containing gas, where the necessary reaction initiating species are produced by flash photolysis,[9] by an electron beam,[10] or by a pulsed discharge.[11-13] Such premixed reactants, if active enough,[14] may present the following problems. 1) Once the mixture is ignited, the combustion front tends to travel upstream and locate itself adjacent to the entry ports. The flame must be extinguished in order to allow unreacted gas mixture to enter the laser cavity. 2) The combustion is apt to transform via a chain reaction into a detonation wave which can cause damage to the device. A simple way to avoid these problems is to replenish the gases between consecutive laser pulses in a non-continuous manner. This can be achieved, for example, by shutting down the fuel flow for a period.[9] However, this method is not conducive to quasi-continuous operation.

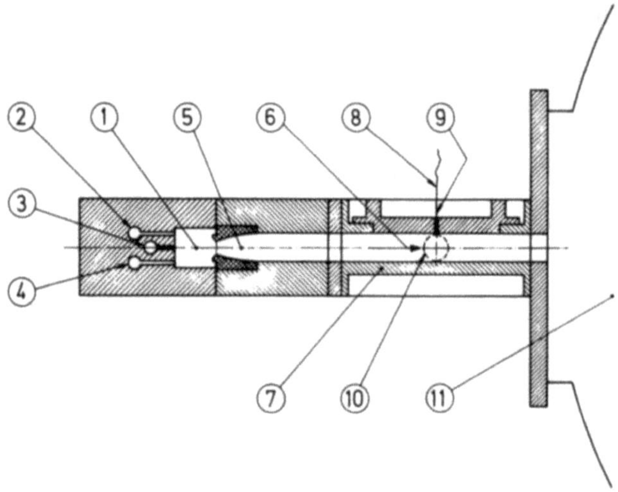

Fig. 1. A sectional side-view of the laser. (1) - 84 ml mixing chamber. (2),(3),(4) - gas inlet jets. (5) - Mach 2.4 nozzle with a throat cross-section of 9 mm × 10 cm. (6) - laser cavity with a height of 2 cm and a length of 10 cm. The laser axis is located 17 cm downstream of the nozzle throat. (7) - aluminum cathode. (8) - high voltage connection. (9) - stainless steel pin anodes. (10) - mirror. (11) - vacuum dump tank.

PULSED SUPERSONIC HF LASER

We here present a supersonic laser which produces high repetition rate pulsed output from premixed gases.[15] The reactants are mixed "on the fly" and then expanded through a supersonic nozzle so that they can be convected through the laser cavity at a velocity greater than the combustion velocity. A pulsed electric discharge is applied at the desired rate, initiating the chemical reaction which produces the lasing species. The spent gas is continuously convected out of the laser cavity, while a new unreacted gas is convected into the cavity at supersonic speed. We have performed experiments with sulfur-hexfluoride/hydrogen or nitrogen-trifluoride/hydrogen mixtures, with or without diluents, and suggest that this system can be used most profitably with fluorine/hydrogen mixtures.

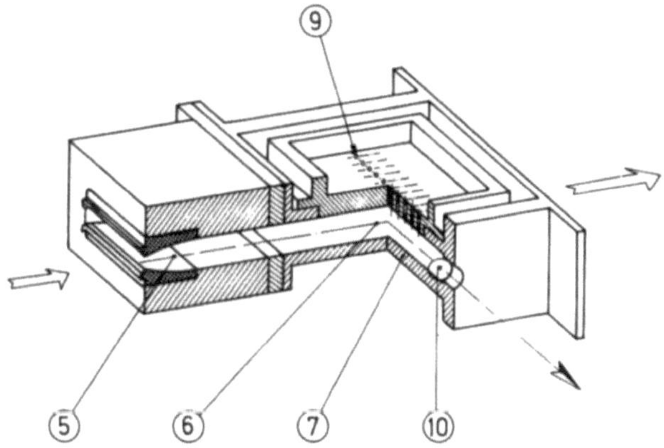

Fig. 2. Isometric drawing of the laser, cut away for clarity. The numbering is the same as in figure 1.

DESCRIPTION OF THE APPARATUS

A sectional side view of the laser is depicted in figure 1; an isometric drawing of the laser, cut away for clarity, is shown in figure 2. The laser comprises in combination a mixing chamber (1), provided with inlet jets (2), (3), and (4) for the gaseous species, which pass via the supersonic nozzle (5) into the 10 cm long laser cavity. High repetition rate electrical pulses are applied between ten stainless steel pin anodes (9), which are evenly spaced in the central 40% of the cavity length, and an aluminum cathode (7). The anodes are connected through connections (8) to a pulse forming network, and the cathode to a ground return. (10) represents the laser mirrors and (11) a vacuum dump tank connected to an evacuation pump.

The laser cavity is composed of a 1 meter radius of curvature dielectric enhanced gold on glass full reflector, and a 1 meter radius of curvature dielectric coated zinc selenide mirror having a 92% reflectivity around 2.8 microns. The mirrors are 10 cm apart, and have effective apertures of 2 cm.

The pulse forming network (figure 3) is comprised of a 10 nFarad capacitor which is resonantly charged through a 220 mHenry air core inductor. The capacitor is discharged through the laser by means of an EEV CX1154 deuterium filled ceramic thyratron, which is attached to the 10 pin anodes through 10 separate, 18 ohm impedance each, coaxial cables of 2 meter length. These cables thus serve as ballast, effectively masking the pins from each other for a period of 20 nseconds. They also serve as peaking capacitors.

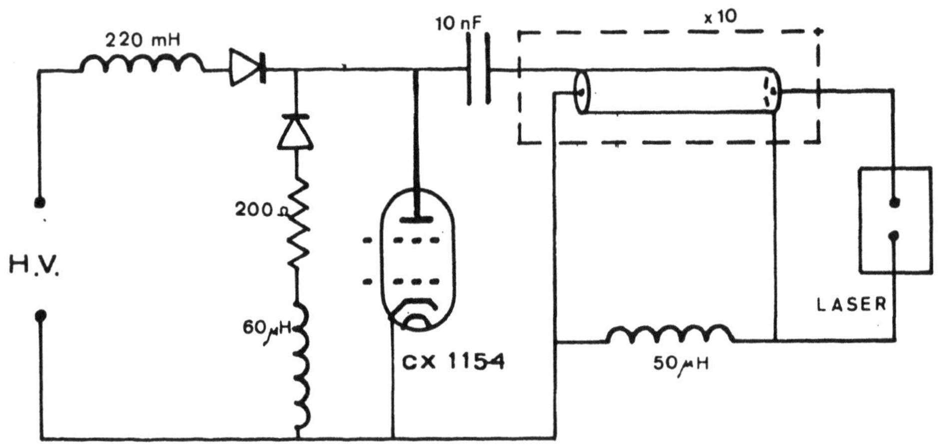

Fig. 3. Schematical diagram of the pulse forming network. Only one of the ten coaxial cable/anode pins is drawn.

EXPERIMENTAL RESULTS

The system was operated with different gas compositions, charging voltages, repetition rates, and pulse forming networks. The best performance achieved to date with the above mentioned system was 12 watts. This was obtained with a gas mixture comprised of 8% sulfur-hexafluoride, 4% hydrogen, and 88% helium (by volume), at a stagnation pressure of 200 torr. The gaseous mixture passed through a Mach 2.4 supersonic nozzle at a flow rate of 63 NTP-liters per second into the discharge chamber. The pressure and temperature of the gas in the laser cavity were 14 Torr and 111 K respectively. Electrical pulses of 2 Joules were applied at a frequency of 6500 Hertz.

CONCLUSIONS

Our laser system has so far been utilized for sulfur-hexafluoride/hydrogen and nitrogen-trifluoride/hydrogen mixtures. Since the gas flow velocity can be greater than the velocity of a fluorine/hydrogen flame, such a device can be used with premixed, oxygen inhibited, diluted, fluorine and hydrogen mixtures without suffering detonation problems. Also, if the device is operated at a pulse repetition rate in the tens of kilohertz range,[16] the supersonic gas flow can be efficiently utilized. Furthermore, in such a system the gas density of the supersonic flow through the laser cavity can be kept low enough to facilitate pulsed discharge, while on the other hand the gas stagnation pressure can be made high enough to allow pressure recovery and gas exhaust to the atmosphere without requiring evacuation pumps. Work is now in progress towards achieving these goals.

ACKNOWLEDGMENTS

We thank I. Smilanski and S. Yatsiv for helpful discussions. The technical assistance of A. Benhenisti, A. Kerman, A. Rotstein, and M. Cohen is gratefully acknowledged.

REFERENCES

1. R.W.F. Gross and J.F. Bott, ed., "Handbook of Chemical Lasers," (Wiley, New York, 1976).
2. L. Bertrand, J.-P. Monchalin, R. Pitre, M.L. Meyer, J.M. Gagne and M. Moisan, Rev. Sci. Instrum. 50:708 (1979).
3. M. Gastaud and P. Bousselet, J. Quantum Electron. 16:246 (1980).
4. Reference 1, pp. 250 ff.
5. D. Proch, H. Pummer, K.L. Kompa and J. Wanner, Rev. Sci. Instrum. 46:1101 (1975).

6. Reference 1, pp. 215 ff.
7. Reference 1, pp. 266 ff.
8. J.A. Shirley, R.N. Sileo, R.R. Stephens and T.A. Cool, AIAA paper #71-27 (1971).
9. R. Limpaecher and J. Woodroffe, AIAA J. 15:1612 (1977).
10. R.A. Gerber and E.L. Patterson, J. Appl. Phys. 47:3524 (1976).
11. G. Wlodarczyk, J. Quantum Electron. 14:768 (1978).
12. K. Midorikawa, S. Sumida, Y. Sato, M. Obara and T. Fujioka, J. Quantum Electron. 15:190 (1979).
13. J.E. Brandelik and R.F. Paulson, J. Quantum Electron. 13:933 (1977).
14. H.L. Chen, J.D. Daugherty and W. Fyfe, J. Quantum Electron. 11:648 (1975).
15. S. Rosenwaks and D. Chuchem, Patent pending.
16. Z. Karny, A. Rotstein, D. Chuchem and I. Smilanski, Rev. Sci. Instrum. 51:1426 (1980).

SATURATION PROCESSES IN DOPPLER-BROADENED HF VIBRATIONAL TRANSITION*

Rolf W. F. Gross, and John G. Coffer

The Aerospace Corporation

Los Angeles, CA. 90009

ABSTRACT

We have experimentally determined the frequency dependent shapes of Doppler-broadened HF vibrational transitions in the presence of a frequency stabilized, single-mode, saturating radiation field. We find that the transitions saturate homogeneously with no effective hole-burning observable.

INTRODUCTION

In a previous investigation,[1] we determined the absorption line-profiles and the anomalous dispersion of HF gas at pressures where the vibrational transitions were Doppler broadened. We were able to show that both quantities corresponded to values predicted by standard Doppler-line theory. The present investigation was originally an attempt to extend these measurements to the gain lines of an active, vibrationally inverted medium of a supersonic HF laser. In the study, an additional attempt was also made to measure gain and anomalous dispersion while the transitions were saturated.

*This work was supported in part by the Defense Research Projects Agency (DARPA) under U.S. Air Force Space Division (AFSD) Contract F04701-82-C-0083.

To permit a clear interpretation of the experimental results, it was decided to measure the gain-line profiles when they were saturated by an external, single-mode radiation field that passed the medium in one direction only, i.e., the active medium was used as a single-pass amplifier. The pressure of the active medium was maintained at 6 Torr, a value at which standard theory predicts that the lines would be Doppler broadened and pressure broadening would be an order of magnitude smaller.[2]

The investigation concentrated on the $P_2(8)$ line for technical reasons: negligible atmospheric absorption, relatively high gain both in the supersonic laser medium and in the saturating laser oscillator, and no absorption by expended ground-state HF molecules that are always present in the amplifier medium.[3]

As a result, the magnitude of the anomalous dispersion was much smaller than expected, and in fact, it proved impossible to extract meaningful dispersion profiles as a function of frequency from the relatively high noise produced by the amplifier medium. Because we suspected that the reason for this result was connected with the saturation behavior of the transition, we concentrated our investigation on a study of saturated gain-line profiles.

The surprising result was that the investigated HF transition saturated homogeneously under the operational conditions of our supersonic diffusion laser. The amplifier gain-line was found to be broadened, which we attribute to source flow effects.[4] Because of this large line broadening we were not able to investigate the line wings and answer the question as to whether the line-profile was of Doppler or Lorentzian shape.

EXPERIMENTAL APPARATUS

Figure 1 indicates a schematic of the apparatus used in our experiments. The investigated active medium was produced by the Aerospace supersonic HF cw laser, which was driven by an arc heater and equipped with a standard 36-slit nozzle.[3] The nozzle produced an active medium of 18 cm width and 2.5 cm height normal to the flow direction. The laser, its flow field, and operating parameters have been exhaustively described elsewhere.[3] We used the laser medium exclusively as an amplifier, so that it was possible to replace the Brewster windows of its oscillator configuration with flat, uncoated calciumfluoride window plates. The windows were not flushed with an inert gas, which permitted operation of the test section at the reduced pressure of 6 Torr.

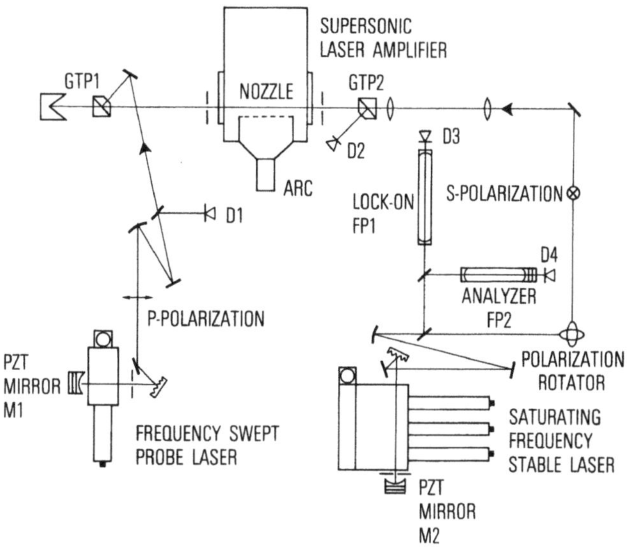

Fig. 1. Optical layout of apparatus

The amplifier was illuminated by two cw HF lasers that probed the active medium from opposite sides, 4 mm downstream of the slit nozzle exit. The beams from the two probe lasers were arranged concentrically in the amplifying medium, so that both interacted with the same molecules of the medium. In one of the two probe lasers a power density sufficient for saturation was produced; the other laser, of much lower power, which was swept in frequency, served to determine the frequency dependent gain-profile of the investigated transition. The amplification experienced by the low-power probe beam was measured by two uncooled, high-frequency InAs-detectors, one before and one after the beam passed the test section.

In order to distinguish the low-power gain signal from the colinear high-power saturating beam, the two beams were polarized orthogonally to each other, which enabled us to separate them with the aid of Glan-Thompson prisms before and after the test section. The prisms were made to our specifications by Karl Lambrecht[5] from single crystals of Rutile (TiO_2). Their rejection ratio, discriminating between S and P polarization, was at least 1/1000 at 2.8 μm.

The laser used as the source of the saturating power was of the type described in Reference 6. It operated on SF_6, He, O_2 mixtures dissociated by a dc discharge in three parallel discharge tubes. H_2 was injected through wall holes into a nearly sonic throat. Equipped with a grating and a 2-m radius gold-coated mirror, this laser produced up to 35 W of continuous, multimode power on the $P_2(8)$ line of HF. In our experiments, the cavity length was reduced to 40 cm. The resultant longitudinal mode separation, after correcting for mode-pulling, was about 320 MHz. This mode separation is of the order of the gain-line width of this laser, and makes its operation on a single longitudinal mode possible.

The laser cavity was frequency stabilized by locking it to a fixed, stable Fabry-Perot interferometer 25 cm long, corresponding to a free spectral-range of 300 MHz. As depicted in Fig. 1, this reference Fabry-Perot interferometer was arranged externally to the laser cavity.

The total reflector of the laser cavity was mounted on a piezoelectric transducer, to which a sinusoidal signal of 600 Hz and a few tenths of a volt amplitude were applied. This signal provided a small dither signal, which was sensed by a detector mounted behind the Fabry-Perot interferometer. The error signal, produced by a frequency drift of the laser mode with respect to the Fabry-Perot, was amplified and rectified by a phase-lock amplifier and then used to correct the position of the cavity mirror, closing the control loop. A second Fabry-Perot interferometer of 750 MHz free spectral-range was used to monitor the mode structure and stability of the laser cavity. In this manner it was possible to produce up to 30 W in a single, stable mode on the HF $P_2(8)$ line with the freedom of locating this mode at any frequency of the gain-width of this line. Figure 2 indicates typical oscilloscope records of the power output and the mode of this oscillator. The upper trace was obtained by analyzing the laser output with a confocal Fabry-Perot interferometer of 750 MHz free spectral-range while oscillating the total reflector of the laser cavity with a large voltage at 600 Hz. The lower trace is a record of the stabilized single-line output integrated over an extended time. Note that the width of the mode increased dramatically in single-line operation; its FWHM is about 17 MHz. After

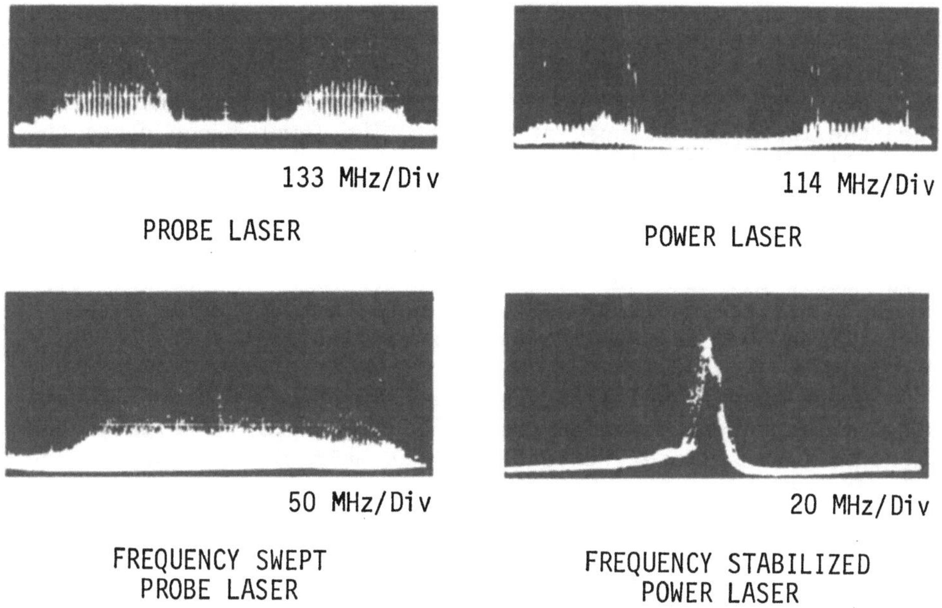

Fig. 2. Frequency analysis of laser oscillators used in experiments. Free spectral range of Fabry-Perot: 750 MHz.

passing several relay mirrors, a simple beam collimator, a polarization rotator, an inverted telescope of magnification 2, the Glan-Thompson prism, and the window of the test section, a Gaussian beam with an average maximum power of 8 W across a diameter of 3.5 mm could be produced in the amplifier medium. By attenuating the beam power, densities between 1 and 85 W/cm^2 were available for saturating the medium.

The low-power probe laser, used for the gain measurements, was of the same construction as the saturating laser[6], except that it was equipped with only one discharge tube. The laser cavity consisted of a 300-l/mm grating blazed at 3 µm and a 2-m radius gold-coated mirror. Power was extracted on the zeroth order beam. The cavity was mechanically stabilized inside four invar rods. The cavity was 30 cm long to assure single mode operation when the total reflector was swept across the line-width of the gain-medium by a piezoelectric translator, which was driven by the sawtooth voltage of an oscilloscope. The duration of a scan could be varied between 20 and 100 msec. The same oscillosope was used to display the signals from the two detectors monitoring the probe laser beam. The free spectral-range of the probe laser, corrected for mode-pulling, was 420 MHz. Figure 2 indicates the laser output spectrally dispersed by a 750 MHz Fabry-Perot interferometer when the laser was scanned at 600 Hz. The lower trace shows the power line-shape expanded to its full width. Note that the mode of this oscillator was at least an order of magnitude narrower than that of the saturating laser during scanned operation. Also, the width in frequency of the power output was about 400 MHz. This record provided the frequency calibration for the gain scans of the amplifier medium. The pronounced Lamb dip defined the center frequency of the line under investigation.

Under these conditions, this laser produced 0.3 W of total, single-mode output power on the $P_2(8)$ line of HF. The laser beam passed a beam splitter, diverting about half of its power to the InAs detector D1. The detector D1 was used to measure the input power I_0 to the test section. After passing several relay mirrors, the probe beam was superimposed on the saturating beam in Glan-Thompson prism GTP1. As depicted in Fig. 1, the probe beam entered the test section in a direction opposite to that of the saturating beam. The diameter of the probe beam was controlled by an inverted telescope of power 2 and to an aperture of 2.5 mm, resulting in a power density of less than 1 W/cm^2 in the amplifier. After the test section, the probe signal was separated from the power beam in Glan-Thompson prism GTP2. The amplified probe laser signal I_1 was finally determined by InAs detector D2.

The signals from the two detectors were amplified and then recorded digitally by an on-line PDP-11/10 computer. The computer was programmed to average 32 individual frequency scans of the two signals to reduce noise and data scatter. The program digitally recorded the two averaged intensities into 2 × 100 channels for each scan along the time/frequency coordinate. From the averaged data, the ratio of I_1/I_0 and, subsequently, the gain-length product $GL = Ln(I_1/I_0)$ was calculated as a function of frequency point by point.

EXPERIMENTAL RESULTS AND DISCUSSION

In Fig. 3, typical averaged intensity records of 32 gain scans as recorded by the two detectors D1 and D2 as a function of frequency are depicted. The saturated intensity signal was measured when a single-mode beam of 60 W/cm^2 illuminated the amplifier medium, 4 mm from the nozzle exit. The saturating laser mode was positioned at the left maximum of the power output curve of the saturating oscillator, about 70 MHz off-line-center, see Fig. 2.

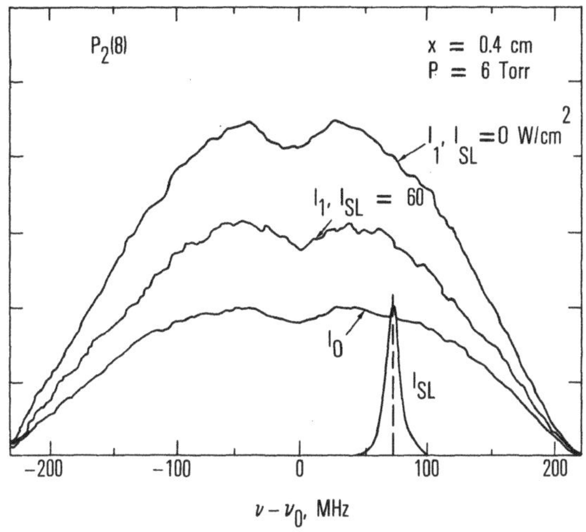

Fig. 3. Intensity of probe laser signal before and after amplifier section with and without saturating laser present.

Line-center is given by the Lamb dip of the probe laser. The ordinate is calibrated by the free spectral-range and the power line-width of the probe laser shown in Fig. 2. The Lamb dips also demonstrate that the frequency resolution of our experiment was sufficient to distinguish any significant hole burning in the amplifier medium, if such were present. The amplification produced by the medium is clearly seen in these records as is the decrease of amplification when the saturating laser field is present. No hole burning or unilateral depression of the gain-profiles is discernible near the position of the saturating laser mode; the depression is uniform across the entire gain-profile.

Fully homogeneous saturation is also shown by the amplifier gain-profiles obtained from these records shown in Fig. 4. The intensity depressions of Fig. 3 produced by the Lamb dip in the probe laser power profile have disappeared in calculating the gain-length GL = $Ln(I_1/I_0)$, and the gain-profiles at various saturating power densities are smooth and self-similar to the unsaturated gain-profile.

Mirels,[7] using the theory of Reference 2, calculated the gain-profile to be expected in this medium under saturation by a single-line laser mode of varying power density. The resulting line-shapes are shown in Fig. 5, normalized with the parameters of

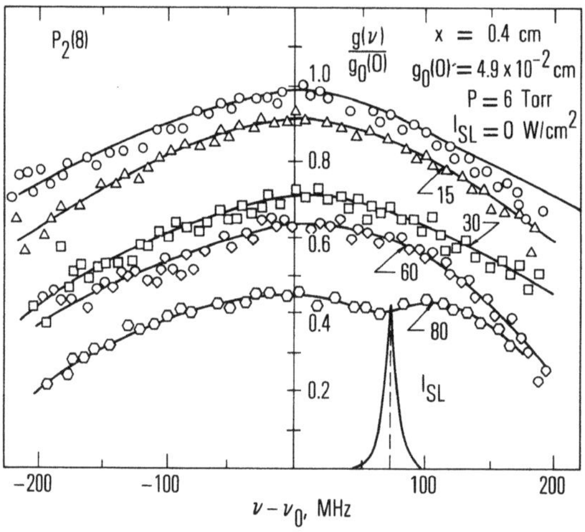

Fig. 4. Measured line shapes of unsaturated and saturated $P_2(8)$ transition in The Aerospace Corporation supersonic HF diffusion laser.[2]

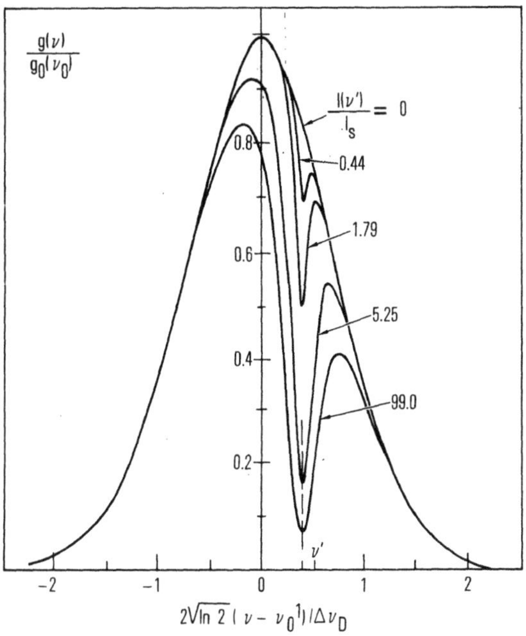

Fig. 5. Hole burning effects predicted by Reference 2 for a supersonic diffusion laser amplifier at P = 3 Torr.

the theory[2]. At low power densities the depth of the hole burned into the Doppler line-profile is small. It is, therefore, possible that at the power densities of our experiments, that hole burning could not be observed, particularly since the theory of Reference 2 may underestimate the cross-relaxation between the various Doppler-shifted molecules. Nonetheless, this theory does not predict any substantial uniform, "homogeneous" depression of the gain-profile as is observed in our experiments.

A similar observation, although under substantially different experimental conditions, was reported in Reference 8 for a CO_2 laser. The authors find a uniform, "homogeneous" depression of the line-profile of their medium when gain is present. This observation was explained by the increased homogeneous relaxation of the investigated transition by higher lying, excited vibrational levels of the CO_2 molecule. It is interesting that they observe inverse hole burning in absorption under otherwise identical conditions in their medium.

With these experimental results we feel justified to calculate a "homogeneous saturation" at line-center, as shown in Fig. 6, a process that would not be meaningful if the lines were saturated truly inhomogeneously. We also plot a curve through the experimental points given by the homogeneous saturation relationship $(g/g_0) = 1/(1 + I/I_s)$ that describes our results well for power densities between 1 and 85 W/cm^2. From this curve we calculate a homogeneous saturation density at line-center of $I_s = 80$ W/cm^2.

The gain-profiles of the arc-laser medium indicated in Fig. 4 are also seen to be considerably broader than those of the probe laser. By extrapolating the measured unsaturated gain-profile to higher and lower frequencies we can estimate a FWHM of over 500 MHz. If one assumes that the line is Doppler broadened, this line-width would correspond to a gas temperature in the laser medium of more than 1050 K, a value that appears to be very high, even when the hot gas in the boundary layers of the nozzle is taken into consideration. A much more comfortable explanation is offered by assuming that the nozzles are overexpanded so that the flow field is similar to the source flows investigated by Mirels[4]. The deformation of Doppler profile and the increase in its FWHM caused by such source flow effects is depicted in Fig. 7. The line broadening is substantial, and the profile loses its characteristic Doppler shape. In order to explain our experimental result of a FWHM of 500 MHz, we see that we have to invoke only a relatively moderate flow velocity component along the laser beam direction of Mach number 1.3. In a flow field of Mach number 4, such as would be expected near the exit of our nozzle, this corresponds to an expansion half-angle of less than 20 degree, which appears entirely plausible.

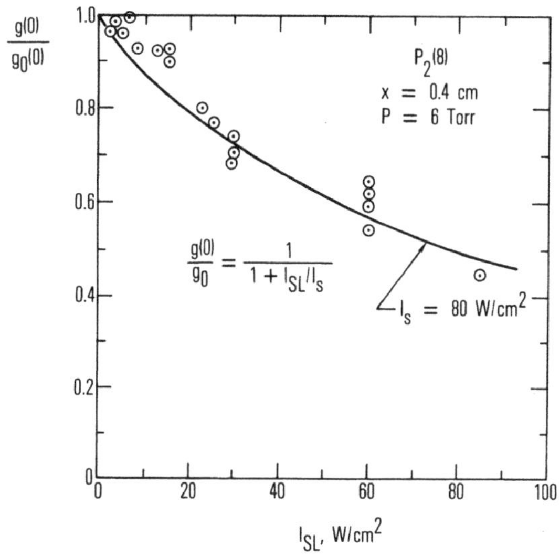

Fig. 6. Measured gain at line center of $P_2(8)$ transition as a function of saturating laser power density I_{SL}.

In conclusion, we have shown that at a pressure of 6 Torr, the $P_2(8)$ transition in the gain-medium of a supersonic mixing-laser is saturated homogeneously even if irradiated by a single-mode, single-line laser of quite moderate power density. We determined a homogeneous saturation intensity at line-center of 80 W/cm^2 for this line at a position 4 mm downstream of the multi-slit nozzle of our laser. This observation is not explained by present theory[2] assuming a Doppler broadened line in the medium of this laser under our conditions. However, a similar observation has been reported for a CO_2 laser[8], although under sufficiently different experimental conditions to warrant a re-examination of

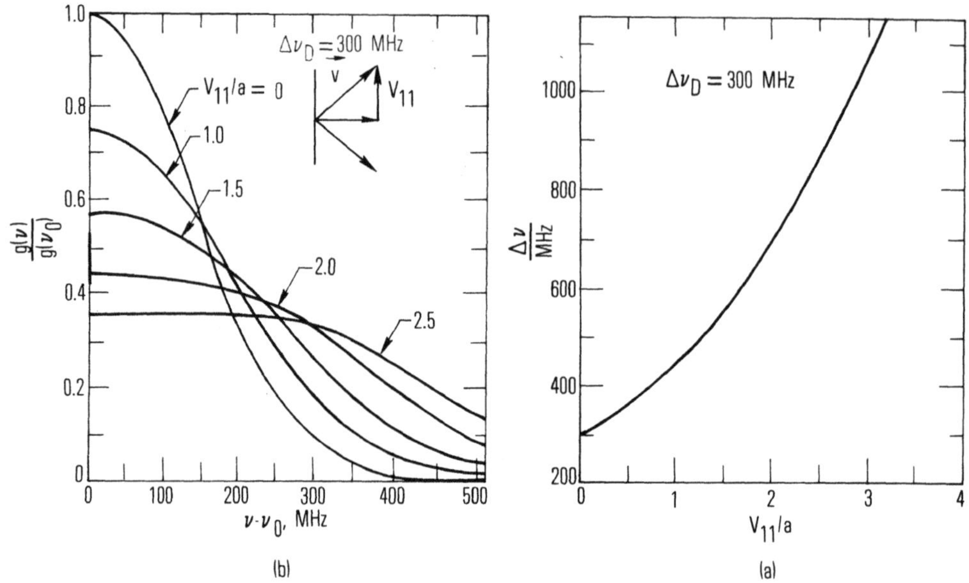

Fig. 7. Source-flow effects on Doppler line shape.[4] (a) Distortion of Doppler line shape caused by source flow, V_{11} = flow velocity parallel to laser beam, a = speed of sound. (b) Full width at half maximum of distorted Doppler line.

present line-shape theory for the HF laser. We also observe a considerable broadening of the gain line that could be explained only by assuming an unreasonably high Doppler temperature of over 1000 K. We propose to explain this observation by a laterally expanding flow similar to the source flows investigated theoretically by Mirels[4].

REFERENCES

1. Gross, R. W. F., Chodzko, R. A., Turner, E. B., and Coffer, J. G., "Measurements of the Anomalous Dispersion of HF in Absorption." J. Quant. Electron. QE-16, 795, 1980.
2. Mirels, H., "Inhomogeneous Broadening Effects in cw Chemical Lasers," AIAA Journal 17, 478, 1979.
3. Gross, R. W. F., Bott, J. F. (eds), "Handbook of Chemical Lasers," Wiley-Interscience, New York, 1976.

4. Mirels, H., "Multimode Low Pressure cw Chemical Laser Performance Including Source Flow Effets," The Aerospace Corporation Technical Report TR-0082(2764)-1, 25 March 1982.
5. Karl Lambrecht Corporation, Chicago, Illinois.
6. Spencer, D. J., Beggs, J. A., and Mirels, H., "Small Scale cw HF(DF) Chemical Lasers," J. Appl. Phys. 48, 1206, 1977.
7. Mirels, H., private communication.
8. Granek, H., Freed, C., and Haus, H. A., "Experiment on Cross Relaxation in CO_2" J. Quant. Electron. QE-8, 404, 1972.

ENHANCED MIXING IN CO-FLOWING LAMINAR STREAMS

L. Manfriani, J. Sandford, & M. Carbonaro

von Karman Institute for Fluid Dynamics
Chaussée de Waterloo, 72
B - 1640 Rhode Saint Genèse, Belgium

INTRODUCTION

In chemical lasers a mixture of F, F_2, He and DF in a thermo-chemical equilibrium state at a total temperature between 1000 and 2000°K and a total pressure between 3 and 5 bars is expanded rapidly through a bank of small supersonic nozzles into the laser cavity. The fluorine-carrying gas exits at a Mach number of about 5 and with a static pressure of about 2000 Pa. Due to the extremely small dimensions of the nozzles and the low density of the gases, the Reynolds numbers based on nozzle width is of the order of 1000 and the flow is commonly laminar.

Lasing efficiency depends on rapid mixing in the cavity. The fundamental mechanism of mixing is one of laminar diffusion but because of the importance of rapid mixing a number of devices have been employed to enhance mixing, in some cases with considerable success but in other cases to little effect.

A particularly useful device has proved to be a "trip-jet" arrangement in which a relatively small amount of an inert gas is injected approximately normal to the streams just upstream of the cavity. The mechanism through which enhanced mixing is obtained with such a device is poorly understood, however. For example, it is not possible to explain why the device is much more effective in some configurations than in others, and indeed why simpler turbulence producing devices, such as vortex generators, have proved to be almost totally ineffective. There is a clear need therefore for a systematic study of mixing devices, designed to investigate the mechanism of mixing and the dependence of mixing rate on geometrical and other parameters[1,2].

Although the mixing process in the cavity will clearly be influenced by compressibility effects and by heat addition and reaction between the various gases, there is evidence to suggest that Reynolds number is a parameter of primary importance. It was therefore felt that useful preliminary experiments could be done at low speeds with air as the working fluid and with Reynolds number simulation, and this paper reports on the results obtained from tests of this type. Such tests are relatively simple and inexpensive to perform and the results will be used as a guideline for further investigation at higher speeds.

THE TEST FACILITY

A special test facility was designed to provide two low speed (up to 1 m/s mean velocity) co-flowing airstreams in rectangular ducts which mix in a cavity downstream of the ducts. Since the flow in the nozzles of a chemical laser is likely to be fully viscous, or at least to have very thick wall boundary layers, duct dimensions were selected in order to achieve a completely established laminar viscous flow in the ducts before the cavity. The cross-section of each duct has an aspect ratio of 10 (200 mm × 20 mm) so that the flow in the inner part of the duct closely resembles a two-dimensional one, and this is the region in which mixing was investigated. The ducts are independently fed from a compressed air supply with particular attention being given to the inlet design to avoid the creation of unacceptable disturbances which would degrade the downstream flow. The volume flow of air is measured by rotameters.

Figure 1 is a sketch of the cavity inlet showing one of the trip-jet configurations used to enhance mixing. The last section of the splitter plate and the duct walls at the corresponding location can be changed, thereby providing the possibility of testing different configurations of jets and indeed other mixing devices.

The jets are fed by lateral tubes, as shown in figure 1, and their volume flow, which varied between 3% and 11% of the volume flow in one duct, is measured by a rotameter and controlled by six small needle valves to achieve a good flow distribution. The diameter of the holes is 1 mm in the configurations tested.

To investigate the clean flow in the ducts, a hot wire anemometer was used to determine mean velocity profiles. The results showed that the flow was indeed sensibly two-dimensional and laminar over most of the cross-section. The RMS turbulence intensity was certainly low but impossible to quantify at the very low flow velocities used because it was masked by the electronic noise of the anemometer.

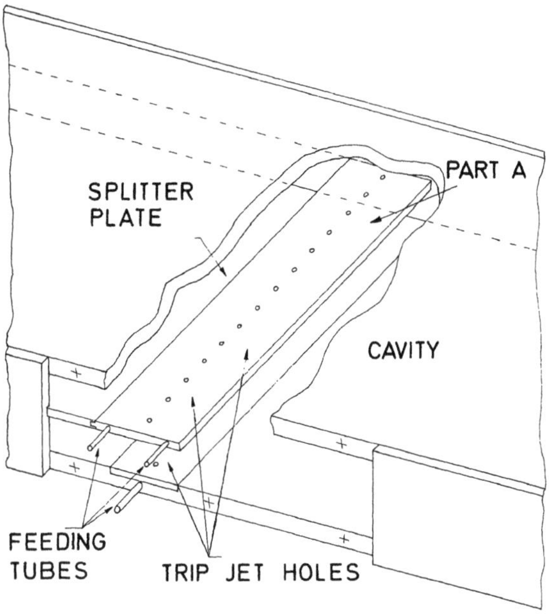

FIG. 1 - Sketch of the Cavity Inlet

TECHNIQUES TO INVESTIGATE MIXING

Two different techniques were employed to assess the degree of mixing, both providing essentially qualitative estimates.

On the one hand, flow visualization by means of the laser sheet technique was used. In this technique the laser is used merely as a high intensity light source. The beam from the laser is spread into a two-dimensional sheet by means of a cylindrical lens, so that by suitable positioning of the components a plane in the flow can be illuminated, this plane usually being selected normal to the axis of the ducts. With smoke injected in the trip-jets the trace of the jets in the plane can be visualized; alternatively, with smoke injected in one of the ducts an appreciation of the mixing that has occurred at the chosen plane and of the flow structures that have contributed to it, can be obtained.

On the other hand, concentration measurements of a foreign gas, injected initially into one of the ducts, were made in a number of cross-sections of the cavity using a hot wire concentration probe. The probe has been described by Colin et al[3]. It is in fact sensitive to the thermal conductivity of the gas mixture; the heat transfer from a hot wire to the fluid is measured in constant flow conditions due to the incorporation of a sonic orifice in the probe downstream of the wire and connection of a vacuum pump to the downstream end of the probe. It is not an absolute concentration sensor and as such requires calibration for the gases used in the experiment; the nature of the gases must clearly remain unchanged during the measurements.

In previous applications of the probe either helium or argon has been used as the foreign gas. For the present experiments the use of either of these gases separately was found to induce a secondary swirling flow in the cavity because of buoyancy effects. The mixing was thereby dramatically affected and the results obtained unacceptable. For this reason a mixture of 64% argon and 36% helium by volume, giving the same density as air, was used. Although the sensitivity of the probe was much diminished by the use of such a mixture it was still found to be high enough for the technique to be employed.

Because of the low speed of flow in the ducts it was not possible to operate the probe isokinetically, and as noted earlier the concentration profiles obtained by traversing the probe should be interpreted qualitatively only. Measurements made using two different sizes of sonic orifice, however, gave almost identical results, with about 10% of the gas mixture injected in one of the ducts.

Figure 2 shows the set-up used for the concentration measurements.

DISCUSSION OF RESULTS

Four configurations of trip-jets were investigated, and these are shown in figure 3. It can be seen that they differ in respect of

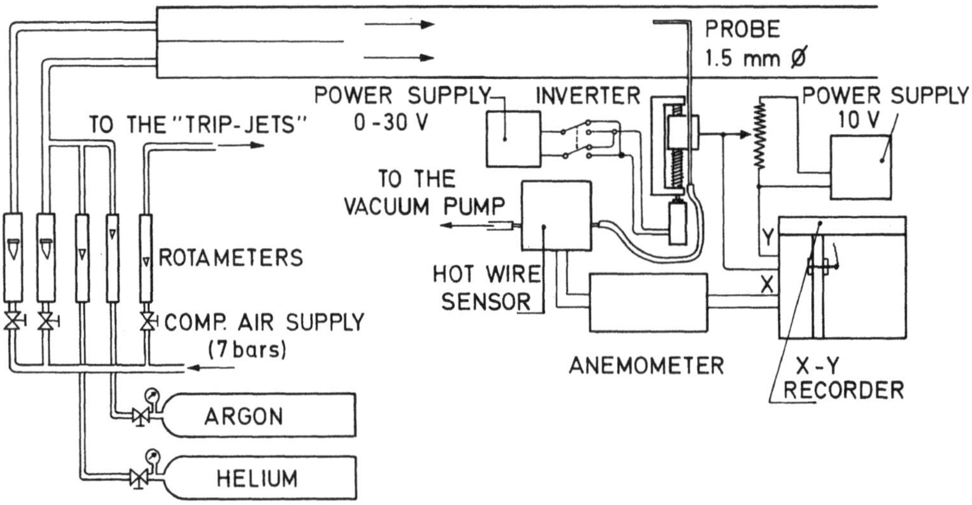

FIG. 2 - Concentration Measurements With a Hot Wire Sensor

jet spacing, in the relative position of opposing jets and in the number of surfaces from which the jets are discharged.

Configuration C4 differs from configuration C3 in that the last part of the splitter plate separating the ducts (part A in Fig. 1) is removed. Mean speeds were always the same in both ducts and varied between 0.25 m/s and 0.5 m/s.

Flow visualization with smoke injected into the trip-jets showed clearly that, at a sufficiently high jet speed, the jets in configurations C1 and C2 interact with each other and develop a turbulent motion dominated by large scale structures. At the time it seemed reasonable to suppose that this motion would enhance the mixing and a series of concentration profiles was taken. The results showed, contrary to expectation, that mixing was very little affected by the trip-jets in these configurations, leading to the conclusion that mere production of turbulence does not of itself guarantee the effectiveness of a mixing device. In fact at the highest duct flow speeds used the jets had an adverse effect on mixing.

In the belief that the jets discharged from the splitter plate were tending to push the flows in the two ducts apart, configuration C3 was then investigated. There was indeed an improvement in mixing at the lower speeds of flow in the ducts, compared with configurations C1 and C2 and the clean flow, but again a deterioration in mixing at higher speeds compared with the clean flow.

Comparison of concentration profiles obtained from the different trip-jet configurations for the same ratio of the trip-jet speed to duct flow speed showed a close similarity between them, suggesting that the mechanism of mixing was the same for a constant value of this ratio, irrespective of the absolute value of duct flow speed.

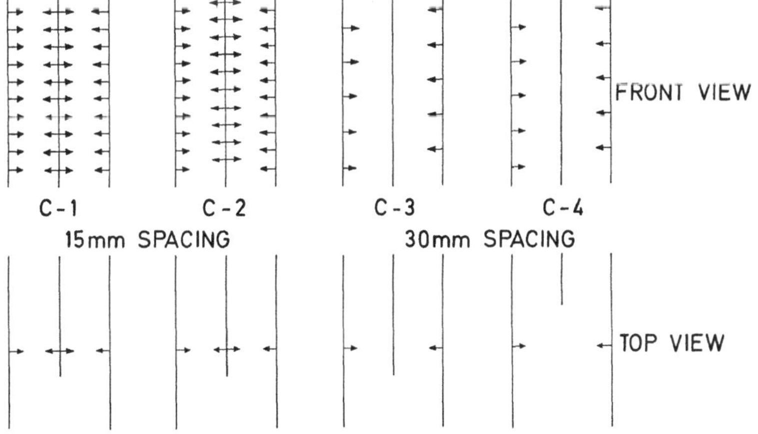

FIG. 3 - Scheme of the Trip-Jet Configurations Tested

This hypothesis was supported by flow visualization which revealed similar flow patterns when this ratio was maintained constant.

For the clean flow, however, the situation is different. At the lower speeds the mixing layer separating the duct flows in the cavity is stable and mixing is slow; at the higher speeds instability waves which roll up into vortices form in the mixing layer and mixing is much improved, probably because of entrainment by the coherent vortices. It then appears that when trip-jets are used at the higher duct flow speeds they tend to stabilize the mixing layer and thus impair mixing.

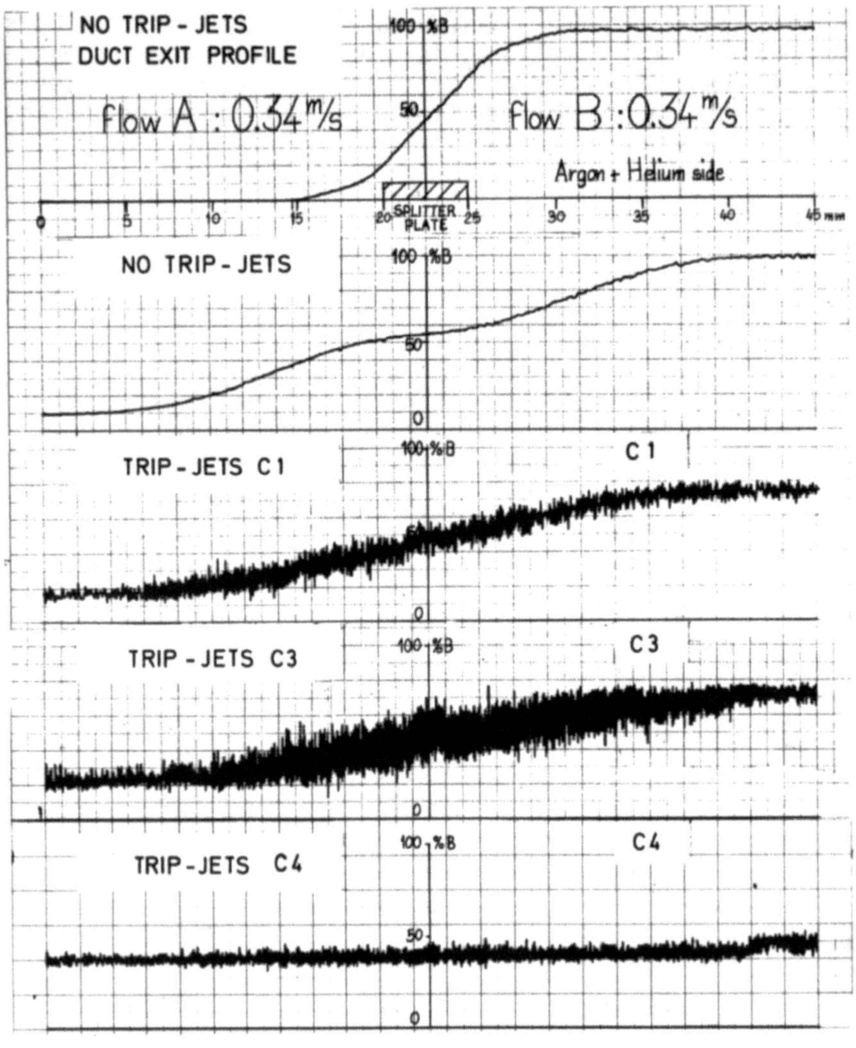

FIG. 4 - Comparison of Concentration Profiles 250 mm (12.5 Duct Widths) Downstream of the Duct Exit. Volume Flow Through Trip-Jets Constant.

Flow visualization also showed that the trip-jets impinged on the splitter plate and were partially reflected in configuration C3. To avoid this the last part of the splitter plate was removed to give configuration C4.

With configuration C4 the trip-jets provide a dramatic improvement in mixing, almost complete mixing being achieved as close as 100 mm (5 duct widths) to the (virtual) end of the splitter plate at sufficiently high jet speeds. Flow visualization confirmed this result.

A comparison of the concentration profiles obtained for configurations C1, C3 and C4 for the case of a particular constant volume flow through the trip-jets is shown in figure 4, and it is clear that the increased turbulence in configurations C1 and C3 is doing little to enhance the mixing. It should be emphasized that the case depicted, for which the total volume flow through the trip-jets is 10.3% of the volume flow in one duct, corresponds to relatively high values of jet velocity (jet velocity = 40 × mean duct velocity for configurations C3 and C4) and cannot therefore be considered representative of conditions in an actual laser cavity even within the limitations of the present investigation, but work to cover more realistic cases is proceeding.

Figure 5 shows cross sections of the cavity at the same location with smoke injected into one of the ducts. Figure 5(a) is for configuration C3 and figure 5(b) for configuration C4 and the enhanced mixing in the latter case is evident.

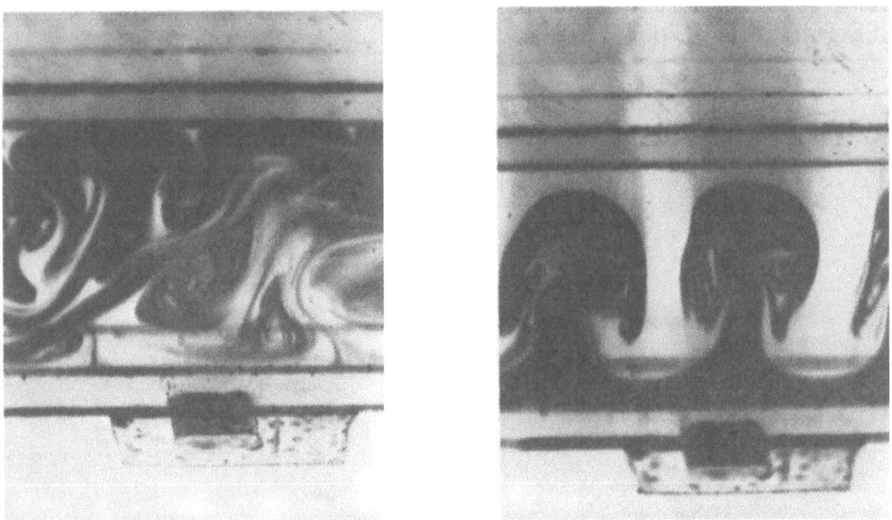

(a) Trip-jets C3 (b) Trip-jets C4

FIG. 5 - Cross Section of the Cavity with Smoke in One of the Ducts

CONCLUSIONS

It has been shown that it is possible to use trip-jets to enhance the mixing of two laminar streams, but the geometry of the arrangement is critical. Use of a poor configuration can make mixing worse rather than better. This can be explained in terms of the necessity to produce an ordered structure which will give rise to the entrainment and transport of mass required to give mixing enhancement.

It is not clear at this point to what extent the results obtained are applicable to chemical lasers, and further investigation is required, as already mentioned, of more realistic cases even within the limitations of the present low speed study, before proceeding to high speed experiments. But the results obtained are interesting enough to warrant this further work which is now proceeding.

This research was sponsored by the US Air Force Weapons Laboratory/AFSC under grant AFOSR 81-0231.

REFERENCES

1. S.W. Zelazny, R.J. Driscoll, J.W. Raymonda, J.A. Blauer and W.C. Solomon; Modelling DF/HF CW lasers : an examination of key assumptions, AIAA J., 16:4 (1978)
2. L.E. Wilson; Deuterium fluoride CW chemical lasers, invited lecture at 3rd G.C.L., Marseille (1980).
3. P. Colin and D. Olivari; Three applications of hot wire techniques for fluid dynamic measurements, ICIASF '71 Record, IEEE Pub., IEEE Pub. 71-C-33 AES, 1971.

EXPERIMENTAL STUDY OF THE SPECTRAL OUTPUT OF A DF-LASER

Juergen H. Massig

DFVLR-Lampoldshausen
Institut für Chemische Antriebe und Verfahrenstechnik
Langer Grund, 7101 Hardthausen, FRG

ABSTRACT

As is well known, rotational equilibrium is not fully established among the lasing molecules of stationary DF-lasers under lasing conditions. In our experiments a shifting of the laser lines to higher J-values was observed when the flow temperature was increased, which indicates that deviations from equilibrium are small. Fluorescence spectroscopy and gain measurements showed no measurable deviation from rotational equilibrium. The rotational numbers of the dominating laser lines were calculated from measured densities and temperatures and are compared to experimental laser spectra.

INTRODUCTION

It is well known that in stationary HF- and DF-lasers rotational equilibrium is not fully accomplished under lasing conditions. This effects the appearence of a multitude of laser lines per vibrational transition and causes a slight loss of efficiency compared to the hypothetical case of infinitely fast rotational relaxation (RR). Several authors[1,2,3] calculated laser output spectra for the case of limited RR and compared the results to experimentally obtained spectra. Hall[3] refers to a DF-laser with the resonator axis extremely close to the exit plane of the laser nozzles where rotational nonequilibrium effects are largest. However, this is not representative for normally operated DF-lasers. The spectra calculated by Yang[2] show lasing at significantly lower J-values than observed experimentally. The author explains the discrepancy by vibrational-to-rotational energy transfer which was not yet included in the model. Sentman[1,4] calculated an HF-laser spectrum peaked at J=9 for rotational equilibrium which is in discrepancy to the experimental spectrum peaked at J=6 whereas the results obtained for limited RR coincided much more with the experimental findings. In all these calculations the DF-particle densities and flow temperatures affecting the spectrum were obtained from computer calculations. Here we report laser spectra and spectroscopically measured temperatures and particle densities. The spectra are compared to the rotational numbers of laser lines calculated from the spectroscopical data assuming rotational equilibrium.

EXPERIMENTS AND RESULTS

The experiments were conducted with a combustion driven DF-laser of moderate size. In these tests the combustion chamber was operated with fluorine and hydrogen. Nitrogen was used as diluent and supplied to both combustor and secondary nozzles. The cavity pressures were in the range of 10 to 20 mbar. Fig. 1 shows a typical laser output spectrum. The fraction of the total power is plotted for each line. No similarity to the pumping distribution of the rotational levels could be observed. This would imply significantly lower J-values for the higher vibrational transitions than for the lowest transition. Instead of this the lines of all

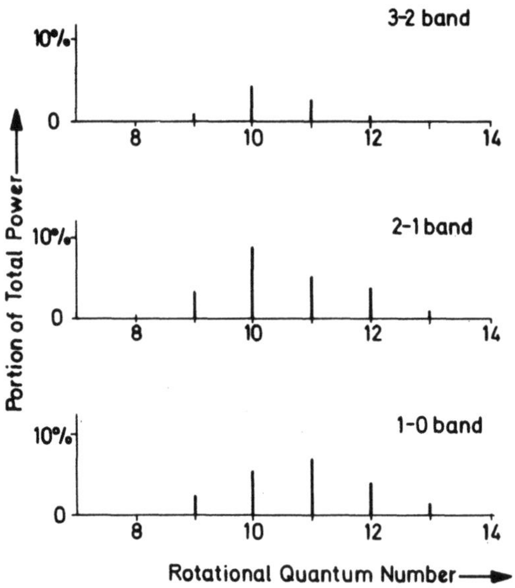

Fig. 1. Example of measured laser output spectra

vibrational bands are typically centered around the same mean rotational number J. In a test series the portion of fluorine available for lasing was varied whereas the plenum pressure and temperature were kept constant. The mean rotational numbers of the spectra recorded in this series are shown in Fig. 2. Clearly, the spectrum is shifted to higher rotational numbers for higher fluorine fractions, which indicates a correlation of laser spectrum and flow temperature.

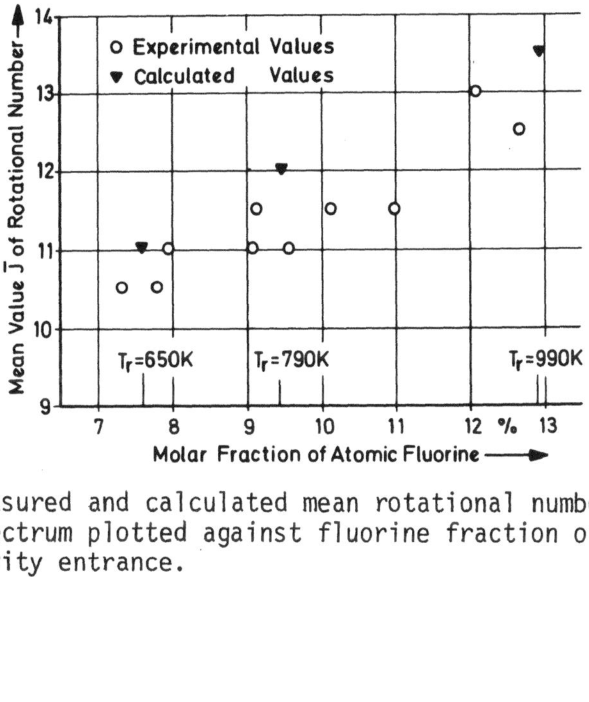

Fig. 2. Measured and calculated mean rotational numbers of laser spectrum plotted against fluorine fraction of gases at cavity entrance.

In order to get further insight DF-particle densities in the vibrational levels v=2, 3 and 4 were measured spectroscopically via the first overtone bands. These measurements were performed under three operating conditions selected from the test series. An example is shown in Fig. 3, where the logarithms of the particle densities divided by the degeneracies are plotted over the rotational energy. No systematic deviation from a Boltzmann distribution appears in these graphs. Therefore a rotational temperature can be assigned which is assumed to be identical to the translational temperature.

Small signal gain coefficients have been measured under the same operating conditions as for the spectroscopic measurements. The optical arrangement is described elsewhere[5]. Fig. 4 shows measured gain coefficients of the $P_2(7)$, $P_2(9)$ and $P_2(11)$ lines and the values calculated for a Boltzmann distribution with a rotational temperature of 650 K as obtained from spectroscopy for these operating conditions. Gain coefficients calculated for 500 K are shown for comparison. Evidently, only minor deviations from rotational equilibrium may exist in the small signal case. Furthermore, small signal gains have been measured on the $P_1(10)$ line. Thus, in combination with the spectroscopic data the DF-particle densities in the v=0 and v=1 vibrational states can also be determined.

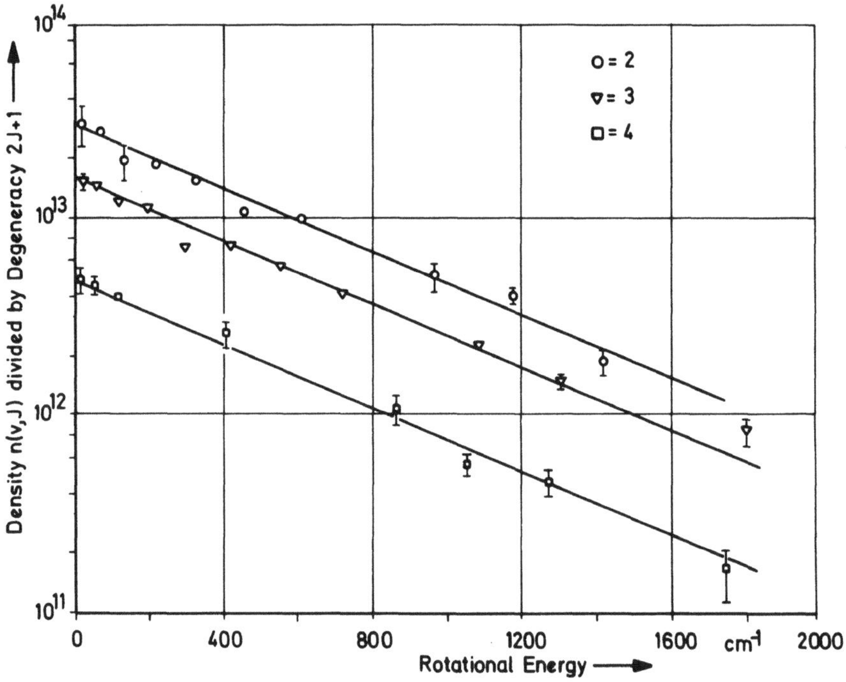

Fig. 3. Example of spectroscopical measurements. DF-densities in vibrational-rotational states divided by degeneracies plotted against rotational energy.

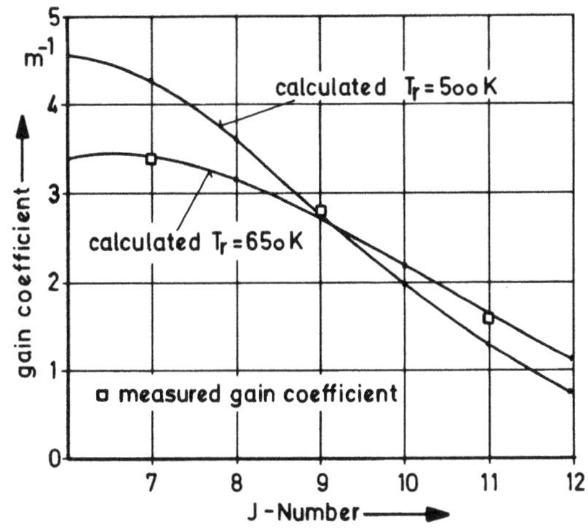

Fig. 4. Measured and calculated small signal gain coefficients of the $P_2(J)$ lines.

CALCULATION

The gain α of the P-lines of an HF- or DF-laser can be expressed as[6]:

$$\alpha = \frac{CJ}{T^{3/2}} \left(n_{v+1}e^{-BJ(J-1)/kT} - n_v e^{-BJ(J+1)/kT}\right) \tag{1}$$

where the rigid rotator was assumed for the rotational energies and the vibration-rotation-interaction has been neglected in the Einstein coefficient. The constant C contains the vibrational matrix element for dipole transitions and is independent of J. In order to determine the rotational numbers of the lasing lines Eq. (1) may be transformed:

$$\frac{n_{v+1}-n_v}{n_{v+1}+n_v} = \frac{\frac{\alpha T^{3/2} e^{BJ^2/kT}}{(n_{v+1}+n_v)C} - J\sinh(\frac{BJ}{kT})}{J\cosh(\frac{BJ}{kT})} \tag{2}$$

In the oscillator case the gain is determined by the resonator and oscillation occurs on that rotational line which permits the largest depletion of the upper vibrational level. Thus, oscillation occurs for that J-value which leads to the minimum of expression (2). The lasing lines were determined in the following way: It was assumed for the oscillator case that the sum of particle densities in the v=0 to v=3 vibrational levels is the same as the sum determined in the corresponding diagnostic experiment. An arbitrary distribution over the vibrational states was assumed and the J-values yielding a minimum of Eq. (2) were determined for each vibrational transition, where noninteger J-values were admitted. As a result of this step Eq. (2) leads to a new distribution over the vibrational levels which is the starting point for the next iteration until finally a self-consistent distribution is obtained. For every data set the resulting J-values were nearly the same for all vibrational transitions. This can be explained by the fact that the vibrational matrix element contained in the constant of Eq. (2) is larger for the higher vibrational transitions whereas the density sum $(n_{v+1}+n_v)$ is lower because only partial inversion exists. Thus, for all vibrational transitions the resulting J-values are nearly the same and can be represented by a single value J.

These calculated values are also shown in Fig. 2 where the measured temperatures are indicated as well. The calculated values are also larger for an increased fluorine fraction which is due to both the higher rotational temperature and the higher DF-molecule density. The values are slightly larger than the mean values of the measured spectra. The agreement is better if a larger resonator loss is inserted in the calculation than the value used here which is derived from the transmission of the coupling mirror.

SUMMARY

For the small signal case the rotational distribution of DF-molecules was found to be in equilibrium. In the oscillator case the laser output spectrum shifted to higher rotational numbers with increasing energy release in the cavity. The oscillating lines were calculated for rotational equilibrium using measured DF-densities and temperatures. These J-values represent quite well the peaks of the measured spectral power distribution.

REFERENCES

1. L.H. Sentman, "Rotational Nonequilibrium Effects in CW Chemical Lasers", Third International Symposium on Gas Flow and Chemical Lasers, Marseille (Sept. 1980)

2. T.T. Yang, "Modeling of CW HF Chemical Lasers with Rotational Nonequilibrium", Third International Symposium on Gas Flow and Chemical Lasers, Marseille (Sept. 1980)

3. R.J. Hall, "Rotational Nonequilibrium and Line Selected Operation in CW DF Chemical Lasers", IEEE J. Quant. Electron., QE 12:453 (1976)

4. L.H. Sentman and W. Brandkamp, "An Efficient Rotational Nonequilibrium Model of a CW Chemical Laser", AAE TR 79-5, Aeronautical and Engineering Dept. University of Illinois, Urbana, IL (July 1979)

5. J.H. Massig, "A Simple Arrangement for Spatially Scanning Gain Measurements in CW Lasers", accepted for publication in Rev. Sci. Instr.

6. G. Emanuel, "Analytical Model for a Continuous Chemical Laser", J. Quant Spectrosc. Radiat. Transfer, 11:1481 (1971)

OPERATION OF A SMALL SCALE, EFFICIENT CHEMICAL OXYGEN-IODINE LASER AND STUDY OF THE REACTION OF $O_2(^1\Delta)$ WITH Pb ATOMS

S. Rosenwaks and J. Bachar

Department of Physics
Ben-Gurion University of the Negev
Beer-Sheva, Israel

ABSTRACT

The operation of a compact, small scale, chemically pumped iodine atom laser is described. More than 40% extractable power efficiency has been achieved in a transverse-flow apparatus. 5W of CW laser emission at 1315 nm has been obtained via energy transfer from chemically generated $O_2(^1\Delta)$ to I atoms in a $10 \times 1\ cm^2$ rectangular flow duct. Simple construction materials, safely handled chemicals and a medium size vacuum pump were used for fabricating and operating the laser. The importance of minimizing quenching of excited species by collision with water molecules or with the wall is demonstrated. The chemical generator has also been used for studying the reaction of $O_2(^1\Delta)$ with Pb atoms. Both excited Pb atoms and excited PbO molecules have been observed and possible excitation mechanisms have been investigated.

The chemical oxygen-iodine laser (COIL) is the only example to date of a CW electronic transition chemical laser.[1-4] The laser operates on the $^2P_{1/2}$ - $^2P_{3/2}$ transition of the iodine atom at 1315 nm which is pumped by energy transfer from chemically generated $O_2(^1\Delta)$. The latter is obtained by reacting $C\ell_2$ gas with basic hydrogen peroxide solution (BHP). I atoms are obtained in the COIL by interaction of $O_2(^1\Delta)$ with I_2 molecules via a process which is not yet fully understood.[5]

Following the first demonstration of this laser by McDermott et al.[1], considerable efforts have been devoted to achieving large-scale, efficient COILs.[2-4] A device delivering $\gtrsim$ 125 W of output power with 30% extractable power efficiency was recently reported.[4] This device, as well as the previous one,[2,3] are high-technology devices, require large vacuum pumps and use large amounts of hazardous chemicals. They are thus not appropriate for use in small laboratories which want a readily available source of chemically generated $O_2(^1\Delta)$ or of coherent radiation in the near infrared.

In the present paper we describe a compact, small scale, efficient COIL which utilizes simple construction materials, small amounts of chemicals and a medium size vacuum pump. 50% H_2O_2 was used rather than the hazardous 90% concentration used in the previous COILs. The apparatus is transparent throughout, permitting convenient monitoring and controlling of the dynamic parameters that govern the laser operation. A schematic drawing of the laser system, including the chemical $O_2(^1\Delta)$ generator, is shown in Fig. 1. Most parts are made of glass. The laser section (plexiglass) is presented in Fig. 2 and the detection system in Fig. 3.

Gas flows are measured with precalibrated ball flowmeters. The I_2 flow rate is measured by weighing the amount of I_2 consumed for a given setting of heating and flow conditions. The production of excited species at various locations of the laser system is monitored by suitable detectors as shown in Figs. 1 and 3. The concentration of the chemically generated $O_2(^1\Delta)$ is monitored relative to that produced by a microwave (MW) generator following the procedure of Benard and Pchelkin,[6] using a cooled intrinsic Ge detector with a 10 nm bandwidth filter centered at 1270 nm. $O_2(^1\Sigma)$, which is formed by the energy pooling reaction of $2\ O_2(^1\Delta)$,[7] is monitored by detecting the $O_2(^1\Sigma-^3\Sigma)$ emission at 762 nm. The $I(^2P_{1/2} - ^2P_{3/2})$ emission in the laser cavity is monitored by a second Ge detector with a 1315 nm filter looking down the laser axis; the side emission from the cavity is monitored by the same detector with the aid of a deflecting mirror.

The operating procedure consists of the following steps. 500 ml of 50% H_2O_2 is pumped into the generator through the reagent feed line while N_2 is continuously flown through the sinter to preclude liquid from entering the gas feed line. After pumping for about 2 hrs, ~ 90% H_2O_2 is obtained. 250 ml of 6N NaOH is then slowly pumped through the reagent feed line and mixed under vacuum with the H_2O_2 for 15 minutes with the aid of the N_2 bubbles. Although the mixing of peroxide and base is an exothermic process, the evaporative cooling due to the pumping off of water vapor maintains the BHP mixture below room temperature. Since water vapor is detrimental to the laser operation,[8] $C\ell_2$ is bubbled through the BHP to generate $O_2(^1\Delta)$ when the temperature of the BHP drops below -10°C and is discontinued when the temperature rises above -8°C. The combination of low BHP termperature and a 195 K trap before the laser ensures low H_2O concentration in the lasing gas mixture.

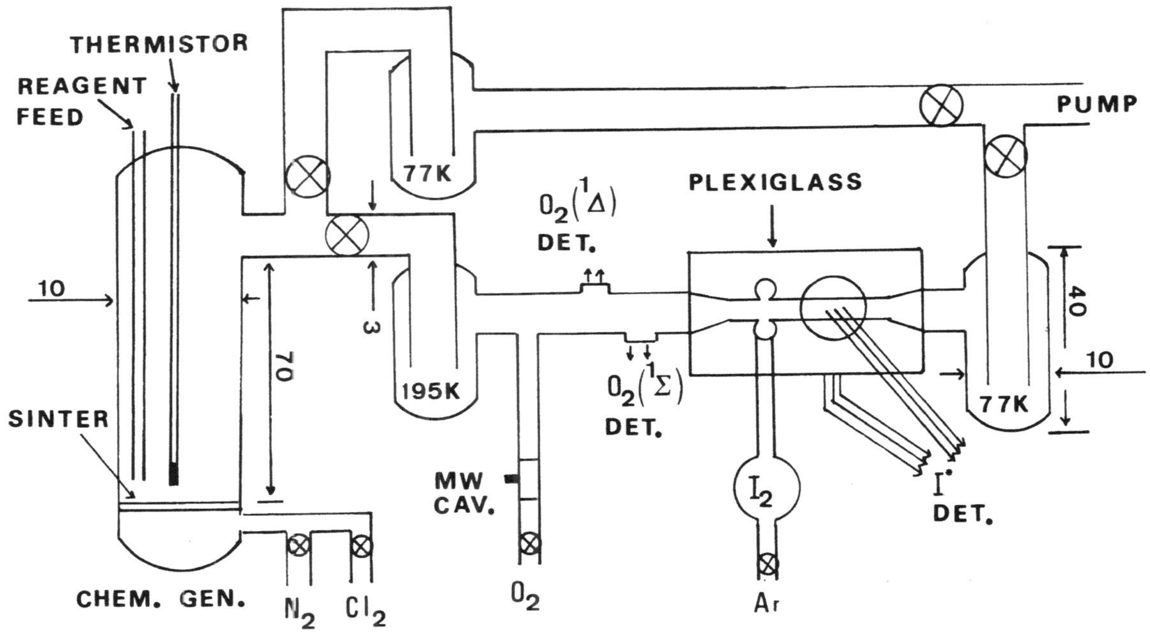

Fig. 1. Schematic drawing of the laser system. All measures are in cm. The design of the generator is a scaled down version of the one described by Benard et al.[2] the main difference being that the $C\ell_2$ gas is bubbled through a 10-20 micron pores sintered glass which occupies the entire cross section of the generator. Two vacuum lines connect the generator to the pump. The first line, which by-passes the laser, is used for preparing and concentrating the BHP solution and includes a 77 K cold trap for water vapor removal. Once the solution is ready the first line is shut and the second line is used for transferring the chemically produced $O_2(^1\Delta)$ from the generator to the laser; it includes a 195 K cold trap before the laser for water vapor removal without significant quenching of $O_2(^1\Delta)$. The effluent from the laser is pumped through a LN_2 trap by a 140 ℓ/s Roots pump.

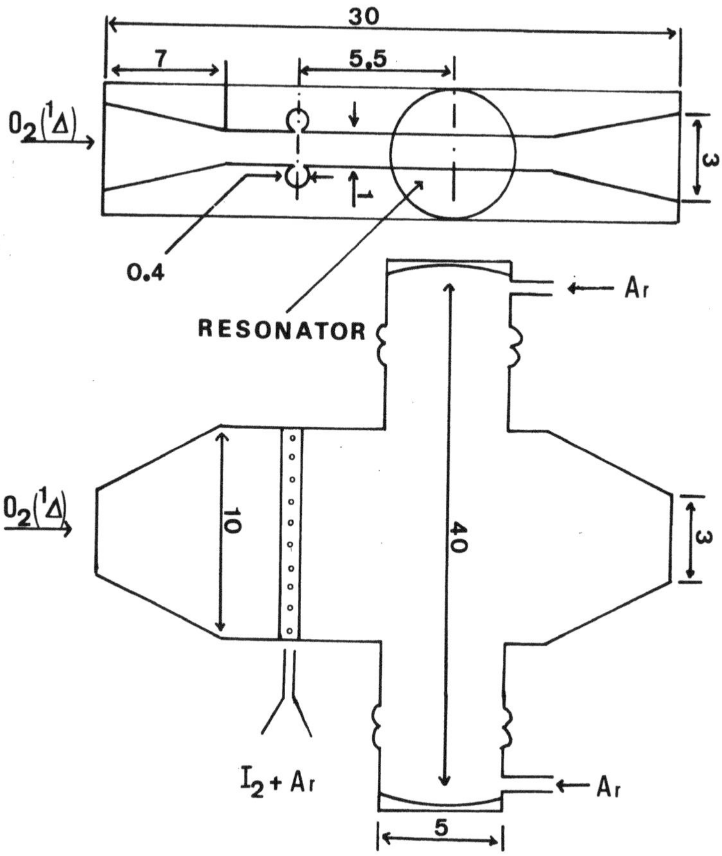

Fig. 2. Diagram of the laser section. It consists of a gradually expanding duct from a 3-cm-diam pipe to a 10 × 1 cm^2 flow duct which is terminated in a 3-cm-diam pipe. The inner surface of the plexiglass duct downstream of the I_2 injectors is coated with thin, transparent, teflon sheet to minimize $O_2(^1\Delta)$ and excited I (I^*) wall deactivation and I atom recombination. The 40 cm long optical cavity consists of high-reflectivity (> 99.9% and > 99.8% at 1315 nm) spherical (r = 2 and 5 m, respectively) internally mounted mirrors. The optical arms are purged with Ar during operation of the laser. Molecular iodine from a radiatively heated I_2 source is carried with a flow of Ar and injected transverse to the oxygen flow. The injectors are located 5.5 cm upstream of the cavity axis in both sides of the duct and are flush with the wall. Each injector is a 4 mm i.d., stainless steel tube with 10 0.5 mm holes evenly distributed along its length.

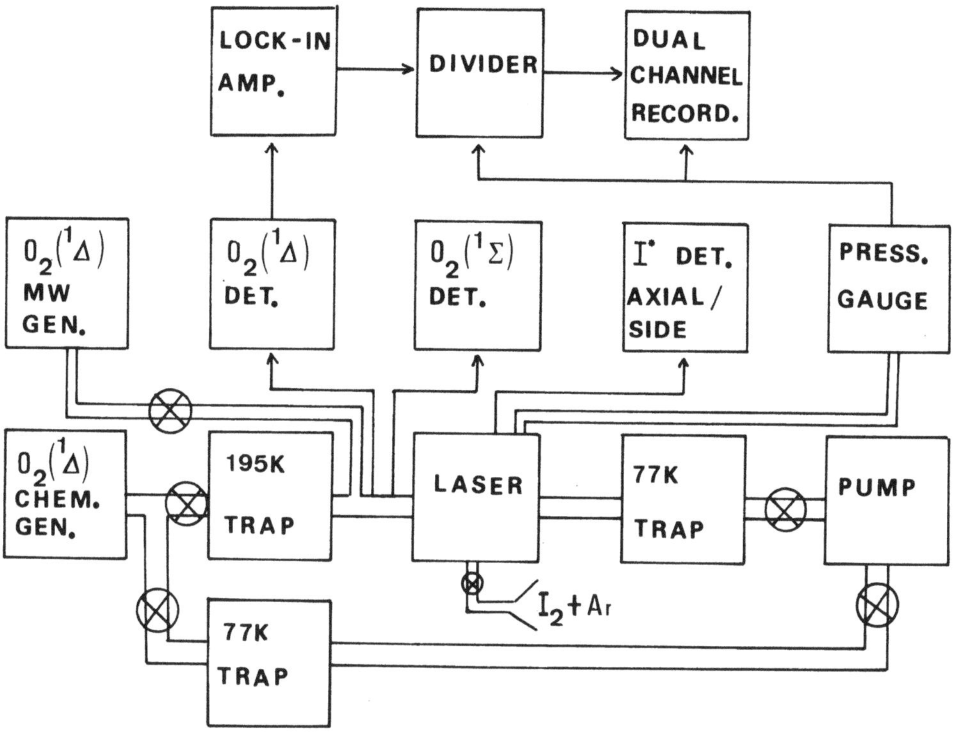

Fig. 3. Block diagram of the laser + detection system.

With $C\ell_2$ flows of 2-5 × 10^{-4} mole/s, a total O_2 pressure of 0.3-0.4 Torr is obtained (when the $C\ell_2$ flow is high, part of it does not react with the BHP and is frozen in the traps). The $O_2(^1\Delta)$ concentration is then 40-48% at the entrance to the laser section. The calculated linear flow velocity in the laser cavity is around 20 m/s. Since $O_2(^1\Sigma)$, which is formed from 2 $O_2(^2\Delta)$, is efficiently quenched by H_2O,[9] the emission intensity ratio $R = Int^2(^1\Delta)/Int(^1\Sigma)$ at a given pressure is a measure of the H_2O concentration. Since I^* is efficiently quenched by H_2O,[8] R must be as low as possible for a desired $[O_2(^1\Delta)]$ before operating the laser. Utilizing R as a monitor, a procedure for qualitatively checking $[H_2O]$ is adopted. R is first measured for MW produced $O_2(^1\Delta)$ and then compared to that measured for chemically produced $O_2(^1\Delta)$ at the same total pressure. The highest result that has been obtained for R(MW)/R(chem), ~ 0.6, is considered as an indication that the amount of H_2O in the system is low enough for efficient laser operation.

With $C\ell_2$ flow of 3×10^{-4} mole/s, stable operation of the chemical generator has been obtained for more than 35 minutes, with 44% $O_2(^1\Delta)$. When I_2 is injected transverse to the $O_2(^1\Delta)$ flow, a bright yellow flame, due to I_2(B-X) emission, is observed in the interaction zone. The flame dies off within a few cm of the point of injection when the I_2 is completely dissociated. The flame's length and shape is easily controlled by viewing it through the transparent plexiglass walls of the laser and adjusting the I_2 heating and the carrier and purge gas flows.

The best laser performance is obtained when the flame fills the entire 10×1 cm^2 cross section of the rectangular flow duct and dies off just before reaching the optical cavity zone, 3 cm down stream of the point of the I_2 injection. ~ 5 W of laser output power (3W from the front and 2W from the back mirror) was obtained with a $C\ell_2$ flow of 4×10^{-4} mole/s, I flow of 1.4×10^{-6} mole/s and 45% $O_2(^1\Delta)$. The total O_2 pressure in the cavity was 0.32 Torr and the carrier and purge gas added 0.15 Torr. The laser beam is a 1 cm square spot near the mirrors. The apparent divergence half angle is three miliradians. The output power of the laser under these conditions is ~ 30% of the power conducted in the $O_2(^1\Delta)$ flow, assuming that the total O_2 production is stoichiometric with the $C\ell_2$ flow. Since $[O_2(^1\Delta)] \gtrsim 15\%$ is required to maintain gain,[3] the extractable power efficiency under our conditions is $\gtrsim 45\%$. Allowing for uncertainties in the $[O_2(^1\Delta)]$ measurements,[6] the minimum efficiency is 40%.

The efficiency of the laser decreases drastically if $[H_2O]$ is increased to a level that decreases the $\mathrm{Int}(^1\Sigma)/\mathrm{Int}^2(^1\Delta)$ ratio. Also, the laser does not operate when I_2 accumulates on the teflon coating of the plexiglass walls. This causes a yellow glow near the surface due to $I^* + I$ wall recombination.

The chemical and the MW generators have also been used for studying the reaction of $O_2(^1\Delta)$ with Pb atoms. For this study the laser section in the system described in Fig. 1 is replaced by a metal evaporation furnace.[10,11] The Pb atoms evaporated from the furnace are entrained in argon stream and carried to the Pb/O_2 mixing zone. Typical O_2 and Ar pressures are 0.15 and 0.2 Torr, respectively. Typical working temperatures are 700 K inside the furnace (measured by a thermocouple) and 400 K in the mixing zone (measured by a thermistor). The concentration of $O_2(^1\Delta)$ is monitored using the procedure described above. The emission of $O_2(^1\Sigma)$ and of excited PbO (PbO*) emerging from the mixing zone is measured by a monocromator/photomultiplier/picoammeter/recorder system. Ground state Pb atoms and excited (metastable) Pb atoms (Pb*) are monitored by atomic absorption spectroscopy using a Pb hollow cathode lamp.

Mixing of ground state O_2 with Pb atoms did not result in any visible reaction or measurable amount of excited species. On the other hand, the reaction of $O_2(^1\Delta)$ with Pb resulted in a bluish, diffuse flame, and both Pb^* and PbO^* were observed. In a search for the Pb states present in the flame when operating the MW generator, the ground state, $^3P_0$, and the two lowest excited states, $^3P_1$ and $^3P_2$, were detected, their concentration ratio being 12/4/1; the next higher states, $^1D_2$ and $^1S_0$, were not observed. The absolute concentration of Pb atoms was evaluated to be $\sim 10^{10}$ cm^{-3}. The use of the chemical generator for [Pb] evaluations is not convenient since for reliable absorption measurements the $O_2(^1\Delta)$ has to be switched on and off without changing the flow conditions.

The $Pb(^3P_1)$ is formed, most probably, by the highly resonant energy transfer process

$$Pb(^3P_0) + O_2(^1\Delta) \rightarrow Pb(^3P_1) + O_2(^3\Sigma) + 0.01 \text{ eV} . \qquad (1)$$

The $Pb(^3P_2)$ can be formed by reactions of $Pb(^3P_1)$ with $O_2(^1\Delta)$ and of $Pb(^3P_{0,1})$ with $O_2(^1\Sigma)$.

Analysis of the PbO^* spectrum revealed the seven lowest excited states of PbO: a,b,A,B,C,C' and D, all emitting to the ground state, X. The emission intensity increased with $[O_2(^1\Delta)]$; the increase of the emission from the highest observed state, PbO(D), was twice as that from the other states.

Production of PbO^* by reactions of $Pb(^3P_{0,1,2})$ with $O_2(^3\Sigma,^1\Delta,^1\Sigma)$ is thermochemically precluded. On the other hand, PbO can be formed by reactions of $Pb(^3P_2)$ with $O_2(^3\Sigma,^1\Delta,^1\Sigma)$, of $Pb(^3P_1)$ with $O_2(^1\Delta,^1\Sigma)$ and of $Pb(^3P_0)$ with $O_2(^1\Sigma)$. The formation of PbO^* can thus proceed via excitation of PbO by energy transfer from excited O_2 or Pb. Taking into account both the concentrations of the excited species and the exothermicity of the possible reactions between them, the main route for production of PbO^* is expected to follow the sequence

$$Pb(^3P_2) + O_2(^1\Delta) \rightarrow PbO(X,v) + O(^3P) , \qquad (2)$$

$$PbO(X,v) + O_2(^1\Delta) \rightarrow PbO(a,b) + O_2(^3\Sigma) , \qquad (3)$$

higher states being formed via further excitation by $O_2(^1\Delta)$. The above mentioned behavior of PbO(D) supports this mechanism.

SUMMARY

Highly efficient COIL operation in a small scale apparatus is described. The efficient operation is obtained when due care is taken to ensure good mixing of $C\ell_2$ with BHP, low $[H_2O]$ in the lasing gas mixture, inert laser walls and adequate control of the flow conditions. The use of the combination of a metal evaporation furnace and a chemical $O_2(^1\Delta)$ generator for studying the reaction of Pb atoms with $O_2(^1\Delta)$ molecules is also described and the potential of the generator as a means of electronic excitation of metal atoms and oxides is demonstrated.

REFERENCES

1. W.E. McDermott, N.R. Pchelkin, D.J. Benard and R.R. Bousek, Appl. Phys. Lett. 32:469 (1978).
2. D.J. Benard, W.E. McDermott, N.R. Pchelkin and R.R. Bousek, Appl. Phys. Lett. 34:40 (1979).
3. R.J. Richardson and C.E. Wiswall, Appl. Phys. Lett. 35:138 (1979).
4. C.E. Wiswall, R.J. Richardson and K.V. Reddy, Paper No. ThS12, presented at CLEO 81 (June 10-12, 1981).
5. R.F. Heidner III, C.E. Gardner, T.M. El-Sayed and G.I. Segal, Chem. Phys. Lett. 81:142 (1981).
6. D.J. Benard and N.R. Pchelkin, Rev. Sci. Instrum. 49:794 (1978).
7. a) R.G. Derwent, D.R. Kearns and B.A. Thrush, Chem. Phys. Lett. 6:115 (1970);
   b) R.F. Heidner III, C.E. Gardner, T.M. El-Sayed, G.I. Segal and J.V.V. Kasper, J. Chem. Phys. 74:5618 (1981).
8. A.J. Grimley and P.L. Houston, J. Chem. Phys. 69:2339 (1978)
9. R.G. Aviles, D.F. Muller and P.L. Houston, Appl. Phys. Lett. 37:358 (1980).
10. J.B. West, R.S. Bradford Jr., J.D. Eversole and C.R. Jones, Rev. Sci. Instr. 46:164 (1975)
11. I. Nadler and S. Rosenwaks, Chem. Phys. Lett. 69:266 (1980).

A NEW NUMERICAL METHOD FOR TURBULENT MIXING OF SUPERSONIC REACTING FLOWS IN CHEMICAL LASER CAVITIES

René Schmitt

Office National d'Etudes et de Recherches Aérospatiales

92320 Châtillon (FRANCE)

INTRODUCTION

The first studies on supersonic diffusion chemical lasers involved the hypothesis of one-dimensional premixed flow and the use of an approximate flame-sheet model[1]. Then, more detailed fluid-dynamics calculations[2-5] have been developed from the two-dimensional boundary-layer equations. More recently the full Navier-Stokes equations have been used, for the prediction of chemical laser performances, in two-dimensional calculations[6,7] and even in a three-dimensional approach[8]. But these methods have the disadvantage of being time consuming. Thus it is interesting to develop in parallel other methods less general but taking short computer times.

The new numerical method proposed hereafter allows the calculation of laminar or turbulent mixing, with boundary-layer approximations, of two supersonic flows of different chemical composition and takes into account lateral pressure gradient, chemical reactions and vibrational relaxations effects.

The method is applied to the study of a DF combustion chemical laser cavity, where the chemical reactions are highly exothermic. It allows to determine the detailed variations of velocity, pressure, temperature fields as well as chemical species concentrations, DF vibrational populations and laser gain throughout the region shown in figure 1.

* Research work partly sponsored by DRET (Direction des Recherches, Etudes et Techniques de l'Armement)

In order to validate the method, a HF chemical laser configuration is also calculated and then compared with the finite-difference Navier-Stokes calculations of Kothari et al.[6]. It appears that the numerical results obtained through the two approaches are in very good agreement.

1 - PHYSICAL PROBLEM

Let us consider the two-dimensional mixing of two parallel streams, one is composed of D_2, the other is a mixture of F, F_2, HF and helium as shown in figure 1, where the concentrations are given in molar fractions. The computational domain covers the region downstream the nozzle exits and is limited by the symmetry axes of the two streams.

In the mixing region the cold reaction between D_2 and F creates vibrationally excited DF subsequently deactivated by collisions. The chemical reactions included in the present calculation are those mentioned by Gross and Bott[9], with reaction constants brought up to date. Among these, the most important are :

$$F + D_2 \rightleftarrows DF(v) + D \qquad \text{(1-a)}$$

$$DF(v) + M \longrightarrow DF(v-1) + M \qquad M = D_2, DF, HF \qquad \text{(1-b)}$$

$$DF(v) + DF(v') \rightleftarrows DF(v-1) + DF(v'+1) \qquad \text{(1-c)}$$

Each of the nine DF vibrational levels is treated as a separate species, which leads to 16 species conservation equations. The integrated small-signal gain is calculated at the line center in the same way as in LAMP computer program[3] for the transitions 1-0, 2-1 and 3-2.

2 - GOVERNING EQUATIONS

The flows considered here are assumed to obey the boundary-layer approximations. Recirculating regions at the nozzle exits are not treated in the model. The conservation equations of mass, longitudinal momentum, total enthalpy and species are written :

$$\frac{\partial \rho u}{\partial x} + \frac{\partial \rho v}{\partial y} = 0 \qquad (2)$$

$$\rho u \frac{\partial u}{\partial x} + \rho v \frac{\partial u}{\partial y} + \frac{\partial p}{\partial x} = \frac{\partial}{\partial y}\left[(\mu + \rho\varepsilon)\frac{\partial u}{\partial y}\right] \qquad (3)$$

$$\rho u \frac{\partial H}{\partial x} + \rho v \frac{\partial H}{\partial y} = \frac{\partial}{\partial y}\left[\left(\frac{\mu}{P_{r_L}} + \frac{\rho\varepsilon}{P_{r_T}}\right)\frac{\partial H}{\partial y}\right] + \frac{\partial}{\partial y}\left\{\left[\mu\left(1 - \frac{1}{P_{r_L}}\right) + \rho\varepsilon\left(1 - \frac{1}{P_{r_T}}\right)\right] u \frac{\partial u}{\partial y}\right\} + Q \qquad (4)$$

$$\rho u \frac{\partial c_i}{\partial x} + \rho v \frac{\partial c_i}{\partial y} = \frac{\partial}{\partial y}\left[\left(\frac{\mu}{S_{c_{i_L}}} + \frac{\rho\varepsilon}{S_{c_{i_T}}}\right)\frac{\partial c_i}{\partial y}\right] + \dot{w}_i \qquad i = 1, \cdots, s = 16 \qquad (5)$$

with the state equation :

(6) $$p = \rho r T \qquad \text{where} \quad r = R \sum_{i=1}^{s} \frac{c_i}{m_i}$$

The mixture total enthalpy is given by :

(7) $$H = \int_0^T \sum_{i=1}^{s} (c_i c_{p_i})\, dT + \sum_{i=1}^{s} q_i c_i + \frac{u^2}{2} = \sum_{i=1}^{s} c_i h_i + \frac{u^2}{2}$$

The specific heat c_{p_i} and the heats of formation q_i of species i are assumed independent of temperature. The quantity Q, which represents the rate of heat addition to the gas from chemical reactions and vibrational deactivation, is written :

(8) $$Q = \frac{\partial}{\partial y}\left[\left(\frac{\mu}{P_{r_L}} + \frac{\rho\varepsilon}{P_{r_T}}\right)\sum_{i=1}^{s} h_i \left(\frac{P_{r_L}}{S_{c_{iL}}} - 1\right)\frac{\partial c_i}{\partial y}\right]$$

The species production rate terms $\dot{w}_i$ are calculated explicitly from the elementary reactions, including chemical pumping, dissociation and V-T and V-V vibrational transfers. The laminar Schmidt numbers are calculated from the multicomponent diffusion coefficients, where the 256 (16 x 16) binary diffusion coefficients[10] occur. The laminar and turbulent Prandtl numbers and the turbulent Schmidt number are assumed constants : $P_{r_L} = 0.75$ and $P_{r_T} = S_{c_{iT}} = 0.9$. The laminar mixture viscosity is obtained by using the Wilke estimation method for gases at low pressure[11].

To allow a description of turbulent flows, the model of two transport equations by Wilcox and Traci[12] is used. These equations, which take into account turbulence compressibility effects, are written :

(9) $$\rho u \frac{\partial e}{\partial x} + \rho v \frac{\partial e}{\partial y} = \left(\alpha^* \rho \left|\frac{\partial u}{\partial y}\right| - \beta^* w\right) e + \frac{\partial}{\partial y}\left[\left(\mu + \rho\sigma^*\varepsilon\right)\frac{\partial e}{\partial y}\right]$$

(10) $$\rho u \frac{\partial w^2}{\partial x} + \rho v \frac{\partial w^2}{\partial y} = \left\{\alpha \rho \left|\frac{\partial u}{\partial y}\right| - \left[\beta + 2\sigma \left(\frac{\partial L}{\partial y}\right)^2\right] w\right\} w^2 + \frac{\partial}{\partial y}\left[\left(\mu + \rho\sigma\varepsilon\right)\frac{\partial w^2}{\partial y}\right]$$

where α , β , σ , α^*, β^* and σ^* are constants.

Knowing the turbulent kinetic energy e and turbulent dissipation rate w, the eddy viscosity ε is determined by the relation :

(11) $$\varepsilon = \frac{\rho e}{w}$$

This turbulence modelisation has only an influence on the mean velocity field and does not take into account temperature and concentrations fluctuations.

The boundary conditions on the symmetry axes (see fig. 1) are : $v = 0$ and $\partial/\partial y = 0$ for all the other variables.

3 - NUMERICAL METHOD

To solve the x-momentum equation (3), the longitudinal pressure gradient $\partial p/\partial x$ must be known. However in supersonic flows, the latter can show large variations in the transverse direction y. Therefore it is determined from an inviscid flow calculation.

The originality of the present method lies in solving the inviscid flow equations and the viscous ones by two distinct numerical methods in the same computational domain. These two calculations are simultaneously carried out and then coupled at each step according to the following scheme :

-<u>stage 1</u> : the pressure field p(x,y) is determined from the resolution of Euler's equations, to which are added as second members the viscous terms calculated in the previous step (see stage 2).

-<u>stage 2</u> : all the other variables (velocity, total enthalpy, concentrations) and dissipative terms are determined by solving the boundary-layer equations which have become parabolic due to the explicit knowledge of the pressure gradient $\partial p/\partial x$ defined from stage 1.

The variables calculated in stage 2 at the step n in the x-direction are then used as initial conditions in stage 1 for the pressure field calculation at the next step n+1. A low level pressure gradient correction (below one per-cent) is then iteratively calculated in stage 2 in order to verify the mass conservation. This pressure correction, independent of y, is also used in the calculation of subsonic regions.

At stage 1, the Euler's equations, written under conservative form and with explicitly known second members, are solved by a predictor-corrector finite difference scheme of second order, analogous to the MacCormach approach[13]. The integration step Δx must verify a stability criteria of CFL type.

At stage 2, the 20 viscous equations are written in the Von Mises coordinates (x, ψ) and are solved with a second order precision by a finite difference scheme of the Crank-Nicholson type. The 16 species conservation equations are iteratively solved with a step about twelve times smaller than the step Δx used for calculating the other variables, the chemical rates of production and destruction being more rapid than the diffusion term.

4 - NUMERICAL RESULTS

Solutions are obtained for the flow described in figure 1. The initial conditions for the velocity u and the turbulent variables e and w are determined from the Chevray and Kovasnay's experiments on the flat plate trailing edge in incompressible turbulent flow. This requires the choice of a very narrow grid in the neighbourhood of the confluence line $y/h = 0.654$. The initial temperature is determined from the velocity profile by using the Crocco's law and the pressure is the same in both flows.

Profiles of velocity, temperature and pressure are shown in figures 2, 3 and 4 respectively. In each figure, results for the longitudinal stations $x/h = 5$ and 10 are given, where the solid and dashed lines represent mixing with and without chemical reactions respectively, the latter case being purely artificial. Note from figure 2 that the velocity profiles downstream of the nozzle exits show only small variations due to chemical reactions. In contrast, the corresponding temperature profiles in figure 3 show more distinctly the effect of exothermic reactions. In figure 4 a strong adverse pressure gradient in the x-direction clearly appears for the reacting flow, with a weak lateral pressure gradient effect.

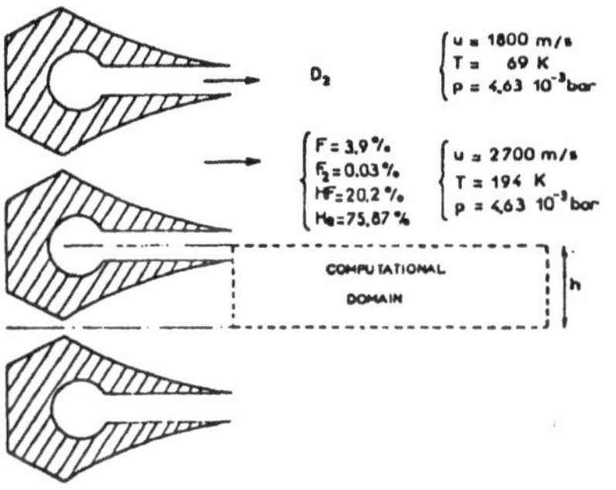

Fig.1 Physical problem and initial conditions

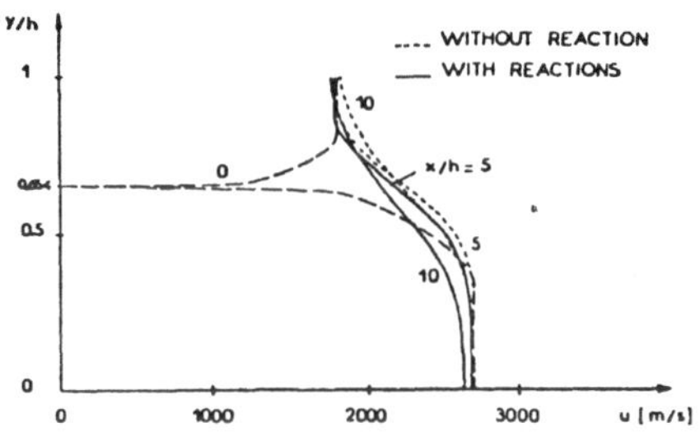

Fig.2 Velocity profiles at various stations

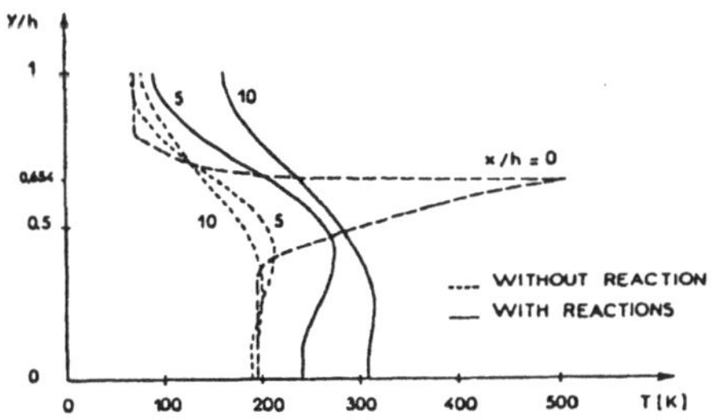

Fig.3 Temperature profiles at various stations

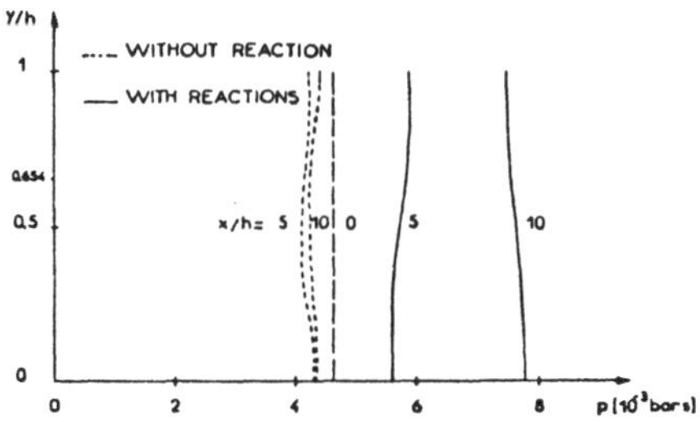

Fig.4 Pressure profiles at various stations

It also appears in figure 2 and 3 that the viscous mixing is very rapid. The laminar Reynolds number being of the order of 1000, the influence of the turbulence is weak. The evolution of the turbulent kinetic energy (see fig.5) is classical for a mixing flow, with a very fast restructuration of the turbulence between x/h = 0 and x/h =1 and with a reponderant diffusion phenomenon in the far flow.

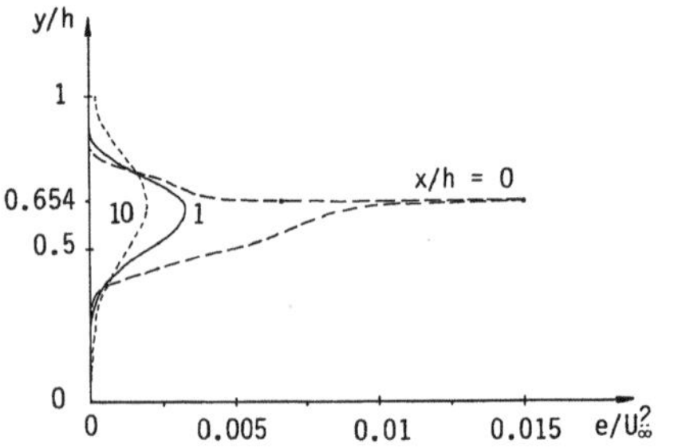

Fig.5 - TURBULENT KINETIC ENERGY PROFILES AT VARIOUS STATIONS

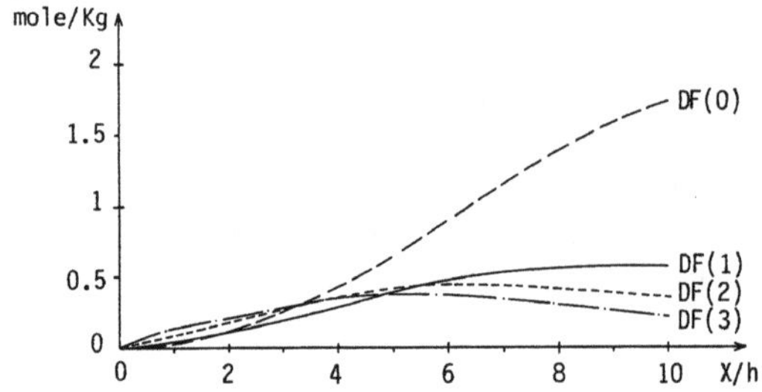

Fig.6 - X-WISE VARIATION OF DF(v) AVERAGE CONCENTRATIONS

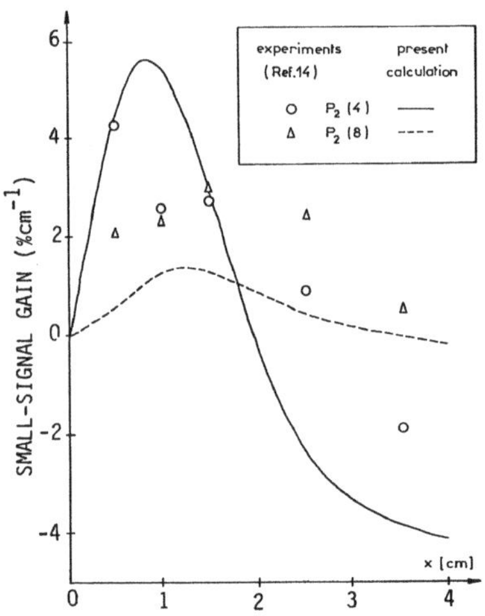

Fig.7 - VARIATION OF SMALL-SIGNAL GAIN WITH THE FLOW DIRECTION ; ROTATIONAL LEVELS $P_2(4)$ & $P_2(8)$

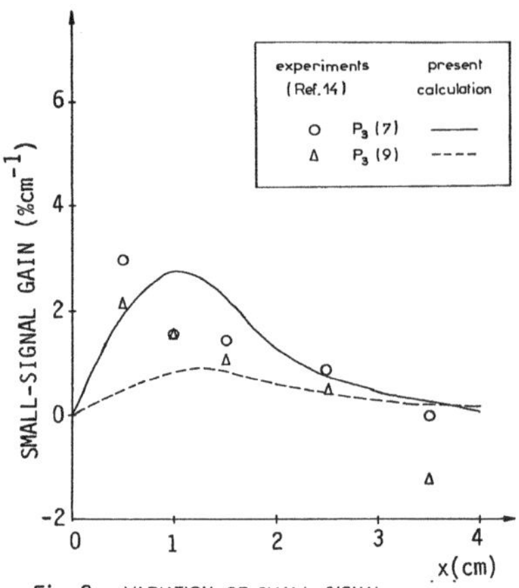

Fig.8 - VARIATION OF SMALL-SIGNAL GAIN WITH THE FLOW DIRECTION ; ROTATIONAL LEVELS $P_3(7)$ AND $P_3(9)$

The existence of partial population inversion and hence laser action is seen in figure 6, where average concentrations of various DF vibrational levels are plotted versus x. The inversions, which exist between the 0-1 and 1-2 levels, yield the small-signal gain. The calculated variations of the integrated small-signal gain with respect to x at the center line for the rotational levels $P_2(4)$ - $P_2(8)$ and $P_3(7)$ - $P_3(9)$ are shown in figures 7 and 8 respectively. These calculated values are compared with measured values of Gastaud et al.[13] in the same configuration.

The time needed on a VAX 11/780 computer for solving the equations for the 16 species, the x-momentum, the total enthalpy and the two turbulent quantities, in a variable transverse grid of 41 points up to x/h = 10, is of 10 min. in the cold case and of 60 min. in the reactive case.

5 - COMPARISON WITH NAVIER-STOKES CALCULATIONS

In order to validate the present numerical method, a HF chemical laser configuration (see fig. 9) with laminar mixing is also calculated and then compared with the finite-difference Navier-Stokes calculations, using a MacCormack method, of Kothari et al.[6]. This comparison, dealing with the purely fluid-dynamic aspects, is carried out in the case where the chemical kinetics are artificially switched off. The results obtained through the two methods are shown in figures 10, 11 and 12. A very good agreement is reached for the velocity, temperature and pressure profiles, since the relative differences are below three per-cent.

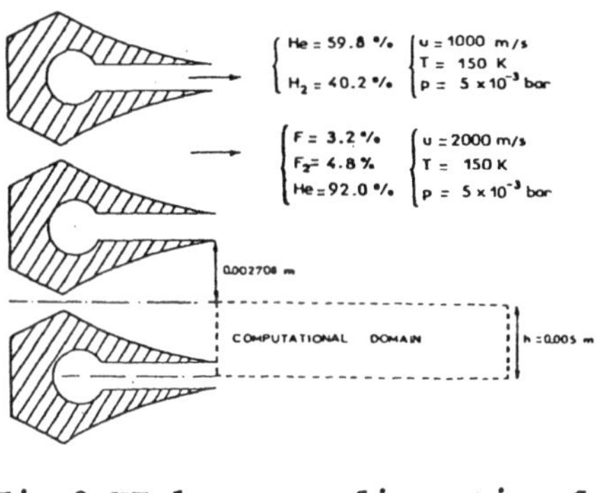

Fig.9 HF laser configuration for compariosn with ref.6

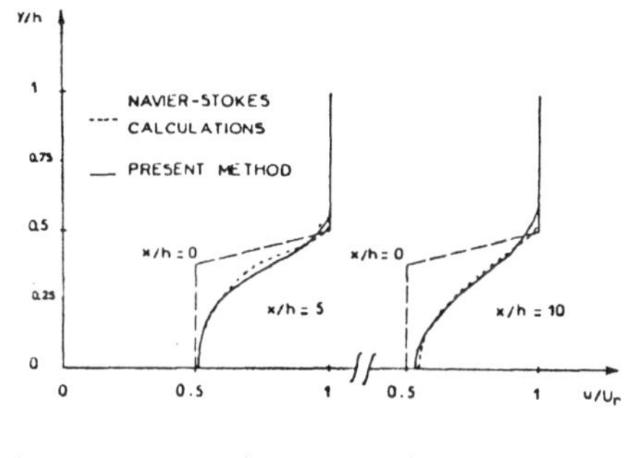

Fig.10 Velocity profiles at various stations

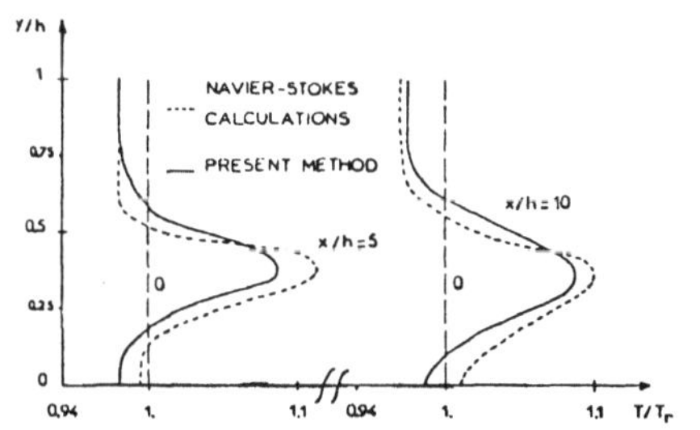

Fig.11 Temperature profiles at various stations

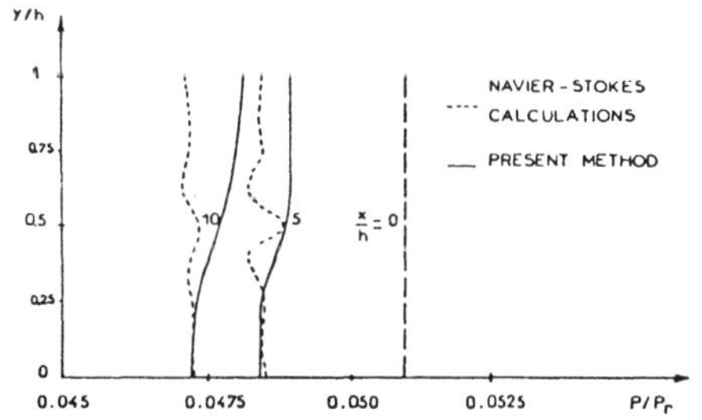

Fig.12 Pressure profiles at various stations

For a grid of 17 points in the y-direction, the computer time on VAX 11/780 in the present calculation is of 6 min. up to x/h = 10, while the Kotheri et al. program needs 30 min. of

UNIVAC 1108 with only 9 points in the y-direction. Another advantage of the present computer code is its weak memory size, which allows the use of a very narrow grid for accurate description of particular phenomena (turbulence for example).

CONCLUSION

The present new numerical method allows the calculation of laminar or turbulent mixing of two supersonic flows of different chemical composition and takes into account lateral pressure gradient effects, chemical reactions and vibrational relaxations. A comparison with Navier-Stokes calculations shows a very good agreement between the results obtained by the two approaches. The advantages of the present method are the calculation rapidity and memory-space saving.

ACKNOWLEDMENT

I want to express my gratitude to Dr. Lucien Torchin from Compagnie Générale d'Electricité (Laboratoires de Marcoussis - France) for his active contribution to this work and most particularly, for carrying out the chemical kinetic programs.

REFERENCES

1. R. Jr. Hofland and H. Mirels, AIAA J., vol. 10, n°4, April 1972, pp. 426-428.
2. W.S. King and H. Mirels, AIAA J., vol. 10, n°12, Dec. 1972, pp. 1647-1654.
3. J. Thoenes and A.W. Ratliff, HREC TR D390222 - vol. I, Lockheed Missiles and Space Cy, Huntsville, Ala., June 1974.
4. A.W. Ratliff and J. Thoenes, AIAA Paper n°74-225, Washington, D.C., Jan. 30-Feb. 1 (1974).
5. R. Tripodi, L.J. Coulter, B.R. Bronfin and L.S. Cohen, AIAA J., vol. 13, June 1975, pp. 776-784.
6. A.P. Kothari, J.D. Anderson and E. Jones, AIAA J., vol. 15, n°1, Jan. 1977, pp. 92-100.
7. T.J. Chung and G.R. Karr, Int. J. for Num. Meth. in Engg., vol. 16, 1980, pp. 1-12.
8. W.L. Hendricks, R.R. Mikatarian, B.J. Gross and N.L. Rapagnani, AIAA Paper n°81-1133, Palo Alto, Ca., June 23-25 (1981).
9. R.W.F. Gross and J.F. Bott, Handbook of chemical lasers, John Wiley and Sons, New-York, 1976.
10. R.C. Reid and T.K. Sherwood, The properties of gases and liquids, Mc Graw-Hill, 1966.
11. R.S. Brokaw, NASA TR-R-81, 1960.
12. D. Wilcox and R.M. Traci, AIAA Paper n°76-351, San Diego, Ca., July 14-16 (1976).
13. R.W. MacCormack, AIAA Paper n°69-354, Cincinnati, Ohio, 1969.
14. M. Gastaud, F. Voignier, P. Bousselet and P. Regnier, Journal de Physique, Suppl. au n°11, tome 41, November 1980.

THEORETICAL AND EXPERIMENTAL STUDY OF cw HF CHEMICAL LASER PERFORMANCE#

L.H. Sentman+, M.H. Nayfeh*, W.O. Mosebach+,
P. Renzoni+, K. Herrick+, K. King*, P. Schmidt+, and
S. Townsend+

University of Illinois
Urbana, IL 61801

INTRODUCTION

The objectives of this research are to measure the frequency and amplitude of the time dependent oscillations which have been predicted to occur in cw chemical lasers employing unstable resonators to extract power[1]. These time dependent fluctuations in cw power, which were shown to be a consequence of rotational non-equilibrium, occur on lines whose saturated gain zone does not fill the unstable resonator[1,2]. Thus, the experimental program depends upon the ability to design an unstable resonator in which the saturated gain zone of one or more of the peak power lines does not reach further into the unstable resonator than the center line. The importance of this point is illustrated by a previous experiment[3] which obtained a null result (no oscillations were observed) because the saturated gain zone of the lasing lines filled the resonator. The initial efforts have thus been directed toward verifying the ability of the computer models[4,5,6] to predict the length of the saturated gain zones of the lasing lines.

CL II PERFORMANCE DATA

The CL II laser is an arc driven subsonic device consisting of two adjacent laser bodies having a variable cross sectional area, Fig. 1. A discharge tube is attached to each laser body into

---

#This study was supported by an AFOSR grant
+Aeronautical and Astronautical Engineering Department
*Physics Department

which He, O_2 and SF_6 are introduced. H_2 is introduced through a series of sonic orifices at the center of the laser body. The gases leave the laser through exhaust tubes that discharge into a vacuum pump. A mercury manometer bank is attached to 10 pressure taps along the channel centerline and 10 off centerline pressure taps in each laser body. Stable resonator multiline optics can be attached to the laser using vacuum mirror mounts or can be detached from the laser body, in which case CaF_2 Brewster windows are used.

To provide a data set for baselining the computer models, stable resonator multi-line power, power spectral distributions (PSD), pressure distributions and beam diameters were measured using vacuum mirror mounts. The feedback mirror was a 2.5 meter radius of curvature ETR (r=0.995+) and the outcoupling mirror was a 2 meter radius of curvature, ZnSe 43% reflector. The laser performance data for six combinations of SF_6 and H_2 flow rates for fixed He and O_2 flow rates are summarized in Table 1. From two burn blocks made at 5 cm and 34.3 cm from the outcoupling mirror, the beam divergence and diameter at the outcoupling mirror were calculated. The measured pressure distribution showed there is no transverse pressure variation and that the pressure is constant throughout the laser bodies. The zero power gain pressure was about 1 torr less than the power on pressure.

The PSD, which are the average of a minimum of two scans for each run, are shown in Fig. 2. The shift in the peak of the power spectral distributions and the loss of the high J $2\rightarrow1$ line and the appearance of the low J $1\rightarrow0$ line[2] as the SF_6 flow rate decreases are indicative of the decrease in temperature which accompanies the reduced heat release due to the reduced F atom flow rate. The same shift toward lower J is observed when the H_2 flow rate decreases for fixed SF_6 flow rate, which is also indicative of a drop in temperature due to a decrease in the formation of HF as a consequence of the decrease in H_2 flow rate.

CL II BASELINE MODELING

The numerical modeling of the CL II laser was performed using the rotational equilibrium Blaze II code[4,5]. After agreement with experimental data was achieved with Blaze II, the rotational nonequilibrium code MNORO[6] was used to calculate the PSD and the saturated gain distributions for all lasing lines. Run 36 was selected as the baseline case for the determination of the four unknowns, the initial temperature of the flow at the H_2 injectors, the % dissociation of the SF_6 in the discharge tubes and the mixing lengths of the flow. Since rotational equilibrium models tend to overpredict the outcoupled power and underpredict the length of the lasing zone,[6] the Blaze II model should overpredict the measured power by about 10% and the lasing cut off should fall in the range 2.5-3.0 mm.

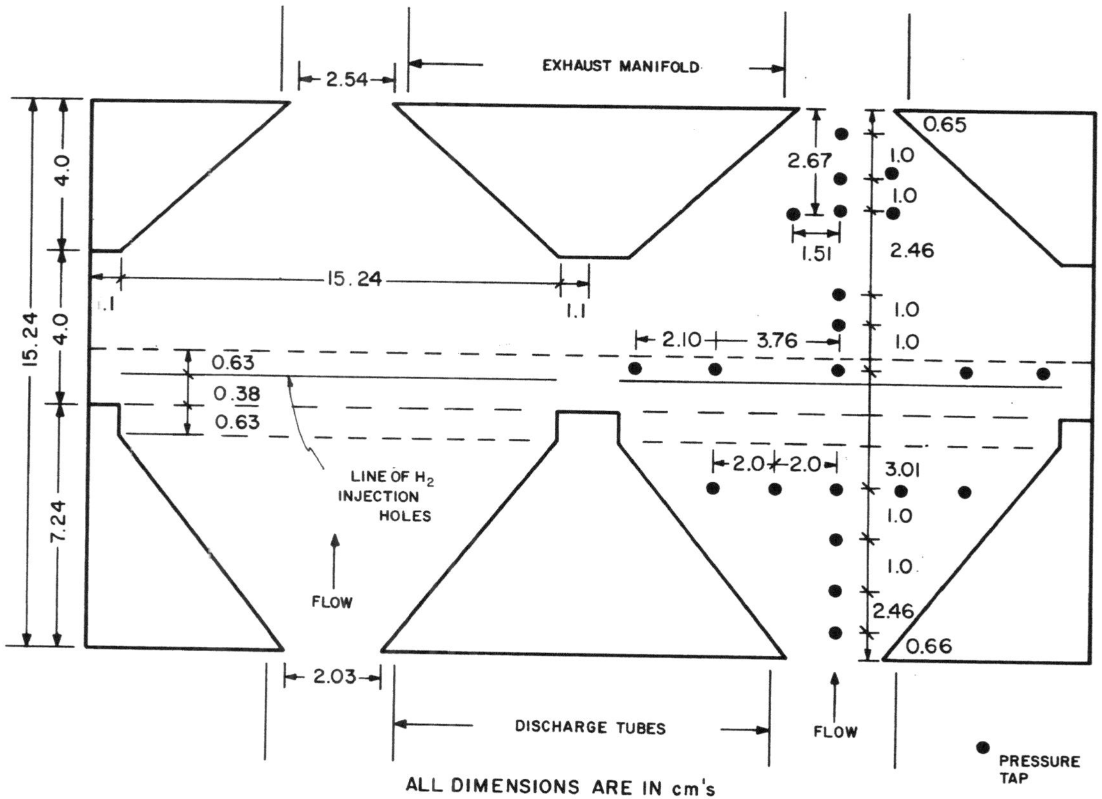

Fig. 1. Geometry of the Helios CL II laser body (not drawn to scale). The H_2 injection holes are 0.0254 cm in diameter on 0.127 cm centers.

A sketch of the hydrogen jets is shown in Fig. 3. To utilize the Blaze II model to treat this complicated, three dimensional mixing, the flow was idealized as a series of alternating primary and secondary slit nozzles. The exit width of the secondary nozzle, which was assumed to be a sonic slit, was taken to be the diameter

Table 1. Comparison of the Model Results with the Data for the Helios CL II Laser. The Primary Mixing Length is 0.001 cm for All Cases

RUN NO.	32	33	34	35	36	37
$\dot{m}_{SF_6}$, gm/sec	1.36	1.01	0.67	1.36	1.01	0.67
$\dot{m}_{H_2}$, gm/sec	0.0375	0.0375	0.0375	0.0545	0.0545	0.0545
P, Torr	12.0	11.0	10.0	12.0	11.0	10.0
% SF_6 Dissoc.	3.3	4.0	4.0	3.3	4.0	4.0
Initial T, °K	500	450	450	500	450	450
H_2 mixing length, cm	3.0	3.5	4.0	3.0	3.5	4.0
P_T, watts						
Data	35.0	32.5	26.0	39.0	36.0	26.0
Blaze II	36.1	32.0	25.4	44.2	39.5	29.4
P_{10}/P_T						
Data	0.475	0.477	0.495	0.491	0.471	0.495
Blaze II	0.537	0.522	0.512	0.523	0.514	0.507
P_{21}, watts						
Data	18.4	17.0	13.1	19.1	19.0	13.1
Blaze II	16.7	15.2	12.4	21.1	19.2	14.5
MNORO	14.8	13.3	10.8	18.9	17.0	12.5
Beam Dia., mm						
Data	2.92	3.03	2.89	2.84	2.86	2.77
Blaze II						
1-0	3.44	3.09	2.88	2.79	2.65	2.31
2-1	2.87	2.75	2.67	2.44	2.40	2.19
MNORO 2-1	2.74	2.70	2.58	2.35	2.32	2.12

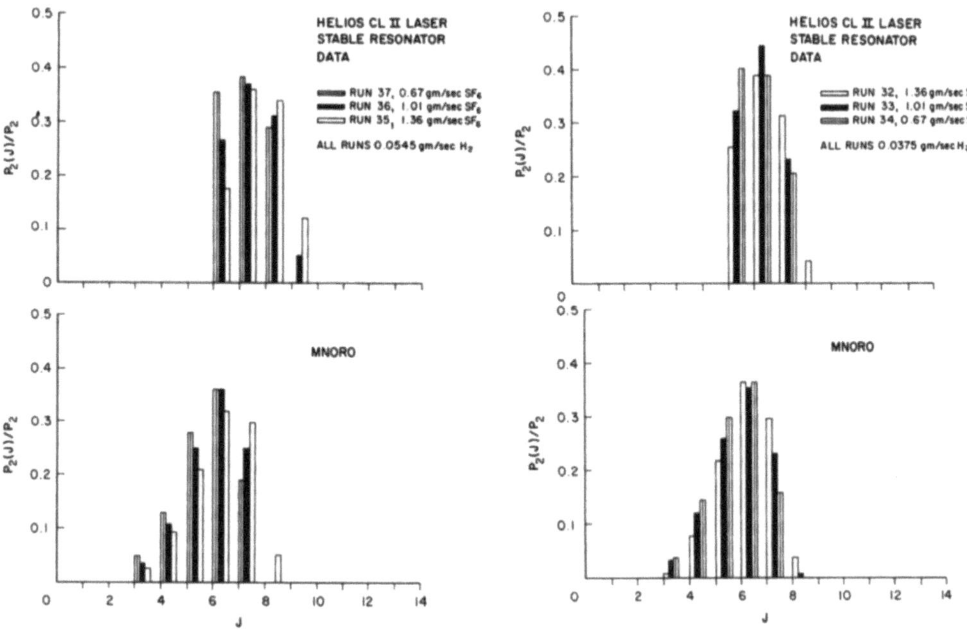

Fig. 2. Experimental and MNORO PSD as a function of SF_6 flow rate. $\dot{m}_{He}$ = 0.078 gm/sec; $\dot{m}_{O_2}$ = 0.28 gm/sec.

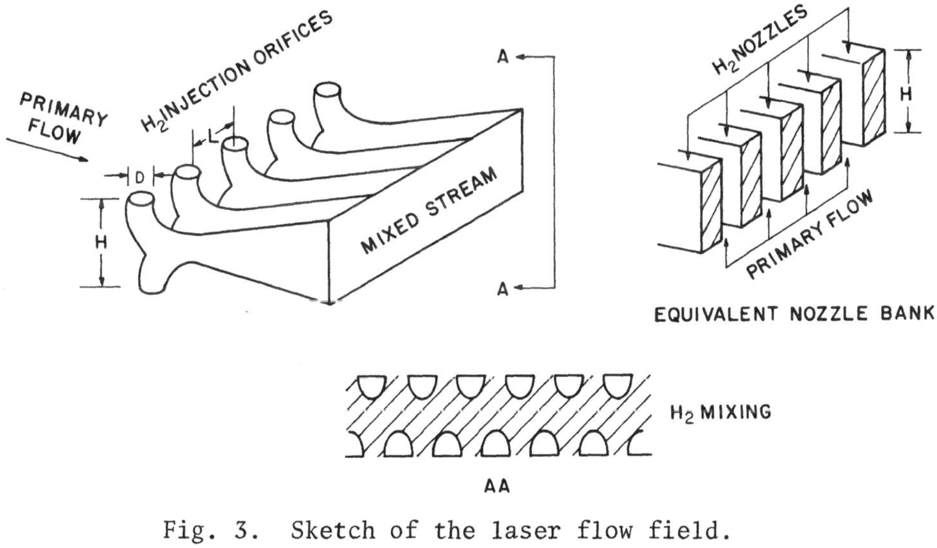

Fig. 3. Sketch of the laser flow field.

of the H_2 orifice. The exit width of the primary nozzles was taken to be the edge to edge spacing between the H_2 orifices, 0.1016 cm. Since each laser body contains 121 H_2 injection hole pairs, the equivalent nozzle bank consisted of 241 H_2 nozzles and 240 primary nozzles. The scheduled mixing option with a separate mixing length for the primary and secondary streams[5] was used. From an area defined run starting at the exit of the discharge tubes, the Mach number of the primary flow at the H_2 injectors was estimated to be 0.2.

Two independent upper bounds on the % dissociation of the SF_6 were obtained, one from a measurement of the F^- ion concentration in the exhaust scrubber for a fixed flow rate of SF_6 for a fixed run time and the other from a measurement of the power absorbed by the arc discharge as a function of SF_6 flow rate. The results for an SF_6 flow rate of 1.188 gm/sec were 13% and 15% respectively.[2] The lasing cut off length was primarily determined by the mixing lengths. The power split between vibrational bands was mainly influenced by the initial temperature. The higher the initial temperature, the greater the fraction of power contained in the 1→0 band. The combination of a mixing length for the primary stream of 2.2 mm, a mixing length for the secondary stream of 6 mm, an initial temperature of 400°K and 7% SF_6 dissociation were found to be the best values of the four unknowns for the baseline case. With this combination of initial parameters, the Blaze II results for the total power, the percentage of power in 1→0 and the lasing cut off were 38.5 watts, 48% and 2.2 mm. A longer cut off could not be obtained.

When this case was run using the MNORO rotational nonequilibrium code,[6] the power and lasing cut off were a factor of 4 smaller than the Blaze II results. Furthermore, MNORO gave an equilibrium initiation sequence and PSD and the lasing species profiles showed an almost Boltzmann distribution over J levels prior to lasing threshold. This clearly showed that the lasing species rotationally relaxed too fast.

The cause of this exceptionally fast rotational relaxation was traced to the HF concentration at threshold which was two orders of magnitude larger in the CL II model than in the CL XI and the Aerospace laser models[2,6]. The only way of decreasing the rotational relaxation was to decrease the rate of formation of HF prior to threshold. When the H_2 mixing length was increased to 4 cm, the H, and hence the HF, concentration at threshold decreased by a factor of 1.7. When the primary mixing length was decreased to a value less than the length at which lasing threshold occurred, the H profiles for the CL II, the CL XI and the Aerospace laser became very similar in shape and magnitude. With these mixing lengths, the MNORO calculations showed a rotational nonequilibrium character for the rotational population distributions, the initiation sequence and the PSD, which was now in reasonable agreement with the data.

With the new mixing lengths, the initial temperature of 400°K and 7% SF_6 dissociation gave too large a value for P_{10}/P_T. The combination of all three parameters which gave good agreement with the data is given in Table 1, together with the data and the predicted laser performance. This baseline model was run with MNORO. The agreement between the Blaze II and MNORO results and the data was very good, Table 1. MNORO predicts about 10% less power in the 2→1 band than Blaze II, as expected from a rotational nonequilibrium code. The lasing cut offs predicted by Blaze II and MNORO for the 2→1 band differ by less than 5%. The computed PSD compares well with the data, Fig. 2, with the peak at one J lower than the data.

EXTENSION OF THE BASELINE MODEL TO OTHER FLOW RATES

To determine the validity of the model which has been developed, it was applied to the flow rates corresponding to the other five sets of experimental data given in Table 1. Since the percent dissociation of the SF_6 in the arc changes as the SF_6 flow rate changes,[2] the model was first applied to Runs 35 and 37 which have the same H_2 flow rate as the baseline case. When a lower SF_6 flow rate is used, the momentum of the primary stream that meets the sonic H_2 jets is less and therefore less penetration of the F atoms into the H_2 stream takes place. Hence, the secondary mixing length should increase for the lower SF_6 flow rate and decrease for the higher SF_6 flow rate. An estimate for the % SF_6 dissociation as the SF_6 flow rate changes was obtained from the % SF_6 dissociation determined from the modeling exercise and the % SF_6 dissociation calculated from the power absorbed in the discharge tubes as a function of SF_6 flow rate[2]. Since the % SF_6 dissociation decreases as the SF_6 flow rate increases, and the discharge tubes are run at a constant current, there is an increase in the power absorbed by the flow and thus the initial temperature of the flow at the H_2 injectors increases slightly. With these changes, the models gave good agreement with the data including the observed shift in the peak of the PSD as SF_6 flow rate changed, Fig. 2.

The validity of the model was checked by comparing the Blaze II and MNORO results to the data when the H_2 flow rate was reduced. The only input parameter changed for these calculations was the H_2 flow rate. As shown in Table 1, good agreement with the data was obtained in all three cases, Runs 32 through 34. The MNORO PSD are in reasonable agreement with the data, and give the observed shift from lower to higher J's as the SF_6 flow rate increases, Fig. 2.

To further check the model, its predictions were compared to the data obtained with a 39% and a 56% reflective mirror. The only input parameter changed for these calculations was the mirror reflectivity. From Fig. 4 it is seen that for all six flow rate combinations, the model and the data are in good agreement as the mirror reflectivity varies. Fig. 4 shows that the general shift in the PSD toward higher J as the mirror reflectivity increased is reproduced by the model although the model does not quite shift the peak one J higher as occurs in the data.

An unexpected result was the appearance of the lines $P_2(10)$, $P_2(11)$ and $P_2(12)$ with a minimum at $P_2(11)$. These lines were observed for the high and middle SF_6 flow rates for both H_2 flows for the 56% mirror. The corresponding $P_1(J)$ lines observed were J=7 through 10 with the peak at J=9. The occurrence of these lines is probably related to the lower losses associated with the 56% reflective outcoupler and the upstream-downstream coupling of the stable resonator. However, a detailed understanding of the reason for the occurrence of these lines and the minimum at $P_2(11)$ is lacking at present.

CONCLUDING REMARKS

The Helios CL II stable resonator multi-line performance with vacuum mirror mounts was measured as a function of the SF_6 and H_2 flow rates and mirror reflectivity. Through a comparison of the measured and the predicted power, power split between $2\rightarrow1$ and $1\rightarrow0$, beam diameter, pressure distribution and PSD, a set of values for the % SF_6 dissociation in the arc, the initial temperature of the flow at the H_2 injectors and the mixing lengths for the primary and secondary streams which gives good agreement with the data was determined. The physical description of the mixing which has emerged and is supported by visual observation of the flame indicates that the hydrogen jets remain intact with minimal breakup and mixing with the primary stream. The reaction appears to occur on the surface of the hydrogen jets. Since these H_2 jets are similar to the trip jets introduced to promote mixing in trip nozzles, these results suggest that the enhanced mixing which results from the trip jets does not occur due to a break up of those jets but from some other fluid dynamical effect which they induce in the flow. When additional fuel was introduced through the trip jets, no increase in performance occurred because the minimal mixing of the trip jets with the primary nozzle flow prevented the formation of a significant amount of additional lasing species.

During this modeling exercise, it was found that the rotational nonequilibrium effects, in particular the PSD, were significantly affected by the mixing rate. This implies that for situations in which the fluid dynamics of the mixing process are so complicated that they are represented by a very approximate model, the determination of the mixing parameters using a rotational equilibrium kinetics model could lead to serious errors. The mixing parameters cannot be considered fixed until they have been used in a rotational nonequilibrium model and the correct PSD obtained.

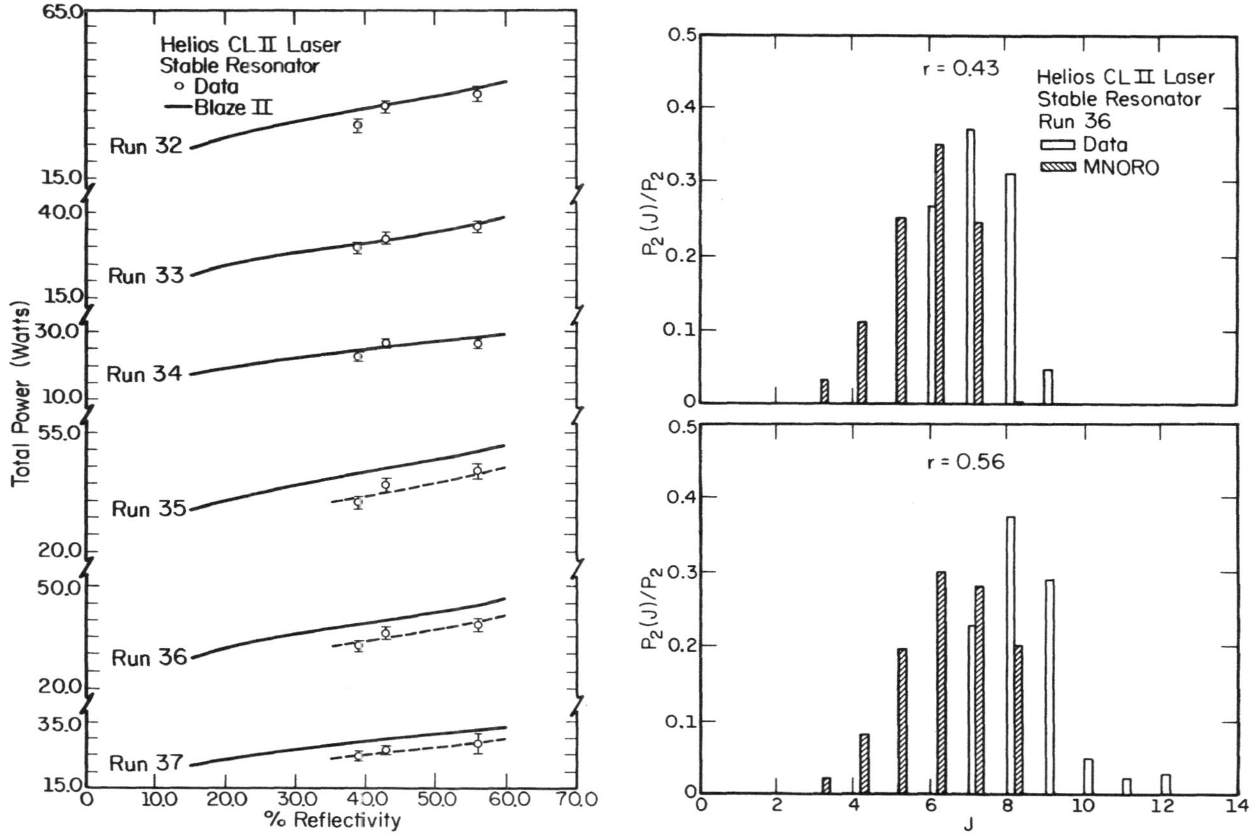

Fig. 4. Experimental and predicted laser performance versus mirror reflectivity.

The model was tested by comparing its predictions with the experimental data for different SF_6 and H_2 flow rates and mirror reflectivities. As would be expected, the % dissociation of the SF_6 and the mixing length changed as the SF_6 flow rate changed. For fixed SF_6 flow rate, the model gave good agreement with the experimental data as the H_2 flow rate or the mirror reflectivity changed without requiring any other changes in the input parameters.

While the model gives reasonable agreement with the measured PSD and reproduces the observed shift toward higher J's as the SF_6 flow rate increases, the model generally predicts more lines than are observed. This is probably a consequence of the Fabry-Perot resonator employed in the calculations. If the calculations were repeated with a stable resonator model which would include the upstream-downstream coupling, the calculated PSD would come into better agreement with the data.

REFERENCES

1. L. H. Sentman, Chemical laser power spectral performance: a coupled fluid dynamic, kinetic and physical optics model, Applied Optics, 17, 2244 (1978).
2. L. H. Sentman, W. O. Mosebach and P. Renzoni, A Theoretical and Experimental Study of cw HF Chemical Laser Performance, TR AAE 81-8, UILU Eng. 81-0508, Aeronautical and Astronautical Engineering Dept., University of Illinois, Urbana, IL, 1981.
3. E. B. Turner, R. A. Chodzko and H. Mirels, Temporal stability of single line cw HF chemical laser with unstable resonator, J. of Appl. Phys., 48, 1163 (1977).
4. L. H. Sentman, M. Subbiah and S. W. Zelazny, Blaze II, a Chemical Laser Simulation Program, TR H-CR-77-8, Bell Aerospace Textron, Buffalo, NY, 1977.
5. J. W. Raymonda, M. Subbiah, J. T. Schimke, S. W. Zelazny and L. H. Sentman, Advanced HF and DF Chemical Laser Performance Modeling, Vol. I, The CNCDE/BLAZE Rotational Equilibrium Code, TR DRCPM-HEL-CR-79-7, Bell Aerospace Textron, Buffalo, NY, 1979.
6. L. H. Sentman and W. Rushmore, Computationally efficient, rotational nonequilibrium cw chemical laser model, AIAA J., 19, 1323 (1981).

PROPAGATION VELOCITY OF THE IONIZATION FRONT DRIVEN BY A HIGH-CURRENT RELATIVISTIC E-BEAM THROUGH THE ACTIVE LASER MEDIA OF INCREASED PRESSURE

G. V. Sholin, A. V. Demura, J. V. Karasev,
G. P. Maksimov, G. L. Nedoseev, V. D. Rusanov,
A. M. Spector, and V. L. Shiryaevsky

I. V. Kurchatov Institute of Atomic Energy
123182 Moscow, USSR

INTRODUCTION

Relativistic electron beams (REB) are widely used for the creation of an active laser medium. They provide homogeneous excitation of large volumes in gas-flow and chemical lasers at low and intermediate pressures. On transition to the increased pressures of gaseous mixtures, the dynamics of excitation starts to depend on the propagation velocity of the ionization wave driven by REB. The fact that the REB-front velocity can evidently differ from the initial electron velocity and be determined by the motion of the ionization front has been discovered by measurements at low and intermediate pressures in electrically neutral gases [1-3]. With a rise in pressure the front velocity seems to approach the initial velocity of the beam electrons, $V_{bo} \cong C$, since the time of electrical neutralization is inversely proportional to the gas concentration. Indeed, such an increase in velocity has been observed in the pressure region from $P \cong 0.1$ torr up to $P \cong 1$ torr [4]. With a further pressure rise to $P > 100$ torr, and especially at the transition to electrically-negative gases, the velocity of the beam current does not remain constant but drops with pressure, as the experimental results given below show. Therefore, the lower limit of the radiation pulse duration of the chemical laser initiated by REB rises with a rise in the pressure of the gas mixture, and this fact should be taken into account in extrapolation of the experimental data to the range of higher gas pressures and REB currents [5].

The theory of the propagation of an ionization wave driven by a high-current REB through dense gases and the experimental results derived from initiation of a chemical laser by this REB are presented below.

THEORY

Let us consider an E-beam with such a high particle density, n_{bo}, and, consequently, with such a high frequency, $\omega_{pb} = (4\pi n_b e^2/m)^{\frac{1}{2}}$, that the corresponding scale of charge separation, $r \cong V_{bo}/\omega_{pb} \cong$ $\cong c/\omega_{pb}$, becomes small in comparison with the characteristic dimensions of the system, the distance from the beam boundary to the wall of the chamber included. Under these conditions, a physical picture of the beam transport through a dense neutral gas may be represented as follows. The pulse high-current electron accelerator is known to be a diode with a thin foil as an anode. Electrons of the beam penetrate through it, at the time of injection, into a drift chamber filled with a working gas. Before injection, the diode gap is a flat capacitor with charges -Q and +Q at its plates. Electrons driven from a cathode transfer part of the charge, q, beyond the anode foil and a new capacitor is produced with one positively-charged central plate (anode) and two negatively-charged plates surrounding the anode: a cathode with charge -(Q-q) and a virtual cathode in the drift gap with a charge -q. The distance from the foil to the virtual cathode is found from the ratio between the initial velocity of the beam electrons, $V_{bo} \cong C$, and their plasma frequency, ω_{pb} [6]. The characteristic electric field strength between the virtual cathode and the anode foil is equal to $E_o = \gamma mc\omega_{pb}/e$. In this case, the field steadily decreases from $E = E_o$ to $E = 0$, and is directed from the anode foil to the head of the beam. The followed breakdown results in the formation near the anode foil of a plasma with high conductivity, that is similar to the growth of the anode foil thickness. In this case, the double layer between the anode and the virtual cathode flies away with the velocity of an ionization wave driven by REB, which the current coming from the cathode as a relativistic beam is compensated by the current of plasma electrons at the anode. Stationary wave propagation occurs when the electrostatic energy stored in the accelerator is high in comparison with the energy transmitted to the beam, and the length of the drift chamber does not exceed the penetration path for relativistic electrons. Since the characteristic time for the translation of the main gradient of electric potential is determined by the conductivity, $\sigma = e^2 n_e/$ $/(m\nu_{eff})$, of the plasma produced on break-down, and since the distance, along with the displaced double layer, is determined by the scale of charge separation, $\Delta x = c/\omega_{pb}$, the propagation velocity can be estimated:

$$v_{REB} = \frac{\Delta x}{(4\pi\sigma)^{-1}} \cong \frac{4\pi ce^2 n_e}{\omega_{pb}\nu_{eff} m}.$$

A geometric mean value of the Townsend ionization frequency,

$$\nu^T_{ion} = n_o < \sigma^T_{ion} V_e >$$

and of the frequency of elastic collisions,

$$\nu_{el} = n_o < \sigma_{el} V_e >$$

between plasma electrons and molecules of the gas is considered to be an effective frequency of collisions. When electron density is calculated from the balance between recombination and beam ionization:

$$n_e^2 \alpha_r = n_o n_b < \sigma^b_{ion} V_b >,$$

a dependence for the ionization wave velocity on the concentration of a gas, $v_{REB} \sim n_o^{-\frac{1}{2}}$ is obtained.

If the electron density is controlled by the processes of attachment and of beam ionization: $n_e n_o < \sigma_{att} V > = n_o n_b < \sigma^b_{ion} V_b >$, the wave velocity decreases linearly with the rise in the concentration of the gas; it also depends on the concentration of electrons in the beam: $v_{REB} \sim n_b^{\frac{1}{2}}/n_o$.

This qualitative model is also true when the beam current is considerably lower than the Alfven one and when a rather high external longitudinal magnetic field provides one-dimensional motion. This can be confirmed by analysis of a corresponding set of equations describing both electron and plasma driven beams:

$$\frac{\partial n_b}{\partial t} + \frac{\partial}{\partial z} (n_b V_b) = 0, \tag{1}$$

$$\frac{\partial V_b}{\partial t} + V_b \frac{\partial V_b}{\partial z} = - \frac{eE}{m}, \tag{2}$$

$$\frac{\partial}{\partial t} (n_e - n_i) + \frac{\partial}{\partial z} (V_e n_e - V_i n_i) = 0, \tag{3}$$

$$\frac{\partial n_e}{\partial t} + \frac{\partial}{\partial z} (n_e V_e) = n_e n_o < \sigma^T_{ion} V_e > + $$
$$+ n_b n_o < \sigma^b_{ion} V_b > - n_e n_i \alpha_r - n_e n_o < \sigma_{att} V_e >, \tag{4}$$

$$m V_e n_o < \sigma_{el} V_e > = -eE, \tag{5}$$

$$\frac{\partial E}{\partial z} = 4\pi e (n_i - n_e - n_b). \tag{6}$$

Here, n_b, n_e, V_b, V_e are concentration and velocity of electrons in the beam and in the plasma, respectively; n_i, V_i are concentration and velocity of ions in the plasma which are considered later to be immobile (V_i=0); $\nu_{el} = n_o < \sigma_{el} V_e >$ is the frequency of elastic collision between electrons and molecules in the plasma, i.e., collisions which are accompanied by a loss of impulse; $\nu^b_{ion} = n_o <$ $< \sigma^b_{ion} V_b >$ is the frequency of ionization by the beam; $\alpha_{att} = <$ $< \sigma_{att} V_e >$ is the rate coefficient of attachment of electrons and

α_r is that of the electron-ion recombination. Restricting ourselves to the solution of the wave-type, dependent on the variable $\xi = Z - ut$, and introducing some dimensionless quantities $t = \omega_{pb}t'$; $x = \omega_{pb}x'/c$; $\beta = u/c$; $V_\alpha = V'_\alpha/c$; $n_\alpha = n'_\alpha/n_{bo}$; $E = eE'/mc\omega_{pb}$, where $\omega_{pb}^2 = 4\pi n_{bo}e^2/m$ and n_{bo} is the concentration of electrons in the beam near the anode foil, we find:

$$n_b(V_b-\beta) = \text{const}, \qquad (1')$$

$$\varepsilon_b + \int E d\xi = \text{const}, \qquad (2')$$

$$\beta(n_i-n_e-n_b) + n_eV_e + n_bV_b = \text{const}, \qquad (3')$$

$$\frac{\partial}{\partial\xi} n_e(V_e-\beta) = n_e\nu_{ion}^T + n_b\nu_{ion}^b - n_en_i\alpha_r - n_on_e\alpha_{att}, \qquad (4')$$

$$V_e\nu_{el} = -E, \qquad (5')$$

$$\frac{\partial E}{\partial\xi} = n_i - n_e - n_b. \qquad (6')$$

Here ε_b is the beam energy in a moving coordinate system. In order to arrive at a solution, the following conditions are necessary: the boundary conditions $V_b \to \beta$, $E \to 0$, $n_e = n_i \to 0$ at $\xi = \xi_o$ and the stationary condition $(\partial/\partial\xi = 0)$ at $\xi \to -\infty$, i.e., in the region near to the anode foil:

$$n_i = n_e + n_b;\ \nu_{el}n_bV_b = n_eE;\ n_en_i\alpha_r + n_on_e\alpha_{att} = n_b\nu_{ion}^b.$$

Relativism is not very important here. Townsend ionization occurs only in a rather high electric field. As the electric field changes rapidly up to a value $E \cong 10^6$ v/cm, ν_{ion}^T can be approximated by a step function:

$$\nu_{ion}^T = \text{const at } E \gg E_{break};\ \nu_{ion}^T = 0 \text{ at } E < E_{break}.$$

Clearly there are three characteristic ranges where E and n are changed. First is the so-called, head of the beam, whose behavior is determined by the dynamics of deceleration of relativistic electrons and where $n_b \gg n_e \cong n_i$. Plasma concentration is dependent here on the direct ionization by electrons of the beam only $n_b\nu_{ion}^b \gg n_e\nu_{ion}^T$ and the following relationship is true:

$$\frac{\partial n_e}{\partial E} = \frac{\nu_{ion}^b}{\beta}. \qquad (I)$$

A region of break-down, where ν_{ion}^T prevails and the effect of beam ionization may be neglected, follows the head. Here, the following relationship is true:

$$\frac{\partial n_e}{\partial E} = \frac{n_e \nu_{ion}^T}{\nu_{el}(V_{bo}-\beta+n_b\beta) - n_e E} \; . \qquad \text{(II)}$$

Finally, after passage of the field E through its maximal value at

$$E_{max} \cong \frac{n_b}{n_e} \beta\nu_{el},$$

the system enters the region where the field, E, owing to the translation of charges by conductivity currents, drops down to a magnitude at which break-down can terminate, but the plasma density is already so high that

$$En_e >> \nu_{el}(V_{bo}-\beta+n_b\beta).$$

In this region, the third one, the following relationship is true:

$$\frac{\partial n_e}{\partial E} = -\frac{\nu_{el}\nu_{ion}^T}{E} - \frac{n_e\nu_{el}}{\beta} \; . \qquad \text{(III)}$$

From the transition from region I to regions II and III, we obtain an algebraic set:

$$\frac{\beta}{\frac{b}{\nu_{ion}}} = \frac{V_{bo}-\beta+n_i\beta}{n_1\nu_{ion}^T} - \frac{E^*}{\nu_{ion}^T\nu_{el}} \qquad \text{(IVa)}$$

$$\frac{1}{\frac{\nu_{ion}^T\nu_{el}}{E^*} - \frac{n_2}{\beta\nu_{el}}} = \frac{\nu_{ion}^T(V_{bo}-\beta+n_b\beta)}{n_2} - \frac{E^*}{\nu_{ion}^T\nu_{el}}. \qquad \text{(IVb)}$$

Out of this set, choosing the points of joining of the solution: n_1 and n_2 at the same field value $E = E^*$, we obtain the dependence of the wave velocity on the pressure of a gas and on the plasma concentration in conformity with the qualitative description developed above.

$$v_{REB} = \beta c \sim cn_e/\sqrt{\nu_{ion}^T\nu_{el}}.$$

The following assumption natural for rather high fields is used in this case:

$$n_b\beta > V_{bo} - \beta \text{ and } E^* \sim \nu_{el}.$$

We find that solutions of the flying double layer type do really exist at the definite velocities of the ionization wave driven by REB.

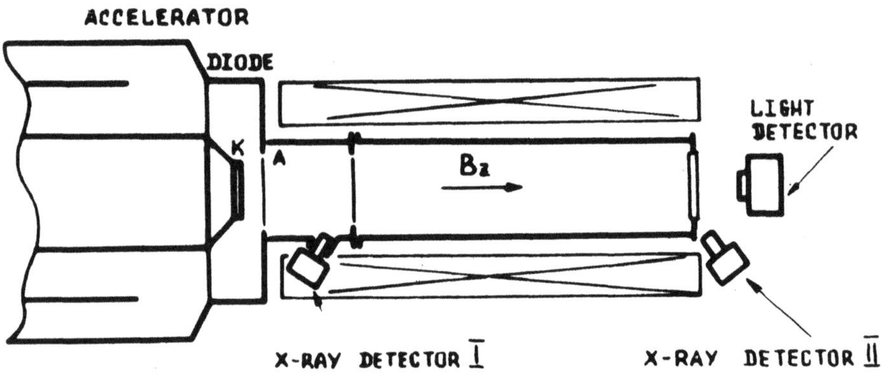

Fig. 1. The scheme of the experiment: K - the cathode of the electron accelerator; A - anode foil; B_Z - axial magnetic field.

EXPERIMENT

A scheme of the experiment is given in Figure 1. The operation chamber is a hollow cylinder of stainless steel, 2 m long, with an internal diameter of 20 cm. A drift tube with the pressure of air $P \sim 1$ torr sustained during the experiments is located between the operating chamber and the outlet window of the electron accelerator. The drift tube is separated from the operating chamber with a Ni-foil, 50 μm thick, attached to a supporting frame. The Ni-foil and its supporting frame determine the losses and the angular spread of electrons in REB at the entrance into the operating chamber. A geometrical transparency of the frame is ≅ 60%. A titanium foil, 50 μm thick, is used as an anode. The accelerator diode and the operation chamber are within a longitudinal magnetic field $B_Z \cong 4$ kG. The REB-parameters in the operation chamber are: current $I \cong 30$ kA, maximal electron energy 1 meV, FWHM of the pulse, $\tau_{\frac{1}{2}} \cong 50$ ns; beam energy, $W_b \cong 1.3$ kJ; diameter, d = 8 cm. The REB-front propagation velocity is determined by a time-shift between signals from two X-ray probes located at input and output of the operation chamber. The time of REB-arrival is determined by the level of 0.1 of the probe signal height at the leading edge of the Bremsstrahlung pulse (Figure 2). The Bremsstrahlung from the plate of steel, 3 mm thick, which closes the output end of the operating chamber, is registered at its output. The time resolution of the probes is not longer than 2 ns. Signals from the probes are registered with the 6 ПОР-04 oscillograph. The REB-current front propagation velocity, which is assumed to be kept constant, is given in Figures 3 and 4 as a function of the gas pressure. An analysis of the results shows that a change in the time of the current propagation with a rise in the gas pressure cannot be explained only by peaking of the REB-current front owing to an effective deceleration of low energy-electrons within it. For example, in N_2 within the pressure range 20-380 torr, no current rise-time reduction (from 0.1 up to 0.9 of magnitude) has been observed. In this case, the propagation time of the current front

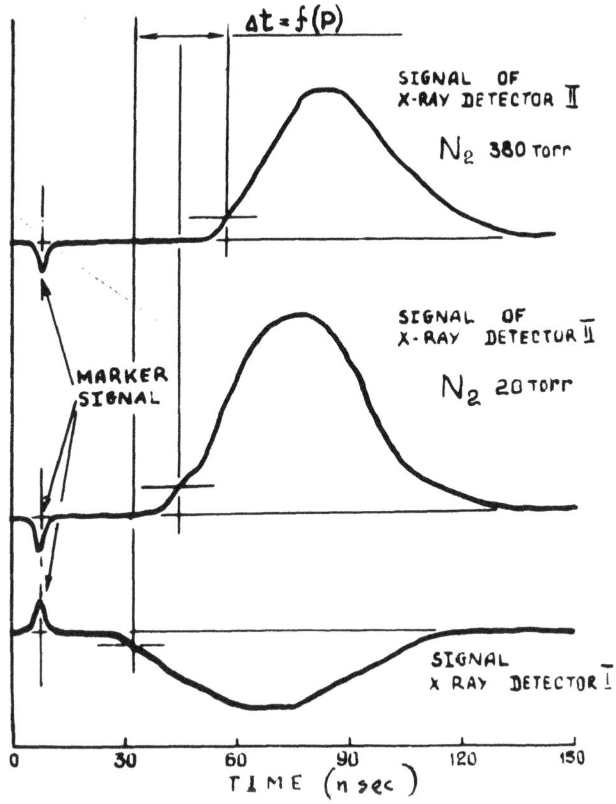

Fig. 2. Correlated signals of the X-ray probe placed at the input and at the output of the operation chamber.

through the operating chamber is changed from 15 ns at 20 torr, up to 25 ns at 380 torr. This effect is more effective for SF_6 gas: at the same mass density, the front velocity in SF_6 gas is lower than that in nitrogen.

The REB-parameters at the entrance into the operation chamber, current and energy passage along the chamber are measured with a Faraday cup combined with a calorimeter.

Experimental results on the REB-energy propagation in N_2 and in F_2 are given in Figure 5. These results show that REB is decelerated more effectively in F_2 than in N_2 (at the same mass density).

A sapphire window, 130 mm in diameter, has been installed at the output end of the operation chamber during the measurements of the laser parameters. An energy and a lasering time vs. partial concentration of the mixture SF_6 + RH, where RH is H_2, CH_4, C_3H_8 have been determined. Lasering time is determined with a pyro-detector. The time resolution in the experiment is not worse than 10 ns. The drive pulse durations vs. pressure of the hydrogen donors are given in Figure 6. The pressure of SF_6 is fixed, and provides

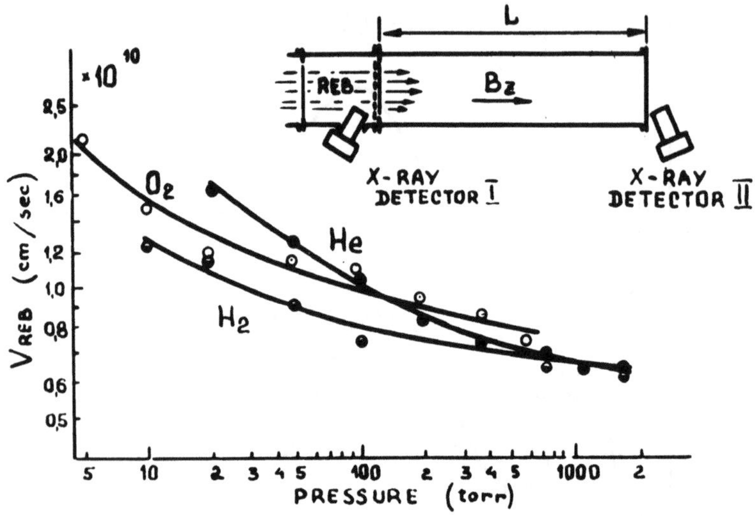

Fig. 3. Propagation velocity of the REB-current front vs. gas pressure in O_2, H_2 and in H_e.

stopping of REB along the operation chamber. It is known that the REB-electrons produce second and tertiary electrons in a medium; those electrons, in their turn, produce free atoms of F in the reaction: $SF_6 + e \rightarrow SF_5 + F + e$.

Atoms of F enter the pumping reaction: $F + RH \rightarrow HF^* + R$. Obviously, at the high rate of this reaction, an accumulation of

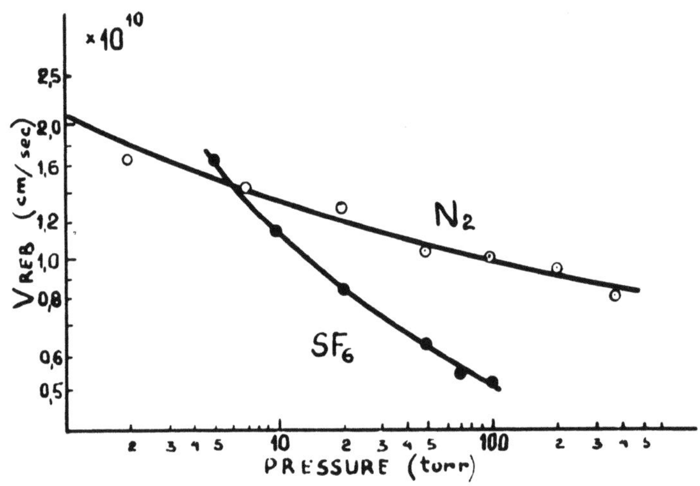

Fig. 4. Propagation velocity of the REB- current front vs. gas pressure in N_2 and in SF_6.

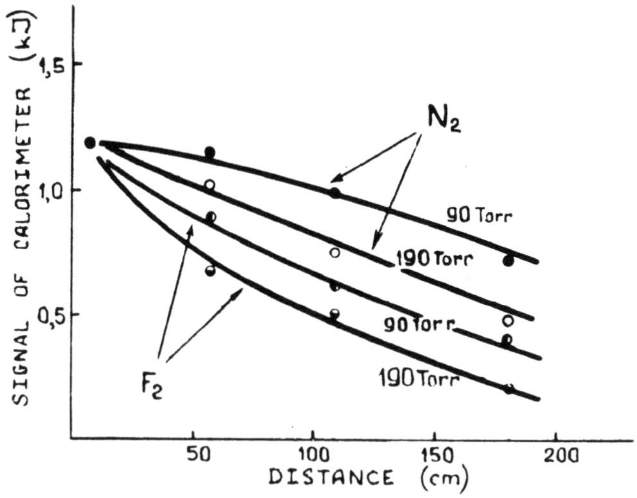

Fig. 5. REB energy transfer in gases N_2 and F_2.

HF* will follow an accumulation of atoms F, i.e., it will follow the REB-current. On the other hand, some estimations of the lifetime for a molecule HF* in the resonator show that it is of the order of a few nanoseconds. Therefore, at an instantaneous spatial initiation of chemical reactions within the operation volume, the

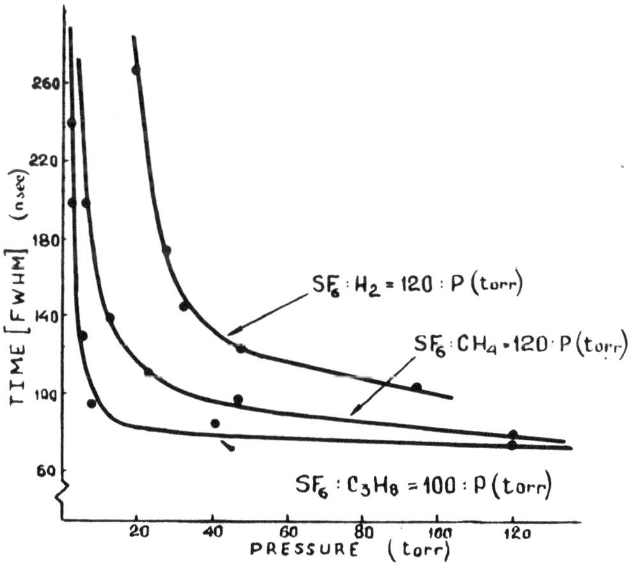

Fig. 6. Dependence of the laser pulse duration on the gas pressure and on the type of hydrogen donor.

initiation time should correspond to the REB duration. The long lasering time we have used in the limiting case, can be related to the value of the velocity of REB propagation through the operation chamber.

CONCLUSION

At the initiation of a chemical laser with a high-current-REB, the ionization wave and, consequently, the pumping wave of an active medium propagate with a velocity which decreases with a rise in the gas pressure. This phenomenon can restrict the pressure range in the operating mixture in systems with a lasering duration of the order of 10 ns, which is attractive for laser fusion. The REB-propagation through a dense gas is likely to occur with a velocity close to the initial one, when the beam continuously touches the metallic walls of the chamber. In that case, radial electric fields prevail and a virtual cathode cannot be produced because the reverse currents to the anode foil pass along the metallic walls of the chamber.

REFERENCES

1. J. V. Tkach, J. J. Magda, G. V. Skachek, S. S. Pushkarev and J. P. Panchenko, J. Techn. Phys. (Russian), 44:658 (1974).
2. B. N. Brejzman, V. S. Burmasov, V. A. Kornilov, E. P. Kruglyakov and V. N. Lukyanov, Proc. XIII ICPIG., Berlin, GDR (1977); Contributed Papers, Part 2 pp. 907-908, Leipzig (1977).
3. S. G. Arutyunyan, O. V. Bogdankevich, I. F. Bondar et al., J. Quantum Electronics (Russian), 9,N2:234 (1982).
4. L. S. Levin et al., J. Appl. Phys., 42:1863 (1971).
5. G. W. Cooper, IEEE., J. Quant. Electr., QE-17,N9:1840 (1981).
6. D. D. Ryutov and G. V. Stupakov, Plasma Physics (Russian), 2:566 (1976).

TIME RESOLVED SPECTRA AND SMALL SIGNAL GAIN IN HF: AN EXPERIMENTAL AND THEORETICAL INVESTIGATION*

P. E. Sojka, W. K. Jaul, and R. L. Kerber

Michigan State University
East Lansing, MI 48824 USA

I. INTRODUCTION

Evaluation of the performance of chemical lasers requires an understanding of the chemical, radiative and relaxation kinetic mechanisms that pump and deactivate the energy levels associated with lasing. The detailed representation of the mechanisms necessary for accurate prediction of laser performance requires a mix of experimental and computer modeling investigations. Several researchers have conducted diagnostic studies of small signal gain[1-2], others have reported experimental studies of P-branch time resolved spectra,[2-7] and still others have compared observed time resolved spectra with computer simulations[8-13].

This paper presents our efforts to compare experimental observations with computer simulations. To that end, we have performed experiments to fully characterize a pulsed H_2+F_2 flash photolysis laser at selected, well controlled operating conditions. An experimental study was conducted where the time-resolved small signal gain on eleven lines and total P-branch laser emission were measured for an H_2+F_2 laser. These results were compared with a very comprehensive computer model including level-to-level kinetic mechanisms and P-branch and R-R lasing. This is the first time such comprehensive diagnostics and computer simulations have been compared.

II. EXPERIMENTAL APPARATUS

The five liter laser cavity consisted of a 100 cm long aluminum

*This work was supported by AFOSR Grant No. 80003.

cavity of 10 cm x 10 cm cross section. Two 1.6 cm diameter inlet ports were located in one cavity sidewall 10 cm from one cavity end and two 1.6 cm diameter outlet ports were symmetrically located in the opposite cavity sidewall 10 cm from the opposite cavity end. The gas mixture, fed in through the inlets, flowed 80 cm longitudinally through the cavity and exhausted through the outlets. A 10 cm diameter dump port was opened after each run to rapidly exhaust the cavity of combusted gases.

The ultraviolet light necessary for initiation of the gas mixture was coupled into the cavity through 11 cm x 11 cm x 2 cm synthetic quartz windows (Suprasil II). The window configuration consisted of five quartz windows mounted on the top of the cavity and five on the bottom. The quartz windows were sealed to the cavity with viton O-rings and clamped in place by aluminum brackets.

A high voltage discharge drove the four flashlamps (ILC: 60 cm long, 0.9 cm diameter, filled with 50 torr Xe). The lamps were mounted in pairs directly above and below the cavity, just outside the quartz windows. The high voltage pulse was produced by charging four capacitors (0.7 microfarads) to approximately 30 kV and discharging two capacitors through each pair of flashlamps using spark gaps.

A CaF_2 window at the brewster angle was attached to each end of the cavity by an aluminum mount. Internal brewster mounts were used to hold the 5.08 cm diameter CaF_2 windows in place. The window spacing was 81 cm and there was no window path to purge. All window to brewster mount and brewster mount to cavity surfaces were sealed with viton O-rings.

The optical resonator was of a stable configuration and external to the cavity. The maximum reflector was a 15.2 cm diameter 500 cm radius of curvature copper mirror. An output coupler with maximum 81% reflectivity was used. The mirror spacing was 121 cm.

The entire cavity-quartz window-brewster assembly was passivated by exposing it to increasing concentrations of fluorine (10% to 50% F_2 in He, total pressure 400 torr) for periods of one half hour.

The gases used were: oxygen (99.98%), fluorine (98.2%), helium and hydrogen (98%). Batch analysis of the fluorine showed the following impurities: HF < 0.4 molar percent, Air 1.69 molar percent, CO_2 443 molar ppm, CF_4 301 molar ppm, and SF_6 10 molar ppm. The helium, oxygen and hydrogen flows were monitored using rotameter flow meters (Matheson: 7400 series). The fluorine flow was monitored using a linear mass flowmeter (Matheson: 8611). A hydrogen fluoride trap (Matheson), using NaF_2 pellets, was installed in the fluorine delivery line to minimize the initial cavity HF concentration. The helium, oxygen and fluorine flows were mixed in a 5.1 cm diameter 76.2 cm long stainless steel tube. This mixture was routed to the cavity in two 1.9 cm OD stainless steel lines. The hydrogen flow was injected into the helium-oxygen-fluorine mixture 20 cm upstream of the laser cavity using a radial "sting" arrangement. The complete helium-oxygen-fluorine-hydrogen mixture then flowed into the cavity where lasing occurred.

The cavity pressure was measured using a Bourdon tube gauge (Heise: Cu-Be) and was controlled by two orifices located 20 cm downstream of the cavity exits in the two 1.9 cm OD exhaust lines. The combusted product gases were exhausted through these two lines to a 15.2 cm OD vacuum duct and then to a triplex pump (Kinney: KT500B, 500 cfm).

Experiments reported here were run at 101.7 torr with a gas mixture of: $He:O_2:F_2:H_2$ = 20.8:1.0:4.6:1.2. The cavity Reynolds number was calculated to be 23,000. Thus, the flow was mostly turbulent.

III. DIAGNOSTIC MEASUREMENTS

Measurements of the Time Resolved Spectra (TRS) of P-branch and R-R (pure rotational) lasing, and the time history of P-branch Small Signal Gain (SSG) were performed.

In the TRS experiments, the laser beam left the output coupler and entered an enclosed, dry nitrogen purged optical path to an electromagnetic interference (EMI) shielded room. The beam then was steered through a CaF_2 lens and focused onto the entrance slit of a monochromator (GCA-McPherson: 218, 0.3m). The monochromator diffraction grating (GCA-McPherson: 150 grooves/mm, 4.0 micron blaze) dispersed the desired component of the total beam through the monochromator exit slit and onto a biased, liquid nitrogen cooled Ge:Au detector (Raytheon: QKN-1568). The change in detector bias current was displayed on an EMI shielded oscilloscope (Tektronix: 7844) with fifty ohm vertical amplifier plugins (Tektronix: 7A19). A photograph was taken of each data run; a sample is shown in Figure 1b.

TRS measurements of pure rotational lasing intensities were also performed. The experimental apparatus was exactly that of the P-branch lasing case with three exceptions: (1) the CaF_2 brewster windows were replaced with NaCl brewster windows, (2) the 150 grooves/mm, 4.0 micron blaze monochromator diffraction grating was replaced by a 75 grooves/mm, 16.0 micron blaze monochromator diffraction grating, and (3) the Ge:Au detector and biasing system was replaced by a liquid helium cooled Ge:Cu detector (Raytheon: QKN 1546) or a liquid nitrogen cooled HgCdTe detector (SBRC: 40742) and applicable biasing circuit.

A slightly different system was used during the SSG experiments. The five liter laser was modified by removing the optical resonator. A commercially available continuous wave (CW) HF laser (Helios: CLI) was employed to probe the laser medium. This laser utilized an electric discharge in a mixture of SF_6, O_2, He and H_2 to produce laser emission. The probe laser resonator consisted of a gold coated maximum reflecting mirror and a grating output coupler which allowed single line operation. Transitions available for probing from this laser were: $P_1(3)$, $P_1(4)$, $P_1(5)$, $P_1(6)$, $P_1(7)$, $P_1(8)$, $P_1(10)$, $P_2(5)$, $P_2(6)$, $P_2(7)$, and $P_2(8)$.

The small signal gain diagnostics were performed by tuning the probe laser to a specific transition. The output was then chopped and steered onto a beam splitter. One beam traversed a path through the laser medium colinear with the path of the laser emission described in the TRS section. The time dependent chemistry occurring within the laser medium perturbed this beam. A reference beam was steered into the EMI shielded room and focused by a CaF_2 lens into a monochromator-diffraction grating-detector-bias circuit system identical to that of the perturbed beam's path. A dual beam, EMI shielded oscilloscope (Tektronix: 7844) monitored the reference and signal beam intensities simultaneously. Photographs were taken of each data run; a sample is shown in Figure 1a.

Flashlamp intensities were experimentally characterized vs. time at 0.1 micron intervals in the region 0.25 microns to 0.40 microns. A sample is shown in Figure 2. A schematic diagram of the experimental apparatus is shown in Figure 3.

III. RESULTS AND DISCUSSION

The time history of the observed small signal gain is shown in Figure 4 along with that predicted by the theoretical model. A description of the model used here is given in references 7 and 8 and will not be repeated here. The kinetic rate coefficients used in the model were updated to reflect the most recent measurements. The values used are given in reference 13. The model predictions of these time histories closely approximate the experiment. A value of 0.005 was used for the F/F_2 dissociated by the lamp pulse. The magnitude of the predicted SSG was approximately four times larger than that observed by experiment. Several scenarios were considered to enhance the comparison of the magnitude of experiment and theory such as larger prereaction and uncertainty in initiation. These corrections lead to discrepancies in the comparison of time evolution or relative magnitude of 2-1 to 1-0 SSG. Hence, improvement in this comparison will require an additional modeling effort.

The time resolved spectra from the experiment is compared with the model in Figure 5. The length of the line represents the duration of lasing on the transition with lower level V,J and the dot indicates the location of peak power on the line. The model and experiment predict strong line overlap and sequential J-shifting in a given band. Of course, these are characteristics of rotational nonequilibrium. The model predicts more lasing transitions than experiment. This trend is consistent with the over prediction of SSG by the model. The model predictions of temporal features of both SSG and TRS agree very well with the experiment.

No significant lasing was observed on rotational-rotational lines. The model also predicted that rotational lasing should not be observed.

In summary, an extensive investigation of a well-diagnosed experimental H_2+F_2 laser has been made and compared with model predictions. The agreement of theory and experiment is quite good. It is expected that some refinements in the model will improve the agreement. Additional work is currently in process to remove the observed discrepancies.

(a) (b)

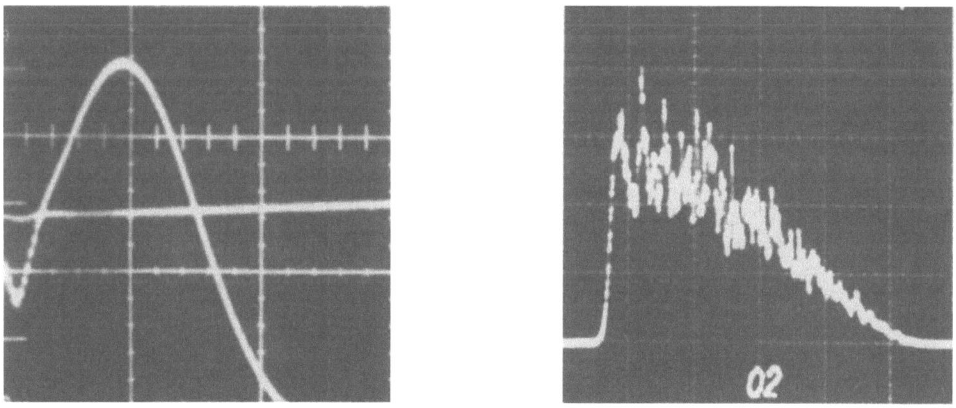

Figure 1. (a) Typical time history of observed gain; 100 m volts/div and 10 μsec/div.

(b) Typical time history of observed laser emission; 100 m volts/div and 5 μsec/div.

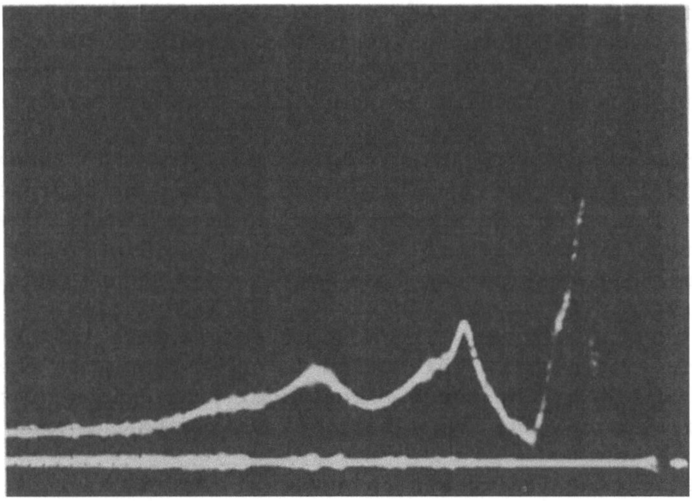

Figure 2. Time history of flash lamp output; 50 m volts/div and 1 μsec/div.

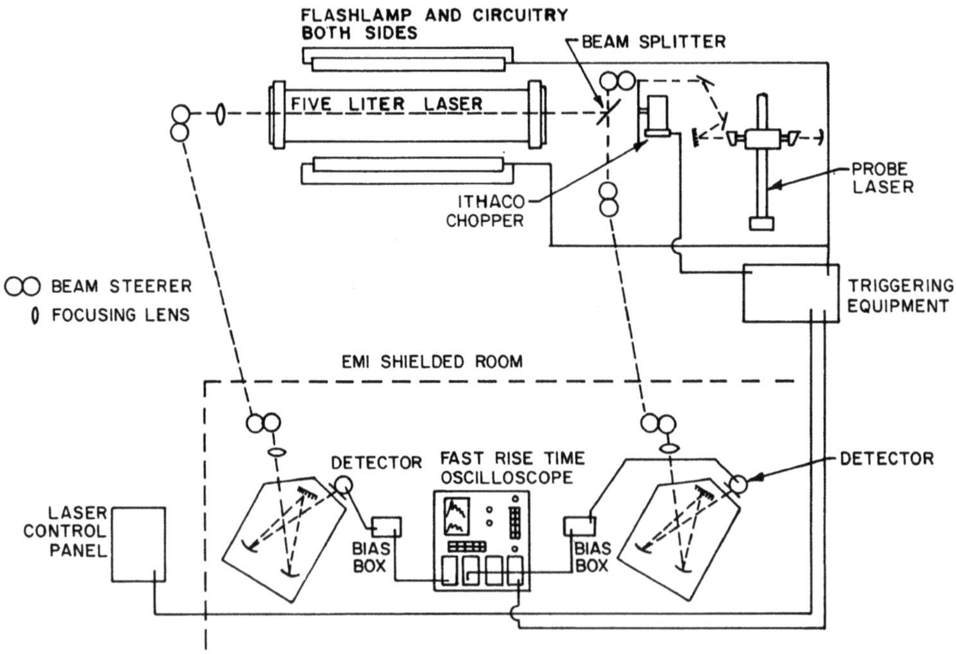

Figure 3. Schematic diagram of laser apparatus.

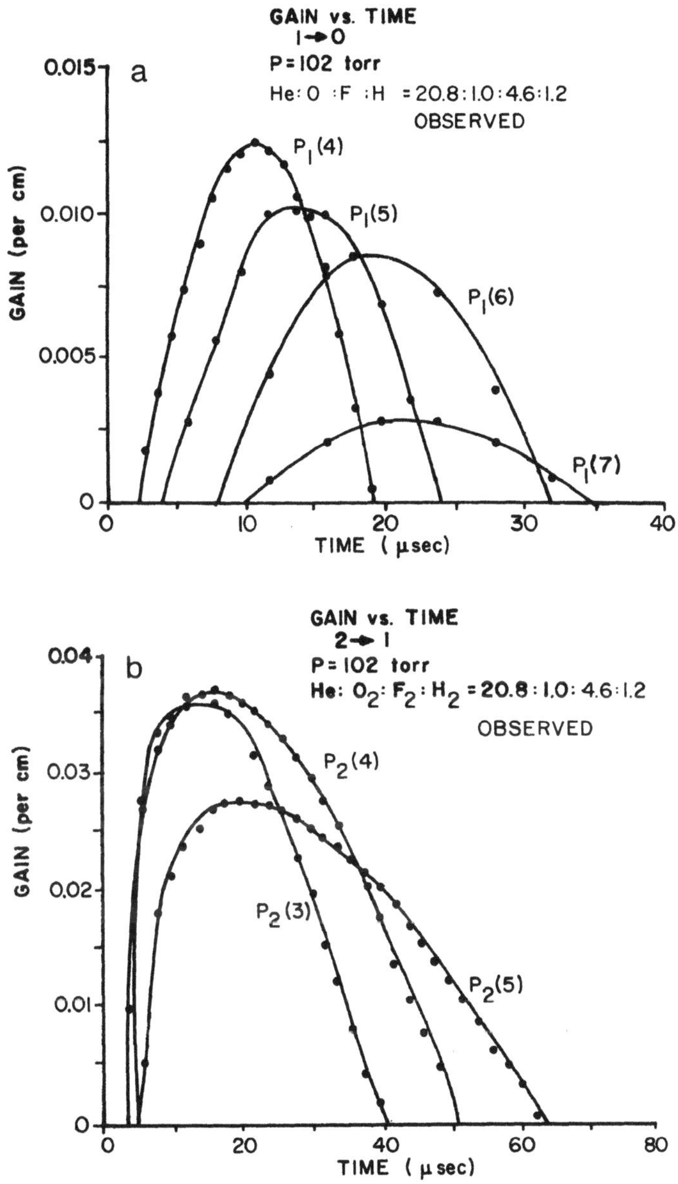

Figure 4. Time histories of small signal gain.
(a) observed 1→0 (b) observed 2→1

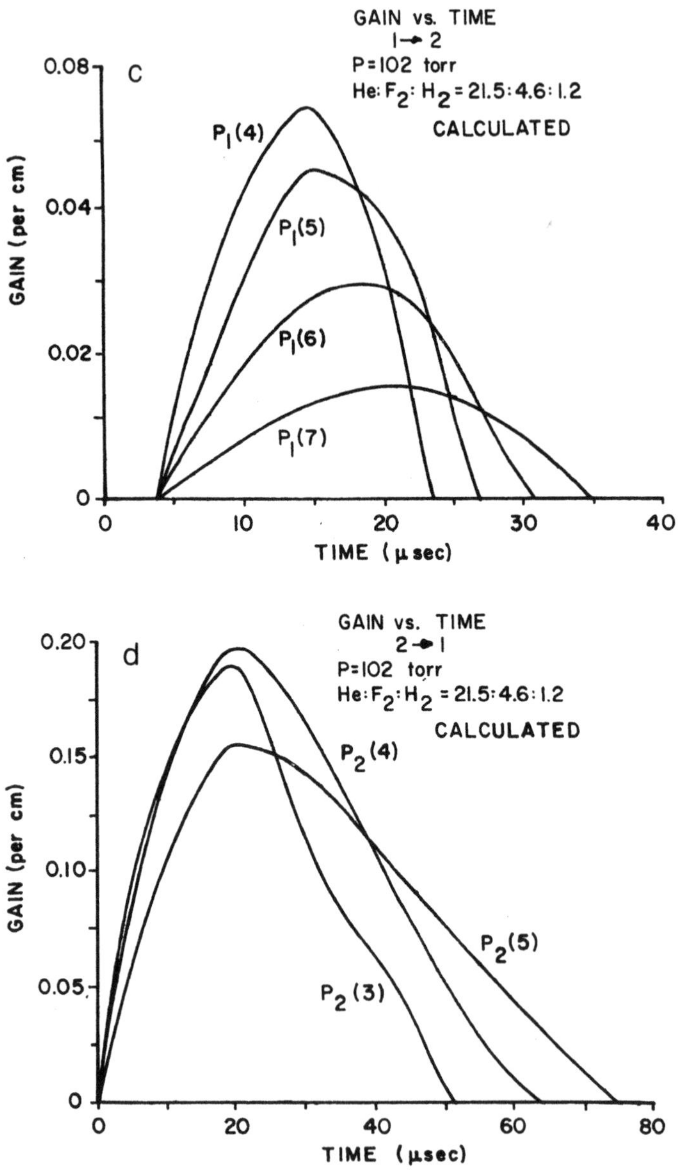

Figure 4. Time histories of small signal gain. (c) calculated 1→0 (d) calculated 2→1

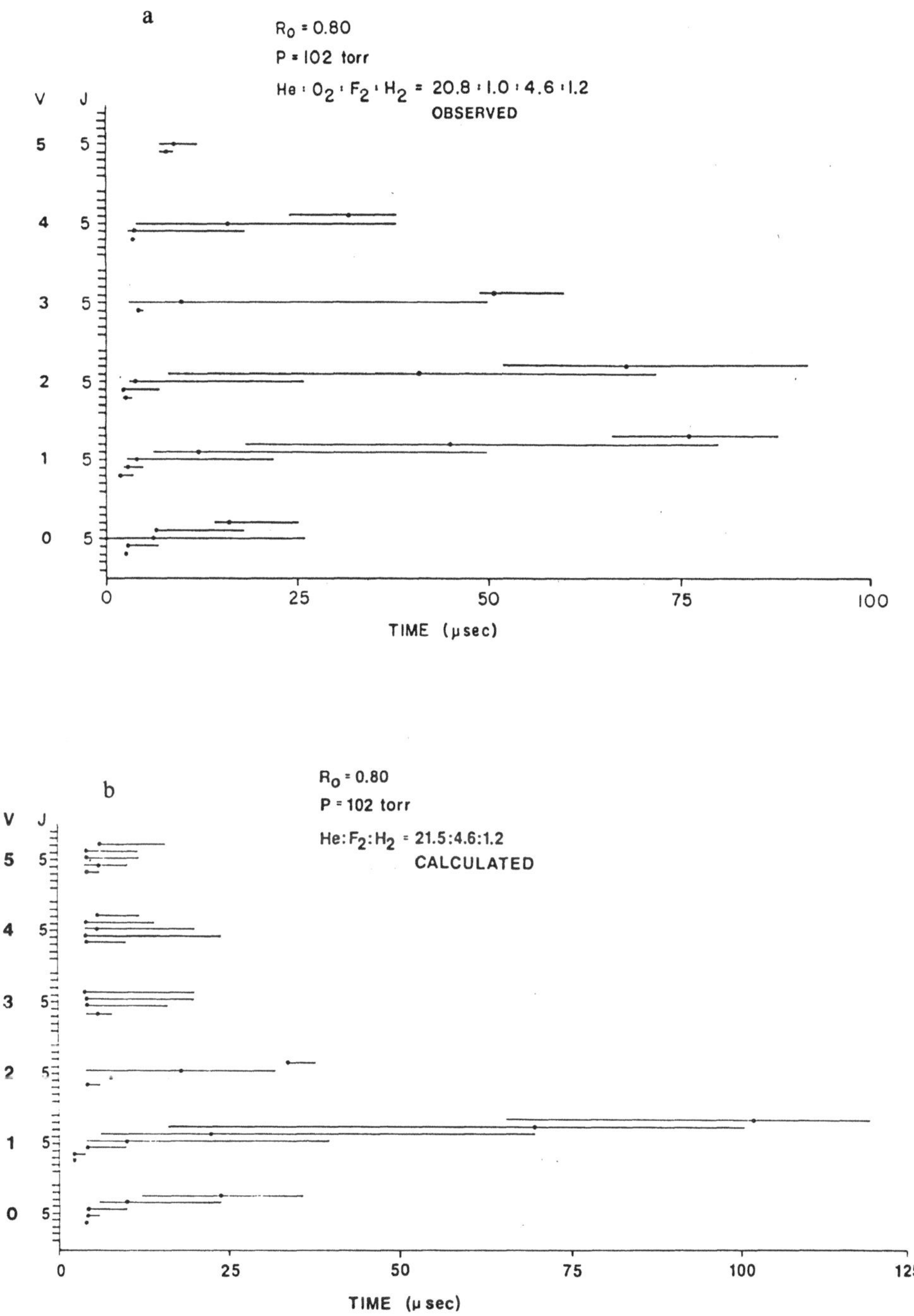

Figure 5. Time resolved spectra.
(a) experiment
(b) calculated

IV. REFERENCES

1. S. Marcus and R. J. Carbone, "Gain and Relaxation Studies in Transversely Excited HF Lasers," IEEE J. Quan. Elec. QE-8 (7), 651-5 (1972).
2. S. N. Suchard, "Small-Signal Gain and Time-Resolved Spectroscopy of the D_2-F_2/CO_2 Pulsed Chemical Transfer Laser System," IEEE J. Quant. Elec. QE-10 (1), 87-9 (1974).
3. J. V. Parker and R. R. Stephens, "Pulsed HF Chemical Laser with High Electrical Efficiency," Appl. Phys. Lett. 22 (9), 450-2 (1973).
4. N. G. Basov, V. T. Galochkin, V. I. Igoshin, L. V. Kulakov, E. P. Martin, A. I. Kikitin, and A. N. Oraevsky, "Spectra of Stimulated Emission in the Hydrogen-Fluorine Reaction Process and Energy Transfer from DF to CO_2," Appl. Opt. 10 (8), 1814-20.
5. N. R. Greiner, "Submicrosecond Pulses from a Hydrogen-Fluorine Laser with High Energy Density and Quantum Efficiency," IEEE J. Quant. Elec. QE-8 (12), 872-6 (1972).
6. G. G. Dolgov-Savel'ev, V. F. Zharov, Uy. S. Neganov, and G. M. Chumak, "Vibrational-Rotational Transitions in an H_2+F_2 Chemical Laser," Sov. Phys. JETP 34 (1), 34-7 (1972).
7. R. L. Kerber, R. C. Brown, and K. A. Emery, "Rotational Equilibrium Mechanisms in Pulsed H_2+F_2 Chain Reaction Lasers. 2: Effect of VR Energy Exchange," Appl. Opt. 19 (2), 293-300 (1980).
8. R. C. Brown, "A Theoretical Assessment of Vibrational to Rotational Energy Exchange in the Hydrogen Fluoride Chemical Laser," Ph.D. Thesis (1980).
9. J. J. T. Hough, "A Theoretical and Experimental Investigation of the Mechanisms of the Hydrogen Fluoride Pulsed Chemical Laser," Ph.D. Thesis (1975).
10. S. N. Suchard, R. L. Kerber, G. Emanuel, and J. S. Whittier, "Effect of H_2 Pressure on Pulsed H_2+F_2 Experiment and Theory," J. Chem. Phys. 57 (12), 5065-75 (1972).
11. D. B. Nichols, K. H. Wolstad, and J. D. McClure, "Time Resolved Spectroscopy of a Pulsed H_2-F_2 Laser with Well-Defined Initial Conditions," J. Appl. Phys. 45 (12), 5360-6 (1974).
12. H. L. Chen, R. L. Taylor, J. Wilson, P. Lewis, and W. Frye, "Atmospheric Pressure Pulsed HF Chemical Laser," J. Chem. Phys. 61 (1), 306-18 (1974).
13. W. K. Jaul and R. L. Kerber, "A Theoretical Assessment of the Kinetic Limitations of the Chemically Pumped HF Optical Resonant Transfer Laser," Michigan State University Technical Report MSU-ENG-82-018.

QUASI ONE-DIMENSIONAL MODEL OF A RING DIFFUSION

CW CHEMICAL LASER AND SOME OF ITS APPLICATIONS

A. A. Stepanov and V. A. Scheglov

Lebedev Physics Institute
Moscow

INTRODUCTION

The theoretical study of a ring model diffusion cw HF chemical laser (HF-CL) has shown the advantages of using such a system [1,2]. In contrast with the traditional (plane) model, the ring laser (Figure 1) includes a cylindrical nozzle unit consisting of a large number of small coaxial nozzles with alternating jets of oxidizer (F, F_2) diluted by helium and fuel (H_2). The chemically active jets run radially outward and mix in the direction of the z-axis.

The quasi one-dimensional model of a cw CL is based on the sheet flame concept [3,4]. The quasi one-dimensional model described allows us to show the multilevel character of active molecule radiaction and thermal and gas-dynamic effects in the chemically active flows of diffusion lasers. The theory in References 3-5 concerns a nozzle unit of plane construction.

The aim of this work is to extend the quasi one-dimensional approach to the case of a nozzle unit of cylindrical construction. The energy characteristics of a ring HF laser with an unstable telescopic resonator consisting of torroidal mirrors will be calculated as an example.

THEORY

The Main Equations of the Quasi One-Dimensional Model

The active medium of a supersonic chemical HF laser with a cylindrical nozzle unit consists of a large number of periodically

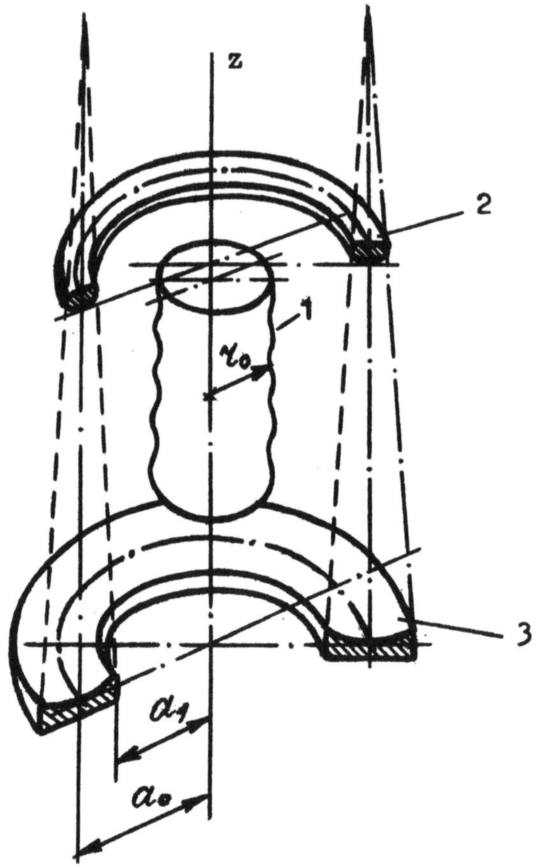

Fig. 1. The principal scheme of ring diffusion HF laser with the unstable telescopic resonator: 1) the cylindrical nozzle unit; 2) and 3) resonator mirrors.

placed ring-form reaction zones. These zones result from fuel (H_2) diffusion into the oxidizer flow (F+He); fluorine diffusion (the heaviest component) into the H_2 flow may be neglected. In accordance with the sheet flame concept the reaction zones are divided from the injection zones of F + He flows by the flame surfaces $z_f(r)$, defined by the stoichiometry of fuel and oxidizer (Figure 2). The sheet flame height $z_f(r)$ defines the depth of reagent mixing in section r. Taking a phenomonological approach, we assume that the surface form $z_f(r)$ is known. The main equations of the quasi one-dimensional model of cw CL with the ring form of the active medium may be formulated using axial symmetry [5]; in this work, however, the cylindrical expansion of flow will be considered. Let r be the coordinate along the gas direction, z, along the system axis and $z_f(r)$, the sheet flame height. As the stoichiometric correlation

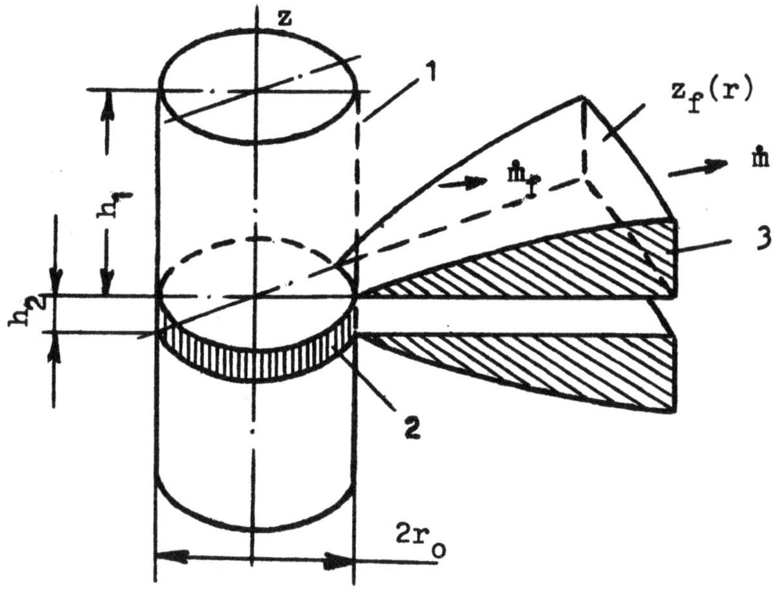

Fig. 2. The mixing model in the quasi one-dimensional approach: 1) the elementary nozzle of oxidizer (F+He); 2) fuel nozzle (H_2); 3) active medium.

of oxidizer and fuel particles is a characteristic of the flame surface, the reagent mixing process may be described as if the F + + H_2 particle flow, mixed in the stoichiometric correlation, but beginning to react only on surface $z_f(r)$, collides with the stationary sheet flame.

Let the gas dynamic parameters (pressure p, density ρ, flow speed u and temperature T) take the subscript "f" at the surface z_f. These parameters will depend on the value r, owing to the cylindrical character of the expansion. Let the gas expenditure mass in a cylindrical tube of current of unit height (along the z-axis) be marked as $\dot m = \rho ur$; specific enthalpy of mixture, h ($h = \sum_i C_i h_i$; h_i, the partial enthalpy of i component; $C_i = \rho_i/\rho$, its mass part in the mixture) and let heat content be H ($H = h + u^2/2$).

Taking into account both the chemical, kinetic and radiation transformations in the reaction zone, and the particle flow through the mixing surface in the quasi one-dimensional approach, the main equations of gas-dynamics may be written thus:

$$\frac{d}{dr}(\dot m z_f) = \dot m_f \frac{dz_f}{dr}; \quad \frac{d}{dr}(\dot m u z_f) = \dot m_f u_f \frac{dz_f}{dr} - z_f r \frac{dp}{dr}; \tag{1}$$

$$\frac{d}{dr}(\dot{m}Hz_f) = \dot{m}_f H_f \frac{dz_f}{dr} - z_f r \cdot \sum_v g^j_{v,v-1} I_{v,v-1};$$

$$p = \rho RT/W; \ \frac{d}{dr}(\dot{m}_i z_f) = \dot{m}_{f_i} \frac{dz_f}{dr} + z_f r u \left(\frac{d\rho_i}{dr}\right)_{chem},$$

where $\dot{m}_f = \rho_f u_f r$; $\dot{m}_{f_i} = \rho_{f_i} u_f r$; $H_f = h_f + u_f^2/2$; where R is the universal gas constant; W, the molecular weight of mixture; $(d\rho_i/dr)_{chem}$, the velocity of particle production in the unit volume owing to the chemical, kinetic and radiation transformations; $g^j_{v,v-1}$, local value of active medium gain in the reaction zone in vibration-rotation transition v, j-1 → v-1, j and $I_{v,v-1}$, radiation intensity in this transition.

Equation (1) is not algebraically closed. The usual equations of isentropy cylindrical flow should be used for the definition of $\dot{m}_f$, $\dot{m}_{f_i}$, u_f and H_f. Chemical and kinetic transformations in the flow at the flame surface $z_f(r)$ are considered to be absent:

$$\rho_f u_f r = \text{const}; \ \rho_f u_f \frac{du_f}{dr} = -\frac{dp_f}{dr}; \tag{2}$$

$$\rho_f u_f \frac{dh_f}{dr} = u_f \frac{dp_f}{dr}; \ p_f = \rho_f RT_f/W_f;$$

$$\rho_{f_i} u_f r = \text{const},$$

where W_f is the molecular weight of the flow.

Equations (1) and (2) should be supplemented by the initial conditions at the cylindrical surface $r = r_o$ matching the nozzle unit section.

Some Applications

We shall examine the ring model of an autonomous cw chemical HF diffusion type laser [1,2]. In this model, production of active centres (fluorine atoms), in the gas generator unit is usually due to the combustion process of molecular deuterium (D_2) in superfluous F atmosphere diluted by helium: $D_2 + \alpha F_2 \rightarrow 2DF + (\alpha-1)F_2$. The heat (Q) extracted in the reaction leads to the molecular fluorine dissociation $F_2 + Q \rightarrow 2F$, so that, at the exit, the oxidizer flow mainly consists of molecules DF and of particles F and He. The fuel flow consists of molecular hydrogen (H_2).

In the presence of a cylindrical nozzle unit, reaction zones occurring on H_2 diffusion into oxidizer flow have the form shown in Figure 2. In the case of laminar flow the following equation may be used for the flame surface:

$$z_f(r) = \frac{h_1}{2} \cdot \begin{cases} (x/x_D)^{1/2}, & x \leqslant x_D \\ 1 & , x > x_D, \end{cases}$$

where h_1 is the height of oxidizer jet; x_D, the character diffusion length and $x = r - r_o$.

In the quasi one-dimensional approach for an HF laser with a single excitation mechanism, $F + H_2 \rightarrow HF(v) + H$, the kinetic equations written for the mole mass concentrations are analogous [5].

Resonator with plane mirrors of a ring form. This is one of the simplest cases which can be considered in the quasi one-dimensional model. For a resonator with plane mirrors the energy characteristics calculation of a diffusion HF laser is made using one-dimensional equations. The quasi stationary approach may be used for radiation intensity definition in the resonator, assuming that in generation the active medium average gain equals its threshold value

$$\dot{G}^j_{v,v-1}(r) = \frac{z_f}{h_*} \dot{g}^j_{v,v-1}(r) = G_{thr} ,$$

where $h_* = (h_1+h_2)/2$; $G_{thr} = L_a^{-1} \ln(R_o R_L)^{-1/2}$ - value gain threshold; L_a is the active medium length along the optical axis in the resonator and R_o, R_L, reflectivity coefficients of mirrors. It should be noted that at each flow point the rotation quantum number j of the radiative transition is usually chosen from the condition $g^j_{v,v-1}$ gain maximum.

The numerical calculations are based on the following data: for simplicity, temperature, speed and pressure in the flows of oxidizer and fuel are taken to be the same; their values are: $T_o = 250°K$, $u_o = 3$ km/s, $p_o = 5$ torr. It is considered that particle correlation in the oxidizer flow equals F : DF : He = 1 : 1 : 5, and that the fuel flow consists of molecular hydrogen. The ratio of height of oxidizer and fuel jets is taken as high as $h_1/h_2 = 1$, the halfperiod width of nozzle grating, $h_* = 0.4$ cm, the characteristic length of diffusion, $x_D = 20$ cm; some of the parameters have been changed.

Figure 3 compares calculations made via the quasi one-dimensional approach (curve 1), and via two-dimensional equations of the boundary layer (curve 2). The calculations have been made from the same initial data and at the same gain threshold $G_{thr} = 5.5 \cdot 10^{-3}$ cm^{-1}. In this and other figures r_{opt} is the value of nozzle unit radius corresponding to the chemical laser efficiency maximum η_{chem} for the two-dimensional approach.

Unstable telescopic resonator. A ring flowing laser with an unstable telescopic resonator is shown schematically in Figure 1.

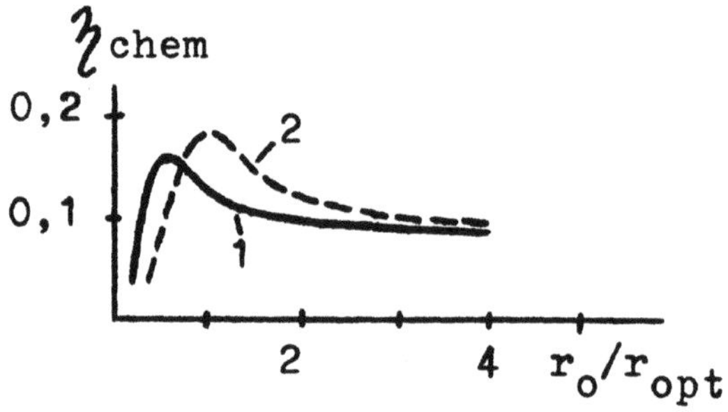

Fig. 3. The dependence of chemical laser efficiency on nozzle radius r_o.

The calculation for a ring HF laser with telescopic resonator has been made by the iterative method (the resonator field is described by geometrical optics). It is considered that the inner diameter of the large mirror equals the nozzle unit diameter ($a_1=r_o$), and that external diameter is defined by the laser zone width Δr_{las} in the absense of radiation (Figure 4a). The position of the ring resonator axis (a_o) is defined by the position of maximum active medium gain along the flow (Figure 4b).

The rotation quantum number j is considered to be fixed in the calculations (j=5). At the nozzle section the initial conditions are assumed to be the same as for the resonator with the plane mirrors.

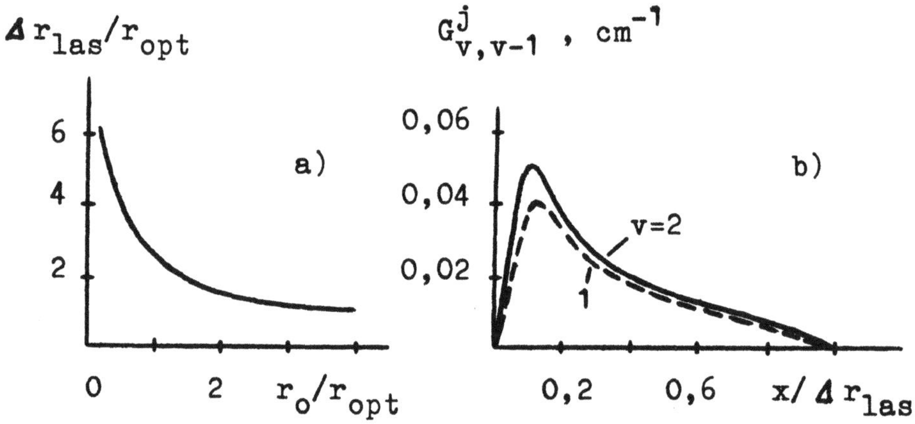

Fig. 4a. The dependence of the laser zone width Δr_{las} (in absense of radiation) on nozzle radius r.

Fig. 4b. The distribution of gain coefficients $G^{j}_{v,v-1}$ along the flow:
$r_o/r_{opt} = 1$.

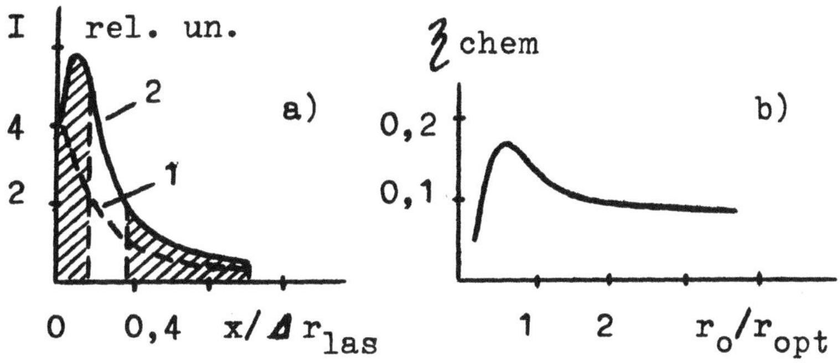

Fig. 5. The distribution character of summary intensity of radiation (a) 1 - small mirrors; 2 - large mirrors; (b) Dependence of chemical efficiency on the nozzle radius r_0 for the telescopic resonator.

Figure 5a demonstrates (by two bands) the whole distribution of the summary radiation intensity at the resonator mirrors. The results are derived from M = 3 and $r_0/r_{opt} = 1$ (M is the magnification coefficient). Radiation from the resonator is shaded in the Figure. The dependence of η_{chem} on the nozzle radius r_0 (when $x_D = 20$ cm is fixed) is shown in Figure 5b. The dependence is rather similar to the dependence found for plane mirrors.

CONCLUSION

This work has shown that use of the quasi one-dimensional approach for the description of the active medium of ring diffusion cw CL leads to results that agree satisfactorily with the results for a two-dimensional boundary layer model. It should be noted that use of the quasi one-dimensional approach is convenient when it is difficult to use the boundary layer model. In particular, the quasi one-dimensional approach is very effective for investigating amplification regimes; for calculations of generation regimes and amplification of a diffusion laser in which multiple telescopic schemes are used; and it is very effective for diffraction calculations of concrete laser systems.

REFERENCES

1. A. A. Stepanov and V. A. Scheglov, Preprint FIAN., M., No. 182 (1976); No. 269 (1978); Kvantov. Electronica., 6:1476 (1979); JTF., 51:1273 (1981).
2. A. A. Stepanov and V. A. Scheglov, Preprint FIAN., M., No. 59 (1979); JTF., 50:557 (1980).

3. H. Mirels, R. Hofland and W. S. King, AIAA J., 11:156 (1973); AIAA Paper, No. 145 (1972).
4. J. E. Broadwell, Appl. Optics., 13:962 (1974).
5. A. A. Stepanov and V. A. Scheglov, Preprint FIAN., M., No. 134 (1975); JTF., 46:563 (1976).

FLOWFIELD DIAGNOSTICS OF SUPERSONIC CW HF CHEMICAL LASER BY USING IMPROVED LIF AND CHEMILUMINESCENCE METHODS

Zhuang Qi, Xie Xingbin, Sang Fengting,
Huang Ruiping, Sun Siyuan, and Zhang Cunhao
Dalian Institute of Chemical Physics
Chinese Academy of Sciences
Dalian, China

With a view to upgrading the diagnostics of supersonic mixing cw chemical laser flowfields, two improved methods have been developed. The improved LIF method is noted for its capability to display the whole cold flowfield and to obtain distinct shock wave patterns with a spatial resolution normal to the incident laser beam higher than 0.5 mm. The second method deals with measuring the spatial intensity profile of the HF overtone emission in a "hot" reactive flow by using an OMA system. The results should help in elucidating the nature of diffusion between fuel and oxidizer streams in a supersonic chemical laser.

INTRODUCTION

In a cw HF chemical laser, the inversion is produced by supersonically mixing initially separate oxidizer and fuel streams and subsequent reaction inside the optical cavity. The process of mixing is a controlling step, on which the efficiency strongly depends. The investigation of the flowfield, therefore, is an important aspect for studying this type of laser. As a matter of fact, extensive theoretical and experimental studies have already been directed on it. Nevertheless, due to the low spatial resolution of existing methods, these methods are capable of visualizing neither the shock wave pattern within the flowfield nor the fine structure of the mixing streams. By using an improved LIF technique, we have succeeded in displaying the whole flowfield of cw supersonic chemical lasers. A salient feature of this method is its high spatial resolution in the direction normal to the incident laser beam. Owing to the fluorescence quenching at the shock front, distinct shock patterns become observable in the LIF photograph. These are subject to quantitative data processing to obtain the local Mach numbers, and to investigate the mechanism of the trip enhancement on the mixing of two supersonic gas streams. In addition, a chemiluminescence method is developed for the "hot" flowfield diagnostics of cw chemical lasers in operation, which has the merits of both a shortened measuring time and a higher spatial resolution, as compared with the infrared optical scan method.

I. LIF METHOD

Rapagnani and Davis reported in 1979 a laser induced fluorescence method for the flow diagnostics of cw chemical lasers, where one of the two gas streams was seeded with iodine and irradiated with a focused Ar^+ laser beam (5145A) at the immediate nozzle exit. However, their method suffers from being unable to measure a wide field of view, because the I_2 excited state is short-lived.

By using a similar apparatus with, however, an improved focusing optical system, as shown in Fig.1, which consists of two column reflectors and a cylindrical lens, we have succeeded in displaying the whole flowfield, e.g., 5 cm in the direction of flow. The fluorescence patterns are recorded with conventional photograph. The filter in front of the camara is used to interrupt the strongly scattered light.

It is fortuitous that, owing to the quenching action by the shock front on the fluorescence, distinct shock patterns become observable in the photograph,as shown in Fig.2.

In order to measure the local Mach numbers, a wedge with an angle A is placed in the flowfield with its centerline coincident with the direction of flow. A distinct shock wave appears in the photo with a shock angle B (Fig.3). We can calculate the Mach

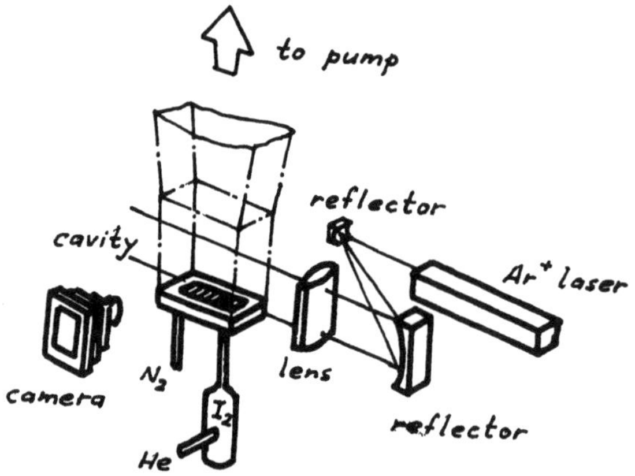

Fig.1 Apparatus schematic of LIF

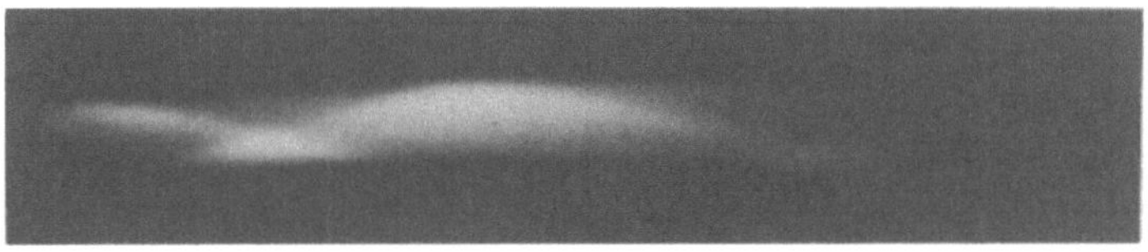

Fig.2 Fluorescence photograph

number from these two angles and the specific heat ratio.

Besides the method presented above, we can also compute the local Mach number under the same set of conditions by measuring the gas pressure both in the plenum and in the cavity.

The computed local Mach number by these two methods for one set of experimental conditions are compared as follows:

Method	Revelant data	Computed Mach No.
I. Shock angle (LIF)	$A= 15.5^{o}$ $B= 26.5^{o}$	M= 5.2
II. Pressure measurment	P= 2470 torr P= 7 torr	M= 5.3

From the data listed in the table, the deviation of Mach number measured by the LIF method from that calculated by the nozzle flow formula is about 2 percent.

We expect that, with some improvements, this method may be applied to supersonic flowfield studies in very small wind tunnels.

Although the phenomenon that the nozzle trip may enhance the gas mixing has been pointed out by several authors, its mechanism remains unclarified as yet. In the present set of experiments, 3-dimensional nozzles are used. A small amount of He diluent, approximately 2 percent of the main flow, is ejected from tiny holes at different angles, 0°, 45°, 90°, into one of the main

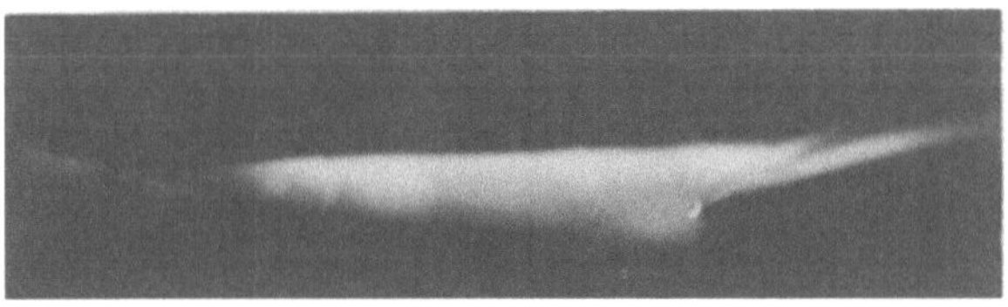

Fig.3 Wedge and shock angle

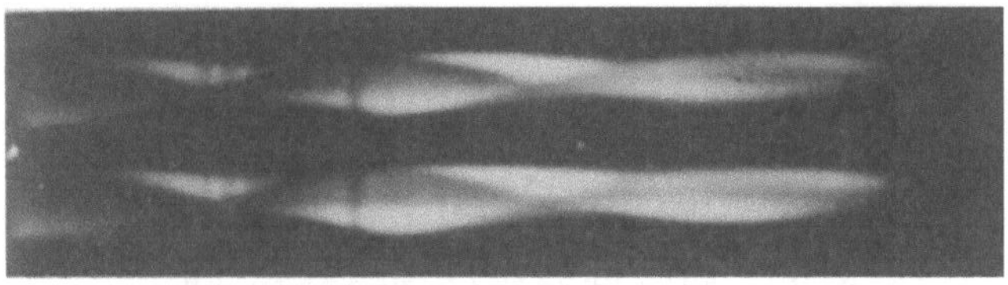

Fig.4 Untripped flowfield

streams. The untripped and tripped flowfields were compared, which are shown in Fig. 4,5,6,7. Fig.4 is an untripped flowfield.

It is evident from these photos that tripping is primarily to enhance the rate of spreading of two main streams, the enhancement being most pronounced for the trip with 45° angle. In this case, the two main streams are broadened at 1 cm downstream from nozzle exit. At 3 cm downstream they start to merge with each other, but the laminar flow profile is unaffected (Fig.5).

With 90° angle tripping, the spreading takes place at 5 mm downstream as shown in Fig.6. However, the boundary between the two streams does not disappear all the way downstream. For trips parallel to main flows, the expansion of flows is likewise observed, as evidenced by the closer gap between the two streams. However, there seems no marked effect on mixing (Fig.7).

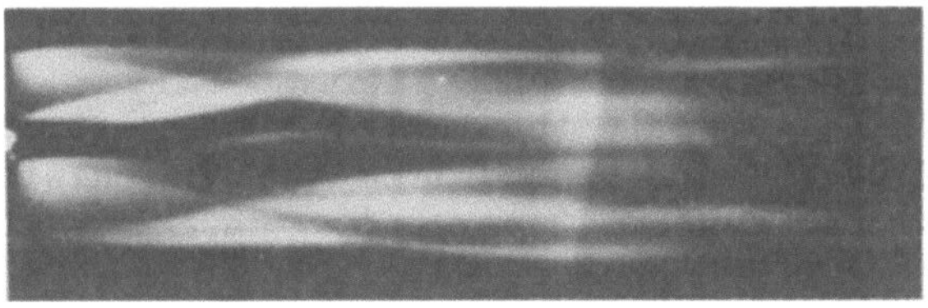

Fig.5 45° trip angle

We imagine that tripping at an angle may be reflected back and forth between the two streams, resulting in broadening of the main streams, which, in turn, enhances the mixing.

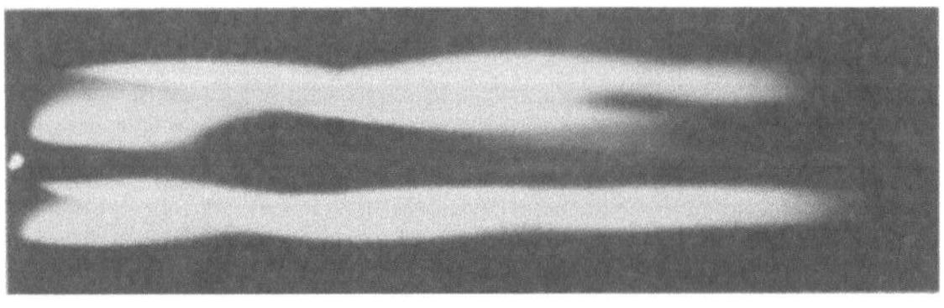

Fig.6 90° trip angle

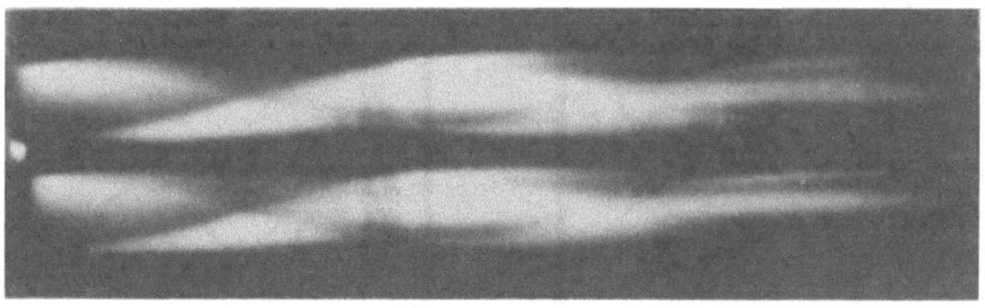

Fig.7 Parallel trip

So far the study of the gas dynamic process of the interaction of a weak trip with the main flows is inconceivable by any other method. It has become possible with the present LIF method owing to its high spatial resolution and its capability to display the whole flowfield.

II. CHEMILUMINESCENCE METHOD

For the flow diagnostics of a cw HF chemical laser, use is made of the spontaneous overtone emission of excited HF species produced by the F_2/H_2 chain pumping reaction in the cavity. The spontaneous emission of HF was imaged on the silicon target of a SIT vidicon through a lens. The filter in front of the vidicon was used to select one band from several overtone bands, for example, V=4-0 (at 6686A), or V=3-0 (at 8666A), etc. Then the spatial intensity profile of the selected band was recorded by using an OMA system. The schematic of apparatus is shown in Fig.8.

Although the chemiluminescence method as described above is basically a 2-dimensional diagnostic technique*, it may be applied with caution to the study of 3-dimensional flowfields. For example the 3-dimensional behavior could be minimized by injecting a large excess of hydrogen and this was exactly what we did in the following experiments.

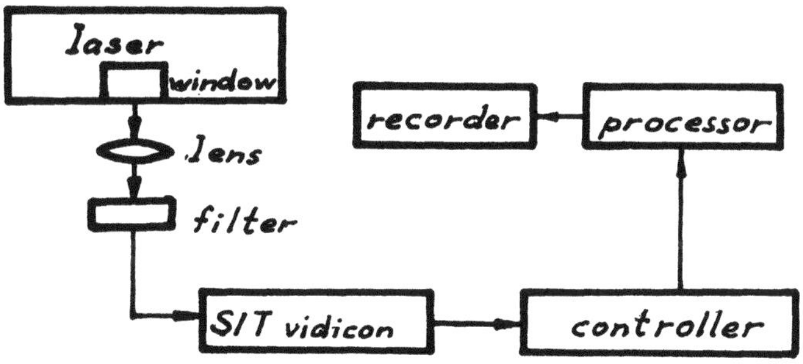

Fig.8 Schematic of apparatus for "hot" flow diagnostics

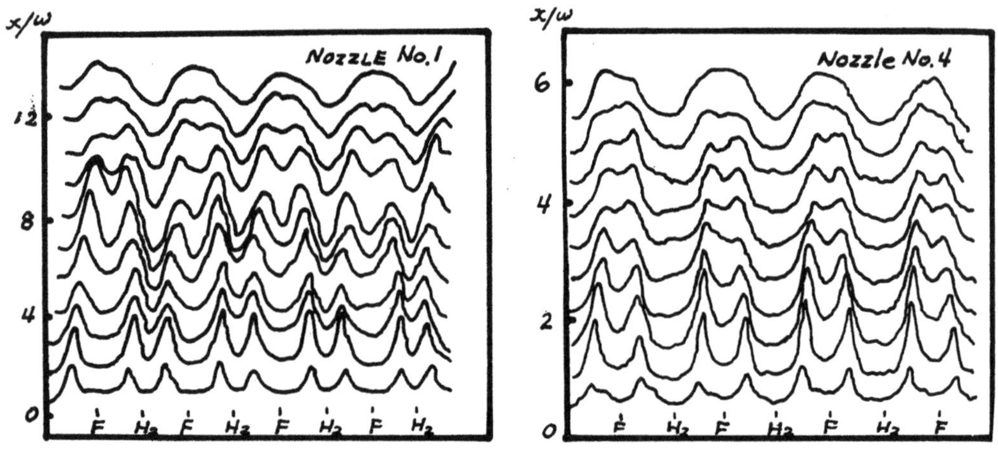

Fig.9,10 Intensity profiles of the excited HF overtone emission (V=4–0)

The typical results of V=4–0 emission are shown in Fig.9 for the nozzle bank No.1 with throat diameter 1.0 mm and in Fig.10 for the nozzle bank No.4 with a smaller throat (0.5 mm in diameter). The location of F and H_2 nozzles have been designated in these figures. As the distance from nozzle exit increases, the reaction zone becomes merged completely, as expected.

It may be noted that the merging for the smaller throat nozzle (No.4) takes place earlier than that for the nozzle No.1. After merging, the cold- reaction lasing would cease, since no more F is entrained into the mixing layer. This is in essential agreement with earlier measurements of small-signal gain distributions in our laboratory.

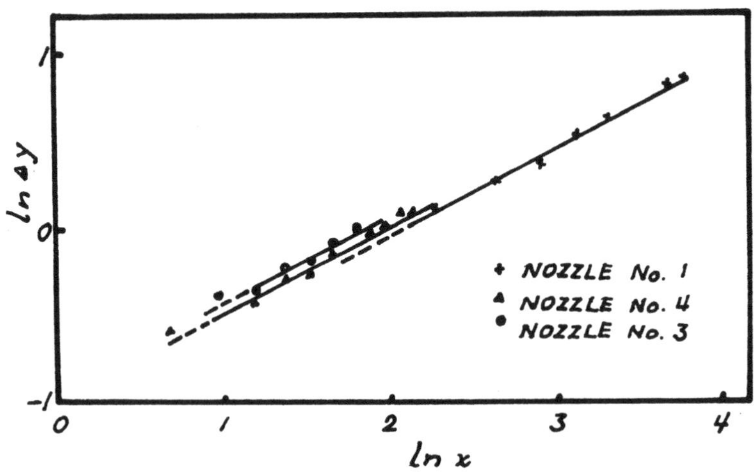

Fig.11 Variation of the mixing half width with the distance X

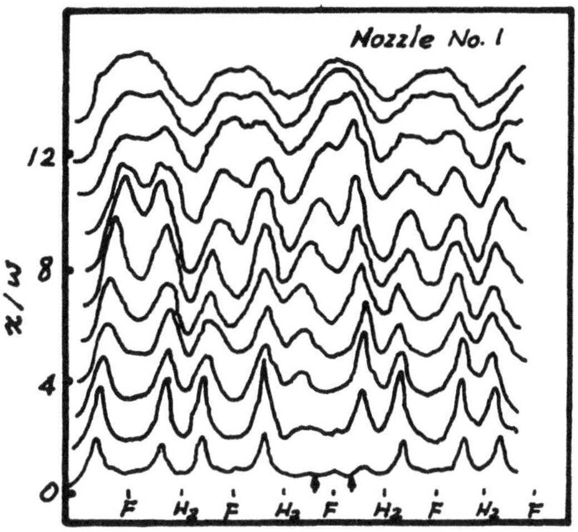

Fig.12 The influence of the trip on the flowfield

The relation between the mixing half width and the distance X from the nozzle exit has been correlated in a log-log plot (Fig.11). The straight line with slope 1/2 implies that laminar diffusion prevails under the present set of conditions. Since the cavity pressure P in our tests is about 3-5 torr, this conclusion is in compliance with the semiempirical law, which states that laminar diffusion would be dominant for $PX < 20$ torr-cm.

The influence of the trip on the flowfield has been observed.

Comparing with the infrared optical scan method, the present technique has the merits of both a shortened measuring time and a higher spatial resolution. Presumably, it will become a potential tool for the flowfield diagnostics of cw HF chemical lasers in operation.

* We appreciate comments from Drs. S.Davis & L.Rapagnani of AFWL, USA on this point.

References

1. R.W.F.Gross and J.F.Bott, "Handbook of Chemical Lasers" Wiley, New York, 1976
2. N.L.Rapagnani and S.J.Davis, AIAA J.17, 1402 (1979)
3. R.P.Huang, Dalian Inst. Chem. Physics, Dalian, China Rept. No. 79101813
4. W.L.Shackleford, A.B.Witte, J.E.Broadwell, J.E.Trost, and T.A.Jacobs, AIAA J. 12, 1009 (1974)

RAPID EXPANSION NOZZLES FOR GASDYNAMIC LASERS

D. Chuchem

Laser Unit - Nuclear Research Centre-Negev

P.O.Box 9001, Beer-Sheva, Israel

ABSTRACT

We here present an array nozzle having axial symmetry, where large area ratios can be attained without suffering from the limitations of conventional gasdynamic laser array nozzles. This facilitates operation of gasdynamic lasers at higher than usual stagnation temperatures and pressures to enable attainment of higher efficiencies. The design of such nozzles is detailed.

INTRODUCTION

In the gas dynamic laser (GDL), the role of the supersonic nozzle is to convert thermal energy from the rotational-translational degrees of freedom to the ordered kinetic energy of a supersonic flow, and thus facilitate the removal of population from the final state laser levels. In the evolution of the GDL, output and efficiency have been increased by raising the stagnation temperature and stagnation pressure[1,2]. As the pressure is scaled up, it is necessary to make the nozzle "faster", i.e., the nozzle must accelerate the gas in a shorter time. This can be accomplished by decreasing the nozzle throat area. For multi-array two-dimensional slit nozzles generally used in GDLs, increasing the area ratio and decreasing the nozzle dimensions are limited by: mechanical strength considerations--the difficulty in building nozzles with long but very narrow throats, and increasing viscous effects in narrow slit throats.

CONVENTIONAL NOZZLES

In figure 1 is depicted an axially symmetric deLaval nozzle. This nozzle has favorable properties regarding area ratio, fast opening, and mechanical strength. It has been used successfully in solid explosive fueled GDLs with very high stagnation temperatures[3]. However it cannot be stacked in an array to give the long and narrow active medium desireable for laser resonator cavities. In this case, the dead spaces between adjoining nozzle elements cause wakes in the supersonic flow which cause optical refraction and degradation of the laser cavity[4]. On the other hand, the two dimensional nozzle drawn in figure 2 can produce a flow field of unlimited length parallel to

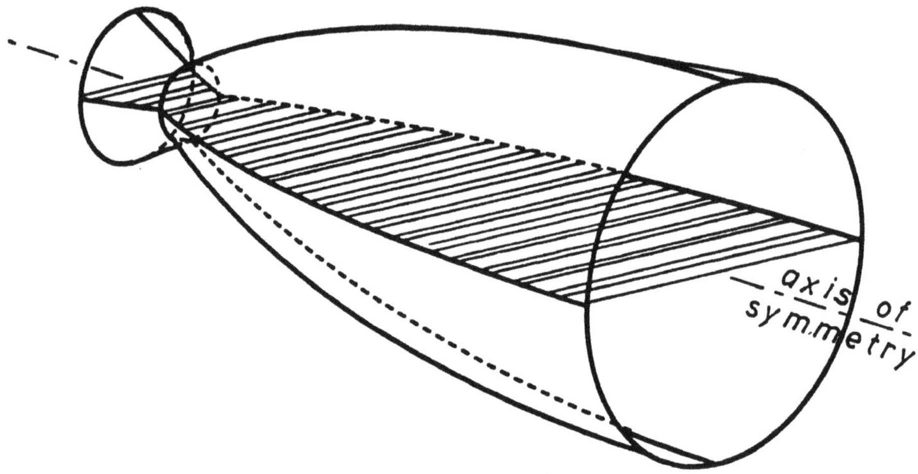

Figure 1. Axisymmetric deLaval nozzle

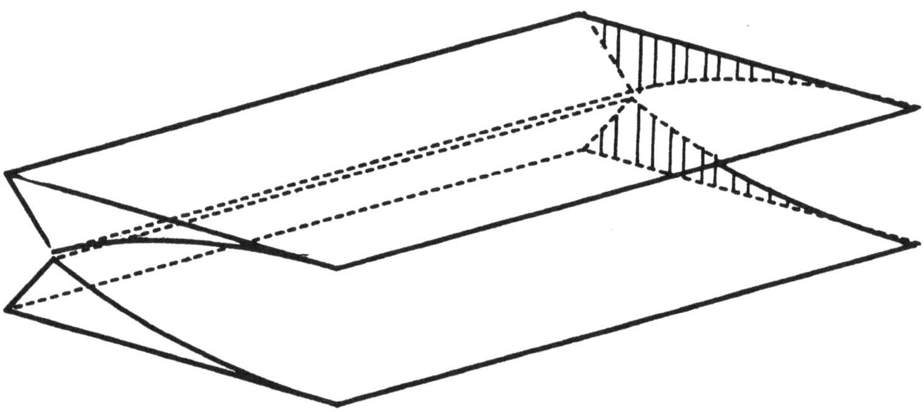

Figure 2. Two dimensional slit nozzle

the nozzle throat. Unfortunately it is not as fast opening as the deLaval nozzle. To achieve fast opening, the 2-D nozzle must be scaled down. To maintain desireable laser cavity height, the 2-D nozzle is cut up and packed in an array as shown in figure 3. In the 2-D array nozzle, wakes emanate from the ends of the array elements, but they do not seriously degrade the optical cavity because the inhomogenieties are perpendicular to the optical path and their effect can be made to cancel out[5]. However, the 2-D array nozzle has the limitations described in the preceding paragraph, and it is to overcome these limitations that we have developed the nozzle described in the following.

TRANSVERSE AXIALLY SYMMETRIC NOZZLE

In figure 4 we see that in a horizontal cross-section parallel to the gas flow, the nozzle is similar to the 2-D array nozzle. The vertical cross-section parallel to the flow shows that it differs from the 2-D nozzle in that in this plane the nozzle has axial symmetry. In the deLaval nozzle the axis of symmetry is parallel to the flow direction. In our nozzle, the axis of rotational symmetry is transverse to the gas flow direction. We can see that this nozzle combines some of the desireable properties of both the deLaval and the 2-D nozzle. Like the deLaval nozzle, the gas flow diverges in both the horizontal and vertical directions; like the 2-D nozzle, the elements of the TAS nozzle can be stacked in an array enabling the production of an optical gain medium of sufficient length for the efficient laser operation, and the wakes of the individual elements

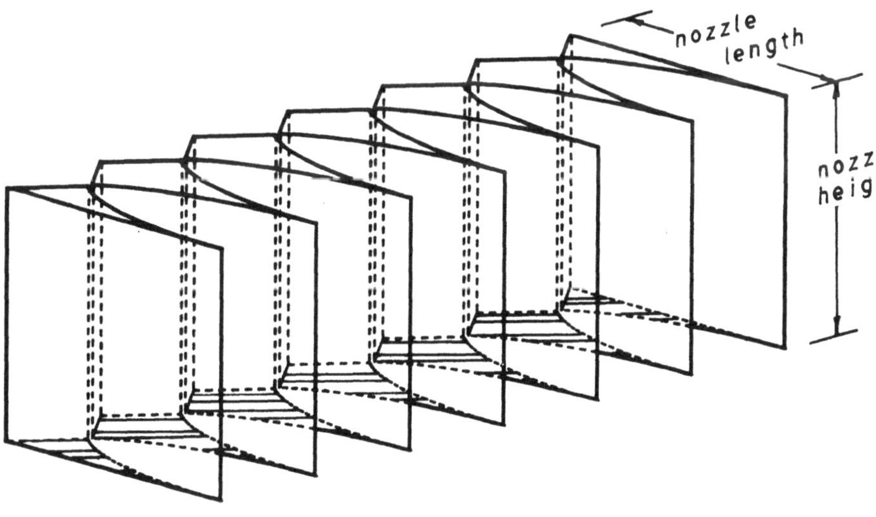

Figure 3. Two-dimensional array nozzle

are perpendicular to the radiation direction and thus do not overly degrade the optical cavity.

We have thus seen that the transverse axially symmetric (TAS) nozzle is similar to the deLaval and the two-dimensional nozzle. In fact, all three nozzles belong to the same family, in that to describe them, it is sufficient to describe a two-dimensional cross-section and then stipulate the corresponding symmetry operation which generates the nozzle in three dimensions from its two-dimensional cross-section. Thus the deLaval nozzle is generated by rotating its cross-section (see figure 1) about its central axis, the "two-dimensional" nozzle is generated by translating its cross-section perpendicular to its plane, and the transverse axially symmetric nozzle is generated by rotating the cross-section about the upstream transverse axis. Therefore the calculation of the contours of these nozzles is simplified in that it can be accomplished in two dimensions.

NOZZLE DESIGN

To further simplify the calculation in two dimensions, we assume steady-state irrotational inviscid flow with negligible heat conductivity, and express the gas flow equations in the natural curvilinear coordinate system. In this system, the s-coordinate measures distance along the streamline in the flow direction, the n-coordinate measures distance along the normal to the streamline, and θ is the angle of inclination of the streamline relative to our reference direction, which we choose as the final flow direction. Since we assumed irrotational flow, the velocity vector has non-zero components

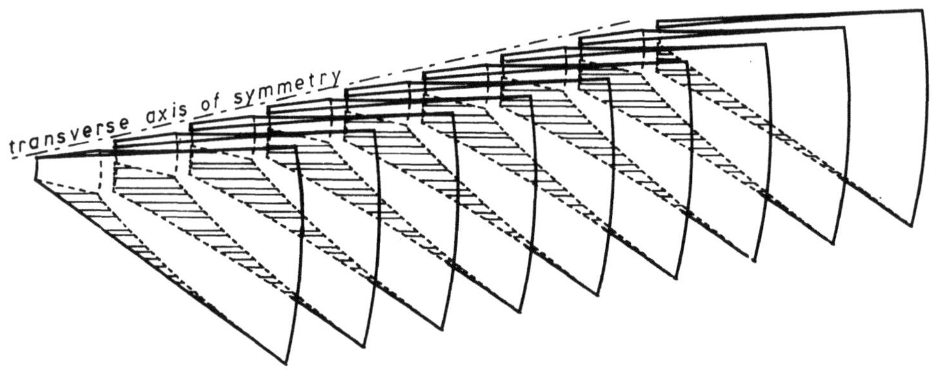

Figure 4. Transverse axially symmetric nozzle.
The cross-hatched section is parallel to the gas flow.

only in the s-direction. The equations of motion of the gas are thus considerably simplified.

However since the s,n coordinates are not Euclidian and inertial, we must include a pseudo-force in the n-direction in the momentum equations, and note that the divergence in the continuity equation must also be expressed in the proper curvilinear form (see figure 5).

BASIC EQUATIONS IN THE NATURAL COORDINATE SYSTEM

The equations describing the gas flow thus are:

energy conservation,

$$v \frac{\partial v}{\partial s} + \frac{\partial h}{\partial s} = 0 \tag{1}$$

s-direction momentum,

$$v \frac{\partial v}{\partial s} + \frac{1}{\rho} \frac{\partial P}{\partial s} = 0 \tag{2}$$

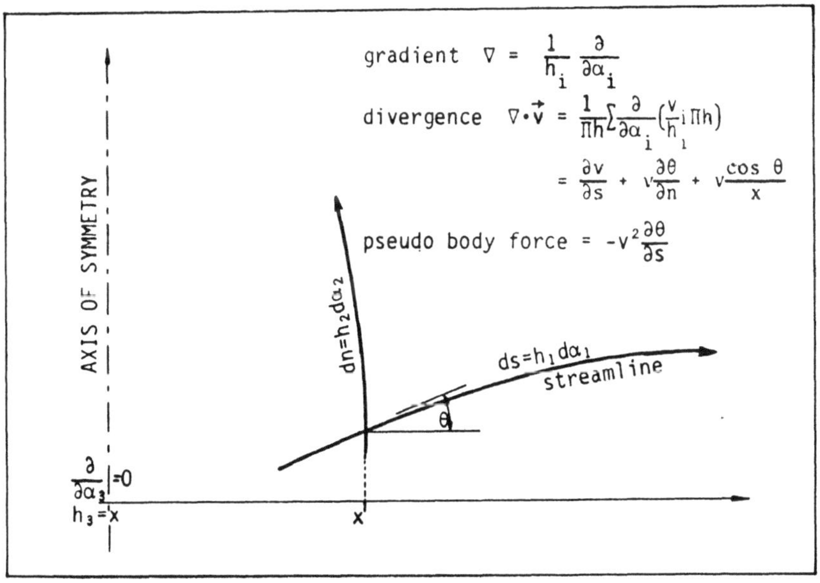

Figure 5. Curvilinear coordinate system.
Detailed for the transverse axially symmetric case.

n--direction momentum,

$$v^2 \frac{\partial\theta}{\partial s} + \frac{1}{\rho}\frac{\partial P}{\partial n} = 0 \qquad (3)$$

and continuity,

$$\frac{1}{\rho v}\frac{\partial(\rho v)}{\partial s} + \frac{\partial\theta}{\partial n} + J = 0 \qquad (4)$$

where

$$J = \begin{cases} \sin\theta/y & \text{for deLaval nozzle} \\ 0 & \text{for 2-D nozzle} \\ \cos\theta/x & \text{for TAS nozzle} \end{cases}$$

and the x-axis is along the center streamline, while the y-axis is along the transverse axis of symmetry.

CHARACTERISTIC DIRECTIONS AND COMPATIBILITY RELATIONS

We note that having chosen the streamline coordinate system has enabled equations (1) and (2) to be expressed immediately as ordinary differential equations. Thus we surmise the obvious fact that the streamline is a characteristic direction and equations (1) and (2) express the corresponding compatibility relations along this characteristic.

The remaining equations, (3) and (4), are independent partial differential equations. We must find the characteristic directions along which these equations reduce to ordinary differential form. Now we notice that the velocity can be eliminated from equations (1) and (2); if the resulting expression is compared with the definition of the enthalpy, we see that we have proved that the entropy is conserved along the streamlines, and thus we may write

$$\partial P/\partial s = c^2\, \partial\rho/\partial s \quad \text{where} \quad c^2 = (\partial P/\partial\rho)\Big|_{\text{const. entropy}}$$

is the velocity of sound propagation. We may thus replace the derivatives of the density and velocity in equation (4) in terms of the pressure derivative. The resulting equation can then be combined with equation (3) multiplied by an arbitrary constant, λ, to obtain:

$$\left[\left(\frac{1}{\rho c^2} - \frac{1}{\rho v^2}\right)\frac{\partial}{\partial s} + \frac{\lambda}{\rho v^2}\frac{\partial}{\partial n}\right] P + \left[\lambda\frac{\partial}{\partial s} + \frac{\partial}{\partial n}\right] \theta + J = 0$$

If $\{1/(\rho c^2)-1/(\rho v^2)\} : \lambda/(\rho v^2) = \lambda:1$ then we see that we have in fact obtained ordinary differential equations from equations (3) and (4). That is, we have derived two characteristic directions, $\chi_\pm$, defined by

$$\frac{\partial}{\partial \chi_\pm} = \{(v^2/c^2)-1\}^{\frac{1}{2}} \frac{\partial}{\partial s} \pm \frac{\partial}{\partial n} \tag{5}$$

along which exist the compatibility relations

$$\frac{\{v^2/c^2-1\}^{\frac{1}{2}}}{\rho v^2} \frac{\partial P}{\partial \chi_\pm} \pm \frac{\partial \theta}{\partial \chi_\pm} + J = 0 \tag{6}$$

Now using equation (5), we can express the equations for the characteristics in x-y coordinates:

$$\left.\frac{dy}{dx}\right|_{\text{along } \chi_\pm} = \frac{(\partial y/\partial \chi_\pm)}{(\partial x/\partial \chi_\pm)} = \frac{\{(v^2/c^2)-1\}^{\frac{1}{2}} \sin\theta \pm \cos\theta}{\{(v^2/c^2)-1\}^{\frac{1}{2}} \cos\theta \pm \sin\theta} \tag{7}$$

Next, in order to simplify the compatibility relations, notice that the conditions for irrotational flow, $U(\partial\theta/\partial S) = \partial u/\partial n$ substituted into equation (3) gives us the same form for the momentum equation in the n-direction as in the s-direction. Since we may take any linear combination of these two, we may say that $dp/\rho = -du^2/2$ in all directions.

We have already seen that as a direct consequence of equations (1) and (2), entropy is constant along the streamlines. If we now stipulate that the entropy is also the same on neighbouring streamlines, then we may write that $dp/\rho = dh = c_p dT = dc^2/(\gamma-1)$ in all directions. Thus we see that $c^2 d(u^2/c^2) = -{}^{P}\{2+(v^2/c^2)(\gamma-1)\}dp/\rho$ and thus the first term in equation (6) may be rewritten using:

$$\frac{\{(v^2/c^2)-1\}^{\frac{1}{2}}}{\rho v^2} dp = -\frac{\{(v^2/c^2)-1\}^{\frac{1}{2}}}{\{(v^2/c^2)(\gamma-1)+2\}} d \ln(v^2/c^2)$$

Futhermore noting that: $d\chi_\pm = dx/(\partial x/\partial \chi_\pm) = dx/\{\{(v^2/c^2)-1\}^{\frac{1}{2}}\cos\theta \mp \sin\theta\}$ then we may rewrite equation (6) for the TAS case:-

$$d\theta \mp \frac{\{(v^2/c^2)-1\}^{\frac{1}{2}} d(v^2/c^2)}{(v^2/c^2)\{(v^2/c^2)(\gamma-1)+2\}} \pm \frac{dx/x}{\{(v^2/c^2)-1\}^{\frac{1}{2}} \mp \tan\theta} = 0 \tag{8}$$

We now immediately see that equations (7) and (8) may be made more amenable to numerical computation if we remove the radical and transcendental function by defining $\beta = \{(v^2/c^2)-1\}^{\frac{1}{2}}$ and $\zeta = \tan\theta$. Thus along the characteristics described by:-

$$dy = \frac{\beta\zeta \pm 1}{\beta \mp \zeta} dx \tag{9}$$

We have the compatibility relations:-

$$\frac{d\zeta}{\zeta^2+1} \mp \frac{2\beta^2 d\beta}{(\beta^2+1)\{\beta^2(\gamma-1)+\gamma+1\}} \pm \frac{dx/x}{\beta \mp \zeta} = 0 \tag{10}$$

CONCLUSIONS

We thus see that we have derived a set of equations from which a flow field which in both the vertical and horizontal directions can be calculated. The four equations, (9) and (10), can be employed to calculate a net of positive and negative going characteristics at whose vertices we can solve for the four unknowns, x, y, β, ζ, which respectively describe the vertex position (x,y), the flow Mach number (through β), and the flow direction (through ζ). Since these expressions do not contain transcendental functions, they are especially suited for iterative numerical solution by digital computer. The shape of the nozzle contour can be found by tracing a streamline through the flow field.

By choosing small enough values of x for the throat position, one can attain relatively high exit Mach numbers with nozzle dimensions in the flow direction considerably shorter for the TAS case than for the 2-D case. A TAS type array nozzle composed of elements of the same length as a 2-D nozzle would thus be composed of fewer elements and therefore introduce fewer wakes into the downstream flow field than would a 2-D type array nozzle. Furthermore, since the array elements are wider, they have greater mechanical strength and are more amenable to the installation of coolant channels.

ACKNOWLEDGEMENTS

We gratefully acknowledge the help of Prof. S. Yatsiv of the Hebrew University of Jerusalem in useful discussions during the progress of this work.

REFERENCES

1. Anderson,J.D.,Jr. and Harris,E.L.; "Modern Advances in the Physics of Gasdynamic Lasers", AIAA Paper No. 72-143 (1972).
2. Christiansen,W.H.; "Pulsed Gasdynamic Lasers", AIAA Paper No. 71-572 (1971).
3. Robinson,C.P. and Sullivan,J.A.; "Condensed Explosives Laser", American Physical Society Paper No. DI-10 (1972).
4. Simons,G.A.; "The Effect of Boundary Layers on GDL Medium Homogeneity", AIAA Paper No. 72-709 (1972).
5. Hoffman,A.L. and Jones,T.G.; "Geometrical Methods for Improving the Optical Quality of Gasdynamic Lasers", AIAA Paper No. 72-217 (1972).

TURBULENCE MEASUREMENT IN THE FLOW OF A CO_2 LEDA* TYPE LASER

M. Duchet, J-P. Crançon, and J. Solmon

Laboratoires de Marcoussis
Centre de Recherches de la C.G.E
F - 91460 Marcoussis

Abstract

Optical index turbulence was measured in a LEDA* type, CO_2, 10.6 μm amplifier medium. The first method using the broadening of a focused probe laser beam appeared inefficient because of disturbance due to the quasi continuous component of the index gradient. However, this turbulence was measured using the fluctuation spectrum of the quasi pin-point intensity of a collimated probe laser beam [1].

This method made it possible to measure the structure index constant of the medium at different heights, and then to calculate the phase correlation radius of a plane wave propagating over some range in the medium. Knowing the beam dimension of the laser to be amplified, the quality factor of the output beam may be obtained by measuring the turbulence.

EXPERIMENT AND RESULTS

In this study we give some results of optical index turbulence of a LEDA* type, CO_2, 10.6 μm amplifier medium. In such a laser amplifier with sonic or supersonic flow, nitrogen is electrically excited in discharge tubes before being mixed with CO_2 and helium. For this measurement, we use a 0.63 μm He-Ne probe laser beam. The He-Ne beam perturbations are compared to Tatarski's theoretical results [1]. We can then estimate the turbulence effect on the

* LEDA: Laser à Excitation Electrique et Detente Adiabatique (Electrical Excitation/Gas Dynamic Laser).

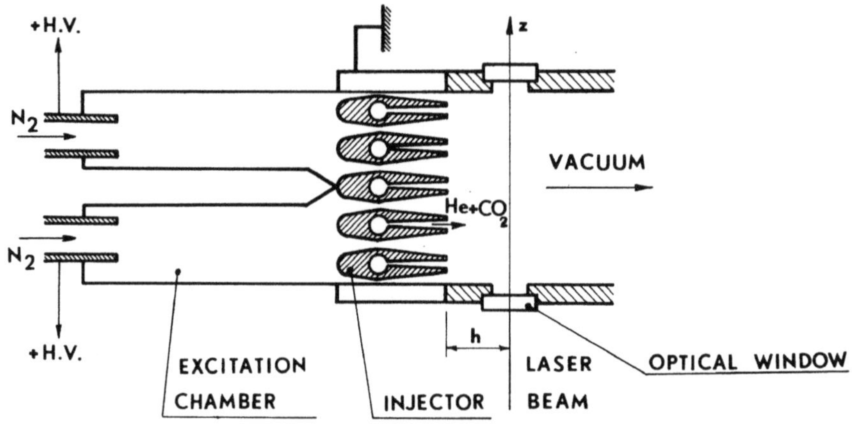

Fig. 1. LEDA configuration.

quality factor of an amplified CO_2 laser beam. We characterize the turbulence medium by its structure index constant C_n, assumed to be homogeneous, isotropic and stationary along a path parallel to the injector plane and transverse to the flow plane (see Figure 1). For a Gaussian focused beam, and for the turbulence profile shown in Figure 2, we observe a long-term broadening Δ such that [2]:

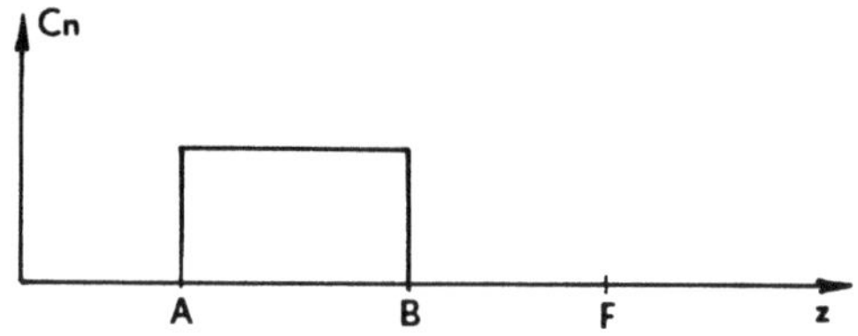

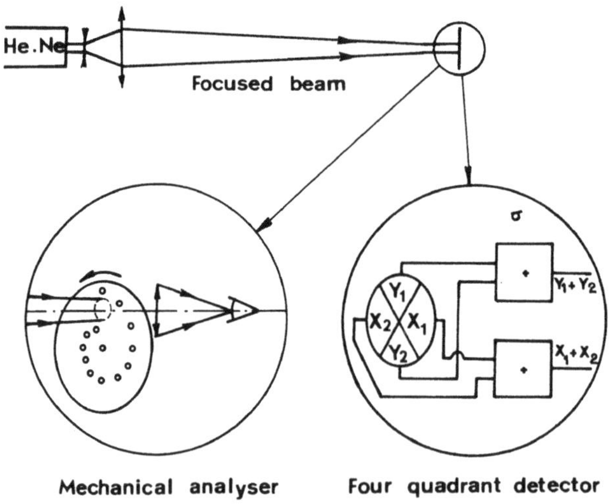

Fig. 2. Beam broadening measurement.

$$\Delta^2 = \rho^2(F) - \rho_o^2(F) = 6.25\ F^2 k^{2/5} \left[\int_o^F C_n^2(z) \left(\frac{z}{F}\right)^{5/3} \right]^{6/5},$$

where $P_o(F)$ is the e^{-1} intensity beam focal radius with the turbulence effect and $P_o(F)$ is the same radius without turbulence effect. F is the focal length, λ the wavelength, and $k = 2\pi/\lambda$.

In the particular case shown in Figure 2, we have:

$$\Delta^2 = 1.93\ k^{2/5} C_n^{12/5} \left[Z_B^{8/3} - Z_A^{8/3} \right]^{6/5}.$$

Then the C_n structure index constant may be defined by a broadening measurement.

The theory leads us to expect a long-term Gaussian beam. A first experiment using such a configuration shows that the focal image after examination with a mechanical image analyzer is not circular but quasi elliptical. This phenomenon is tested in order to know if we have a quasi continuous effect due to medium turbulence. We use a four-quadrant silicon cell to analyze the focused He-Ne beam with a high frequency response, giving the ΣX and ΣY signals (see Figure 3). The theorectical results of R(K) are also given in this Figure.

In Figure 4, the oscillogram shows that the focused beam is flattened, in accordance with the theoretical results. Therefore, this lateral beam compression is a quasi continuous phenomenon which is due not to turbulence but to the index gradient.

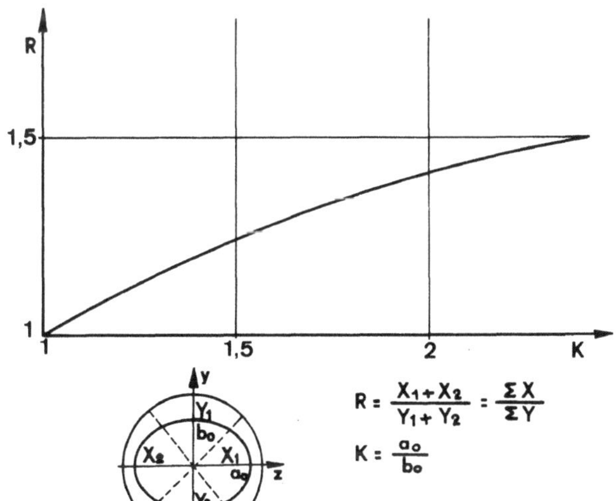

Fig. 3.

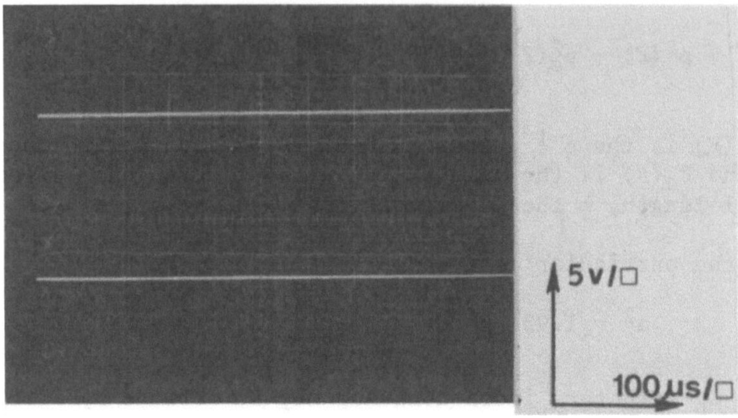

Fig. 4. ΣX and ΣY signals.

We then use another measurement method based on the spectrum analysis of quasi pin-point intensity in the case of a plane wave, this wave being approximated by a collimated He-Ne beam (Figure 5).

$W_\ell(f)$ being the power spectrum of normalized fluctuations logarithm (ℓ), $\bar{I}$ being the mean intensity of the circular detector, we have:

$$W_\ell(f) = \frac{\Delta P_\ell}{\Delta f},$$

P_ℓ being the ℓ power between f and f + Δf.

In this case, according to Tatarski [1], we have:

$$F(\Omega,\rho) = \frac{fW_\ell(f)}{\sigma_\ell^2(o)} = 1.35 \int_o^\infty \left(1 - \frac{\sin X}{X}\right) X^{11/6} \left(\frac{2J\ (\rho\sqrt{X})}{\rho\sqrt{X}}\right)^2 dt;$$

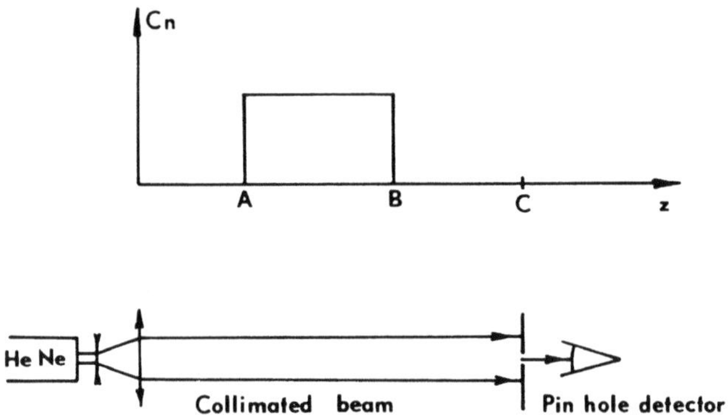

Fig. 5.

with $X = t^2 + \Omega^2$, $\Omega = f/f_o$, $\rho = R\sqrt{\frac{2\pi}{\lambda L}}$, $f_o = \frac{V}{\sqrt{2\pi\lambda L}}$; L being the propagation distance in the medium, V the transverse wind and R the detector radius; $\sigma_\ell^2(o)$ is the intensity logarithm variance when R goes to 0.

From Figure 5, we have:

$$\frac{\sigma_{\ell o}^2}{4} = \sigma_x^2 = 0.56\ k^{7/6}C_n^2 \times \frac{6}{11}\left[(F-z_A)^{11/6} - (F-z_B)^{11/6}\right].$$

With the small signal hypothesis we have, with $I = \bar{I} + \tilde{I}$,

$$\ell = \log\frac{\tilde{I}}{\bar{I}} = \log(1+\frac{\tilde{I}}{\bar{I}}) \# \frac{\tilde{I}}{\bar{I}}\ .$$

$W_{\tilde{I}}$ being the $\tilde{I}$ power spectrum, measured with a Δf frequency width, we have:

$$W_{\tilde{I}} = W_\ell(f)\bar{I}^2\Delta f,$$

then

$$\sigma_\ell^2(o) = \frac{W_{\tilde{I}}f}{\bar{I}^2\Delta f F(\Omega,\rho)}\ .$$

With $G(\Omega,\rho) = F(\Omega,\rho)/\Omega$ we have:

$$C_n^2 = \frac{W_{\tilde{I}}f_o}{\bar{I}^2\Delta f G(\Omega,\rho) \times 1.22\ k^{7/6}\left[(F-z_A)^{11/6} - (F-z_B)^{11/6}\right]}\ .$$

The $W_{\tilde{I}}$ function is directly obtained with a spectrum analyzer in dB/Hz.

We have calculated the 10 log $G(\Omega,\rho)$ function with $\rho = 0.5376$ (see Figure 6).

We may then compare the theoretical curve to the experimental curve (Figure 7) and measure the C_n.

The Table in Figure 8 gives some values of estimated C_n.

With a CO_2 laser beam amplifier in this medium, having a a_oe^{-1} intensity radius, the phase correlation radius after a L_A propagation range in the medium is approximately given by:

$$\rho_o = (1.5\ k^2C_n^2L_A)^{-3/5}.$$

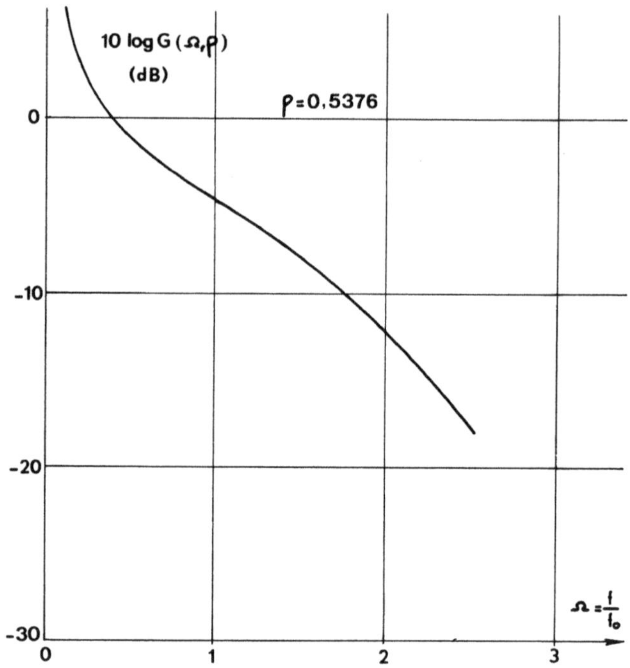

Fig. 6. $G(\Omega,\rho)$ function with $\rho = 0.5376$.

If the initial CO_2 beam is at the diffraction limit, we have at the output a K factor quality given by:

$$K \# \sqrt{1 + \frac{4a_o^2}{\rho_o^2}} \ .$$

The Table in Figure 8 gives some values of ρ_o and K for different values of C_n and L_A, with $a_o = 1.77$ cm.

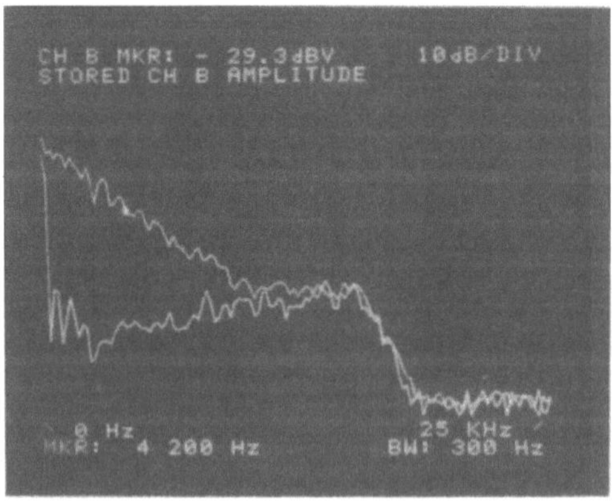

Fig. 7. Fluctuation spectrum.

Range from injector h (cm)	1.5	6	10.5
Cn $(10^{-6} m^{-1/3})$	6.27	2.7	2.46
ρ_0 (LA = 1m)	16.2cm	44.6cm	49.8cm
ρ_0 (LA = 10m)	4.07cm	11.2cm	12.5cm
K (LA = 1m, a_0 = 1.77cm)	1.023	1.003	1.0025
K (LA = 10m, a_0 = 1.77cm)	1.32	1.05	1.04

Fig. 8. Results.

We find that the only turbulence effect is very small, excepting $C_n = 6.27 \cdot 10^{-6}\ m^{-1/3}$ and $L_A = 10$ m.

REFERENCES

1. V. I. Tatarski, "Wave propagation in a turbulent medium", McGraw Hill Book Company (1961).
2. H. T. Yura, J. Applied Optics, Vol. 10, No. 12, p. 2772, Dec. (1971).

ACKNOWLEDGMENT

This work was supported by the Direction des Recherches, Etudes et Techniques (D.R.E.T.).

SUBSONIC TE N_2-CO MIXING LASER

Mitsuhiro Iyoda, Kazumi Sugino, Katsuhiko Komatsu, and Tomoo Fujioka

Department of Electrical Engineering, Keio University
3-14-1, Hiyoshi, Kohoku-ku, Yokohama-shi, 223, Japan

INTRODUCTION

CO lasers have the advantages of high power and high efficiency. The electrical conversion efficiency is remarkably high and the value of 47% was achieved with the coaxial flow CO laser at a wall temperature of 77 K [1]. CO lasers are, however, not commercially used to date because a temperature below 100 K is necessary in order to achieve such high efficiency. Although the operation of CO lasers at room temperature is also possible, the conversion efficiency is as low as 15%. It is well known that the addition of N_2 is effective to increase the output performance of the CO laser operating near room temperature [1,2]. For the coaxial flow lasers the concentration of N_2 is limited. And high N_2 concentration is achieved in supersonic flow and pulsed CO lasers. The latter two types operate for short durations, and the effect of the addition of N_2 is supposed to be the optimization of E/N values of the discharge, rather than the vibration to vibration (V-V) energy transfer between N_2 to CO, because the N_2-CO V-V transfer rate is twice as small in magnitude as the CO-CO V-V transfer.

The higher operation temperature of the CO laser is better for its practical use. Under these circumstances the authors first applied transverse excitation (TE) configuration to the subsonic flow CO laser [3]. As a result, specific laser power of more than 1 W/cm^3 and electrical conversion efficiency of more than 20% are achieved with a mixture of CO/N_2/He = 1.4/11.2/22.4 Torr at an inlet gas temperature of 185 K. By the use of the TE configuration, uniform self-sustained discharge was obtained in a mixture containing more than 30% of N_2. The input power density was increased to 4 W/cm^3. Effective vibrational energy storage in N_2 and V-V energy transfer from N_2 to CO in the subsonic flow were supposed to contribute to the novel laser performance. But based on the experiment results, we expect laser operation occurred only by V-V energy transfer, and we developed the subsonic flow TE N_2-CO mixing laser.

EXPERIMENTAL SETUP

The fundamental construction of the experimental setup is the same as our previous TE CO laser, in order to compare laser performance under the same conditions of gas temperature and input power. For the N_2-CO mixing laser, only N_2 is electrically excited. The experimental setup is shown in Fig. 1. The cross section of the flow channel has the same dimension of $12^H \times 270^W$ mm^2 as our previous apparatus. Bakelite is used for most of the walls. The left half of the figure is the discharge region, where the pre-cooled mixture of N_2 and He is introduced and electrically excited. The detailed configuration is shown in Fig. 2. The electrode consists of hollow cathodes and a flat-plate anode. The 49 tube-cathodes are located in a row with 5 mm of separation. The cathodes, are brass tubes with outer diameters of 3 mm and inner diameters of 2 mm, and are set 3 mm inside the insulating material made of ceramics. The electrode gap is therefore 15 mm. For the stabilization of the discharge, a ballasted resistance of 150 kΩ is connected to each cathode. The anode is made of SUS-304, with a cross section of $25^L \times 250^W$ mm^2. The mixing region is the right half of Fig. 1. A pre-cooled mixture of CO and He is injected at 82 mm downstream of the discharge region through a row of 27 holes, with diameters of 1 mm. Distance between the injectors is 1 cm. For a uniform injection, the CO mixture passes through a buffer tank with a diameter of 50 mm and length of 270 mm. By aerodynamical calculation, the mixing is estimated to be completed within 1 mm downstream of each injector center. For uniform mixing between injectors, however, more length along the flow is necessary.

A sufficient density of vibrationally excited N_2 should be maintained in transitting from the discharge region to the mixing region. The vibration to rotation and transration (V-R/T) energy transfer of N_2 is dominated by He. In the experiment, the gas temperature downstream of the discharge region is less than 230 K. From the value of $p\tau = 60$ Torr·sec for the N_2-He V-R/T energy transfer at 230 K [4], and the partial pressure of He less than 15 Torr, the relaxation time is estimated at $\tau = 4$ sec. The flow velocity is at least 6 m/s, so the relaxation of N_2 is negligible. The optical axis is transverse to the flow and located downstream of the mixing region. In the same way described above, the V-V energy transfer time is estimated in order to decide the position of the optical axis. For the transfer

$$N_2(1) + CO(0) \longrightarrow N_2(0) + CO(1) + 188\ cm^{-1},$$

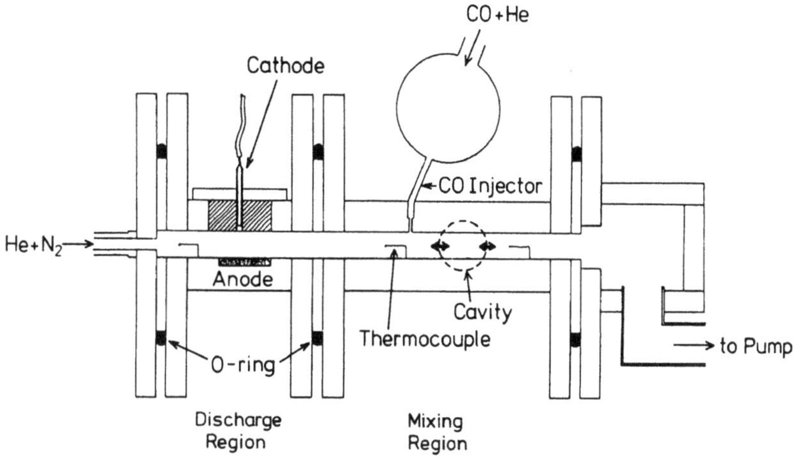

Fig. 1 Schematic of the experimental setup.

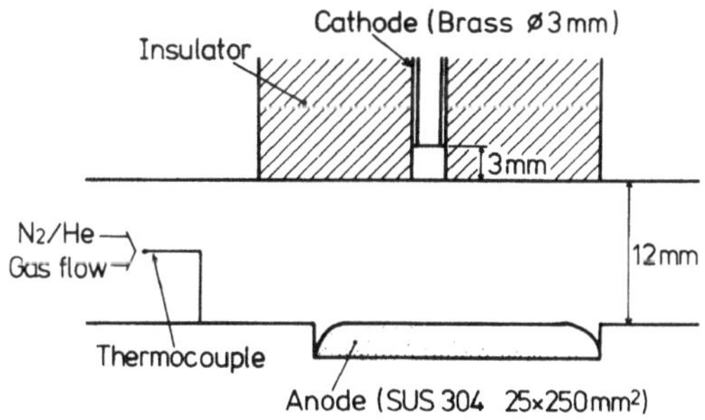

Fig. 2 Cross-sectional view of the discharge region.

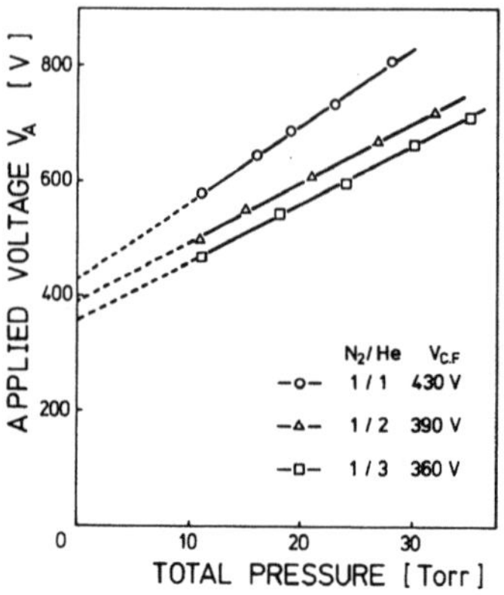

Fig. 3 Dependence of applied voltage V_A on total pressure. Discharge current is 400 mA, total gas flow rate is 50 l/min (7 m/s), and the gas temperature in the discharge region is 230 K.

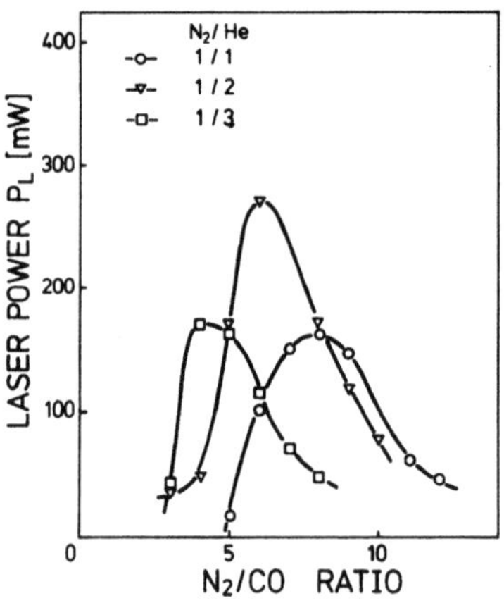

Fig. 4 Effect of N_2/CO mixture ratio on laser power P_L. Discharge current is 400 mA, total gas pressure is 28 Torr, N_2-He flow rate is 70 l/min, and the gas temperature at the cavity is 210 K.

$p\tau$ is 2.4×10^{-3} Torr·sec at 200 K [5]. Under the CO partial pressure of 1-5 Torr, τ is estimated 0.5-2.5 msec. The actual transfer time is expected to be smaller than the calculated value considering the existence of resonant V-V transfer of

$$N_2(v+6) + CO(v) \longrightarrow N_2(v+5) + CO(v+1).$$

For the CO-CO V-V transfer, $p\tau$ is 5×10^{-6} Torr·sec [5], which is more than twice as small as for N_2-CO. With the calculated transfer time and the total flow velocity of 10 m/s, the optimum position of the optical axis is estimated at around 20mm downstream of the mixing region. Based on these results, the position of the optical axis is designed to be variable from 0-50 mm downstream of the injectors. Two inner mirrors with the reflectivity of 99.5%(R = 4 m) and 99%(R = 1 m) comprise the optical cavity, whose length is 640 mm. The gas temperature, which gives significant influence on CO laser performance, should be monitored. The temperature at the inlet of the discharge region, the inlet of the mixing region, the outlet of the optical axis, and the CO injector were meaured with a Chromel-Alumel thermocouple. The laser power was monitored with a Coherent-radiation Model 210 power meter.

EXPERIMENTAL RESULTS AND DISCUSSION

In the experiment the volumetric flow rate of the electrically excited main stream, that is the mixture of N_2 and He, was fixed to 70 l/min. The N_2/He mixing ratio was set at 1/1, 1/2, and 1/3. The discharge characteristics are shown in Fig. 3. The cathode fall voltage was estimated at 430, 390, and 360 V for N_2/He ratio 1/1, 1/2, and 1/3, respectively. The discharge voltage was kept constant against the variation of the discharge current from 250 to 450 mA. Figure 4 shows the dependence of the laser power on the N_2/CO mixing ratio. The optical axis was located 20 mm from the mixing region. The total gas pressure was kept to 28 Torr. As shown in Fig. 4 the optimum N_2/CO mixing ratio was established. The optimum value was 8, 6, and 4 for N_2/He ratios 1/1, 1/2, and 1/3, respectively. Although the flow rate of the main stream was kept constant, the total flow rate changed depending upon the N_2/He ratio. In order to analyze the experimental results, Fig. 4 is converted into Fig. 5, where one horizontal axis represents the injected CO-He flow rate and the other horizontal axis represents the transit time of the total mixture from the mixing region to the optical axis. According to the figure, all the optimum values exist around the same CO-He flow rate of 20 l/min at a transit time of 1.85 msec. Here, the partial mixture ratio and partial pressure of CO are 4.4% and 1.24 Torr, respectively. The calculated V-V transfer time between N_2 and CO is around 1.9 msec and nearly equal to the transit time. The experimental results are thus attributed to the sufficient V-V energy transfer between N_2 and CO.

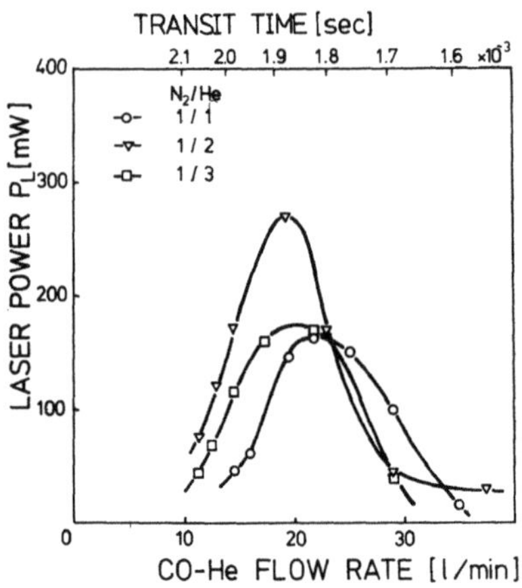

Fig. 5 Effect of CO-He flow rate on laser power P_L. Discharge current is 400 mA, total gas pressure is 28 Torr, N_2-He flow rate is 70 l/min, and the gas temperature at the cavity is 210 K.

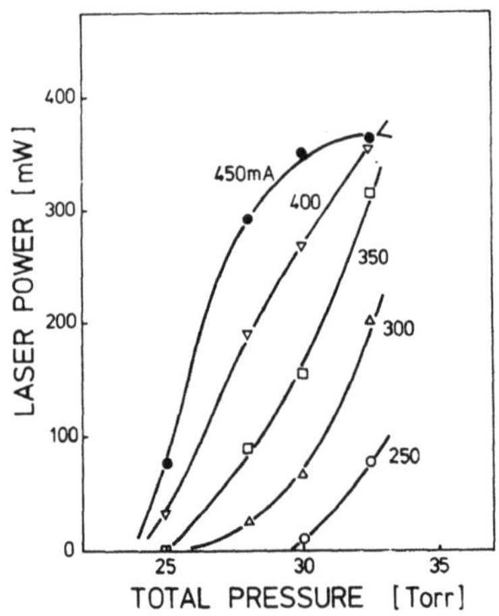

Fig. 6 Dependence of laser power P_L on total pressure. N_2-He flow rate is 70 l/min.

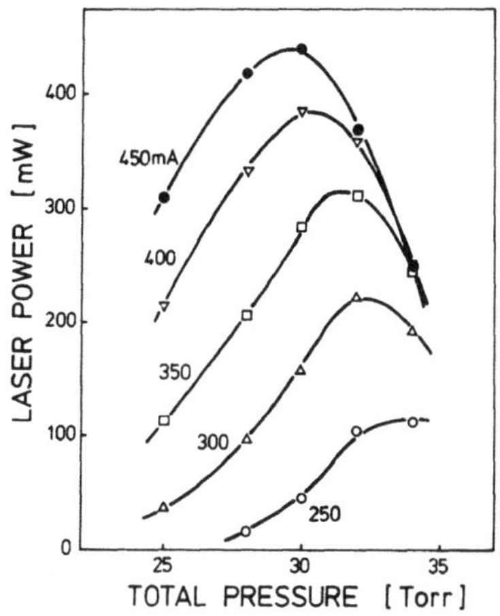

Fig. 7 Dependence of laser power P_L on total pressure. N_2-He flow rate is 70 l/min.

The maximum laser power was obtained where the N_2/He mixture ratio was 1/2. In this case, the excitation of N_2 is supposed to be the most efficient. In addition, the total mixture ratio, CO/N_2/He = 4.4/25.6/70(%) was close to the optimum ratio, CO/N_2/He = 4/32/64 (%), of our previous TE CO laser. Figure 6 and 7 show the dependence of laser power on total pressure and discharge current. N_2/He mixture ratios are 1/3 and 1/1.5 for Figs. 6 and 7, respectively. The N_2/CO ratio is fixed at 7. Comparing Fig. 6 and Fig. 7 under the same pressure and current, the laser power in the case of Fig. 7 is greater than that of Fig. 6 by at least 20% in the pressure region of less than 30 Torr. This is considered to result from the high concentration of excited N_2 and CO. In Fig. 7, however, a decrease of laser power is observed in the high pressure region of greater than 30 Torr. With a high concentration of N_2, the condition of discharge is not supposed to be uniform nor suitable to efficiently excite N_2. The observed laser lines under the optimum experimental condition are regarded as $P_{9-8}(15)$, $P_{10-9}(14)$, $P_{10-9}(15)$, $P_{11-10}(14)$, $P_{12-11}(13)$, $P_{12-11}(15)$, and $P_{13-12}(13)$. The line distribution varies little from our previous TE CO laser. This result predicts that V-V energy transfer took place in our previous CO laser as well as in the N_2-CO laser.

CONCLUSION

By the first successful operation of TE N_2-CO laser, the important role of N_2, that is, the energy storage in vibrational mode of N_2 and V-V energy transfer from N_2 to CO, was clarified. Considering the good scale of the results of the TE configuration, the experiment predicts the possibility of high power TE N_2-CO laser operation.

ACKNOWLEDGEMENT

The authors are grateful to Dr. M. Obara for the useful suggestion. We are thankful also to Mr. S. Sato, T. Abe, and T. Isagoda for their cooperative work in the experiment.

REFERENCES

1. M. L. Bhaumik, W. B. Lacina, and M. M. Mann, "Characteristics of a CO laser," IEEE J. Quantum Electron. QE-8:150 (1972).
2. R. M. Osgood,Jr., W. C. Eppers,Jr., and E. R. Nichols, "An Investigation of the High-Power CO Laser," IEEE J. Quantum Electron. QE-6:145 (1970).
3. T. Arai, K. Sugino, M. Obara, and T. Fujioka, "Experimental Investigation of the Optimum Specific Input Energy on a Subsonic CO EDL," Proceedings of the International Conference on LASERS '81, Dec. 14-18, 1981, New Orleans, Louisiana, USA.
4. D. J. Miller and R. C. Millikan, "Vibrational Energy Exchange between N_2 and CO. The Vibrational Relaxation of Nitrogen," J. Chem. Phys. 39:98 (1963).
5. J. C. Stephenson and E. R. Mosburg,Jr., "Vibrational energy transfer in CO from 100 to 300 K," J. Chem. Phys. 69:3562 (1974).

NUMERICAL ESTIMATIONS OF 16 μm CO_2 GASDYNAMIC LASERS WITH VARIOUS CONTOURS OF 2-DIMENSIONAL NOZZLES

Koichi Kasuya and Kazuhiko Horioka

Department of Energy Sciences, The Graduate School at Nagatusta, Tokyo Institute of Technology, Nagatusta, Midori-ku, Yokohama, JAPAN

INTRODUCTION

The temperature and the pressure of the mixed gas in the CO_2 gasdynamic laser (GDL) cavity can be easily changed by the nozzle contour or the stagnations, so that we can easily select the optimum conditions for the maximum population inversion of the 16 μm band. The mean value of the laser output is also large, because of the inherent high speed flow across the laser cavity.

The effects of the stagnation temperature and pressure to the 16 μm CO_2 GDL performances are numerically estimated in another article[1], using only one kind of 2-dimensional nozzle. On the contrary, we discuss in this article the effect of nozzle contours to the laser performances. Under an assumption of a simple model, numerical estimations of 16 μm CO_2 GDLs are performed, covering a wide range of parameter changes.

BASIC EQUATIONS AND NUMERICAL METHODS

Energy levels of the CO_2-N_2 GDL system are shown in Fig. 1. After we obtain a population inversion between (00^01) and (02^00) levels, a laser pulse of 9.4 μm band is injected into the laser medium to get a population inversion between (02^00) and (01^10) levels.

The reactions concerned are as follows.

$$CO_2\ (00^01) + h\nu\ (\lambda=9.4\ \mu m) \rightarrow CO_2\ (02^00) + 2h\nu. \quad (1)$$
$$CO_2\ (02^00) + M \rightleftarrows CO_2\ (10^00) + M - 102.8\ cm^{-1}. \quad (2)$$
$$CO_2\ (02^00) + M \rightleftarrows CO_2\ (02^20) + M - 49.9\ cm^{-1}. \quad (3)$$

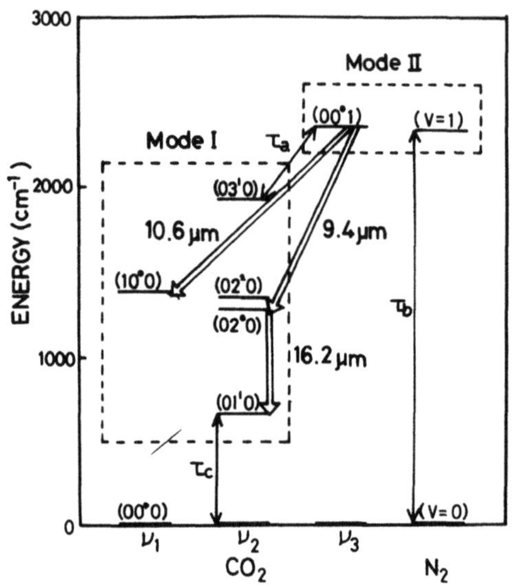

Fig. 1 Vibrational energy levels and the simplified 2-mode kinetic model of CO_2-N_2 system.

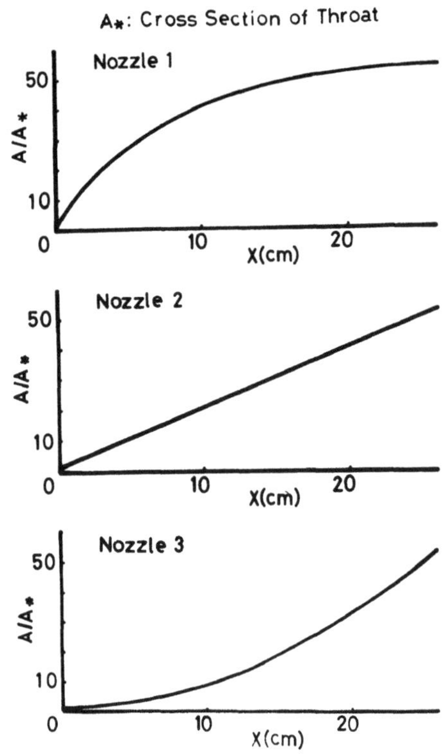

Fig. 2 Supersonic portions of the 2-dimensional nozzles used in this article.

$$CO_2\ (02^00) + CO_2\ (000) \rightleftharpoons 2CO_2\ (01^10) - 50.1\ cm^{-1}. \qquad (4)$$

Here M is the collision partner (CO_2, N_2, and He).

Before the input of the 9.4 μm laser pulse, the flow field is calculated with the so called 2-level model of Anderson[2]. The basic equations of the quasi-one dimensional non-stationary flows are as follows,

$$\frac{\partial \rho}{\partial t} = -\frac{1}{A}\frac{\partial(\rho u A)}{\partial x}. \qquad (5) \qquad \frac{\partial u}{\partial t} = -[\frac{1}{\rho}\frac{\partial p}{\partial x} + u\frac{\partial u}{\partial x}]. \qquad (6)$$

$$\frac{\partial E}{\partial t} = -[\frac{p}{\rho}(\frac{\partial u}{\partial x} + \frac{\partial \ln A}{\partial x}) + u\frac{\partial E}{\partial x}]. \qquad (7)$$

$$E = e + e_{vib,I} + e_{vib,II}. \qquad (8) \qquad e = \sum_j [(C_v)_j T] X_j. \qquad (9)$$

$$\frac{\partial e_{vib,i}}{\partial t} = \frac{e^{eq}_{vib,i} - e_{vib,i}}{\tau_i} - u\frac{\partial e_{vib,i}}{\partial x}. \qquad (10) \qquad p = \rho RT, \qquad (11)$$

where the definitions of various symbols are shown in the last part of the text.

After we calculate the flow field along the nozzle axis x by the time dependent method[2], the upper level population of 16 μm band under the input of 9 μm laser pulse is estimated with the following equation[1].

$$N'_{02^00} = (N_{02^00} + N_{00^01}) / 2. \qquad (12)$$

The small signal gain coefficient is also estimated as follows.

$$G = \frac{\lambda^2}{4\pi\tau_s\nu_c}(N_u - N_1)\frac{2B(2J+1)hc}{kT}\exp(-\frac{hc}{kT}BJ(J+1)). \qquad (13)$$

(In the following calculations, J = 20, without any specification.) Another example of estimation is the available output power density P_{16} as follows.

$$P_{16} = \frac{hc}{\lambda}(\frac{N'_{02^00} - N_{01^10}}{2})u. \qquad (14)$$

RESULTS AND DISCUSSIONS

4 kinds of 2-dimensional nozzle contours are considered here, 3 of which are shown in Fig. 2. Only the supersonic parts of the nozzles are shown in this figure. Nozzle-1 is a Busemann nozzle obtained with the characteristics method[3], nozzle-2 is a wedge nozzle, and nozzle-3 is a cone nozzle. These 3 nozzles breathe

stagnated gases through the subsonic parts and throats. Nozzle-4 has the same contour as the nozzle-1, but it breathes non-stagnated gas behind a bow shock wave[4]. The designed mach number of the nozzle exits is 6 (γ = 1.4, A/A^* = 53.2, nozzle length = 26 cm). All of the subsonic parts have shapes of direct lines, and 3 kinds of declining angles are considered (θ_{sub} = 30°, 45° and 60°). The mixing ratio of the laser gas is fixed at $X_{CO_2} : X_{N_2} : X_{He}$ = 0.2 : 0.1 : 0.7.

An example of numerical results is shown in Fig. 3, where the small signal gain coefficient and the temperature are shown along the flow axis.

In next 3 figures are shown (1) the small signal gain coefficient of 16 μm band G_{16} (Fig. 4), (2) the available power density P_{16} (Fig. 5), and (3) the maximum values of $G_{10.6}^{max}$, G_{16}^{max} and P_{16}^{max} (Fig. 6). The optimum value of the stagnation pressure seems to be about 10 atm in these 3 figures.

The effects of the nozzle contours are shown in Fig. 7 for the case of P_0 = 10 atm and θ_{sub} = 45°. Higher stagnation temperature is necessary to make the available upper level energy larger. On the contrary, lower temperature is necessary to make the cavity gas temperature lower and also to get the higher G_{16} or P_{16}. This compromise corresponds to the optimum conditions shown in Fig. 7.

G versus x is shown in Fig. 8 for the case of P_0=10 atm, T_0= 1750 K and θ_{sub}=45°. The nozzle contour effect to G_{16} is much more than the one to $G_{10.6}$. The same kind of nozzle size effect is shown in Fig. 9. Similar contours of 1/1, 1/2 or 1/4-size nozzle (nozzle-1) are compared in this figure. On the other hand, the contour difference in the subsonic part does not effect much, as shown in Fig. 10.

An example of respective population density is shown in Fig. 11. Solid lines are the results for thermal-equilibrium-Boltzman distribution. Other 2 lines are the results for the case of $T_{vib,II}$ =1500 K and 1000 K.

Only one result for the nozzle-4 is shown in Fig. 12. G_{16} of P(13) branch versus x is shown together with the parameter P_1, fixing M_1=5. A large value of G_{16} is obtained to be about 6 m^{-1} under the condition of M_1=5 and P_1=60 torr.

CONCLUSIONS

Numerical estimations of 16 μm CO_2 gasdynamic lasers are performed and the effects of nozzle contours are discussed in detail. The nozzle contour effects to the 16 μm band is much more than the one to the 10 μm band, and the lower translational temperature is

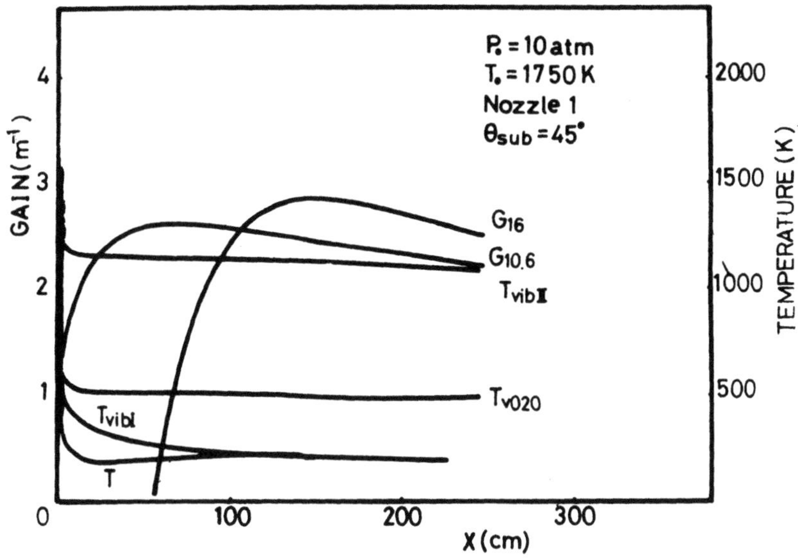

Fig. 3 Typical profiles of small signal gain coefficients and vibrational temperatures along the nozzle axis (nozzle 1).

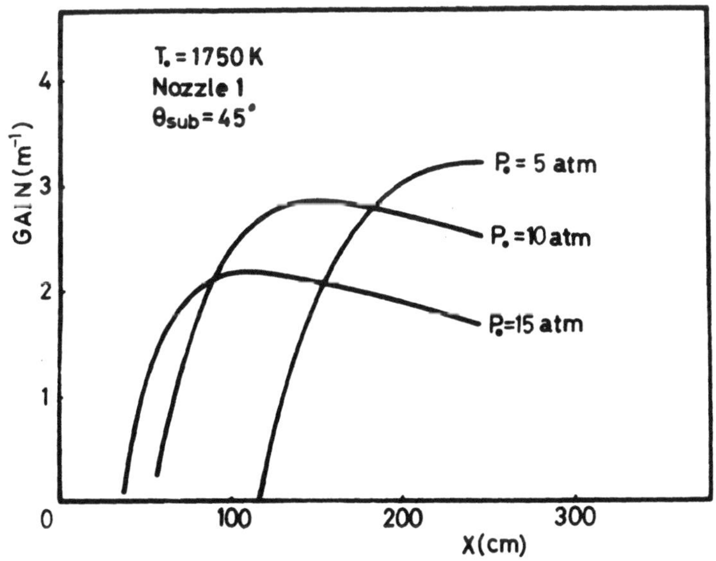

Fig. 4 G_{16} profiles for various P_0 values.

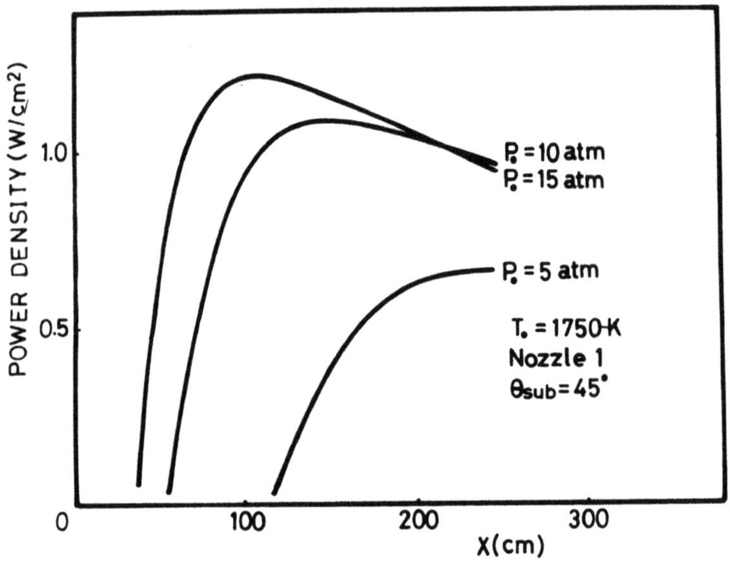

Fig. 5 P_{16} profiles for various P_0 values.

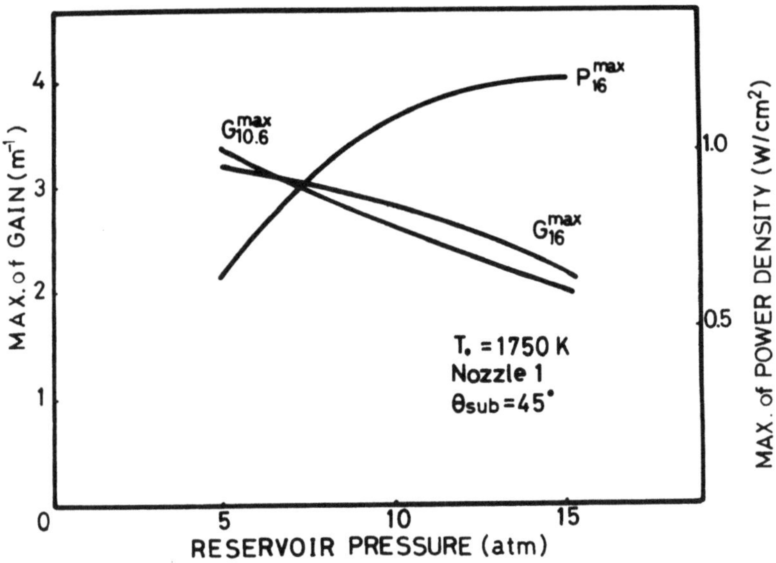

Fig. 6 Effect of P_0 value on maximum small signal gain coefficients and maximum available power density with nozzle 1, T=1750K.

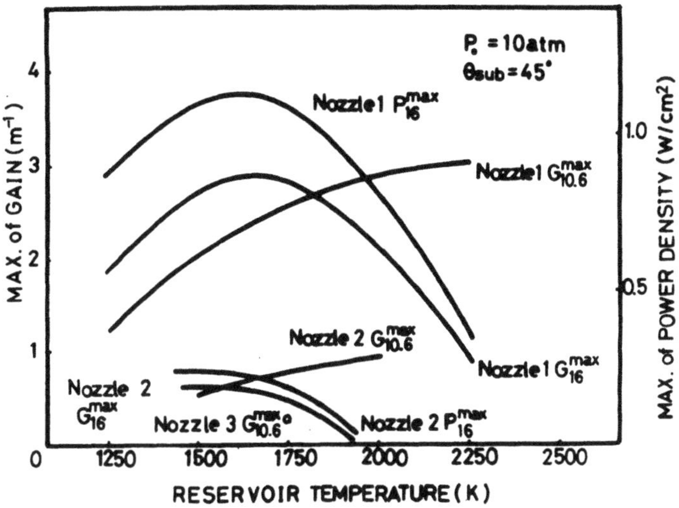

Fig. 7 Variations of maximum small signal gain coefficients with reservoir temperature of the three nozzles (nozzles 1,2,3).

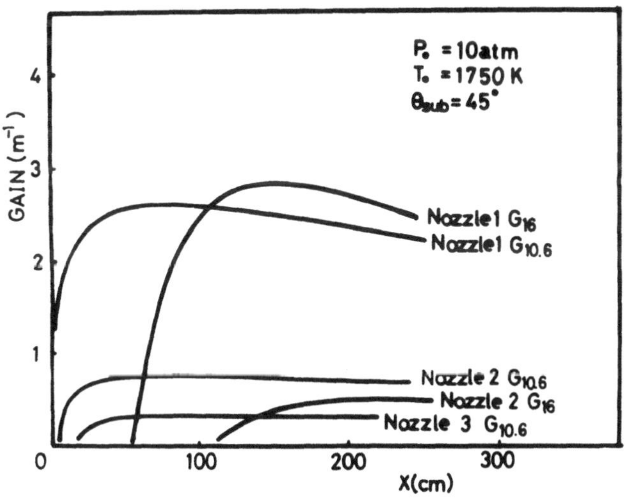

Fig. 8 Various small signal gain coefficients along the nozzle axis; reservoir and nozzle conditions are fixed.

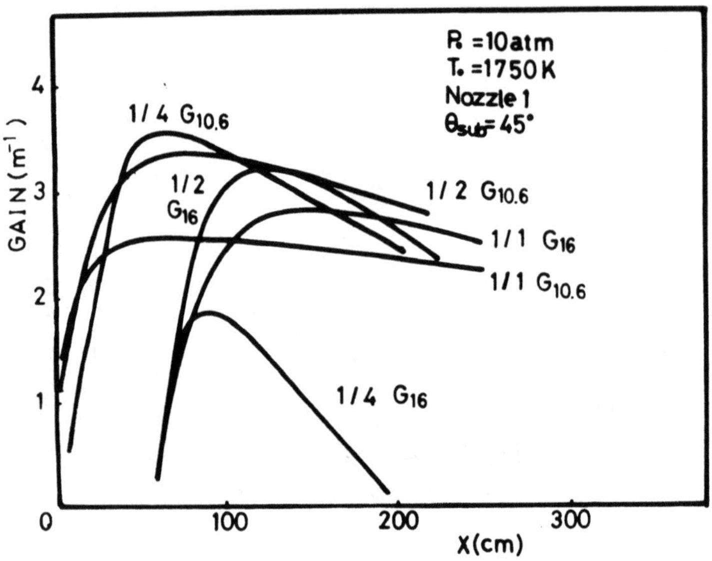

Fig. 9 Effect of nozzle size on the small signal gain coefficient profile.

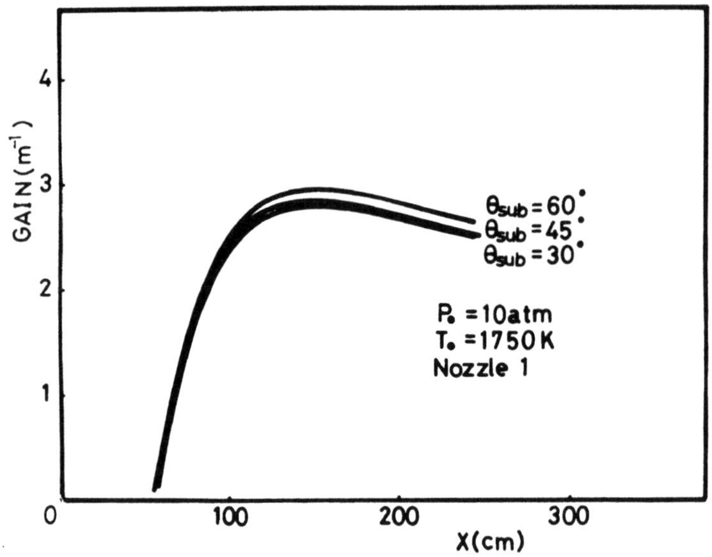

Fig.10 Effect of subsonic entrance angle on the small signal gain coefficient profile G_{16}.

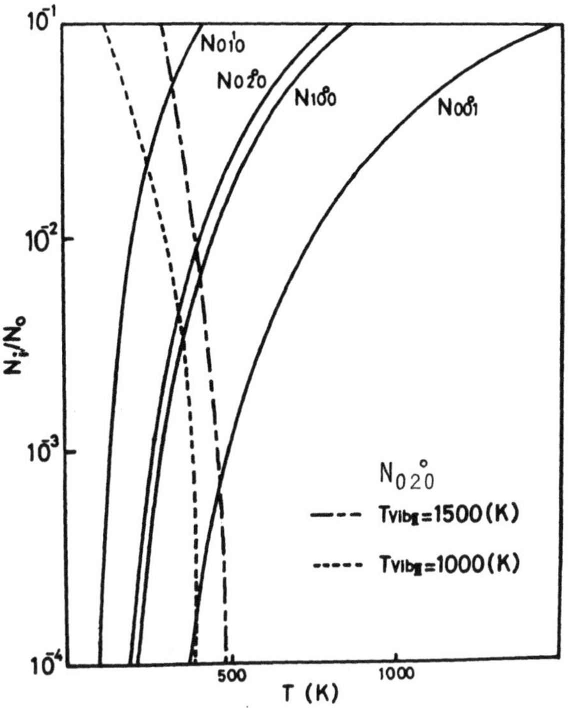

Fig.11 Population distributions as a function of the gas temperature and effect of the vibrational temperature on the $N^{1}_{02^{0}0}$ values.

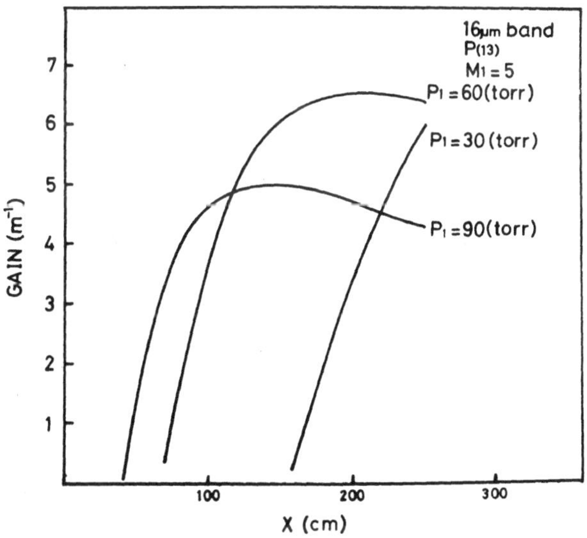

Fig Fig.12 Small signal gain coefficient profiles for various P_1 values (nozzle 4).

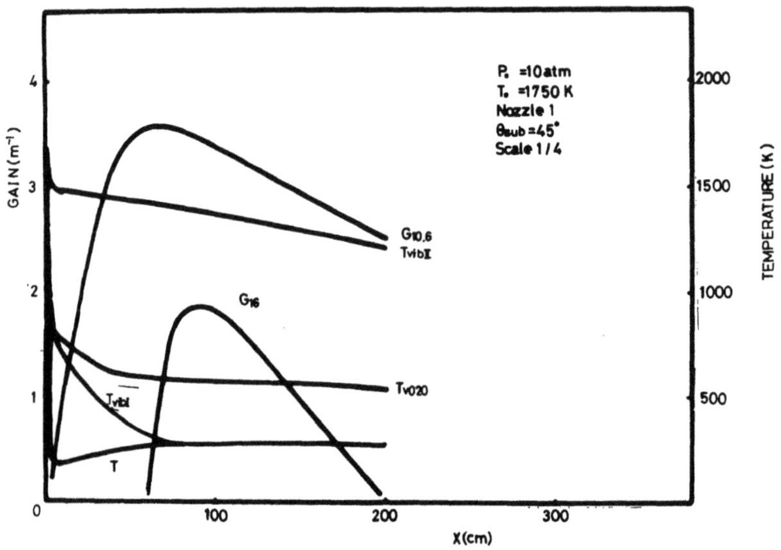

Fig.13 Profiles of small signal gain coefficients and temperature along the nozzle axis of 1/4 length nozzle.

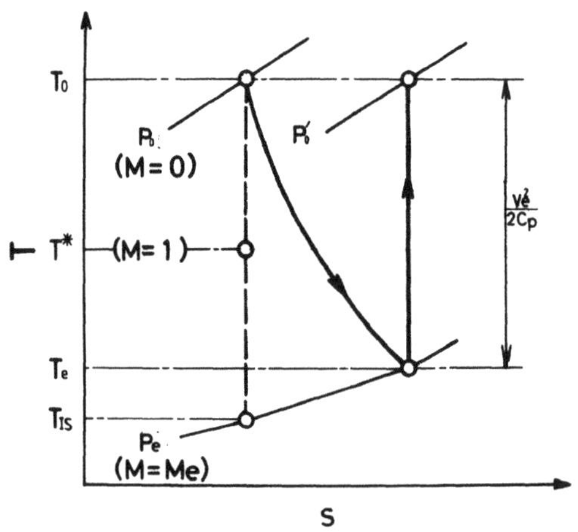

Fig.14 Flow processes on T-s (temperature- entropy) diagram, (a) isentropic expansion (broken line), (b) irreversible expansion (thick solid line).

necessary for the higher 16 μm laser performances. The exit mach number of the nozzle in this article is fixed at 6, but a higher mach number is expected to get higher laser efficiency[5].

Examples of our results shown here are as follows. (1) Nozzle-1 (length 133 mm), P_0=10 atm, T_0=1750 K, θ_{sub}=45°, G_{16}^{max}=3.2 m^{-1}, P_{16}^{max}=1.3 w/cm^2, u=2400 m/s. (2) Nozzle-4 (non-stagnant), M_1=5, P_1= 60 torr, G_{16}^{max}(P(13))=6 m^{-1}.

ACKNOWLEDGEMENT

The authors thank Mr. H. Kanazawa for his assistances. The work was supported by the Ministry of Education and Culture (the grant-in-Aid for Research Work etc.). Numerical calculations were performed with large computers in Tokyo Institute of technology and Osaka University.

REFERENCES

1. K. Suzuki, S. Saito, M. Obara and T. Fujioka, Theoretical study for a 16-μm CO_2 gasdynamic laser, J. Appl. Phys. 51:4003 (1980), S. Saito, M. Obara and T. Fujioka, Computer simulation of cw 16-μm CO_2 gasdynamic laser, Appl. Opt. 20:2838 (1981).
2. J. D. Anderson, Jr., "Gasdynamic Laser, An Introduction," Academic Press, New York (1976).
3. H. Shames and F. L. Seashore, Design data for graphical construction of two-dimensional sharp-edge-throat supersonic nozzles, NACA RM No.-E8J12 (1948).
4. K. Kasuya, Y. Minami, K. Horioka and K. Niu, Small signal gain coefficient and power output calculations of high temperature CO_2 gasdynamic laser without flow stagnation , J. de Physique 41:C9-175 (1980).
5. D. Kuehn, Importance of nozzle geometry to high-pressure gasdynamic lasers, Appl. Phys. Lett. 21:112 (1972).

SYMBOLS

P,p	:pressure
P_{16}	:available laser output power density (16 μm band)
ρ	:density
u	:flow speed
A	:cross section of flow
T	:temperature (without suffix-translational temperature)
t	:time
E	:internal energy
e	:sum of translational and rotational energy
e_{vib}, e_v	:vibrational energy
x	:distance along flow axis
R	:gas constant
X_j	:mole-fraction of j-specie
C_V	:specific heat under constant volume
I, II	:vibrational mode
N	:population density
N_{02^00}, N_{00^01}	:population density before 9.4 μm laser input
N'_{02^00}	:population density after 9.4 μm laser input
G	:small signal gain coefficient
λ	:wave length
ν	:oscillation frequency
ν_c	:collision frequency
τ_s	:life time for spontaneous emission
B	:rotational constant
J	:rotational quantum number
h	:Planck's constant
k	:Boltzman's constant
c	:speed of light
γ	:ratio of specific heat
θ	:declining angle of subsonic part (nozzle)
M_1	:Mach number of incident shock wave

SUFFIXES OR SUPERSCRIPTS

eq	:equilibrium condition
u	:upper laser level
l	:lower laser level
*	:throat condition
sub	:subsonic part
16	:16 μm band
10.6	:10.6 μm band
0	:stagnation condition
1	:condition ahead of incident shock wave
max	:maximum value
i	:mode I or II

APPENDIX

A smaller size (1/4) nozzle of nozzle-1 is considered under the same conditions of Fig. 3. A result of the same kinds of calculations is shown in Fig. 13. Although $T_{vib,II}$ is maintained higher in the nozzle than before, G_{16}^{max} remains smaller than before. This is because the energy dissipation results in increase of T.

The entropy increase ds in the flow is described as follows.

$$ds = \sum_i \left(\frac{1}{T_{vib,i}} - \frac{1}{T} \right) de_{vib,i}. \qquad (15)$$

The entropy increase is zero only under the perfect equilibrium ($T_{vib,i}$=T) or the perfect frozen ($de_{vib,i}$=0) processes. The larger the difference between $T_{vib,i}$ and T, the larger the entropy increase is. The large ds brings the increase of T, which results in the decrease of G_{16} in Fig. 13.

The temperature-entropy diagram of the vibrational relaxation processes is shown in Fig. 14. Along the isentropic line (broken) the stagnation conditions P_0, T_0 expands into the nozzle exit conditions P_e and T_{IS}, while P_0, T_0 into P_e, T_e along the entropy increase process (thick solid line). In the latter case, the corresponding stagnation pressure P_0', after the expansion is smaller than P_0. In Fig. 14, V_e and C_p are the exit flow speed and the specific heat under constant pressure. The above kind of temperature increase can not be neglected to estimate the 16 μm-GDL performances.

ON SIMPLE ESTIMATION OF CHARACTERISTICS

FOR CO_2/N_2 MIXING GASDYNAMIC LASER

K. Maeno

Department of Mechanical Engineering
Muroran Institute of Technology
27-1 Mizumoto Muroran, Hokkaido, Japan

ABSTRACT

A simple numerical estimation is performed for CO_2/N_2 mixing gasdynamic laser with screen nozzle. Steady quasi-onedimensional flow, instantaneous mixing, and three-mode model are assumed to estimate the characteristics of laser medium. Also the unsteady onedimensional analysis is conducted to clarify the laser intensity variation in cavity. The application of this estimation to the experiments by Hugel et al. shows qualitatively reasonable agreement. Some degree of under-estimation, however, is obtained quantitatively, which is to be the limitation of instantaneous mixing analysis. Optimal operation range for the screen nozzle can be obtained by this analysis.

INTRODUCTION

The efficient energy transfer in CO_2/N_2 gasdynamic mixing laser has an advantageous possibility getting over the limitation of conventional premixed CO_2 gasdynamic lasers. Of particular, screen nozzle has indicated most favorable mixing efficiency and well-mixed flow conditions of effectively excited N_2 and laser medium CO_2.[1] As regards the gasdynamic mixing process of vibrationally nonequilibrium flows (laminar or turbulent), elaborated numerical researches have been reported[2,3,4] for several CO_2 mixing techniques. In practical use the estimation based on instantaneous mixing assumption[5] seems to be appropriate for the application to screen nozzle, because of its analytical simplicity and admitting of the discontinuity on cross sectional area and flow parameters, peculiar to the conical screen exit.

The purpose of this paper is to estimate the performance of CO_2/N_2 mixing gasdynamic laser of screen nozzle type by instantaneous mixing assumption.[6,7] Steady quasi-onedimensional inviscid equations along with the relaxation equations by three-mode model in CO_2-N_2 system are used for the calculation of flow characteristics, and for the variation of laser intensity, improved unsteady equations with temperature expression involving additional laser extraction terms are employed to the numerical calculation with explicit MacCormack method.[2] Comparison of the estimated results with the experimental data obtained at DFVLR by screen nozzle[8] is mainly discussed, and applicability of the analysis is investigated.

FORMULATION

The equations for the unsteady, quasi-onedimensional, inviscid and vibrational nonequilibrium flow involving laser radiation terms can be written as follows:

$$\frac{\partial \rho}{\partial t} = -\rho\frac{\partial u}{\partial x} - u\frac{\partial \rho}{\partial x} - \rho u \frac{\partial \ln A}{\partial x}, \quad (1) \qquad \frac{\partial u}{\partial t} = -\left[\frac{1}{\rho}\frac{\partial(\rho R T)}{\partial x} + u\frac{\partial u}{\partial x} \right], \quad (2)$$

$$\frac{\partial T}{\partial t} = -u\frac{\partial T}{\partial x} + \frac{1}{C_v}\left[-\frac{P}{\rho}\left(\frac{\partial u}{\partial x} + u\frac{\partial \ln A}{\partial x} \right) + C_c R_c \left[-\theta_2 \Delta T_2^v + (\theta_3 - \theta_1)\Delta T_{31}^v - (\theta_3 - \theta_N)\Delta T_{N3}^v \right] \right], \quad (3)$$

$$\frac{\partial}{\partial t}\left(\frac{T_{12}}{\theta_2}\right) = -u\frac{\partial}{\partial x}\left(\frac{T_{12}}{\theta_2}\right) + \left[\frac{\exp(\frac{\theta_1}{T_{12}})}{(\frac{T_{12}}{\theta_1})^2(\exp(\frac{\theta_1}{T_{12}}) - 1)^2} + \frac{2\exp(\frac{\theta_2}{T_{12}})}{(\frac{T_{12}}{\theta_2})^2(\exp(\frac{\theta_2}{T_{12}}) - 1)^2} \right]^{-1}$$

$$\times \left[\Delta T_2^v + \frac{\theta_1}{\theta_2}\Delta T_{31}^v + \frac{\theta_1}{\theta_3 - \theta_1}\frac{GI}{C_c \rho R_c \theta_2} \right], \quad (4)$$

$$\frac{\partial}{\partial t}\left(\frac{T_3}{\theta_3}\right) = -u\frac{\partial}{\partial x}\left(\frac{T_3}{\theta_3}\right) + \frac{(\frac{T_3}{\theta_3})^2(\exp(\frac{\theta_3}{T_3}) - 1)^2}{\exp(\frac{\theta_3}{T_3})}\left[-\Delta T_{31}^v + \Delta T_{N3}^v - \frac{\theta_3}{\theta_3 - \theta_1}\frac{GI}{C_c \rho R_c \theta_3} \right], \quad (5)$$

$$\frac{\partial}{\partial t}\left(\frac{T_N}{\theta_N}\right) = -u\frac{\partial}{\partial x}\left(\frac{T_N}{\theta_N}\right) - \frac{(\frac{T_N}{\theta_3})^2(\exp(\frac{\theta_N}{T_N}) - 1)^2}{\exp(\frac{\theta_N}{T_N})}\frac{\Psi_c}{\Psi_N}\Delta T_{N3}^v, \quad (6)$$

where

$$\Delta T_2^v = \frac{2}{\tau}\left(\frac{1}{\exp(\frac{\theta_2}{T}) - 1} - \frac{1}{\exp(\frac{\theta_2}{T_{12}}) - 1} \right),$$

$$\Delta T^v_{31} = \frac{2}{\tau_{3,12}} \frac{\exp(\frac{\theta_1}{T_{12}}) - \exp(\frac{\theta_3}{T_3})\exp(\frac{\theta_1-\theta_3}{T})}{(\exp(\frac{\theta_1}{T_{12}}) - 1)(\exp(\frac{\theta_3}{T_3}) - 1)} ,$$

$$\Delta T^v_{N3} = \frac{\Psi_N}{\tau_{3,N}} \frac{\exp(\frac{\theta_3}{T_3})\exp(\frac{\theta_N-\theta_3}{T}) - \exp(\frac{\theta_N}{T_N})}{(\exp(\frac{\theta_N}{T_N}) - 1)(\exp(\frac{\theta_3}{T_3}) - 1)} .$$

$$\frac{dI}{dt} = C(G - G_c)I , \tag{7}$$

$$G = \frac{\lambda^2}{4\pi\tau_s\nu_c}(N_{001} - N_{100})(\frac{45.6}{T})\exp(- \frac{234}{T}) . \tag{8}$$

T_i is vibrational temperature, C_i and R_i are mass fraction and gas constant of i-th gas, Ψ_i is molar fraction, I is the laser radiation intensity in cavity, and $\tau_{i,j}$ is relaxation time of energy transfer between i-th and j-th mode. Equations (4),(5), and (6) are vibrational relaxation equations in explicit temperature expression, which are improved from the relaxation equations[7] by making use of the fact that each vibrational energy depends on single vibrational temperature. These equations are somewhat different from those by Lee[9] in respect of intramolecular V-V energy transfer. The relaxation equations by Lee[9] or Anderson[2] can, of course, be rearranged in direct temperature expression,[10] which have the advantage for time-dependent analysis. Vibrational energy expression can be also obtained. Equation (3) has the temperature form too, which is deduced from energy equation with laser intensity term.[9] The variation of laser intensity in the cavity is specified by Eq.(7), and small signal gain coefficient G is calculated by Eq.(8).[11]

NUMERICAL CALCULATION

To obtain the characteristics of flow without lasing, Eqs.(1)-(6) are modified in the steady form with no lasing term. For the nozzle flow of CO_2(+He) or N_2 dealt with separately, the set of governing equations can be integrated from throat to the exit by Runge-Kutta method.[6] Assuming the instantaneous mixing at nozzle exit, the flow properties after mixing are determined by algebraic conservation equations of total mass, momentum, and energy, with three vibrational temperatures unchanged across the mixing position.[7] Using these properties, the cavity flow can be further calculated.

In order to compare the estimation with the experiment by Hügel et al.,stagnation temperature and mass flow rate are nearly matched to the data. To avoid the great discrepancy in pressures and velocities at both nozzle exits, a crude assumption is also made that CO_2 flow expands rapidly in the assumed cones with the length of inner diameter of CO_2-bore,which is negligible to change the mixing position as compared to the cavity length. Thus CO_2(+He) is expanded in these cones to fill up the cross sectional discontinuity to the neighbouring N_2 nozzle wall before the mixing process occurs. This procedure is performed mainly to approximate to measured pressure distribution in laser channel.

For the estimation of laser intensity, Eqs.(1)-(8) are calculated by explicit MacCormack method.[2] Onedimensional cavity meshes are assumed at t=0, where the transverse laser intensity is amplified from monochromatic spontaneous radiation with T. In order to avoid the singularity at the leading edge of cavity, the threshold G_c is assumed to have the following loss distribution,

$$G_c = G_c^* + (G_0 - G_c^*)\left(1-\sin\frac{\pi}{L}\left(\frac{L}{2}-x\right)\right), \quad (9) \qquad G_c^* = \frac{1}{L_c}\left(\frac{1-r_1}{1+r_1} + \frac{1-r_2}{1+r_2}\right), \quad (10)$$

where G_0 is the gain at leading edge, x is the distance from cavity center, L is the cavity width (L/2 = R), and L_c is cavity length.

RESULTS AND DISCUSSION

Estimated results for the experiments of DFVLR[8] are presented to investigate the performance of screen nozzle. The results for the experiment by nondiaphragm shock tubes were already reported.[6,7] As a compared result, are represented in Fig.1 the distributions of calculated and measured small signal gain G downstream from nozzle exit for pure CO_2 injection, together with calculated vibrational temperature distributions. The estimated and measured pressures are also plotted. Stagnation conditions T^s and p^s, mass flux, and estimated flow parameters at each exit (superscript e) are supplemented. Subscripts denote the component gas (or mixture). From this figure it is shown that the estimation shows qualitatively reasonable agreement to measured data. Quantitatively, however, it underestimates gain distribution with faster decay, and the position of peak gain becomes forward on account of instantaneous mixing.

Figure 2 shows a result for CO_2/He mixture. The effect of He can be observed in higher peak value and slower decline of gain than in Fig.1, mainly because that He increases CO_2-side exit velocity and consequently lowers the temperature T in mixed flow, together with catalytic effect as usual. The degree of quantitative agreement is improved in Fig.2. The same trend as Figs.1 and 2 was obtained for other flow conditions. Owing to the instantaneous mixing, however, calculated data generally have greater quantitative discrepancies in the case of high stagnation pressure (high mass flux) of CO_2(+He) or of high N_2 stagnation temperature,i.e. the case when real fluid dy-

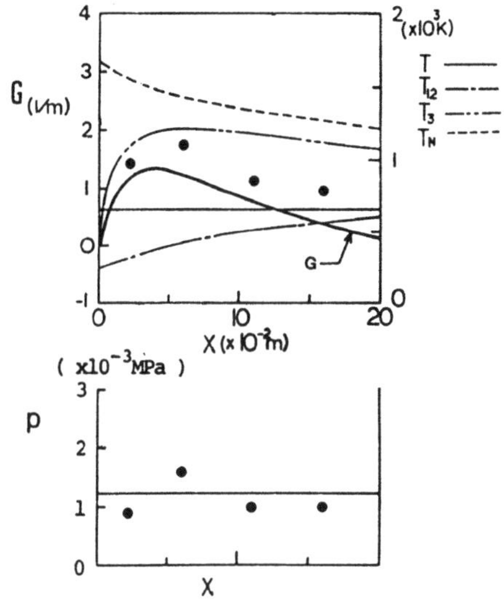

Fig.1 Distributions of gain, temperatures, and pressure. (Pure CO_2)
T_N^s=2000K, p_N^s=0.248MPa, $\dot{m}_N$=12.6g/s.
u^e=1900m/s, T^e=330K, p^e=3.90x10^{-4}MPa.
T_C^s=300K, p_C^s=0.086MPa, $\dot{m}_C$=9.47g/s,
u^e=507m/s, T^e=118K, p^e=2.65x10^{-3}MPa.

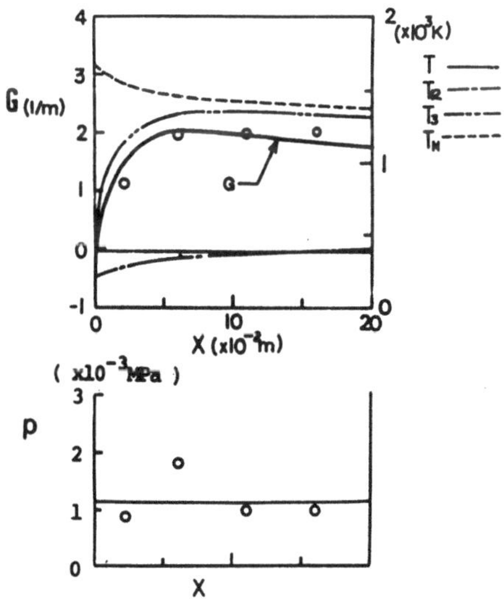

Fig.2 Distributions of gain, temperatures, and pressure. (CO_2+ He)

N_2-side: same conditions as Fig.1

T_{CH}^s=300K, p_{CH}^s=0.161MPa, $\dot{m}_{CH}$=11.3g/s,
u^e=803m/s, T^e=80K, p^e=3.65x10^{-3}MPa,
31CO_2/69He.

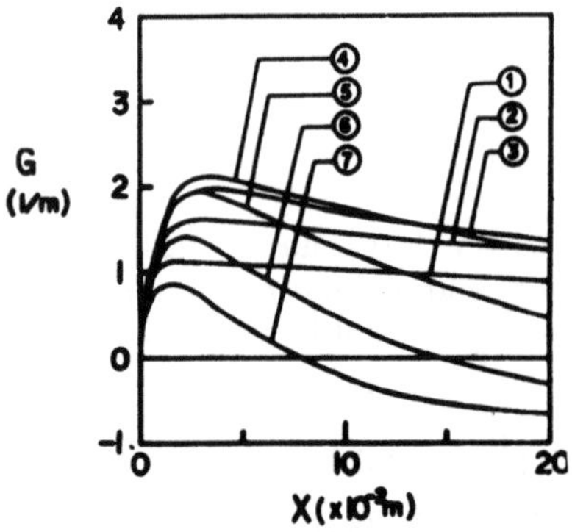

Fig.3 Dependence of gain distributions on N_2 stagnation conditions. $\dot{m}_N$=12.6g/s.

T^s_{CH}= 300K

p^s_{CH}= 0.253MPa

$\dot{m}_{CH}$= 20.78g/s

47.4CO_2/52.6He

No.	T^s_N (K)	p^s_N (MPa)
1	1200	0.193
2	1400	0.208
3	1600	0.222
4	1800	0.235
5	2200	0.260
6	2600	0.283
7	3000	0.304

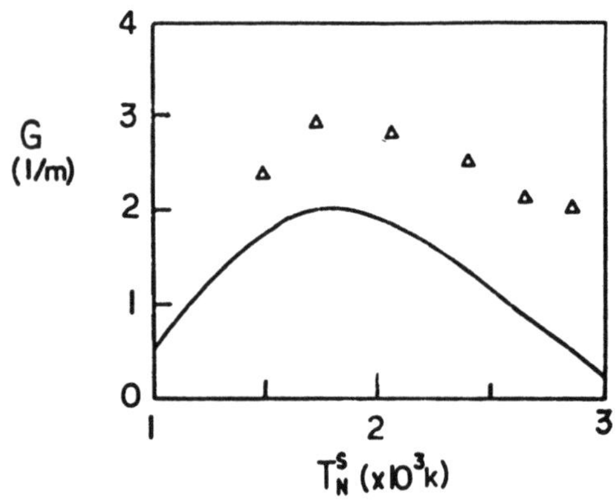

Fig.4 Dependence of gain on N_2 stagnation conditions at 6cm downstream from nozzle exit.

T^s_{CH}=300K, p^s_{CH}=0.253MPa, $\dot{m}_{CH}$=20.8g/s, 47.4CO_2/52.6He.

namical process has severer influence on laser. With the exception of these points, reasonable agreements can be qualitatively obtained by our simple estimation.

In Fig.3 is shown the calculated dependence of G on N_2 stagnation conditions for fixed CO_2/He conditions and approximately fixed N_2 mass flux. Also the dependence of G on N_2 stagnation conditions at a fixed position is represented in Fig.4 with experimental data. From these figures it is noticed that the optimum stagnation range exists if the geometry of screen nozzle is prescribed, and that the optimal range of N_2 stagnation temperature for gain G is restricted to relatively low values near 2000K.

Time variations of laser intensity I and gain G in cavity are shown in Fig.5 at fixed position. The first peak of I corresponds to giant pulse, and the minimum is bound by spontaneous radiation. The oscillation frequency is about 500kHz, and I tends toward a steady state in 30μs. Figure 6 indicates the distributions of I in cavity. The shape similar to a giant pulse is denoted in Fig.6-(a), and the quasi-steady distribution is shown in Fig.6-(b), which is shifted upstream of cavity owing to supersonic inflow. Total output P_t approximating experimental conditions is plotted in Fig.7. This case corresponds to Fig.4, and the discrepancy is still great according to the difference in gain distribution. It is noted that T_N^s giving the maximum P_t becomes 2200K, above the value for maximum gain. So the higher power extraction is realized in the range higher than the gain peak temperature. Experimental result shows much higher values.

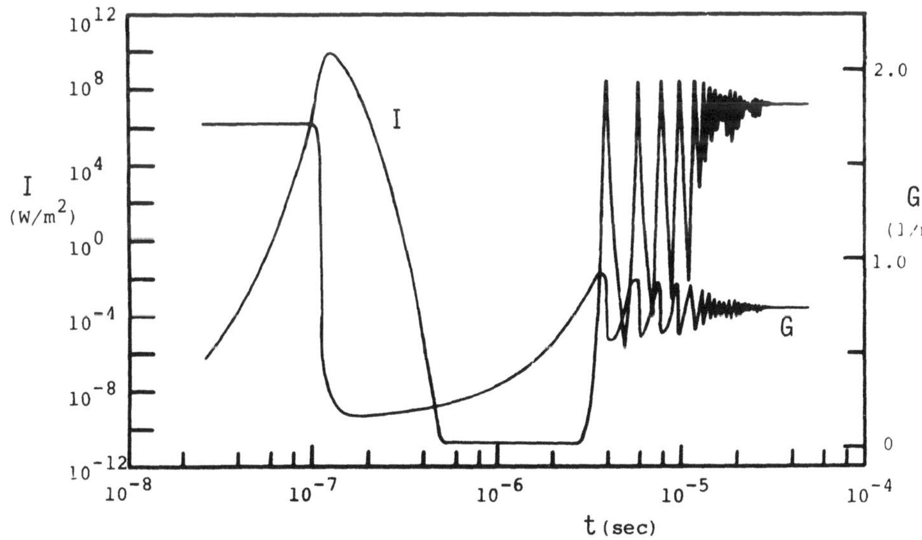

Fig.5 Time variations of laser intensity and gain in cavity. (X = 4mm from the center of cavity) T_N^s=2200K, p_N^s=0.26MPa, $\dot{m}_N$=12.6g/s, T_{CH}^s=300K, p_{CH}^s=0.253MPa, $\dot{m}_{CH}$=20.8g/s.

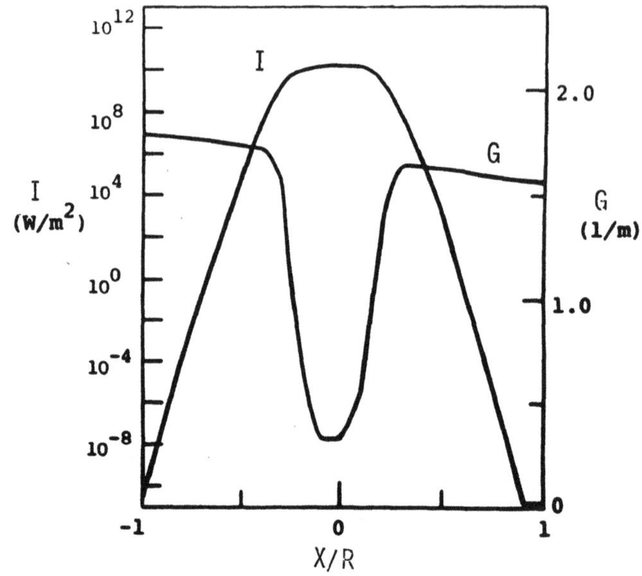

(a) t = 0.13μs (giant pulse shape)

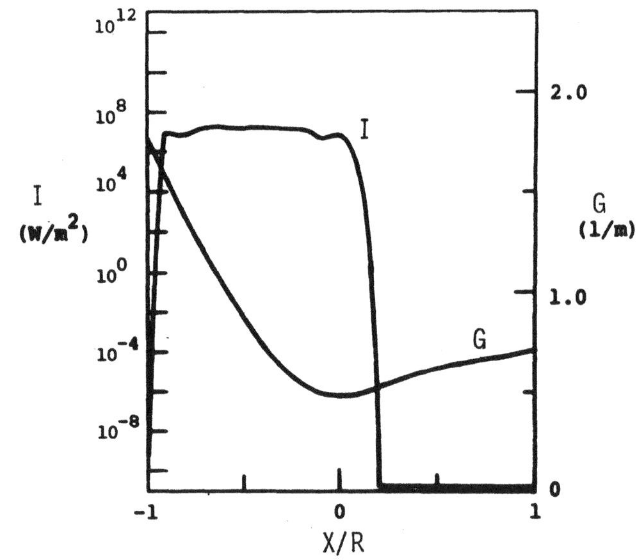

(b) t = 46μs (quasi-steady state)

Fig.6 Distributions of laser intensity and gain in cavity. T_N^s=2200K, p_N^s=0.26MPa, $\dot{m}_N$=12.6g/s, T_{CH}^s=300K, p_{CH}^s=0.253MPa, $\dot{m}_{CH}$=20.8g/s.

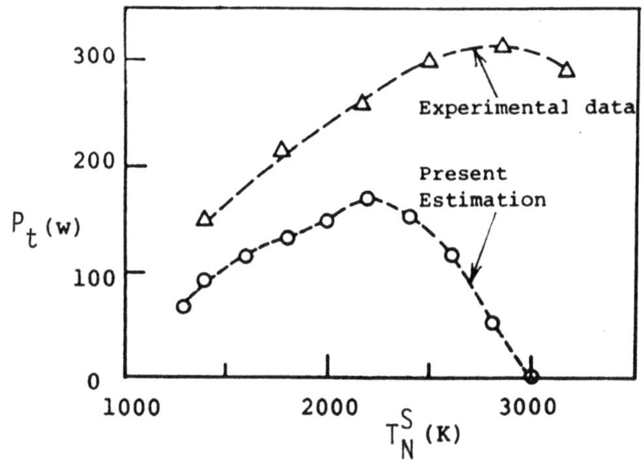

Fig.7 Dependence of total laser output on N_2 stagnation. 6cm downstream, T_{CH}^s=300K, p_{CH}^s=0.253MPa, $\dot{m}_{CH}$=20.8g/s, 47.4CO_2/52.6He.

In conclusion, the applicability and limitation of instantaneous mixing analysis to CO_2/N_2 mixing gasdynamic laser by screen nozzle are clarified. Qualitatively reasonable agreement with experiments is obtained, and among the results the existence of optimal range for prescribed nozzle configuration is made clear, which can be estimated simply by this analysis. Laser intensity can also be investigated.

ACKNOWLEDGEMENT

The author is grateful to express his sincere thanks to Dr.H. Hügel of DFVLR for his kind discussions about their experiments. Also wishes to express his thanks to Prof. H.Oguchi of ISAS for his cordial suggestions, and to Mr.N.Yamaguchi, Muroran Inst. of Tech., for his useful discussions. Numerical calculation was performed by HITAC M-200H of Hokkaido Univ. Computer Center.

REFERENCES

1. P. E. Cassady, in:"Gas-Flow and Chemical Lasers; Pro. 2nd Int. Symp.on Gas-Flow and Chemical Lasers," Hemisphere Publishing Corp., Belgium (1979) p.95.
2. J. D. Anderson,Jr., in:"Gas-Flow and Chemical Lasers," Belgium (1979) p.3.
3. J. P. E. Taran, M. Charpenel, and R. Borghi, AIAA Paper,No.73-622 (1973).
4. N. A. Fomin, R.I. Soloukhin, V.I. Golovichev,and S. A. Munjee, in:"Combustion in Reactive Systems,"Progress in Astron. and Aeron.(AIAA),Vol.76 (1981) p.46.
5. R. I. Soloukhin, in:"Shock Tube and Shock Wave Research; Proc. 11th Int.Symp. on Shock Tubes and Waves,"Univ.of Washington Press, Seattle (1978) p.629.
6. K. Maeno and H. Oguchi, J. de Physique(C9, Suppl. No.11; Proc. 3rd Int.Symp.on Gas-Flow and Chemical Lasers),Vol.41 (1980) p.C9-209.
7. K. Maeno, K.Funabiki,and H. Oguchi, ISAS Rept.(No.593),Vol.46, No.5 (1981) p.175.
8. H. Hügel, W. Schall, and P. Hoffmann, J.de Physique (C9, Suppl. No.11) p.C9-335.
9. G. Lee, The Physics of Fluids, Vol.17, No.3 (1974) p.644.
10. K. Maeno, (to be published by Muroran Inst. of Techn.),Memoirs of Muroran Inst. of Techn.;Science and Engineering, Vol.10, No.4 (1982).
11. J. D. Anderson,Jr., "Gasdynamic Lasers: An Introduction," Academic Press, New York (1976).

AERODYNAMIC DISTURBANCES FROM SUPERSONIC NOZZLE ARRAYS*

David A. Russell and Yung K. Chu

Aerospace and Energetics Research Program
University of Washington
Seattle, WA 98195

INTRODUCTION

Many high-power lasers require expansion of gases through an array of supersonic nozzles. Early studies[1,2] of such arrays for wind-tunnel application established that they produce a decaying system of waves and wakes as depicted on Fig. 1. The waves are caused by truncated walls, use of simple nozzle contours, or the effects of base regions; while the wakes are formed from the nozzle boundary layers and base losses. Motivated by their re-appearance in gasdynamic lasers, an analysis for 2-D waves was

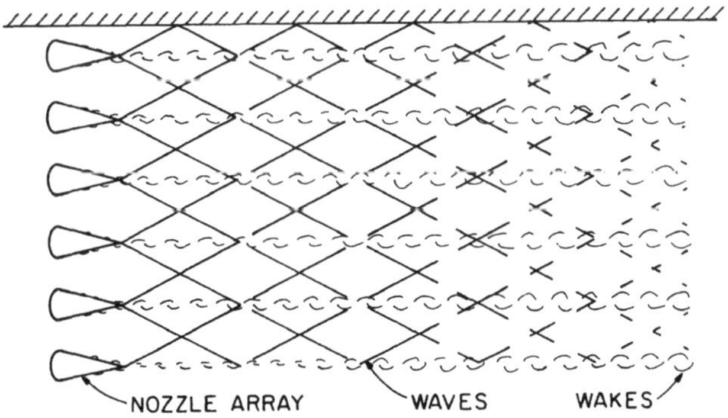

Fig. 1. Nonuniformities generated by nozzle array.

*Research partially supported by U.S. Air Force of Scientific Research, Contract F49620-79-C-0020. The authors wish to acknowledge the contributions of C.H. Lee, now with Applied Technology, Inc., Sunnyvale, CA.

developed from elements of sonic boom theory,[3] and 2-D cavity wakes studied. The effect on kinetic performance was at first found to be minimal, and satisfactory optical quality could be maintained by aligning the optical axis perpendicular to the nozzle trailing edges. However, evolution to simple fine-scale wedge or conical nozzles,[4] high-performance mixing nozzles,[5] and the nozzles of the modern chemical laser have made it important to study the effects of shocks and wakes from real expansions further.

Modern lasers use a variety of nozzle/injector configurations, frequently with multiple scales and gases. The flow downstream of each individual expansion is quickly confined to a channel by the neighboring flows, and this typically requires turning shocks that initiate the wave system. This may be reinforced, often in highly coupled fashion, by the displacement effect of heat release in reaction layers growing out of the wakes. A first step to understanding the aerodynamics of these flows is to study the case of a single gas issuing from an array of identical nozzles. A 3-D close packed geometry is here chosen as representative; other packing and 2-D configurations being expected to be derivative. This geometry introduces hexagonal symmetry, the basic element of which is flow in a 30° right-triangular cross section. The following section presents analyses of the waves and wakes to be expected. This is followed by a description of experiments carried out to measure the disturbances from supersonic nozzle arrays in the near field, a section comparing these with model predictions, and a conclusion.

ANALYSIS

Shock Wave Model

The inviscid flow from an array of identical conical supersonic sources followed by a parallel channel is considered. The assumption of initial radial flow can be accurate above moderate exit Mach number, M_e, and method-of-characteristics calculations can be used to assign an equivalent source angle for low M_e or contoured nozzle designs. Figure 2 illustrates the wave system recently worked out for a close-packed array with the assumption that the flow from each nozzle expands instantly to fill all the space available to it.[6] The basic element then becomes source flow into a 30-degree right-triangle channel as shown. The shock sheets are the results of the velocity tangency condition on the top and side planes of symmetry, and the cavity flow is composed of these systems packed together.

To analyze the flow, the variables were expressed as perturbation series in the source flow half-angle, ε, and the first-order equations solved with careful application of the boundary conditions. This information was then used to systematically solve the second order equations, as the decay of the flow disturbances is

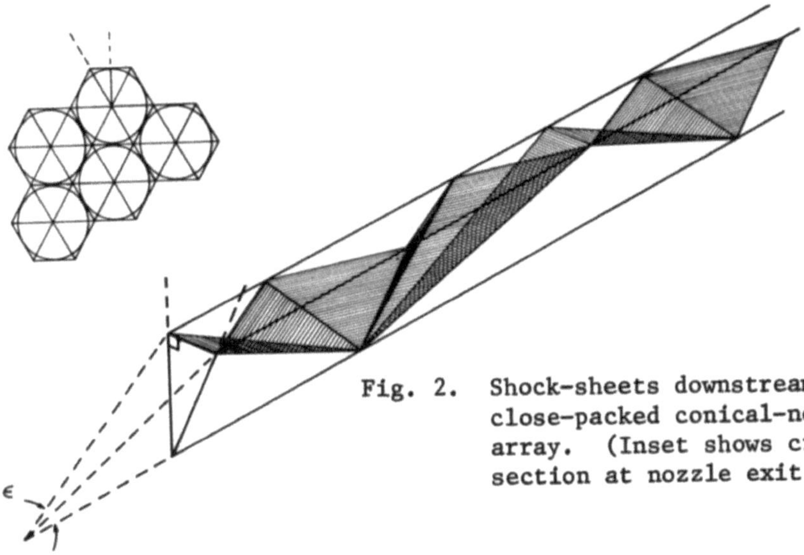

Fig. 2. Shock-sheets downstream of close-packed conical-nozzle array. (Inset shows cross-section at nozzle exit plane)

realized only at this order. A nontrivial subtlety of the analysis was the requirement that the independent variables also be expanded in perturbation series (strained coordinate technique) in order to assure validity of the solution in the far field.

The solution yields velocity, u, and density, ρ, given by

$$(u-u_e)/u_e = (M_e^2-1)^{-1/2}\,\varepsilon[1+\varepsilon\alpha_1(n-1)]^{-1}\,F(\chi)$$

$$(\rho-\rho_e)/\rho_e = -M_e^2\,(M_e^2-1)^{-1/2}\,\varepsilon[1+\varepsilon\alpha_1(n-1)]^{-1}\,F(\chi) \qquad (1)$$

$$\text{where } \chi = 2(M_e^2-1)^{-1/2}\,\tilde{x}/d_e$$

$$\text{and } \alpha_1 = (5\gamma M_e^4 - 2M_e^4 + M_e^2 + 6)\,(M_e^2-1)^{-3/2}/4$$

Here subscript e conditions are at the nozzle exit wall, with the exception of M_e which is at the exit center line. γ is the ratio of specific heats. n is the number of the cell bounded by the main shock; n=1 upstream of the first shock, and cells with velocities on the nozzle axis have even numbers. $\tilde{x}$ is the position within each cell, measured from its midpoint; χ is a dimensionless $\tilde{x}$ with value between -1 and +1. Actual position downstream = $(n-1)(M_e^2-1)^{1/2}(d_e/2)+\tilde{x}$ where d_e is the nozzle exit dimension.

Figure 3 shows slices through Fig. 2 at constant χ for n even. The lines within the triangles are the shocks which separate regions of uniform flow. As χ increases, the main shock between regions 1 and 2 advances towards the nozzle axis, and the shock that satisfies the slant-wall boundary condition moves out from the corner to subsequently reflect off the line of vertical symmetry. The

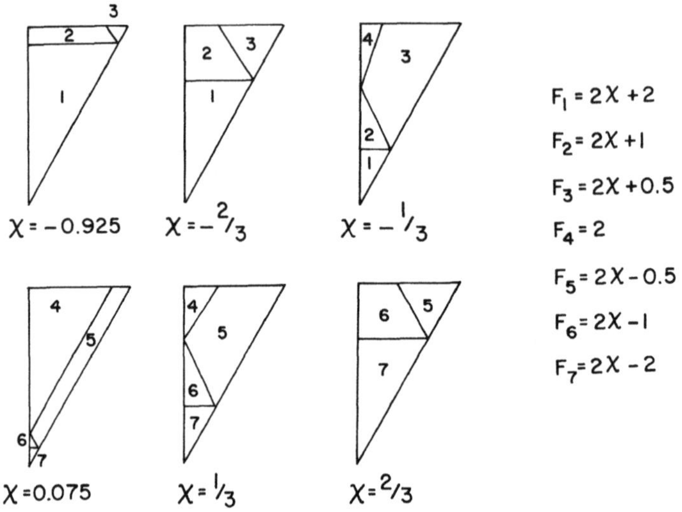

Fig. 3. Shock geometry at constant χ and values of $F(\chi)$ (n even)

value of $F(\chi)$ in the various regions is indicated to the side of the figure. The perturbed pattern repeats itself every time the main shock reflects back to the outer boundary, i.e., every 2χ. imposed on this is a general decay occurring as $(n-1)$ increases. Note that Eqs. (1) predict a direct dependence on ε. They also show that the decay is equivalent to a reduction in effective source angle for a cell to $\varepsilon[1+\varepsilon\alpha_1(n-1)]^{-1}$, with the origin shifting alternately between the nozzle and wake axis.

Wake Model

The wakes are treated as individually axisymmetric and centered at the vertices of the hexagons as sketched on Fig. 4. The wake defect can be related to the drag by

$$C_D(\rho_e u_e^2/2)A_b = \int_0^\infty 2\pi r\rho u(u_e - u)\,dr \qquad (2)$$

where $C_D(A_b/d_e^2) \approx (2/\gamma M_e^2)(1-P_b/P_e)(A_e/d_e^2) + \pi(\theta_e/d_e)(A_e/d_e^2)$.

The first term on the right side of the second equation is the non-isentropic form drag produced by an equivalent axisymmetric body, while the second is the friction drag associated with the boundary layer momentum loss over the circumferential distance of $\pi d_e/2$ assignable to each base. The base area, A_b, comes from the nozzle geometry; the normalized base pressure, P_b/P_e, must be given, and the boundary-layer momentum thickness, θ_e, calculated. Integral laminar boundary-layer calculations for supersonic N_2 expansions from M=1 with straight walls at a temperature of 300°K give

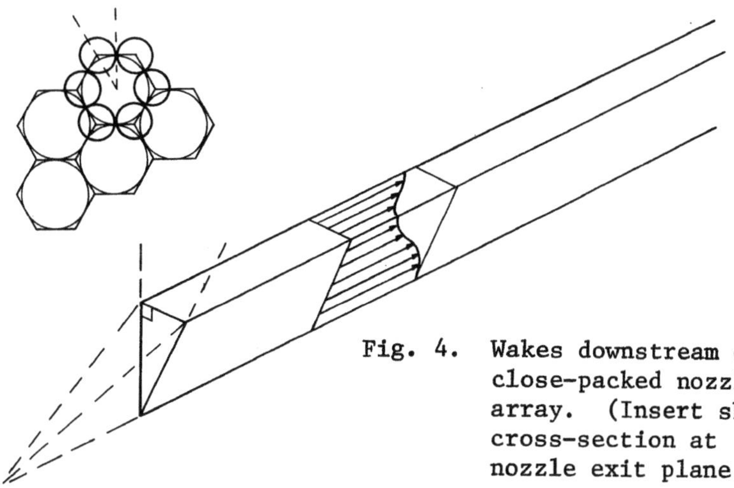

Fig. 4. Wakes downstream of close-packed nozzle array. (Insert shows cross-section at nozzle exit plane.)

$\theta_e/d_e \approx 4.5\,(10)^{-3}\,(P_o d^* \sin\varepsilon)^{-1/2}(T_o/300)$, where P_o and T_o are total values in atm and in °K respectively, and d* is the throat dimension in cm. This applies for both 2 and 3-D nozzles, and is independent of γ or area ratio as long as the latter $\gtrsim 10$. Extensions to other gases may be made by dividing by the square root of the stagnation kinematic viscosity to the stagnation speed of sound ratioed to the values for N2.

The wakes are assumed to be turbulent with a Gaussian velocity-defect profile

$$(u_e - u)/u_e = a(x)e^{-2(\tilde{r}/b)^2}$$

$$\text{where } \tilde{r} = 2\int_o^r (\rho/\rho_e)rdr \qquad (3)$$

$$\text{and } a = \pi^{-1/3}(R_T/6)^{2/3}\, C_D(A_b/d_e^2)(x/d_e)^{-2/3}$$

$$b/d_e \approx [2/\pi a)C_D(A_b/d_e^2)]^{1/2}$$

$\tilde{r}$ is a Dorodnitsyn variable absorbing compressibility effects. It becomes approximately equal to the radius, r, as the wake fills in. The expression for a was obtained from supersonic wake measurements for which the turbulent Reynolds number, R_T, was found to be 12.8.[7] The wake width parameter, b, is readily related to a by inserting the first two of Eqs. (3) into the first of Eqs. (2) and integrating.

Finally, if the temperature defect profile is also assumed to be Gaussian with width $(P_{r_t})^{1/2}$ times that of the velocity defect, conservation of momentum and enthalpy flux for an ideal gas axisymmetric wake leads to

$$(\rho_e-\rho)/\rho_e \approx P_{r_t}[2(1-T_e/T_o) - (\varepsilon_1/\theta)_e](T_e/T_o)^{-1}\,(u_e-u)/u_e \qquad (4)$$

where P_{r_t} is the turbulent Prandtl number and ε_1 the boundary-layer total-enthalpy-flux thickness. $\varepsilon_1/\theta \approx 0.2[(T_o/300)-1]$ with T_o again in °K.

EXPERIMENT

The experiments were carried out in the Ludwieg-tube wind-tunnel depicted on Fig. 5. The nozzle array was mounted just upstream of the 10×7.5 cm test section, and the flow viewed through flush-fitting plexiglass windows located in all four sides at the first station. The array used for most of the present studies was made by drilling an Al plate with conical holes having the cross-section shown on the lower right. The lower left is a schematic view from downstream, showing the close-packed nozzle array with minimum base area calculated = $0.040\ d_e^2$. The half nozzles were formed by machining slots through full nozzles and using inserts. The nozzle plate, test-section and drive tube could be rotated about its axis to change α, the angle between the horizontal and the bottom nozzle row.

A ruby-laser holographic interferometer was arranged with horizontal optical axis across the test section. The laser was Q-switched

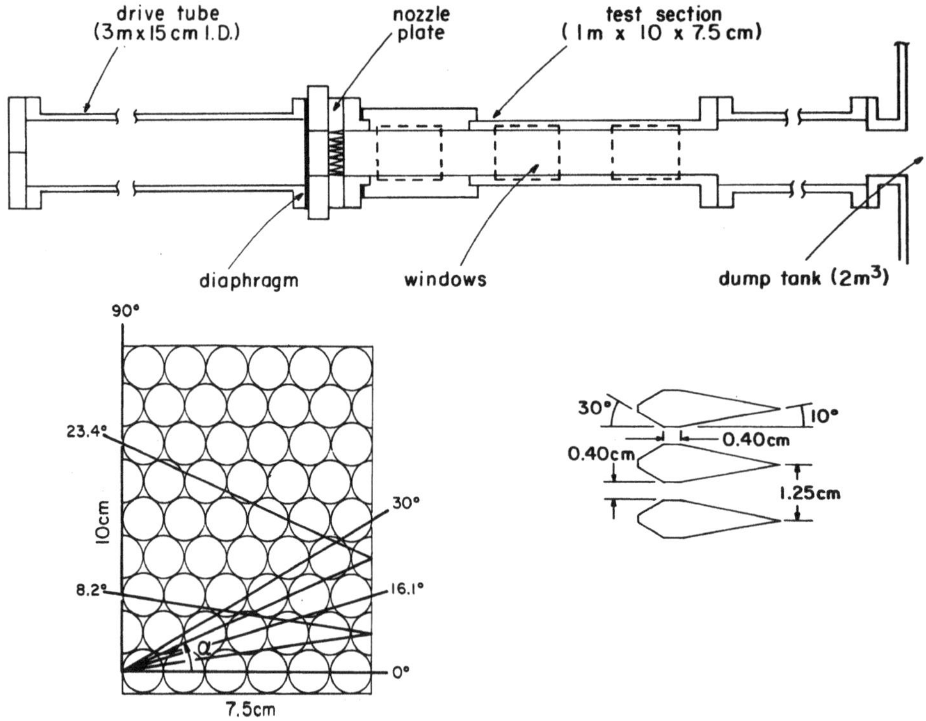

Fig. 5. Schematic of Ludwieg tube and nozzle plate geometry.

to effectively freeze the flow. Its beam was expanded and split to pass both over and through the test section, and it was subsequently recombined on a photographic plate. An exposure was made before each run with a He-filled wedge inserted into one optical path to set fringe spacing. The number of fringe shifts is given as $N = \beta(L/\lambda)(\rho_e/\rho_r)\overline{\Delta\rho}/\rho_e$ where $\overline{\Delta\rho} = L^{-1}\int_0^L(\rho_e-\rho)dl$. Here β is the Gladstone-Dale constant in the index of refraction λ the laser wavelength (694.3 nm), ρ_r the reference atmospheric density, dl the elemental distance along the optical path, and L the total distance. L was 7.5 cm for $\alpha=0°$ and 10 cm for $\alpha=90°$, unless it was doubled by operating the interferometer in a double-pass configuration. L for intermediate α was determined by the geometry of Fig. 5. Values of α were used that provided even-multiple repetitive patterns along ray paths through the nozzles such as shown on the figure. $\alpha=30°$ was replaced by 90° in order to obtain larger and more sensitive interferograms.

The Ludwieg tube was charged with N_2 at 5-20 atm, and prescribed soft Al diaphragms pressure broken to initiate the flow. A delayed accelerometer signal fired the laser $\approx 10(10)^{-3}$sec into the available $\approx 20(10)^{-3}$sec of steady flow. Fringe information for x/d_e between 5-15 (n=3-10) was obtained by a combination of location selection on the photographic plate and translation of the optical axis. Fourteen successful runs were made with the nozzle plate of Fig. 5, a few with smaller d_e plates.

RESULTS AND DISCUSSION

Figure 6 presents representative measured interferograms in the left column, and computer-generated interferograms using the new models in the remaining columns. The top row is for $\alpha=0°$, the middle for 16.1°, and the bottom for 90°. x/d_e and P_o are nominally the same for each α. Corresponding interferometric sensitivities were $\overline{\Delta\rho}/\rho_e$ = 4.6, 5.4 and 4.4% per fringe. All of the interferograms cover a distance in the flow direction of $(M_e^2-1)^{\frac{1}{2}}d_e$, corresponding to a shock cell. The $\alpha=0°$ and 90° experiments show the expected transverse periodicity on a scale of $|d_e\cos(\alpha+30°)|$, while the 16.1° case shows no large amplitudes or regularity in fringe structure. The latter is due to the averaging effect along the optical path, see Fig. 5. The effect of α on apparent optical quality has been noted before.[8]

The fringe shifts calculated by integrating ρ through a field of hexagonal cells using Eqs. (1) with $F(\chi)$ from Fig. 3 are shown in the second column for the conditions of the experiments. The agreement in amplitude and pattern with the measured interferogram for $\alpha=0$ is striking. The fringe direction reversal is basic to a diamond wave system. Equations (3) and (4) were used to calculate the wakes with C_D obtained from Eqs. (2) with $P_b/P_e = 0$. Pr_t was taken as 1.0. The predictions are shown in the third column. Note the agreement

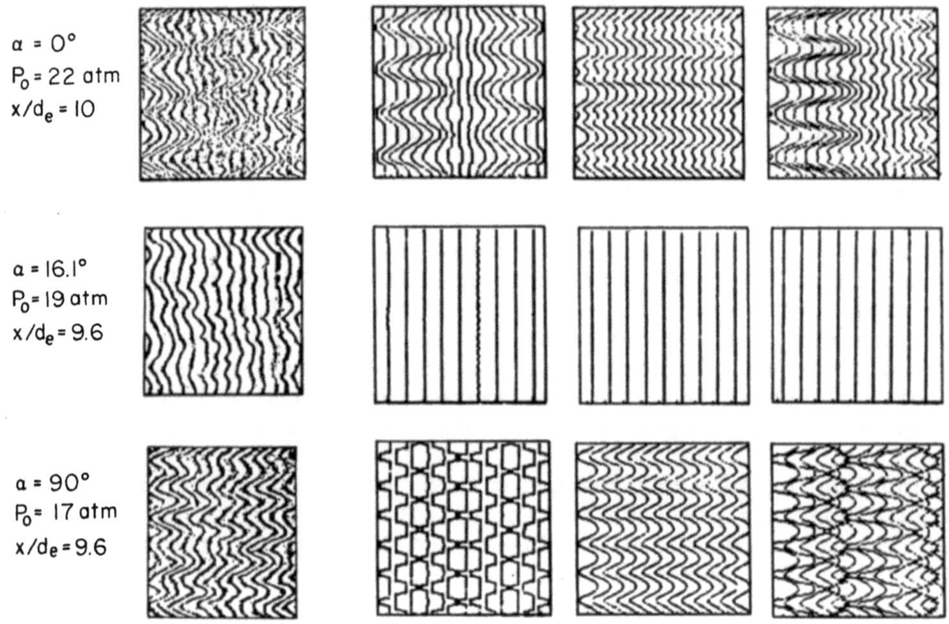

Fig. 6. Interferograms of flow (measurements on left, theoretical models on right).

in amplitude and pattern between prediction and experiment for α=90°. Both show no fringe direction reversals.

The final column of Fig. 6 is the cumulative prediction, for which the results of the second and third columns were superimposed. While the peak amplitudes generally agree with experiment, the pattern for α=0° and 90° is less satisfactory. As noted above, a wave-only prediction would be best for α=0°, and a wake-only for α=90°. For superposition to work, the models would need to contain at 90° and 0° respectively a large repetitive distance similar to the 16.1° case, or feature little variation transverse to the optical axis. No rational way of adding such features has been found.

Figure 7 shows the maximum difference in $\overline{\Delta\rho}/\rho_e$ plotted versus x/d_e. The black symbols are from the wave model with x/d_e at the center of sequential cells (X=0), and the ordinate evaluated at X=±0.5. The solid lines are for the wake model. While the disturbance decay rate is initially slower than that for the waves, the wakes begin to merge at $x/d_e \gtrsim 10$. Each wake profile is cut off halfway to its neighbor and this speeds up the decay, starting first for α=90° as shown. Incorporation of profile overlap would speed the process further; at larger x/d_e it may be necessary to develop models starting with continuous trigonometric function representation at fixed x.[8] Both wave and wake models predict much reduced ordinate values for intermediate α as expected from Fig. 6.

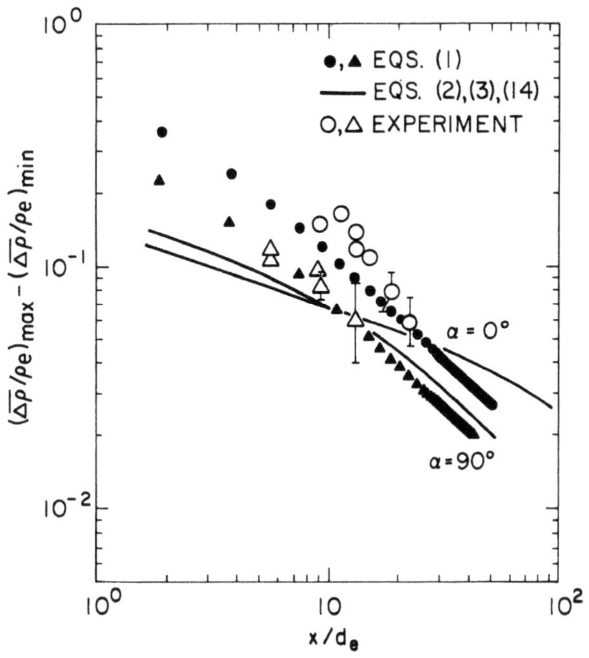

Fig. 7. Maximum difference in optical path vs. x/d_e.

The open symbols on Fig. 7 were obtained from the maximum fringe excursion on the measured interferograms, i.e., Fig. 6. There is a factor of ≈2 reduction for α=90° over 0°, and values for 16.1° showed a factor of ≈3 further reduction. The experiments are seen to generally agree with the wave model predictions, as did earlier α=0° measurements.[6] There is also less satisfactory agreement with the wake model at 90°. Either model may be used to estimate diffractive and refractive effects of array generated disturbances for the simplest cases. While these can be significantly controlled through selection of α, Fig. 7 indicates a maximum $\overline{\Delta\rho}/\rho_e$ difference still in excess of 10% at $x/d_e = 10$. The wave model predicts that local ρ perturbations in a cell vary up to ±2 times more,[6] with P and T variations further multiplied by γ and (γ-1) respectively. Gas comes from different parts of this perturbed field to mix and react in the shear layers of the chemical laser, and the effect on rate-sensitive processes can be substantial.

CONCLUSIONS

Closed-form 3-D models are presented for both the waves and wakes from nozzle arrays. Interferograms taken in the near field of a close-packed array show that either may dominate, and that strong cancellation effects occur with optical axis rotation. The measurements show good agreement with the dominant model, which may be used to estimate flow optical quality and the effect of systematic flow nonuniformity on fast rate processes.

REFERENCES

1. R.G. Pankhurst and D.W. Holder, "Wind Tunnel Technique," Pittman and Sons, London (1965).
2. J.K. Royle, A.G. Bowling, and J. Lukasiewicz, Calibration of Two-Dimensional and Conical Supersonic Multi-Nozzles, RAE AERO 2221, S.D. 23 (1947).
3. G.A. Simons, Decay of a Diamond Shock Pattern, AIAA J 10:1037 (1972).
4. D.A. Russell, S.E. Neice, and P.H. Rose, Screen Nozzles for Gasdynamic Lasers, AIAA J 13:593 (1975).
5. P.E. Cassady, A Survey of Advanced Gasdynamic Laser Concepts, in "Gas Flow and Chemical Lasers," J.F. Wendt, ed., Hemisphere (1979).
6. T.S. Vaidynathan, Wave Decay Downstream of a Nozzle Array, Ph.D. Thesis, University of Washington (1979). To be submitted for publication.
7. A. Demetriades, Mean-Flow Measurements in an Axisymmetric Compressible Turbulent Wake, AIAA J 6:432 (1968).
8. B.S. Masson, U.S. Air Force Weapons Laboratory, private communication.

MODELING OF 16-µm CO_2 GASDYNAMIC LASER IN INTRACAVITY CASCADING MODE

Shigeki Saito, Minoru Obara, and Tomoo Fujioka

Department of Electrical Engineering, Keio University

3-14-1, Hiyoshi, Kohoku-ku, Yokohama-shi, 223, Japan

INTRODUCTION

Although 16-µm lasers have been developed as light sources for isotope separation of UF_6 gas [1,4], these lasers operate in a pulsed mode except the continuous operation reported by Wexler et al. [5,6] using an after-mixing-type electric discharge gasdynamic laser (EDGDL). Since a high-average-power laser is required for laser isotope separation, not only high-energy operation but also high-repetition-rate operation are important problems.

In a CO_2 GDL, the vibrational population inversion for the 9.4-µm transition is formed downstream from the nozzle throat without the injection of the 9.4-µm laser beam. Therefore, the 9.4-µm stimulating beam can be generated internally and, subsequently, a 16-µm emission can be produced in a cascade manner without using a separate CO_2 laser. Therefore, the apparatus becomes simpler and the overall efficiency becomes very high.

Here, we would like to report the result of the analysis of the 16-µm lasing in the intracavity cascading mode with a conventional CO_2 GDL so as to obtain feasibility data on 16-µm cascade lasing in CO_2 GDL. It has been found that this system is able to produce high average power output levels such as are applicable to industrial isotope separation.

SIMULATION MODEL

Carbon dioxide 16-µm lasers use vibrational-rotational transition between the (02^00) and (01^10) levels of CO_2. To produce a population inversion between these vibrational levels in a GDL, first the (001) level is thermally pumped and the (01^10) level is depopulated by adiabatic expansion. Then the (02^00) level must be pumped with a 9.4-μm laser to depopulate the (001) level.

In the simulation of our previous model [8,9], the intensity of the 9.4-μm saturating beam was assumed to be Gaussian. When we analyzed continuous operation using this 9.4-μm beam, a static small signal gain of 16 μm was obtained in the center of the beam spot. As the intensity of the 9.4-μm is small in the foot of the upstream side of the Gaussian distribution, only the population of the (001) level is reduced; there is no population inversion between the (02^00) and the (01^10) levels. In such conditions each population flows to the center of the Gaussian distribution where the intensity is highest. Therefore, in the center of the beam, a large inversion for the 16-μm transition cannot be obtained.

Here, as shown in Fig. 1, we assume that the intensity distribution of the 9.4-μm beam is uniform. Using such a 9.4-μm beam, each level's population collides abruptly with the intense part of the 9.4-μm beam in the upstream side of the spot. Therefore, a high static gain of 16 μm is expected. Such a 9.4-μm beam with uniform intensity distribution is easily obtained by multitransverse-mode oscillation. It is assumed that the cross section of this 9.4-μm beam is circular, and its radius is called the spot size.

The simulation model used in this paper is almost the same as the previous model in which the pulse operation was analyzed [8,10]. The eight vibrational levels of CO_2 up to the (001) level including the (110) and (030) levels ((03^10) and (03^30) are combined), and the v = 1 level of N_2, and thirty-one reactions involving V-T reactions are taken into account. We solved a set of rate equations with the population of each vibrational level on the assumption that the rotational population always forms a Boltzmann distribution defined by a given translational temperature. For cascade operation, not only the rate equation with a 16-μm photon flux, but also the equation with a 9.4-μm photon flux is added to a set of rate equations.

Therefore, mirrors having transmissions for both 9.4 μm and 16 μm must be used in a cavity. But it is in fact not easy to produce the mirror with the desirable transmission characteristics, and additional cavity engineering studies will have to be carried out in an experimental stage. This model was based on the assumption that the transmission fraction of 9.4 and 16 μm is decided independently without considering cavity engineering.

For initial orientation, the values calculated by the previous simulation were used as shown in Fig. 1. These parameters turned out to be optimum values for a small signal gain at 16 μm. The cavity was set transversely 15 cm down stream from the nozzle throat.

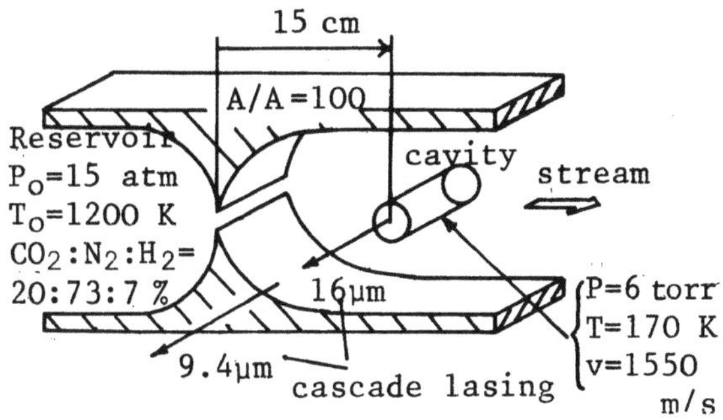

Fig.1 Schematic view of the 16-μm CO_2 gasdynamic laser. The parameters in the reservoir are optimized previously and shown in the figure. The active length (width of the stream) is set to 0.1 m.

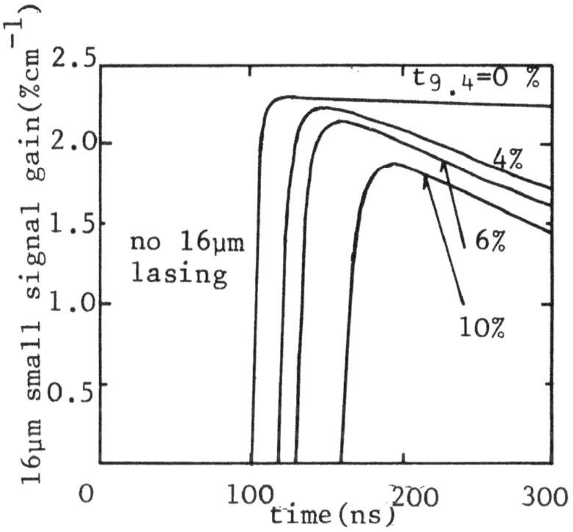

Fig.2 16-μm small-signal gain with time after setting the laser cavity as a number of 9.4-μm transmission fractions.

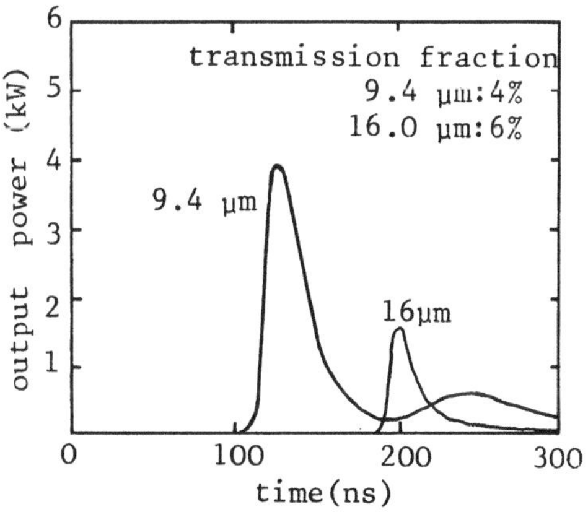

Fig.3 Typical pulse profiles of 16- and 9.4-μm lasings in a cascade lasing system.

At the U.S. Naval Research Laboratory 16-μm lasing has been demonstrated in experiments using an aftermixing-type GDL [5,7]. However, no detailed experimental parameters have yet been reported. To show that our theoretical model can predict the experimental results, we performed the numerical analysis of Wexler et al.[5] using our model. It was found that a 16-μm output of several hundred μJ/pulse is obtainable [9], while Wexler et al. obtained about 190 μJ/pulse. For continuous operation, a 16-μm gain of 1%/cm has been obtained [7], and cw 16-μm lasing has been demonstrated in experiments using EDGDL [5 6]. In our previous study we obtained the maximum gain of 1.1%/cm [9]. Therefore, we believe that our model can precisely predict experimental results.

THEORETICAL RESULTS AND DISCUSSION

Pulsed Cascade Operation

The analysis of the cascade operation is based on the assumption that the optical cavity was set transversely to the gas stream at t = 0 sec. Therefore, the pulse cascade operation corresponds to the first pulse in the Q-switched operation.

We examined how the population inversion of the 16-μm transition is produced by oscillating only 9.4 μm. In the pulsed operation, the population of each level, photon number, and gain form a uniform distribution inside the beam spot up to about 300 nsec, because the effect of the stream is almost nil.

Figure 2 shows a 16-μm small signal gain with time after setting the laser cavity as a number of the 9.4-μm transmission fraction. The 16-μm gain starts to increase suddenly at about 100 nsec and is maximized within 40 nsec from its rising. With a 9.4-μm transmission fraction of 0%, when the 9.4-μm laser energy is not extracted externally, a maximum 16-μm gain of 2.3%/cm is obtainable. Even with a 9.4-μm transmission fraction of 10%, a 16-μm maximum gain of 1.9%/cm is obtainable. However, in the case of externally injecting a 9.4-μm beam into the laser cavity, a maximum small-signal gain of 2.5%/cm was obtained. In the cascade operation, the (001) level is only slightly de-excited until a 9.4-μm pulse rises and a 16-μm gain is produced. Therefore, a maximum 16-μm gain in the cascade operation is just a little smaller than that obtained by externally injecting a 9.4-μm beam[10]. However, when the 9.4-μm intensity generated inside the cavity is almost as strong as an external 9.4-μm beam intensity, it is essentially possible to obtain 16-μm emission by cascade lasing.

Figure 3 shows the typical pulse profile of a cascade lasing system. The 9.4- and 16-μm transmission fractions are 4% and 6%, respectively. These parameters are typical. About 80 nsec after

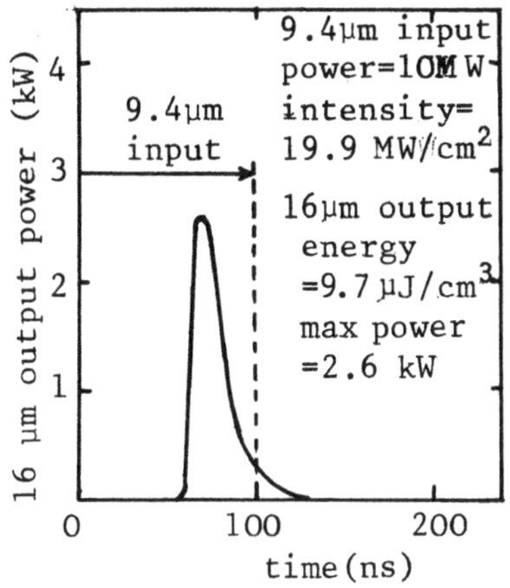

Fig.4 Typical 16-μm pulse profile for the injection of a 9.4-μm laser pulse with a uniform intensity distribution in an 8-mm diam beam spot.

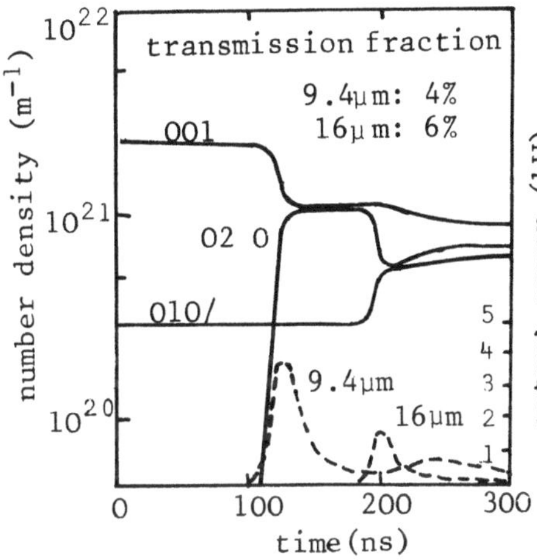

Fig.5 Temporal variation of the population number density of each vibrational level during lasing. The laser pulses (by dotted line) are also shown for reference.

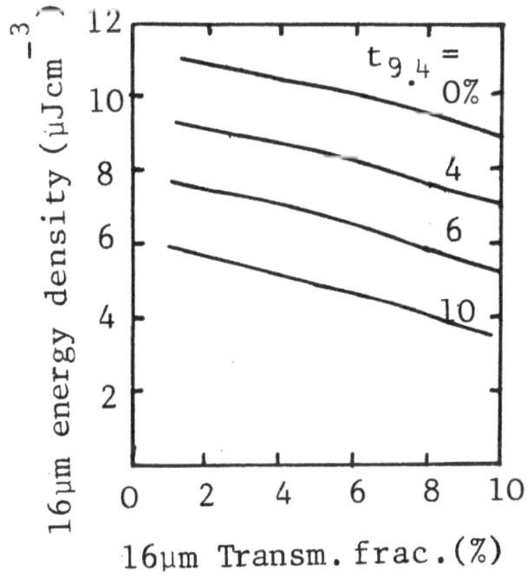

Fig.6 16-μm laser energy density against the 16-μm transmission fraction as a number of 9.4-μm transmission fractions.

9.4-μm lasing a 16-μm pulse is generated, the maximum power of which is 1.6 kW; the energy density is 8.3 $\mu J/cm^3$ with an active length of 0.1 m. However, in the case of externally injecting a 9.4-μm beam whose power is 10 MW, and whose intensity is 19.9 MW/cm^2, the spot size is 8-mm in diameter, and the pulse width is 100 nsec, the 16-μm output energy of 9.7 $\mu J/cm^3$ and the 2.6-kW maximum power were obtainable, as shown in Fig. 4 [10]. Thus, by using cascade lasing one can obtain the 16-μm output energy level which is almost the same as that obtained by externally injecting a 9.4-μm beam. This means that the cascade operation is a very efficient and simple method to produce a 16-μm laser in a CO_2 GDL.

Figure 5 shows the trend of the population density for each level in this case. At the same time as the 9.4-μm pulse rises up at 100 nsec, the population inversion between the (02^00) and the (010) levels is produced. Successively, 16-μm photon starts to be amplified rapidly, resulting in a laser pulse about 70 nsec later. By 16-μm lasing, the (02^00) level population is deactivated abruptly, and the population inversion between the (001) and the (02^00) levels is again generated. Therefore, the 9.4-μm pulse has its second peak at 240 nsec. But the second pulse of the 9.4-μm lasing is not strong enough to produce the second peak of a 16-μm lasing system.

The 16-μm energy density against the 16-μm transmission fraction is shown in Fig. 6 for a number of 9.4-μm transmission fractions. As the 9.4-μm transmission fraction decreases, the 9.4-μm photon flux stored in the cavity becomes strong. Similarly, both the population inversion for 16 μm transition and the 16-μm energy density become large. It is thus desirable to use a mirror with a 9.4-μm transmission fraction of 0%. However, as the 16-μm transmission fraction increases, the 16-μm energy density uniformly decreases and does not have a peak. This is because no detrimental loss in the cavity is considered in the calculation model. The decrease rate of the 16-μm energy density for the 9.4-μm transmission fraction is seen to be larger than that for the 16-μm transmission fraction. It is found that the 9.4-μm transmission fraction has a greater effect on the 16-μm output energy than the 16-μm transmission fraction.

In the CO_2 GDL considered in this analysis, the flow velocity is 1550 m/sec and it takes 5.2 μsec for the gas to pass through the cavity with an 8-mm-diameter spot size. In other words, a new gas is supplied into the cavity every 5.2 μsec. As shown in Fig. 3, a pulse cascade lasing cycle is completed within about 300 nsec after setting the laser cavity. Because this value of 300 nsec is so small in comparison with the value of 5.2 μsec, the distance traversed by the gas flow in 300 nsec is negligible. This means that pulse cascade lasing is theoretically feasible at about 200 kHz. Since the energy of about 8 $\mu J/cm^3$ is obtainable in cascade with the 9.4-μm and 16-μm transmission fractions of 4% and 6%, respectively, a high

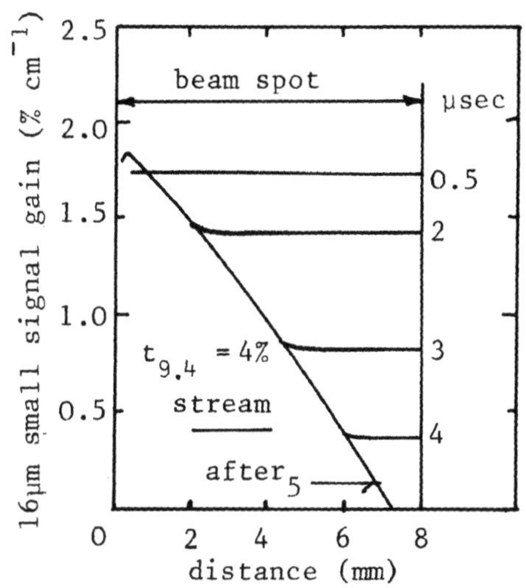

Fig.7 Distribution of the 16-μm small signal gain in the beam spot as a function of time after setting the laser cavity.

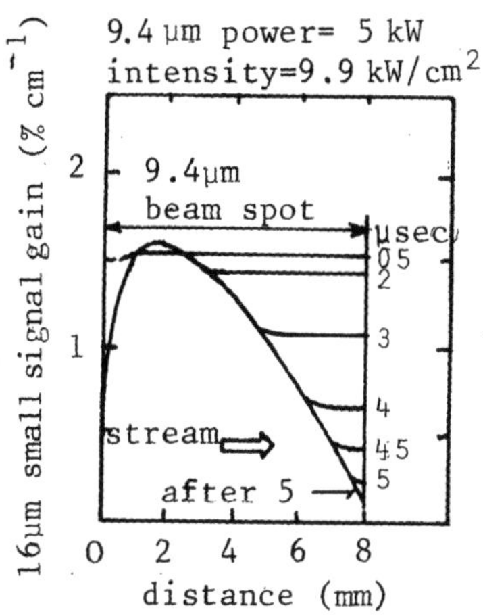

Fig.8 Distribution of the 16-μm small signal gain in the beam spot as a function of time after injection of the 9.4-μm saturating beam.

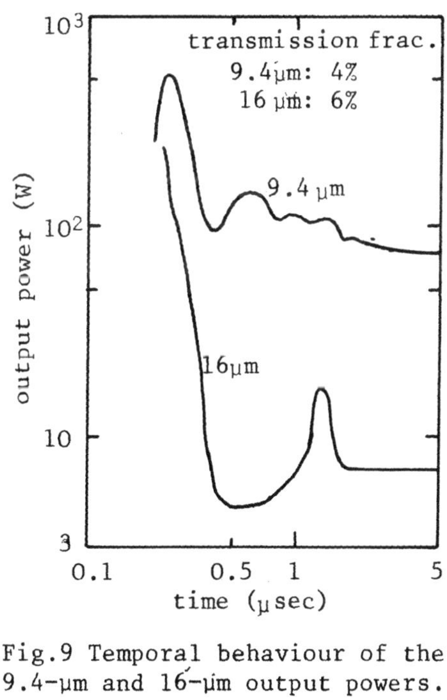

Fig.9 Temporal behaviour of the 9.4-μm and 16-μm output powers.

repetition-rate operation at 200 kHz by using the Q-switch technique can produce the average 16-μm output power of about 1.6 W/cm^3.

Continuous Cascade Operation

The 16-μm small signal gain distribution in the beam spot is shown in Fig. 7 as a number of the time after setting the laser cavity. The 9.4-μm transmission fraction is 4%. Within a short time from setting the cavity, a gain distribution becomes uniform in the beam spot. However, the 16-μm gain decreases downstream and it becomes a steady state after about 5 μsec. While a maximum small signal gain of 1.8%/cm is achievable close to the upper stream end of the beam spot, at the lower stream end the gain is a negative quantity. Such a tendency is very similar to the 16-μm gain distribution shown in the case of injecting continuously an external 9.4-μm beam with a uniform intensity distribution, as shown in Fig. 8. This is due to that the 9.4-μm photons generated inside the cavity are distributed over the beam spot due to the high flow velocity.

Figure 9 shows the temporal behavior of the 9.4- and 16-μm output power. When the 9.4- and 16-μm transmission fractions are 4% and 6%, respectively, a 9.4-μm output power of 73 W and a 16-μm output power of 6.8 W are obtained with an active length of 0.1 m and a spot size of 8 mm in diameter in a steady state. This figure follows Fig. 3 describing the pulse cascade operation. After the pulse cascade lasing, both output powers remain in a steady state from about 3 μsec. Thus, the possibility of continuous cascade operation becomes apparent.

The population density distribution of each level in the beam spot in this steady state is shown in Fig. 10. While the populations of the (001) and the (02^00) levels are distributed uniformly over the beam spot, the (010) level's population increases downstream. The difference between the (001) and the (02^00) level populations indicates a 9.4-μm gain, and the difference between the (02^00) and the (010) level populations contributes to the 16-μm gain. A 16-μm laser oscillates at a distance of only 2 mm when the 16-μm gain is 0.3%/cm. In spite of this, a 16-μm small signal gain is generated over the beam spot, as shown in Fig. 7. The region where a 16-μm laser will oscillate is very restricted. As the (010) level populations which increase with the 16-μm laser emission are de-excited by flowing out of the cavity, the 16-μm lasing is obtainable only at upper stream side.

The 16-μm power density of the 1-1.5 W/cm^3 is obtainable on the average. This value is not small in comporison with the case of injecting an external 9.4-μm beam with a uniform intensity distribution. When we analyzed a 16-μm oscillation by continuously injecting a 9.4-μm beam with 5-kW power, 9.9-kW/cm^2 intensity, and 8-mm diam.

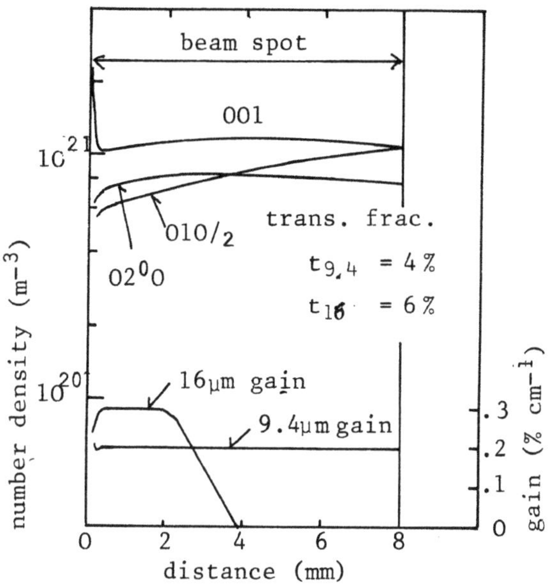

Fig. 10 Population number density of each vibrational level in the beam spot in the steady state when lasing.

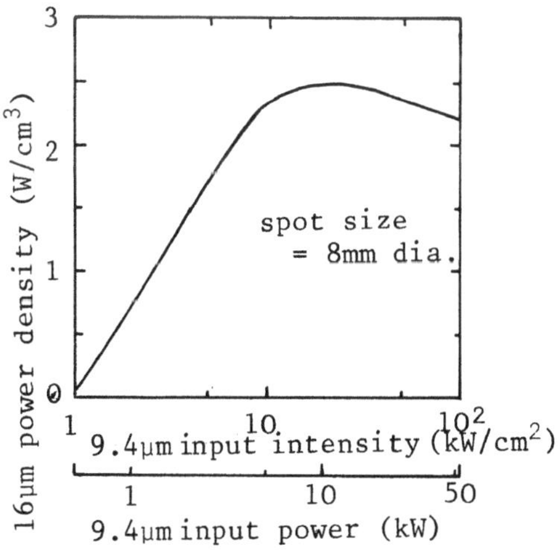

Fig.11 Density of the 16-μm power against intensity and power of the 9.4-μm beam of 8-mm diam spot size.

spot size, a 16-μm continuous oscillation of 11.7 W (2.3 W/cm^3) was obtained, which is shown in Fig. 11.[10] While the ratio of the 16-μm output power to the input power of the 9.4-μm beam is 0.23%, the ratio of the 16-μm output power to the 9.4-μm stimulated emission is 10.8%. This means that, since the stimulating beam is not absorbed but used only for stimulated amplification, one can increase the overall efficiency up to 10.8% when the injected beam is fully utilized in the flow, for example, by using multireflection devices.

CONCLUSION

As a result of our theoretical analysis of a 16-μm cascade operation in a CO_2 GDL with a nozzle of a large area ratio (A/A^*= 100), it has been found that the 16-μm high-power laser is obtainable in the cascading mode without an external 9.4-μm laser.

The 16-μm power density is 1-1.5 W/cm^3 in this case, while by injecting a 9.4-μm uniform saturating beam continuously, the 16-μm small signal gain of over 1%/cm and 16-μm continuous power of over 2.0 W/cm^3 with the cavity were obtained.

Therefore, it has been shown for the first time that CO_2 GDL operation in the cascading mode for which an externally generated 9.4-μm laser is not necessary has great potential for the industrial application of a 16-μm laser.

References

1. R. M. Osgood,Jr.,"1-mJ line-tunable optically pumped 16-μm laser" Appl. Phys. Lett. 32:564 (1978).
2. J. J. Tiee and C. Wittig,"CF_4 and NOCL molecular lasers operating in the 16-μm region" Appl. Phys. Lett. 30:420 (1977).
3. W. H. Kasner and L. D. Pleasance,"Laser emission from the 13.9 μm 10^00-01^10 CO_2 transition pulsed discharges"APL 31:82(1977).
4. P. Rabinowitz, A.Stein, R. Brickman, and A. Kaldor," Efficient tunable H_2 Raman laser" Appl. Phys. Lett. 35:739 (1979).
5. B. L. Wexler and R. W. Waynant ,"High-average power 16-μm gasdynamic CO_2 laser using a multipass cavity" Appl. Phys. Lett. 34:674 (1979).
6. B. L. Wexler, T. J. Manuccia, and R. W. Waynant,"cw and improved pulsed operation of the 14- and 16-μm CO_2 laser" Appl. Phys. Lett. 31:730(1977).
7. T.J.Manuccia, J.A.Stregack, N. W. Harris, and B.L.Wexler,"14- and 16-μ gasdynamic CO_2 laser" Appl. Phys. Lett. 29:360(1976).

8. S.Saito, K. Suzuki, M. Obara, and T. Fujioka,"Theoretical analysis of 16-μm CO_2 gasdynamic laser" Third International Symposium on Gas Flow and Chemical Lasers, Marseille 1980.
9. S. Saito, M. Obara, and T. Fujioka,"Theoretical characteristics of 16-μm CO_2 gasdynamic laser" J. Appl. Phys. 52:4422 (1981).
10. S. Saito, M. Obara, and T. Fujioka,"Computer simulation of cw 16-μm CO_2 gasdynamic laser" Appl. Optics 20:2838 (1981).

DEVELOPMENT OF A CLOSED-CYCLE CW TE CO LASER

OPERATING IN THE TEMPERATURE REGION 100-300 K

Shunichi Sato, Mitsuhiro Iyoda, and Tomoo Fujioka

Department of Electrical Engineering, Keio University

3-14-1, Hiyoshi, Kohoku-ku, Yokohama-shi, 223 Japan

INTRODUCTION

The CO laser has high quantum efficiency of nearly 100%, so operation with high conversion efficiency can be obtained. Its output wavelength of 5 μm is within the atmospheric transmission window, and therefore applications such as medical engineering, optical communications, metalworking, etc., can be anticipated.

Owing to these excellent characteristics, much research has been done on CO lasers. However, because of many technical problems, there are only a few reports on practical high-power CO laser systems to date.

One of the most important problems of CO laser technology is effectively cooling the laser mixture. The main excitation process of the CO laser, that is, the vibration-vibration (V-V) energy exchange rate becomes faster as the operating temperature is reduced. As a result, laser characteristics are very sensitive to gas temperature. For example, with a coaxial flow DC discharge CO laser the conversion efficiency of 47% was obtained at a wall temperature of 77 K, while 12% at 300 K [1]. In this configuration the gas mixture is cooled by liquid N_2 or water. But in high-power CO laser systems, supersonic expansion is also available for gas cooling. In recent years many excellent performance data on supersonic discharge CO lasers have been reported (both pre-excited [2, 3] and post-excited configurations [4, 5, 6, 7] are included). However, these systems have some problems in practical use. First, operating life is limited to less than a few minutes. Secondly, in post-excited configurations it is not so easy to obtain a stable glow discharge in supersonic flow, therefore resultant discharge techniques become rather complicated.

To overcome these disadvantages, another novel approach has been examined in our laboratory. Based on the experimental results in a small-scale test facility [8], a practical CW CO laser with closed-cycle subsonic flow was designed and fabricated. In this system a transversely excited (TE) laser configuration is applied. The main purpose of this paper is to present the design of this novel CO laser system. The results of preliminary experiments using an electrode module are also described.

DESIGN

Experimental results which were obtained at the test facility [8] are summarized in Table 1 compared with other CO laser systems. At the relatively high temperature of 200 K, laser power of 25 W with a conversion efficiency of 24% was obtained. The input power density and the specific laser power (laser power/discharge volume) were estimated at 4 W/cm^3 and 1 W/cm^3, respectively. At a pressure of 35 Torr, the flow velocity of the laser mixture ($CO/N_2/He$ = 1/8/16) is about 5 m/sec. Some useful information can be extracted from this study. First, the TE laser configuration, which could be easily scaled to large volume, was successfully applied to CO laser excitation. Secondly, it was experimentally demonstrated that the hollow cathode produces a stable, self-sustained glow discharge in the gas flow of CO laser mixture. Thirdly, the expensive Xe gas was not added to the laser mixture, while the concentration of N_2 was remarkably high. The effect of the abundance of N_2 is very interesting. In this system, vibrational energy transfer from N_2 is supposed to have a very important role in the excitation of the CO molecule. This problem has been discussed extensively in our other paper [9].

Based upon the results of this study, a 200 W CW TE CO laser having a wide performance range was designed and fabricated. The purpose of this work is to develop engineering information on scaling to larger CO laser systems whose laser power is more than 1 kW.

A schematic and a photograph of the setup are shown in Fig. 1 and Fig. 2, respectively. The system fundamentally consists of a centrifugal blower, two-stage heat exchangers, and laser head.

It is the relationship between the conversion efficiency (laser power/electric power deposited in discharge volume) and gas temperature that we must first understand exactly. As described above, the conversion efficiency of the CO laser increases as the operating temperature is reduced, while the electric power consumed in the heat exchange system also increases. There must be,

Table 1 - CW subsonic CO laser experimental data.

Authors	This work	Bhaumik et al.	Lim et al.	Kan et al.
Laser Configuration	TE	coaxial	coaxial	coaxial
Output Power (W)	25	70	750	400
Specific Output Power ($W \cdot cm^{-3}$)	1	0.16	1.44	0.82
Input Power Density ($W \cdot cm^{-3}$)	4	0.35	5.2	4.1
Conversion Efficiency (%)	24	47	28	20
Flow Velocity ($m \cdot sec^{-1}$)	5	12	96	100
Gas Temperature (K)	200	77 (at wall)	100	80

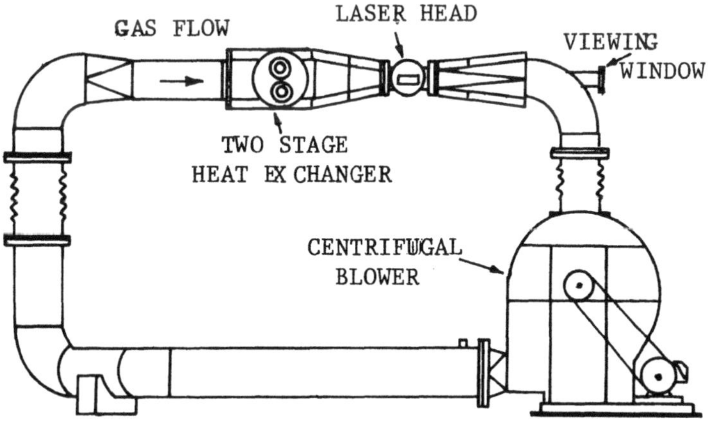

Fig.1 Schematic drawing of the experimental apparatus

therefore, a temperature allowing maximum system efficiency. According to our simple calculation on the electric power balance in the CO laser system, this temperature was predicted to be about 170 K [10]. The prediction must be confirmed experimentally in our setup.

The laser mixture is cooled through two-stage finned-tube heat exchangers operating with water, liquid N_2, or its own vapor. The temperature is controllable in a wide range of 100-300 K by adjusting the coolant flow. A photograph of the one-stage finned-tube, which is inserted through the side wall aperture of the wind tunnel, is shown in Fig. 3. In the setup, twenty thermocouples are attached for gas flow temperature measurements.

The gas mixture is recirculated by a centrifugal blower driven by a 1.5 kW motor. The variable flow rate is as high as 30 m^3/min at 1 atm. In the pressure range of less than 100 Torr, the maximum flow velocity of 30 m/sec is obtained. For shaft sealing, labyrinth seals combined with lip-type oil seals are applied. A small amount of the laser mixture is expelled through the shaft seal.

A sectional view of the discharge section is shown in Fig. 4. A self-sustained DC glow discharge is produced between a hollow cathode array which is individually ballasted with a 50 kΩ resistor and a planar anode. Oxygen-free copper pipes (4 mm OD by 3 mm ID) with equal spacings of 7 mm are used. The anode plate is also made of oxygen-free copper. In this electrode configuration the CO_2 laser mixture can be excited, also. The low temperature characteristic of the CW TE CO_2 laser is another very interesting behavior that must be revealed.

The specifications of this system are summarized in Table 2. The concept is said to be a typical gas circulating system, but severe cryogenic resistance is taken into account in the design. The walls of the wind tunnel and the blower runner are all made of stainless steel 304, and metallic gaskets (solder and oxygen-free copper) are used for joint sealing. The outside walls of the system are covered with a heat insulator (in Figs. 1 and 2 heat insulator is not shown).

ELECTRODE MODULE EXPERIMENTS

The 12-mm discharge gap in the test facility is enlarged to 30 mm in our novel system. In order to reveal engineering problems on this scaling, some preliminary experiments using an electrode module were performed. The dimensions of the electrode module are 30 mm in the discharge gap and 100 mm in the length. Other conditions of the electrode configuration are similar to the practical system. The test section was equipped in another small-scale closed-cycle wind tunnel.

Fig.2 Photograph of the experimental apparatus.

Table 2 Specifications of the experimental apparatus

EXCITATION	:SELF SUSTAINED DC DISCHARGE TRANSVERSE EXCITATION
GAS TEMPERATURE	:100 - 300 K
GAS FLOW RATE	: $< 30\ m\ sec^{-1}$
TOTAL PRESSURE	: < 100 Torr
DISCHARGE VOLUME	: $3cm^{H}$ $3cm^{W}$ $40cm^{L} = 360\ cm^{3}$
OUTPUT POWER	: > 200 W

Fig.3 Photograph of the finned-tube heat exchanger.

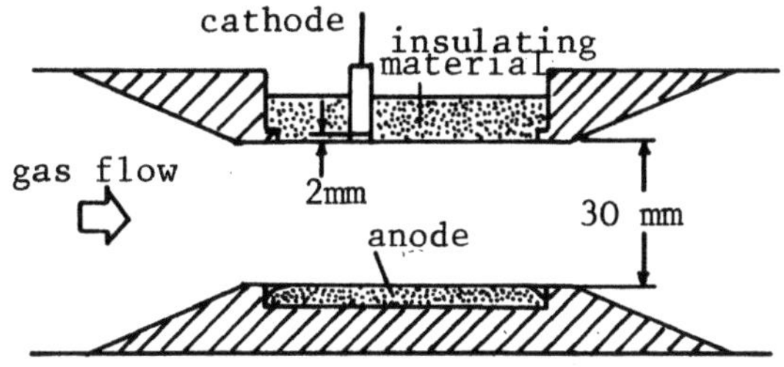

Cathode : hollow cathode array
oxygen-free copper pipes
4mm O.D. 3mm I.D.

Anode : oxygen-free copper plate

Fig.4 Electrode configuration in laser channel

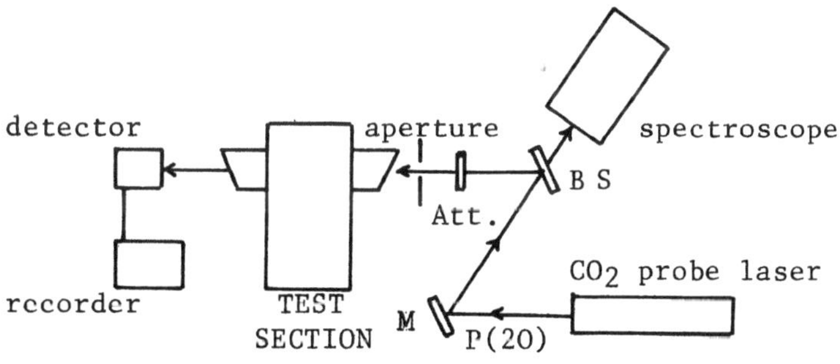

Fig.5 Schematic diagram of the system used to measure gain of CO_2 laser discharge

A. Spatial Distributions of CO_2 Laser Gain

For fundamental understanding of laser excitation in the hollow cathode configuration, small signal gain distributions were measured on CO_2 laser, at room temperature. The experimental setup is schematically shown in Fig. 5. The measurements were made with a tunable CO_2 probe laser operated in the TEM_{00} mode. The discharge region could be probed by moving the beamsplitter and mirror. The dependence of the gain at the P(20) transition upon gas pressure and discharge current were studied. The gas composition ($CO_2/N_2/He$ = 1/1/8) and flow velocity of 20 m/sec were fixed during this experiment.

Fig. 6-(A) shows the measured spatial distribution of small signal gain g_0 at a pressure of 40 Torr. The total discharge current was 275 mA, which is equivalent to 25 mA per cathode pipe. For convenience, the point just below the cathode pipe on the anode surface was taken as the origin of the X-Y coordinate system. It is found that the small signal gain becomes higher in the region near the cathode. A similar trend was observed for other gas pressures and discharge currents. Some papers, however, report the results contrary to our measurements [11, 12]. The gain, in general, strongly depends on both the electron density and gas temperature. The electron density in the region near cathode is high, while the mixture in this region is heated by electron impact and heat conduction from the cathode, therefore, the small signal gain reflects their both conditions. The reason why the gain is high in the region near the cathode must be revealed by further study.

The similar measurements were performed in the structure of two hollow cathode rows, also. Other experimental conditions are same as the former case. The total discharge current of 275 mA, in this case, corresponds to about 13 mA per cathode pipe. The experimental result is shown in Fig. 6-(B). The uniformity of the gain is improved in this double-row structure, compared with single-row configuration. It is supposed that in the excitation configuration of CO laser the multiple hollow cathode structure is also effective to obtain the uniform gain distribution.

B. Xe Effects on CO Laser Discharge

It is well known that the addition of Xe to CO laser mixture, because of its low ionization potential, lowers the average electron energy, and consequentially increases the fractional power transfer to CO vibrational levels [1, 14]. But this beneficial effects of Xe have been experimentally confirmed in only a sealed-off or slowly flowing dicharge tube CO lasers, because Xe is too expensive to use

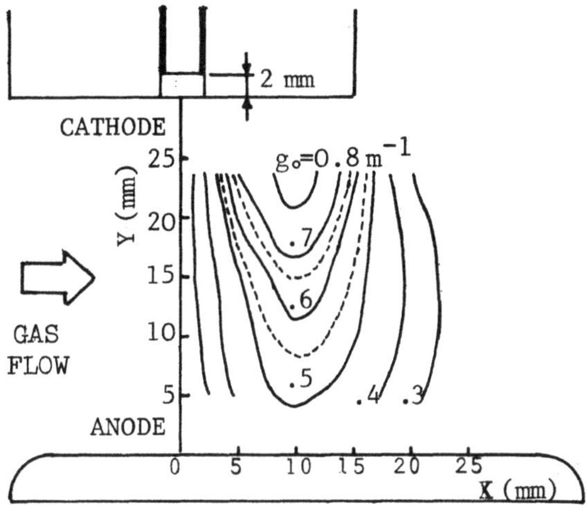

(A)

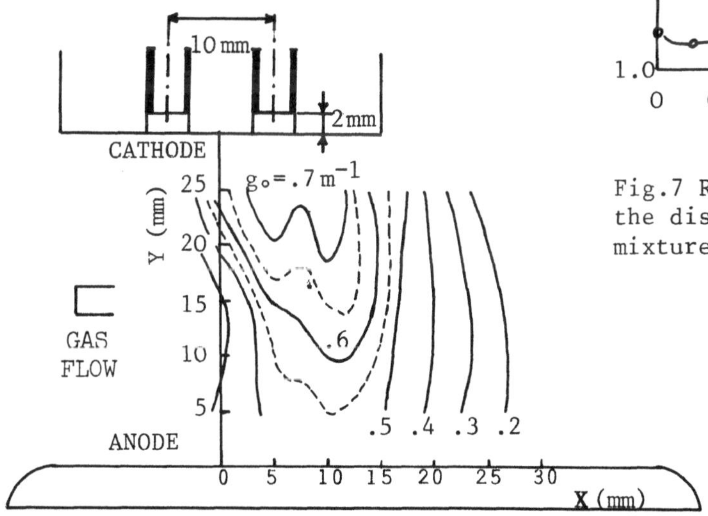

(B)

Fig.6 Spatial distribution of small signal gain measured in (A) single-row structure and (B) double-row structure. The total discharge current and pressure are 275 mA and 40 Torr, respectively.

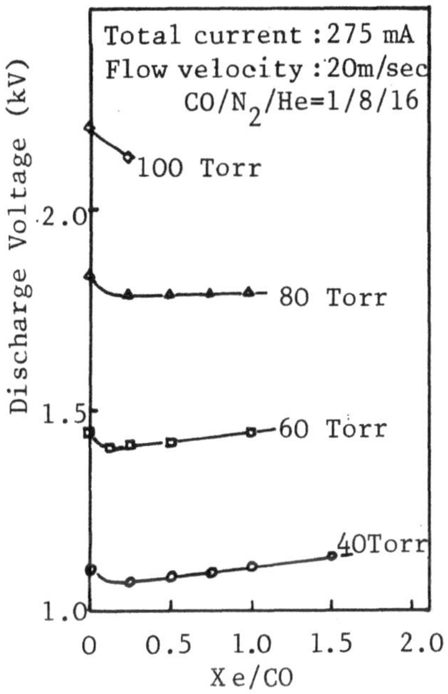

Fig.7 Relationship between the discharge voltage and mixture ratio of Xe/CO

in a high-flow or large systems. The effects of Xe in TE laser configuration with closed-cycle gas flow is another interest in our study on CO lasers.

At a room temperature the V-I characteristics in CO laser mixture with/without Xe were measured in the same test section which is described above (single-row hollow cathode structure). Fig. 7 shows the relationship between the discharge voltage and mixture ratio of Xe/CO. The basis of the mixture was $CO/N_2/He = 1/8/16$, which had been optimized in the test facility. The flow velocity was fixed to 20 m/sec at the pressure range of 40-100 Torr. And the constant discharge current was 275 mA. First, it was found that the maximum voltage drop by the addition of Xe was only a few percent of the total discharge voltage for any pressures were tested, and the minimum discharge voltage was obtained at Xe/CO $\simeq$ 0.25. In the region Xe/CO $\gtrsim$ 0.25 the voltage increased again, and the discharge tended to be unstable. And the excess of Xe resulted in arcing.

In this experiment the maximum input power of 600 W, which corresponds to 2400 W in our novel system, was obtained at the pressure of 100 Torr for the mixture of $CO/N_2/He/Xe = 1/8/16/0$, while the active volume in the test section was estimated at 90 cm^3 from the luminosity of the discharge. Therefore, the input power density could be calculated as 6.7 W/cm^3. At the presumed conversion efficiency of 10% an output power of 240 W was predicted from the discharge volume of 30x30x400 mm^3. In practice, much improved performance appears possible.

CONCLUSION

In order to develop engineering information on scaling to larger CO laser systems for practical use, a 200 W CW TE CO laser with closed-cycle subsonic flow was designed and fabricated. In the preliminary experiments using electrode module, the input power density of 6.7 W/cm^3 was obtained at a pressure of 100 Torr for the mixture of $CO/N_2/He = 1/8/16$. This fundamental result predicts a successful operation of our novel CO laser system.

ACKNOWLEDGEMENTS

The authors gratefully acknowledge the expert technical assistance rendered by Mr. T. Abe, K. Komatsu, and Y. Kato. Particular thanks go to Drs. M. Obara and H. Saito for a private communication and technical discussions.

References

1. M. L. Bhaumik, W. B. Lacina, and M. M. Mann, "Characteristics of a CO Laser," IEEE J. Quantum Electron., QE-8:150 (1972)
2. J. W. Daiber and H. M. Thompson, "Performance of a Large, CW, Preexcited CO Supersonic Laser," IEEE J. Quantum Electron., QE-13:10 (1977)
3. S. Krause and F. Maisenhälder, "Gasdynamic CO Laser with Closed-Cycle Gas Flow," Appl. Phys., 22:421 (1980)
4. D. J. Monson and G. Srinivasan, "A Pulser-Sustainer Carbon Monoxide Electric-Discharge Supersonic Laser," Appl. Phys. Lett., 31:828 (1977)
5. E. L. Klosterman and S. R. Byron, "Electrical and Laser Diagnotics of an 80-kW Supersonic CW CO Electric Laser," J. Appl. Phys., 50:5168 (1978)
6. W. Schock, W. Schall, H.Hügel, and P. Hoffmann, "CW Carbon Monoxide Laser with RF Excitation in the Supersonic Flow," Appl. Phys. Lett., 36:793 (1980)
7. P. Hoffmann, H. Hügel, W. Schall, and W. Schock, "CW Carbon Monoxide Laser with Microwave Excitation in the Supersonic Flow," Appl. Phys. Lett., 37:673 (1980)
8. T. Arai, K. Sugino, M. Obara, and T. Fujioka,"Experimental Investigations of the Optimum Specific Input Energy on a Subsonic CO EDL," Proceedings of the International Conference on LASERS '81, USA (1981)
9. M. Iyoda, K. Sugino, K. Komatsu, and T. Fujioka, "Subsonic TE N_2-CO Mixing Laser," (to be published)
10. M. Iyoda, Y. Imai, S. Sato, T. Fujioka, H. Saito, K. Watanabe, and T. Taira, "High-Power Closed-Cycle CW TE CO Laser at Relatively High Temperature," Appl. Phys., B-28:285 (1982)
11. H. J. Seguin and G. Sedgwick, "Low Voltage Gas Transport TE CO_2 Laser," Appl. Opt., 11:745 (1972)
12. T. Akiba, H. Nagai, and M. Hishii, "Gain Characteristics of an Atmospheric Sealed CW CO_2 Laser," IEEE J. Quantum Electron., QE-15:162 (1979)
13. GA Murray and ALS Smith, "The Efficacy of Xenon as an Additive Gas in Carbon Monoxide Lasers," J. Phys. D: Appl. Phys., 11: 2477 (1978)

MICROWAVE EXCITED GASDYNAMIC CO_2-LASER

W. Schall, P. Hoffmann*, H. Hügel, and W. Schock

DFVLR-Institut für Technische Physik
Pfaffenwaldring 38-40, D-7000 Stuttgart 80

INTRODUCTION AND EXPERIMENTAL SETUP

Microwave energy has been used in several arrangements for the excitation of CO_2 gas lasers. Golovitskii et al.[1] applied microwave discharge pulses as a preionizing source in a pulsed TE laser. The two groups Vasyutinskii et al.[2] and Handy et al.[3] used a pulsed microwave discharge alone to vibrationally pump CO_2 lasers. Joeckle and Stern[4] coupled microwave energy sideways into a supersonically flowing CO_2 laser gas. They were able to improve the stability and to raise the energy deposited by a transversal dc main discharge.

A new technique to excite a cw gasdynamic CO laser with a microwave discharge in the supersonic flow has been described in Refs. 5-7. The same configuration is now used to operate a cw CO_2 laser. In the experiments presented herein the main emphasis is laid on the establishment of the power potential under various conditions.

Fig. 1 shows the principal arrangement of the laser channel: Continuous microwave radiation at 2.45 GHz enters the high pressure plenum in the direction of the flow through a dielectric window. The microwave energy propagates inside a waveguide of 2.5 x 11 cm, coinciding with the inner dimensions of the flow channel. The resulting TE_{10} mode has a cosine E-field distribution across the broad side of the waveguide/flow channel. The laser gas expands from the plenum through a convergent-divergent nozzle, made

*present address: Laser Innovation GmbH, Im Betzengaiern 9,
D-7000 Stuttgart 70

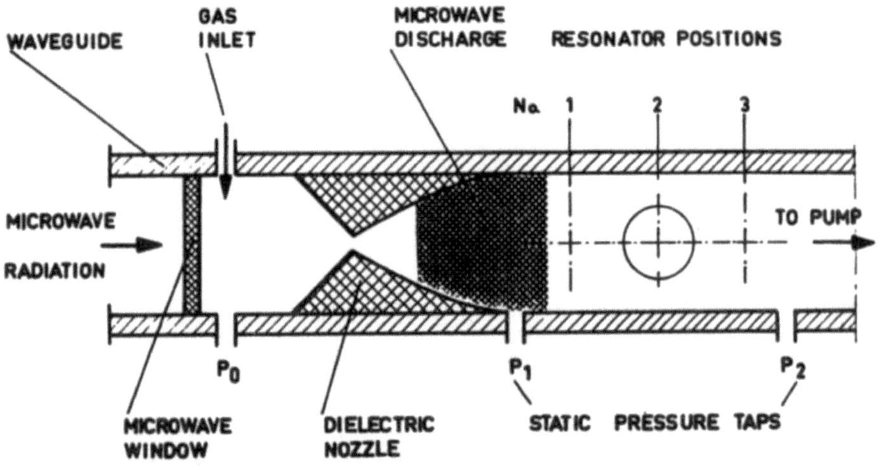

Fig. 1. Scheme of the laser.

of dielectric material with minimum loss angle to the microwave radiation. In the divergent section of the nozzle the breakdown occurs and peak electron densities up to 10^{12} cm^{-3} are produced[7]. The discharge spreads about 5 to 7 cm across the channel width. With proper adjustment of the waveguide system to match the discharge impedance a near 100 % deposition of the microwave power into the core of the supersonic flow is achieved. No further power source is needed to vibrationally pump the laser gas. Laser power is coupled out from a stable resonator located at either one of three ports, 10, 20 and 30 cm downstream of the nozzle throat. The resonator consists of a flat total reflector and a 1.5", 4 m radius partial transmitter with 95 % reflectance. Because discharge heating occurs partly inside the nozzle, the nozzle walls were only roughly contoured to match a minimum length nozzle for an area ratio of 15. Static pressure has been measured in the plenum, at the nozzle exit (still inside the discharge region), and beyond the last resonator port.

The following parameters are varied in the experiments:

- microwave power (up to 4.75 kW)
- nozzle area ratio, by varying the throat height (A/A* from 9 to 35)
- resonator distance from nozzle throat.
- composition and mass flow of the laser gas.

EXPERIMENTAL RESULTS

It should be understood that all results presented are obtained at an optimum mixture ratio and pressure if not stated otherwise. At the range of investigated area ratios the static

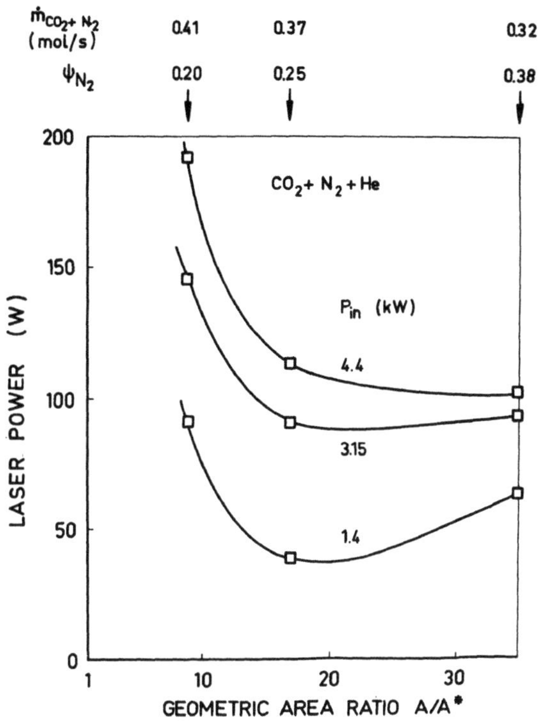

Fig. 2. Laser power versus expansion ratio with microwave power input as parameter.

temperature in an isentropic expansion may drop below 80 K in mixtures of $CO_2/N_2/He$. Despite the possibility of CO_2 condensation the laser can be operated even at large area ratios (Fig. 2). Pressure measurements indicate much reduced expansion ratios due to heat addition from the discharge. Hence, taking into account a supersaturation difference of 40 to 50 K for the very rapid expansion[8], CO_2 condensation may not occur at least for the lowest area ratio of 9. On the other hand, condensation may partly explain the rapid decrease of laser power within the range of the first two expansion ratios. Besides, a significant fraction of the observed power reduction for larger expansion ratios can be attributed simply to a change in laser gas composition, required for optimum laser power. In the example of Fig. 2, the flow rates of N_2 and CO_2 have to be reduced by 25 %, if A/A* is increased from 9 to 35. Consequently, less vibrational energy is storable in the laser gas for equal input powers.

For gas mixtures of $CO_2/N_2/He$ and low expansion ratios the laser power increases proportional to the microwave input power, if the massflow rate is adjusted appropriately (Fig. 3). At high area ratios the necessary high N_2 fractions for attaining optimum laser

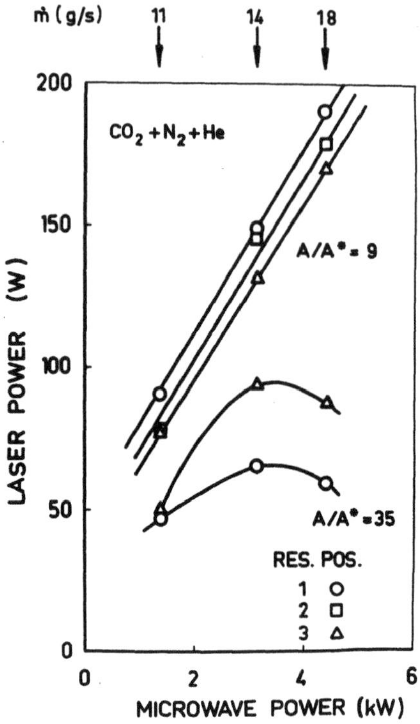

Fig. 3. Laser power versus microwave power for $CO_2/N_2/He$ mixtures.

power impair the discharge homogeneity and provoke reductions of the laser power. The location of the resonator within 30 cm distance from the nozzle is of little importance for the achievable power level. However, it does influence the composition of the gas mixture required for optimum power. The maximum power is reached with half the CO_2 mass flow for resonator position 3, compared to position 1, but the fraction of N_2 need nearly be doubled. This is a clear indication of the CO_2 being a deactivating component to the excited N_2 via collisional losses.

Plenum pressures for optimum operation are reduced from 750 mbar to 350 mbar, if the area ratio is reduced from 35 to 9 and static channel pressures increase from 5 to 15 mbar.

The laser can also be operated with a mixture of CO_2 and laboratory air (Fig. 4). Laser power data scatter strongly and a maximum is found within the range of the available microwave power. Furthermore, for optimum laser power a short distance between discharge and location of the resonator is a prerequisite. Laser operation with an inhomogeneous discharge is also more frequent in the operation with air. Under such conditions, high mass flow rates

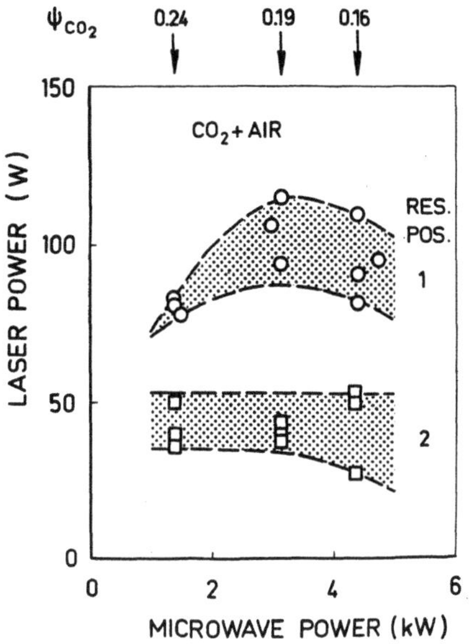

Fig. 4. Laser power versus microwave power for CO_2/air mixtures.

are required to achieve appreciable laser power. This results in correspondingly lower values of the mass specific energy.

Despite the detrimental influence of large amounts of O_2 in laser discharges and its deactivating nature, only about one third of laser power is lost, compared to mixtures with N_2 and He. The strong scatter of the data partly reflects the sensitivity in adjusting the correct flow rate. In addition, it may be attributed to changes of the humidity of the laboratory air.

The admixture of H_2 of 2 % to 10 % to the laser gas produces a substantial increase in laser power (Fig. 5) for both, mixtures with N_2 and He, as with air. Highest values (320 W) were achieved for $CO_2/N_2/H_2$. A saturation of laser power has not been found within the range of available microwave power, even in the case with air. Hence, considerably higher power values are anticipated for the same laser device. The efficacy of H_2 is attributed to two effects: 1) H_2 is seen to improve the homogeneity of the discharge. 2) H_2 is a strong deactivator of the (0,1,0)-level of the CO_2 molecule at the prevalent low static temperatures, where He becomes very inefficient. Both effects have been described for different types of CO_2 discharge lasers[9] or were even exploited to achieve

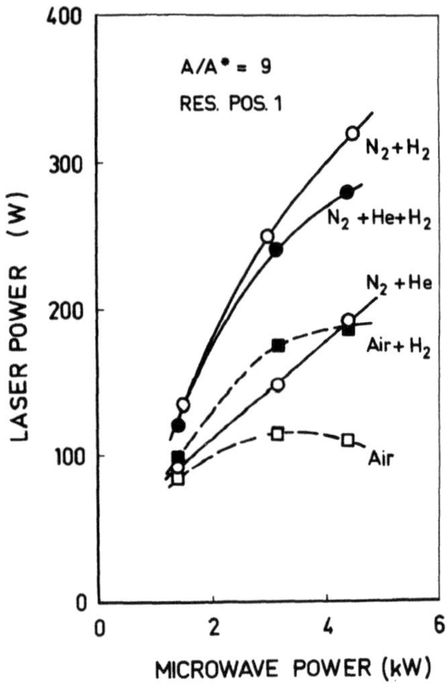

Fig. 5. Effect of hydrogen on laser power.

intramode lasing[10]. In addition, a possible role in controlling chemical reactions has also been discussed[11].

Compared to the operation with CO as the lasing species, the CO_2 laser shows a similar performance, if He is a component of the laser gas mixture (see table). Pressure requirements for operation in a closed cycle gas system appear moderate. Operation with air reduces the laser power by about one third, but with even better specific energy. In this case gas consumption costs are extremely low in a system with exhaust to the atmosphere.

Beside the enhancement of input power, other means to improve the microwave CO_2 laser became apparent from the experiments: A reduction of the expansion ratio to the point where a breakdown still occurs within the nozzle ($A/A^* = 4$-5) not only would ease the requirements for pressure recovery but also promises a further increase in laser power and efficiency. A reduction of unexcited flow regions on either side of the discharge by decoupling the

Table. Comparison of power data for the CO and CO_2 microwave laser. Laser gas mixtures are: a) 6.4 CO/0.3 O_2/93.3 He, b) 12.6 CO_2/20.6 N_2/61.6 He/5.2 H_2, c) 14.7 CO_2/75.3 air/10 He. Coupling mirror diameter 2" for a) and 1.5" for b) and c).

gas mixture	laser power (W)	specific energy (J/g)	effi- ciency (%)	mass flow rate (g/s)	stagn. pressure (bar)	cavity pressure (mbar)
a)	340	11.2	7.1	30.3	2.0	19
b)	280	12.4	6.4	22.6	0.46	16.5
c)	187	13.7	4.3	13.7	0.20	12

waveguide from the flow channel would render considerably higher specific energies. This is provided by dielectric wall inserts. Further improvements would aim at larger active lengths and better coupling characteristics of the resonator arrangement.

Scaling of this laser to a multikilowatt power level is straightforward. A possible upscaled version of the microwave laser, using parallel waveguides is depicted in Fig. 6.

CONCLUSIONS

The microwave discharge arrangement, originally designed for exciting a supersonically flowing CO laser medium, has also been applied successfully to operate a 10.6 µm CO_2 laser. Maximum achieved power values are comparable to values from the CO laser.

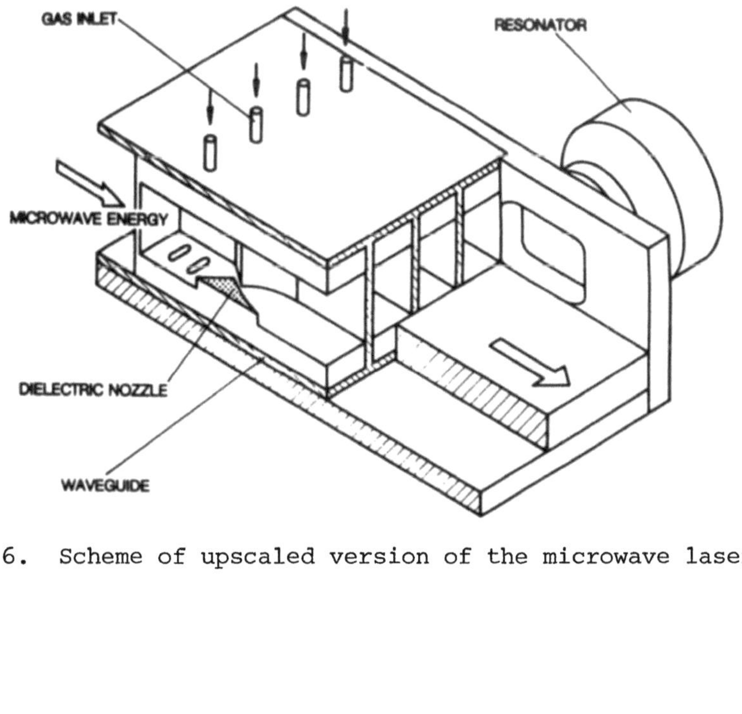

Fig. 6. Scheme of upscaled version of the microwave laser.

However, requirements for total mass flow and stagnation pressures are much relieved. The CO_2 laser may not only be operated with the classical gas composition of $CO_2/N_2/He$, but also with CO_2 and air. With air, maximum achievable laser power is reduced by about one third. Significant increases in laser power were achieved through the addition of H_2 in quantities of up to 10 %. H_2 has a favorable influence on the discharge uniformity, but also participates in the laser kinetics. With H_2, the maximum obtainable laser power in any case was limited by the available generator power only. The successful application of the system to two different laser media promises its usefulness for other lasers as well.

References

1. A. P. Golovitskii, V. A. Kruzhalov, T. M. Perchanok, S. G. Ronanov, "Investigation of the amplifying properties of a CO_2-N_2-He medium with different microwave preionization systems", Sov. J. Quant. El., 11: 535 (1981).
2. O. S. Vasyutinskii, V. A. Kruzhalov, T. M. Perchanok, D. K. Terekhin, S. A. Fridrikhov, "Pulsed microwave discharge as a pump for the CO_2 laser", Sov. Phys. Tech. Phys., 23: 189 (1978).
3. K. G. Handy, J. E. Brandelik, "Laser generation by pulsed 2.45-GHz microwave excitation of CO_2", J. Appl. Phys., 49: 3753 (1981).
4. R. Joeckle, G. Stern, "Création d'énergie vibrationelle par décharge électrique autonome dans un mélange gazeux laser CO_2 en écoulement supersonique", ISL-Report R 111/77 (1977, unpublished).
5. P. Hoffmann, H. Hügel, W. Schall, W. Schock, "cw carbon monoxide laser with microwave excitation in the supersonic flow", Appl.Phys. Lett., 37: 673 (1980).
6. P. Hoffmann, H. Hügel, W. Schall, W. Schock, "cw carbon monoxide laser with microwave excitation in the supersonic flow", J. de physique, Colloque C9, supplément au no. 11, 41: 217 (1980).
7. P. Hoffmann, W. Schall, H. Hügel, W. Schock, "Microwave excited supersonic carbon monoxide laser", Proc. Int. Conf. on Lasers 80, STS Press, McLean (1981).
8. K. M. Duff, P. G. Hills, "Condensation of Carbon Dioxide in Supersonic Nozzles", Proc. 1966 Heat Transfer and Fluid Mechanics Institute, Stanford Univ. Press (1966).
9. T. F. Deutsch, "Effect of Hydrogen on CO_2 TEA Lasers", Appl. Phys. Lett., 20: 315 (1972).
10. B. L. Wexler, T. J. Manuccia, R. W. Waynant, "cw and improved pulsed operation of the 14- and 16 μm CO_2 lasers", Appl. Phys. Lett., 31: 730 (1977).
11. D. S. Stark, P. H. Cross, M. R. Harris, "A sealed uv-preionization CO_2 TEA laser with high peak power output", J. Phys. E : Sci. Instrum., 11: 311 (1978).

SCREEN NOZZLE MIXING LASER CALCULATIONS

W. Schall and H. Hügel

DFVLR-Institut für Technische Physik
Pfaffenwaldring 38-40, D-7000 Stuttgart 80

INTRODUCTION AND BRIEF OUTLINE OF THE MODEL

The gasdynamic CO_2/N_2 mixing laser is generally considered a device to overcome some of the inherent power limitations of classical gasdynamic CO_2 lasers. However, in comparison with expected performance data the experimentally obtained values are almost an order of magnitude lower. Extensive parametric variations of gas composition and stagnation temperature[1] did not reproduce the expected trend of increasing small signal gain and power with increasing stagnation temperature of the N_2. A definite correlation of the experimental results to either a theory implying instantaneous mixing[1] or one with laminar mixing[2] could not be found.

Here, a simple semi-empirical model is suggested that allows for the rapid mixing process in gasdynamic lasers of the screen nozzle type. As a base for the numerical model a kinetic code[3] is transformed into a 1-dimensional space dependent description of the pertinent variables. Since the gas temperature and vibrational excitation temperatures enter this model as the independent variables they have been prescribed as starting and boundary conditions according to the following considerations and assumptions:

- N_2 as the primary gas flow and CO_2 plus either He or H_2O as the secondary flow are expanded isentropically in separate nozzles.

- The vibrational temperature of the N_2 at the exit of the N_2 nozzles is assumed in equilibrium with the N_2 stagnation temperature T_{ON}. The vibrational temperatures of CO_2 are assumed in equilibrium with the exit temperature of the CO_2 stream.

- A 1-dimensional mixing process is assumed to occur over a finite distance z_M in flow direction, according to a free mixing parameter $A_t = h/z_M$, where h is the distance between the axes of adjacent N_2 and CO_2 nozzles (Fig. 1). For typical mixing processes A_t runs between 0.01 and 0.1. Linearly spreading mixing zones are assumed. Only in these zones inversion is produced.

- In mixing processes the final temperature T_M is always higher than the arithmetic mean value T_A of the exit temperatures of the jets to be mixed. From qualitative considerations a linear increase of the gas temperature from T_A to T_M over the mixing length z_M appears reasonable (Fig. 1). With this assumption, the energy equation need not be regarded, in contrast to more rigorous models[4,5].

- In cases where the secondary mass flow enters the mixing region unmatched from sonic orifices, a somewhat higher secondary Machnumber M_C is assumed to account for possible post expansions.

Small signal gain distributions along z are the primary output of the calculations. Upper and lower bounds of the extractable power can also be derived. The upper bound is given, if all remaining vibrational energy stored in the N_2 can be exploited. In general, this requires a broad resonator in flow direction or a multipass arrangement to account for the energy exchange time between N_2 and CO_2. A lower bound is given, if only the energy already transfered to the CO_2 is extracted. This may be achieved in a rather narrow, single pass resonator.

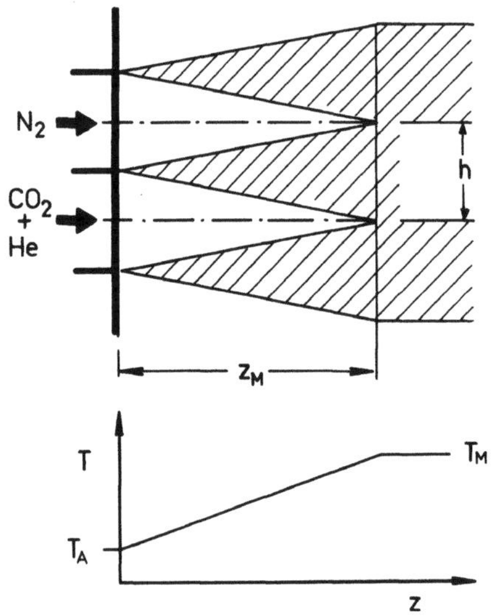

Fig. 1. Scheme of the mixing model.

COMPARISON WITH EXPERIMENTAL DATA

Two free parameters have to be selected for comparison with the experimental results: A_t and M_C. Fortunately, the contributions of both parameters can be clearly discriminated in the results: Inreasing A_t tends to stretch the gain profil in the flow direction, while increasing M_C raises the small signal gain values.

Fig. 2 shows the result of small signal gain calculations with appropriately fitted A_t and M_C for a N_2 stagnation temperature of 2000 K. In general, a selection of A_t = 0.03 and M_C = 1.75 produces results that agree well with the experimental data, irrespective of the presence of helium in the secondary flow. Fig. 3 is an example of the calculated extractable power (lower bound). In experiments, where the expansion ratios and gas composition are not selected for minimum gas temperature T_M, and hence, minimum loss rates, a rather significant amount of vibrational energy is lost by deactivating collisions before the mixing is complete (Fig. 4). In addition, the small signal gain decreases above a certain stagnation temperature of the N_2, T_{ON}, when the gas temperature becomes too high (Fig. 5).

The good agreement of these and many other analyzed examples is believed to establish a sound base for the use of this model to predict the performance of a screen mixing laser, when nozzle geometries, stagnation temperatures and gas compositions of both flows are varied.

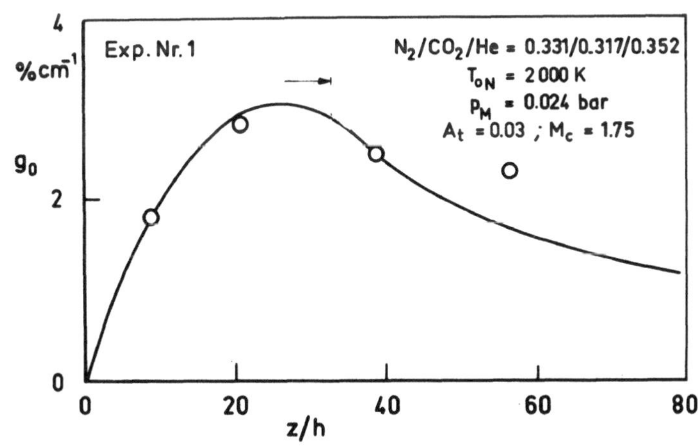

Fig. 2. Comparison of experimental and theoretical small signal gain g_O in normalized flow direction z/h. The arrow indicates the end of the mixing zone.

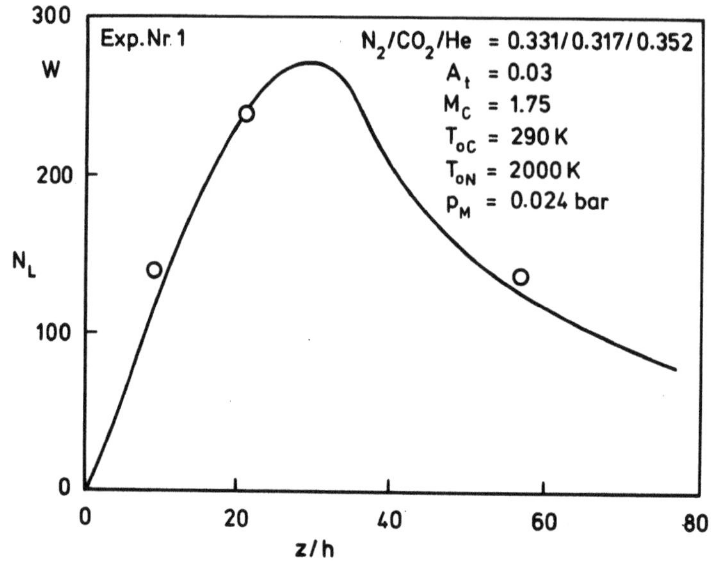

Fig. 3. Comparison of experimental and theoretical laser power N_L.

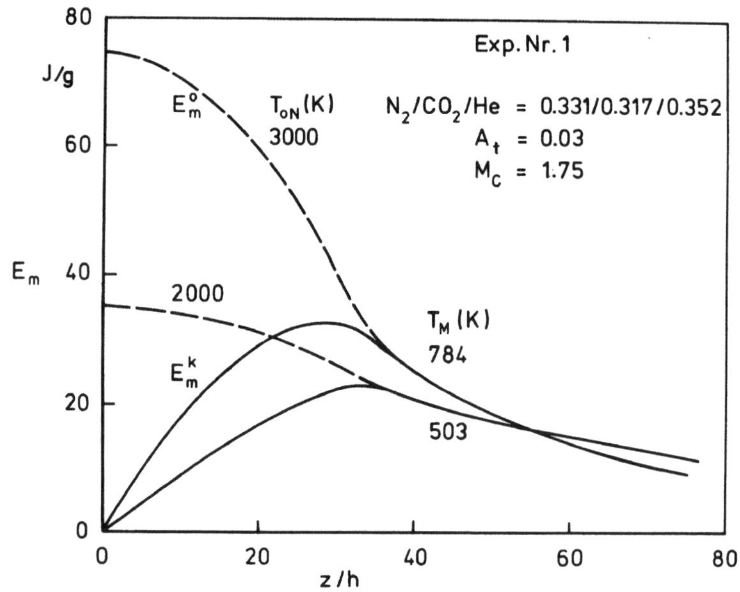

Fig. 4. Variation of total vibrational energy in the flow, E^{o}_{m}, and available vibrational energy in the mixed regions of the flow, E^{k}_{m} .

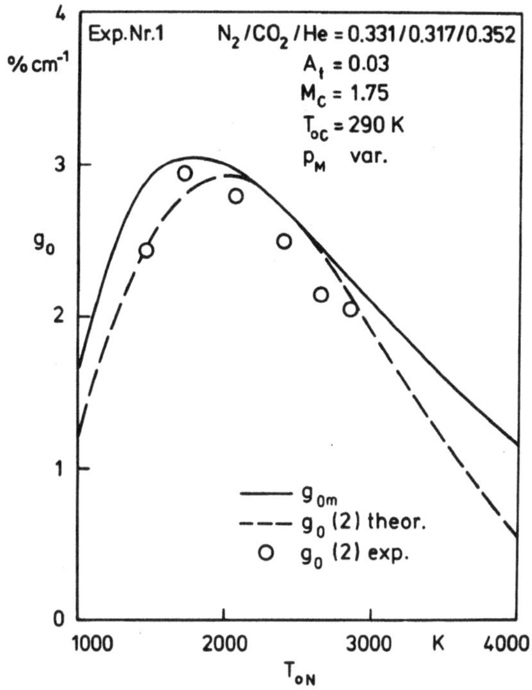

Fig. 5. Variation of small signal gain with N_2 stagnation temperature, T_{ON}: $g_o(2)$ represents values at the second resonator port (z/h = 21); g_{om} is the calculated peak value of the profile $g_o(z)$.

PREDICTIONS AND LIMITATIONS OF A SCREEN MIXING LASER

One way to improve the performance of the CO_2 screen mixing laser is to minimize the crucial parameter, i.e. T_M. Fig. 6 is an example of hypothetically attainable small signal gain values for a mixture optimized for maximum g_{om} (mixtures with a large fraction of He). The nozzle exit temperatures of the gas streams are set to 300 K. From the curve connecting the maxima it is seen, that irrespective of a practical design limit (i.e. by pressure requirements), high values T_{ON} require high stagnation temperatures of the secondary flow, T_{OC}. Due to the low N_2 fraction in the considered mixture a specific energy of 50 J/g will never be exceeded.

There are other ways to improve the laser. On the expense of high small signal gain values, the specific energy can be much improved by significantly increasing the amount of N_2 (Figs. 7a, b). The problem then encountered is the increased length of the resonator in flow direction to allow for the complete transfer of

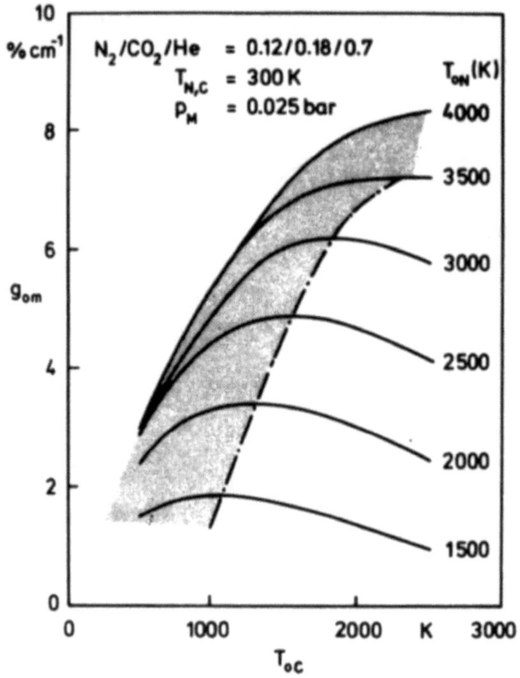

Fig. 6. Dependence of maximum small signal gain on stagnation temperatures.

energy from N_2 to CO_2. For a change in composition, marked by the dots in Figs. 7a,b, the required resonator width has to be stretched in flow direction by about two orders of magnitude.

The short run times of the numerical model on the computer (ca. 7 sec on a Siemens 7551 computer) allowed optimizing the primary input variables in dialogue. From this, ultimate performance values of the power potential within technological limitations could be derived. The main constraint is set by the stagnation pressures, which have been limited arbitrarily to 200 bars. (Only shock wave operated lasers may exceed this value for present day technology). A further, yet minor constraint has been imposed on the calculated resonator width (b_R < 30 cm). The temperature of the CO_2 gas is assumed not to drop below the condensation value. In any case, the cavity pressure is assumed to 0.05 bar, allowing gasdynamic recompression to atmosphere for all but the lowest stagnation temperatures.

The same optimization has been carried out for a gas mixture containing a low percentage of H_2O instead of He. In this case an optimum N_2 temperature was found to be ca. 3000 K. The upper bounds of technically achievable performance values for the investigated

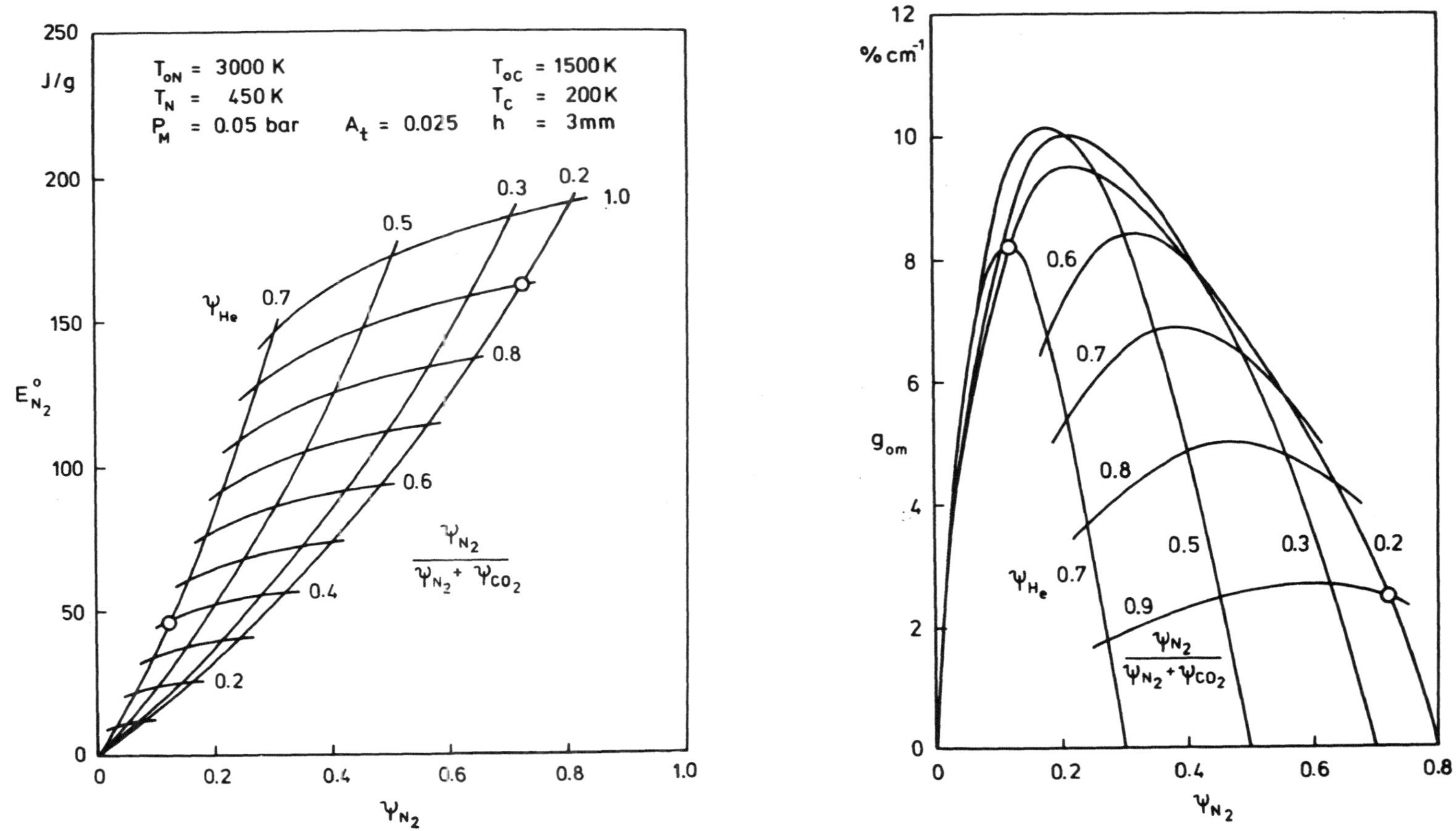

Fig. 7a,b. Variation of initial vibrational energy, $E^o_{N_2}$, and peak small signal gain, g_{om}, with gas composition.

laser are stated in the table. Efficiency is defined as maximum achievable laser power per enthalpy deposited in both gas streams. In the case with He a fraction of 20 % is assumed to promote sufficient lower level depopulation and raise the secondary flow speed. In the second example the gas composition is 10 % CO_2 and 0.5 % H_2O. The small signal gain values stated in the table correspond to the temperature for maximum power, but can be higher for lower values of T_{ON}.

Table. Limits to the performance data of screen nozzle mixing lasers

Laser gas mixture	N_2 stagnation temperature K	specific energy J/g	J/l	power flux kW/cm^2	peak small signal gain $\% \, cm^{-1}$	efficiency %
$N_2/CO_2/He$	4000	230	7.5	1.9	1.3	2.8
$N_2/CO_2/H_2O$	3000	150	6.5	1.3	1.6	3.5

CONCLUSIONS

A simple, semi-empirical model for a gasdynamic mixing laser with screen nozzle mixing has been developed, that

- describes the experimentally found laser performance values with sufficient accuracy

- led to the better understanding of experimentally observed trends and the focussing on the pertinent parameters

- allows predictions for design considerations and gives upper bounds of the achievable laser power

- encompasses the power potential of this laser within physical and technological constraints.

References

1. H. Hügel, W. Schall, P. Hoffmann, "Experimental Investigations of Gasdynamic N_2/CO_2 Mixing Lasers", Journal de Physique, 41: C9-335 (1980).
2. K. N. Parthasarathy, J. D. Anderson Jr., "Downstream Mixing Gasdynamic Lasers - A Numerical Solution", AIAA J. 17: 1208 (1979).
3. R. J. Harrach, T. H. Einwohner, "Four Temperature Kinetic Model for a CO_2 Laser Amplifier", UCRL-Report 51399 (1973).
4. K. Horioka, K. Kasuya, T. Murasaki, "Simplified Model for Cavity Flows of Mixing-Type CO_2 Gasdynamic Lasers", Japan. J. Appl. Phys. 17: 745 (1978).
5. K. Maeno, K. Funabiki, H. Oguchi, "Experimental and Analytical Study of CO_2/N_2 Mixing Gasdynamic Laser", ISAS Report No. 593 46: 175 (1981).
6. W. Schall, H. Hügel, "Modellrechnungen zum CO_2-Mischungslaser mit Siebdüse", to be published as DFVLR-FB Report.

THEORETICAL DESIGN OF MIXING GASDYNAMIC LASERS

Hai-Xing Yan* and Ji-Hua Xu

Institute of Mechanics
Chinese Academy of Sciences
Beijing, People's Republic of China

ABSTRACT

Using theoretical simulation of nonequilibrium flow, the performance of CO_2 mixing gasdynamic lasers (MGDL) has been thoroughly studied. A scheme to premix a nitrogen stream with small amounts of water which is in turn later mixed with a CO_2 stream is presented, and the feasibility for lasing when $T<3000K$ is shown. The effects of velocity, temperature, compositions, etc. on the performance of the lasers have been investigated and approaches for improving performance are considered. The effects of the radiation field are extremely important for CO_2 MGDL. The design of a cavity from which laser energy can be extracted more effectively is presented. On the basis of the mechanism of microscopic dynamics, physical explanation of various relationships are given.

INTRODUCTION

As is well known, gasdynamic laser (GDL) is an important high energy laser device.[1] Intrinsic defects of a premixed GDL are: (1) because of dissociation of CO_2 it can not work at very high temperatures (T_0 <2300K), so thermal efficiency is low; (2) premixing of CO_2, N_2 and catalyst in the reservoir decreases the freezing efficiency of the device. Mixing gasdynamic laser (MGDL) can overcome these defects effectively and has improved efficiency, specific power, gain, etc.[2] This improvement has motivated research on MGDL in different countries for the past twelve years.

*Present address: Department of Physics & Astronomy, The University of Mississippi, University, MS 38677, U.S.A.

In MGDL, a key problem is mixing nonequilibrium gasdynamics. Using numerical simulation, optimum design and operating conditions of devices can be determined. Based on the simulation of a premixed GDL,[3] a Gradually Mixing Mode (GMM)[4] and comparison of different models in simulation of MGDL,[5] the Instantaneously Mixing Model (IMM) has been chosen for numerous calculations because it can yield various important relationships needed in theoretical design with much less computer time.

SCHEME OF PREMIXING WATER IN A NITROGEN STREAM

Premixing CO_2 with small amounts of water and then mixing with a high-speed N_2 stream has been used widely.[2] But, it is believed that accurately and continually adding a small amount of water to the CO_2 stream at low flow and low reservoir temperature has made manufacture and operation of the device more difficult. Furthermore, in order to keep water unfrozen, the CO_2 stream has to be preheated up to >1100K;[6] this leads to additional problems. Inversely, if without water, lower reservoir temperature of CO_2 stream (500-600K) is necessary and lower temperature after expansion is favourable for performance of MGDL (vide post). Therefore, a scheme of premixing water in N_2 stream has been presented and investigated.

The dissociation of water at high temperature will be considered first. The dissociation reactions are

$$H_2 + 1/2\ O_2 \rightleftarrows H_2O(g) + 57{,}107\ \text{Cal/mole} \tag{1}$$

and

$$OH + 1/2\ H_2 \rightleftarrows H_2O(g) + 67{,}107\ \text{Cal/mole.} \tag{2}$$

$$K_1 = \frac{X_{H_2}\sqrt{X_{O_2}}}{X_{H_2O}}\sqrt{p} \tag{3}$$

and

$$K_2 = \frac{X_{OH}\sqrt{X_{H_2}}}{X_{H_2O}}\sqrt{p} \tag{4}$$

where X_i is the equilibrium mole fraction of i and p is the total pressure. The other two equilibrium equations are

$$2X_{H_2} + X_{OH} + 2X_{H_2O} = 2(2X_{O_2} + X_{H_2O} + X_{OH}) \tag{5}$$

and

$$X_{H_2} + X_{OH} + X_{O_2} + X_{H_2O} = 1 \tag{6}$$

From (3)-(6), X_i can be calculated and are presented in Table I, where the K_i have been taken from Ref.[7]. It can be seen from Table I that when T<3000K, the dissociation of water is not large. An increase of pressure is favourable to restrain the dissociation of water.

Table I. Dissociation of Water at High Temperature

T(K)	p = 10 atm				p = 50 atm			
	X_{H_2O}	X_{H_2}	X_{O_2}	X_{OH}	X_{H_2O}	X_{H_2}	X_{O_2}	X_{OH}
2200	0.98767	0.00691	0.00281	0.00261	0.99277	0.00406	0.00164	0.00153
2600	0.94613	0.02869	0.01073	0.01446	0.96801	0.01703	0.00637	0.00859
3000	0.84735	0.07781	0.02692	0.04792	0.90661	0.04760	0.01647	0.02932
3500	0.64137	0.17486	0.05531	0.12846	0.76427	0.11493	0.03636	0.08444

Using the GDL computer program,[3] a scheme of premixing N_2 with water has been checked. The results are given in Table II. It is shown that for two-dimensional and conic nozzles, if <5% water is premixed in N_2, the available energy stored in N_2 only decreases a few percent. The scheme is feasible for lasing when T<3000K. In comparison, in a premixed GDL much more available energy is lost.

Table II. Effect of Premixing H_2O with N_2 (Unit of E_{max} = J/g)

Nozzle	A/A*	h*(cm)	T_0(K)	p_0(atm)							
					X_C	0.00	0.00	0.00	0.00	0.00	0.10
					X_N	1.00	0.00	0.98	0.95	0.90	0.89
					X_H	0.00	0.01	0.02	0.05	0.10	0.01
Two dimen-sional	29.33	0.03	2200	10	T	345.2	349.0	352.8	364.3	383.5	420.0
					T_N	2190.9	2182.0	2173.9	2149.5	2113.7	2037.1
					E_{max}	111.12	109.43	107.82	103.03	95.58	85.54
Conic	60	0.04	2200	10	T	258.6	262.8	265.6	274.7	289.7	291.0
					T_N	2197.3	2188.2	2181.0	2156.2	2121.6	2076.4
					E_{max}	112.88	111.22	109.76	105.11	98.07	91.43
Conic	60	0.04	2200	60	T	260.2	266.6	270.8	284.0	298.8	332.5
					T_N	2191.4	2146.4	2106.2	2004.5	1909.5	1677.8
					E_{max}	112.28	107.05	102.38	90.61	78.81	57.83
Conic	60	0.04	3000	10	T	354.5	360.1	364.9	380.4	407.5	397.2
					T_N	2977.5	2972.4	2963.3	2931.8	2872.9	2824.2
					E_{max}	193.81	191.87	189.52	181.96	168.72	156.26

EFFECTS OF GASDYNAMIC PARAMETERS ON PERFORMANCE OF MGDL

The N_2-H_2O stream is denoted as stream 1 with T_0 = 2200K, p_2 = 10 atm. A conic nozzle (A/A* = 60, h* = 0.04cm) is used for the theoretical design of CO_2 MGDL (See Table II). The pure CO_2 stream is denoted as stream 2 and its reservoir conditions and nozzle parameters can be determined from parameters presented here. The areas of screen nozzle for streams 1 and 2 are 11.9022 and 1.7374 cm^2 respectively. We assume $p_1 = p_2$.

Effects of u_2

It is shown in Fig. 1. that when $u_2 = u_1/0.9$, G_0 is the highest and both larger and smaller u_2 decrease G significantly. When $u_1/u_2 = 1.2, 1, 0.9, 0.8$, $X_C = 0.164, 0.191, 0.208, 0.228$ respectively and G_0 is proportional to X_C. So roughly speaking, when $u_1 \sim u_2$, G_0 is largest.

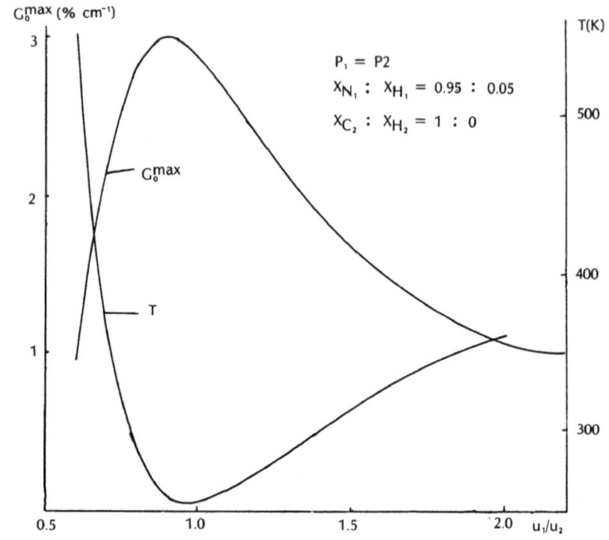

Fig. 1. Effect of u_2 on CO_2 MGDL

The physical explanation is that a difference in velocities for the two streams makes translational temperature T in the mixing region higher (see Fig.1) and higher T is harmful to MGDL. First, relaxation of available vibrational energy is quicker at higher T. Second, originally lower $T\nu_2$ is increased by fast $\nu_2 \leftrightarrow T$ process to approach higher T, and $T\nu_1 \sim T\nu_2$ (due to Fermi resonance between ν_1 and ν_2)[3,8], so that population inversion is smaller. Third, the negative temperature dependence of the $\nu_3 \leftrightarrow \nu_N$ rate makes pumping of the upper laser level so that $T\nu_3$ is lower.

Effect of T_2

Fig.2 shows that when a colder stream 2 is mixed with the same stream 1, a colder mixing stream, so that a higher G_0 is produced. If $u_2 \pm u_1$, similar results are obtained.

A decrease of T_2 is favourable for MGDL. The limitation is that stream 2 can not be frozen. Therefore, without premixing water with CO_2, the lowest T_2 can be changed from $\sim$270K to $\sim$170K so that G_0 changes from 1.8 to 2.8%cm^{-1}.

Effect of Composition

Fig.3 shows that the higher X_{H2}, the lower G_0. This is a unique result for MGDL. Unlike the premixed GDL, CO_2 in the MGDL has low translational and vibrational temperatures. The fast relaxation rates of water is harmful for MGDL, i.e., to increase $T\nu_1$, to relax available vibrational energy. But, in a radiation field, water will

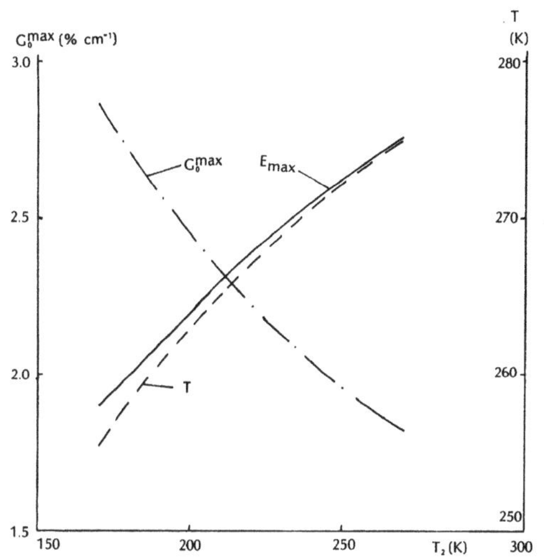

Fig.2. Effect of T_2 on CO_2 MGDL

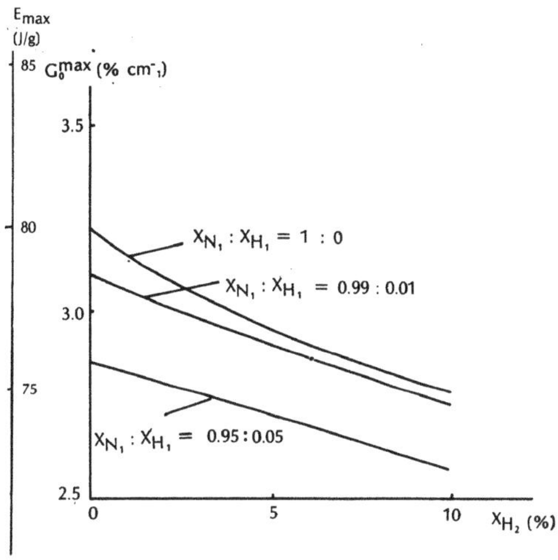

Fig.3. Effect of X_H on CO_2 MGDL

be indispensable (vide post). The decrease G_0 with X_{H1} is because of the relaxation effect of water in the mixing region and the harmful effect of water premixed with N_2 on freezing efficiency of stream 1 (see Table II).

If the design of MGDL is $p_1 \sim p_2$, $u_1 \sim u_2$ and as low T_2 as possible, X_C in the mixing flow will only depend on flow of stream 2, i.e., area of stream 2. Due to the proportionality of G_0 to X_C, G_0 increases with X_C. But, at the same time, E_{max} will decrease significantly because the corresponding decrease of X_N makes available energy mainly stored in N_2 lower and CO_2 can relax more available energy through $\nu_N \leftrightarrow \nu_3$, $\nu_3 \leftrightarrow \nu_1, \nu_2$ and $\nu_2 \leftrightarrow T$. On the other hand, gain in MGDL has been high enough to extract laser energy. Unnecessarily high gain is unfavourable for the device (for example, it can produce parasitic oscillation). Thus, a little lower X_C (<10%) is preferable in MGDL.

EFFECTS OF RADIATION FIELD AND DESIGN OF CAVITY

The final goal of the design process is to obtain the highest laser output. Sometimes, it is not significant that G_0, E_{max}, or both are very high, as is true without water where both G_0 and E_{max} are quite high but little laser output can be obtained. Consideration of the effects of the radiation field is extremely important, especially in CO_2 MGDL.

The effect of the radiation field is considered in a manner similar to Ref.[9]. In another paper,[10] a description of saturated gain variation in CO_2 GDL and the physical explanation were given. It is shown in Fig.4 that the saturated gain of the laser medium

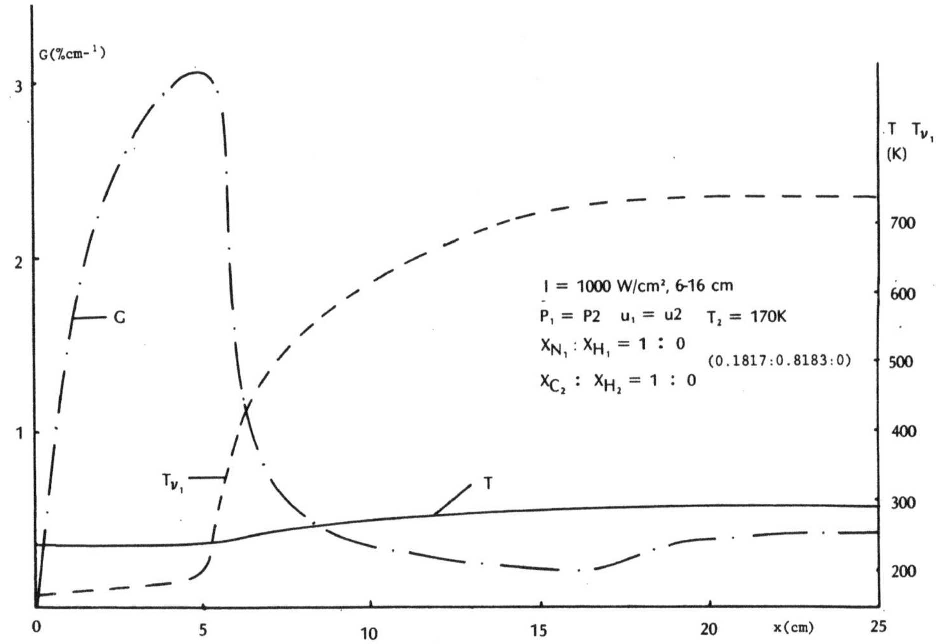

Fig.4. Change of Saturated Gain in Laser Medium Without Water

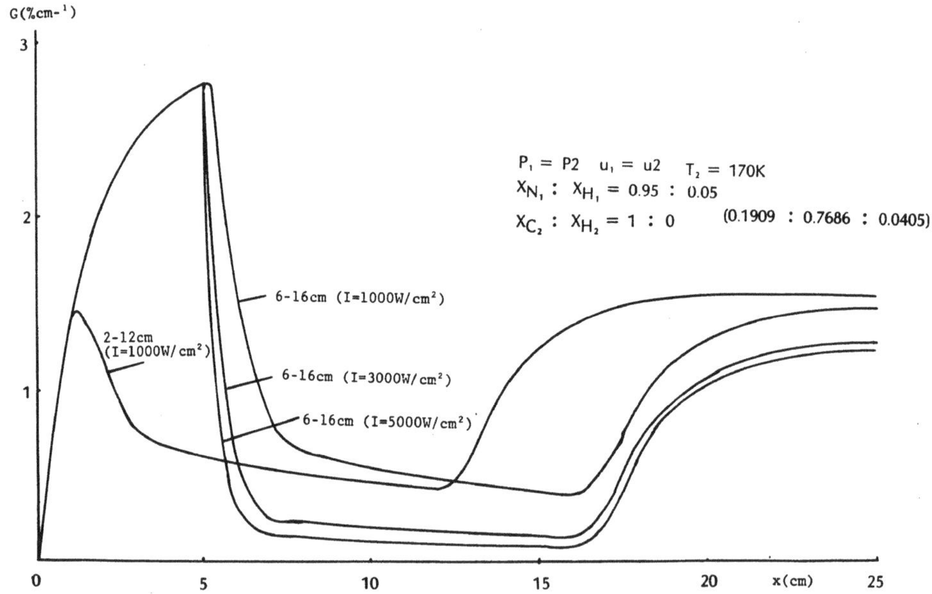

Fig.5. Change of Saturated Gain in Laser Medium With Water

without water is low in the cavity and gain can not rise again outside the cavity. The reason is that without water there is no way to deplete the lower laser level so that radiation yields a rapid accumulation of lower laser level population. Conversely, with water, lower $T_{\nu 1}-T$ and higher saturated gain in the cavity are obtained and gain can be recovered quickly outside the cavity (see Fig.5). It is found that too much water deactivates more available vibrational energy and too little water is not enough to deplete the lower level and recover the gain outside the cavity. A 2-4% mole fraction of water in a CO_2 MGDL seems optimum (see Table III).

Table III. Some Parameters in the Middle of Cavity
($I = 1000 W/cm^2$, length of cavity = 10 cm)

X_H	0	0.0082	0.0182	0.0266	0.0405	0.0595	0.0801
$T_{\nu 1}-T$(K)	362.02	276.51	225.75	184.76	132.13	97.62	67.25
G(cm^{-1})	0.00339	0.00465	0.00493	0.00521	0.00545	0.00486	0.00472
E_{max}(J/g)	60.25	59.22	63.40	60.18	52.18	50.79	42.32

In CO_2 MGDL, the process which pumps the upper laser levels is $\nu_3 \leftrightarrow \nu_N$ which is not extremely rapid. Even if the mixing happened instantaneously, the maximum of G_0 appears at 7-10 cm downstream of the exit nozzle. Considering the limited mixing velocity, it should be further downstream. Additionally, inhomogeneity of the laser medium upstream in a practical device is harmful for beam quality. This is why we calculated the performance of the cavity further downstream. Fig.5 shows that saturated gain in the cavity of 6-16 cm is higher than that of 2-12 cm. Using the cavity of 6-16 cm, higher laser output can be obtained. The steady oscillation condition is[9]

$$\int_0^M (\ln r_1 r_2 + 2GL)\,dx = 0 \tag{7}$$

where r_i are the reflectivities of the mirrors, M is the length in the flow direction, L is the distance between cavity mirrors and G is the saturated gain. The larger G, the higher the output fraction is, i.e., the higher the laser output is when radiation intensity I is same. But, too downstream cavity is not only unnecessary but also unfavourable because of the deactivation of available energy in mixing flow.

Fig.5 shows that higher I yields a very rapid decrease of G, that is, output fraction should be lower so that laser output would not be much higher. And, too high I will yield too rapid decrease of G to keep I constant in the cavity. We have not taken this inhomogeneity of I into account. Similarly, simply to lengthen the cavity in the flow direction x is not effective. A new cavity design suitable for CO_2 MGDL in which mirrors are divided into several groups in x direction and laser energy can be extracted more effectively is presented. A configuration of every section of mirror 5-10 cm in length and interval of about 5 cm in x direction is suggested. Physically, this is because relaxation rates both to pump the upper laser levels and deplete the lower levels in the medium are not rapid enough when I is quite high[10] and in the interval the gain can recover to a higher value.

CONCLUSION

1. The scheme to premix stream N_2 with small amount of H_2O, which is, in turn, later mixed with a stream of CO_2 has been shown feasible for lasing when T<3000K. Some difficulties of manufacture and operation can be avoided and the possibility of decreasing the temperature of stream 2 is favourable for improvement of performance in CO_2 MGDL.

3. Consideration of the effects of the radiation field is extremely important, especially in CO_2 MGDL. Small amounts of water plays a very important role in depleting lower laser levels in and downstream of the cavity to maintain and recover the gain. Either simply to enhance I or lengthen the cavity in the flow direction are not effective. Using a new cavity design in which mirrors are divided into several groups in x direction with intervals, laser energy can be extracted more effectively. The beginning of the first group of mirrors should be set up properly downstream.

3. A u_2 approximately equal to u_1 and as low a T_2 as possible are favourable for improvement of performance of MGDL. 2-4% mole fraction of water in mixing flow seems optimum. Too much CO_2 (>10%) is not only unnecessary but also unfavourable.

4. The concept of downstream mixing really is a powerful approach to imrpove performance of GDL.

REFERENCES

1. J.D. Anderson, Jr., Gasdynamic Lasers: An Introduction. Academic, New York (1976).
2. P.E. Cassady, A survey of gasdynamic laser concepts, J. Energy, 4:145 (1980).
3. Hai-Xing Yan and Li-Yin Chen, Calculation of nonequilibrium flow in gasdynamic lasers, Acta Mechanica Sinica, No. 4:174 (1978). (in Chinese)

4. Hai-Xing Yan, Gradually mixing model used to calculate mixing nonequilibrium flow, in Proceedings of the 4th International Conference on Lasers (LASERS '81), New Orleans, December 14-16, 1981.
5. Hai-Xing Yan and Ji-Hua Xu, Two simplified models and their application to performance study of mixing gasdynamic lasers, in Proceedings of the 4th International Symposium on Gas Flow and Chemical Lasers, Stresa, Italy, September 13-17, 1982.
6. P.E. Cassady, J. Newton and P. Rose, A new mixing gasdynamic laser, AIAA J., 16:305 (1978).
7. B. Lewis and G. vonElbe, Combustion, Flames and Explosions of Gases, Academic, New York (1951), 742.
8. Hai-Xing Yan, The vibrational relaxation processes in CO_2-N_2-H_2O laser system: Rates data, Laser Journal, 8(6):1 (1981). (in Chinese)
9. G. Lee, Quasi-one-dimensional solution for the power of CO_2 gasdynamic lasers, Phys. Fluids, 17:644 (1974).
10. Zhong-Xiang Wu and Hai-Xing Yan, Basic rules of saturated gain variation in a flowing laser cavity, Laser Journal, 7(3):5 (1980). (in Chinese)

TWO SIMPLIFIED MODELS AND THEIR APPLICATION TO PERFORMANCE STUDY OF MIXING GASDYNAMIC LASERS

Hai-Xing Yan* and Ji-Hua Xu

Institute of Mechanics
Chinese Academy of Sciences
Beijing, People's Republic of China

ABSTRACT

Two simplified models, Gradually Mixing Model (GMM) and Instantaneously Mixing Model (IMM), have been used to study the performance of CO_2 mixing gasdynamic lasers (MGDL). The relaxation rates and, more importantly, the profiles of the mixing region have considerable effect on computational results of mixing nonequilibrium flow. The computational results of GMM can be compared satisfactorily with solutions of the two-dimensional Navier-Stokes (N-S) equations. It is pointed out that Anderson et al. used an incorrect gain equation which gave about two times the correct gain. In CO_2 MGDL, the effects of mixing rates are not as serious as in chemical lasers. IMM can yield various relationships needed in theoretical design.

INTRODUCTION

As is well known, mixing gasdynamic lasers (MGDL) can overcome some of the defects of a premixed GDL and have improved efficiency, specific power, gain, laser output, etc.[1]

Supersonic diffusion chemical lasers (CL) are very similar to MGDL in models used for numerical simulation, design and operation of the devices, and techniques of measurement. In both CL and MGDL, a key problem is mixing nonequilibrium gasdynamics. Mixing models which give sufficiently accurate results with less computer time

---

*Present address: Department of Physics and Astronomy, The University of Mississippi, University, MS 38677, U.S.A.

are important for theoretical simulation and design.

Anderson and his colleagues have published a two-dimensional solution for the Navier-Stokes (N-S) equations in a CO_2 MGDL.[2] But, the effect of boundary layer losses of nozzles was not included[1] and, more importantly, the numerical calculation is so tedious that it is not practical to select optimum operation conditions by simulating many different conditions. A new Gradually Mixing Model (GMM) has been presented to overcome some of the defects of present computational techniques.[3] In this paper, a simpler model - Instantaneously Mixing Model (IMM) is presented. These three models are compared to select the most suitable one for theoretical design of CO_2 MGDL.

OUTLINE OF GMM AND IMM

In an array configration, streams G and J inject into mixing stream M gradually. X_G and X_J are the mixing lengths. The nozzle exit is taken as $x = 0$ (see Fig.1). A quasi-one-dimensional flow is assumed, i.e., all parameters within a stream tube are uniform and only change with x. Discontinuity of parameters between adjacent stream tubes is allowed.

The mixing region (in stream M, see Fig.1) is able to be divided into several segments. At grid point 1, stream G with area G_1 is instantaneously mixed with stream J with area J_1. Then, stream M develops up to point 2 as a non-equilibrium flow with constant area $G_1 + J_1$ and is instantaneously mixed with G_2 and J_2 at point 2. The resultant continually develops as a nonequilibrium flow with constant area $G_1+G_2+J_1+J_2...$ When $x > X_G$, the flow is a nonequilibrium one with constant area.

All flow parameters after instantaneous mixing at grid points can be obtained from solving the integral formulation of the conservation equations in a steady flow

$$M_0 = \sum_i \rho_i u_i A_i = \rho u A \qquad (1)$$

$$L_0 = \sum_i (\rho_i A_i R_i T_i + \rho_i u_i^2 A_i) = \rho ART + \rho u^2 A = M_0(\frac{RT}{u} + u) \qquad (2)$$

$$H_0 = \sum_i \rho_i u_i A_i (\frac{u_i^2}{2} + H_i) = \rho u A (\frac{u^2}{2} + H) = M_0(\frac{u^2}{2} + H) \qquad (3)$$

For the sake of simplicity, a nonessential assumption is introduced,

$$A = \sum_i A_i \qquad (4)$$

where subscript i = G,J,M. The ethalpy is given by

$$H = RT + E_t + E_r + E_v. \qquad (5)$$

For the CO_2 MGDL, it is assumed that

$$E_v^j = \sum_i E_{v,i}^j \tag{6}$$

where superscript j refers to the vibrational mode. The values of T, u, ρ after instantaneous mixing at grid points can be solved for numerically from Eqs. (1) - (6) and

$$p = \rho RT. \tag{7}$$

For a CO_2 MGDL, the one-dimensional steady nonequilibrium flow equations with constant area can be transformed into a very simple form.[3] Coupled with vibrational relaxation equations obtained from a relaxation model with three modes and four vibrational temperatures,[4] they can be solved numerically using a Runge-Kutta method.

Combining solution of instantaneous mixing at grid points and that of nonequilibrium flow with constant area, the solution for the entire mixing region is obtained in Gradually Mixing Model (GMM).

In the Instantaneously Mixing Model (IMM), it is assumed that at the exit of the nozzle streams G and J are homogeneously mixed instantaneously. That is, the mixing lengths X_i (see Fig.1) in GMM are taken as zero. All flow parameters after mixing can be obtained from solving Eqs. (1) - (6) only once, then the flow is a nonequilibrium one with constant area.

FACTORS AFFECTING COMPUTATIONAL RESULTS OF GMM

Because of gradual mixing, the width of stream M, in which the medium has a gain, is different at different positions in the x direction. In order to describe the inhomogeneous laser medium, Ref.[2] suggested the use of integrated gain. For a quasi-one-dimensional model,

$$G_I = G \frac{h}{h_C + h_N} \tag{8}$$

where G is the gain in the mixing region (its height is h) at x, and h_C and h_N are the heights of streams CO_2 and N_2 at that point, respectively. In Ref.[2], h_G and h_J are 0.0815 and 0.0451 (the unit of length in this paper is cm), respectively.

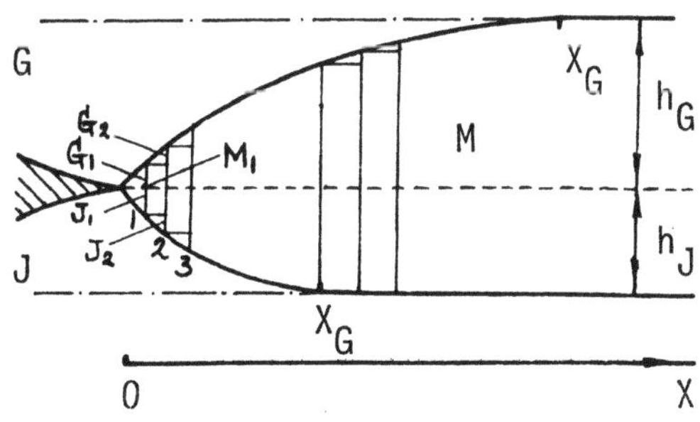

Fig. 1. Mixing Flow Pattern

In GMM, the most indeterminate factor is the profile of the mixing region. The profile must be determined by a different theoretical simulation to specifically investigate the mixing processes or experimental measurements. Ours have been determined from the results of Ref.[2]. The first effect of profile is that because the injection rates of streams G and J are different, the distribution of compositions, temperature, etc. in the mixing region will be different, which makes the effects of relaxation processes different, so that G(x) and $G_I(x)$ [through Eq.(8)] are different. The second one is that the different distribution of h(x) means different distribution of $G_I(x)$ directly.

In principle, any type of profile can be used in GMM. But, in fact, we only used linear profile and two kinds of quadratic curves. One of the quadratic curves is called rear-flat profile in which

$$h = ax^2 + bx. \quad (9)$$

At point X_i the curve is parallel with x and $h(X_i) = h_i (i=G,J)$ (see curve 1 in Fig.2). In another profiel (curve 2 of Fig.2)

$$h = \sqrt{ax^2 + bx.} \quad (10)$$

The point x = 0 is the vertex and the curve passes through another point (x', 0.02). a and b are obtained from these requirements. It is called an elliptic or hyperbolic profile and denoted as $X_G(x'_G)$-$X_J(x'_J)$. Also, it is required that $-b/2a \geq X$, otherwise, a maximum will lie between 0 and X (curve 3 in Fig.2).

The computational results for linear and rear-flat profiles are quite similar, the results of rear-flat profiles at small x are much lower than those of Ref.[2] and comparison of results of elliptic or hyperbolic profiles with those of Ref.[2] is much better (see Fig.3). The change of profiles has had small effect on the evolution of the gain (the gain outside the mixing region). However, the distribution of integrated gain in the mixing region is quite sensitive to the profiles.

The relaxation rates are another important factor which affect computational results. We have used relaxation rates surveyed by us,[5] using these gave good agreement with experimental small-signal gain of premixed CO_2 GDL.[4] The rates used in Ref.[2] were taken from Ref.[6] and have been denoted as A new rates. These rates and our own $\nu_3 \leftrightarrow \nu_N$ rate have

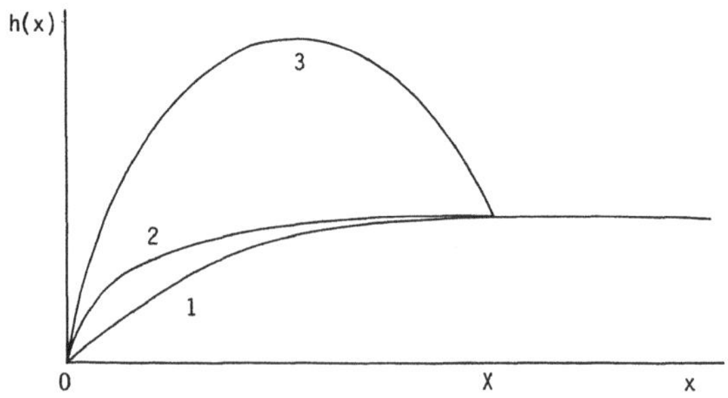

Fig. 2. Profiles of Mixing Region

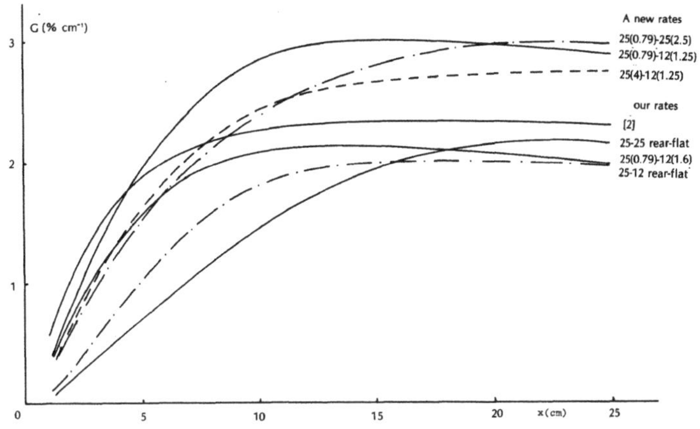

Fig. 3. Effects of Profiles on Results of Simulation

been used in the caluclation. For the process $CO_2(\nu_3 \leftrightarrow 3\nu_2)$, Anderson's $p\tau_a$ used in the linearized relaxation equation should be multiplied by $[1-\exp(-\Theta_2/T)]^3/[1-\exp(-\Theta_3/T)]$ when used in our non-linearized equation.[7]

Fig.4 shows that a higher $\nu_2 \leftrightarrow \nu_N$ rate yields a higher gain, expecially in the initial part of the mixing region. Using A new rates, much higher gain has been obtained when the medium includes water. It mainly comes from the difference in $CO_2(\nu_2 \leftrightarrow T)$-$H_2O$ rates (see Table I).

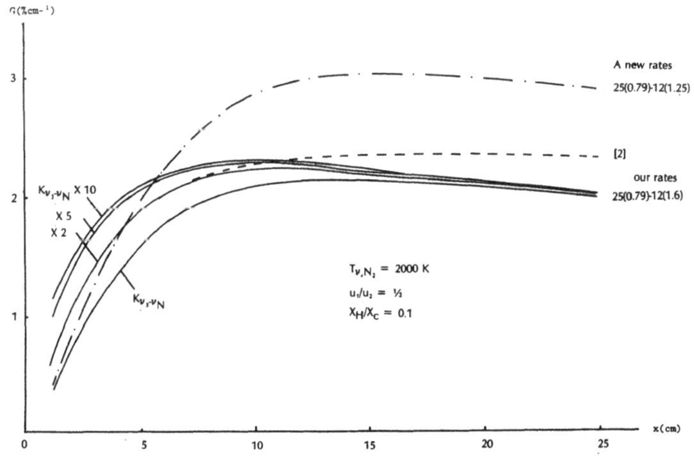

Fig. 4. Effect of Relaxation Rates on Results of Simulation

Table I. Difference Between our Relaxation Rates and A New Rates ($p\tau$: μs atm)

Process		200K	300K	400K	500K	600K
$CO_2(\nu_2\leftrightarrow T)-CO_2$	Our	10.44	7.610	5.059	3.686	2.499
	A new	17.21	7.095	4.054	2.721	2.006
$CO_2(\nu_2\leftrightarrow T)-N_2$	Our	25.70	12.08	6.659	4.225	2.965
	A new	34.42	14.19	8.108	5.442	4.012
$CO_2(\nu_2\leftrightarrow T)-H_2O$	Our	0.00080	0.00253	0.00471	0.00699	0.00922
	A new	0.01672	0.01672	0.01672	0.01672	0.01672
$N_2(\nu_N\leftrightarrow T)-H_2O$	Our	25.59	9.477	5.475	3.858	3.027
	A new	30.66	7.742	3.245	1.747	1.088
$CO_2(\nu_3\leftrightarrow 3\nu_2)-CO_2$	Our	5.497	4.016	2.541	1.600	1.031
	A new	6.123	3.157	1.974	1.371	1.018
$CO_2(\nu_3\leftrightarrow 3\nu_2)-N_2$	Our	13.42	11.44	7.256	4.376	2.660
	A new	22.68	11.69	7.310	5.077	3.770
$CO_2(\nu_3\leftrightarrow 3\nu_2)-H_2O$	Our	0.0616	0.0448	0.0444	0.0475	0.0518
	A new	0.055	0.055	0.055	0.055	0.055

COMPARISON OF GMM RESULTS WITH TWO-DIMENSIONAL N-S SOLUTIONS

The gain of the laser medium is

$$G = \left(N_2^{(v)} - \frac{g_2}{g_1} N_1^{(v)}\right) \frac{c^2}{8\pi\nu_0^2\tau} g(\nu), \tag{11}$$

where superscript (v) refers to particles per unit volume, subscripts 1 and 2 refer to lower and upper laser levels, respectively, g_i is degeneracy, c is speed of light, and τ is the spontaneous radiation lifetime of laser levels. The line shape profile function with pressure-broadening and Doppler-broadening at line center ν_0 is[8]

$$g(\nu_0) = \frac{2}{\pi\Delta\nu} B \exp(B^2) \left(\sqrt{\pi} - 2\int_0^B \exp(-\xi^2)d\xi\right), \tag{12}$$

where
$$B = \frac{\Delta\nu\sqrt{\ln 2}}{\Delta\nu_D}, \tag{13}$$

$\Delta\nu$ and $\Delta\nu_D$ are half width for pressure-broadening and Doppler-broadening, respectively. At very low pressure, $B\ll 1$ and

$$g_D(\nu_0) = \frac{2\sqrt{\ln 2}}{\sqrt{\pi}\,\Delta\nu_D}. \tag{14}$$

At very high pressure, $B\gg 1$ and

$$g_P(\nu_0) = \frac{2}{\pi\Delta\nu} \tag{15}$$

Our calculation showed that in CO_2 MGDL, the pressure is a few torr,

$B \sim 0.5$, so that it is necessary to consider the two kinds of broadening simultaneously. In this case $\int_0^B \exp(-\xi^2)d\xi \sim 0.4\text{-}0.5$ and only considering pressure-broadening [using Eq.(15)], the gain obtained is about twice that found using Eq.(12).

We think that Eq.(15) was improperly used in Ref.[2]. In Ref.[2], $T_N(0)$ = 2000K and $T(0)$ = 300K. It is necessary that in mixing flow $T_3<T_N<T_N(0)$, $T_1 \geq T > T(0)$. Using Eq.(12), even with the limiting condition (T_3 = 2000K, T_1 = 300K), a gain of 2.3% cm^{-1} as given by Ref.[2] cannot be obtained. Furthermore, vibrational temperatures calculated from vibrational energies given by Ref.[2] can very easily be compared with those from GMM by which about half of the gain given by Ref.[2] can be obtained. In the present paper, all comparisons with Ref.[2] are using Eq.(15).

From Fig.5 it is shown that using A new rates, GMM results without water are a little higher than those of Ref.[2], but when X_H/X_C = 0.1, and even more when X_H/X_C = 0.5, the GMM results are much higher than Ref.[2]. If a larger value of X in the medium with more water is assumed, the comparison is better. Using our rates, with no water, the results are almost the same as those using A new rates, however, the more water, the larger difference.

In our calculation, the gasdynamic losses in mixing processes have not been considered, although, in principle, GMM can include them. So, our higher integrated gain is understandable.

Maybe, the profiles used in our calculations are too simple to describe the mixing process very well.

Considering the large difference between the two models and our lack of information about the profiles which significantly affect computational results, the GMM results agree nicely with two-dimensional N-S solutions.

COMPARISON OF GMM RESULTS WITH IMM RESULTS

Fig.6 shows that when the mixing length of one stream is constant, the integrated gain approaches a limiting value as the mixing length of another stream decreases. The results approach those of IMM as the mixing lengths of both streams approach zero. The mixing lengths have a very small effect on the evolution of the gain. And IMM results represent the limiting approach to the highest gain most rapidly.

Physically, this is because the deactivation processes of vibrational energy in CO_2 MGDL are not very fast. In the limit of no relaxation processes in the medium, the mixing rates should not affect parameters of the mixing flow. Therefore, the effects of

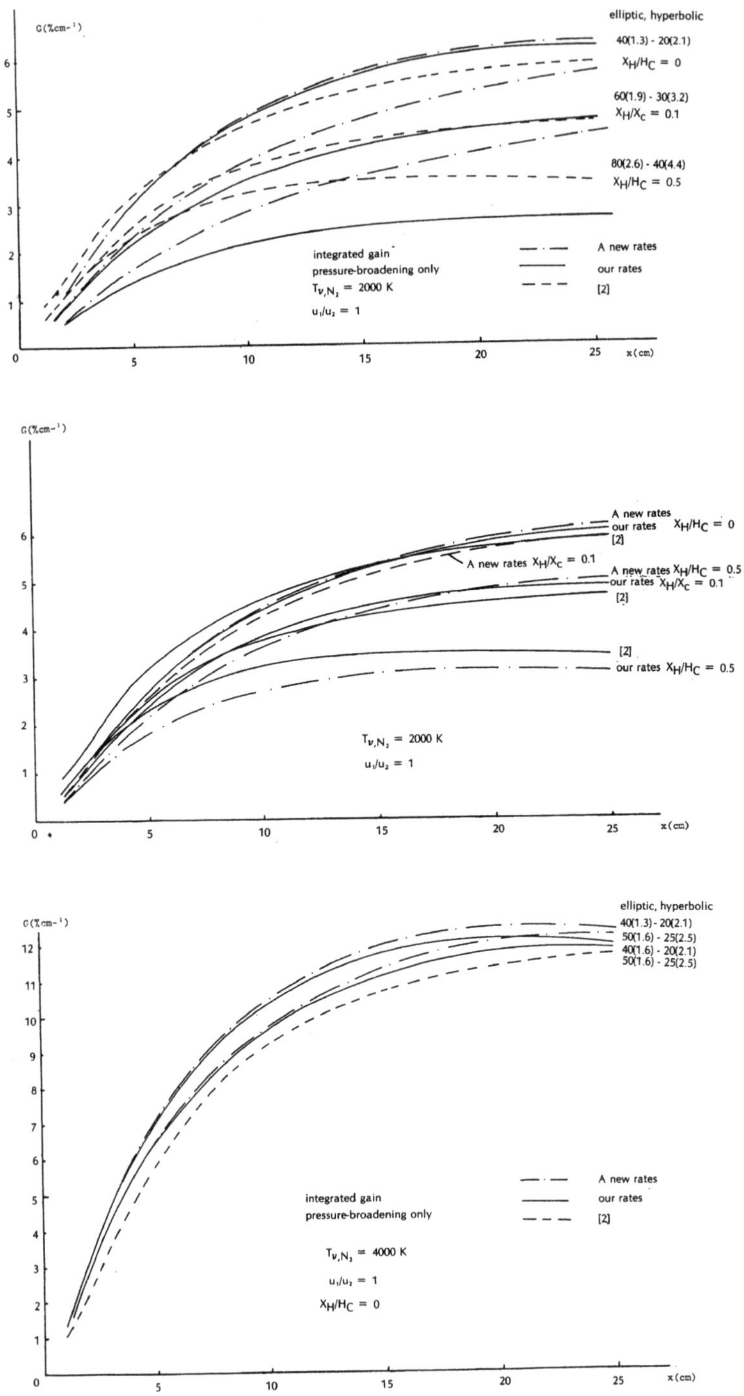

Fig. 5. Comparison of GMM results with Ref.[2]

mixing profiles on the integrated gain are due largely to the different h(x). The computational results show that for quite long mixing lengths or very much water, the effects of mixing rates are more obvious. This is understandable physically.

It can be said that the effects of mixing rates in CO_2 MGDL are not as serious and essential as in chemical lasers. The IMM can yield various relationships needed in theoretical design of CO_2 MGDL[9] and save a lot of computer time (about one order of magnitude). It is known from two-dimensional N-S solutions[2] that in CO_2 MGDL, the mixing processes are basically realized in very short distance (only tens characteristic lengths). In practical screen nozzles, the mixing processes are faster.[10] And we are primarily interested in gain outside the mixing region. We believe that IMM is a good approximation for CO_2 MGDL.

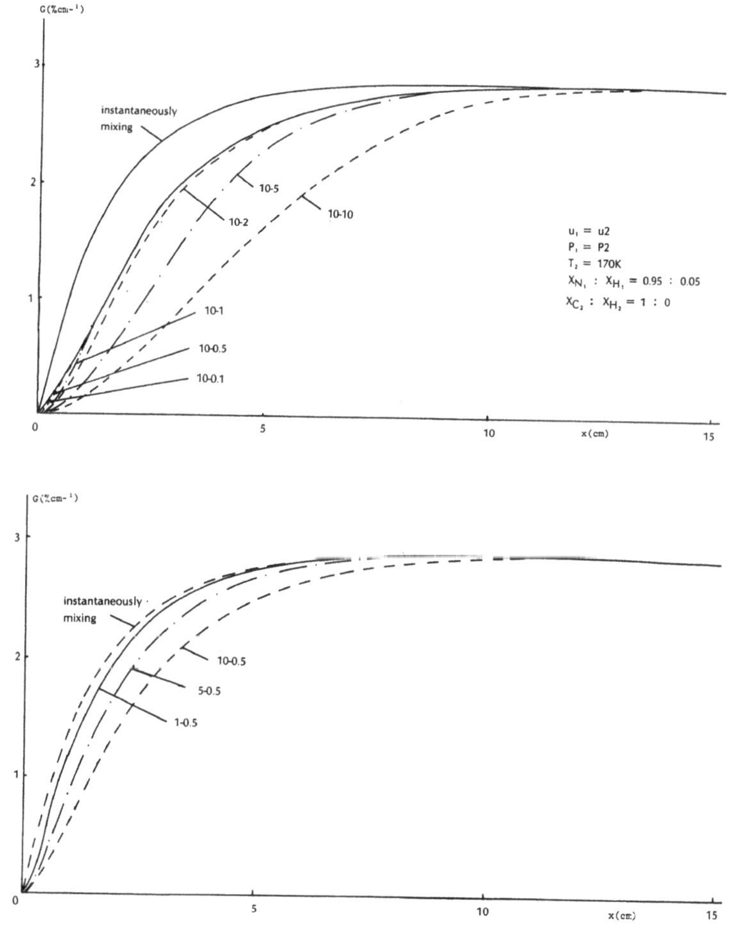

Fig. 6. Comparison of GMM results with IMM results

CONCLUSION

1. Two simplified models (GMM and IMM) can be used to study mixing nonequilibrium flow.

2. The relaxation rates and, especially, the profiles of the mixing region which we do not know a lot about have considerable effects on computational results of GMM.

3. Anderson et al. did not consider the Doppler-broadening effect in their gain formula so that about two times the correct gains were obtained.

4. Using the relaxation rates and gain formula of Ref.[2] and appropriate mixing profiles, the computational results of GMM agree nicely with the two-dimensional N-S solutions.

5. The effects of mixing rates in CO_2 MGDL are not as serious as in chemical lasers. The IMM can yield various relationships needed in theoretical design with much less computer time.

REFERENCES

1. P.E. Cassady, A survey of advanced gasdynamic laser concepts, J. Energy, 4:145 (1980).

2. K.N. Parthasarathy, J.D. Anderson, Jr. and E. Jones, Downstream mixing gasdynamic lasers - a numerical solution, AIAA J., 17:1208 (1979).

3. Hai-Xing Yan, Gradually mixing model used to calculate mixing nonequilibrium flow, in Proceedings of the Fourth International Conference on Lasers (LASERS '81), New Orleans, December 14-18, 1981.

4. Hai-Xing Yan and Li-Yin Chen, Calculation of nonequilibrium flow in gasdynamic lasers, Acta Mechanica Sinica, No.4: 274 (1978). (in Chinese)

5. Hai-Xing Yan, The vibrational relaxation processes in CO_2-N_2-H_2O laser system: Rates data, Laser Journal, 8(6):1 (1981). (in Chinese)

6. J.D. Anderson, Jr., Effect of kinetic rate uncertainties on gasdynamic laser gain and energy predictions, AJAA J., 12:1699 (1974).

7. Hai-Xing Yan, Determination of vibrational relaxation rates from decay constants, Acta Mechanica Sinica, No.3: 319 (1980). (in Chinese)

8. G. Birnbanm, Optical Masers, Academic, New York (1964), 18-19.

9. Hai-Xing Yan and Ji-Hua Xu, Theoretical design of mixing gasdynamic lasers, in Proceedings of the 4th International Symposium on Gas Flow and Chemical Lasers, Stresa, Italy, September 13-17, 1982.

10. D.A. Russell, S.E. Neice and P.H. Rose, Screen nozzles for gasdynamic lasers, AIAA J., 13:593 (1975).

PHOTOIONIZATION OF CO_2 AND CO LASER MEDIA CONTAINING LOW-IONIZATION-POTENTIAL $NO-NO_2$ ADMIXTURES

G. V. Abrosimov, S. A. Akhmanov, K. S. Klopovsky,
E. A. Muratov, T. S. Pulinets, A. T. Rakhimov, and
V. B. Saenko

Institute of Nuclear Physics
Moscow State University, Moscow 117234, USSR

Photoionization as a means for sustaining gas discharges has been extensively investigated in connection with the excitation of dense CO_2 laser media [1]. Plasma produced from spark discharges have been used as a photon source; their UV-radiation acts on a low-ionization-potential admixture intentionally added to the working gas mixture. These admixtures have been organic liquid vapors with the ionization potential E_i = 7.1 to 7.5 eV. The subsequent use of molecular gases possessing a low ionization potential, particularly NO, for which E_i = 9.25 eV [2,3] has been associated with attempts at expanding the application field of photoionization-sustained discharges to include low-temperature gas flow laser systems (CO_2 lasers, plasma chemical reactors, etc.).

The present report shows that the action of a high-repetition-rate UV-radiation on NO molecules added in optimal concentrations to CO_2 and CO-based mixtures makes it possible to sustain successfully a quasi stationary gas discharge with performance characteristics no worse than those of the cw E-beam controlled discharges [4]. In addition, it deals with the plasma chemical processes affecting the electrical and energy characteristics of photoionization-sustained gas discharges.

EXPERIMENTAL RESULTS

Figure 1 illustrates the experimental set up used. The set up enables us to simulate conditions realized in cw gas flow systems. The gas chamber is 10 liters in volume and contains two 3 x 11 cm plane parallel discharge electrodes spaced 2 cm apart. The discharge

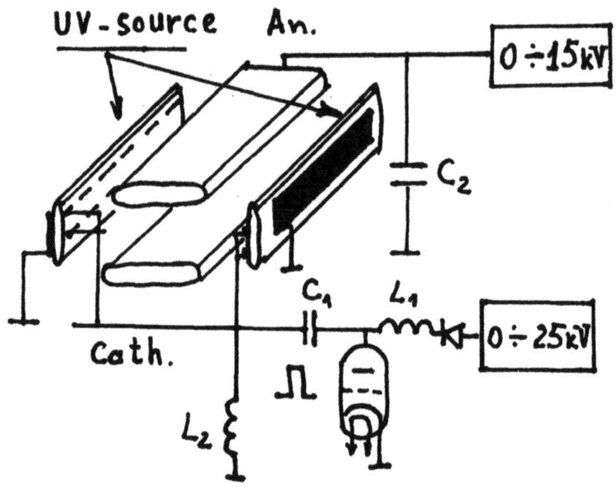

Fig. 1. Schematic diagram of UV sustained discharge apparatus.

electrodes are placed between two surface-glow UV-sources, the distance between the electrodes and the sources being 2 cm. The discharge circuit of the UV-sources is of LCR-type in series with a hydrogen thyratron. With C_1 = 2.5 nF, U_1 = 20 kV, and L = 1 μH, the ionization pulses are stable at frequencies up to 25 kHz. The discharge electrodes are connected across a capacitor C_2 = 0.5 μF, charged to U_2 = 0 - 15 kV. Discharge is initiated in N_2-NO and N_2-CO_2 (CO)-NO mixtures at a pressure of 0.5 to 1 atm. The NO gas used as an admixture has been obtained in laboratory conditions via the reaction

$$3KNO_2 + KNO_3 + Cr_2O_3 \rightleftarrows 2K_2CrO_4 + 4NO$$

that required heating the reagents to 500°K. The partial pressure of the admixture varies in the range 0 - 5 Torr.

In the course of the experiments we have recorded the voltage drop across the capacitor C_2, current pulse oscillograms for the main and ionizing discharges, and the intensity of UV-radiation in the range 110 to 200 nm. Typical current pulse oscillograms are presented in Figure 2 and the current of the non-self-sustained discharge as dependent on the NO concentration in Figure 3.

The photoionization efficiency of the auxiliary discharges are compared when made to glow directly in the working mixture and on the surface of various hydrogen-containing dielectric plates, and we find the intensity of emission of photons with an energy in excess of E_{NO} to be much higher in the latter case. Apparently, this is due to the excitation of the L_α line of atomic hydrogen.

Spark current
(300 A/Div)

Sustainer current
(10 A/Div)

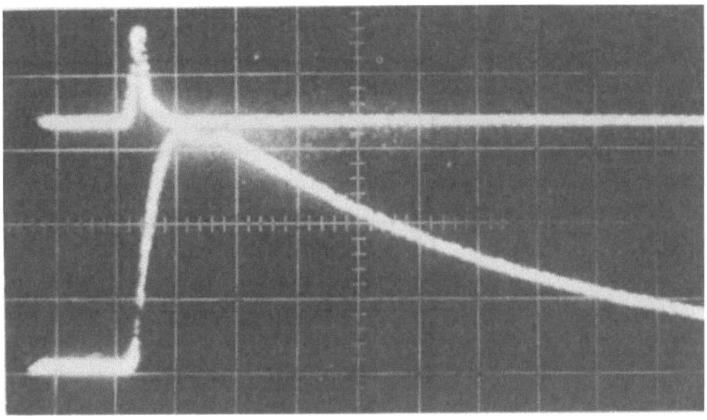

Time (1 mcS/Div)

Fig. 2a. Sustainer and spark array currents for N_2:NO (0.3%) mixture at $E/N = 2 \times 10^{-16}$ V cm^2; C_1 = 5 nF, U_1 = 20 kV.

Spark Array current
(300 A/Div)

Sustainer current
(10 A/Div)

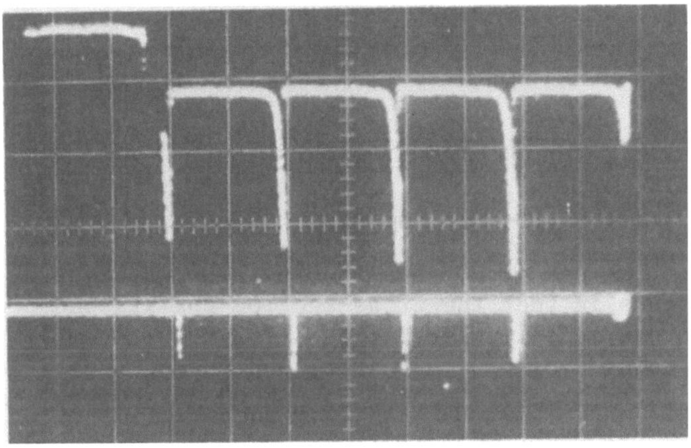

Time (100 mcS/Div)

Fig. 2b. Impulse-periodic discharge $E/N = 4 \times 10^{-16}$ V cm^2; P = 400 wt/cm^3 atm; C_2 = 0.5 mcF.

The auxiliary discharge circuit used in our experiments enables us to operate the photoionization-sustained discharge in a quasi stationary mode. In this case the discharge is sustained during 1 ms at E/P = 7.5 kV/cm atm in an N_2-NO mixture at a pressure of 1 atm. The repetition period of the main discharge is 150 μs. The maximum specific power input in the discharge attained in the experiment is around 500 W/cm^3 atm, the power consumed by the external photoionizer being 25 percent.

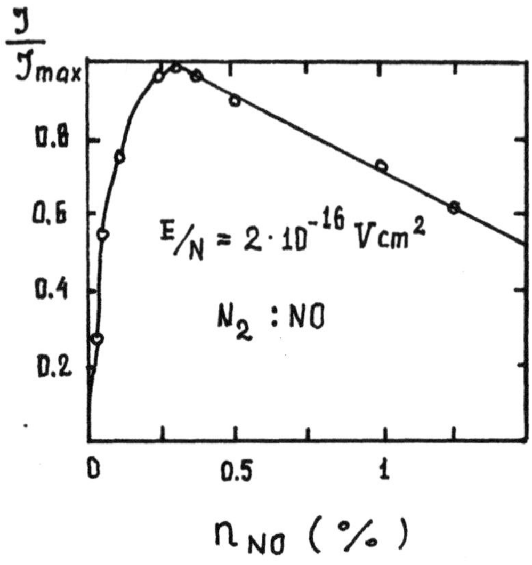

Fig. 3. Sustainer current J/J_{max} as a function of density NO (total pressure 0.5 atm).

DISCUSSION

The use of NO as a low-ionization potential admixture is advantageous in that the cross-section for its photoionization by L_{α} photons is faily large (σ_{NO} = 2.2 x 10^{-17} cm^2), the quantum yield reaching 0.8 [5]. Moreover, photons of such a wavelength are but slightly absorbed by the N_2 and CO molecules. As to the CO_2 molecules, the cross-section for the absorption of the photons by them is about 10^{-19} cm^2 [6]. The NO molecule has a low affinity to the electron (V_{NO} = 0.031 eV). Therefore, in the discharge plasma there occur the fast processes of electron detachment from NO^-:

$$NO^- + M \rightarrow NO + M + e; \; K = 10^{-11} - 10^{-12} \; cm^3s^{-1}. \qquad (1)$$

With high NO concentrations, the NO^- ions rapidly convert to NO_2^- by the following reactions [7]:

$$\begin{array}{c} e + NO + M \rightarrow NO^- + M \\ NO^- + 2NO \rightarrow NO_2^- + N_2O. \end{array} \qquad (2)$$

In CO_2-based laser mixtures, another mechanism of conversion of NO^- is possible [6]:

$$NO^- + 2CO_2 \rightarrow NO^- \times CO_2 + CO_2; \; K = 7.5 \times 10^{-29} \; cm^6s^{-1}. \qquad (3)$$

The balance equation for the electron concentration n_e in the discharge plasma has the form

$$dn_e/dt = Q - \eta \nu_a n_e - \beta n_e n_+, \tag{4}$$

where Q is the photoionization rate of the low-ionization-potential admixture, η is a coefficient correcting for the effective electron detachment, ν_a is the electron attachment frequency, n_+ is the number density of the positive ions, and β is the electron-ion recombination coefficient.

If $\eta \ll 1$ is much less than unity (in N_2(CO)-NO mixtures), this means that the attachment process in (2) may be neglected, the behavior of the electron number density will then be determined by the electron-ion recombination. In this case, the decay of plasma in the afterglow period following the discharge must be of recombination character:

$$J = J_{max}/(1 + n_+\beta t),$$

where J is the current density.

When $\eta\nu_a \gg \beta n_+$, the plasma decay is governed by the exponential law $J = J_{max}\exp(-\nu_a t)$.

Figure 4 shows $\ell_n(J_{max}/J)$ as a function of time in the afterglow period following the de-energizing of the UV-source. It is readily apparent that even in mixtures containing no CO_2, decay of the discharge plasma is determined by the electron attachment processes. Also, as shown by our experiments, with the discharge operating in a high-repetition-rate mode, the discharge current decreases somewhat from pulse to pulse.

As noted earlier, the plasma chemical processes in N_2(CO)-NO mixtures are such as to exclude electron attachment. It is therefore only logical to attribute the attachment behavior of the plasma decay to the presence of some impurity. In this connection, it is of interest to note the following circumstance. It is a well-known fact that a long-term storage of CO in steel bottles gives rise to the formation of iron pentacarbonyl. The molecules of this compound have a high affinity to the electron and therefore their presence in the discharge plasma greatly intensifies the loss of electrons through attachment. On the other hand, these molecules may be photoionized by the L_α line, since their ionization potential is about 8 eV [8]. In fact, a number of our experiments with CO containing NO admixtures show a noticeable discharge current sustained by UV ionization.

It should also be noted that in the process of manufacture NO may become contaminated with an admixture of oxygen, and so some NO will be oxidized to NO_2 by way of the reaction:

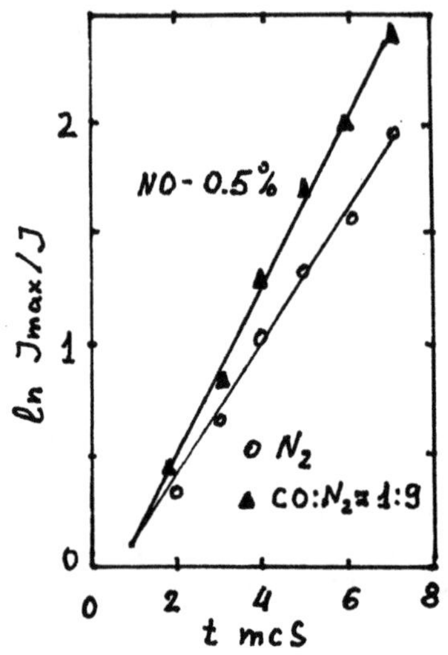

Fig. 4. $\ell_n(J_{max}/J)$ as a function of time.

$$2NO + O_2 \rightarrow 2NO_2.$$

Besides, some NO molecules may dissociate under the effect of UV-radiation and therefore NO_2 may be formed by the reaction:

$$O + NO + M \rightarrow NO_2 + M$$

during time intervals between the discharge current pulses.

Note that the NO_2 molecule, whose ionization potential is 9.78 eV, may be either ionized or decomposed to NO by UV-radiation [5]. Hence it follows that NO_2 is a suitable low-ionization-potential admixture for realization of photoionization-controlled discharge in CO_2 laser mixtures.

However, the high electron affinity of NO_2 ($U \sim 2.43$ eV) will make the electron attachment processes essential. By varying the concentration of NO_2, we may, as in the case of NO, optimize the mixture composition.

REFERENCES

1. I. S. Levin and A. Javan, Appl. Phys. Letts., 22:55 (1973).
2. D. R. Suhre and S. A. Wurtke, J. Appl. Phys., 52:3858 (1981).
3. Yu. K. Zemtsov, K. S. Klopovsky, E. A. Muratov, A. A. Ozerenko, T. S. Pulinets, A. T. Rakhimov and V. B. Saenko, Proc. ICPIG-15, part II, p. 875 (1981).

4. E. P. Velikov, V. D. Pismennyi and A. T. Rakhimov, Uspeki Fiziologicheskih Nauk, 122:c. 419 (1977).
5. H. Massey, Negative ions, Cambridge Univ. Press, Cambridge (1976).
6. D. A. Parkes, J. Chem. Soc. (London), 68:2103 (1973).
7. D. A. Parkes and T. M. Sugden, J. C. S. Farad., Trans. II, 68: 600 (1973).
8. F. I. Vilesov and V. L. Kurbatov, Dokl. Akad. Nauk, SSSR., 140: c. 1361 (1961).

HIGH POWER BLACKBODY PUMPED CO_2 LASERS*

Walter H. Christiansen and R.J. Insuik

Aerospace and Energetics Research Program
University of Washington
Seattle, Wash. 98195, U.S.A.

INTRODUCTION

Direct conversion of sunlight into laser light without the need for a complicated intermediate step of energy conversion is an important possibility for high power cw space-based lasers. The principle of broadband optical pumping using flashlamps to achieve a working laser is well known and has been put into practice with the ruby laser, Nd:YAG, and dye lasers.[1] Broadband pumping of gas lasers has not been so extensively studied.[2] Recently, however, an optically pumped iodine laser has been demonstrated using a solar simulator.[3]

Solar pumping of bound-bound absorption transitions, typical of infrared lasers, is also a possibility. This approach normally would be very inefficient, and therefore of little interest, due to the fact that the absorption bandwidths of many potential laser media are small in relation to the effective bandwidth of the solar spectrum. Consequently, only a very small fraction of the solar energy can be absorbed and converted into laser light. A concept for efficient optical pumping of an infrared laser medium has evolved using blackbody cavity radiation which has the potential of making solar pumped infrared lasers very efficient.[4]

Because of the potential of this idea, an experiment to demonstrate a blackbody-pumped laser is being carried out. This paper describes the physics of the blackbody-pumped CO_2 laser and gives the preliminary results of a non-steady lasing experiment. The laser experiment uses thermal radiation produced by an electrically

*This work is supported by the National Aeronautics and Space Administration.

heated cavity as the pumping source. While it is too early to experimentally evaluate the energetics of this idea, lasing has been demonstrated over a wide range of conditions and the first results are presented herein. Certain differences between these results and early predictions of gain are inferred from the observations. A description and extrapolation of the blackbody approach to a high power solar pumped laser is discussed at the end of the paper.

BACKGROUND IDEAS

The conventional method of solar pumping is to use a collector system which focuses the solar radiation directly on the laser medium. Assuming the medium has been placed in a transparent vessel, the radiation is selectively absorbed wherever there is an absorption band or line. For IR laser mediums, the overall efficiency is extremely poor due to their narrow absorption spectrum in relation to the radiation spectrum. In addition, many gas lasants have infrared absorption bands which do not lie near the region of peak radiant energy density of the solar spectrum. These basic problems imply an overall low system efficiency for gases with line absorption in the IR.

To overcome the severe limitations of direct solar pumped or broadband pumped lasers, an approach has been suggested wherein concentrated solar radiation is absorbed and reradiated via an intermediate blackbody. This body would be heated to a high temperature by focused sunlight and must be engineered to have small heat losses. The blackbody is then used as the optical pumping source. The basic idea is shown in Fig. 1, where an insulated blackbody provides the intermediate 4π steradian radiation field. In a static system, cooled laser tubes would be placed within the cavity. Alternatively, the lasant gas could be flowed through the blackbody cavity into a laser cavity where it would then be caused to lase via the introduction of catalysts and/or resonant transfer of energy to other lasants. In either case, blackbody temperatures (T_B) of 2000°K to 3000°K seem possible.

In order to maintain a blackbody spectrum, radiation will be continuously re-emitted at any frequency which has been depleted by the selective absorption of the lasant gas. Any radiation not absorbed by the lasant will be re-absorbed by the cavity walls and then re-emitted covering the entire spectrum, rather than being wasted as in the direct absorption case. The hot walls permit only small deviations from the equilibrium radiation field of the blackbody (see Fig. 2). In this way the pumping radiation of this type is continuously replenished and all of the energy source is utilized. The actual efficiency of this method of excitation is primarily a function of the ratio of the laser medium absorption to the

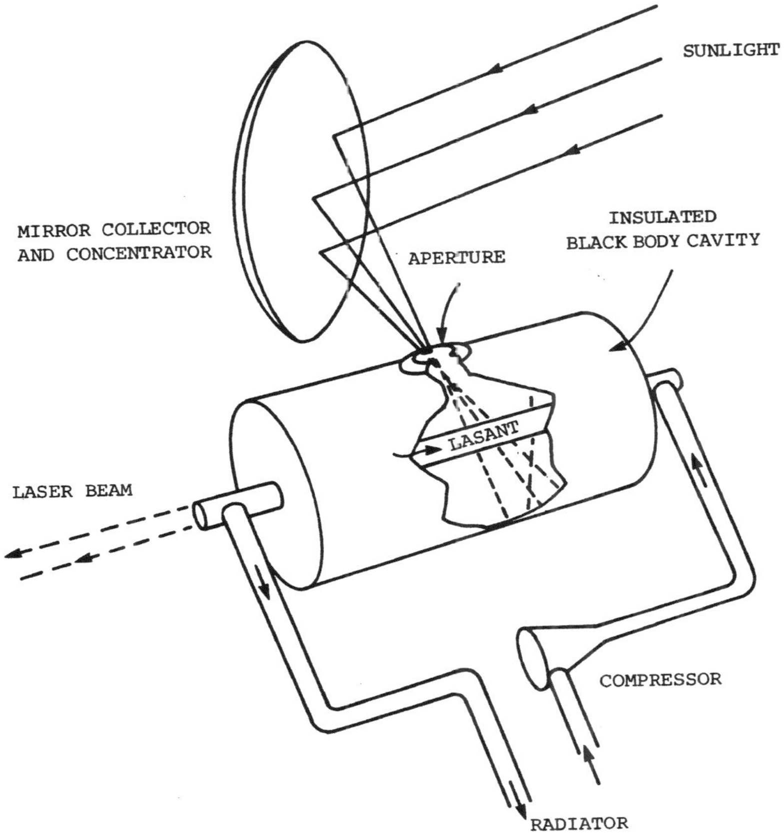

Fig. 1. A solar pumped laser concept.

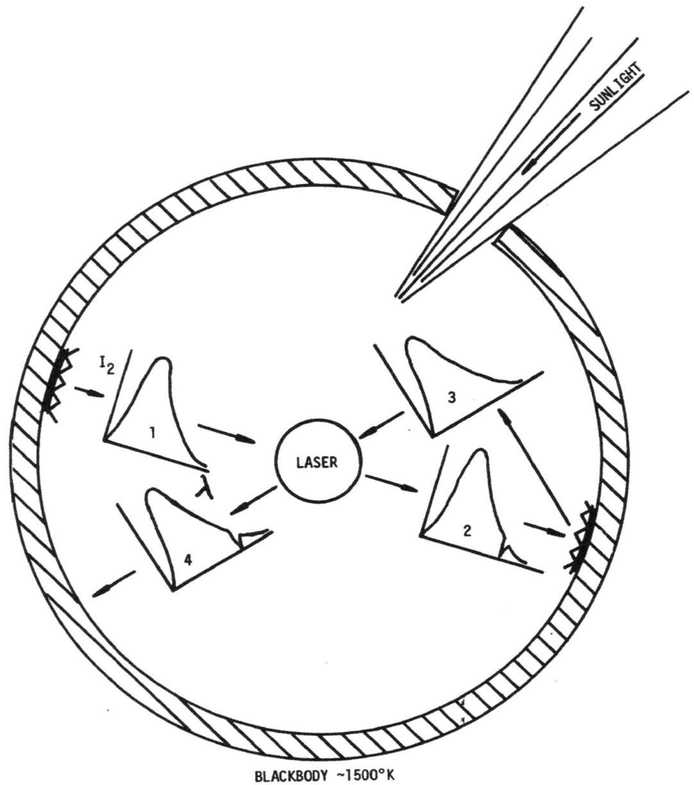

Fig. 2. The thermalization of a nonequilibrium radiation field by a blackbody.

blackbody heat losses,[4] and could in principle be many times that of a directly pumped solar laser.

This laser system seems well suited for space applications, particularly in view of new concepts of waste heat rejection.[5] Utilizing the pumping concept, scaled versions of a laser have been made to the megawatt level,[6] where the overall efficiency of conversion of solar radiation to laser radiation was shown to be 10-20%. It may be noted that the cavity may be heated in a number of ways without affecting the laser performance if the appropriate blackbody temperature is maintained. Any high temperature heat (perhaps waste heat from nuclear reactor or chemical source) could be used to heat the cavity. Thus, blackbody pumped lasers may offer a wide range of application.

Optically-pumped laser media can be chosen from those systems already lased by other means, such as the electric discharge method. The most direct approach is to find a lasing gas which has a spectral absorption band connecting the upper laser level to the ground state. It is also feasible to use other atomic or molecular gases which can contribute their absorbed energy via collisional transfer. Because of the relatively weak specific intensity of broadband thermal sources (only 1.6 kW/cm^2 for concentrated sunlight) and the low pumping rate within a band, the best laser candidates are those systems with low threshold requirements.

Laser candidates may be pumped using bound-bound transitions such as atomic or molecular line spectra or bound-free transitions such as photodissociation. The possibilities of the first case extend over the whole spectrum whereas the latter case tends toward the visible part of the spectrum because of typical bond strengths. Some examples of molecular laser systems with IR absorption bands are given in Ref. 6. All of these molecules have produced laser emission in the range between 4.5μ and 10.8μ. It is possible that mixtures of these gases and their isotopes may be used as a laser medium. While a range of potential candidates exist, CO_2 seems to be the best choice for demonstrating the physics of blackbody lasers. Furthermore, the kinetics of CO_2 and the theoretical aspects of modeling this molecule are well known. Previous calculations have estimated the levels of gain and confirmed that threshold conditions should be reached fairly easily at translational temperatures on the order of 300°K and quite modest blackbody temperatures. In this paper we confine our discussion only to the CO_2 laser system.

CURRENT RESULTS

Experimental Apparatus

The lasing experiment is conducted in a non-steady manner, that is, an uncooled laser tube is exposed only briefly to

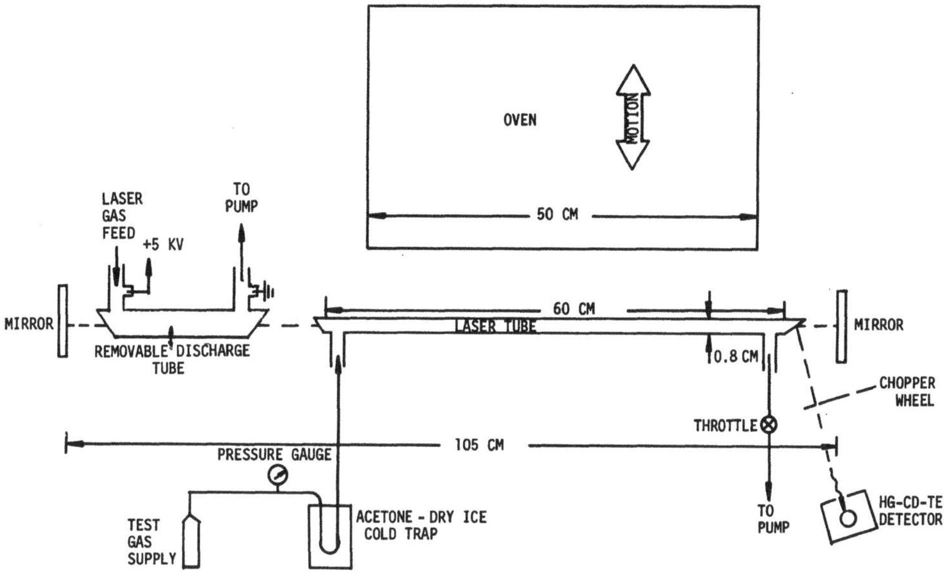

Fig. 3. Experimental setup

radiation. Lasing should occur until the laser tube and the gas within become hot. Actively cooling the laser tube should permit cw lasing, but this has not yet been tried. The radiation field is produced by an electrically heated oven capable of achieving 1500°K. The oven consists of two half shell electrical heaters and is well insulated so that the surroundings remain near room temperature. This ensures that the laser tube (which sits adjacent to and outside the heater) will not be preheated.

The details of the setup are shown in Fig. 3. The laser tube is a sapphire tube which is 60 cm long and has an 8 mm internal diameter. It is illuminated only over the center 50 cm portion of its length. Sapphire is used because it transmits very well (85%) out to wavelengths of 5-6 μm and can withstand very high temperatures. The optics and the laser tube are mounted on a breadboard which is vibrationally isolated from the oven and its supports so that the motion of the oven does not disturb the laser cavity or its alignment. The laser cavity is nearly confocal with a one meter optical path between two reflecting spherical mirrors. The laser output is taken from the reflection off one of the Brewster windows using a Hg-Cd-Te detector. The alignment of the cavity is optimized using a small electric (glow) discharge laser operated with a CO_2 laser mixture. Care is taken to ensure that the mirrors are properly adjusted so that the optical path of the discharge laser is centered in the sapphire tube. As shown, the discharge apparatus can be removed from the optical system when desired without disturbing the alignment of the cavity.

Once the proper oven temperature is achieved, in the neighborhood of 1200°K - 1500°K, a CO_2-He-Ar mixture is introduced into the laser tube and the oven is opened and moved around the laser tube to start pumping action. Data is taken for a few seconds and the heater is then removed from around the laser tube, thus completing the cycle. The tube is allowed to return to room temperature before the cycle is repeated. In this experiment the gas mixture flows through a dry ice acetone cold trap just upstream of the laser tube. This removes any contaminants which may have been introduced within the lines and precools the gas somewhat. The gas flow through the tube is controlled by a throttle placed downstream of the laser tube which ensures a flow transit time many times longer than the optical pumping time for CO_2.

Preliminary Results

Preliminary tests were carried out to measure the heating rate of the laser tube. By sealing off the laser tube and monitoring the gas pressure within the tube a type of gas thermometry was achieved. The first set of experiments was conducted using an I.R. quartz laser tube which has similar physical properties to sapphire. It was found that the gas within the laser tube reached temperatures of 700-750°K after 45 seconds with oven temperatures ranging from 1000-1275°K. This is in agreement with expectations. Thus there is sufficient time (10 sec.) for a lasing experiment before the laser mixture becomes too hot.

Actual lasing tests were then undertaken. Figure 4 shows an oscilloscope trace of a typical laser signal from this apparatus at a gas pressure of 10 torr. In this figure time is the horizontal axis at 1 sec/cm and the vertical axis is calibrated to be 0.08 mw/cm. The laser signal is about 0.2 mw and lasts 3.7 sec. The signal is obtained with a Hg-Cd-Te detector measuring the weak laser output from a Brewster window, at approximately 0.025% output coupling. Lasing has not yet been observed with a 5% output coupler which implies that threshold conditions have barely been reached. The laser signal is chopped at 30 cps thus giving a hashy look to the trace. However, the upper envelope which is used to measure the laser signal is clearly defined. It is known that the oven is opened just before the laser turns on and is closed over the laser tube at about two seconds into the run. The heating of the tube and laser mixture terminates the signal. A two humped signal, seen many times, is believed to be due to rapid cooling off and reheating of the blackbody radiation field as the heater is opened and closed around the laser tube.

A series of experiments was conducted in which the laser output

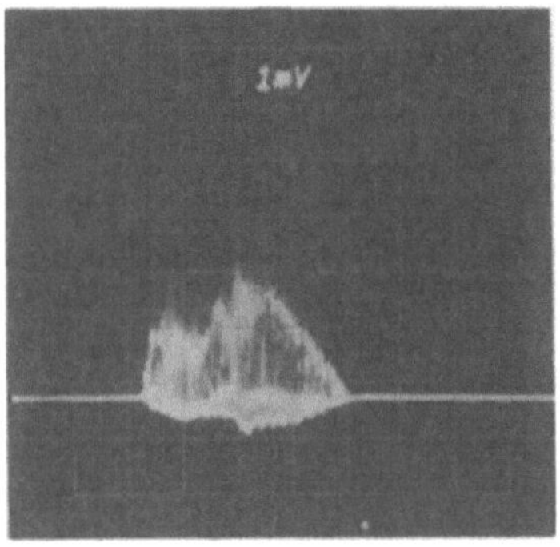

Fig. 4. A CO_2 laser output; vertical axis 0.08 mw/cm, horizontal axis 1 sec/cm.

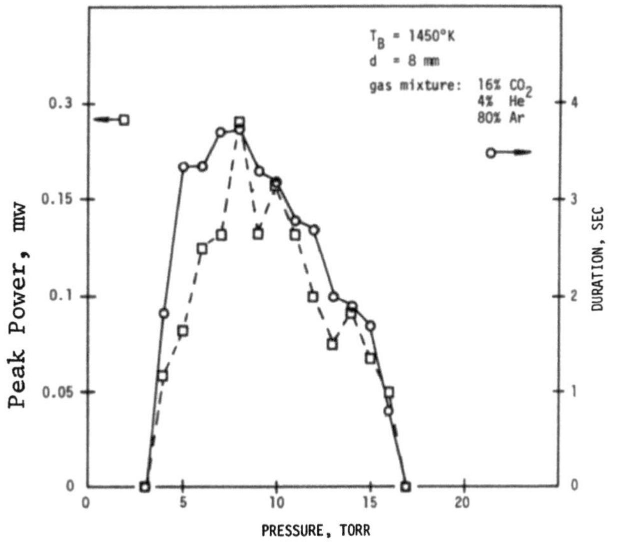

Fig. 5. Peak power and lasing duration vs pressure for an estimated output coupling of 0.025%.

and the duration of lasing was determined for various gas mixtures over a range of pressures. The results for the optimum mixture, 16% CO_2-4% He-80% Ar, are plotted in Fig. 5. These results show a maximum laser signal and signal duration at 8 torr. No laser signal was obtained at pressures less than 4 torr or more than 16 torr. The results have been very repeatable. The time duration of the signal is consistent with expectations based upon the results of the heating rate measurements.

In its simplest terms, the kinetics of an optically pumped laser system are little different from other forms of the laser and can be treated thusly once the pumping has been specified. The level of small signal gain and its relationship to the blackbody temperature have been calculated using a simple model for the case of an optically thin gas.[7] Mixture compositions were chosen so that the optical thickness of the mixture was appropriate for an 8 mm tube. The calculations assumed a transmission coefficient of 85% through the sapphire tube at 4.3μ based on the reflection coefficient estimates. The model included inhomogeneous and homogeneous broadening effects on the spectral line. As shown in Fig. 6, this model (case D=0) predicts high gains at low pressures (e.g., a 1 torr mixture of 0.8 torr CO_2 and 0.2 torr He) and decreasing gain as argon is added to increase the pressure. These calculations imply that lasing should be easy to achieve at low pressure conditions because the gain is so high there. The experimental results do not

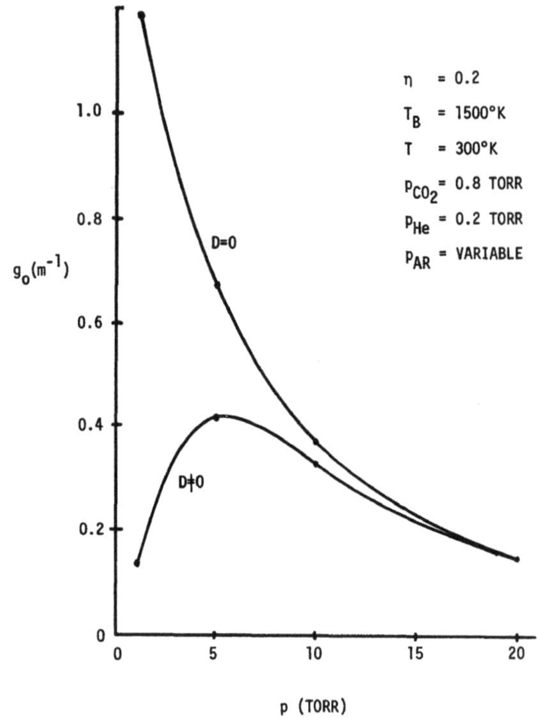

Fig. 6. Calculated small signal gain on ℄ with and without diffusor.

suggest this trend as no lasing is seen at low pressure for any gas mixture.

It is believed that the discrepancy is due to catalytic wall deactivation of excited states. It is known that wall deactivation of some excited gases such as N_2* is quite small, but measurements of CO_2* on many materials[8] indicate significant deactivation probability, η, O(20%). Because the laser tube has such a small diameter, diffusion of excited states to the wall and possible deactivation would seem to be a significant effect. The gain model is now being revised to include diffusion and wall deactivation losses. This model while not completely formulated is a linearized version based on excited state concentrations rather than a mode calculation as done in Ref. 7. Some preliminary results with diffusion present (case D≠0) are also shown in Fig. 6. The results show sharply reduced gain at low pressures if the wall deactivation probability is indeed 20%. Gain is not affected as much at pressures above 10 torr as argon inhibits the diffusion mechanism. The overall trend shows that peak gain occurs near 8 torr in fair agreement with the experimental results. Further modeling and experimental proof is required.

A CONCEPTUAL LASER

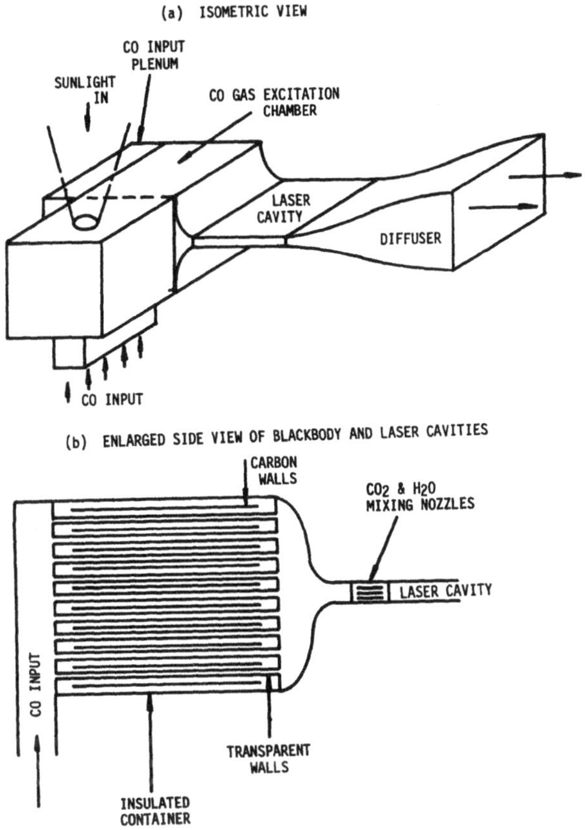

Fig. 7. Conceptual CO-CO_2 mixing laser with blackbody pump.

There are a number of possible designs which may incorporate the blackbody concept. Here we discuss one which uses rapid mixing of CO_2-H_2O with excited CO as the laser system. A possible cw mixing gas laser system is shown in Fig. 7. In this system CO gas is excited by flowing it through the blackbody radiation field. After emerging from the blackbody at high subsonic speeds, the gas is mixed with CO_2 and H_2O catalyst, the excitation energy is transferred to the CO_2, and the gas mixture is then lased. After lasing the gas must be cooled and the CO_2 and H_2O catalyst must be removed from the CO gas before the CO can begin the cycle anew; otherwise collisional processes will deactivate CO during the pumping process.

The blackbody cavity in a solar pumped laser system must satisfy constraints imposed by the pumping time (radiation lifetime of the pumped state) of the laser gas, and the temperature requirements of the laser gas. The excitation of CO gas by blackbody radiation has been examined in an arrangement of sets of two parallel windows through which the CO gas would flow. The layer approach would appear to be easy to construct, involving no frames or struts for support, and may not require any cooling other than that provided by the flowing CO gas itself as depicted in Fig. 7. In this arrangement the CO pumping gas is exposed to the blackbody radiation emanating from the walls of the cavity and the carbon sheets between the tube layers. The CO is then rapidly transferred into a laser cavity and ultimately recycled for mixing with CO_2. In this case the pressure of the lasant can be considerably higher (may be 1/2 atm) because of the low deactivation rates.

Adjusting the number of tubes or size of the blackbody chamber scales the size of the laser. Megawatt sizes of these have been examined in Ref. 6. For continuous closed-loop operation, necessary auxiliary equipment is required to condition the gas properly for lasing, and stable chemistry is needed for long duration operation. Stability can be quite good with photon pumped systems in contrast to electrically pumped systems; very few high energy photons are available for bond breaking, and these may be filtered out if present. To close the laser gas loop, heat exchangers to remove waste heat from the system are required and a pump will be required to circulate the laser medium. Large cw lasers with efficiencies on the order of 10% may be possible, based on the above concept.

REFERENCES

1. A. Yariv, Introduction to Optical Electronics, 2nd ed., Holt, Rinehart, and Winston, Inc., New York (1976), pp. 157-197.
2. A. Bokhan, "An Optical Pumping of a Molecular Laser by Blackbody Radiation," Optics of Spectroscopy 26:423 (1969).
3. W.R. Weaver and J.H. Lee, "A Solar Simulator Pumped Gas Laser," Laser Focus, p. 16 (March 1981).
4. W.H. Christiansen, "A New Concept for Solar Pumped Lasers," Progress in Aeronautics and Astronautics, Vol. 16, Radiation Energy Conversion in Space, K. Billman, ed., American Institute of Aeronautics & Astronautics, New York (1978), pp. 346-356.
5. A.T. Mattick and A. Hertzberg, "Liquid Droplet Radiators for Heat Rejection in Space," J. Energy 5:387 (1981).
6. R. Taussig, C. Bruzzone, L. Nelson, D. Quimby and W. Christiansen, "Solar-Pumped Lasers for Space Power Transmission," AIAA Paper No. 79-1015, AIAA Terrestrial Energy Systems Conf., Orlando, Florida, 1979.
7. O. Yesil and W.H. Christiansen, "Optically Pumped CO_2 Laser Mixtures, J. Energy 3:315 (1979).
8. M. Kovacs, D.R. Rao and A. Javan, "Study of Diffusion and Wall De-Excitation Probability of 00°1 State in CO_2, J. Chem. Phys. 48:3339 (1968).

A NOVEL COMMAND CHARGING SYSTEM FOR HIGH REPETITION-RATE PULSED DISCHARGE LASERS

D. Chuchem and E. Margalith

Laser Unit - Nuclear Research Centre-Negev

P.O.Box 9001, Beer-Sheva, Israel

ABSTRACT

The proper excitation of repetitively-pulsed avalanche-discharge lasers at high pressures requires thyratrons capable of high rates of rise of current and low inductance. The use of a "special for lasers" thyratron employed as part of a pulse forming network for the operation of an electric-discharge-initiated pulsed supersonic chemical laser is described. Different charging configurations were employed. A novel method of high-repetition-rate pulse charging by means of a grounded-grid thyratron was found. The rationale for and efficacy of using the grounded-grid thyratron is described. An example of the performances of the different charging configurations in igniting such a laser is given.

INTRODUCTION

The pulse length and efficiency of avalanche-discharge lasers are limited by the formation of plasma arcs: the desired E/N can not be maintained, and the homogeneous discharge is constricted to streamers. The transition of evenly distributed "glow" discharges to localised arcs makes it impossible to utilise all of the laser volume, while the intense heating in the narrow plasma arc columns produces gas inhomogeneities which can disrupt the optical cavity. It is thus desireable to limit the discharge pulse widths to times less than that required for an arc to develop (typically less than a few tens of nanoseconds [1]).

Obviously, in order to efficiently utilise supersonically flowing medium, one would like to increase the repetition rate of the laser as much as possible. For example. for a flow velocity of 1 km/sec (Mach 3) and a 1 cm wide active area, the gas transit time would be 10 microseconds. Thus a pulse repetition rate approaching 100 kHz would be desireable to enable utilization of the gas flow.

To enable short pulse discharges, hydrogen thyratrons capable of very high rates of current rise have been produced commercially specially for use in repetitively pulsed lasers [2,3]. However, conventional charging methods are not adequate for the full realization of these tubes' capabilities. In the most common scheme, resonant charging, the immediate reapplication of the high voltage across the tube following its discharge, interferes with the tube recovery process and limits the repetition rate. Command pulse charging (where the passive charging diode is replaced with an active switch) allows the reapplication of voltage and current to the thyratron after firing to be delayed, thus allowing the thyratron to achieve recovery under optimum conditions. Additionally, since the high voltage dwell time is considerably shortened, the thyratron can successfully hold off higher voltages without spontaneously firing through.

Vaccum tubes and conventional thyratrons employed as command charge switches [4] suffer from an inherent limitation, which is described below. We have employed a grounded grid thyratron to overcome this limitation. The rationale for and efficacy of using the grounded-grid thyratron is described. An example of the performances of the different charging configurations in pulsing an electrically initiated supersonic chemical laser is given.

BASIC PULSED DISCHARGE CIRCUIT

The most common discharge circuit used with repetitively pulsed lasers is presented schematically in figure 1. A capacitor C is alternately charged from a high voltage DC supply via the inductors L_1 and L_2 through the diode D, and discharged through the laser by firing the thyratron switch S. The repetition rate of such a circuit is limited by the charging time:

$$T = \pi\sqrt{LC} \qquad (1)$$

It would seem, that for fixed capacitance, by lowering the inductance of the charging circuit one could decrease the charging time to any desired value, and thus enable an increase in the repetition rate. But due to the characteristics of the thyratron tube there is a minimum acceptable inductance: after a thyratron is fired it needs time to recover before it can hold off high voltage again. This "recovery time" depends on the extent to which the current flowing through the tube is held below I_{hold} (I_{hold} is the minimum current for which the tube's plasma will sustain itself). During the recovery process, either the voltage across the tube has to be kept below its conduction voltage, or the current through the tube must be below I_{hold}. However in the resonant charging scheme the charging current starts growing as soon as the pulse forming network is discharged. It develops according to the following relation:

$$I(t) = V_o\sqrt{C/L} * SIN(t/\sqrt{L*C}) \qquad (2)$$

At the same time, the voltage across the tube is determined by the charging rate of the capacitor:

$$Vc = Vc_o + 1/C\int I*dt = Vc_o + V_o*t^2/2LC*COS(t/\sqrt{LC}) \qquad (3)$$

where Vc_o is the capacitor's voltage immediatly after the discharge and V_o is the supply voltage.

Thus in order to keep the tube current far enough below I_{hold} and its voltage below the "tube drop" value (to enable the achievement of short recovery times), one must use an inductance high enough to limit the rate of rise of the current.

(Another approach to increase the repetition rate is by decreasing the capacitance, but it is assumed here to be determined by laser requirements).

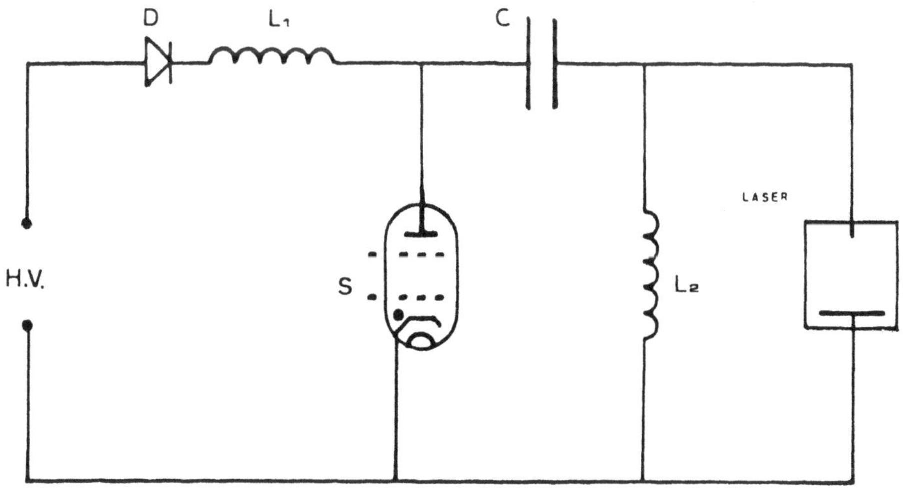

Figure 1 Conventional resonant charging ciruit. This circuit uses passive elements only.

CONVENTIONAL COMMAND PULSE CHARGING

Nevertheless the above mentioned repetition rate limit can be surpassed by employing a command charging scheme as follows, (see figure 2): A second thyratron, S_2, is introduced into the circuit. It replaces the diode in the previous scheme. S_2 is triggered in order to charge the PFN. When the charging current approaches zero on the completion of charging, the thyratron S_2 can recover quickly. Following the recovery of S_2, S_1 is fired and the capacitor discharges through the laser. Since S_2 is now nonconducting, S_1 can now recover in minimum time. The recovery time for thyratrons is of the order of 10 microseconds, and thus the repetition rate could theoretically approach the hundred kiloHertz range compared to a few kiloHertz in the previous scheme.

This circuit suffers from a major drawback: when the main thyratron fires, the cathode of the charging thyratron sees a large negative going pulse whose magnitude is proportional to the charging voltage. Since the grid potential will unavoidably lag after the falling cathode potential, a positive going pulse will appear between the grid and cathode. Thus for charging voltages above some threshold value, the charging thyratron will fire simultaneously with the triggering of the main thyratron. In which case the entire pulse charging scheme is rendered uneffective.

Heavy capacitive coupling between the grid and the cathode of

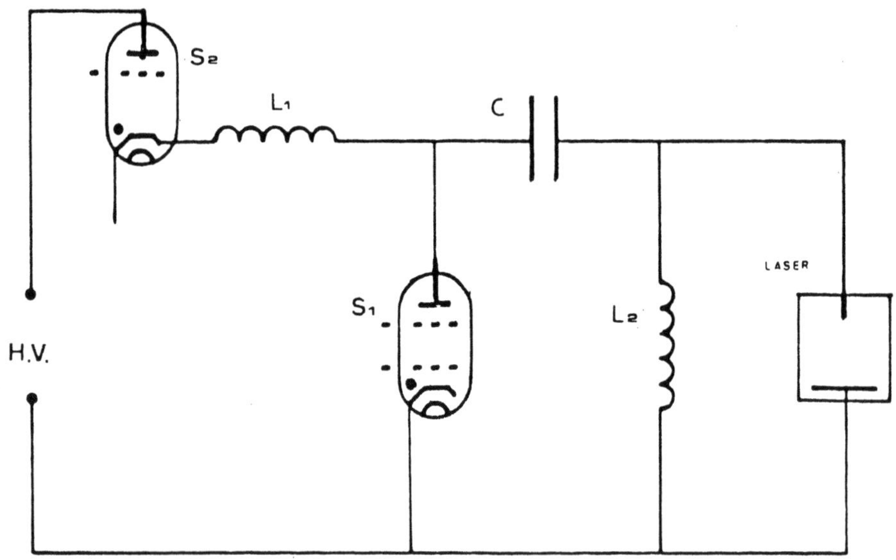

Figure 2 Conventional pulse charging scheme. The charging current flows through the cathode of the charging thyratron.

the charging thyratron may be introduced in order to overcome this problem but then very large pulses are required from the pulser to fire the thyratron when desired, and as the charging voltages are increased such a solution becomes impractical.

PULSE CHARGING WITH A GROUNDED-GRID THYRATRON

A simple method for solving the above mentioned problem, is to replace the conventional charging thyratron with a grounded-grid tube. In a grounded-grid tube [5] the anode current is conducted out through the grid structure, (see figure 3) while the cathode serves only to produce the electons necessary for initiating commutation of the grid-anode gap. Thus it is the thyratron grid which is connected to the pulse forming network, and when the main thyratron is fired, the negative going pulse is returned to the grid rather than to the cathode. Since the negative pulse appears at the grid and not at the cathode as in the above case, the charging thyratron is not triggered spuriously as before. Thus, it is the grid which is connected to the PFN, and the cathode is grounded. The negative pulse which follows the firing of the main thyratron (S_1), lowers the potential of the grid with respect to the cathode (in contrast to the previous case) and prevents its spurious triggering. In fact, the large negative pulse which arrives at the grid of the charging thyratron at the same time the laser is fired, ensures that the charging thyratron remains immune to any other noise generated by the laser circuit.

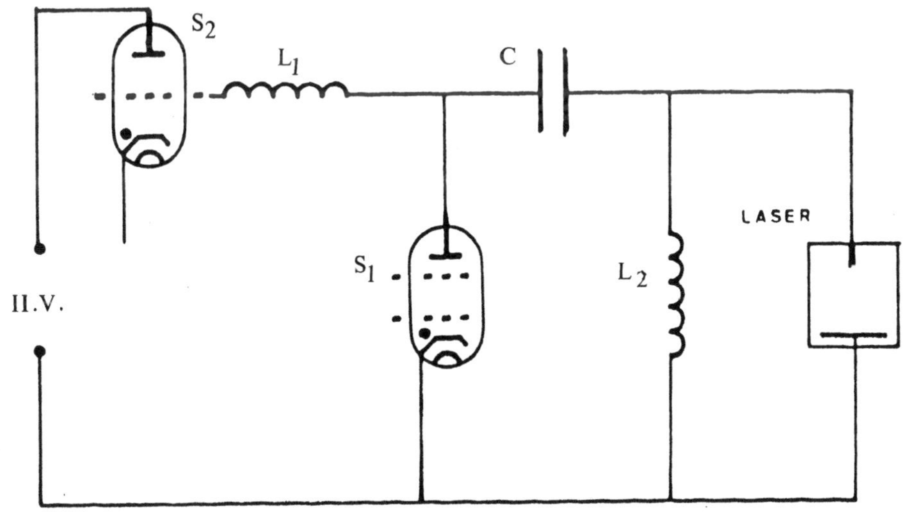

Figure 3. Pulse charging scheme employing a grounded-grid thyratron. In this scheme the charging current flows out through the grid of the charging thyratron.

In its normal operation mode, the grounded-grid thyratron is not as long lived as the conventional thyratron due to depletion of the grid by arc formation. However, at the low peak currents found in the charging circuit, the thyratron will operate in the glow mode and thus normal life can be expected for the tube [6].

EXAMPLE

We have used a HY-3013 EG&G thyratron to pulse an electric-discharge-initiated pulsed chemical laser. The pulse forming network consisted of an 11 nanoFarad capacitor which was charged through an inductance of 330 milliHenries.

With this system we could achieve full resonant charging of the capacitor at repetition rates up to 5.2 kHz (equation 1). At this repetition rate charging voltages up to 5.5 kV were attained. At higher charging voltages the thyratron could not achieve recovery.

We incorporated a conventional pulse charging circuit using an EG&G 1802 tube as the charging switch. With this scheme and a charging inductance of 150 milliHenry we could operate at 6 kHz with charging voltages up to 7 kV. However, replacing the 1802 charging thyratron with its grounded-grid equivalent, the HY-3202, we were able to increase the operating frequency up to 16 kHz at charging voltages of 7 kV. The charging inductance for this case was 15 milliHenry, the dead time before triggering the HY-3202 was 10 microseconds, and the dwell time at high voltage for this case was approximately 12 microseconds.

CONCLUSIONS

The advantage in using a grounded grid thyratron for pulse charging is explained and clearly demonstrated. A comparison between the performance of the conventional grounded cathode 1802 with the otherwise equivalent grounded grid HY-3202 shows an improvement by a factor of 3 in the pulse repetition rate of the circuit. The performance of the charging circuit with each one of the tubes in the lower frequency range was identical.

Introducing the grounded-grid pulsed charging system into the original passive resonant circuit led to an increase of almost a factor of six in the energy delivered to the laser by the pulse forming network.

REFERENCES

1. G. A. Mesyats, Yu. I. Bychkov, and V. V. Kremnev, "Pulsed Nanosecond Electric Discharges in Gases", Soviet Physics Uspekhi, Vol. 15, pp. 282-297, 1972.
2. D. Turnquist, "Thyratron Development for High Repetition Rate Gas Lasers", International Conference on Lasers, Orlando, Florida, 1979.
3. R. F. Caristi, S. Friedman, S. S. Merz, D. V. Turnquist, "New developments in Super-Fast, High-Power, Hydrogen Thyratron Switching", IEEE Trans. on Electron Devices, Vol. ED-26, pp. 1427-38, (1979).
4. "Hydrogen Thyratron Preamble", English Electric Valve Company Limited, 1972.
5. D. Turnquist, S. Merz, and R. Plante, "Grounded Grid Thyratrons", Twelfth Modulator Symposium, New-York, 1976.
6. S. Merz, E.G.&G. Inc., private communication.

FLOW EFFECTS ON THE LASER POWER OF TRANSVERSE FLOW TE LASERS

V. Fantini, V. Donati , L. Garifo, and G. Incerti

CISE S.p.A., P.O.B. 12081
20134 Milano, Italy

A. Borghese

Istituto Chimica Industriale, Università, P.le Tecchio
80100 Napoli, Italy

INTRODUCTION

The purpose of this work is to investigate experimentally the influence of the motion of active laser medium on the laser output power in a convectively cooled electrically sustained CW CO_2 laser.

The analysis will be confined to laser systems in which electric field direction, flow direction and optical axis are mutually orthogonal.

For calculating the output laser power it is used in the literature[1] the Rigrod model[2] which strictly applies to non flowing active medium lasers.

In predicting the output optical power by this procedure, at a given coordinate $\bar{x}$ in the gas flow direction the effects due to laser radiation oscillating into the cavity upstream $\bar{x}$ are not taken into account.

Several authors have considered flow effects. De Maria[3] gives an expression for the saturation intensity I_s.

*Work performed under CNR/CISE Contract n. 79.01579.98

I_s turns out to depend on the gas transit time τ_F across the modal or interaction volume.

When τ_F is much shorter than the lifetime of the upper and lower laser levels, I_s is inversely proportional to τ_F.

The small signal gain g_o is considered independent on the gas flow velocity.

By ad-hoc assumptions, Siegman et al.[4] have obtained, in the case of gasdynamic lasers, an approximate expression for the gain g of the active medium in presence of oscillating radiation in the laser cavity. g is given in terms of the small signal gain g_o as a function of the x coordinate along the flow direction. In this expression the saturation intensity of the laser gas mixture is kept constant along x.

Cool[5] has elaborated a model for transversal flow gasdynamic lasers. Using a four levels scheme of the active medium, he gives an expression for the small signal gain g_o and the saturation intensity I_s, keeping into account the flow effects and the oscillating radiation. The conclusion of the work of Cool is that it is possible to use a Rigrod model for flowing active media , but it is necessary to replace g_o and I_s with equivalent expressions in which flow effects are taken into account.

More recently Armandillo et al.[6] have worked out a computer code for studing a transversal flow electrically excited CW CO_2 laser.

The behaviour along the gas flow coordinate of the various parameters characterizing the active laser medium has been computed.

In particular the saturation intensity behaves as that obtained by Cool in his analytic treatment.

In the present work some experimental results are reported, showing both the importance of flow effects in estimating the available laser power and the dependence of saturation intensity on the gas flow coordinate.

EXPERIMENTAL RESULTS

In our investigations we have used a self sustained transversally electrically excited CW CO_2 laser. The gas mixture, flowing across the discharge channel at a velocity of 40 m/s, is a $CO_2/N_2/He$: 2/8/30 torr.

The excited volume of gas is about 1000 cm^3, and the typical electrical power density inside the discharge is 8.5 W/cm^3.

In Fig. 1 the schematic experimental setup for measuring the influence of the flow effects on the extractable laser power P is shown. We have used optical cavities with toroidal mirrors[7,8] to obtain a rectangular cross section of the volume of the oscillating modes. The geometrical optical coupling of the cavities is t = 15%.

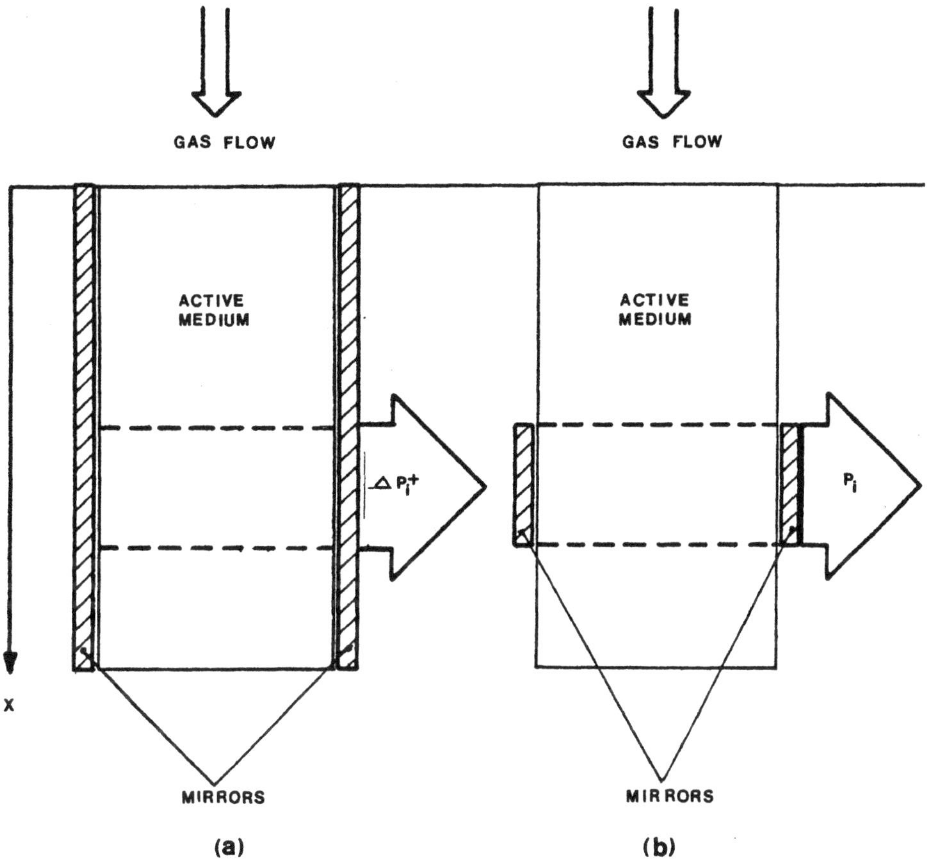

Fig. 1. Schematic experimental setup for measuring the influence of the flow effects on the extractable power P.

ΔP_i^+ and P_i are the powers extracted from a region of the active medium, when laser radiation is respectively oscillating or not oscillating upstream this region. In particular P_i has been measured with a cavity of 20 mm fixed dimension movable inside the discharge volume along the flow direction x (Fig. 1b). Powers ΔP_i^+ are instead obtained using a cavity of variable dimension along x (Fig. 1a). The cavity dimension has been successively taken equal to 20, 40, 60, 80 mm. ΔP_i^+ is computed as a difference between two power values corresponding to two different consecutive cavity dimensions.

In Fig. 2 we report the values of P_i and ΔP_i^+ as a function of x. P_i and ΔP_i^+ are assigned to the central points A,B,C,D of the four considered regions of the active medium. At x = A we have obviously $\Delta P_i^+ = P_i$, because upstream this region there isn't oscillating radiation, while for the other regions $P_i > \Delta P_i^+$ and $P_i - \Delta P_i^+$ is increasing vs.x.

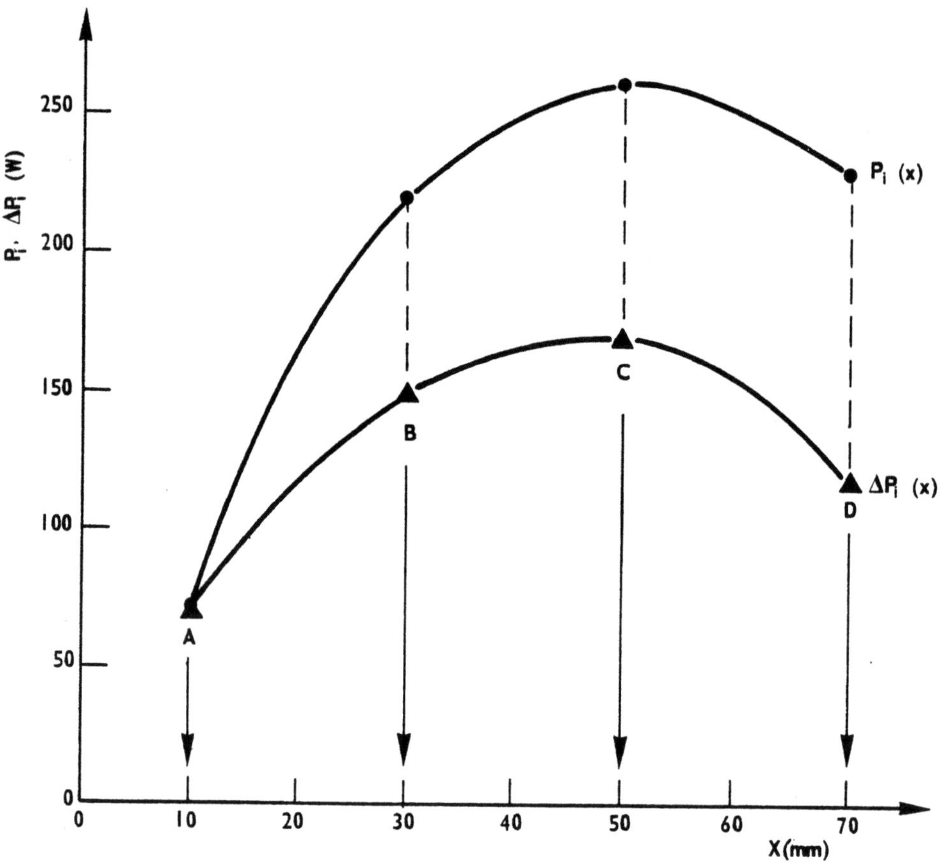

Fig. 2. Values of the powers extracted with different active medium volumes as a function of x coordinate along flow direction.

This can be explained as due to the fact that when oscillation is occuring upstream a given region, the oscillating radiation partially exploits the active medium. The extractable power from that region is therefore smaller than that which can be obtained from the same volume without upstream oscillating radiation. In our experimental conditions the total laser output power P extracted from the active medium, given by $P = \Sigma \ \Delta P_i^+$, is 510 W, while the total power predicted by the sum of the powers P_i is 780 W, i.e. a value 1.5 greater than that effectively measured.

In Fig. 3 another experimental setup to measure the influence of the flow effects on the extractable laser power P is shown.

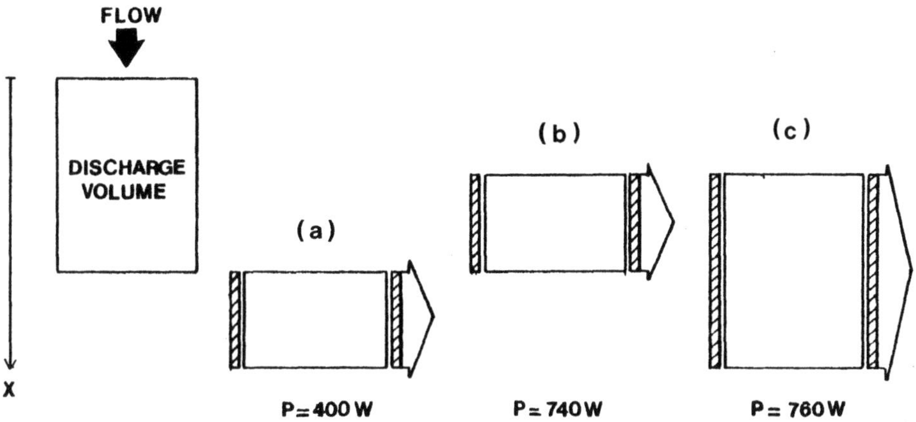

Fig. 3. Scheme of an other experimental setup to measure the influence of the flow effects on the extractable power P.

When the active medium flows the small signal gain has a tail downstream the discharge volume and positioning the optical cavity in this region it is possible to extract some laser power. In our experimental conditions the power extracted downstream the discharge volume is 400 W without upstream oscillating radiation (Fig. 3a), while the contribution of the same region when there is upstream laser radiation is negligible, as can be inferred from a comparison of Fig. 3b and 3c. We therefore can conclude that the presence of oscillating radiation into the cavity inside the discharge volume depresses gain tail downstream the discharge.

Another relevant effect due to the motion of the active medium is that saturation intensity I_s is not constant as in the case of stationary active medium, but varies along flow direction, being higher at the discharge volume entrance than at the exit[5,6].

To investigate the behaviour of I_s along flow direction we have used the experimental setup shown in Fig. 4. A stable probe cavity with t = 15% and rectangular cross section is positioned inside the discharge volume at different positions along flow direction x. A proper aperture inside the cavity limits the dimension of the modal volume along x to 5 mm.

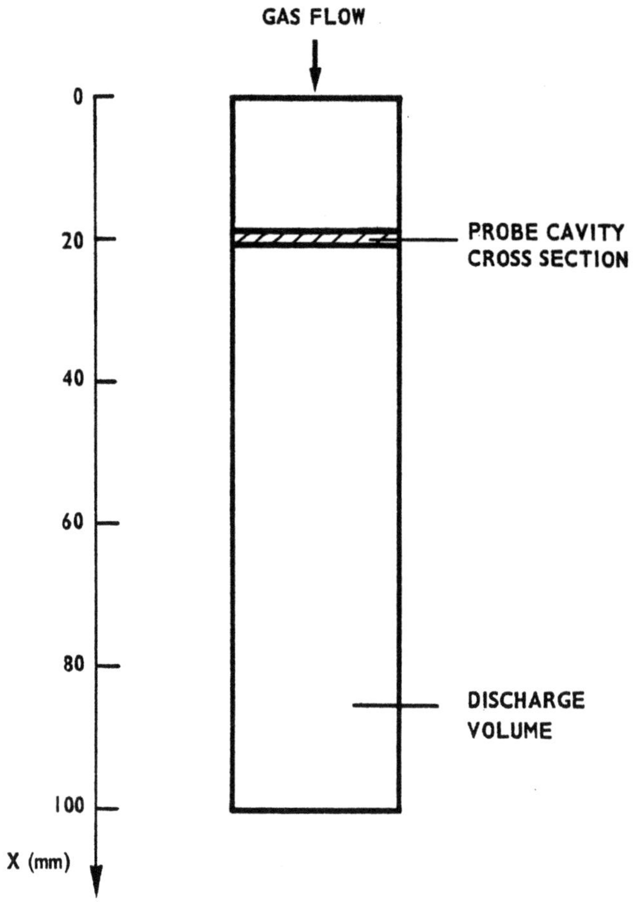

Fig. 4. Schematic experimental setup to show the influence of flow effects on the saturation intensity I_s.

The values of I_s at each x position are obtained from Rigrod equation[2], using the measured values of the laser output power and small signal gain at that x position. The use of Rigrod model is now correct, because the dimension of the probe cavity along the flow direction has been chosen sufficiently small that flow effects on small signal gain can be neglected. Fig. 5 shows the behaviour of I_s vs.x. As previously said I_s is very high at the entrance of the discharge volume and, as x increases, I_s decreases reaching a constant value of about 550 W/cm^2.

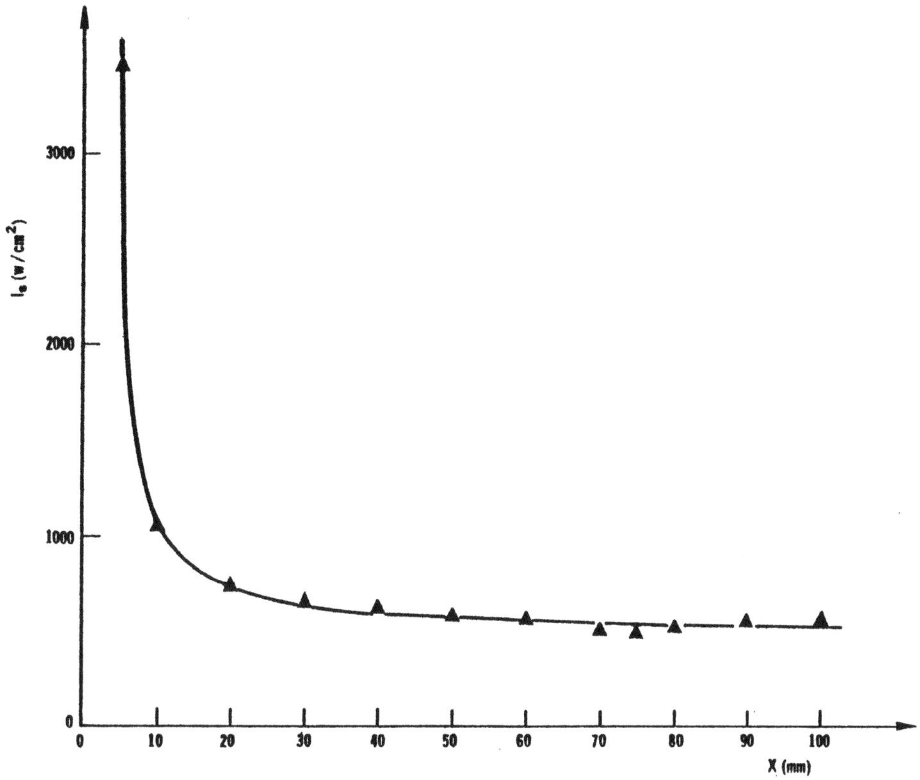

Fig. 5. Behaviour of I_s vs. x for the laser mixture $CO_2/N_2/He$: 2/8/30 torr flowing with v = 50 m/s.

This experimental behaviour of the saturation intensity agrees very well with the theoretical results obtained by Cool[5] for a gasdynamic laser and with the results of the computer code carried out by Armandillo et al.[6] to describe transverse flow CW CO_2 lasers.

CONCLUSIONS

The effects of the active medium flow on the laser power and on the saturation intensity have been experimentally investigated. For laser system design the most important result is that not taking into account flow effects can lead to a considerable overestimation of the extractable laser power.

Moreover experimental evidence of the behaviour of the saturation intensity along flow direction has been given.

These data agree satisfactorely with theoretical results

obtained by other authors[5,6]. As a concluding remark we point out that results on flow effects imply the necessity of using a model where small signal gain and saturation intensity are modified to take them into account, so that the obtainable laser power from a given active medium can be realistically predicted.

REFERENCES

1. E. Hoag, H. Pease, J. Staal and J. Zar, Performance Characteristics of 10-kW Industrial CO_2, Appl. Opt., 13, 1959 (1974).
2. W.W. Rigrod, Saturation effects in High-Gain Lasers, J. Appl. Phys., 36; 2487 (1965).
3. A.J. De Maria, Review of CW High-Power CO_2 Lasers, Proc. IEEE, 61; 731 (1973).
4. A.E. Siegman, E.A. Sziklas, Mode Calculations in Unstable Resonators with Flowing Saturable Gain. 1: Hermite-Gaussian Expansion. Appl. Opt., 13; 2775 (1974).
5. T.A. Cool, Power and Gain Characteristics of High Speed Flow Lasers, J. Appl. Phys., 40; 3563 (1969).
6. E. Armandillo and A.S. Kaye, Modelling of transverse-flow CO_2 lasers: theory and experiment, J. Phys. D: Appl. Phys., 13; 321 (1980).
7. A. Borghese, R. Canevari, L. Garifo and F. Pandarese, Toroidal Unstable Resonant Cavities in Fast-Flow CW CO_2 Lasers. CLEOS/ICF '80, Feb. 26-28, 1980, San Diego, California.
8. A. Borghese, R. Canevari, V. Donati and L. Garifo, Unstable-stable resonators with toroidal mirrors, Appl. Opt., 20; 3547 (1981).

FLOW CHARACTERISTICS OF CLOSED LOOP HIGH PRF E-BEAM OR X-RAY-ASSISTED DISCHARGE EXCIPLEX LASER*

S. M. Fournier, M. L. Sentis, B. M. Forestier, and B. L. Fontaine

Institute of Fluid Mechanics of Marseille
Aix-Marseille University, 1 Rue Honnorat
13003 Marseille

INTRODUCTION

The potential applications of high-average power short-wavelength lasers have led to intensive research in this field [1], and a study of methods for achieving such a laser using wellknown exciplexes capacities has been undertaken at I.M.F.M. The method chosen here is to excite a flowing active medium by means of a high-repetition-rate long-pulse avalanche discharge stabilized either by a low-density electron or X-ray beam. Specifically this paper is devoted to the numerical study of the transient effects of electrical excitation on flow behavior for typical conditions of exciplex laser at high loading ($Q \geqslant 100$ J/ℓ). The experimental study is under development on a new device, "LUX", described in a companion paper [5].

In this paper the well-known characteristics method applied to simulate the transient effects induced by electical excitation in the base flow will be briefly presented as well as predicted waves diagrams. At high-repetition-rates the transient phenomena due to the rapid deposition of energy in the laser medium cause development of large acoustic perturbations which may be strongly attenuated in order to achieve a new laser pulse. The laser beam optical quality is qualitatively related to the flow density perturbations [2] as follows:

$$\frac{\Delta I}{I_o} \simeq \delta\phi^2_{rms}$$

* Work supported by D.R.E.T. and C.E.A., France.

$$\delta\phi_{rms} = \frac{2\pi \ell}{\lambda} \beta \frac{\rho}{\rho_s} (\frac{\Delta\rho}{\rho})_{rms},$$

with: I_o, diffraction-limited far-field central spot intensity; ΔI, variation of the extracted intensity; ℓ, optical path length; λ, laser wavelength; β, Glastone-Dale constant; ρ, laser medium density; ρ_s, density at S.T.P. conditions and $(\frac{\Delta\rho}{\rho})_{rms}$, rms fractional density fluctuations.

The criteria established in terms of the maximum tolerable value of $\Delta\rho/\rho$ for a short wavelength laser are based on beam quality requirements. They are very restrictive owing to the wavelength dependence and more severe than in infrared pulsed chemical lasers. For the XeCl laser under investigation at I.M.F.M., $\Delta\rho/\rho$ in the range of 10^{-5} must be challenged.

The density inhomogeneities in the laser cavity may be due to three main processes:

(a) initial turbulence of the baseline flow,
(b) temperature fluctuations,
(c) pressure waves.

The first process (a) has already been studied experimentally on the "LUX" device, and results are given in a companion paper [5]. Simple energy considerations seem to show that control of temperature fluctuations (b) can be achieved by classical thermal exchangers, notably in the case of a burst mode working of the "LUX" device.

Contrary to the first two processes the $\Delta\rho/\rho$ fluctuations induced by pressure waves (c) are very strong for an atmospheric gas laser with an input energy of about 200 J/ℓ and may be greater than 20 percent. These fluctuations must be attenuated and more carefully studied.

EXPERIMENTAL

Numerical Study of Pressure Waves

The classical method of characteristics is used to analyse the transient phenomena induced by the electrical excitation in the gas flowing in the laser cavity.

The energy deposition, in a few microseconds, causes a quasi-instantaneous rise of pressure and temperature in the excitation zone and generates shock waves. The propagation and reflection upstream and downstream of these shock waves and associated expansion

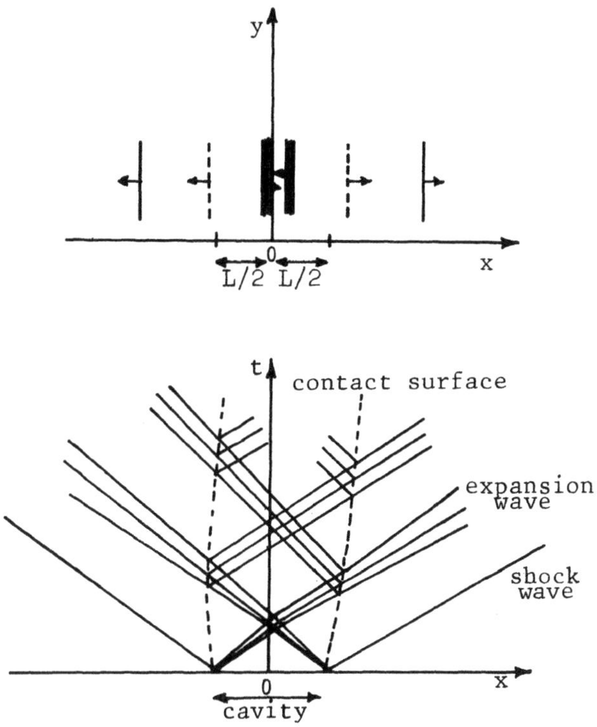

Fig. 1. Wave-front motion.

waves and contact surfaces between heated and cold gases are at the origin of the main thermoacoustic phenomena to be damped before a subsequent pulse.

A schematical representation of phenomena is given on Figure 1 for an idealized infinite duct.

The hypothesis used to establish the numerical model are as follows:

- unidimensional flow,
- the gas follows the ideal gas law and the specific heat are constant,
- body and dissipative forces are neglected,
- the initial conditions are calculated by use of a constant volume heating law (justified by the shortness of the excitation pulse),
- the cross-section of the duct is rectangular and constant,
- no flow and no thermal exchanges across sidewalls.

The two final points will subsequently be modified to take into account the effects on the one hand of cross-sectional area changes and screens and on the other hand of sidewall attenuators.

The classical mass and momentum equations are written using Riemann invariants following a characteristic path, so that

$$\frac{\partial N}{\partial \tau} + (U+A)\frac{\partial N}{\partial \xi} = 0 \quad N = \frac{2}{\gamma-1}A + U$$

$$\frac{\partial Q}{\partial \tau} + (U-A)\frac{\partial Q}{\partial \xi} = 0 \quad Q = \frac{2}{\gamma-1}A - U,$$

with the following non-dimensional parameters

$$\xi = \frac{2x}{L}; \ \tau = \frac{2a_o t}{L}$$

$$A = \frac{a}{a_o}, \ U = \frac{u}{a_o}, \ P = \frac{p}{p_o}, \ R = \frac{\rho}{\rho_o},$$

where L is the excitation zone length (2 cm) and (o) index means relative to gas initial parameters. Interactions between expansion waves are treated by means of these equations.

When an expansion wave reaches a shock wave, the classical Rankine-Hugoniot equations are associated to the Riemann relations to determine the reflected and transmitted pressure waves and contact surface which are created.

When an expansion wave reaches a contact surface the calculations are classically made [4] taking into account the relation between entropy and sound speed.

In Figure 2, we see an x-t diagram for an energy input of 150 J/ℓ in a typical mixture Ne/Xe/HCl (5300/190/10) flowing at an initial velocity of 50 m/s. Two quasi-parallel contact surfaces isolate the heated gas slug flowing downstream. Two shock waves and two expansion fans are generated at t = 0 at each end of the excitation zone. We see also reflected and transmitted expansion waves which are created. Interactions between expansion and shock waves are very weak. For the next step, the numerical program must be capable of handling interaction of pressure waves with screens, cross-sectional area changes and effects of successive pulses.

The four diagrams of Figure 3 represent variations of non-dimensional flow parameters P, U, A, R versus x at two given times 240 μs (strong lines) and 380 μs (dashed lines) for the same conditions as those of Figure 2. Contact surfaces isolating the heated gas slug are visible in Figure 3c and 3d (A and R profiles). Figure 3d underlines as strong the induced phenomena with $\Delta\rho/\rho$ ranging between ± 25%. Moreover, these gradients decrease quite slowly

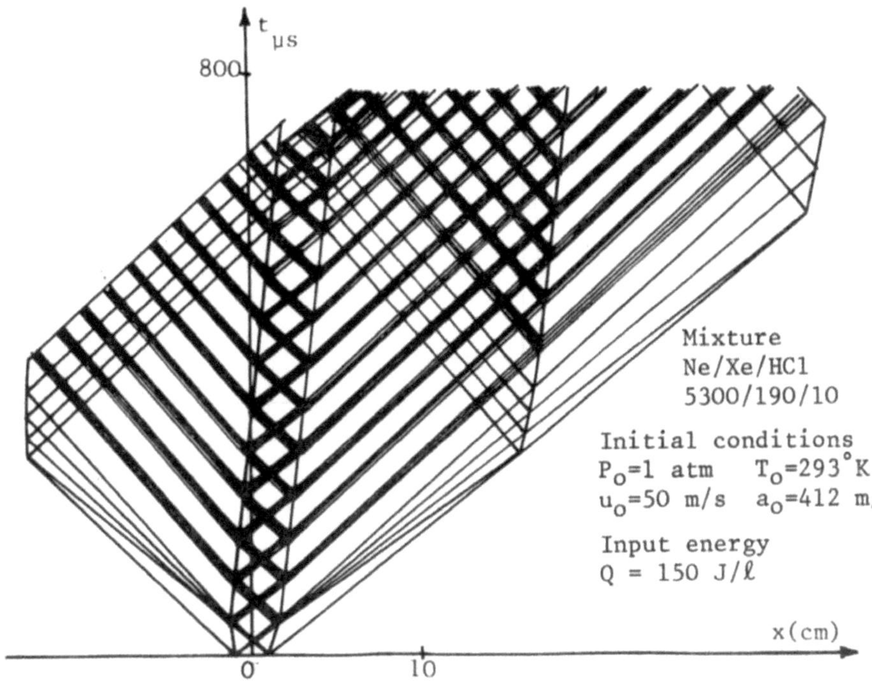

Fig. 2. Induced aerodynamic phenomena in flowing active medium.

from t = 240 μs to t = 380 μs. Variations induced by downstream and upstream shock waves are off the graph at t = 380 μs.

At the loading considered here, the time when the first reflected wave again reaches the excitation zone is about 550 μs. With a 100 J/ℓ input energy, this time is about 750 μs. At 1 ms, the heated gas slug is swept away from the excitation zone but if a new energy pulse would be added, the corresponding shock waves very soon would interact with contact surfaces. This kind of interaction must be considered and a program with this aim is under development.

Effect of Screens on Pressure Waves

Screens have been used as pressure waves attenuators in open cycle CO_2 laser systems [6]. In closed cycle laser systems, this type of attenuator is less convenient owing to friction losses. However, a model has been established [3]. It has been used here and will be associated to the characteristics program. The model is greatly simplified by the following considerations: as mentioned in Figure 4a, screens are characterized by an area reduction on a

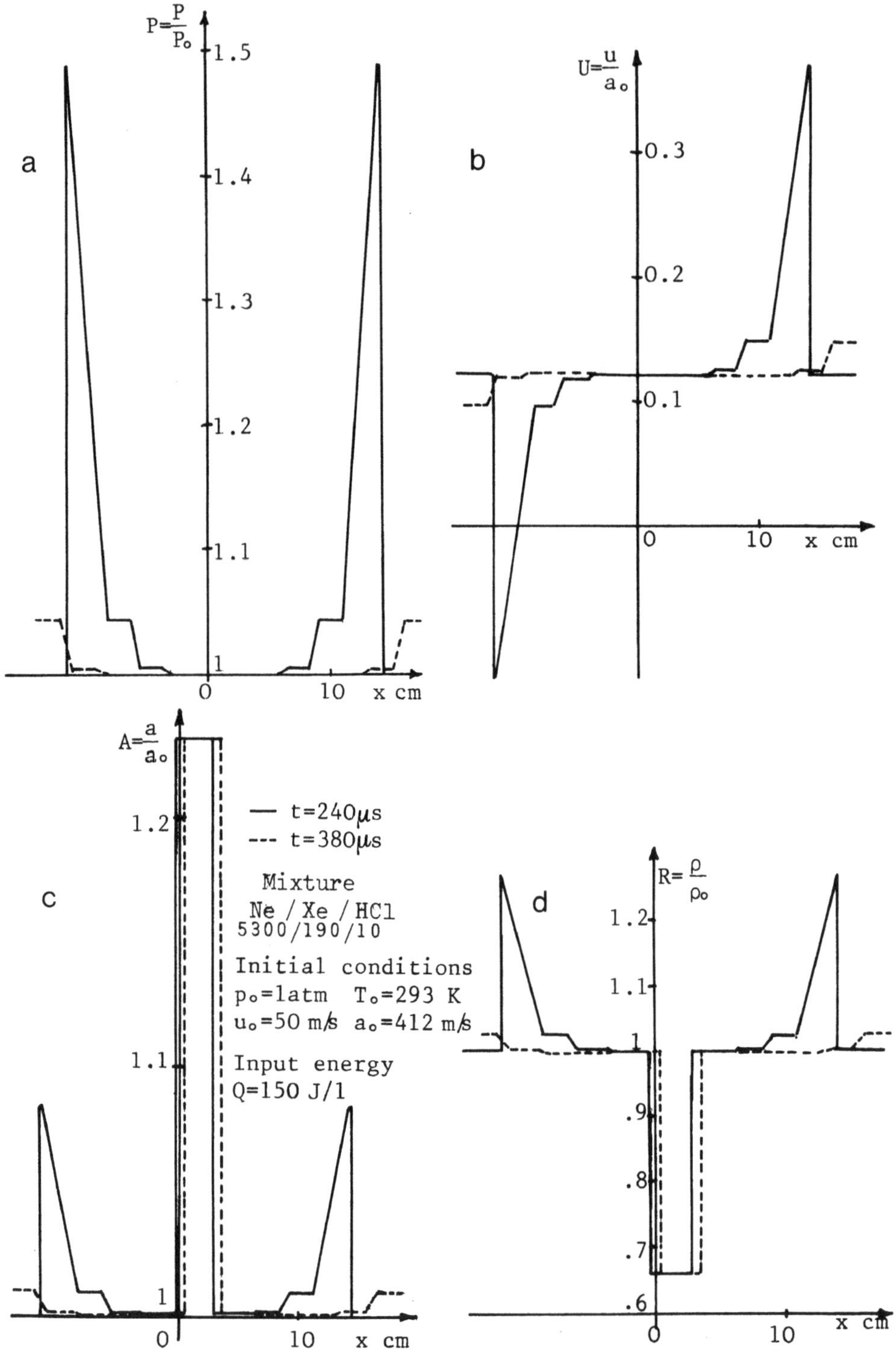

Fig. 3. Flow parameters evolution versus x for two given times: a) pression; b) velocity; c) sound speed; d) density.

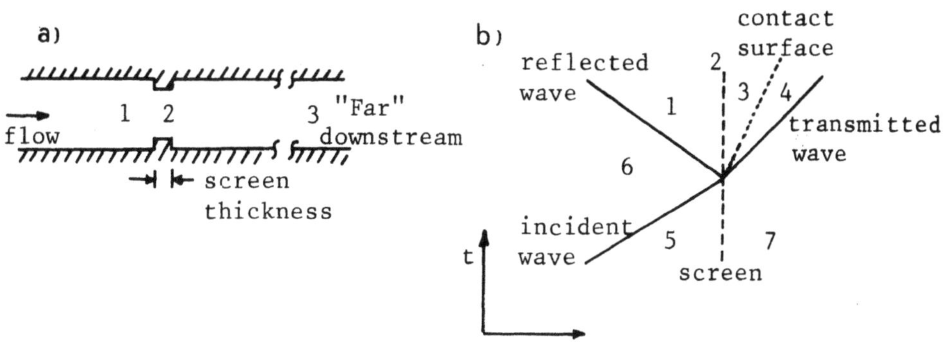

Fig. 4. Screen modelization: a) Flow only; b) Flow and pressure wave.

distance equal to the wire diameter followed by a zone of turbulent mixing on a distance equal to 100 - 200 times the mesh spacing. The wave diagram corresponding to the interaction of a pressure wave with a screen is shown in Figure 4b with the various zones to be considered. Screen and turbulent mixing zone are considered as a simple discontinuity. The main hypothesis [7] are:

- isentropic and quasi one-dimensional flow between zones 1 and 2,
- entropy increase between zones 2 and 3 is related to the stagnation pressure ratio by an experimentally derived formula [8],
- heat exchange between screen and flow is neglected.

As an example in Figure 5, variations of density gradients in zones 1 and 4 are given versus density gradient induced by the incident pressure wave for two screen solidities (ratio of solid area to total frontal area). Where screens are very efficient in attenuating initial turbulence in the flow, the curves of Figure 5 show up their low efficiency as pressure wave dampers.

Calculations presented here will be extended to the case of porous sidewall dampers. Results of calculations taking into account effects due to successive energy pulses, screens, cross-sectional area changes and porous sidewall dampers will be compared with an experimental study currently in progress.

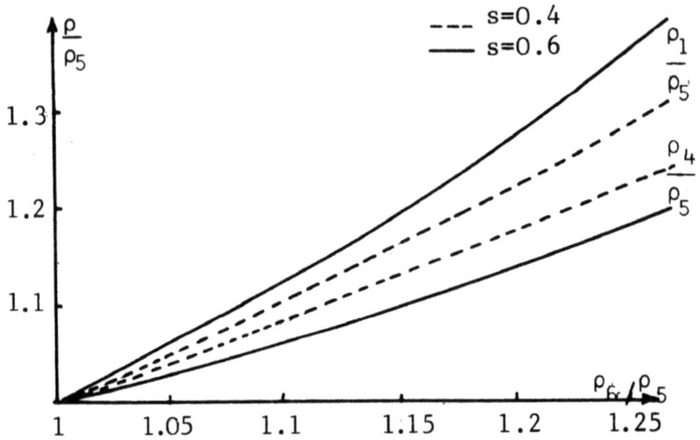

Fig. 5. Density variations in zones 1 and 4 for two screen solidities versus initial density variation (due to a shock wave).
Mixture: Ne/Xe/HCl, 5300/190/10.
Initial conditions: u_5=50 m/s, a_5=412 m/s, T_5=293°K and γ=1.675.

REFERENCES

1. C. K. Rhodes, Excimer lasers, in: "Topics in Applied physics," Vol. 30 Springer-Verlag (1979).
2. C. C. Shih and C. M. Cason, Gas dynamic and acoustic management for visible wavelength lasers, in: "Advances in Laser Technology," Vol. 138, Proceedings of the S.P.I.E., D. Finkleman and J. Stregack (1978),
3. M. J. Pechersky, Shock waves in pulsed gas lasers, in: "Laser Focus," Vol. 17, No. 12, J.A. Camuso (December 1981).
4. G. Rudinger, "Wave Diagrams for Non-steady Flow in Ducts," D. Van Nostrand Co. NY (1955).
5. M. L. Sentis, B. L. Fontaine, and B. M. Forestier, High repetition rate long pulse discharge XeCl laser preionized by a low density electron or X-rays beam, 4th GCL, Stresa (September 1982).
6. B. N. Srivastava, C. J. Knight, and O. Zappa, Acoustic suppression in a pulsed laser system, AIAA Journal, Vol. 18, no. 5 (May 1980).
7. K. O. Tong, C. J. Knight, and B. N. Srivastava, Interaction of weak shock waves with screens and honeycombs, AIAA Journal, Vol. 18, no. 11 (November 1980).
8. A. A. Adler, Variation with Mach number of static and total pressures through various screens, NACA Wartime Report L-23 (Originally Confidential Bulletin L5F28, February 1946).

A NEW 10kW INDUSTRIAL CARBON DIOXIDE LASER

A.S. Kaye, A. Delph, E. Hanley, & C. Nicholson

UKAEA Culham Laboratory,
Abingdon,
Oxon OX14 3DB, UK

INTRODUCTION

Multikilowatt carbon dioxide lasers designed specifically for industrial applications are becoming available at steadily increasing output powers[1-4]. These lasers all utilise electric discharge excitation of rapidly flowing gas, the gas flow being either parallel or normal to the optic axis. Both multi-mode and near-diffraction limited resonators have been used. The former are primarily aimed at the surface treatment range of applications (transformation hardening, surface cladding etc.) which require modest but highly uniform power densities on the surface. Precision hardening by laser has become increasingly established as a viable industrial process. The near diffraction limited lasers, usually fitted with unstable resonators, are in general aimed at welding and cutting applications which require much higher peak intensities. Cutting in particular demands good beam quality, but may prove less interest at high multikilowatt powers. Industrial welding by multikilowatt lasers has been much slower in becoming accepted as a production process due to the more demanding quality control requirements of both the laser and the process. Focal spot size appears to be a less critical parameter in welding; on the other hand, lasers combining very high power with good spot quality have not been generally available for developing this process.

This paper outlines the design and operating characteristics of a new industrial 10kW laser (designated CL10). In the design of this laser, emphasis was placed on achieving good beam quality at the highest power, aiming essentially at applications in deep-

Figure 1 The CL5 Laser at Springfields Nuclear Laboratory

penetration welding of thick section metals. Initial results on the welding performance of this laser are also described.

Laser Design Features

A five kilowatt industrial laser, designated CL5, was designed and built at Culham in 1976-77, cf. Fig 1. This design has since been continuously refined, and a further two lasers built for use within the Authority. Very extensive operational experience has been accumulated, totalling in excess of 13,000 running hours. This 5kW module is a compact, transverse flow, transverse current device, excited by a self-sustained glow discharge (Fig 2). The high specific power input (18kW/litre) of the discharge is obtained by dividing the anode into a large number of independently ballasted segments. An $M = 2$ unstable cavity is used to extract the power with 14% discharge efficiency. The gas consumption is $0.15m^3$/hr of a 6:3:1 mix of $He:N_2:CO_2$. A small amount of oxygen ($\sim$ 2%) is added to reduce the dissociation of CO_2. Typical operating pressure is 50mBar. At maximum input power, the small signal gain is typically 0.8% cm^{-1} and the saturation intensity $400W/cm^2$, both these parameters varying with distance through the discharge[5].

Two of these modules have now been coupled together to form a 10kW laser (Fig. 3). This approach takes advantage of the small size of the CL5 module and minimises additional development. The modules are mounted within a single unstable optical cavity, giving several advantages. The increased single pass gain allows the output coupling to be increased, thus improving both the efficiency of extraction and the beam quality. The increased cavity length also reduces the longitudinal mode spacing and removes the small power modulation detected on CL5. Furthermore, the tolerance to mirror misalignment is improved.

The effect of magnification on beam quality is illustrated in Figure 4, where the calculated intensity distribution and enclosed power at the focal spot are plotted for various values of magnification, assuming uniform illumination and a plane wavefront; it is observed that for $M = 3$, the peak intensity is close to the optimum, whilst the Airy disc includes more than 60% of the total power.

The new laser is also fitted with a catalytic loop to reduce the dissociation of CO_2. With this loop in operation, the dissociation is about halved and the maximum output power increased by about ½kW. At low power, the improvement is more pronounced.

The output window can be a critical component on lasers of this power. The optimum design is a trade-off between thermal distortion and lifetime. Scaling from long experience on CL5 suggests that service times of 500 hours can be achieved. Calculated thermal distortion of a window transmitting an annular beam is shown in Figure 5. A spherical aberration of $\lambda/4$ due to the temperature coefficient of refractive index, and a further $\lambda/12$ from the stress-optic effect is predicted, the latter component being dependent on the beam polarization. This degree of distortion corresponds to the Rayleigh limit and is sufficient to about double the focal spot size.

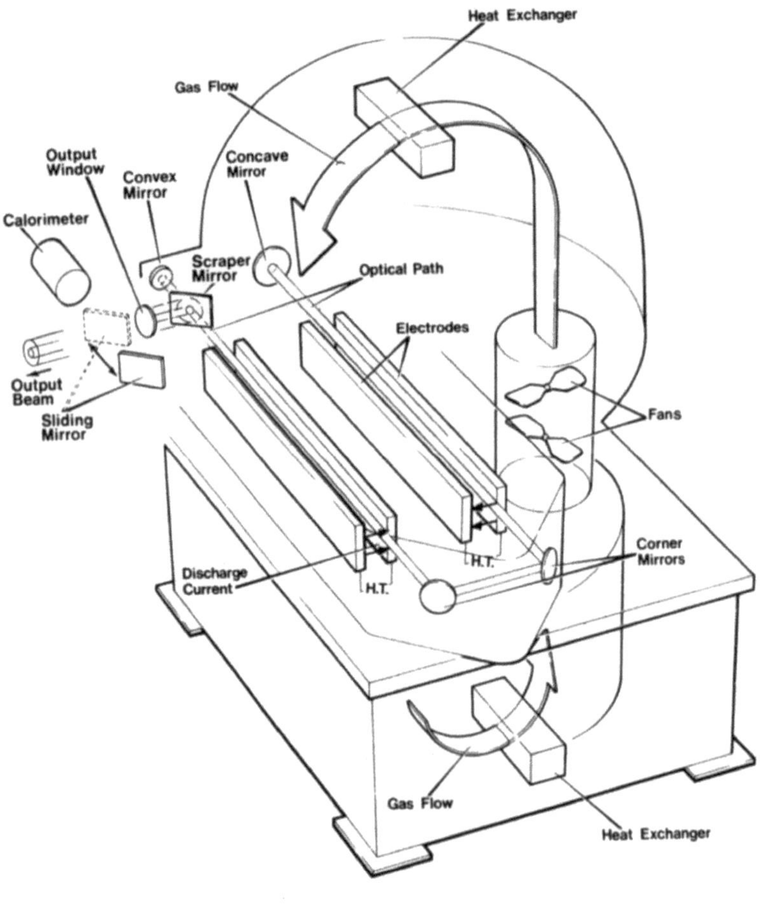

Figure 2 Diagram of the CL5 Laser module

Figure 3 The CL10 Laser

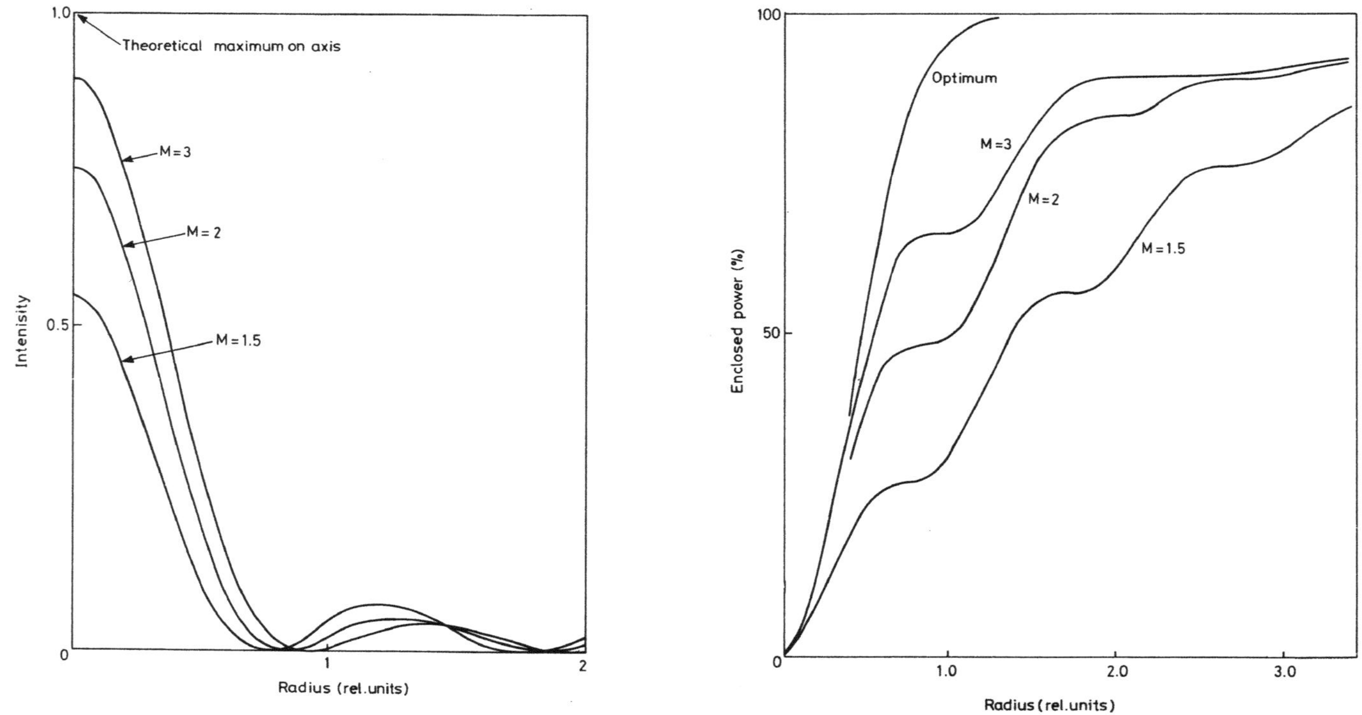

Figure 4 Computed intensity distribution and enclosed power at the focal plane for an annular beam of magnification M

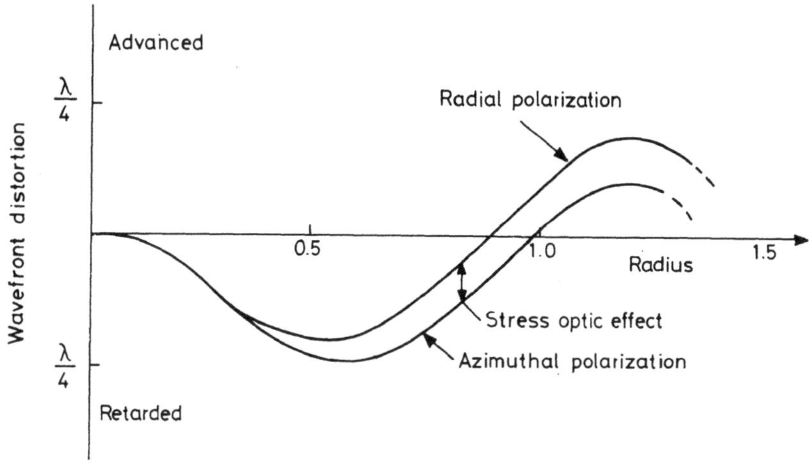

Figure 5 Calculated wavefront distortion due to the output window at 10kW

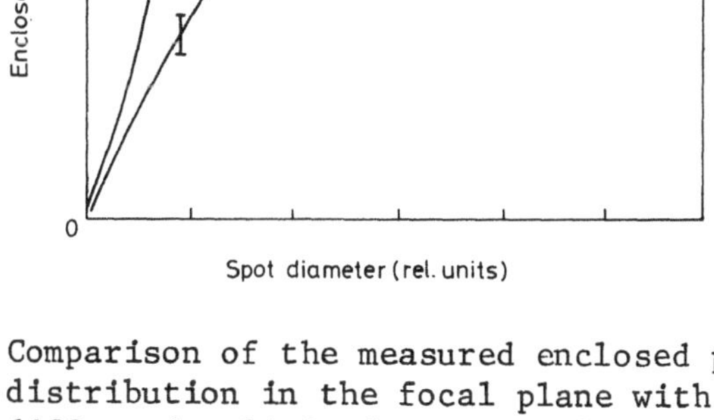

Figure 6 Comparison of the measured enclosed power distribution in the focal plane with the diffraction limit for an M = 3 annular beam

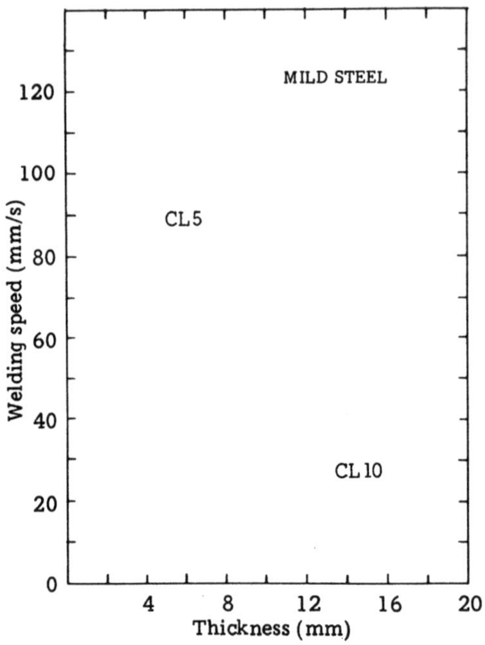

Figure 7 Weld section in 19mm steel

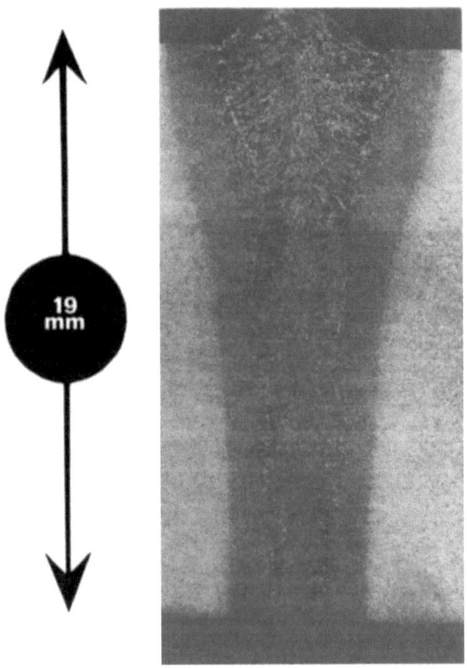

Figure 8
Variation of welding speed with thickness of steel

Laser Performance

Many hundreds of hours of operational experience have shown the laser to be very reliable at the design power. Usefully stable operation is maintained comfortably in excess of the design power, enabling experimental work at up to 12kW on the workpiece, the anticipated scaling in power of better than a factor of 2 from combining the two modules having been realised. No difficulty was found to arise from electrical, optical or gas dynamic interference between the two modules. Total gas consumption on the CL10 laser is less than $0.1m^3$/hour; the laser will operate fully sealed for many hours.

The focal spot intensity distribution has been measured by spatially filtering an f.100 focal spot. The results are shown in Figure 6. The power enclosed in the Airy disc is close to the diffraction limit for the CL10 cavity magnification. The lifetime of the window is typically 500 hours as anticipated.

On the basis of the measured powers and spot sizes, the intensity in the CL10 focal spot will be typically 10^7 watts/cm^2 averaged over the Airy disc at f.8 with aberration-free focusing optics.

Welding Performance

Preliminary welding trials on thick section steel and titanium have been conducted. The maximum penetration in steel is about 19mm at 11kW on the work at 8mm/s, a section of such a weld being shown in Figure 7. The weld is observed to be very narrow in relation to its depth, to be near parallel-sided and to have a weakly defined nail-head. The variation of optimum welding speed with thickness is shown in Figure 8. These results are fully comparable with published data for other lasers[6], reflecting the good beam quality of the CL10 laser.

Conclusion

The main design and performance features of a new 10kW industrial laser have been described. The laser combines high power with good beam quality and low running costs. Initial welding trials show a capability of welding up to 19mm steel.

Acknowledgements

The cooperation of Dr. J.H.P.C. Megaw and Mr. M. Hill in carrying out the welding trials is gratefully acknowledged.

References

1. E.V. Locke, Opt. Eng. 17, 3, 192, 1978.
2. A.V. Artamonov et. al., Sov. J. Quantum Electron. 8, 528, 1978.
3. M. Hichii, H. Nagori, A. Nagori and T. Akiba, J. Appl. Phys., 52, 8, 4953, 1981.
4. E. Haag, H. Pease, J. Stall and J. Zar, Appl. Optics 13, 1959, 1974.
5. E. Armandillo and A.S. Kaye, J. Phys. D: Appl. Phys. 13, 321, 1980.
6. C.M. Banas, Opt. Eng., 17, 3, 210, 1978.

FURTHER DEVELOPMENT OF THE NON-STATIONARY STOCHASTIC MODEL OF DIFFUSE ELECTRIC DISCHARGES IN TURBULENT GAS FLOWS

Yu. Khait

Department of Physics

Ben-Gurion University of the Negev, Beer-Sheva, ISRAEL

A new development of the non-stationary statistical kinetic theory of non-equilibrium transient electronic phenomena in hydrodynamically turbulized gas portions speedily passing during a short time through the discharge zone with electric field E at $eE\bar{\lambda} << W_0$, is presented. Here $\bar{\lambda}$ is the electron mean free path length and W_0 is the ionization potential. The first Townsend ionization coefficient, characteristic times of the formation of avalanches, time-dependent electric concentration and current and other parameters associated with non-equilibrium transient electron processes occuring in turbulent gas portions during a short residence time in the discharge zone, are calculated.

Diffuse electric discharges (e.g. coronas) in hydrodynamically turbulized gas flows are presently of interest due to their applications to electric discharge convection lasers [1-14] and to plasma chemistry [2,8,10-14]. Many differences exist between diffuse discharges under no-flow and under hydrodynamically turbulent conditions [2,10]. These differences lead to directly observable consequences, some of which were considered in Ref. [2,3,6,8-14]. Among them we want to mention the following: (i) A great enhancement of transport coefficients, e.g. of turbulent ambipolar diffus ivity [2,10]

$$D_{At} \simeq \frac{R_e}{R_e^*} \cdot \frac{\bar{\varepsilon}_e}{\bar{\varepsilon}_i} \cdot D_i \quad \text{at} \quad \frac{R_e}{R_e^*} >> 1 \quad \text{and} \quad \frac{\bar{\varepsilon}_e}{\bar{\varepsilon}_i} >> 1 \qquad (1.1)$$

due to high ratios R_e/R_e^* (e.g. $\simeq 3{\cdot}10^3$) of the Reynolds, R_e, and the critical Reynolds, R_e^*, numbers and $\bar{\varepsilon}_e/\bar{\varepsilon}_i$ (e.g. $\simeq 10^2$) of the mean electron, $\bar{\varepsilon}_e$, and ion, $\bar{\varepsilon}_i$, energies, where D_i is the ion diffusivity (e.g. $D_i \simeq 0.2$cm/S). (ii) A sharp increase of power cousumption and electric current at the prearcing mode up to 10^2 to 10^3 of non-flow conditions without losing the discharge stability [2,3]. (iii) Random fluctuations of gas velocity Δv around mean value $\bar{v}$, of pressure ΔP near $\bar{P}$, etc., which can be described in terms of random functions of space and time [15], and, on the other hand, can be characterized by the spectrum of space scales Λ and frequencies Ω, where the largest, ℓ, and the smallest, Λ_0, linear scales and the corresponding frequencies are linked by well-known equations [1,6]

$$\Lambda_0 = (R_e^*/R_e)^{3/4}\ell \text{ and } \Omega_0 = \frac{\Delta v}{\ell}\left(\frac{R_e}{R_e^*}\right)^{3/4} \tag{1.2}$$

used in Ref. [2,10 to 14] for the analysis of electrical discharges in turbulent flows.

At high R_e/R_e^* scales Λ_0 can be rather small and frequencies Ω_0 high enough. For instance, $\Lambda_0 \simeq 10^{-2}_{cm}$ and $\Omega_0 \simeq 10^6 S^{-1}$ at $R_e/R_e^* \simeq 3 \cdot 10^3$, $\ell \simeq 4$cm and $\Delta v \simeq 10^4$cm/S. (iv) A rapid and great enhancement of the electron number density $n(t)$ in gas portions speedily passing through the discharge zone of volume $V \simeq L^3$, during short residence time $\Delta\tau \simeq L/\bar{v}$, that leads to fast and drastic changes of properties of the gas converted non-equilibriumly into weakly ionized plasma [2,10-14]. The transient non-equilibrium phenomena associated with the formation of electron avalanches and plasmas in practically non-ionized gas portions passing over the discharge zone during the short time $\Delta\tau$, we shall briefly term the non-equilibrium gas-to-plasma transitions (GPT's), following the suggestion given in [10-14]. The non-equlibrium non-stationary statistical kinetics of the GPT and the related phenomena in turbulent gas discharges and, in particular, in turbulent coronas, is the main subject of this paper. In this work we face some new classes of physical problems which are different from conventional gas dynamics, from the theory of non-flow gas discharges, e.g. from non-flow coronas [17-19], from the standard equilibrium theory of phase transitions, etc. [2,10-14]. On the other hand, there are similarities between transient phenomena in the turbulent gas portions passing through the discharge zone during $\Delta\tau \simeq L/\bar{v}$ and processes in pulse discharges [2,10-14].

Consider a small gas portion of volume $\Delta V << V$ containing a very small initial electron number density $n^{(0)} = n(t_0)$ (formed by cosmic rays, by the natural radioactivity, etc.) entering at moment t_0 the discharge zone of volume V having steady but spatially non-uniform electric field strength $E_0(r_0) = E_0(x_0, y_0, z_0)$. $E(r_0)$ can be higher near electrode pins or wires and much weaker at some distances from them. If one considers the corresponding electric field strength $E(r,t)=E(x,y,z,t)$ acting on the gas portion in the accompanying coordinate system moving with velocity $v=\bar{v}+\Delta v$ (having components $v_k=\bar{v}_k+\Delta v_k$, $k=1,2,3$), one finds that $E(r,t)$ and ratio $E(r,t)/P$ change with time. Characteristic of such time evolution of $E(r,t)$ depends on $E_0(r_0)$, v and the traectory of the gas portion in the discharge zone. Namely, components $E_i(r,t)$ of $E(r,t)$ change with time according to the following relations

$$\frac{dE_i}{dt}=\sum_{k=1}^{3} v_k \frac{\partial E_{0_i}}{dx_{0_k}}=\overline{\frac{dE_i}{dt}}+\Delta\left(\frac{dE_i}{dt}\right),\qquad i=1,2,3 \tag{1.3}$$

$$E_i(r,t)=\int_{t_0}^{t}(dE_i/dt)\,dt=\bar{E}_i(r,t)+\Delta E_i(r,t) \tag{1.4}$$

where

$$\overline{\frac{dE_i}{dt}}=\sum_{k=1}^{3}\bar{v}_k\left(\frac{\partial E_{0_i}}{\partial x_{0_k}}\right)\quad\text{and}\quad \Delta\left(\frac{dE_i}{dt}\right)=\sum_{k=1}^{3}\Delta v_k\left(\frac{\partial E_{0_i}}{\partial x_{0_k}}\right) \tag{1.5}$$

Here $\Delta E_i(r,t)$ presents random oscillations of components $E_i(r,t)$ of $E(r,t)$, associated with turbulent fluctuations $\Delta v=\{\Delta v_k\}$ of the gas velocity.

Correlation tensor of $\Delta E_i(r,t)$

$$B_{is}(r_1,t_1;r_2,t_2)=\overline{\Delta E_i(r_1,t_1)\cdot\Delta E_s(r_2,t_2)} \tag{1.6}$$

is closely connected with correlation functions of turbulent pulsations of Δv_k, according to Egs (1.3) to (1.6).

If volume ΔV is small enough, one can assume that $E(t)$ does not depend on r within ΔV. In some cases $\bar{E}(t)$ can be described through the triangular approximation

$$\bar{E}(t)=\begin{cases}\dfrac{E_M}{t_M-t_0}(t-t_0) & \text{at}\quad t_0\le t\le t_M\\[2ex] \dfrac{E_M}{t_0+\Delta\tau-t_M}(t_0+\Delta\tau-t) & \text{at}\quad t_M\le t<t_0+\Delta\tau\end{cases} \tag{1.7}$$

where t_M is the moment when $\bar{E}(t)$ reaches its maximum value

$E_M = \bar{E}(t_M)$, for the gas portion moving along a given traectory in the discharge zone. In this case ratio $\bar{E}(t)/\bar{P}$ is also described in the correspoding triangular approximation. Hence one can see that

$$\dot{E} = \frac{E_M}{t_M - t_0} \text{ at } t_0 < t < t_M \text{ and } \dot{E} = -\frac{E_M}{t_0 + \Delta\tau - t_M} \text{ at } t_M < t < t_0 + \Delta\tau \qquad (1.8)$$

For example, a gas portion passing with $\bar{v} \simeq 10^4$cm/S near an electrode pin or wire with maximum field strength $E_{oM} \simeq (5-10)\cdot 10^4$v/cm and strong field localized in the corona region with linear scale $L_a \simeq 0.3$cm, is affected by the pulsed action of electric field characterized by

$$\Delta\tau \simeq 2(t_M - t_0) \simeq 6\cdot 10^{-5}\text{S}; \quad E_M \simeq (5-10)\,10^4\text{v/cm}; \quad |dE/dt| \simeq (1.6-3.2)\,10^9\text{v/cmS}$$

$$E_M/\bar{P} \simeq (66-132)\,\text{v/cm torr and } |d\bar{E}/\bar{P}dt| \simeq (2.1-4.2)\,\text{v/cm}\cdot\text{S}\cdot\text{torr} \qquad (1.9)$$

at $\bar{P} \simeq 760$torr. Here $t_M - t_0$ is the time interval during which the gas portion approaches the pin (wire), and $t_0 + \Delta\tau - t_0$ is the time when the gas portion moves from the pin within the distance L_a.

In this paper we use the concept of the "impulse" action of the electric field on gas portions passing the discharge zone during a short period of time to develop an approximate non-stationary kinetic theory of the transient non-equilibrium bevahviour of electrons during the formation of avalanches or GPT's, using the approach suggested in Ref. [10-14]. Here we consider the low E/P approximation satisfying conditions [10-14]

$$e\bar{E}\bar{\lambda}_{min} \ll W_0 \text{ or } \bar{E}/N \ll (\bar{E}/N)_B = W_0\sigma_{max}/e \qquad (1.10)$$

where $\bar{\lambda} = (N\cdot\sigma_{max})^{-1}$ is the minimum of the electron mean free path length associated with the maximum total cross section σ_{max} of electron-gas particle collisions, $N = \bar{P}/KT$ is the total number of gas particles in 1cm^3, W_0 is the ionization potential.

2. THE DESCRIPTION OF THE MODEL

The initial $\Delta N_e^{(0)} = N^{(0)}\cdot\Delta V$ free electrons that exist in the gas portion entering the electric field $E_0(r_0)$ at t_0 and forming the nil electron generation (nil EG) have low initial mean energy $\bar{\varepsilon}^{(0)}(t_0) \ll W_0$. Therefore the average free electron (AFE) of the nil

EG cannot ionize gas particles and change $n(t)$. The AFE needs finite time $\Delta\theta^{(0)}$ of the order of the mean time between the two successive ionizing collisions at given $E(t)$,P,T, etc. to be accelerated up to $\bar{\varepsilon}^{(0)}(t_0+\Delta\theta^{(0)})\geq W_0$ and ionize a gas particle. The nil EG is characterized by:

(a) the fixed initial electron concentration $n^{(0)}$ which is practically constant during time $\Delta\theta^{(0)}$ (see fig. 1);

(b) the time dependent electron energy distribution $f^{(0)}(\varepsilon^{(0)},t^{(0)})$ shifting systematically towards higher energies during $\Delta\theta^{(j)}$, according to the non-stationary electron kinetic equiation and its solution presented in Ref. [10], and

(c) time dependent AFE energy

$$\bar{\varepsilon}^{(0)}(t^{(0)})=\int\varepsilon^{(0)}\cdot f^{(0)}(\varepsilon^{(0)},t^{(0)})d\varepsilon^{(0)} \qquad (2.1)$$

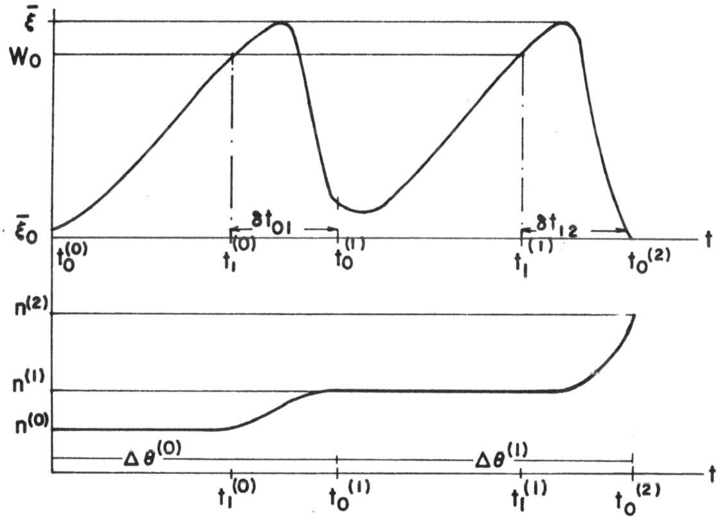

Fig. 1: Schematic Diagrams of the Time Evolutions of the AFE Energy $\bar{\varepsilon}(t)$ and Electron Number Density $n(t)$ of the Nil and First EG's During Lifetimes $\Delta\theta^{(0)}$ and $\Delta\theta^{(1)}$. $n^{(0)}$, $n^{(1)}$ and $n^{(2)}$ are the Electron Number Densities for the Nil, First and Second EG's Starting at $t_0^{(0)}$, $t_0^{(1)}$ and $t_0^{(2)}$.

At moment $t_1^{(0)} = t_0^{(0)} + \Delta t_1^{(0)}$ the AFE energy $\bar{\varepsilon}^{(0)}(t_1^{(0)}) \simeq W_0$ and then AFE with $\bar{\varepsilon}^{(0)}(t^{(0)} > t_1^0)$ needs additional time δt_{01} to ionize a gas particle and then (see fig. 1).

$$\Delta\theta^{(0)} \approx \alpha^{(0)} \langle \bar{\tau}_e^{(0)} \rangle_{\Delta t_1} + \langle N \cdot \sigma_{in}(\bar{\varepsilon}^{(0)}) \cdot \bar{u}^{(0)} \rangle_{\delta t} \tag{2.2}$$

where $\bar{\tau}_e^{(0)} = \bar{\lambda}/\bar{u}^{(0)}$ is the AFE mean free path time, $\alpha^{(0)}$ and $\delta\alpha^{(0)} \approx \sigma(\bar{\varepsilon}^{(0)})/\sigma_{in}(\bar{\varepsilon}^{(0)})$ are the numbers of non-ionizing AFE-gas particles collisions during $\Delta t_1^{(0)}$ and δt_{01}, σ_{in} is the ionization cross section, $\bar{u}^{(0)} = (2\bar{\varepsilon}^{(0)}/m)^{\frac{1}{2}}$, $t_0^{(0)} = t_0$.
At moment $t_0^{(0)} + \Delta\theta^{(0)}$ the AFE's create $\delta n_{01} = n^{(0)}(1 - \chi^{(0)})$ free electrons and

$$n^{(1)} = n^{(0)} + \delta n_{01} = n^{(0)}(2 - \chi^{(0)}) \tag{2.3}$$

The treatment given above for the nil EG can be applied to an arbitrary jth EG, with the obvious replacement o ⟶ j and 1 ⟶ j+1. Then the GPT and an avalanche can be modelled by the non-stationary kinetic chain composed of g >> 1 kinetic segments each of which includes the two parts : (i) The continuous evolution of an approximately fixed number $\Delta N^{(j)} = n^{(j)} \cdot \Delta V$ of free electrons during time $\Delta\theta^{(j)}$ of the order of the mean time between the two successive ionizing collisions of the AFE with gas particles at given $E(t^{(j)}) = \bar{E}(t^{(j)}) + \Delta E(t^{(j)})$, $P = \bar{P} + \Delta P$ and time-dependent electron energy distribution $f^{(j)}(\varepsilon^{(j)}, t^{(j)})$ of the jth EG calculated from the non-stationary electron kinetic equation [10] ; here j=0, 1, 2,..., g. (ii) The discontinuous jth EG-to-(j+1)th EG transition associated with the increase in n(t)

$$\delta n_{j,j+1} = n^{(j+1)} - n^{(j)} = n^{(j)}(1 - \chi^{(j)}); \quad j = 0, 1, \ldots, g \tag{2.4}$$

and the corresponding decrease in the AFE energy . Multiple repetitions of the above processes lead to the formation of electron avalanches, GPT's, etc. The electron concentration

$$n^{(j)} = n^{(0)} + \sum_{k=0}^{j-1} \delta n_{k,k+1} = n^{(0)} \exp\left(\frac{t_0^{(j)} - t_0}{a \cdot \langle \Delta\theta \rangle_j}\right) \tag{2.5}$$

at moment

$$\theta^{(j)} = t_0^{(j)} - t_0 = \sum_{k=0}^{j-1} \Delta\theta^{(k)} = \dot{j} \cdot \langle\Delta\theta\rangle_j \; ; \; j = 0, 1, \ldots, g \tag{2.6}$$

appears as a result of the development of j EG's, where

$$j = a \cdot \ln(n^{(j)}/n^{(0)}) \quad \text{and} \quad a^{-1} = 0.7 + \langle \ln(1 - \int/2) \rangle_j \tag{2.7}$$

and brackets $\langle \ldots \rangle_j$ mean averaging over g EG's.

PARAMETERS OF TURBULENT GAS DISCHARGES AT LOW E/P.

The model described in the previous section leads to the following equations for parameters of the gas speedily passing over the discharge zone.

(i) the time between the two successive ionizing AFE particle collisions averaged over j EG,s is

$$\langle \Delta\theta \rangle_j = \langle \Delta t_1 \cdot \} _1 \rangle_j = \Delta\theta_{og} \cdot \exp(B \cdot P_g / E_g) \tag{3.1}$$

where factors

$$B \approx \left\langle [2\delta(\bar{\varepsilon})]^{\frac{1}{2}} \cdot \frac{\bar{\varepsilon}}{kT} \cdot \frac{\sigma(\bar{\varepsilon})}{e} \right\rangle_g \; ; \; \Delta\theta_{og} = \frac{m \cdot W_0 \cdot \}}{e \cdot E_g \cdot [m \cdot \delta(\bar{\varepsilon}) \cdot \bar{\varepsilon}]^{\frac{1}{2}}} \tag{3.2}$$

$\overline{\Delta W} = \delta(\bar{\varepsilon}) \cdot \bar{\varepsilon}$ is the mean AFE energy losses per one non-ionizing collision, $E_g = \bar{E}_g + \Delta E_g$ and $P_g = \bar{P} + \Delta P_g, \} = 1$. Eqs (3.1) and (3.2) are obtained from the consideration of AFE's with random free path times τ and energy $\delta\varepsilon = e \cdot E(t^{(j)}) \cdot \delta z$ the AFE's gain from E during τ distributed according to

$$\varphi^{(j)}(\tau) = \frac{1}{\bar{\tau}_e^{(j)}} \exp\left(-\frac{\tau}{\bar{\tau}_e^{(j)}}\right) \quad \text{and} \quad \psi^{(j)} = \frac{1}{2 \cdot \delta\bar{\varepsilon}} \exp\left[-\left(\frac{\delta\varepsilon}{\delta\bar{\varepsilon}^{(j)}}\right)^{\frac{1}{2}}\right] \tag{3.3}$$

and satisfying the following conditions

$$\delta e = [\delta\varepsilon - \delta(\bar{\varepsilon}) \cdot \bar{\varepsilon}] > 0 \quad \text{or} \quad \tau > \tilde{\tau}_e = [m \cdot \delta(\bar{\varepsilon}) \cdot \bar{\varepsilon}]^{\frac{1}{2}} \cdot [e \cdot E(t^{(j)})]^{-1} \tag{3.4}$$

Here $\delta z = w \cdot \tau = \mu \cdot E \cdot \tau$ is the random AFE way along $E(t^{(j)})$ during τ, with drift velocity w and mobility μ,

$$\delta\bar{\varepsilon}^{(j)} = e \cdot E(t^{(j)}) \cdot \overline{\delta z} = [e \cdot E(t^{(j)})]^2 \cdot \left\{ 2\bar{\varepsilon}^{(j)} \cdot [N \cdot \sigma(\bar{\varepsilon}^{(j)})]^2 \right\}^{-1} \ll W_0 \tag{3.5}$$

(ii) The mean number of ionizations the AFE makes on 1 cm of its way along the electric field

$$\eta_T = \langle \eta^{(j)} \rangle_g = (\langle w^{(j)} \cdot \Delta\theta^{(j)} \rangle_g)^{-1} = F \cdot P \cdot \exp(-B \frac{P_g}{E_g}) \qquad (3.6)$$

is closely connected with the ionization coefficient

$$\bar{\eta}_T = \bar{F} \cdot P \cdot \exp(-B \frac{\bar{P}}{\bar{E}_g}) \quad \text{with} \quad \bar{F} = \frac{\xi \cdot e \cdot \bar{E}_g}{W_o \cdot \bar{P}} \qquad (3.7)$$

in the gas without turbulent pulsations, where $F = \xi \cdot e \cdot E_g / W_o \cdot \bar{P}$.

(iii) $\bar{\eta}_a$ is close to the first Townsend coefficient $\bar{\alpha}_T \approx \bar{a}^{-1} \eta_T \approx \approx (\langle w \rangle_g \cdot \tau_T)^{-1}$ in non-flowing gases [18-21]. Eq(3.7) complies with the empirical equation for $\bar{\alpha}_T$ [18-21]. Here $\tau_T \approx a \cdot \langle \Delta\theta \rangle$. Value η_T given by Eq(3.6) can be treated as the first ionization coefficient in turbulent gases. The obtained equations lead to an agreement between calculated $\bar{\alpha}_T$ and observed $(\bar{\alpha}_T)_{obs}$, calculated and experimental parameters of single avalanches , etc.

(iv) If one uses the triangular approximation (1.7) for E(t) at $t < t_M$, one can obtain

$$\eta_T \approx \frac{e \cdot E_m}{W_o} \left(\frac{t - t_o}{t_M - t_o}\right) \exp\left(-\frac{\overline{\Delta T_d}}{t - t_o}\right) \qquad (3.8)$$

where

$$\overline{\Delta T_d} = \frac{B \cdot \bar{P}}{E_M}(t_M - t_o) = [2\delta(\bar{\varepsilon})]^{\frac{1}{2}} \cdot \frac{\bar{\varepsilon} \cdot \bar{P} \cdot \sigma(\bar{\varepsilon})}{kT \cdot e \cdot E_M}(t_M - t_o) \qquad (3.9)$$

Hence one can conclude that $\Delta\theta^{(o)} \approx \overline{\Delta T_d} \approx t_M - t_o$ and $\Delta\theta^{(o)} \approx \approx 3.10^{-5}$ s (see Eqs(1.9)), since $B.\bar{P}/E_M \approx 1 - 3$. Therefore for the avalanche composed of $g \gg 1$ EG's one has $\theta^{(g)} > \Delta\theta^{(o)} \approx 3.10^{-5}$ s $> \omega_o \approx \approx 10^{-6}$s. Since durations of avalanche chains usually preceding breakdowns of the discharge are longer than $\theta^{(g)} > \omega_o^{-1}$, turbulent pulsations have time to disperse charged particles over the discharge zone and to increase the discharge stability.

(v) Electron current density $J(t^{(j)})=e.n^{(j)}w^{(j)}$ and conductivity Σ =J/E and other values depending on $n^{(j)}$ are exponential functions of time, $\bar{E}$, gas parameters and turbulent pulsations ΔE_g and ΔP_g, according to Eqs(2.5), (3.1) and (3.2).

REFERENCES.

1. A.C.Eckerbeth and F.S.Owen, Rev.Sci.Instrum. 43,995 (1972).
2. Yu. Khait and O.Biblarz,J.Appl.Phys. 50, 4692 (1979).
3. O.Biblarz and R.E.Nelson,J.Appl.Phys. 45, 633 (1974).
4. J.Shwartz and Y.Lavie, AIAA Jornal 13, 647, 1975.
5. J.L.Barto, M.S.Thesis, US Naval Postgraduate School, Monterey, USA, 1976.
6. O.Biblarz, J.L.Barto and H.A.Post, Israel Jorn. of Technology 15, 59, 1977.
7. H.A.Post, M.S. Thesis, US Naval Postgraduate School, Monterey, USA, 1976.
8. J.L. Barto, Report NPS 67-80-005.
9. C.H.Davis, M.S. Thesis, US Naval Postgraduate School, Monterey, USA, 1980.
10. Yu.Khait, Jorn.de Phys. (Paris) C 9 41, 317 (1980).
11. Yu.Khait, A paper at 4th Int. Symp. on Plasma Chemistry, Zürich (1979).
12. Yu.Khait, Abstracts XIV UPAP Int. Conf. on Statistical Physics, Edmonton, Alberta, Canada (1980).
13. Yu.Khait, Invited Paper, Int.Workshop on Plasma Chemistry in Technology, Ashkelon, Israel (1981).
14. Yu.Khait, Extraprogram Invited Paper, 5th Int.Symp. on Plasma Chemistry, Edinburgh, U.K. (1981).
15. A.S.Monin and A.M.Yaglom, Statistical Hydromechanics, Nauka, Moscow, 1967.
16. L.D.Landau and E.M.Lifshitz, Fluid Mechanics,Addison-Wesley Reading,Mass. (1959), Chap.III.
17. I.W. McAllister, G.C.Crichton and E.Brengsbo,J.Appl.Phys. 50, 6797 (1979).
18. A.von Engel, Ionized Gases, Oxford University Press, 1965.
19. S.C.Brown, Introduction to Electrical Discharges in Gases, Wiley, New York, 1966.
20. H.S.W.Massy, E.H.S.Barhop and H.B.Gilbody, Electronic and Ionic Impact Phenomena, Clarendon Press, Oxford, 1969.
21. H.Raether, Electron Avalanches and Breakdown in Gases, Butterworths, London, 1964.

THE EFFECT OF CHEMICAL PROCESSES ON THE STRATIFICATION OF VOLUME-DOMINATED GAS DISCHARGES

K. S. Klopovsky, A. P. Osipov, I. G. Persiantsev,
A. T. Rakhimov, T. V. Rakhimova, N. V. Suetin, and
M. A. Timofeev

Institute of Nuclear Physics, Moscow State University
Moscow, 117234, U.S.S.R.

Besides the processes of ionization of atoms and molecules, recombination of charged particles, and attachment of electrons that occur in the plasma of volume-dominated gas discharges, there may also take place some slower processes causing changes in the chemical composition of the plasma and in the excitation level of the various degrees of freedom of neutral particles. Such processes may include the formation of easy-to-ionize particles (these may be new molecules formed as a result of plasma chemical reactions, or neutral particles so excited as to reach a metastable state) or of particles conducive to the acceleration of detachment of electrons from the negative ions. The above processes lead to an increase in the plasma conductivity and the acceleration of ionization instabilities[1,2].

This report concentrates on the fact that chemical reactions taking place in the plasma of gas discharges may lead to the formation in the discharge volume of new electronegative species to which electrons attach with higher rates and this may tell materially on the physical properties of the gas-discharge plasma. The formation of ozone molecules in the plasma of a nonself-sustained discharge in air may serve as an example of such a reaction. In fact, according to ref. 3, the loss of electrons as a result of their dissociative attachment to ozone moleucles

$$e + O_3 \rightarrow O^- + O_2 \tag{1}$$

occurs fairly intensively when the electron energy is ~ 1 eV.

It should be noted that the data of ref. 3 were obtained when investigating the loss of electrons in carbon dioxide where the ozone molecules were only a minor admixture. Presented in ref. 3 is the dependence of the rate constant for the dissociative attachment of electrons to ozone molecules on the mean electron energy. With the E/N values in the range 1 Td < E/N < 50 Td, the characteristic electron energy 1 eV < D/μ < 1.2 eV. Comparing the D/μ values for air and CO_2, one can estimate the rate constant for the attachment of electrons to ozone molecules: $K_{O_3} \approx 10^{-9}$ cm^3s^{-1}. Comparison between the rate constants for the attachment of electrons to the oxygen and ozone molecules shows that at the air pressure P = 150 Torr as small a quantity of ozone molecules as 10^{13} - 10^{14} cm^{-3} is sufficient to change substantially the ionization equilibrium of the discharge plasma. As follows from the numerical calculations made in[4] for an O_2-N_2 (1:4) mixture, such concentrations of ozone molecules are quite readily attainable.

To check the above assumptions, we investigated, both experimentally and theoretically, the process of establishment of a quasi-stationary nonself-sustained discharge in air. The experiments were conducted on the setup described in[5]. The discharge was sustained with the aid of an e-beam with a current density 100 $\mu A/cm^2$, the gas pressure in the discharge chamber being 150 Torr. The E/N parameter was varied over a broad range of values corresponding to the nonself-sustained mode of discharge. The experiments revealed a number of features typical of air discharges[1]. In particular, as can be seen from the current oscillogram shown in Fig.1a, the discharge current j at E/N = 31.5 Td starts to decrease smoothly after the lapse of approximately 50 μs required by transient phenomena and continues so far around 700 μs and then being rising again. Streak camera photographs revealed that the increase of the discharge current is due to the formation of highly conductive channels on the electrodes. The presence of the drooping section in the current oscillogram may be explained by the accumulation during the discharge of ozone molecules causing an intensive electron attachment. Note that calculations of ref. 4 do not show any decrease in the conductivity of plasma for time periods up to around 100 μs, and this is apparently due to the disregard of the reaction (1).

To quantitatively explain the experimental data obtained by us, use was made of the model expounded in ref. 4, due regard being taken of the reaction (1). That this process should be taken account of also follows from the fact that electron production rate provided by the external ionizer was selected so as to have in the calculations an electron concentration of up to $n_e \approx 10^{11}\ cm^{-3}$. Since the production rate of ozone molecules depends on the electron concentration n_e, the ozone concentration values obtained in our calculations were higher than the maximum value of $n_e \approx 10^{10}\ cm^{-3}$ indicated in[4].

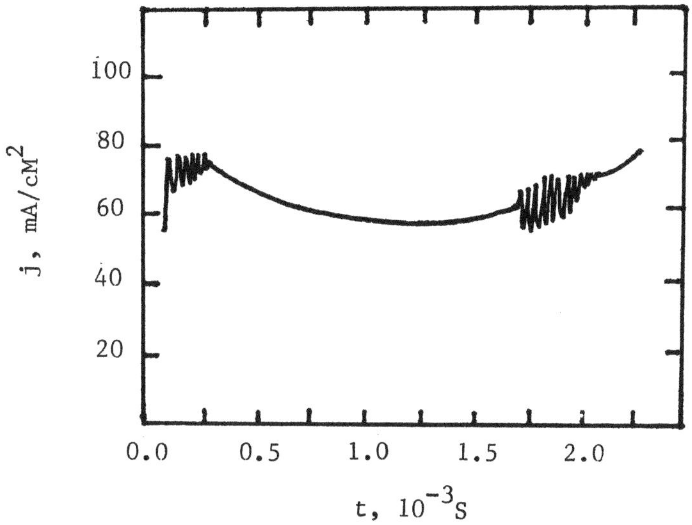

Fig.1a Current oscillogram of nonself-sustained discharge in air.

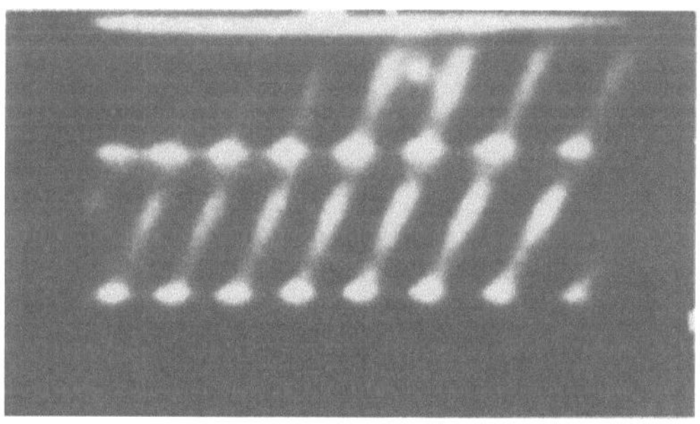

Fig.1b Image-converter-intensifier streak camera photography of discharge in air, track duration-40 mcs, the glow evolved from the cathode (at the bottom) to the anode.

The results of numerical analysis performed within the framework of the above-considered model are presented in Fig.2. As seen from Fig.2a, the inclusion of the process (1) enables one to explain the presence of the drooping section in the discharge current oscillogram (Fig.1a).

In the course of experiments it was also found that for $E/N < 25$Td there was no drooping section in the oscillogram. This effect may be attributed to the strong dependence on E/N of the oxygen dissociation rates:

$$e + O_2 \rightarrow \begin{cases} 2O + e \\ O + O^- \end{cases} \qquad (2)$$

and of the excitation of nitrogen molecules to reach the metastable

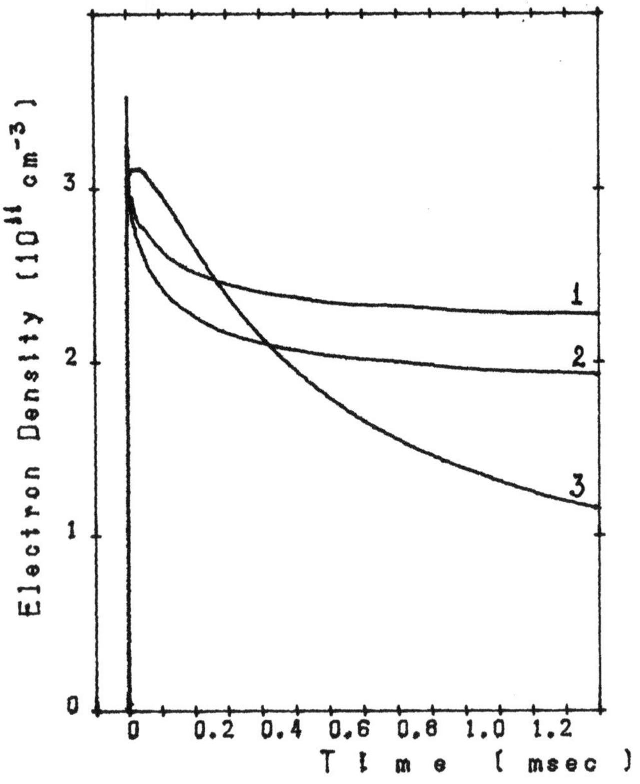

Fig.2a Electron density time evolution at various parameter E/N: 1-E/N = 20 Td, 2-E/N = 31.5 Td, 3-E/N = 50 Td.

state $N_2(A^3\Sigma_u^+)$

$$e + N_2 \rightarrow N_2(A^3\Sigma_u^+) + e \qquad (3)$$

followed by the dissociation of oxygen molecules by the reaction:

$$O_2 + N_2(A^3\,\Sigma_u^+) \rightarrow 2O + N_2 \quad (4)$$

The calculations also showed the reaction (4) to be one of the most important routes of formation of atomic oxygen, hence ozone.To wit, the experimentally observed decrease of the discharge current cannot be explained within the framework of the model used unless the reaction (4) is taken into consideration.

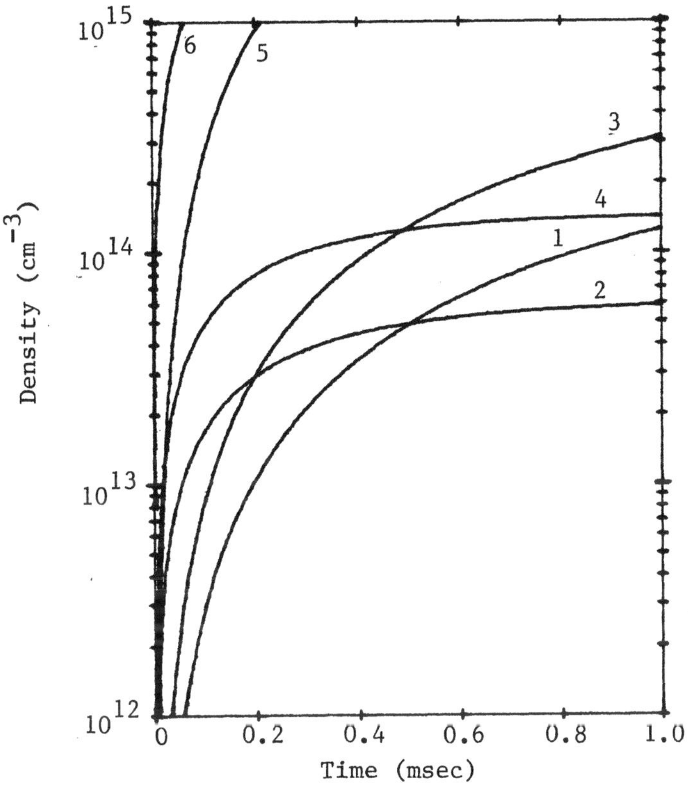

Fig.2b Ozone (1,3,5) and atomic oxygen densities (2,4,6) time evolution at E/N = 20 Td; 31.5 Td, 50 Td, respectively.

As the formation of ozone results from the nonequilibrium dissociation of O_2 in the discharge, the effective electron attachment rate at a certain concentration of O_3 is a sharply rising function of the E/N parameter. The circumstance may cause the development of domain instabilities even when the E/N value is such as conduces to the three-body attachment of electrons to the oxygen molecules.

To find the conditions for the evolution of such an instability, let us write down equations descriptive of the nonstationary behavior of the electron concentration (n_e) and the concentration of electronegative species (n_{a_1}) formed in the plasma of the nonself-sustained discharge:

$$\frac{dn_e}{dt} = S - K_a n_e n_a - K_{a_1} n_e n_{a_1} \qquad (5)$$

$$\frac{dn_{a_1}}{dt} = Q - \frac{n_{a_1}}{\tau} \qquad (6)$$

where S is the specific electron production rate provided by the external ionizer, K_a and K_{a_1} are the rate constants for the attachment of electrons to the electronegative molecules existing in the gas at equilibrium and to those produced as a result of plasmochemical reactions, respectively, and Q is the effective production rate of the latter molecules. For simplicity, the terms accounting for the processes of recombination and detachment of electrons are omitted from equations (5) and (6). It is also assumed that (as is usually the case)

$$\tau \gg (K_a n_a + K_{a_1} n_{a_1})^{-1} \qquad (7)$$

Then, taking into consideration the continuity of the total current, we obtain from equations (5) and (6) the following expression for the evolution increment of *chemical* striations:

$$\gamma = \tau^{-1}\left[\frac{\hat{Q}\, K_{a_1}\, n_{a_1}}{K_a n_a + K_{a_1} n_{a_1}} - 1\right] \qquad (8)$$

where $\hat{Q} = \partial \ln Q/\partial E$.

When deriving expression (8), it was assumed that the attachment rate constants do not depend on the electron temperature,i.e., considered as the case of three-body attachment.

Note that although the attachment of electrons to ozone molecules is of dissociative nature, the rate constant K_{a_1} depends but slightly on E/N in the range of field intensity values used[3].

A similar analysis can also be made for the case where K_a and K_{a_1} strongly depend on the electric field intensity. In such a case, *chemical* striations will exist against the background of attachment-induced domains. Since usually $\tau \gg (K_a n_e)^{-1}$, the evolution increment of chemical domains will be much less than that of attachment-induced instability and the *chemical* striations will be observed against the background of the attachment-induced ones, like in the case where the domains due to the nonmonotonous dependence of the electron drift velocity on the electric field intensity are observed against the background of attachment-induced striations[1]. Apparently, it is exatly this situation that was observed in our experiments under E/N > 30 Td. Fig.1 shows (a) a typical discharge-current oscillogram and (b) the corresponding streak-camera photograph of the discharge-plasma glow at E/N = 31.5 Td. As can be seen from the figure, the region where, the attachment-induced domains originate shifts towards the anode and there is a corresponding dip in the discharge current oscillogram. It may be assumed that both the shift of the striation origin and the dip in the discharge-current oscillogram are due to the development and propagation of a high-potential region in the discharge plasma. The occurrence of this region may be explained by the formation of the *chemical* domain.

In conclusion, it may be stated that plasma chemical reactions leading to the formation of new electro-negative species may have a material effect on the properties of nonself-sustained gas discharges.

REFERENCES

1. E.P. Velikhov, ed., Molecular Gas Lasers, Mir, Moscow.
2. N.L. Aleksandrov, F.I. Vysikailo, R.Sh. Islamov, et al., Teplofizika Vysokikh Temperatur 19:22-27, 1981.
3. D. Stelman, J.L. Moruzzi, and A.V. Phelps, J. Chem. Phys. 56:4138, 1972.
4. N.L. Aleksandrov, F.I. Vysikailo, R.Sh. Islamov, et al., Teplofizika Vysokikh Temperatur 19:485-490, 1981.
5. M.N. Andreeva, I.G. Persiantsev, V.D. Pis'mennyi, et al., Fizika plazmy 3:1384-1389, 1977.

A STUDY OF PROPAGATION OF A CONDUCTING CHANNEL IN THE PLASMA OF A NON-SELF-SUSTAINED DISCHARGE

A. S. Kovalev, A. M. Popov, and A. T. Rakhimov

Institute of Nuclear Physics
Moscow State University
Moscow, 117234, USSR

INTRODUCTION

In References 1-4 it has been experimentally shown that the breakdown of a discharge gap occurs as a result of the development in the electrode regions of highly conducting channels which propagate towards the opposite electrodes. An attempt was made [4] to explain qualitatively the propagation of a channel with a subsonic velocity by the Joule heating of the gas and its expulsion from the region of increased field intensity near the channel tip. The plasma conductivity out of the channel is considered to be constant. This assumption permits the use of electrostatic theory in order to calculate the electric field and current density values near the channel tip. However, the increase in electron concentration owing to the self-ionization processes over the channel tip must lead to a decrease in the electric field value in this region in comparison with one calculated by the electrostatics approach [5,6]. Therefore, we examine the process of highly conductive channel propagation, taking into consideration the region of self-sustained discharge over the channel tip. For simplicity we propose that input energy instantaneously goes into the translational degrees of freedom of gas molecules; we also neglect the step-ionization process and effects connected with electron attachment.

THEORY

Assuming that the discharge plasma is quasi-neutral and that the electron current in plasma is much higher than the ion current, the set of equations describing the electrodynamic processes which occur in the gas will be:

$$\operatorname{div} \vec{j} = 0 \quad \operatorname{rot} \vec{E} = 0$$

$$Q + \alpha(E/N)v_e(E/N)n_e = \beta n_e^2. \tag{1}$$

Here $\vec{j} = e\, n_e\, \vec{v}_e$ is the current density, n_e and v_e are the concentration and the drift velocity of electrons, respectively, $\vec{E}$ is the electric field intensity, N is the concentration of gas molecules, β is the recombination coefficient, Q is the specific electron production rate provided by an external ionizer, and $\alpha = AN\exp(-BN/E)$ is the first Townsend coefficient.

The set of equations (1) must be supplemented with the equation for the gas temperature, which for the case p = const has the form

$$\frac{\gamma}{\gamma-1}\frac{dT}{dt} = j\,\frac{E}{N}, \tag{2}$$

where γ is the specific heat ratio.

Since the characteristic times for electrodynamic processes are substantially shorter than those for gas-dynamic ones, equations (1) may be solved independently of (2), assuming that $N(\vec{r},t)$ is a slowly changed parameter. We shall assume that at the temperature T^* the plasma is thermally ionized so that its conductivity is high. Then, it is safe to consider the equation $T(r,t) = T^*$ to be that of the surface of the highly conductive channel at the moment t, and its propagation velocity, is defined as a rate of isotherm movement $T = T^*$.

We approximate the conducting channel by a semi-ellipsoid with semi-axes $\ell(t)$ and R $(\ell(t)>R)$. It is evident that the problem of calculating the stationary current density distribution in a plasma of varying conductivity is equivalent to that of finding the electric induction vector in a medium of varying permittivity, dealt with in electrostatics. It is also evident that the supplementary concentration of the electric current at the channel tip is a result of the existence of a self-sustained discharge region near the tip. However, if the channel length ℓ is much greater than its thickness, i.e., $\ell \gg R$, the size of the self-sustained discharge zone is substantially less than the channel length, and the self-ionization region has practically no effect on current distribution near the tip of the channel. For this reason, the problem of calculation of the electric current density distribution near the channel tip is equivalent to that of finding an electric induction vector in the medium of constant permittivity near the conducting ellipsoid in a uniform electric field. Using the well-known solution to this problem [7], we get the following approximate expression for the current density distribution along the channel axis:

$$j(z,\ell) = j_o \frac{1}{\ell n(2\ell/R)-1} \frac{\ell^2}{R^2} \left(1 + 2\frac{\ell(z-\ell)}{R^2}\right)^{-1} . \quad (3)$$

Here, j_o is the non-self-sustained discharge current density; coordinate z is measured from the channel tip ($z \geqslant \ell$).

Using the formula (3) it is easy to evaluate $E/N(z,\ell)$. There are regions of different dependence $E/N(j)$ in the vicinity of the channel tip. The dependence $E/N(j)$ in the non-self-sustained discharge zone $\{\alpha(E/N)v_e(E/N) < \sqrt{Q\beta}\}$ is the following:

$$\frac{E}{N}(z,\ell) = (E/N)_o \frac{j(z,\ell)}{j_o} . \quad (4)$$

As we approach the channel tip the E/N parameter grows proportionally to the electric current density and at last reaches a value when the self-ionization process begins to dominate, i.e., $\alpha(E/N)v_e(E/N) \geqslant \sqrt{Q\beta}$. The expression for $E/N(j)$ dependence in this region is

$$\frac{E}{N}(z,\ell) \cong B\left(\ell n \frac{Aj_o}{eQ} \frac{j_o}{j(z,\ell)}\right)^{-1} . \quad (5)$$

Equation (5) shows that E/N varies only slightly in the vicinity of the channel tip. Taking into account that the gas concentration considerably decreases in this region, we find that electric field intensity may increase, but may even drop. Nevertheless, it follows from expression (3), that the electric current density continues to increase as far as the channel tip.

Let the region of thermally ionized plasma exist at the time moment $t = t_o$; let this region have a semi-ellipsoid form with semi-axes ℓ_o and R. We propose, the temperature distribution outside this region is given as follows:

$$T(z,\ell_o) = T^* \frac{Q(z,\ell_o)}{Q(z=\ell_o,\ell_o)} . \quad (6)$$

Here, $Q(z,\ell) = j(z,\ell) \times \frac{E}{N}(z,\ell)$.

As mentioned above, the relaxation time for the calculated distribution $j(z,\ell)$ and $E/N(z,\ell)$ is considerably shorter than that for rising temperature. Then, assuming that during the process of propagation the channel keeps the ellipsoid form with constant small semi-axes, we find from equation (2) the plasma temperature distribution at the moment t along the channel axis:

$$T(z,t) = T^* \frac{Q(z,\ell_o)}{Q(\ell_o,\ell_o)} + \frac{\gamma}{\gamma-1} \int_{t_o}^{t} Q(z,\ell(t'))dt' . \quad (7)$$

Recalling that the movement of the channel tip is described by the equation $T(z=\ell,t) = T^*$, we have:

$$\int_{t_o}^{t} Q(\ell(t),\ell(t'))dt' = \frac{\gamma}{\gamma-1} T^* \left(1 - \frac{Q(\ell(t),\ell_o)}{Q(\ell_o,\ell_o)}\right). \tag{8}$$

The above expression is the Volterra integral equation in the unknown $\ell(t)$.

Following [8-10], we substitute in equation (4) $dt' = dx/V(x)$, where $V(x)$ is the unknown propagation velocity of the channel tip with length x, to get

$$\int_{\ell_o}^{\ell} \frac{Q(\ell,x)}{V(x)} dx = \frac{\gamma}{\gamma-1} T^* \left(1 - \frac{Q(\ell,\ell_o)}{Q(\ell_o,\ell_o)}\right). \tag{9}$$

Substituting expressions for $j(z,\ell)$ and $E/N(z,\ell)$ from (3), (5) and also neglecting slight logarithmical dependences in equation (9), we get

$$\int_{\ell_o}^{\ell} \frac{x^2/R}{1+2x(\ell-x)/R^2} \frac{dx}{V(x)} = 10T^* \frac{\gamma}{\gamma-1} (j_oB)^{-1}\left(1 - \frac{1}{1+2\ell_o(\ell-\ell_o)/R^2}\right). \tag{10}$$

An approximate analytical solution of equation (10) may be obtained by the iteration method (see, for example [11]) in the form:

$$V(\ell) \cong \frac{\gamma-1}{\gamma} \frac{j_oE_o}{\rho} \left(0.1 \frac{T_o}{T^*} \frac{B}{(E/N)_o}\right) \frac{\ell^2}{R} . \tag{11}$$

It should be noted that the value of the channel tip velocity given by equation (11) is the value measured for relatively motionless gas. In reality, gas movement is the result of its heating and expulsion from the region of thermally ionized plasma. The estimate of gas movement influence on the experimentally measured channel velocity for three-dimensional gas flow has already been made [12]. According to the results of this paper the experimentally measured channel tip velocity is connected with the value $V(\ell)$ by means of relation

$$V_v = V \cdot \sqrt{\rho_o/\rho_{th}} , \tag{12}$$

where ρ_o, ρ_{th} are the gas densities of volume dominated non-self-sustained discharge and thermally ionized plasma respectively. The relation (12) shows that visible channel propagation velocity is approximately 3 - 5 times higher than the value found from formula (11).

Taking into consideration that $V(\ell) = d\ell/dt$, we obtain the length of the propagating channel as a function of time

$$\ell(t) = \frac{\ell_o}{1-t/\tau_H} , \tag{13}$$

where

$$\tau_H = \frac{\gamma}{\gamma-1}\frac{\rho}{j_o E_o}\left[10T^* \frac{(E/N)_o}{B}(\rho_{th}/\rho_o)^{\frac{1}{2}}\right]\frac{R}{\ell_o}.$$

It should be noted that the above expressions for the propagation velocity of the channel and for its length are valid only in the case of subsonic channel propagation velocities. As the channel grows in length its propagation velocity rises and, once the sonic velocity is reached, there occurs a change in the propagation mechanism. However, the breakdown time of the discharge gap is governed by the initial stage of channel propagation and, as seen from equation (13), it is τ_H. This time has great dependence upon the initial inhomogeneity shape ℓ_o/R, which has formed near the cathode.

Let us present some numerical estimates. With typical non-self-sustained discharge parameters (electric current $j_o \cong 0.1$ A/cm^2, electric field intensity $E \sim 10^3$ V/cm, gas pressure $p \sim 150$ torr) for channel length $\ell \sim 1$ cm and its radius $R \sim 0.1$ cm, we have $V \sim$ $\sim 10^4$ cm/s. This value of channel velocity qualitatively corresponds to experimental data [4,13,14].

CONCLUSION

The region over the propagating channel is that of the volume instability initiation of the non-self-sustained discharge. As a result, the conducting channel propagates in the region of more rare gas and more high parameter E/N value in comparison with bulk gas volume. For this reason the channel propagation velocity may increase. On the other hand, the volume instability will also speed up owing to the discharge gap reduction and a rise in electric field intensity.

Such mutual acceleration of both mechanisms may decrease the time of the uniform discharge glow operation and limit the input energy in the volume of the non-self-sustained discharge.

REFERENCES

1. E. P. Velikhov, S. A. Golubev, I. G. Persiantsev, V. D. Pis'mennyi, A. T. Rakhimov and T. V. Rakhimova, "Stationar Non-Self-Sustained Discharge in Molecular Gases", Fiz. Plasmy., 1:847 (1975).
2. Yu. I. Bychkov, S. A. Genkin, Yu. D. Korolev, G. A. Mesyats, V. G. Rabotkin and A. G. Filonov, "The Change of the Volume Non-Self-Sustained Discharge to the Spark One", Izv. VUZov Fizika, Iss. 10:146 (1978).

3. D. B. Gurevich, M. A. Kanatenko and I. V. Podmoshenskyi, "Development of the Breakdown of Volume Non-Self-Sustained Discharge", Fiz. Plasmy, 5:1359 (1979).
4. A. S. Kovalev, I. G. Persiantsev, V. M. Polushkin, A. T. Rakhimov, N. V. Suetin and M. A. Timofeev, "On the Mechanism of the Non-Self-Sustained Discharge Breakdown", Zh. Tekh. Fiz., Pis'ma Red., 6:743 (1980).
5. G. G. Gladush and A. A. Samokhin, "On the Mechanism of a Conductive Channel Propagation in the Non-Self-Sustained Discharge", Preprint IAE., No. 3406/6 (1981).
6. A. S. Kovalev, A. M. Popov and A. T. Rakhimov, "A Study of Propagation of a Conducting Channel in the Plasma of a Non-Self-Sustained Discharge", Zh. Tekh. Fiz., Pis'ma Red., 8:561 (1982).
7. L. D. Landau and E. M. Lifshits, "Electrodynamics of Continuous Media", Gostekhizdat, Moscow (1957).
8. M. M. Kekez and P. Savic, "A Hypersonic Interpretation of the Development of the Spark Channel in Gases", J. Phys. D: Appl. Phys., 7:620 (1974).
9. M. M. Kekez and P. Savic, "Fast Mode of Spark Channel Development", Proc. XII Int. Conf. on Phenomena in Ionized Gases, Eindhoven (1975).
10. M. M. Kekez and P. Savic, "Further Support for the Hypersonic and Volterra Models of the Spark Channel Development", Proc. IV Int. Conf. on Gas Discharges, p. 129, Swansea (1976).
11. G. Korn and T. Korn, "Mathematical Handbook", McGraw-Hill Book Company, New York (1968).
12. I. A. Bufetov, A. M. Prokhorov, V. B. Fedorov and V. K. Fomin, "Propagation of the LSC-Wave in the Nd-Laser Beam", Sov. Quantum Electronics, 8:751 (1981).
13. V. N. Babichev, S. A. Golubev, A. S. Kovalev, V. D. Pis'mennyi, and A. T. Rakhimov, "The Near Cathode Region of the Non-Self-Sustained Discharge in the Gas Flow", Fiz. Plasmy, 6:195 (1980).
14. I. G. Persiantsev, V. M. Polushkin, A. T. Rakhimov, S. P. Rebrik and M. A. Timofeev, "Investigation of the Instability of the Non-Self-Sustained Discharge in Nitrogen", Proc. XV Int. Conf. on Phenomena in Ionized Gases, Minsk (1981).

SUPPRESSING ATTACHMENT-INDUCED INSTABILITIES

BY OPERATING GAS DISCHARGES IN AN HF MODE

A.S. Kovalev, A.T. Rakhimov, N.V. Suetin, and
V.A. Feoktistov

Institute of Nuclear Physics, Moscow State University
Moscow 117234, USSR

High-pressure glow discharges have found wide application in laser physics and plasma chemistry[1-4]. A special feature of both self-sustained and non-self-sustained discharges is their inherent instability[5-6]. One of the methods for improving the stability of glow discharges is to use high-frequency electric fields[7-8].

It is shown in this report that the development of attachment-induced instabilities in high-frequency glow discharges in electronegative gases can be suppressed on account of the shunting action of the bias current. The suppression mechanism can be understood within the framework of the perturbation theory.

Using the Maxwell equations, we get the following conservation equation for the total current for the case where ionic currents can be disregarded:

$$e n_e \mu_e E + \frac{1}{4\pi} \frac{\partial E}{\partial t} = j_0 e^{i\omega t} \qquad (1)$$

where e, n_e, μ_e are the electronic charge, electron concentration, and electron mobility, respectively, E is the electric field intensity, j_0 is the peak discharge current provided by the external circuit, and ω is the discharge-current frequency.

For high-pressure discharge plasma, the diffusive drift of electrons out of the discharge volume can be neglected. In this case, the number density of electrons and negative ions will be governed by volume-dominated processes and the balance equations for the charged particles in the self-sustained discharge plasma will be

$$\frac{\partial n_e}{\partial t} = \nu_i n_e - \nu_a n_e + \nu_d n_- - \beta_e n_e (n_e + n_-) \tag{2}$$

$$\frac{\partial n_-}{\partial t} = \nu_a n_e - \nu_d n_- - \beta_i n_- (n_e + n_-) \tag{3}$$

where ν_i is the frequency of the gas ionization by electron impact, ν_a is the frequency of dissociative attachment of electrons, ν_d is the frequency of detachment of electrons from negative ions, β_e, β_i are the electron-ion and ion-ion recombination coefficients.

If the characteristic time for the establishment of equilibrium electron concentration is much longer than the discharge-current period, i.e., if $\nu_a \ll \omega$, the concentration of charged particles at any moment of time will not be determined by the electric field intensity at the same moment, but will settle at a certain level depending solely on the peak field intensity. Then, the quantity n_e in Eq.(1) can be considered to be constant and, after the lapse of the Maxwellian time $\tau_M = 1/4\pi\sigma$, the solution of Eq.(1), accurate to within ν_a/ω, will be

$$E = \frac{4\pi j_0}{\omega^2 + (4\pi\sigma)^2} (i\omega + 4\pi\sigma) e^{i\omega t} \tag{4}$$

where $\sigma = e n_e \mu_e$. Hence we find the peak field intensity

$$|E| = \frac{4\pi j_0}{\{\omega^2 + (4\pi\sigma)^2\}^{1/2}} \tag{5}$$

Since $\tau_M \ll \nu_a^{-1}$, then, linearizing the set of Eqs.(2),(3),(5), we get the following expression for the evolution increment of instability:

$$\Omega = -a/2 \pm \sqrt{a^2/4 - b} \tag{6}$$

$$a = \eta\nu_i \hat{\nu}_i - \eta\nu_a \hat{\nu}_a + \nu_d \frac{n_-}{n_e} + \nu_a \frac{n_e}{n_-} + \nu_i \frac{n_e}{n_+} \tag{7}$$

$$b = \eta\nu_i \hat{\nu}_i \left(\beta_i n_- + \nu_a \frac{n_e}{n_-}\right) - \eta \hat{\nu}_a \nu_a \left(\nu_a \frac{n_e}{n_-} - \nu_d + \nu_i \frac{n_e}{n_+}\right) + + \nu_d \beta_i \frac{n_- n_+}{n_e} + \beta_e \nu_a \frac{n_e n_+}{n_-} \tag{8}$$

where

$$\eta = (4\pi\sigma)^2/\{\omega^2 + (4\pi\sigma)^2\} \; ; \; \hat{\nu}_a = \frac{\partial \ln \nu_a}{\partial \ln|E|} \; ; \; \hat{\nu}_i = \frac{\partial \ln \nu_i}{\partial \ln|E|}$$

As in Refs. 5,6, we assume that ν_d is constant, because the accumulation of detaching species takes time much in excess of ν_a^{-1}. It is seen from Eqs.(6)-(8) that when $\omega = 0$, the evolution increment of attachment-induced instability corresponds to the values obtained in[5]. It is also evident that for any discharge parameter, there is a certain frequency ω_k that marks the limit above which the instability cannot develop, i.e., for $\Omega < 0$.

In the case of a non-self-sustained discharge in an electronegative gas, where the concentration of electrons is governed by their dissociative attachment to the gas molecules, domain instabilities can develop even when no detachement processes occur. Then, the electron balance equation will have the form

$$dn_e/dt = S - \nu_a n_e \tag{9}$$

where S is the electron production rate provided by the external ionizer, and from Eqs.(5),(9), the evolution increment of attachment-induced instability will be

$$\Omega = \eta \, \hat{\nu}_a \, \nu_{\bar{a}} - \nu_a \tag{10}$$

As in the case of a self-sustained discharge, the instability will be suppressed if $\omega > \omega_k$, where $\omega_k^2 = (4\pi\sigma)^2(\hat{\nu}_a - 1)$.

Thus, it is shown that attachment-induced instabilities of both self- and non-self-sustained volume-dominated discharges will be suppressed if $\omega \geq 4\pi\sigma$. It should be noted that the suppression of domain instabilities due to the nonmonotonic dependence of the electron drift velocity on the electric field intensity[6] can be analyzed in a similar way.

Numerical calculations for a high-frequency non-self-sustained discharge in an N_2-O_2 (4:1) mixture allowed finding the spatial and temporal distributions of the charged particle concentrations and electric field intensity in the discharge gap, investigating the current-flow mechanism in the boundary layers of the volume-dominated discharge, and analyzing, by comparison with similar calculations for a d.c. non-self-sustained discharge, the specifics of the evolution of attachment-induced instabilities in the high-frequency discharge, with due regard for the instabilities developing in the electrode regions.

Let an alternating voltage of the form $U(t) = U_0 \cos \omega t$ be applied across a plane-parallel discharge gap filled with an N_2-O_2(4:1) mixture. At the moment $t = 0$ an external ionizer is put into operation, and the specific electron production rate $S(t)$ provided by it first rises linearly up to $t = 10\ \mu s$ and then remains constant. The basic set of equations has the form

$$\frac{\partial n_e}{\partial t} + \frac{\partial j_e}{\partial x} = S(t) + \nu_i(E/P)n_e - \{\nu_a(E/P) + \nu_a'\}n_e - \beta_e n_e n_+$$

$$\frac{\partial n_+}{\partial t} + \frac{\partial j_+}{\partial x} = S(t) + \nu_i(E/P)n_e - \beta_e n_e n_+ - \beta_i n_+ n_-$$

$$\frac{\partial n_-}{\partial t} + \frac{\partial j_-}{\partial x} = \{\nu_a' + \nu_a(E/P)\}n_e - \beta_i n_- n_+ \qquad (11)$$

$$\frac{\partial E}{\partial x} = 4\pi e(n_+ - n_e - n_-); \quad \int_0^d E(x,t)dx = -U_0 \cos \omega t$$

$$j_e = -\mu_e n_e (E/P); \quad j_- = -\mu_- n_- (E/P); \quad j_+ = \mu_+ n_+ (E/P)$$

where j_e, j_+ and j_- are the respective flow densities of the electrons and the positive and negative ions, ν_a' is the frequency of the three-body attachment of electrons to oxygen molecules, and μ_+ and μ_- are the mobilities of the positive and negative ions, respectively.

The magnitude of the coefficient β_e was taken at $10^{-7} cm^3 s^{-1}$. The magnitude of the coefficient β_i was estimated from the formula $\beta_i = 10^{-25} N$, N being the concentration of neutral molecules. The rate constant for the three-body attachment of electrons to oxygen molecules was taken at $4 \times 10^{-31} cm^6 s^{-1}$ (Ref.9). The relation between the frequency of dissociative attachment of electrons and the ratio E/P was established to approximate the calculation results in Ref. 10.

The set of Eqs.(11) was supplemented with the following initial and boundary conditions:

$$n_e(t=0) = n_+(t=0) = n_-(t=0) = 0$$

$$\text{when } x = 0 \begin{cases} j_e = -\gamma j_+, \; j_- = 0 & (E(0,t) < 0) \\ j_+ = 0 & (E(0,t) > 0) \end{cases}$$

$$\text{when } x = d \begin{cases} j_+ = 0 & (E(d,t) < 0) \\ j_e = -\gamma j_+ , \; j_- = 0 & (E(d,t) > 0) \end{cases}$$

These conditions reflect the situation where there is no secondary emission of positive and negative ions from the electrodes, and the electron emission occurs as a result of ion bombardment at the rate γj_+. The magnitude of the coefficient γ was taken at 0.01. In all our calculations, the gas pressure P, interelectrode distance d, and field frequency were set at 100 Torr, 3 cm, and 7.4 MHz, respectively.

The fact that the dependences of ν_i and ν_a on E in the set (11) are only valid for a constant field presupposes that in our case of a HF field the electron energy distribution function at any instantaneous field intensity has enough time to settle, which is true if $\omega << \delta\nu$, where ν is the electron-neutral-molecule collision frequency and δ is the electron energy portion lost upon collision. At a pressure of around 100 Torr and a field frequency of the order of 10 MHz this conditions is more than satisfied.

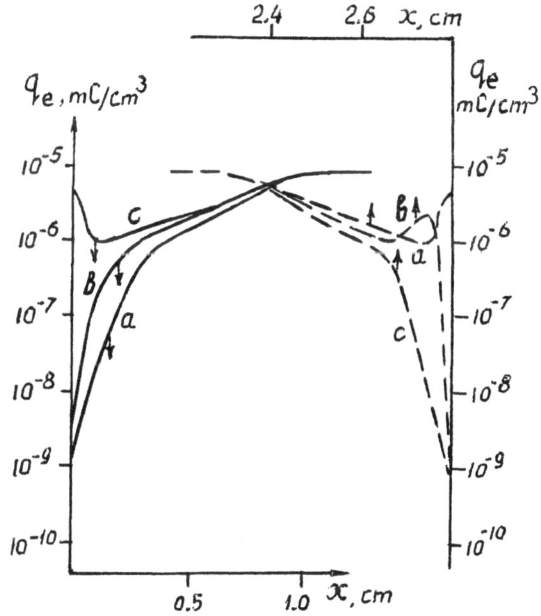

Fig.1

Fig.1 shows the electron distribution near the electrodes for U_0/Pd = 13 V/cm Torr, eS = 1.6 mC/cm^{3}s (solid lines - left-hand electrode, dashed lines - right-hand electrode). The curve refer to the moments of time indicated by the corresponding letters on curve 1 of Fig.3.

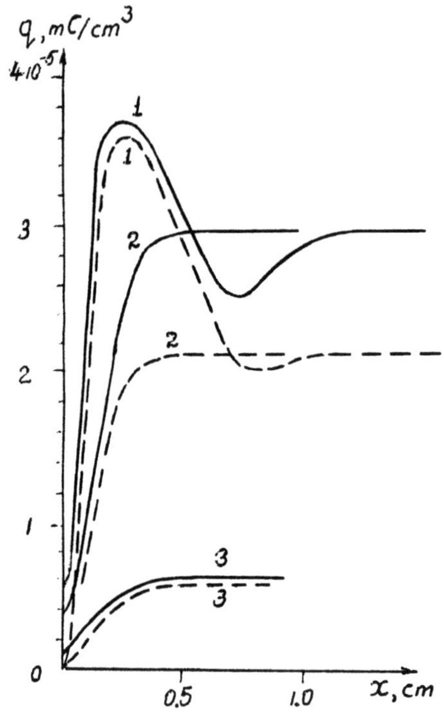

Fig.2

As the distance from the electrodes grows larger, all the curves reach one and the same level corresponding to the homogeneous and quasi-stationary distribution of electrons in the discharge volume. The quasi-stationarity of this distribution is due to the fact that the characteristic times defining the balance of electrons (i.e., ν_i^{-1}, ν_a^{-1}, $\hat{\nu}_a^{-1}$, $(\beta_e n_+)^{-1}$) are much longer than the field frequency period. Curves a and c of Fig. 1 show the electron distribution at the moments the electron column occupies, respectively, the extreme right and left-hand positions. Curve b refers to an intermediate position. It is seen from the figure that the electron column oscillates not as a single whole, but suffers deformation in the boundary regions as a result of the nonuniform distribution of the positive and negative ions near the electrodes. Fig.2 shows the distribution of the positive ions (solid curves) and negative ions (dashed curves) near one of the electrodes [U_0/Pd = 13 V $cm^{-1}Torr^{-1}$ (curve 1) U_0/Pd = 5(2), U_0/Pd = 16(3); eS = 1.6 mC $cm^{-3}s^{-1}$ (curves 1,2), eS = 0.16(3)]. The sections of the curves that run parallel to the x-axis represent the ion distribution in the discharge volume, which, like that of electrons, is both homogeneous and quasi-stationary. As to the electrode regions, the distributions of the positive and negative ions there may, in contrast to that of the electrons which, being much more mobile than the ions, oscillate at the field frequency, become quasi-stationary if $\langle\nu_i\rangle$, $\langle\nu_a\rangle$, $\langle\nu_a'\rangle \ll \omega$. In this case, the oscillation amplitude of the mean ion concentration is insignificant.

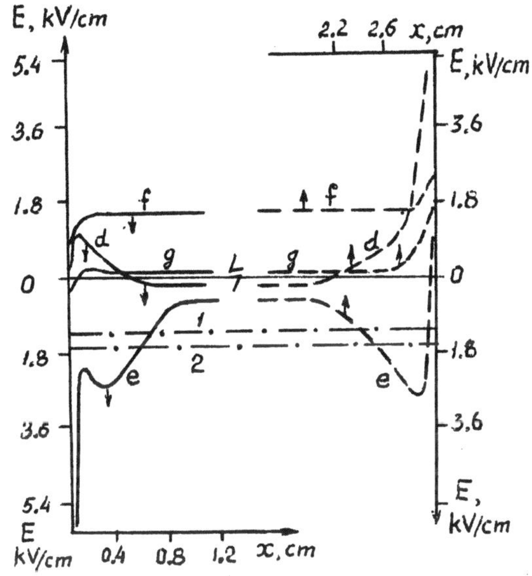

Fig.3

The character of the positive and negative ion distributions at the discharge boundaries mainly depends on the values of U_0 and S. With these values being not too high, these distributions drop monotonically in the direction towards the electrodes (curves 2 and 3 of Fig.2). The decrease in the positive ion concentration is, in particular, due to their escape on to the electrodes, for the averaged (over a period) intensity of the nonuniform electric field in the electrode regions is such that the ions drift the faster the nearer they are to the electrodes. The averaged field intensity in the electrode regions rises with increasing U_0 or S, and the contribution of self-ionization to the balance of the positive ions and, correspondingly, that of dissociative attachment to the balance of the negative ions become comparable with the contributions from other processes. This is precisely this fact that explains the occurrence of humps on curves 1 of Fig.2.

Fig. 3 illustrates the behavior of the discharge current during a period of a steady HF discharge (the discharge parameters are the same as in Fig.2). The points d, e, f and g on curves 1 and 3 indicate the moments of time for which the field distribution curves of Fig.4 were calculated. The capacitive nature of the reactances due to the electrode layers causes a phase shift between the discharge current and the applied voltage (curves 1 and 2 of Fig.3); for curve 3 this shift is barely evident. Comparison between the curves of Fig.1 with curve 1 of Fig.3 shows that the discharge current passes through zero at the moment the electrons are displaced the closest to one of the electrodes.

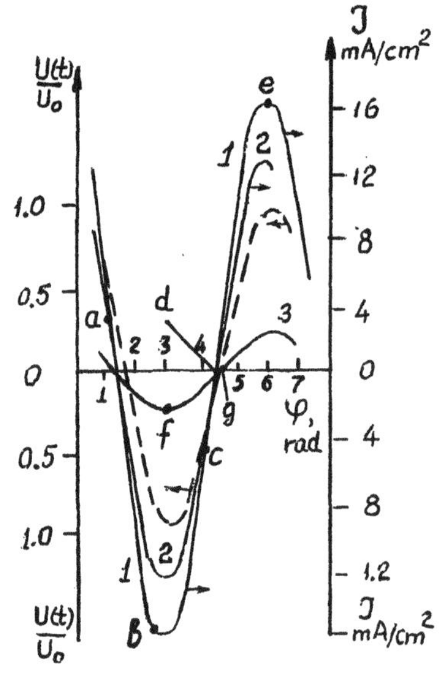

Fig.4

The field distribution curves shown in Fig.4 refer to the moments the discharge current passes through a maximum and zero. The field amplitude in the discharge volume depends on the extent to which the field is being shielded by the electrode layers. As can be seen from a comparison of curves e and f of Fig. 4, both the thickness of these layers and the voltage drop across them increase with increasing field amplitude and S . The nonmonotonic variation of the electric field near the electrodes is due to the electrons oscillating against the background of the nonuniform distribution of the positive and negative ions at the discharge boundaries.

It would be of interest to compare the characteristics of a HF non-self-sustained discharge with those of its d.c. counterpart in the field intensity region where the d.c.discharge develops instability as a result od electron attachment processes. In[11] the attachment-induced instability in air was analyzed at U_0/Pd = 13 V/cm Torr , eS = 1.6 mC/cm^{3}S, P = 100 Torr, d = 3 cm. This instability evolves as follows. As the rate constant for the dissociative attachment of electrons grows sharply with increasing field intensity, near the cathode there forms a region with a high negative ion concentration and alow electron concentration. The mobility of the negative ions being substantially lower than that of the electrons, the field intensity in this region considerably exceeds the mean intensity in the discharge gap. The region moves from the cathode towards the anode. Our calculations for a HF discharge under similar conditions (see Figs.2, 3 and 4) revealed the discharge to be free from attachment-induced instabilities. A numerical analysis shown that in this case the suppression of this kind of instability, which owes to the shunting action of the bias currents, takes place mainly in the electrode regions that give birth to domains.

Thus, our calculations showed the domain instability to be absent even when the field intensity is so high that the loss of electrons is obviously governed by their dissociative attachment to electronegative molecules. Note that Fig.4 (curves g and f) is illustrative not only of the absence of domain instabilities and the uniformity of the electric field in the discharge volume, but also of the fact that the increase in the field intensity and the drop of potential near the electrodes are insignificant even when the discharge current density in a HF discharge is not very high.

REFERENCES

1. E.P. Velikhov, V.D. Pis'mennyi, and A.T. Rakhimov, Uspekhi Fiz. Nauk 122:419 (1977).

2. "Principles of Laser Plasmas," edited by George Bekefi, New York (1976).
3. V.D. Rusanov, A.A. Fridman, and G.V. Sholin, Uspekhi Fiz. Nauk 134:185 (1981).
4. V.M. Donnelly and D.L. Flamm, Solid State Technol. 24(4):161 (1981).
5. R.A. Haas, Phys. Rev A8:1017 (1973).
6. A.P. Napartovich and A.N. Starostin, "Plasma Chemistry," Vol. 6, p. 153, Atomizdat, Moscow (1979).
7. T.V. Rakhimova and A.T. Rakhimov, Teplofiz. Vys. Temp. 15:1313 (1976).
8. A.N. Vasil'eva, I.A. Grishina, A.S. Kovalev, V.I. Ktitorov, and A.T. Rakhimov, Plasma Physics 5:1135 (1979).
8. A.N. Vasil'eva, I.A. Grishina, A.S. Kovalev, V.I. Ktitorov, A.T. Rakhimov, and S.E. Rozanov, Pis'ma Zh. Tekh. Fiz. 6:551 (1980).
9. A.N. Vasil'eva, I.A. Grishina, V.I. Ktitorov, A.S. Kovalev, and A.T. Rakhimov, Plasma Physics 5:1135 (1979).
10. N.L. Aleksandrov, F.I. Vysikailo, R.Sh. Islamov, I.V. Kochetov, A.P. Napartovich, and V.G. Pevgov, Teplofiz. Vys. Temp. 19:20 (1981).
11. A.S. Kovalev, A.T. Rakhimov, and V.A. Feoktistov, Plasma Physics 8(5) (1982).

LIMITING ENERGY INPUTS IN THE NON-SELF-SUSTAINED DISCHARGE OF A CO_2-FLOW LASER

G.B. Lopantseva, A.P. Napartovich, A.F. Pal',
A.F. Perevoznov, I.G. Persiantsev, and A.N. Starostin

Institute of Nuclear Physics, Moscow State University
Moscow 117234, USSR

At present the non-self-sustained discharge supported by a beam of fast electrons is being widely used to excite laser media at high pressures[1,2]. The principal characteristics of the discharge (the limiting value of the E/N, the limiting discharge current density j) significantly depend on the composition of gas mixtures and the discharge-burning duration. In particular, the limiting values of E/N abruptly decrease with the increased duration of a pumping pulse τ_ρ (in a gas-flow discharge, τ_ρ is the time in which the gas passes through the discharge region), reaching, at technologically convenient values $\sim$ 1 ms, such magnitudes that the amount of energy spent to excite the upper laser levels decreases[3]. In this case, the discharge instability substantially reduces the possibility of optimization of the device parameters.

The physics of the processes that determine the conductivity and composition of the discharge plasma for mixtures filling the CO_2-laser has been studied rather well. Many mechanisms of instabilities leading to a rapid increase in plasma conductivity are also known (see e.g., Refs. 2,4). For a number of specific mixtures numerical calculations have been made. In these, the linear theory was used to determine increments of the developing instability caused by the overheating and ion-molecular mechanisms[5-7]. To our knowledge, there have so far been no quantitative comparisons of the calculated discharge instability in a laser mixture with experiment (one should mention the report[8] comparing some experimental dependences with approximate theoretical ones). The chief purpose of the present communication is to test the theoretical model chosen to describe the experimentally observed development of instability of the longitudinal non-self-sustained discharge in mixtures $CO_2:N_2:He:O_2$.

It is more convenient to compare a calculation with experimental results obtained in the longitudinal discharge, because in this case the effect of the near-electrode values is reduced[9]. We have experimentally investigated the non-self-sustained discharge at atmospheric pressure in the laser mixture $CO_2:N_2:He = 1:7:8$, using the installation described in Ref. 10, in which a beam of fast electrons with an energy $\mathcal{E}$ of 120 KeV and a current density j_b to 50 $\mu A/cm^2$ was injected into the discharge gap towards the gas flow through a grid cathode. A similar grid electrode served as the anode for the discharge gap. The area of the electrodes was 2.4 cm^2, the distance between them 1 cm, the flow speed 10 m/s. The experimental setup is shown in Fig.1. The gas mixture under study was prepared of commercially pure gases.

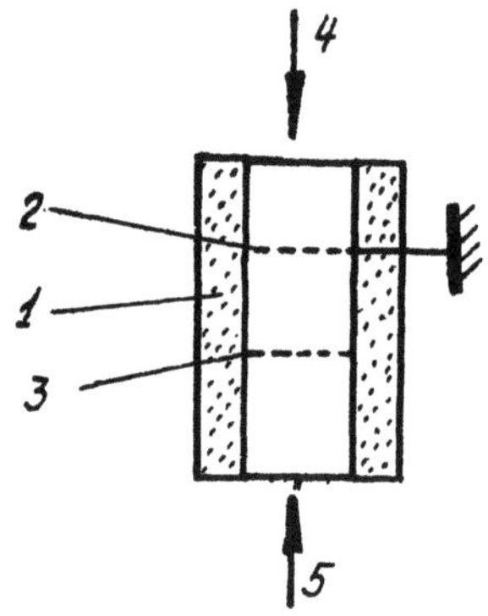

Fig.1 Experimental setup: 1 - discharge chamber; 2 - cathode; 3 - anode; 4 - electron beam; 5 - gas flow.

We have used the homogeneous discharge model, which takes into account all the major processes in the laser medium in the absence of lasing action: ion-molecular and radical reactions, processes of V-T relaxation, excitation, de-excitation, ionization of various particles in a discharge, gas rarefaction due to heating and gas-dynamical expansion. Use was made of the simplified condition of a constant pressure in the discharge region. The beam-current density and the energy of fast electrons were assumed known. The model takes care of the change with time in the concentrations of neutrals ($CO, O, N, NO, NO_2, N_2O, NO_3, O_3, N_2O_5$), charged particles ($He^+, He_2^+, O^+, N^+, N_2^+, CO^+, C^+, N_2O_2^+, CO_2^+$,

$N_4^+, C_2O_4^+, C_2O_3^+, C_2O_2^+, NO_2^+, N_3^+, O_2^+, NO^+, O_4^+, N_2O^+, O^-, O_2^-, He(2^1S, 2^3S), O_3^-, N_2O^-, CO_3^-, CO_4^-, NO_2^-, NO_3^-$) and excited molecules $N_2(A^3\Sigma_u^+)$, $N_2(v)$, $CO_2(v_1, v_2, v_3)$, where the symbol v stands for the vibrational excitation. It is assumed that the above particles could be formed by the action of both fast beam electrons and discharge electrons. The dependences of the chemical transformation rates on the gas temperature and the average electron energy (E/N) , if known from the literature, have also been taken into account in the calculations.

The vibrational kinetics in the mixture was treated within the conventional *temperature* model[11] with three temperatures: T_2 for the symmetrical and deformational mode, T_3 for the asymmetrical mode of CO_2, and T_4 for the vibrational temperature of nitrogen. Account was taken of the vibrational energy relaxation on both the initial mixture components and the reaction products in the discharge. In gas heating, both the *fast* channel (due to direct heating and the excitation of molecular rotations) and that of the V-T relaxation were considered. Due to gas heating and expansion during a discharge, the parameter E/N, which determines for the given mixture the average electron energy and a number of rate constants for processes involving electrons, changes (increases). The initial value of E/N also was widely varied ($0.8 \leqq E/N \times 10^{16}$ V cm $\leqq 2.5$). The rate constants for excitation, dissociation, ionization by discharge electrons have been calculated as a functions of E/N for particular mixtures, according to the procedure of[12] by I.V.Kochetov and V.G.Pevgov who kindly supplied us with these results. The rate of stepwise ionization of nitrogen has been recalculated by us from the data of Ref. 13 by introducing the average electron temperature T_e.

The set of equations describing the change in the above quantities was solved numerically on a BESM-6 computer using the Runge-Kutta method.

The change of all the sought-for quantities with time has been found. Both the ionic composition and the shape of oscillogram have been found to depend significantly on the molecular oxygen admixture. In the absence of oxygen the principal ion that exists by the moment of instability onset is $C_2O_2^+$. The calculated current oscillogram is then nonmonotonic. Fig.2 compares the experimental and the calculated current oscillograms obtained at three concentrations of O_2. The discharge occured in the mixture $CO_2:N_2:He=1:7:8$ at P = 1 atm, E=3.1 kV/cm, j_b = 35 μA/cm^2. For the same experimental conditions and O_2 =

= 0.05%N Figs.3 and 4 present the calculated time dependences of concentrations of the principal ionic components, the effective recombination coefficient β_{eff}, the temperatures of the gas and the electrons, the vibrational temperatures and the specific energy input, W. as can be seen from Fig.2, at O_2 = 0.05%N the calculated current oscillogram agrees with the experimental both average current and in the time of instability development. It should be noted that since commercially pure gases were used in the experiment, there is a spread in the values of averaged current, instability development time τ_{in} and W, which is apparently due to composition fluctuations. The average spread is about 20%. as the beam current density and electric field intensity are changed the agreement between the calculation and the experiment is retained.

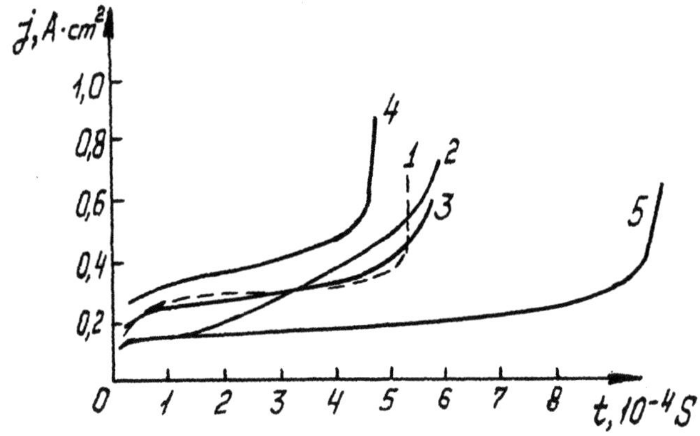

Fig.2 Time dependence of discharge current density at j_b= 35 mcA/cm^2 and E = 3.1 KV/cm (allowing for a cathode fall of 0.7 KV); 1 - experimental; 2,3,4 - calculated at O_2 = 0.1% N, 0.05% N and 0.01% N respectively; 5- corresponds to curve 2 with the processes (4) left out.

An analysis of the obtained results enables one to formulate a simplified model retaining the most important processes which determine the development of instability. Since under the conditions assumed in the calculation, the V-T relaxation time turns out much shorter than that of instability development, the heating of a gas can be described by the simple equation:

$$\dot{\theta} = \nu_T \, n_e/n_{eo} \tag{1}$$

where $\theta = (T-T_0)/T_0$, T_0 being the initial gas temperature, $\nu_T = (\gamma-1)/\gamma \times \gamma_0 \times E/P$, γ the adiabat index, j_0 and n_{eo} are the quasistationary values of the current and electron densities in the absence of secondary processes involving excited particles. The evolution of the current oscillograms with changing O_2 concentration (curves 2-4 in Fig.2) is explained by the following processes. The effective recombination coefficient is determined by the dominant species of positive ions which as can be seen from Fig.3 may change with time ($O_4^+ \rightarrow O_2^+$), mainly due to such reactions as:

$$O_2^+ + O_2 + He \xrightarrow{K_2} O_4^+ + He \qquad (2)$$

$$O_4^+ + O \xrightarrow{K_3} O_2 + O_3 \qquad (3)$$

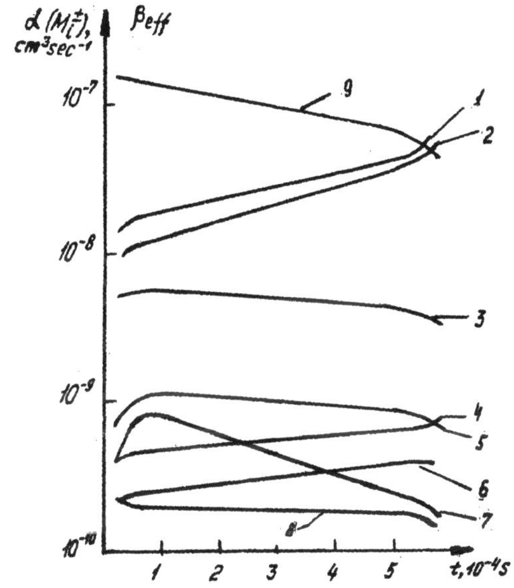

Fig.3 Calculated time dependences of the relative ionic composition $\alpha(M_i^{\pm}) = M_i^{\pm}/N$: 1 - n_e, 2-O_2^+ ; 3-O_4^+ ; 4-NO^+ ; 5-NO_2^+ ; 6-N_3^+ ; 7-CO_4^- ; 8-N_2O^+ and β_{eff} - 9 for the conditions of Fig.1(3).

The main source of oxygen atoms in the conditions adopted for the calculation is the reaction

$$N_2(A^3\Sigma_u^+) + O_2 \rightarrow 2O + N_2(X) \qquad (4)$$

concerning which there is no reliable information in the literature.

We have assumed that the quenching of the state $A^3\Sigma_u^+$ by oxygen molecules proceeds entirely through the reaction channel (4). The importance of the processes described above has been confirmed by a calculation in which all the remaining plasma chemical processes (except for the direct and stepwise ionization) were discarded. The variation in discharge characteristics did not then exceed 15%. We note that leaving out the stepwise ionization channel we have τ_{in} not much greater, only by $\sim$10%, the average current remaining unchanged. On the other hand at $[O_2]$ = 0.1% N leaving out the process (4) alone has markedly changed the current oscillogram (curve 5 in Fig.1) and led to τ_{in} becoming greater by a factor 1.6.

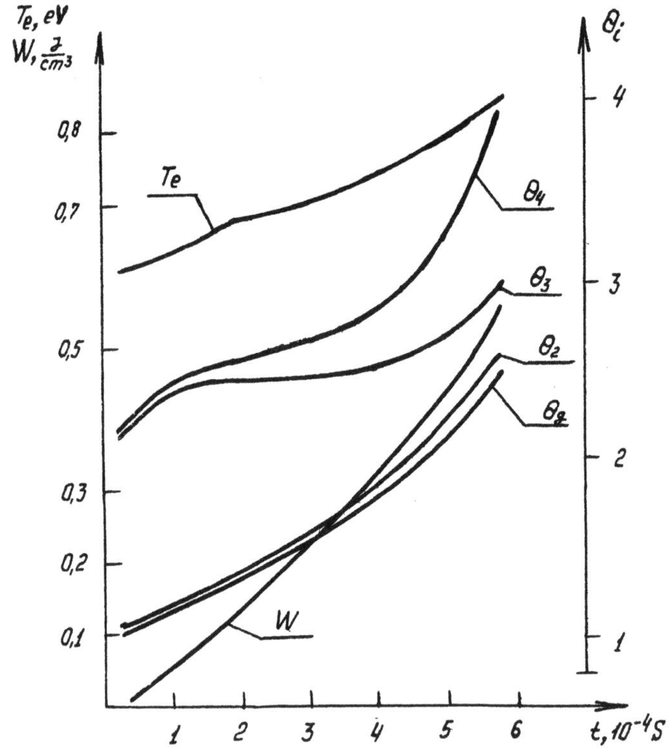

Fig.4 Calculated time dependences of the reduced gas ($\theta_g = T_g^\circ/300$) and vibrational ($\theta_i = T_i^\circ/300$) temperatures, electron temperature T_e and specific energy W for the conditions of Fig.1 (curve 3)

Bearing in mind the above-listed processes, we have

$$n_e = n_{eo}\{1 + \lambda \exp(B\theta)\}(1+K)^{\frac{1}{2}}/\{1 + (\beta_4/\beta_2)K\}^{\frac{1}{2}} \tag{5}$$

where $n_{eo} = \sqrt{q/\beta_2}$, q being the rate of ionization by a fast electron beam, β_2, β_4 are the recombination coefficients for the ions O_2^+ and O_4^+, $K = [O_4^+]/[O_2^+] = K_2[O_2][He]/K_3[O]$, $\lambda \ll 1$ the relative fraction of ionization due to electrons of discharge, the factor $\exp(B\theta)$ allows for the increase in the rate of such ionization with increasing $E/N \sim \theta$.

The concentration of oxygen atoms increases in accordance with the equation

$$[\dot{O}] = n_e[N_2]K_b\, e^{A\theta} \tag{6}$$

where $K_b\, e^{A\theta}$ is the constant of the excitation rate of N_2 to the level $A^3\Sigma_u^+$ by discharge electrons. Using Eqs.(1),(5) and (6) one can find the time of instability development

$$\tau_{in} = \frac{1}{\nu_T}\int_0^{\infty} \frac{d\theta}{(1+\lambda\, e^{B\theta})} \cdot \frac{\{1+K(\theta)\}^{\frac{1}{2}}}{\{1+(\beta_4/\beta_2)K(0)\}^{\frac{1}{2}}} \tag{7}$$

where

$$K(\theta) = \frac{K_2[O_2]\,[He]\,\nu_T}{K_3[N_2]\,n_{eo}K_b} \quad \frac{A}{1 - \exp(-A\theta)} \quad .$$

The expression (7) may be simplified in a number of limiting cases. In particular for a low concentration of O_2 when the process of conversion (2) is insignificant $([O_2] \ll \{K_3K_bn_{eo}[N_2]\}/\{A\,\nu_TK_2[He]\})$ the expression (7) reduces to the known expression (see[2,4]):

$$\tau_{in} = (1/B\,\nu_T) \times \ln(1/\lambda)$$

Thus we have outlined a model that allows for the known processes in the active medium of a CO_2-laser which makes it possible to calculate the limiting characteristics of the non-self-sustained discharge for a wide range of conditions. The comparison between the results of the calculation and experiment shows a good quantitative agreement which suggests that inhomogenities in the discharge play an insignificant role. For the experimental conditions described above it is possible to construct a crude model which allows one to describe more vividly the development of an instability.

REFERENCES

1. N.G.Basov, E.M.Belenov, V.A.Danilychev, A.F.Suchkov, Usp.Fiz. Nauk.,114:213 (1974)
2. E.P.Velikhov, V.D.Pismennyi, A.T.Rakhimov, Usp.Fiz.Nauk.,122: 419 (1977)
3. L.A.Kostylev,Ya.I.Londer, A.P.Terentiev,K.N.Ul'yanov,ZhTF,47: 2293 (1977)
4. A.P.Napartovich, A.N.Starostin, in:"Khimiya plasmy" Part 6:153, Atomizdat, Moscow,1979
5. W.L.Nighan,Phys.Rev.15A:1701 (1977)
6. W.L.Nighan,Phys.Rev.16A:1209 (1977)
7. R.Bruzzese,S.Solimeno,I.L.Braglia,S.Marbelucci,Nuovo Cimento, 60B:113 (1980)
8. N.A.Blinov,L.A.Vasiliev,A.M.Seryging et al.,Proc.of XV Intern. Conf.Phenom.Ion.Gases,Minsk,p.227 (1981)
9. M.I.Yoder,H.H.Legner,I.H.Iacob,D.K.Ahouse,J.Appl.Phys.,49:3171 (1978)
10. A.F.Pal',A.F.Perevoznov,I.G.Persiantsev,A.T.Rakhimov,Pisma v ZhETF, part 21 7:1286 (1981)
11. B.F.Gordiets,A.I.Osipov,E.V.Stupochenko,L.A.Shelepin,Usp.Fiz. Nauk,108:655 (1972)
12. N.V.Karlov,Yu.B.Konev,I.V.Kochetov,V.G.Pevgov,Preprint n.91, (1976)
13. N.L.Aleksandrov,A.M.Konchakov,E.E.Son,Fiz.Plasmy,4:169 (1978)

SUBSONIC FLOW CO_2 LASER WITH TRANSVERSE RF EXCITATION

W. Schock and H. Hügel

DFVLR-Institut für Technische Physik
Pfaffenwaldring 38-40,
D-7000 Stuttgart 80.

INTRODUCTION

In an effort to increase the performance of fast flow CO_2 lasers, many techniques have been investigated and are still being studied. Among these, the application of rf discharges to enhance the stability of dc glow discharges and thereby to raise the electrical power density has been well-known for more than a decade [1]. The use of a rf discharge alone as the energy source for producing laser action has also been suggested [2]. Accordingly, successful operation of a cw carbon monoxide laser with rf excitation in a supersonic flow has been reported [3]. In the past, many experiments have proved the merits of rf discharges for waveguide and small-scale lasers (e.g., Ref. 4). More recently, growing interest in this technique for the excitation of large volume molecular gas lasers can be observed, particularly in the Soviet literature [5,6]. In this contribution, experimental work is reported which is aimed at the application of a transverse rf discharge as an efficient excitation means for fast flow CO_2 lasers.

As compared to dc discharges, rf discharges offer several advantages which result from the physical phenomena encountered.

Two different discharge modes occur during a given period depending on the time variation of the voltage and field strength. In the short phases near the voltage maximum the reduced field strength, E/n, is sufficient to maintain a self-sustained discharge. During the much longer phases with lower values of E/n no substantial ionization can occur, a situation characteristic of a non-self-sustained discharge. Since for typical operating conditions the recombination time is about three orders of magnitude longer than the rf period, a time constant value of the electron density is established. Hence, the rf discharge may be regarded as a non-self-sustained one in a recombining plasma. The achievable high values of electron density at relatively low effective mean values of E/n thus provide the premise for an efficient vibrational excitation of the laser gas.

A rapid change of polarity avoids the growth of thermal instabilities emanating from electrodes, which, in the case of dc discharges is observed to occur in the cathode fall region. This change also gives rise to a discharge with a symmetrical distribution of plasma properties and hence yields a symmetrical gain profile.

Theoretically the electrical energy may be coupled into the plasma by means either of metallic or of dielectric electrodes. At relatively high pressures (40 to 200 mbar) and current densities of about 60 mA/cm^2 and frequencies of the order of 10 MHz, a rather low voltage drop in front of the electrodes is found to be present [6]. It occurs across a sheath the thickness of which is a few tenths of a millimeter. The current transport within the sheaths is estimated to ensue from displacement currents. Thus, heat production in the electrode region is very low and consequently very low effort is needed for cooling the electrodes. Furthermore, to achieve homogeneous current distribution, no segmentation of the electrodes is necessary (or at most a coarse one when employing metallic electrodes). Even in the latter case no external ohmic stabilization need be provided because of the positive current-voltage characteristic of the rf discharge. By this the energy consumption may be reduced by up to 30% compared with externally stabilized dc discharges.

As a result, low expenditure in design and construction of the laser device is possible and high acievable values of loading and electrical efficiency may be anticipated. In addition, using a master-excited rf generator, the discharge can be modulated and/or switched within rise-times of the order of microseconds thereby meeting the requirements of various laser applications in a particularly simple manner.

EXPERIMENTAL

This work continues the investigations into the development of efficient excitation techniques [3,7].

The generator at our disposal was originally designed for other investigations so that in the present experiments not all requirements could be met. In particular, we have a frequency of 6 MHz and maximum power of 13 kW. The experiments are performed in "open" operation with typical run times of 10 s.

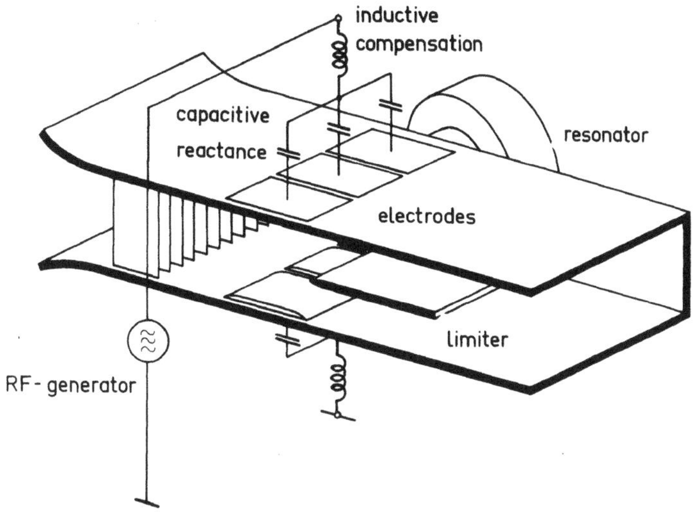

Fig. 1. Scheme of the laser.

A scheme of the apparatus is shown in Figure 1. The cross-section of the laser channel made of Plexi-glass is 48 x 4 cm^2, but it may be reduced to 48 x 2 cm^2. By appropriate installations care is taken to achieve a uniform flow field at the discharge entrance. Depending on the various operation parameters the gas velocity ranges between 35 and 90 m/s at pressures from 30 to 160 mbar. The results presented herein are obtained with a gas mixture of $CO_2/N_2/$ /He = 5/18/77. No substantial changes in the laser performance are observed when the gas composition is altered in a wide range, a reduction of the helium fraction to 0.5 included.

The discharge region is designed so that various electrode configurations may be tested by exchanging a completely mounted electrode support. The suitability of both metallic and dielectric electrodes is studied especially. In any case, the extension of the electrodes in the flow direction is kept constant at 3.8 cm.

Figure 1 shows a version of the apparatus with metallic electrodes arranged in a single row. They are made of sandblasted aluminium and are equipped with an air cooling system (which, however, has proved to be unnecessary for the short run times in the "open" operation). The six individual segments on each side of the channel are losslessly decoupled by capacitive reactances and are connected to a joint inductive compensation. Thus the rf generator has a pure ohmic termination which arises from the resistance of the discharge plasma.

A further version with two rows of (smaller) segments permits the variation of rf input power along the flow direction.

A rather simple construction is involved applying dielectric electrodes. Here, no segmentation or cooling system is needed. In addition, the homogeneous dielectric material of the electrode itself makes use of capacitive reactances superfluous.

With both metallic and dielectric electrode systems completely uniform, stable and arc-free discharges are obtained. The highest value of electrical power density so far achieved is 35 W/cm^3. This figure seems to be limited only by the generator in present use. From the measured current densities, typically of around 60 mA/cm^2, the values for the electron density may be deduced to be of the order of $5 \cdot 10^{10}$ cm^{-3}.

A single pass resonator is directly superimposed onto the discharge region. The results presented herein are obtained with a stable resonator. Its length is 82 cm and the diameter of the mirrors is 2". The resonator axis may be shifted along the flow direction. Maximum laser power is measured at a position 1 to 2 cm downstream of the discharge middle. Burn patterns in the near field give an indication of the extreme uniformity of the laser medium. They show a completely symmetrical plane ring structure with about 40 concentric rings. Preliminary investigations are performed with an unstable resonator having a magnification of 1.2. The measured laser power is approximately 30% lower compared to the results obtained with the stable resonator at a transmittance of 10%.

Laser oscillation is found to occur in the P20 line. Hence, measurement of small signal gain is concentrated on that transition.

The dependence of laser power on electrical input power is plotted in Figure 2 for two pressure levels. The results are obtained with metallic electrodes and a transmittance of 10%. Owing to constraints imposed by the rf generator the loading varies around a typical value of 350 J/g. The efficiency achieved ($\eta = P_L/P_\ell$) is in the range of 13 to 20%.

The effect of varying the power distribution along the flow direction becomes apparent on comparing the two different values of P_L obtained with an input power of 8 kW: in the experiment characterized by the full triangle a higher fraction of power is coupled into the discharge through the upstream electrode row.

The development of a small signal gain along the flow direction within the discharge is seen from Figure 3. The electrode configuration is the same as for the power measurements. Here again the full symbol characterizes the changed power distribution. Two features are evident from Figure 3: (1) it is preferable to have a higher power density at the upstream region of the discharge, and (2) there is some, although minor, down-blowing effect of the flow upon the discharge (which, however, is not discernible by eye).

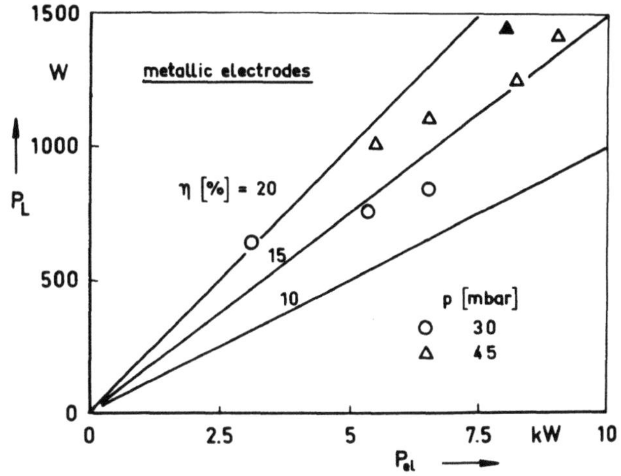

Fig. 2. Laser power as function of rf power and pressure.

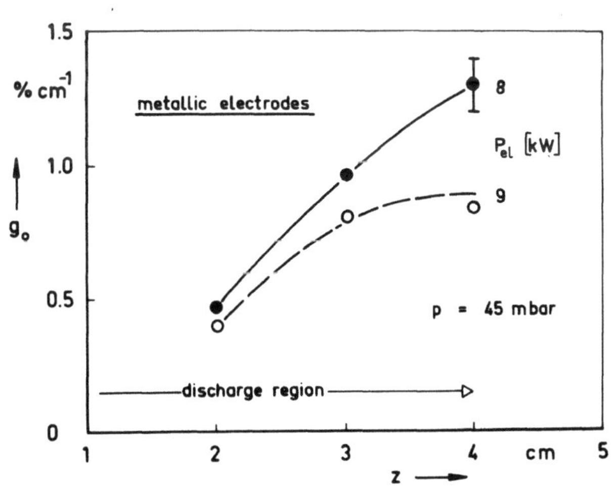

Fig. 3. Small signal gain in the middle of the channel height; different symbols indicate different power distribution in flow direction.

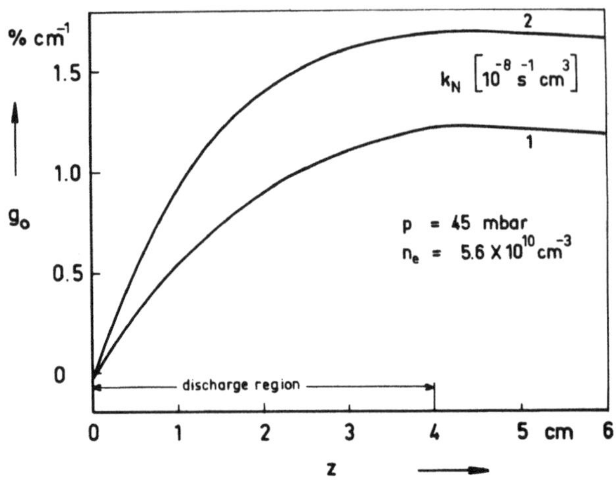

Fig. 4. Calculated small signal gain.

Support for the second observation is given by the theoretical results presented in Figure 4. They are obtained from a simple code that has been developed using the kinetic four-temperature model [8]. The parameter k_N, the coefficient of vibrational excitation of nitrogen by electrons, is the most sensitive term influencing the results. Its value is taken from the reduced field strengths E/n, encountered in the experiments. Altogether, the theoretically predicted distribution of small signal gain is in good agreement with the measured one.

The dependence of laser power on the transmittance of the coupling mirror is depicted in Figure 5. The circles represent measured values at experimental conditions of 45 mbar pressure and 8 kW electrical input using metallic electrodes. Two resonator specific curves are included: I/I_s is the ratio of the intensity near the coupling mirror inside the resonator to the saturation intensity; η is an efficiency defined as internal available power per cm^3 divided by the maximum available laser power per cm^3, $I_s \cdot g_o$. The calculations are performed according to Reference 9 and are based upon the experimental data of an active gain length of 48 cm, a reasonable averaged value of small signal gain of 0.8% cm^{-1}, and a loss on each side of the resonator of 2%. The agreement of the results with the experimental data is very good, which, among other things, is seen from the coincidence of the theoretical curve corresponding to a saturation intensity of 507 W/cm^2 with the experimental results.

Figure 6 gives some results of the optical power density (defined by measured laser power per cm^3 discharge volume) achieved with dielectric electrodes. The maximum value of approximately

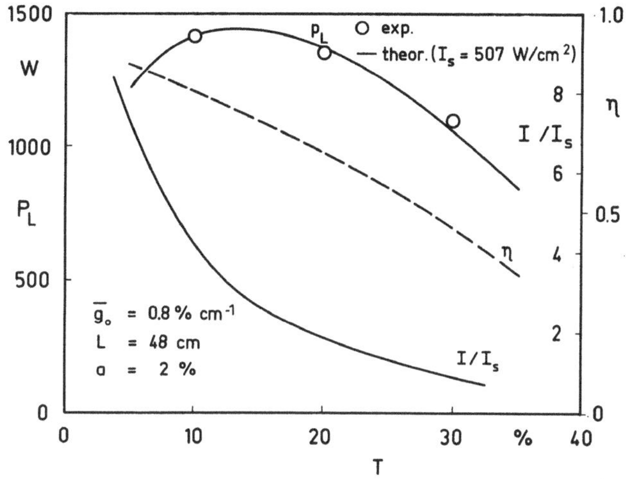

Fig. 5. Resonator characteristics (description see text).

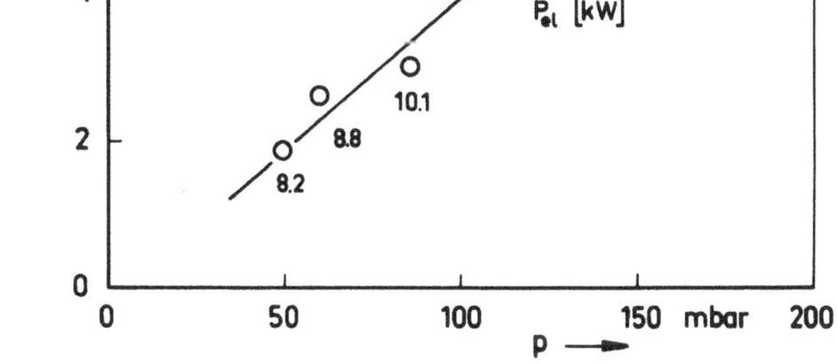

Fig. 6. Optical power density vs. pressure.

7 W/cm^3 is attained with an efficiency of 20% at an electrical power density of 35 W/cm^3.

CONCLUSIONS

The experimental results presented in this contribution have clearly demonstrated the merits of rf discharges as an efficient excitation means for large volume CO_2 lasers. High values of optical power density can be achieved at high efficiency. This results in the construction of compact laser devices utilizing single pass resonators. Furthermore, the implications of this technique with respect to the design of the apparatus itself are most advantageous. In particular, they yield but a low effort concerning the construction of the elctrodes and the operation of the discharge.

REFERENCES

1. A. C. Eckreth and J. W. Davis, "Rf-Augmentation in CO_2 Closed-Cycle Electric Discharge Convection Lasers", Appl. Phys. Lett., 21:25 (1972).
2. E. R. Pugh, "Radio Frequency Electrically Excited Flowing Gas Laser", US Patent 3748594 (1973).
3. W. Schock, W. Schall, H. Hügel and P. Hoffmann, "CW Carbon Monoxide Laser with Rf Excitation in the Supersonic Flow", Appl. Phys. Lett., 36:793 (1980).
4. C. P. Christensen, F. X. Powell and N. Djeu, "Tranverse Electrodeless Rf Discharge Excitation of High-Pressure Laser Gas Mixtures", IEEE J. Quant. Electron., QE-16:949 (1980).
5. V. D. Gavrilyuk, A. F. Glova, V. S. Golubev, A. B. Kusnetsov, F. V. Lebedev and V. A. Feofilaktov, "Characteristics of a CO_2 Laser Excited by a Capacitive Ac Discharge", Sov. J. Quant. Electron., 9:326 (1979).
6. V. I. Myshenkov and N. A. Yatsenko, "Prospects of Using High-Frequency Capacitive Discharges in Lasers", Sov. J. Quant. Electron., 11:1297 (1981).
7. W. Schock, H. Hügel and P. Hoffmann, "Ein neues Laserkonzept: Quergeströmter CO_2 Laser mit Hochfrequenzanregung", LASER+ Elektro-Optik, 2:76 (1981).
8. R. J. Harrach and T. H. Einwohner, "Four-Temperature Kinetic Model for a CO_2 Laser Amplifier", UCRL-51399 (1973).
9. W. W. Rigrod, "Saturation Effects in High-Gain Lasers", J. Appl. Phys., 36:2487 (1965).

HIGH-REPETITION-RATE LONG-PULSE-DISCHARGE XeCl LASER PREIONIZED BY A LOW-DENSITY ELECTRON OR X-RAY BEAM*

Marc L. Sentis, Bernard L. Fontaine, and
Bernard M. Forestier

Institute of Fluid Mechanics of Marseille
Aix-Marseille University, 1 Rue Honnorat
13003 Marseille, France

INTRODUCTION

High average power short wavelength laser systems and particularly exciplex lasers are a very rapidly growing research field owing to their numerous potential applications: energetics, long range communications, material processing, etc. [1]. For further development of these systems, numerous problems remain to be solved. Among the most important of these problems are:

- Laser discharge stability when long pulse, large volume, and high pulse repetition frequency (PRF) are considered;
- deposition at very high PRF of the necessary high specific power and energy [2,6];
- laser medium homogeneity at very high PRF [3,4].

Low density electron beam and X-ray preionization schemes proposed and demonstrated recently [6,9] represent a very promising alternative to the well developed U.V. preionization scheme for many applications requiring high energy per pulse and high PRF. The advantages of these new preionization schemes lie mainly in the homogeneous preionization of large volumes of high-pressure laser mixtures at high PRF and in maintaining stable discharge during several hundreds of nanoseconds. It is noteworthy that several potential applications of high energy U.V. laser radiation could be favored by a relatively long laser pulse duration (> 100 ns).

* Work supported by D.R.E.T. and C.E.A., France.

A program has been undertaken at I.M.F.M. to study the high-repetition-rate long-pulse avalanche discharge exciplex lasers preionized by either a low-density E-beam or a low-density X-ray beam. This study is related to the development of large-volume, high-repetition-rate, high-average-power U.V. laser systems.

In the first part of this paper are described the most recent experimental results obtained at I.M.F.M. on an XeCl laser (λ = 308 nm) preionized either by an E-beam or an X-ray beam for no-flow low-PRF conditions with two experimental devices with different gain length. In the second part a new fast-flow subsonic closed-loop laser system called LUX (for "laser ultraviolet préionisé par rayons X") is described in detail, with experimental results on the flow characteristics of the loop. A conclusion is presented which includes planned Lux laser characteristics.

EXPERIMENTAL

Long Pulse XeCl Laser Pumped by an Electron Beam or X-ray-assisted Discharge (No Flow)

The experimental apparatus [10] has been slightly modified and consists mainly of a laser head directly fitted to an electron or X-ray gun. Two permaglass (fiber glass and epoxy) laser heads A and B with different gain length and useful active medium volume are used. Head A is 3 cm high and 16 cm wide and allows a working pressure up to 3 atmospheres. Head B is 2.5 cm high and 30 cm wide and allows a working pressure up to 2 atmospheres. The static gas mixture is excited at low PRF by an avalanche discharge assisted either by an E-beam and an X-ray or only by X-rays. Relatively long beam pulses of low intensity are used for preionization. A cold cathode electron-gun energized by a Marx generator permits us to preionize the gas by means of electrons (E $\leqslant$ 300 KV) with a density of 20 A cm^{-2} to 1.6 A cm^{-2}. The time duration of the E-beam may be varied between 0.4 and 2 μs. The preionization beam cross-section (14 x 2 cm^2 or 28 x 2 cm^2) is adapted to the length of the laser head by modification of the gun cathode length. The E-beam reaches active medium through a 25 μm thick titanium window and a tungsten grid which acts as a cathode for the main discharge. Where preionization is made by means of X-rays only, the E-beam is totally absorbed by a 1.5 mm thick aluminium plate before entering the laser head. For these experiments, a tantalium foil (Z = 73), 12.5 μm thick is set inside the gun immediately in front of the titanium window in order to enhance X-ray emission. The laser head is fitted with an anode which faces the discharge cathode. The A laser head is fitted with a copper plate (2 x 14 cm^2) anode with smoothed edges while the B laser head is fitted with an aluminium (3 x 28 cm^2) contoured anode. A voltage up to 20 KV can be applied to the electrodes through a triggered spark gap by means of a simple capacitor

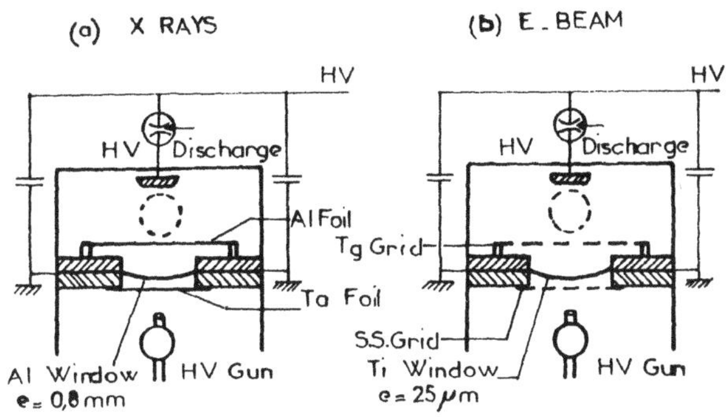

Fig. 1. Schemes of E-beam and X-ray preionized discharge circuits.

(C = 0.5 μF). The circuit inductance of 100 nH controls the current rise-time. Figures 1a and 1b show respectively the X-ray and E-beam preionized avalanche discharge laser schemes.

The laser cavity is formed of MDL hard coating quartz concave mirrors (R = 2 m); one has a transmission coefficient of 10^{-3} at λ = 3080 Å; the other may vary between 10^{-3} and 0.66 at the same wavelength. The useful cavity volume, defined as a cylinder whose cross-section is the laser beam area at the extracting mirror and whose length is the distance between side channel walls, is equal to 11 cm^3 and 100 cm^3 respectively for the A and B laser heads.

Main discharge voltage and current time behavior is recorded on scopes by means of a carbon resistance voltage divider and a Pierson current transformer. The following optical parameters are generally recorded during each experiment: (a) time variation of laser or fluorescence intensity by means of a photomultiplier, or photodiodes (< 30 ns rise-time) and narrow band with interference filters or holographic grating monochromator; (b) ($4 < \Delta\lambda < 40$ Å) time integrated spectrum of laser or fluorescence emission by means of home-made grating spectrographs and an optical multichannel analyzer coupled to a PDP11/03 minicomputer; (c) total extracted optical energy by means of ED100 and ED200 Gentech pyroelectric detectors and (d) laser beam pattern and energy profiles by means of photographs.

A first set of experiments with, as the main objective, the behavior of XeCl laser characteristics for conditions of low-density E-beam preionization has been performed with the A laser head. When a room temperature Ne/Xe/HCl (9100/260/14) mixture at a pressure of 2000 torrs is excited by a discharge (V = 11 KV) assisted by an electron beam of 1.6 A cm^{-2} transmitted current density (V = 270 KV) the laser pulse characteristics are, for a coupling gun mirror of 0.20 transmission, the following ones: energy 31 mj (3 j/ℓ), duration 200 ns, peak power 200 KW (20 MW/ℓ). To achieve such laser energy and laser power it is necessary to apply the voltage to the electrodes after E-beam firing and so to obtain initial preionization before discharge. During these experiments, in order to determine conditions favoring long-pulse high-repetition-rate high average power XeCl laser emission, the current density of the transmitted preionization electron beam has been varied from 20 mA cm^{-2} to 1.6 A cm^{-2} by changing the Marx generator voltage. The constant parameters for these experiments are as follows: discharge capacitor charging voltage (10 KV), Ne/Xe/HCl mixture (5300/190/10); pressure (1430 torrs), temperature (300°K), and coupling mirror transmission coefficient 0.20. Figure 2 shows the variation of laser power and energy versus transmitted E-beam current density. A second abscissa scale indicates the corresponding ratio ($R = E_D/E_G$) between main discharge and E-beam energy absorbed by the active medium. This figure shows that conditions exist for an XeCl laser where energy deposited in the active medium originates essentially from the discharge whilst maintaining high laser efficiency. Laser energy only decreases threefold when preionization E-beam energy is decreased by a factor 60. Nearly constant specific energy and peak power of 150 j/ℓ and 500 MW/ℓ are added to the mixture by means of the discharge during these experiments. Intrinsic laser efficiency varies from 2 to 0.7% and seems to be limited mainly by the short gain length of the A laser head. These results compare favorably with those obtained from experiments where the ratio R was lower than 10 [11].

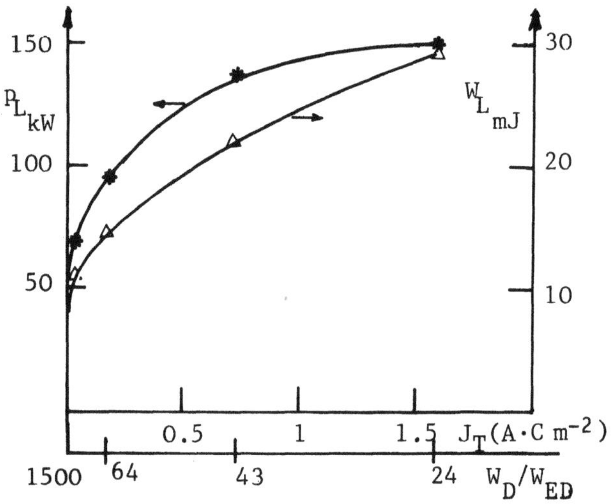

Fig. 2. Laser power and energy versus E-beam transmitted current density (Ne/Xe/HCl 5300/190/10; p = 1430 t; V_D = 10 KV; T_{mirror} = 0.2; L_{gain} = 16 cm).

A second set of experiments with the A laser head is undertaken for conditions where preionization is achieved by means of an X-ray beam produced by the interaction of an E-beam with the gun window. These experiments achieve laser action at λ = 3080 Å from an XeCl excimer when Ne/Xe/HCl mixtures at 1 atmosphere pressure are excited by an X-ray-assisted discharge (V_{gun} = 270 KV). However, extracted laser energy is weaker by an order of magnitude and the laser emission duration is 3 - 4 times shorter than for the case of an E-beam assisted discharge. This difference is attributed to a too short gain length and an unadapted discharge anode profile for X-ray preionization where preionization electron density is much lower.

In an attempt to optimize experimental conditions a third set of experiments uses the B laser head in place of the A head. A careful comparison of XeCl laser performances with E-beam and X-ray preionization for the same energy storage conditions is attempted. Table 1 gives the laser emission characteristics for typical laser conditions, XeCl laser performances with X-ray and E-beam preionization are much more similar than previously.

Results from this set of experiments allow the following observations: (a) when gain length is sufficient and the discharge anode is contoured, long pulse 150 - 200 ns FWHM XeCl emission is achieved following excitation of Ne/Xe/HCl mixtures either by an E-beam or by X-ray-assisted discharges; (b) stable discharge is allowed during several hundred nanoseconds for conditions where discharge energy is several orders of magnitude higher than the absorbed E-beam or X-ray energy and where the energy loading is over 100 j/ℓ atm using a simple triggered spark gap capacitor circuit of relatively high

Table 1. Typical XeCl Laser Characteristics (Ne/Xe/HCl), (5300/190/10).

Preion.	V_{GUN} (KV)	V_D (KV)	Press. (Torrs)	P_{laser} peak (kW)	W_{laser} (mJ)	η intrins. (steady) %	$\eta_t = \frac{W}{W_{in}}$ %	Laser Pulse Length (FWHM) MS
X-ray				500	87	0.5	0.33	140
	210	16						
E-beam				760	105	1.2	0.44	160
X-ray				680	116	0.51	0.31	160
	210	20	800					
E-beam				900	170	0.87	0.4	190
X-ray	210	10		330	55	1.43	0.61	
E-beam	270	20	1134	1470	300	3	0.85	210

inductance; (c) maximum specific laser energy over 3 j/ℓ is obtained for a pressure of 1.5 atm (unoptimized) and a volume of 100 cm^3 with the use of E-beam preionization and is reduced only by a factor of two when using X-ray preionization; (d) the value of extracted laser energy is not strongly dependent on E-beam or X-ray intensity; (e) efficiency $W_{laser}/W_{discharge}$ reaches 1%; (f) today's technology can attain the E-beam or X-ray preionization energy required at a high repetition rate, so favoring X-rays, considering the limitations of gun window heating.

High PRF LUX Exciplex Laser Test Bed

A closed cycle laser test bed called LUX has been built in order to assess carefully issues such as U.V. laser beam quality (damping of aerodynamic disturbances), discharge stability for long pulse laser conditions, and average laser power available with fast transverse flow E-beam or X-ray assisted discharge excitation for conditions of high PRF. The device is shown in Figure 3. The loop has been especially designed to achieve a long "life-time" for the working mixture, very low base-line flow turbulence level, low pressure drop and fast damping of acoustic waves.

The ducts of the 150 ℓ volume flow loop are fabricated from 304 and 316 stainless steel. Gas flow is driven by a nominal 500 ℓ/s 30 torrs ΔP (at 3000 rpm) stainless steel centrifugal fan driven by a direct current 4 KW, 0 - 3000 rpm motor which allows a continuous control of flow speed. This motor is set outside the duct and connected to the fan by use of an HCl compatible "ferrofluidic" seal. Viton "O" rings are used throughout the flow loop and the system reaches a base pressure of 10^{-6} torr when evacuated by a 270 ℓ/s turbomolecular pump. An average flow velocity up to 80 m/s has been measured with a pitot static probe (air, 1 atm) in the laser

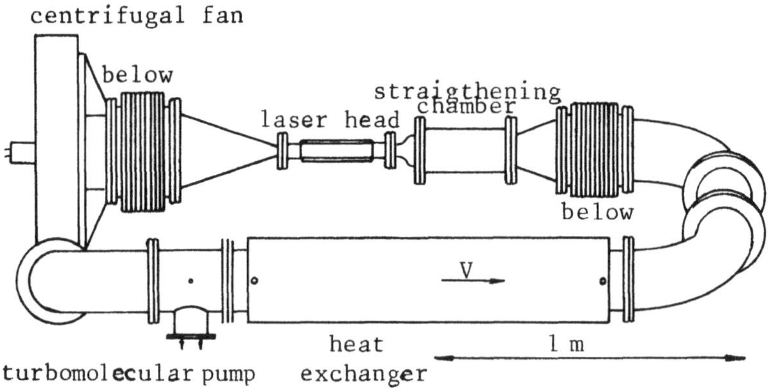

Fig. 3. LUX high average power exciplex laser test bed.

discharge head (2.5 x 30 cm^2 cross-section). The loop includes heat exchangers, convergent and divergent, flow straightening grids and bellows for vibration isolation.

The electrical excitation system consists essentially of a 100 KW average power pulser allowing preionization either by E-beam or X-rays and excitation by the main discharge with 1 KHz PRF in burst mode (≃ 100 pulses), see Figure 4.

A preionization beam is produced by a gun set under the laser head. The beam window cross-section is 2 x 28 cm^2. The gun is driven by a fast pulse transformer type pulser controlled by hydrogen-filled spark gaps (quenchotrons). Energy is stored in a 50 KJ capacitor bank which allows fast charging of the 0.5 μs - 15 KV primary capacitor. A triggered spark gap connects this capacitor to the transformer. A peak capacitor 1 nF - 200 KV and a 200 KV self-triggered spark gap allows a fast rise-time high-voltage pulse (≃ 20 j - 200 KV - 0.5 μs) to be applied to the gun cathode. The main discharge circuit of capacitor type, energized by a 50 KJ storage bank capacitor and controlled by a triggered spark gap delivers 50 j at 30 KV charging voltage (100 - 500 ns) after pre-ionization gun firing.

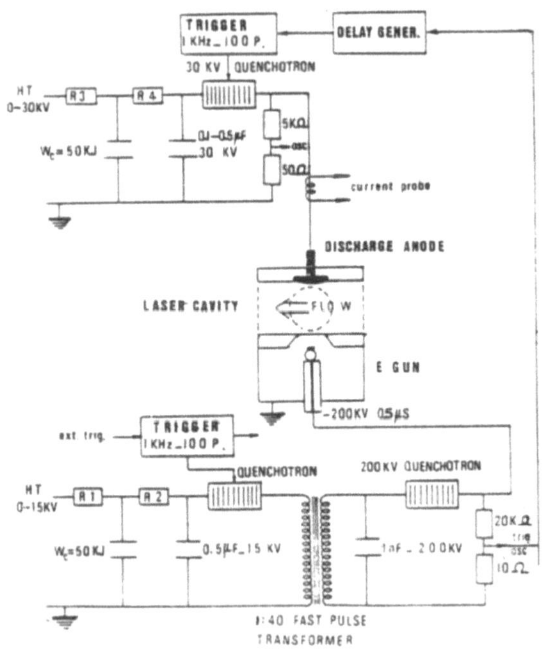

Fig. 4. Scheme of the 100 KW average power 1 KHz 100 pulses electrical excitation system.

An experimental study of the LUX system base-flow characteristics (without electrical excitation) has been fulfilled with air as the working gas for standard conditions of pressure and density and a 50 m/s flow speed. The flow velocity field, as measured by a very small (Ø = 2 mm) pitot probe, is shown to be constant with better than ± 1% outside the boundary layer than that in the two transverse planes. Turbulence intensity (RMS) measured with hot wire anemometry is equal to $1 \cdot 10^{-3}$ when stabilizing grids are present and reaches $3 \cdot 10^{-3}$ without grids. The measured value of boundary layer thickness ($\Delta u/u \simeq 1 \cdot 10^{-2}$) is E = 6 mm on the optical axis direction and E = 4 mm on the other transverse plane.

Relative space and time flow density variations are measured by means of snapshot and streak Michelson interferometry using a laser height source. Interferometer sensibility is $\Delta\rho/\rho = 3 \cdot 10^{-3}$ per frange shift. Measured variations of the base flow density outside the boundary layers is less than measurement uncertainty ($\Delta\rho/\rho \pm 2.5\ 10^{-4}$) and reaches $5 \cdot 10^{-4}$ in the absence of grid when flow speed is increased to 67 m/s^{-1}. These results show that the base flow generated by a LUX device is fully compatible, as far as turbulence and large scale fluctuations are concerned, with high beam quality excimer laser generation. The next step will be to find efficient methods to recover this situation for each pulse for conditions of high PRF excitation (see companion paper).

Work on electrical excitation of the flow with the pulser is well in progress and will be presented in a forthcoming paper.

CONCLUSION

Results obtained both with a low PRF E-beam or X-ray - assisted discharge on XeCl laser performances and with the LUX test bed on base flow characteristics, permit us to expect this system to deliver U.V. laser radiation with several hundred watts of average power in burst mode at high PRF (⩾ 1 KHz) with a good beam quality. Work will be actively pursued at I.M.F.M. in this direction with the aim also to investigate problems which could limit performances of high PRF high energy exciplex lasers.

REFERENCES

1. C. A. Brau, in: Excimer Lasers, C. K. Rhodes ed., Springer-Verlag, Topics in Applied Physics, 30 (1979).
2. J. L. Miller et al., Paper THB2 CLEO 81, Washington DC (1981).
3. S. R. Byron et al., CLEO 82, Pheonix, Arizona (1982).
4. V. Yu. Baranov et al., Sov. J. Quant. Elect., 10:512 (1980).
5. T. S. Fahlen, IEEE JQE., QE16:1260 (1980).
6. J. I. Levatter et al., Appl. Phys. Lett., 39:297 (1979).
7. S. Sumia et al., J. Appl. Phys., 52:2682 (1981).
8. B. Forestier et al., J. Physique Lettres, 42:L211 (1981).
9. H. Shield and A. J. Alcock, Optics Communications, 42:128 (1982).
10. B. Forestier and B. Fontaine, Rev. Sc. Instr., 50:421 (1979).
11. B. Forestier, B. Fontaine and P. Gross, 3rd G.C.L., Marseille, Septembre 1980, proc. in J. de Physique, Coll.C9:41, C9:455 (1980).

A COMPACT 75 WATTS CO_2 LASER AT 750 HERTZ

Robert C. Sze

University of California
Los Alamos National Laboratory
Los Alamos, New Mexico 87545

ABSTRACT

We describe the operation of a compact CO_2 laser that operates to 750 Hz at 100 mJ/pulse in the TEM_{00} mode. The device is line and power tunable ranging from milliwatts to 75-W average power.

We report the development of a compact CO_2 laser system that delivers over 75-W average power in the TEM_{00} mode at pulse repetition rates up to 750 Hz. The device is line tunable and will have the capability of being frequency off-set locked from line center. This laser is intended for use in optical pumping of molecular gases to obtain coherent radiation at longer wavelengths.

Figure 1 shows a schematic of the device. Presently, the laser measures 100 ℓ, which includes all components, except for the charging supply. We have already worked out designs that will reduce the system to one-third its present size or to just a fraction over 35 ℓ. Figure 2 gives the top view of the laser and Fig. 3 gives an overall view of the laser with the cw-laser in the foreground used for injection-locking the pulsed laser. The component that contributes most significantly to the compactness of the laser is that only twelve 1-nF strontium titanate doorknob capacitors are used as the storage element (located on the top of the electrodes directly below the plexiglass plate in Fig. 2) and the decision to use a nearly helium-free gas mix so that the operating E/N is substantially higher than the normal helium-rich mixtures. This allows for sufficient energy storage in the capacitors at high voltage. The peaking capacitor is a folded strip line using a 5-mil Kapton sheet as the dielectric with a value of 0.7 to 0.9 nF. A rectangular quartz tube is located behind the screen electrode and

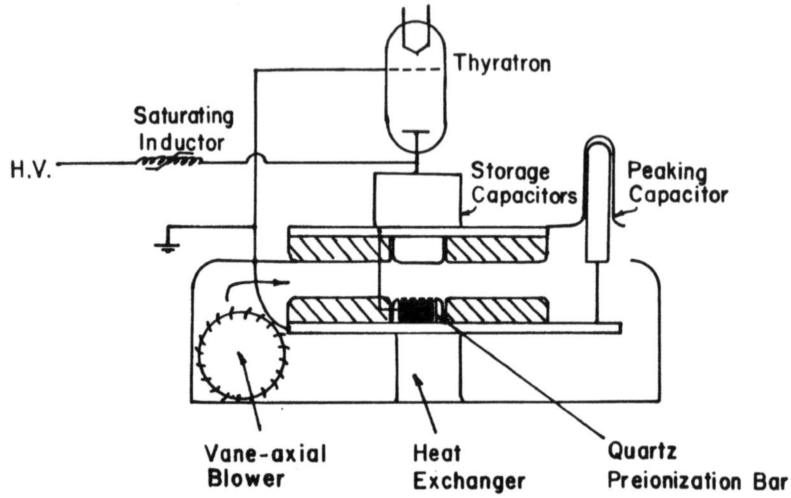

Figure 1

Figure 2

Figure 3

a conductor inside the tube is connected to the opposite electrode, resulting in very dependable preionization of the laser volume. The gas flow is accomplished by a simple inexpensive squirrel cage vane-axial blower used in air-conditioning systems. The 3-in.-diam fan is operated at 3600 rpm. The water-cooled heat exchanger is capable of dissipating 2 kW of power. In our studies we conclude this to be the ultimate limitation to our pulse repetition rate.

The discharge dimensions were determined by the need for TEM_{00} operation and the design decision to limit the energy deposition to approximately 100 J/ℓ atm. The laser, therefore, has a 0.9-cm × 0.9-cm × 60-cm discharge volume with the laser cavity separation at 80 cm.

Performance curves using a gas mixture[1] of (1.0 CO_2/1.2 N_2/ 0.4 He/0.2 H_2) at 350-torr filling pressure for output power versus repetition rate and charging voltage are shown in Figs. 4 and 5. These data used a 5-m radius of curvature total reflector and an 80% R flat output coupler. The output efficiency determined from the stored energy in the capacitor is better than 5%, and the wall-plug efficiency is better than 3.5%. The curve for average power versus pulse repetition frequency shows a faster than linear rise. This is simply caused by the slight gradual droop of the capacitor voltage as a function of time due to back leakage through the charging diode. The average power grows slightly faster than

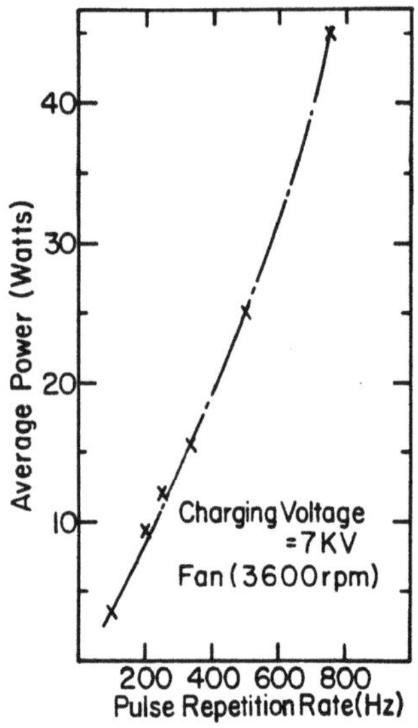

Fig. 4

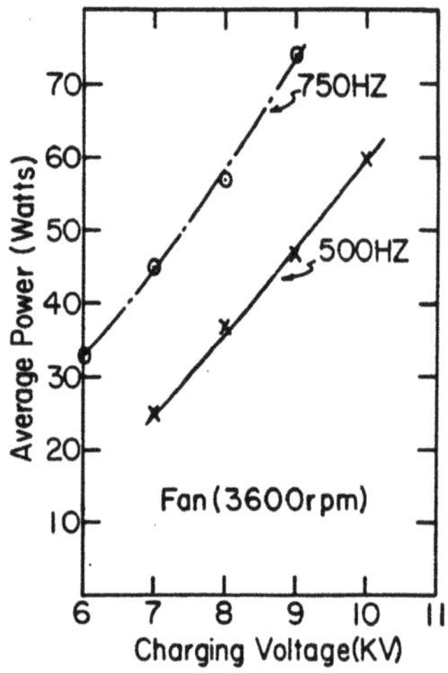

Fig. 5

linear with voltage. For 350-torr filling pressure the upper voltage limit is at 10-kV charging voltage, which gives approximately 20 kV on the capacitors, resulting in an upper limit of energy depositions at 107 J/ℓ atm. A charging voltage of over 11 kV is possible at 500-torr filling pressure.

Line tuning is achieved with a 150 l/mm gold-coated grating and a 5-m radius of curvature, 85% R output coupler. With the laser tuned to the 10-μm P(20) transition, 90% of the energy in the untuned cavity is obtained. Figure 6 shows the energy per pulse we obtain as the laser is tuned through the 9-μm (00°1)-(02°0) and the 10-μm (00°1)-(10°0) R- and P-branch transitions.

The relative energies in the gain-switched spike versus the long tail are dependent on the lasing transition. The lower gain transitions give more energy in the tail of the pulse, as well as exhibit a shorter gain-switched pulse. Figure 7 shows the pulse shapes taken with a photon-drag detector for the (a) 10-μm P(20) transition and (b) the 9-μm R(12) transition.

Frequency off-set locking of the TEA laser is to be accomplished via a cw laser locked onto a molecular absorption. The cw laser is injected via the zero order of the grating into the TEA laser cavity. The cavity is locked to the peak of the cw laser transition. This has been accomplished, however, the system losses lock in the presence of electrical noise from the discharge. We are now addressing this problem.

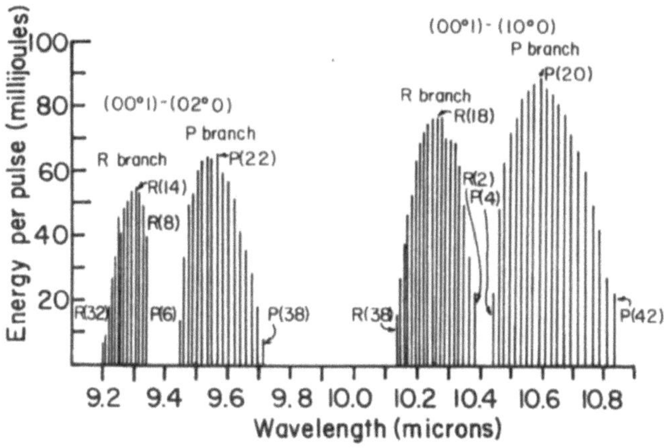

Figure 6

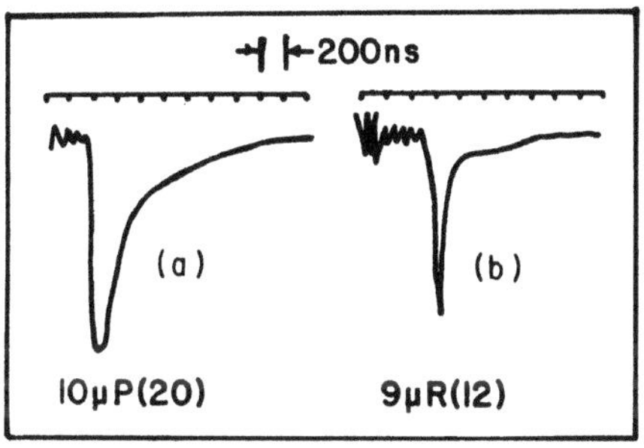

Figure 7

REFERENCE·

1. N. Menyuk, P. F. Moulton, Rev. Sci. Instrum. 51, 216 (1980).

SCALING RELATIONS FOR THE PLASMA RECOMBINATION LASER

W. L. Bohn, and W. Schall

DFVLR-Institut für Technische Physik
Pfaffenwaldring 38-40, D-7000 Stuttgart 80

INTRODUCTION

Although numerous papers have been devoted to the scaling properties of hydrogen-like plasmas and their population densities since the fundamental work of McWhirter[1], little has been published on the scaling of plasma recombination laser properties, especially in the case of non-hydrogen-like ions. The purpose of this paper is to present scaling laws for the limits of electron density, laser power, laser energy and small signal gain for non-hydrogen-like recombination laser transitions covering the spectral range from the IR to the XUV. Furthermore an attempt is made to scale the different cooling mechanisms responsible for the pumping of recombination lasers. As far as available, comparison with experimental data is carried out to support the derived scaling laws.

ELECTRON DENSITY AND LASER POWER OUTPUT LIMITS

Since the plasma recombination rate in the collision dominated regime is strongly dependent on electron density ($\sim n_e^3$) its limiting value in recombination lasers defines a maximum for the pumping mechanism. A good approximation for the upper density limit is obtained from the condition of thermalization by electron collision processes for the corresponding laser transition. By equating collisional deexcitation and radiative decay rates a very simple relation, $n_e \sim \lambda^{-3}$, has been first proposed by Silfvast[2] and Bohn[3]. A more refined approximation which allows for all the radiative decay out of the corresponding level is found by using the equilibrium criterion derived by Griem[4]. In all approaches the laser levels are supposed to be high lying quantum levels and the hydrogenic cross sections to be valid. Thus, the maximum electron density in recombi-

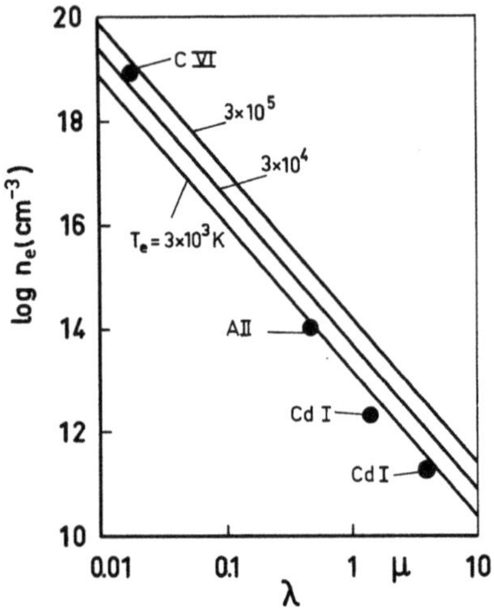

Fig. 1. Electron density limit as a function of laser wavelength (Dots: experimental data).

nation lasers can be written as

$$n_e(\text{cm}^{-3}) = 1.3\ T_e^{1/2}/\lambda^{17/6}$$

where the electron temperature, T_e, is expressed in K and the wavelength of the laser transition, λ, in cm. This relation is plotted in Fig. 1 for three different temperatures and compared to various experimental data (full dots): the CdI-transition[5], the AII-transition[6] and the CVI-transition[7] are in good agreement with the theoretical scaling.

The maximum available power ($P\nu$) and energy density ($E\nu$) for a recombination laser are readily expressed as a function of wavelength using the density relation and remembering that

$$P_\nu = h\nu \quad \alpha\, n^2_{e,max} \quad \text{and} \quad E_\nu = P_\nu \cdot \tau_{rec} = h\nu\ n_{e,max}$$

where α and τ_{rec} are the recombination rate coefficient and the recombination relaxation time, respectively:

$$P_\nu(\text{W/cm}^3) = 10^{-30}/T_e^3\ \lambda^{9.5} \qquad \text{and}$$

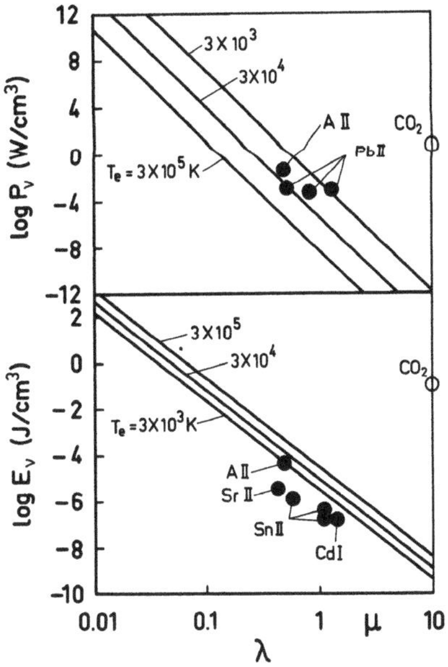

Fig. 2a,b. Laser power and energy density limits in comparison to various experiments.

$$E_\nu (J/cm^3) = 2.6 \; 10^{-23} \; T_e^{1/2}/\lambda^{23/6}$$

Both relations are plotted in Fig. 2a and Fig. 2b for different T_e and completed by several experimental data (dots). The AII-transition fits very well for the energy density, which implies that it does not correspond to a truly cw-operation the CdI-transition reaches 20 % of its theoretical maximum in energy output; the PbII infrared transition[8] fits with its theoretical power density limit. All other transitions[9,10] are found to follow roughly the theoretical scaling within an order of magnitude.

SMALL SIGNAL GAIN SCALING

For a realistic scaling of the gain coefficient, an upper limit of the inversion density, Δn, or at least the upper laser level $\Delta n \sim n(j)$ has to be found. In the quasi-steady state approximation[1] the density of a quantum level in recombining plasma is given by $n(j) = r_o \, n_E(j)$, where, n_E, is the equilibrium value of the density and r_o is a n_e and T_e-dependant factor allowing for departures from thermal equilibrium. Inspection of r_o over a wide range of temperatures and densities shows that, $r_o \leq 1$, is always fulfilled. Therefore an upper limit for the quantum level density may be obtained by setting $n(j) \approx n_E(j)$. After some rearrangements and making use of the electron density scaling relation, one obtains for the gain coefficient (assuming a Doppler dominated line profile) of a recombination pumped laser transition:

$$\frac{g_D}{A\mu_A^{1/2}} = 6.1 \times 10^{-18} \frac{\exp(43.4/T_e \lambda^{2/3})}{X\, T_e \lambda^2}$$

The gain coefficient $g_D(cm^{-1})$ has been normalized to the Einstein coefficient A (sec^{-1}) and the atomic mass number related to the transition involved. The degree of ionization of the plasma is taken into account by X, i.e.: the ratio of electron density to the density of the recombining ion. This relation is shown in Fig. 3 and compares favourably with two available experimental points. The gain of the CVI-transition has been reported in a $T_e = 4 \times 10^5$ K plasma and fits the corresponding theoretical scaling (broken line); the gain of the CaII-transition[10] matches the scaling for $T_e = 10^4$ K which seems to be a realistic approximation for the corresponding experimental temperature. In general, the gain scaling exhibits surprisingly moderate g_D-values as compared to the hydrogenic scaling ($\sim Z^{7.5}$) and may therefore be a more realistic tool for theoretical predictions of short wavelength recombination lasers. Another

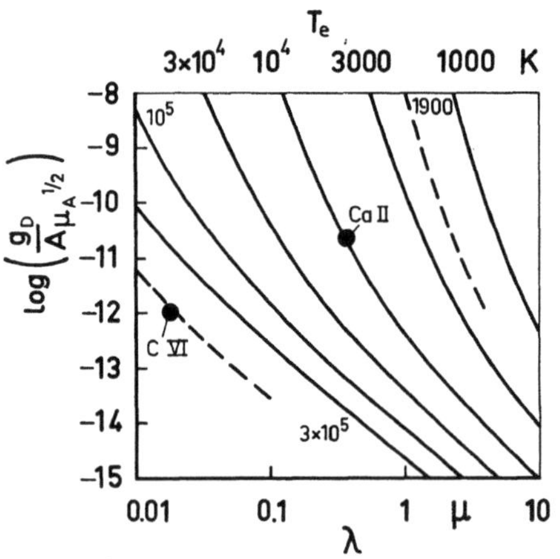

Fig. 3. Gain scaling as a function of laser wavelength.

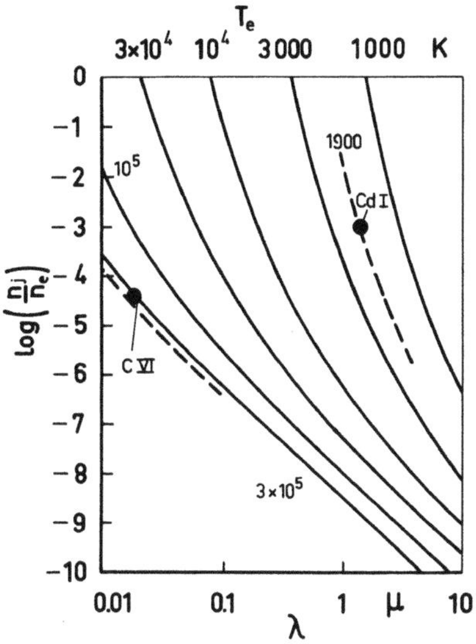

Fig. 4. Ratio n_j/n_e as a function of laser wavelength

measurable plasma parameter which may be useful for scaling purposes is the ratio $n(j)/n_e$. Under similar assumptions than those described above an upper limit can be written as:

$$\frac{n(j)}{n_e} = \frac{1.9\text{x}10^{-12}}{\text{X}\ T_e} \frac{\exp\left(43.4/T_e \lambda^{2/3}\right)}{\lambda^{13/6}}$$

This relation is shown in Fig. 4 for various electron temperatures. A comparison with a CVI-transition in a $T_e \approx 4 \times 10^5$ K plasma shows good agreement with the corresponding theoretical curve (broken line). A CdI-transition reported by Silfvast[2] would imply a temperature of T_e = 1900 K which seems compatible with other experimental parameters; a cross-check may be carried out for the gain scaling of the same transition in Fig. 3 with the corresponnding T_e = 1900 K curve (broken line).

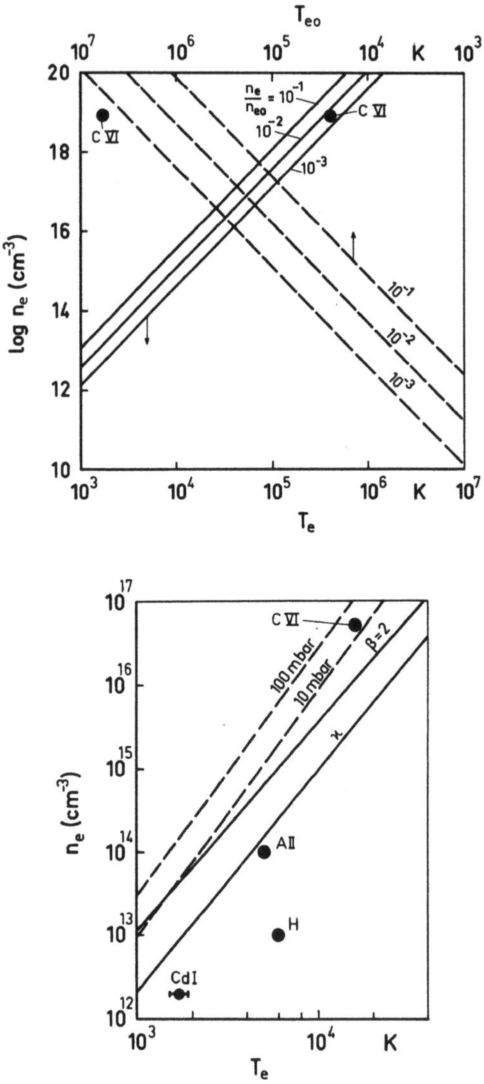

Fig. 5 a) Actual n_e, T_e-values as a function of initial plasma parameters; b) actual n_e, T_e-values for various cooling mechanisms.

SCALING OF THE COOLING RATE

A different quality of scaling property is obtained when looking at the different contributions to the electron energy conservation equation[11].

For a pure adiabatic expansion the initial plasma temperature, T_{eo}, and the temperature, T_e, reached after a time $t = \tau_{rec}$ are related to the actual and initial electron densities by the following relations:

$$T_{eo} = 4.4\ 10^{-3}\ r_o^{0.2}\ \mu_A^{0.1} \left[1-\left(\frac{n_e}{n_{eo}}\right)^{1/\beta}\right]^{1/5} \left(\frac{n_e}{n_{eo}}\right)^{\frac{9}{10}(\gamma-1)-\frac{1}{5\beta}} n_e^{2/5}$$

$$T_e = 4.40\ 10^{-3}\ r_o^{0.2}\ \mu_A^{0.1} \left[1-\left(\frac{n_e}{n_{eo}}\right)^{1/\beta}\right]^{1/5} \left(\frac{n_e}{n_{eo}}\right)^{\frac{1}{10}(\gamma-1)-\frac{1}{5\beta}} n_e^{2/5}$$

where r_o is the initial plasma radius and β a geometric parameter. A comparison of these relations with a CVI-transition is shown in Fig. 5a. Whereas the final temperature is in good agreement with experimental data, the initial plasma temperature claimed by Key[7] is about 3 times higher than predicted by our scaling. A more detailed check with other cooling mechanisms as derived by properly equating the corresponding contributions in the energy equation[11] is shown in Fig. 5b. The adiabatic cooling for cylindrical geometry ($\beta = 2$) is compared to the cooling by elastic collisions of the hot electrons with a helium background gas at pressures of 10 and 100 mbar and with the cooling by thermal conduction of the electrons (κ). A CVI-transition reported by Elton[12] shows good agreement in the case of background gas cooling. The AlI-transition reported from an expansion experiment[6] implies that thermal conduction plays a greater role in the corresponding experiment than adiabatic cooling. This also applies to the CdI-transition. A H-transition reported by Hara[13] does not match any of the predicted cooling mechanisms probably because the plasma is cooled in a 2 step expansion which is not described by our rough cooling approximation.

CONCLUSIONS

A scaling of electron density, gain and laser power/energy over three decades of laser wavelength (from IR to XUV) has been presented and is verified by numerous experimental data. A first attempt to derive scaling properties for the cooling mechanism of recombining plasmas stresses the importance of other cooling schemes than adiabatic expansion.

REFERENCES

1. R. W. P. McWhirter and A. G. Hearn, Quasi Steady State Population Densities in Hydrogenic Ions, Proc. Phys. Soc. Land., 82: 641 (1963).
2. O. R. Wood II and W. T. Silfvast, Electron Density and Energy Output Limits of Plasma-Recombination Lasers, Appl. Phys. Lett. 15: 121 (1982).
3. W. L. Bohn and W. Schall, Scaling Properties of the Plasma Recombination Laser, Conference on Lasers and Electro-Optics, Phoenix Arizona, Technical Digest THT3: 110 (1982).
4. H. R. Griem, "Plasma Spectroscopy", McGraw-Hill, New-York (1964).
5. W. T. Silfvast and O. R. Wood II, Recombination Laser Transitions in Expanding Plasmas of Mg, Ca, Cu, Zn, Ag, Cd, In, Sn, Pb, and Bi, Optics Letters 7: 34 (1982).
6. E. M. Campbell, R. G. Jahn, W. F. von Jaskowsky, and K. E. Clark, Recombination Lasing in a Magnetoplasmadynamic Arcjet, J. Appl. Phys. 51: 109 (1980).
7. M. H. Key, C. L. S. Lewis and M. J. Lamb, Transient Pupulation Inversion at 18.2 nm in a Laser Produced CVI Plasma, Optics Comm. 28: 331 (1979).
8. K. Gnädig, CW Laser Oscillation Observed on Visible and Near-Infrared PbII-Transitions in a Hollow Cathode Discharge Tube, Optics Comm. 35: 427 (1980).
9. K. Gnädig, and Lin FU-Cheng, A Hollow Cathode Tin Laser, Optics Comm. 34 218 (1980).
10. E. L. Latush and M. F. Sém, Stimulated Emission due to Transitions in Alkaline-Earth Metal Ions, Sov. J. Quant. Electron. 3: 216 (1973).
11. P. Wägli, and W. L. Bohn, Population Inversions by the Interaction of an Expanding Plasma with a Background Gas, J. Appl. Phys. 51: 3601 (1980).
12. R. H. Dixon, J. F. Seely, and R. C. Elton, Intensity Inversion in the Balmer Spectrum of C^{5+}, Phys. Rev. Lett. 40: 122 (1978).
13. T. Hara, H. Kodera, M. Hamagaki, Quasi-Steady Laser Oscillation in the Recombining Hydrogen Plasma, Japanese Journal of Appl. Phys. 19: L606 (1980).

KINETIC STUDY OF THE VUV EMISSIONS OF AR-KR MIXTURES

H.Brunet, A.Birot, S.Busquets, J.Galy, P.Millet, and Y.Salamero

Centre de Physique Atomique, Lab. Associe au CNRS n 277
Universite Paul Sabatier, 118 route de Narbonne
31062 Toulouse Cedex, France

ABSTRACT

The far UV spectroscopic study of argon/krypton mixtures demonstrates the very great efficiency of the Ar-Kr transfer processes. Kinetic analysis of the main molecular emissions of the two rare gases shows that the energy exchange essentially involves the molecular states of argon -the Ar_2^* species are destroyed on collision with krypton with an efficient cross-section of roughly 80 Å^2. The metastable $Kr(^3P_2)$ states, which are an indirect result of the transfer, lead, through three-body heteronuclear collisions, to the formation of radiative Kr_2^* species which are themselves responsible for the emission of the continuum at 145 nm.

The study of the VUV emissions of the rare gases and their mixtures is of great interest due to their applications to power lasers[1]. In pure rare gases the VUV continua correspond to transitions between the weakly bound $^1\Sigma_u^+$ and $^3\Sigma_u^+$ molecular states and the dissociative molecular state $^1\Sigma_g^+$. The radiative excited species are created by three-body collisions from resonant and metastable states; previous studies have allowed the kinetic schemas to be defined[3,4]. In rare gas mixtures, the emissions of the heaviest gas are observed even when it is only present in the form of traces. Gedanken[5] reports that molecule atom transfers are very efficient when the atomic state has an energy comparable to that of the transition of the molecular state-use is made of this phenomenon in halogen lasers where the medium is composed of three elements favouring the creation of radiative species with respect to de-excitation processes through collision. We have already carried out very detailed analyses of the Kr-Xe and Ar-Xe systems, the results presented here concern argon-krypton mixtures.

EXPERIMENTAL SET-UP

The equipment was designed to enable the spectroscopic analysis and the kinetic study of mixtures at total pressures of a few hundred torr and at Kr concentrations between a few ppm and several percent. Excitation of the gaseous medium was obtained with a source of alpha particles (^{241}Am). In the experiments the photomultipliers used for the detection were set to detect single photons. The emission spectrum was obtained by counting the anode pulses of a solar blind P.M. (EMR 510 F 09 13) situated behind a MacPherson 218 monochromator. The kinetic study was carried out on the main emissions of the mixture and was based on the reconstitution of the shapes of the light pulses by a time-amplitude conversion method. In this experiment, wavelength selection was obtained with MATRA interference filters. The light pulse shapes, which represent the change with time of the density of the excited species, were numerically analysed and their time constants calculated by a least squares method.

SPECTROSCOPIC ANALYSIS

The spectra of the argon-krypton mixtures were plotted at two different total pressures (100 and 600 torr) over a wavelength range of 100 to 300 nm. The spectrum of pure argon is characterized by the emission of the second continuum at 127 nm which has a mid-height width of 90 Å. As krypton is added to the argon, the spectrum is seen to be greatly modified both by the Kr concentration and the total pressure.

At p_T = 100 torr changes occur as soon as the Kr concentration reaches about 1 000 ppm. Several phenomena are observed simultaneously (Fig. 1):

- decrease of the 127 nm argon continuum,
- appearance at 123.6 nm of the resonance line arising from the $^3P_1$ state of kr,
- appearance at 116.5 nm of the line from the $^1P_1$ state of Kr.

These phenomena become more intense as the Kr concentration increases, at concentrations of the order of 1%, the Ar_2^* continuum at

127 nm completely disappears and only the resonance lines remain. At higher concentrations the 127 nm Kr continuum appears; as the level of Kr reaches 50% the spectrum is similar to that observed in pure Kr, the amplitudes and the mid-height widths are of the same order of magnitude.

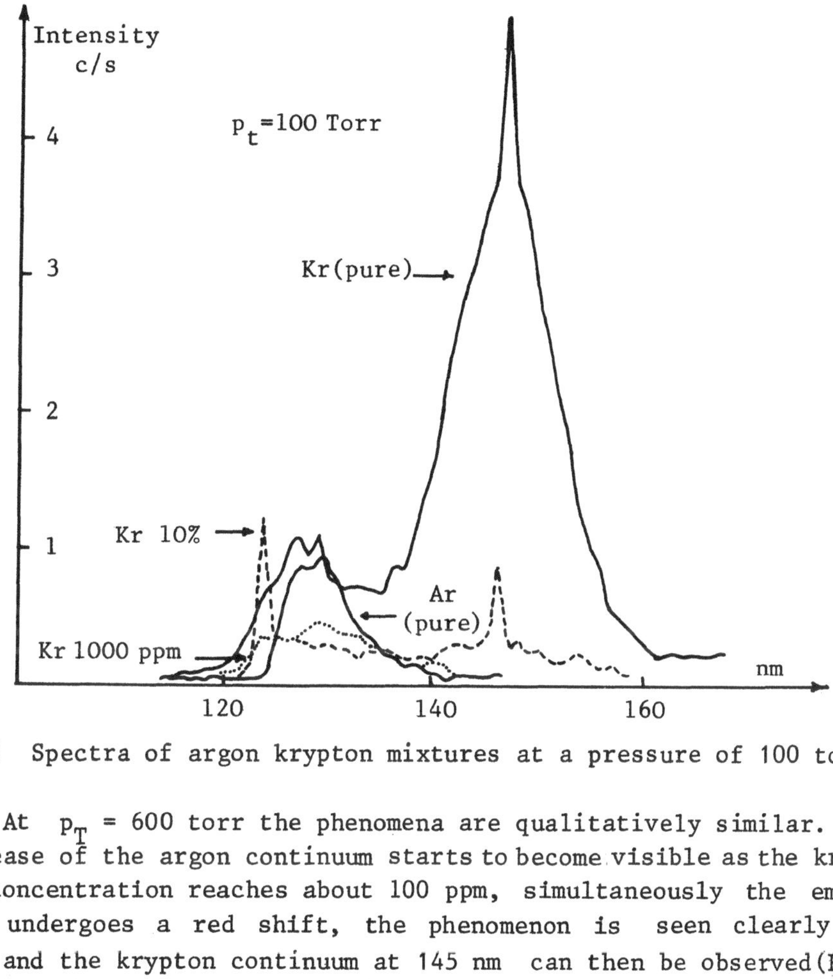

Fig.1 Spectra of argon krypton mixtures at a pressure of 100 torr.

At p_T = 600 torr the phenomena are qualitatively similar. The decrease of the argon continuum starts to become visible as the krypton concentration reaches about 100 ppm, simultaneously the emission undergoes a red shift, the phenomenon is seen clearly at 0.5% and the krypton continuum at 145 nm can then be observed (Fig. 2). When the Kr concentration is greater than 10% only the emission of the second Kr continuum is observed.

Spectroscopic analysis already allows a certain number of hypotheses to be put forward which can be checked later with the kinetic study:

- the disappearance of the Ar_2^* continuum demonstrates the existence of very efficient transfer processes to krypton which could occur from the radiative molecular states or from the precursor states.

- the appearance of molecular Kr emissions, even though the partial pressures of the gas are still relatively low, indicate that heteronuclear type collision processes lead to the formation of excited Kr_2^* molecular states.

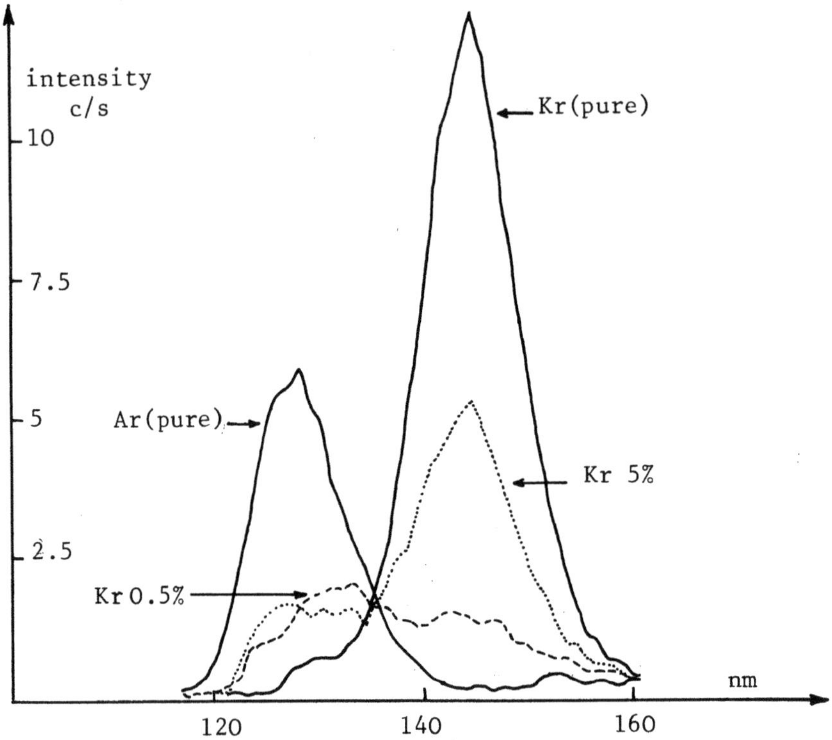

Fig.2 Spectra of argon krypton mixtures at a pressure of 600 torr.

KINETIC ANALYSIS

a) At λ = 127 nm

The wavelength range corresponding to the Ar_2 continuum was selected using a filter centered at λ = 128 nm ($\Delta\lambda$ = 10.5 nm). In pure argon the shapes of the light pulses are described by a law involving three exponential terms. Several measurements were carried out keeping the pressure of one of the gases in the mixture constant in order to study the role of Kr on each of the time constants. Figure 3 gives an example of experimental histograms obtained at 700 torr.

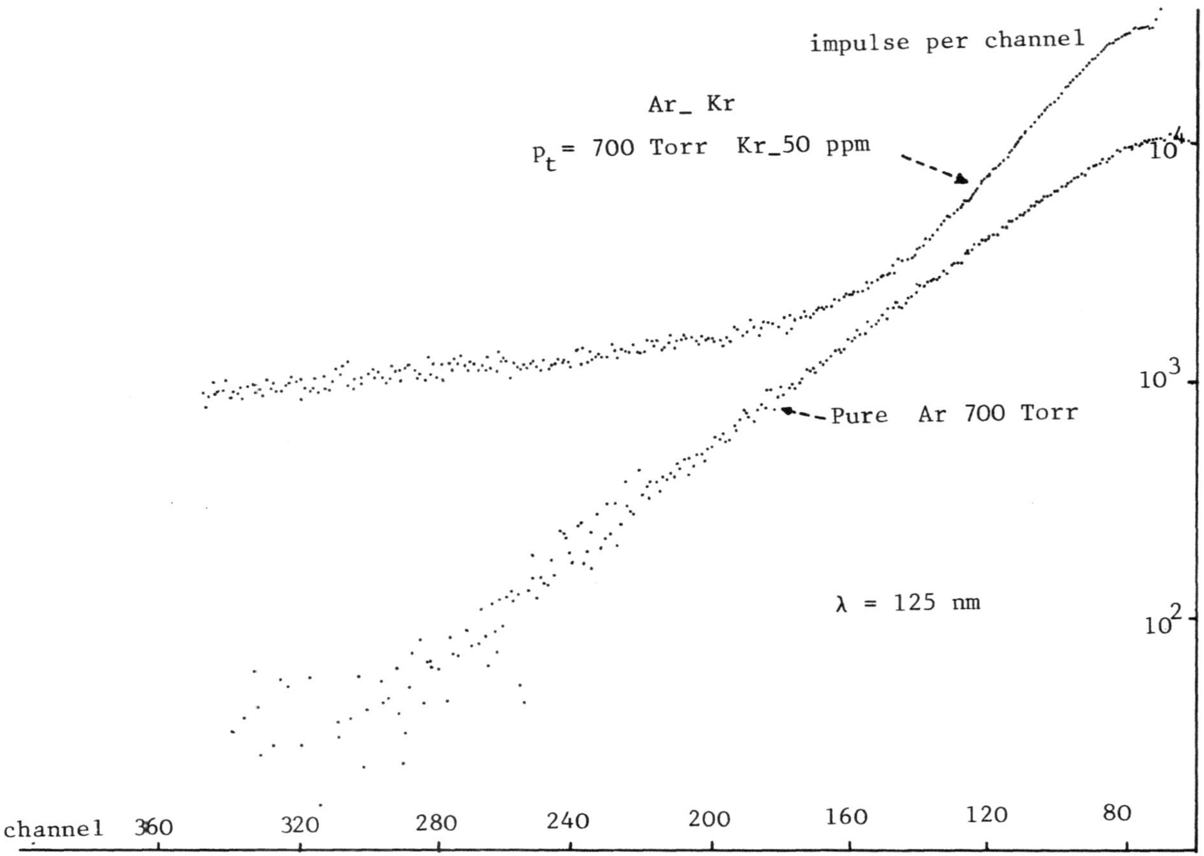

Fig.3 Comparison of the pulse shape obtained at λ = 127 nm for pure argon at 700 torr and for a mixture containing 50 ppm krypton (pt = 700 torr).

As soon as 50 ppm Kr is added, two phenomena occur: the time constant representative of the life time of the Ar_2^* state decreases rapidly and simultaneously an additional time constant, which cannot be correlated to argon, appears: this constant represents a krypton emission which occurs in the same wavelength range.

Complete analysis of the variation of the time constants with the partial pressure allows the various processes to be defined.

- The action of Kr on the time constant representing the life time of excited argon (Ar_2^*) is very efficient:

$$1/\tau_2' = kp_{Kr} + 1/\tau_2$$

$1/\tau_2$ represents the life time of the molecular species $^3\Sigma_u^+$ of Ar_2^* ($\tau_2 \cong 2.8$ μs). k is obtained from the slopes of the plots in Fig.4, $k = 1.1 \times 10^7$ torr^{-1}s^{-1}, the corresponding efficient cross section of the collision process $\sigma = 80$ Å^2.

- The action of Kr on the time constant related to the density changes with time of the $^3P_1$ and $^3P_2$ Ar atomic species is not significant. Unlike the results found in the other mixtures therefore, it would seem that the transfer process onto Kr only involve the Ar_2^* radiative molecular state.

The new time constant which appears for the higher Kr pressures can be correlated to either the $^3P_1$ state or the $^3P_2$ state of this gas. The results obtained at various pressures of the two gases allow us to opt for the metastable state; the findings are confirmed by the kinetic analysis at 145 nm.

b) <u>At λ = 145 nm</u>

These measurements were carried out using a filter centered at 145 nm ($\Delta\lambda$ = 17.5 nm) and mainly concern the emission of the second krypton continuum. Two partial pressure ranges can be considered.

- For the higher Kr partial pressures it is principally the action of argon on the constants related to the modification of the Kr continuum which is observed. In effect, in this range the transfer process only plays a minor role since the excitation of argon is less probable. At krypton pressures fixed at 500 and 600 torr it was checked that the molecular species

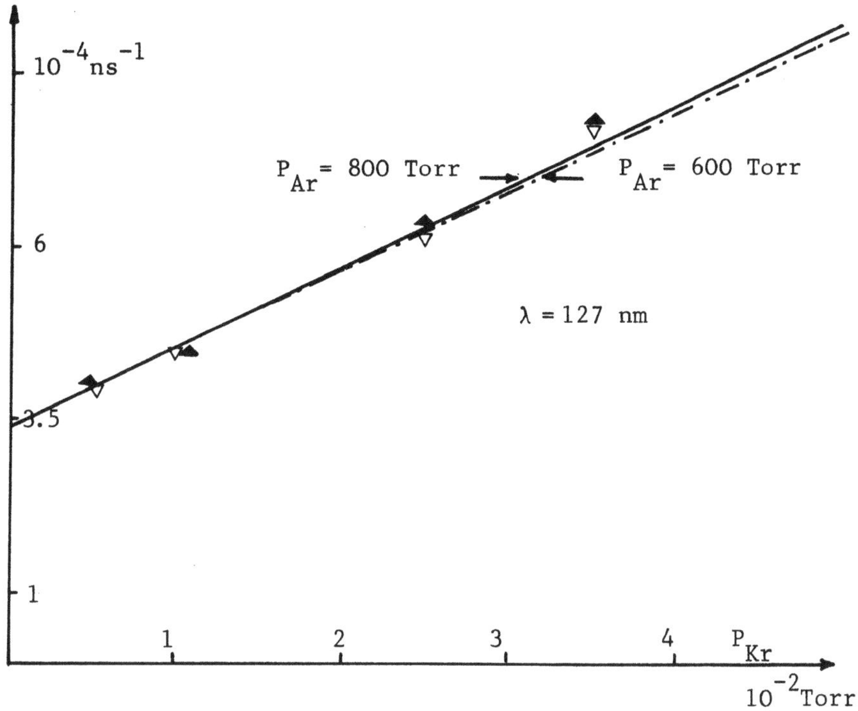

Fig.4 Linear variation of $1/\tau_2$ with the pressure of krypton. τ_2 represents the life time of the molecular species $Ar_2^*(^3\Sigma_u^+)$.

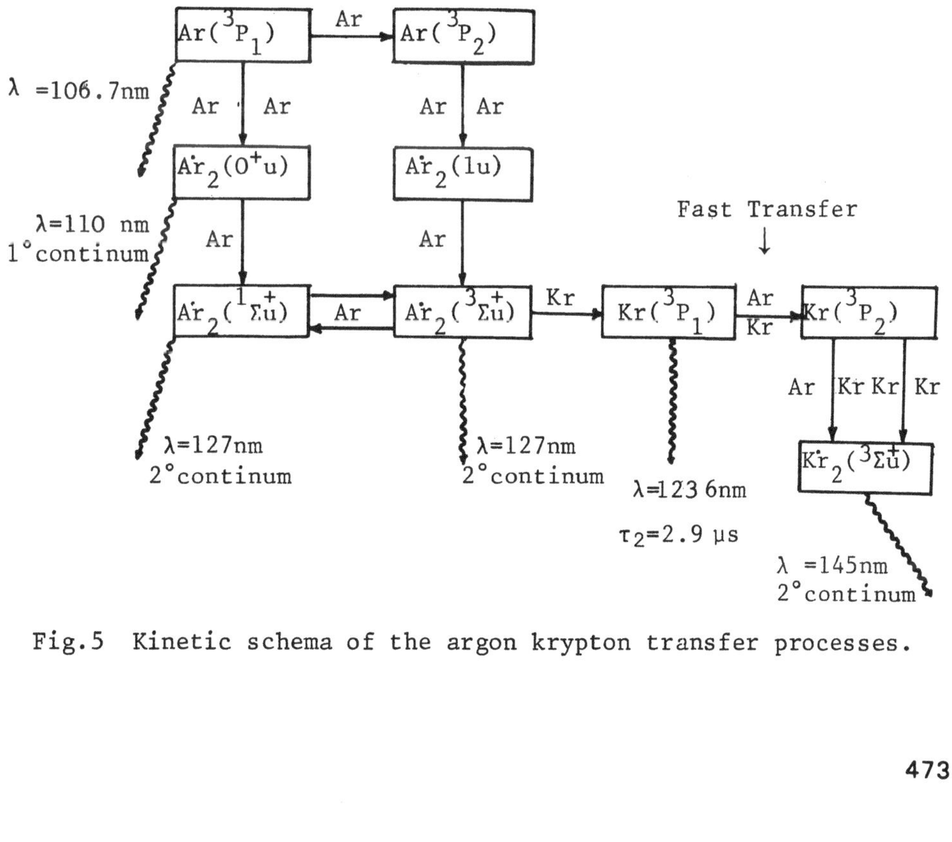

Fig.5 Kinetic schema of the argon krypton transfer processes.

time constants were not significantly modified by the argon pressure.

- Over the range of low Kr partial pressures, the Ar-Kr transfer phenomenon must increase in importance. This can be checked ed by comparing the shapes of the light pulses obtained at 127 and 145 nm in a mixture at 700 torr Ar and 50 torr Kr: the results are similar. The emission of the second Kr_2^* continuum is linked to processes involving excited species which also emit the second Ar_2^* continuum. In this concentration range it was possible to define the law governing the variation of constant τ_3 with the concentration of $^3P_2$ states of krypton. The $^3P_2$ states are mainly destroyed through heteronuclear collisions to create molecular Kr_2^* species: the constant of the process is $k = 12\ torr^{-2}s^{-1}$.

Finally, the kinetic schema can be established according to the diagram represented in Figure 5. Like Gedanken[5], we think that the disappearance of the Ar_2^* states leads to the preferential formation of the $^3P_1$ state. The $^3P_1 \rightarrow {}^3P_2$ transfer must be rapid since it escapes detection and also the formation of the molecular species involves a heteronuclear process.

REFERENCES

1. Lorents, D.C., Physica (Utrecht) 82, 19 (1976)
2. Tanaka, Y., J.Opt.Soc.Am., 45, 344 (1955)
3. Chen, C.H.; Judish, J.P.; Payne, M.G., J.Phys.B, 11, 2189 (1978)
4. Galy, J.; Birot, A.; Brunet, H.; Dijols, H.; Millet, P.; Salamero, Y., Journal de Physique, 41, C9-881 (1980)
5. Gedanken, A.; Jortner, J.; Raz, B.; Szöke, A., J.Chem.Phys., 57, 3456, (1972)

CW ION-ARGON-LASERS WITH THE SUPERSONIC TRANSVERSE GAS FLOW

V.I. Donin and G.N. Alferov

Institute of Automation and Electrometry
Siberian Branch of the USSR Academy of Sciences
Novosibirsk 630090, USSR

INTRODUCTION

Investigations of plasmas of powerful continuous-wave ion-argon-lasers[1] show that as the pumping current J in large-current discharges of reduced pressure grows, the number of neutral atoms N_a decreases. The decrease is caused by plasma heating and atomic ionization in the discharge. This decrease is so strong that it has a great influence on behaviour of all plasma characteristics. In particular, the electron number n_e stops to increase with J due to the reduction of N_a for sufficiently high discharge currents, and the generation power W on the ArII transitions reaches its maximum (saturates) in J. Instabilities resulting in the destruction of a tube develop in the discharge if the discharge currents are close to these required for the generation saturation[1]. These instabilities can also be caused by the decrease of N_a on the discharge axis[2,3]. In discharge tubes with the diameter >20 mm, these instabilities occur (when J increases) before the development of the generation maximum on the ArII transitions and prevent the achievement of the saturation regime of W in pumping current in such tubes. The current saturation of generation on the ArIII transitions in continuous regime was not observed at all, because it should occur for higher J which can not be achieved due to the development of the mentioned instabilities.

Earlier[1] an idea has been suggested that it was advantageous to use the supersonic gas pumping across the tube of a large-current ion laser since it could enhance the flow of atoms into the discharge due to the increase of flow velocity without growth of N_a (i.e. without reduction of the optimal electron temperature). An experimental realization of the laser with supersonic "injection" of atoms into the discharge has demonstrated[4] a possibility of the considerable rise of ion laser output parameters, in particular, on the UV ArIII transitions. In this paper the results of experimental investigations of some plasma parameters and output characteristics of the ion lasers with the supersonic injection of atoms into the discharge are presented. The influence of the supersonic pumping on the plasma instabilities, radial variations of plasma spectral parameters, and the output power of the lasers is considered.

EXPERIMENTAL RESULTS

The investigations were carried out with the large-current construction of the discharge tube of the 16 cm active length and the 7 mm diameter. The active length consisted of water-cooled aluminum sections covered with an Al_2O_3 film. Each of these sections in the direction perpendicularly to the discharge axis had a slit supersonic nozzle on its one side and the pumping-out hole on its opposite side. With the aid of the supersonic nozzles and a vacuum pump of the 120 l/s power, an optically homogeneous argon flow with the Mach number $M \simeq 3$ could have been formed in the active discharge zone. The excitation of the longitudinal continuous discharge was produced with the aid of the electrode system that has been used earlier in high-power ion lasers without the supersonic gas pumping[5]. All the results presented below were obtained at the initial (i.e. before the discharge ignition) concentration of argon atoms $N_a \simeq 1.7 \cdot 10^{16}$ cm^{-3}. The control of J was carried out with the help of the ballast resistor.

In Fig. 1 the scheme of the measurement of plasma spectral characteristics is shown. The spontaneous emission passed through diaphragms 2 of 0.6 mm diameter, an interferometer with lenses, and a monochromator was registered with the aid of a photomultiplier and an XY-recorder. The interferometer scanning was carried out with the help of a piezoceramic fed from a voltage generator. By using mirrors 9-13, the lateral radiation from the tube could have been fed to the interferometer. To extract this radiation there were two opposite lateral quarz windows in one of the sections. Such arrangement permited to record the spectral line contours in the directions along and opposite with respect to the gas flow. The doppler shift of atomic lines enabled one to calculate the gas flow velocity. It proved to be $(4.5 \pm 0.4) \cdot 10^4$ cm/s which was close to the calculated value of $4.8 \cdot 10^4$ cm/s. By removing the mirror 13, one was able to record the spectral line contours along the discharge axis. In Figs. 2,3 the atomic and ionic temperatures, which were found from the line width, are presented. Removing the interferometer and shifting the discharge tube in parallels with its axis with the help of a special stage, one was able to record the radial distribution of the spectral line intensities (Figs. 4,5). With the aid of high-reflectivity dielectric mirrors of the optical resonator, the employed discharge tube was able to operate in the generator

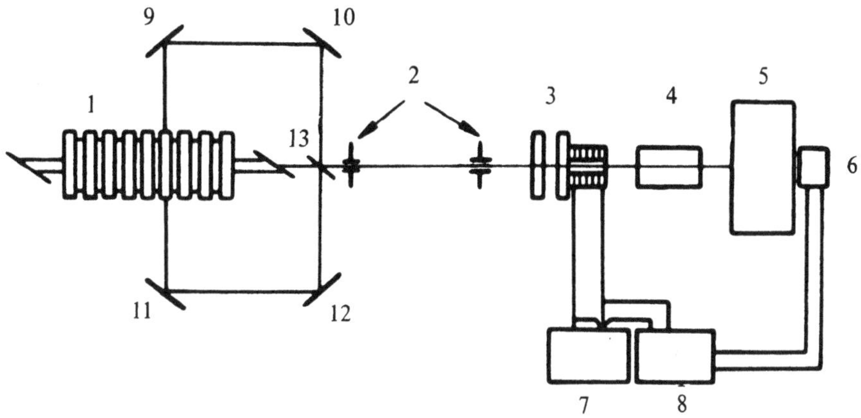

Fig. 1. Setup for the measurement of spectral characteristics. 1 - discharge tube, 2 - diaphragms, 3 - scanned Fabry-Perot interferometer, 4 - condenser, 5 - monochromator, 6 - photomultiplier, 7 - voltage generator, 8 - recorder, 9-13 - turning mirrors.

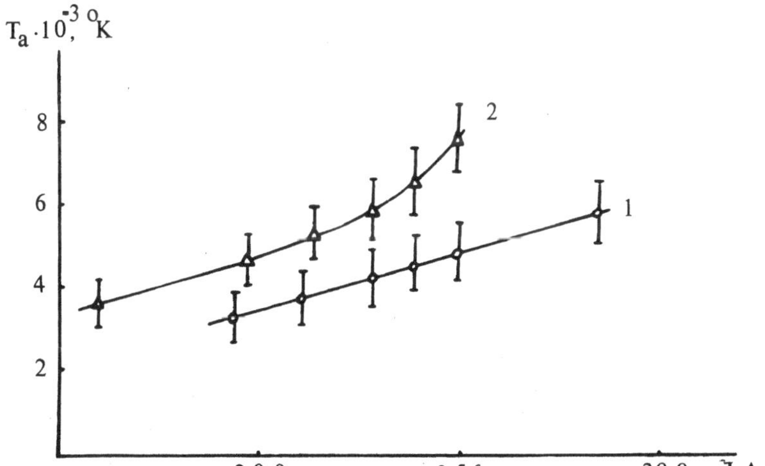

Fig. 2. Values of the atom temperature T_a from the ArI 4198 Å line on the discharge axis as a function of the discharge current J. Curve 1 - with the pumping (TP), 2 - without the pumping (NP).

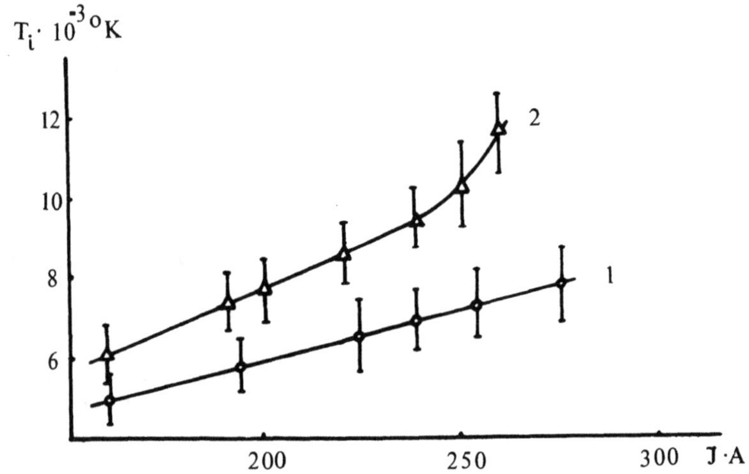

Fig. 3. Values of the ion temperature T_i from the ArII 4880 Å line on the discharge axis as function of J. 1 - TP, 2 - NP.

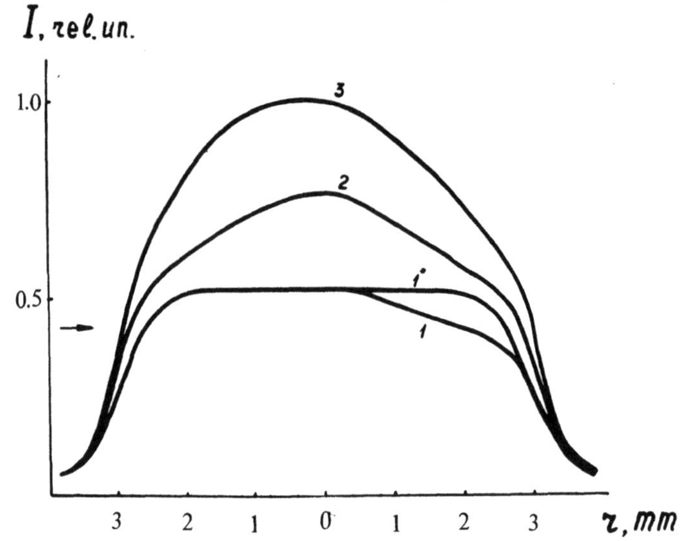

Fig. 4. Spontaneous-emission intensity I for the ArI 4198 Å line as a function of the distance to the tube axis r. 1, 2 and 3 correspond to J=150, 180, and 210 A TP (the arrow indicates the direction of the gas flow); 1* corresponds to J=150 A NP.

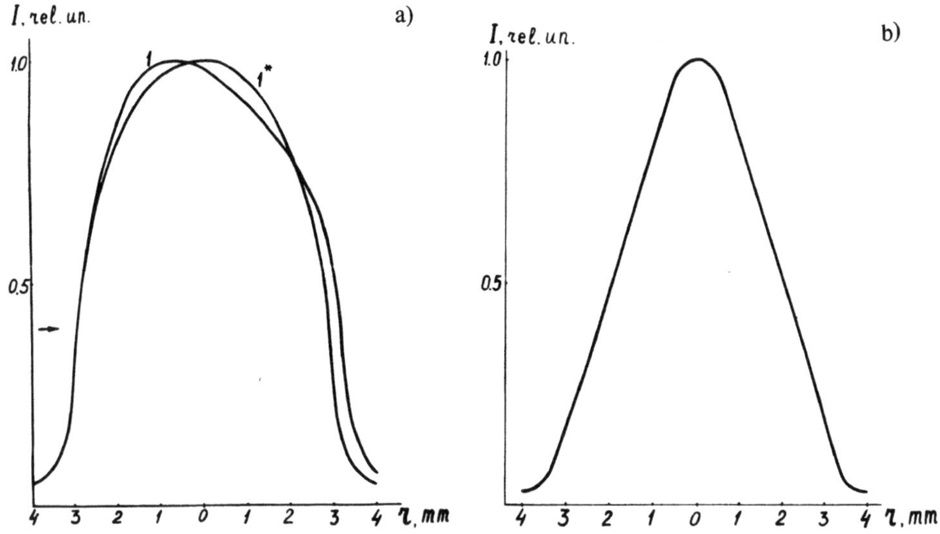

Fig. 5. Spontaneous-emission intensity I for the ion lines as a function of r by J=200 A.
a - the ArII 4880 Å line (1 - TP, 1* - NP); b - the ArIII 3638 Å line TP.

Fig. 6. Dependence of the generation power W for the ArIII lines (3511 and 3638 Å) on J. 1 - TP, 2 - NP (the arrow marks onset of high-frequence plasma instabilities).

mode. The dependence W on J is displayed in Fig. 6. The plasma discharge instabilities were recorded with the spectrum analyzer that was fed by AC component of the potential difference between two tube sections. In the absence of pumping two characteristic frequency bands appeared, namely, the low-frequency one at ≈270 kHz and high-frequence one at ≈15 MHz. The low-frequency oscillations are non-threshold. The high-frequence oscillations possess sharply pronounced threshold in the current. For instance, the exceeding of the threshold current by a few per cents provides the increase of the high-frequence oscillation amplitude by two-three orders of magnitude. In the latter case the oscillation spectrum becomes noise-like in the all recorded band (up to 50 MHz) and the walls of the discharge tube are being quickly demolished. In the supersonic pumping discharge no high-frequence oscillations were observed.

DISCUSSION OF THE RESULTS

The transverse supersonic pumping increases the flow of neutral atoms into the discharge. Therefore, it permits one to eliminate to some extent the sharp decrease of N_a in a vicinity of the discharge axis that takes place with increasing J in the discharges without the pumping. The elimination of the sharp drop of N_a in the discharge has to change behaviour of all plasma characteristics. In particular, the consideration V•A - characteristics of the discharge with the pumping indicates[4] a monotonous change of n_e over the whole realized range of J. Therefore, taking into account the high ionization degree on the discharge axis and the principle of conservation of the total number of heavy particles, one can expect an essential increase of multiple ionization of atoms. The latter fact is especially important from the point of view of further increasing the output power on the transitions of multy-ionized atoms. One can see from Fig. 6 that the use of the supersonic pumping allows one to produce, at least, the 6-fold increase of the specific UV output power on the ArIII ions. As the detailed analysis shows[3,6], the high-frequency instabilities which destroy the tube are coupled with the excitation of an ionic-acoustic instability in central part of the discharge. The elimination of the sharp decrease of N_a in a vicinity of the discharge axis by the pumping leads to suppression of this instability. Besides the transverse pumping produces a conventional pulling-down of plasma instabilities. The latter factor can explain the observed experimentally drastic (~100 times) fall of the low-frequence instability amplitude in comparison with the case without the pumping at the same N_a and J. The low-frequence instabilities are apparently coupled with drift oscillations of ions in the own magnetic field of the discharge[6].

The results of Figs. 2,3 show the temperatures of atoms and ions are noticeably decreased in the discharge with the pumping. This indicates an appreciable role of the flow in cooling of the heavy particles. The noticeable asimmetry of the ArI, ArII lines in Figs. 4,5a can be explained by heterogeneity along the flow of the gas temperature, N_a, n_e (and possibly T_e too). The enhancement of the ArI line intensity in the tube centre indicates[1], in indirect way, an increase of T_e with J in the discharge with the pumping. More definite conclusions on the quantitative changes of the electrical characteristics of the plasma with the supersonic flow and on physical phenomena happening there can be made only from direct measurements of the electrical parameters of the plasma.

In conclusion one can emphasize an interesting result of Fig. 5b. As distinct from the ArI and ArII lines, the radial distribution of the ArIII line within the experimental errors prooved to be simmetrical. This fact is important from the practical view-point since it allows one to exclude influence of the radial distribution asimmetry of the plasma parameters when calculating the output beam characteristics in powerful UV lasers on multiplex ions with the supersonic pumping.

REFERENCES

1. V.I.Donin, "Output Power Saturation with a Discharge Current in Powerful Continuous Argon Laser", Sov. Phys. JETP, 35:858 (1972).
2. J.E.Allen and P.C.Thonemann, "Current Limitation in the Low-Pressure Mercury Arc", Proc. Phys. Soc., B67:768 (1954).
3. G.N.Alferov, V.I.Donin, G.I.Smirnov and D.A.Shapiro, "On Ion Laser Plasma Instability", Kvant. Electr., 8:13 (1981).
4. G.N.Alferov and V.I.Donin, "Ion Laser with the Supersonic Gas Pumping", Izvest. AN USSR, ser. fiz., 44:2079 (1980).
5. V.I.Donin, A.F.Shipilov and V.A.Grigoriev, "High-Power CW Ion Lasers with Longer Service Life", Kvant. Electr., 6:359 (1979).
6. A.A.Apolonskiĭ, V.I.Donin, V.P.Drachov, and D.A.Shapiro, "On Spectra of the Ion Laser Plasma Oscillators", All-Union III Conf. "Optics of Lasers", Leningrad (1982).

ROVIBRATIONAL STATE POPULATION DISTRIBUTIONS CALCULATED FROM SPONTANEOUS FLUORESCENCE MEASUREMENTS FOR CO ($v \lesssim 4$, $J \lesssim 10$) IN HIGHLY HEATED SUPERSONIC FREE JETS OF CO IN N_2 AND Ar

M. A. Gaveau, J. Rousseau, A. Lebéhot, and R. Campargue

Laboratoire des Jets Moléculaires
Département de Physico-Chimie, CEN Saclay, France

INTRODUCTION

Internal energy lags are predicted in free jet expansions of polyatomic molecules, due both to the continuous decrease of the collision frequency as the gas expands in the jet, and to the very different collision numbers required for relaxation of the various modes of motion (much larger for vibrations than for rotations, and larger for rotations than for translations). Consequently, the successive freezings of the vibrational, rotational and translational temperatures ($T_{vib} \gg T_{rot} > T_{trans}$) are expected along the flow axis. This is only a very simplified description because one single rotational or vibrational temperature cannot characterize a non-Boltzmann distribution of the molecules on the internal states, as observed experimentally in a number of laboratories.

During the last few years, the spatial distributions of vibrational-rotational state populations have been measured in free jets for a variety of gases : N_2, CO_2, CO, NO, H_2, D_2, CH_4, C_2H_2, SF_6, Na_2, etc. As reported in our recent paper[1] (see references herein), different spectroscopic techniques have been used to provide insight into the internal energy states of the free jet molecules : electron beam induced fluorescence, one-photon laser induced fluorescence, spontaneous fluorescence, spontaneous Raman scattering, coherent anti-Stokes Raman spectroscopy (CARS), multiphoton induced fluorescence or ionization (MPI). Also many diagnostics have been made on molecular beams skimmed from free jets, in particular by using the time-of-flight, bolometric, or MPI techniques.

Many discrepancies between existing experimental data are not resolved. This is due to the difficulties in finding an appropriate model of the excitation-emission process, for relating the measured intensity distribution (for instance in the rotational structure of the fluorescence spectra) to the initial population distribution. Also, complications may result from additional effects associated for example to collisions (quenching), self-absorption, secondary electrons, laser power, etc.

This fundamental problem of internal energy relaxation in free jets, is very important in the fields of gas dynamic or chemical lasers, molecule or isotope separation, molecular spectroscopy, molecular beam applications, etc. Contrary to the results reported previously on this topic, this paper deals with highly heated free jets[2] produced with stagnation conditions similar to those used in a gas dynamic laser, i.e. $T_o \simeq 2000$ K, $P_o \simeq 10$-30 bars, for 20 % to 50 % CO in nitrogen or argon. Such a mixing, in N_2 or Ar, offers the possibility of varying the rapidity of the expansion.

EXPERIMENTAL

The heated free jets are produced by means of a heating furnace for bringing a high pressure circulating gas stream ($\lesssim 50$ bars) to a very high temperature ($\lesssim 3000$ K). Heating is obtained by the Joule effect through small tubular ducts driving the flow and discharging the hot gases just upstream of a nozzle throat. Such a device[3] makes it possible to produce supersonic expansions from a very high stagnation enthalpy achieved in a thermodynamic equilibrium state.

In the present work, the measurements are performed on free jets of 20 % CO in N_2, or Ar, produced through a sonic nozzle ($D^* = 0.22$ mm in diameter) operated at stagnation pressure $P_o \simeq 30$ bars and temperatures T_o close to 1800 K and 2500 K, for the N_2-CO and Ar-CO mixtures, respectively. In these conditions, the background pressure, $P_1 \simeq 0.15$ Torr, is maintained in the nozzle chamber by means of mechanical Roots pumps. Consequently, the Mach disk is located at about 58 mm from the nozzle throat and the supersonic flow inside the shock barrel is similar to that expanding into a perfect vacuum (zone of silence), at least in the first stages of the jet, now investigated ($X/D^* \lesssim 36$).

The rovibrational relaxation of CO, mixed and expanded with N_2 or Ar in such heated free jets, is studied by measuring infrared fundamental ($\Delta v = 1$) spontaneous fluorescence emission of the CO jet molecules (4.6 to 5 μm). The radiation is collected by an optical system described previously,[1] mechanically chopped at 1000 Hz and spectrally resolved with a scanning grating monochromator. The spectral bandwidth is $\Delta\sigma = 0.5$ cm^{-1} and the spatial resolution corresponds to $\Delta X = 0.2$ mm on the jet axis. The radiation is detected by a liquid nitrogen cooled InSb photovoltaic detector. Then the signal is enhanced through a lock-in amplifier and finally displayed on a (X,T) chart recorder. The position of the nozzle can be adjusted in order to collect the IR emission of the jet at different distances X/D^* from the throat.

RESULTS AND DISCUSSION

Experimental Data

Fluorescence spectra have been recorded for the (v,v-1) systems going from (1,0) to (4,3) at locations $X/D^* = 18$ and 36 in the N_2-CO free jets, and $X/D^* = 18$ in the Ar-CO free jets. In the plots of $\mathrm{Log}[I_{J',J}^{v',v}/(\nu_{J',J}^{v',v})^4 S_{J',J}]$ versus $J'(J'+1)$ (with $I_{J',J}^{v',v}$ = intensity of the $(v',J') \to (v,J)$ transition, $\nu_{J',J}^{v',v}$ = frequency of the transition, $S_{J'J}$ = transition matrix element for rotation) a Boltzmann rotational distribution appears as a straight line, the slope of which is $-B_{v'}/kT_R$ (with $B_{v'}$ = rotational constant for the v' vibrational level). Nevertheless, it is possible to obtain deviations which are not due to nonequilibrium effects, but result for instance from self-absorption phenomena, nonhomogeneity in temperature and density, etc. The departures from linearity, observed in the present results, seem to be due mainly to emission and self-absorption phenomena occurring in the free jet as well as in the background gas. The spontaneous fluorescence of the background gas has been measured by recording the spectrum emitted only from the gas surrounding a N_2-CO free jet. In this case, from the intensity ratio of the (1,0) and (2,1) bands, the vibrational temperature was $T_{BG}^{V} \simeq 1000$ K ; the rotational temperature deduced from the slope of the curve was $T_{BG}^{R} \simeq 300$ K.

Table 1. Rotational and Vibrational Temperatures in Free Jet and Background Gas, Measured for 20 % CO in N_2, or Ar, in Different Experiments Performed with $P_o = 30$ bars.

Gas Mixture	T_o	X/D^*	T_{FJ}^{R}	$T_{FJ}^{V}(4,3)$	T_{BG}^{R}	T_{BG}^{V}
N_2-CO (Fig. 1)	1800 K	18	70 K	1600 K	300 K	1000 K
N_2-CO (Fig. 2)	1800 K	36	45 K	1500 K	380 K	1000 K
Ar-CO (Fig. 3)	2500 K	18	15 K	2000 K	450 K	1200 K

From the experimental results shown in Figs. 1,2 and 3, three main features can be pointed out :

(i) the data concerning the (1,0) and (2,1) bands for $J'(J'+1) \gtrsim 100$, i.e. $J' \gtrsim 10$ (or $J' \gtrsim 7$ in Fig.3) appear to be due to the fluorescence of the background gas, with negligible contribution from the jet,

(ii) the data concerning the (3,2) and (4,3) bands seem to characterize a pure free jet emission (i.e. without any self-absorption produced by the jet molecules, or the background gas) as indicated by the linearity of the plots showing a Boltzmann rotational equilibrium, at least for the low J numbers, with a rotational temperature T^R_{FJ} given by the slope of the straight lines. Also a two-level (4,3) vibrational temperature T^V_{FJ} (4,3) can be deduced from the ratio $I^{4,3}_{J-1,J}/I^{3,2}_{J-1,J}$

(iii) the data concerning the (1,0) and (2,1) bands, for $J' \lesssim 10$, result from the superimposition of the emission and the self-absorption occurring in the free jet and the background gas.

The various temperatures defined above and determined for three different experiments are reported in Table I.

Model of the Emission Self-Absorption Process

The entire spectrum recorded for CO can be calculated by taking into account both emission and absorption in the free jet and the background gas. The intensity of the radiation is integrated over the successive media encountered by an optical path, using absorption and emission constants deduced from Einstein coefficients for each transition.[4] Thus, a good agreement with the experimental data has been found by assuming Boltzmann vibrational and rotational distributions in the free jet and the background gas (Figs. 1,2 and 3). Nevertheless, it cannot be asserted that this model is the only way for interpretation.

The calculation is performed by using the temperatures deduced from the experimental data (Table I). The vibrational temperature T^V_{FJ} is taken as equal to T^V_{FJ} (4,3). The local number density, n, introduced in the calculation, is obtained by reporting in the classical gas dynamic equations the stagnation conditions (n_o deduced from P_o and T_o) and the Mach number distribution $\mathcal{M}(X/D^*, \gamma)$ of Ashkenas and Sherman[5] (with γ = specific heat ratio). Because the temperature is not measurable in the heated nozzle, T_o has been evaluated from the measured value of T^R_{FJ} by using a flow model with thermodynamic equilibrium upstream of the nozzle throat ($T_{vib} = T_{rot} = T_{trans}$) and only vibrational freezing dowstream of the throat.

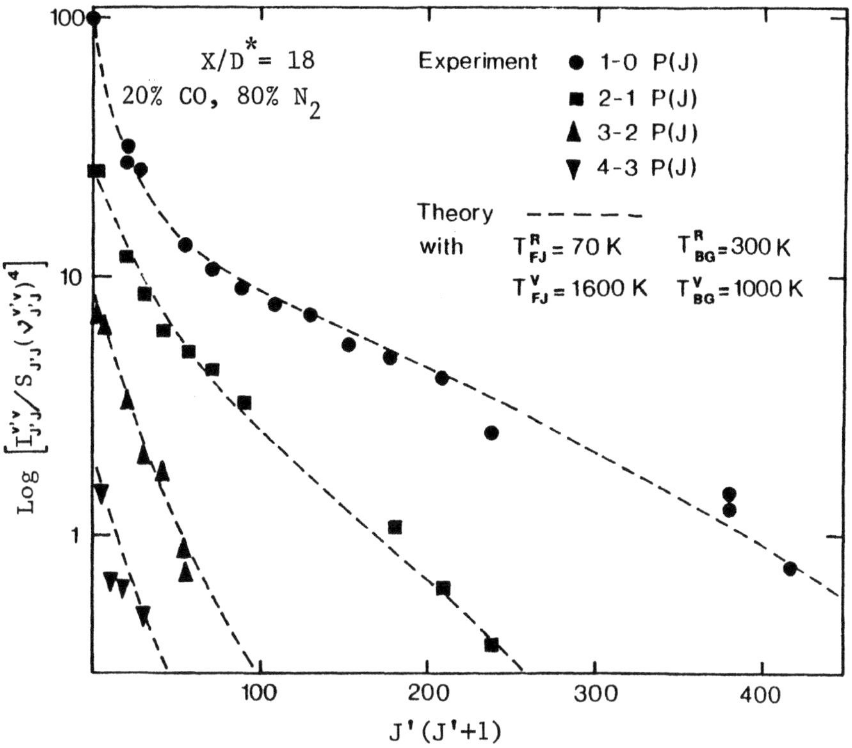

Fig. 1. Measured and calculated line intensities vs J'(J'/1) for N_2-CO free jet (P_o = 30 bars, $T_o \simeq$ 1800 K, X/D* = 18).

With this model used for 20% CO in N_2 (Figs. 1 and 2) the effective values of γ upstream and downstream of the throat are 9/7 and 7/5, respectively. In these conditions, it is found, from the rotational temperature T^R_{FJ} : $T_O \simeq$ 1800 K and $T^* \simeq$ 1580 K at the nozzle throat. Furthermore, it is seen that $T^* = T^V_{FJ}$ and this is consistent with the vibrational freezing defined above for the model. It should be emphasized that the two temperatures $T^R_{FJ} \simeq$ 70 K and 45 K, measured at X/D* = 18 and 36, respectively, lead to the same stagnation temperature $T_o \simeq$ 1800 K.

The interpretation is not so simple for the results concerning 20% CO in Ar (Fig. 3). With the modelled values of γ for this mixtures (1.53 and 1.59 upstream and downstream of the nozzle throat, respectively) the stagnation temperature T_0 deduced from $T^R_{FJ} \simeq$ 15 K (measured at X/D* = 18) seems to be too low for obtaining a vibrational temperature $T^* = T^V_{FJ}$ = 2000 K. This vibrational temperature would be consistent with T_0 = 2500 K and $T^R_{FJ} \simeq$ 25 K at X/D* = 18, instead of T^R_{FJ} = 15 K measured for the mixture and achieved theoretically, at the same X/D* , only for

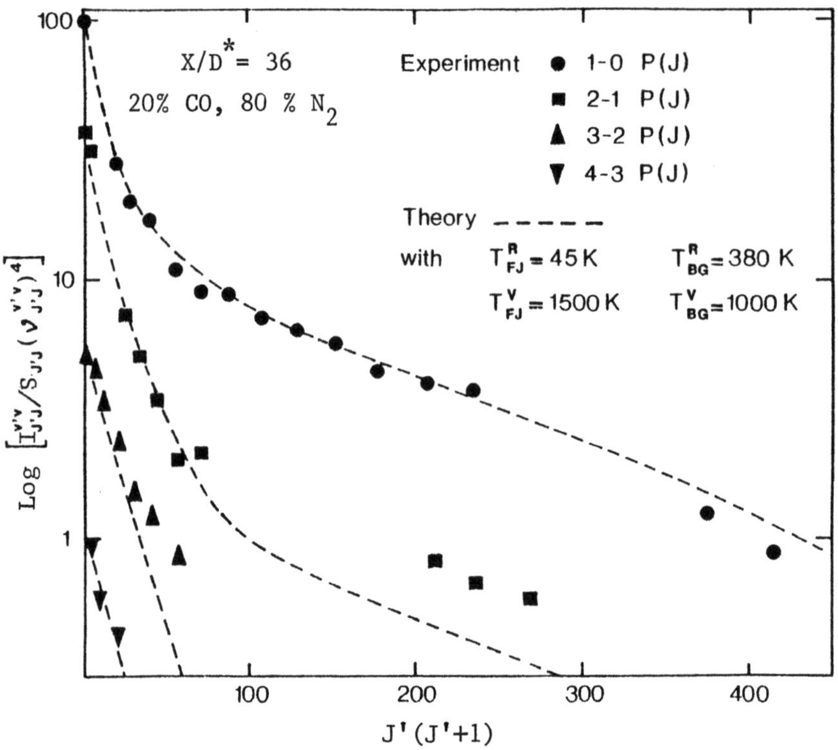

Fig. 2. Same caption as for Fig. 1 for $X/D^* = 36$

monatomic gases ($\gamma = 5/3$). In any case, the deduction from T^V_{FJ} are better than those from T^R_{FJ} and also in agreement with the characteristics of the heated nozzle operated with argon.

The line intensities computed by using the present model, in the same conditions as in the experiments, are reported as dashed lines in Figs. 1,2, and 3. A good agreement with the experimental data is obtained in the three cases.

These results could appear in disagreement with those obtained in supersonic nozzles,[6] where a vibrational temperature cannot be defined because the high vibrational levels are overpopulated compared to a Boltzmann distribution.[7] On the contrary, it seems that T^V_{FJ} is meaningful along the free jet (produced from a sonic nozzle). This is due to the fact that the expansion is then so fast that the V-V exchanges between excited CO molecules are not possible. Indeed, the number of collisions necessary for vibrational relaxation is of the order of a few thousands,[8] whilst only a few collisions per μsec occur in the free jet.

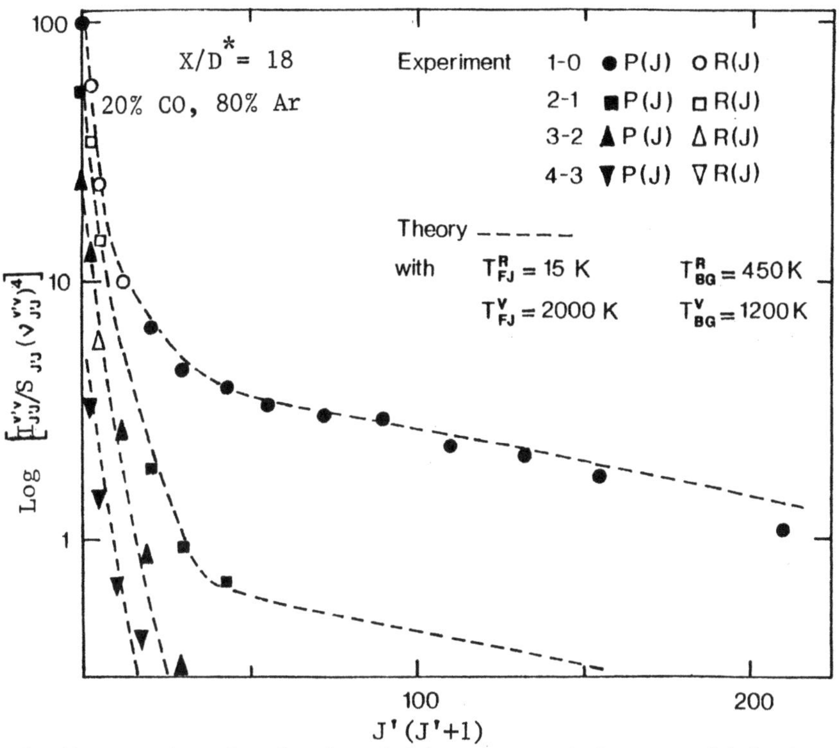

Fig. 3. Measured and calculated line intensities vs J'(J'+1) for Ar-CO free jet (P_o = 30 bars, T_o ≃ 2500 K, X/D* = 18)

CONCLUSION

The spontaneous fluorescence measured in free jets of 20% CO in N_2, or Ar, generated with stagnation conditions similar to those used in gas dynamic lasers (Table I), may be interpreted on the basis of a Boltzmann equilibrium for each internal mode, at least at distances X/D* $\lesssim$ 36 from the nozzle and for the lowest rotational levels. Due to the sudden expansion in the free jet, the vibrational energy is frozen and T^V_{FJ} remains nearly equal to the translational temperature at the nozzle throat. Since the expansion is more rapid with Ar than N_2, the lowest translational and rotational temperatures and the strongest nonequilibrium effects between vibrational and rotational modes are observed in the Ar-CO free jet, where, for the conditions used, positive gains have been calculated for the v+1, v P (J) lines with J $\geqslant$ 5. However, this phenomenon was not detected due to the absorption by the background gas. Also, it is possible that nonequilibrium distributions appear on higher rotational levels (J $\gtrsim$ 10) and/or at higher values of X/D* . Experiments with a better sensitivity are now under way for detecting such transitions.

REFERENCES

1. M. A. Gaveau, J. Rousseau, A. Lebéhot, R. Campargue, and J. P. Martin, Rovibrational state population distributions of CO (v $\lesssim$ 4, J $\lesssim$ 10) in highly heated supersonic free jets of $CO-N_2$ mixtures, 13th International Symposium on Rarefied Gas Dynamics, Novosibirsk, USSR (July 1982), to be published.
2. R. Campargue and A. Lebéhot, High intensity supersonic molecular beams with extremely narrow energy spreads in the 0-37 eV range, in "Rarefied Gas Dynamics," M. Becker and M. Fiebig, eds., DFVLR Press, Porz Wahn, West Germany (1974) ; R. Campargue, A. Lebéhot, J. C. Lemonnier and D. Marette, Measured, very narrow velocity distributions for heated, Xe and Ar-seeded nozzle-type molecular beams of He and H_2 skimmed from free jet zones of silence ; Xe energies up to 30 eV, in "Rarefied Gas Dynamics," S. S. Fisher, ed., AIAA, New York (1981).
3. R. Campargue, J. Bouffenie, and A. Recule, A furnace for heating a circulating gas stream and especially for producing molecular beams, Patent 2,179,045 filed July 1974 (France) and Foreign Patents : 4,063,067 (USA), etc ; R. Campargue, M. A. Gaveau, A. Lebéhot, J.C. Lemonnier, and D. Marette, High enthalpy generator ($P_0 \lesssim$ 50 bars, $T_0 \lesssim$ 3000 K) for supersonic free jets and nozzle beams, 7th International Symposium on Molecular Beams, Riva del Garda, Italy (May 1979).
4. S. S. Penner, "Quantitative Molecular Spectroscopy and Gas Emissivities," Addison-Wesley, Reading, London (1959).
5. H. Ashkenas and F. S. Sherman, The structure and utilization of supersonic free jets in low density wind tunnels, in "Rarefied Gas Dynamics," J. H. de Leeuw, ed., Academic Press, New York, London (1966).
6. J. P. Martin, Thesis "Etude de la relaxation vibrationnelle de CO en écoulement supersonique et derrière une onde de choc par la technique du laser sonde," Université Paris-Sud (1979) ; P. J. Bender, M. Mitchner, and C. H. Kruger, Measurement of vibrational population distributions in a supersonic expansion of carbon monoxide, Phys. Fluids 21 : 1073 (1978).
7. C. E. Treanor, J. W. Rich, and R. G. Rehm, Vibrational relaxation of anharmonic oscillators with exchange-dominated collisions, J. Chem. Phys. 48 : 1798 (1968) ; R. E. Center and G. E. Caledonia, Anharmonic effects on the rate of relaxation of vibrational energy in rapidly expanding flows, J. Chem. Phys. 57 : 3763 (1972).
8. V. F. Gavrikov, A. P. Dronov, V. K. Orlov, A. K. Piskunov, and V. L. Shikanov, Vibrational relaxation of carbon monoxide in supersonic streams, Sov. J. Quantum Electron., 6 : 803 (1976).

A XeF-EXCIMER LASER EXCITED IN A LONGITUDINAL DISCHARGE

T. Gerber, W. Lüthy, and R. Schmiele

Institut für angewandte Physik
Sidlerstrasse 5, CH-3012 Bern

Abstract

We report about the properties of a rare-gas-halide (RGH) laser excited in a longitudinal discharge. The upper laser level formation kinetics in RGH-lasers do not necessarily require high pressure. Therefore, also discharges in a quartz tube at pressures in the order of a few hundred mbar can provide RGH excimer densities that are high enough for laser action. Homogeneity of the longitudinal discharge and beam quality of the XeF laser are discussed.

Raregas-halide excimer lasers are up to now the most efficient UV-lasers to produce high power output. Though these lasers are commercially available, some of their properties may be further improved.

The RGH excimers exhibit a homogeneously broadened emission due to the transition of the excited molecules to the dissociative ground state [1]. This band emission has in general a width of a few nanometers. Therefore, the cross section for stimulated emission of RGH excimers is rather small ($10^{-16} cm^2$) compared to the one observed in other gas lasers oscillating on bound-bound transitions. Hence, high excimer densities in the order of $10^{15} cm^{-3}$ are necessary to obtain a practicable gain The pump power to obtain such a high excimer population amounts to about 1 MW/cm^3 [1].

The RGH excimer formation kinetics involve two- and three-body collisions [1,2]. In the case of XeF* the reactions are:

$$Xe^{*} + F_2 \longrightarrow XeF^{*} + F$$

$$Xe^{+} + F^{-} + M \longrightarrow XeF^{*} + M$$

M denotes those particles which are added as a buffer gas. As the rate for three body collisions increases with the square of pressure, conventional RGH-lasers are operated with gas fills at a pressure of one to several atmospheres.

The problems which arise in RGH-lasers by the requirements of high gas pressure and high power deposition can be solved by the use of preionized transverse discharges.

In order to produce efficiently excited and ionized rare gas atoms in a discharge high electric field strengths are necessary. Transverse discharges with a high electric field, however, tend to instabilities, especially in gases at high pressures of >1 bar, and can be maintained homogeneous only for a periode of a few ten nanoseconds. Laser emission, therefore, has to occur after a short amplification time corresponding to a few resonator roundtrips.

Therefore, most of the realized transversely excited lasers exhibit an unsatisfactory beam quality. The beam quality can be improved by injection locking [3] or by the use of unstable resonators[4].

An alternative approach is the use of a longitudinal instead of a transverse discharge. The main advantage of a longitudinal discharge is the cylindrically symmetric gain volume, which favors oscillation on modes of low order. With a simple passive preionization technique, homogeneous discharges with a pulse duration of about 100 ns can easily be obtained, whereas in transverse discharges such a long pulse can only be achieved with careful optimization of the active preionization and the pulse forming network [5]. Moreover the set up for a longitudinal discharge involves simpler technical means than a transverse discharge device.

In our experiment we investigated discharges in mixtures of Ar, Kr and Xe with F_2 diluted in He at a concentration of 3%. The discharge was driven in a 4 mm bore quartz capillary with 125 mm electrode spacing. The electrodes consisted of stainless steel tubes allowing a free optical path coaxial with the discharge. The capillary was O-ring sealed to high quality quartz Brewster windows and to a a gas handling system. To avoid impurities, the laser system was evacuated to 10^{-5} mbar before each filling.

The resonator is formed by two dielectric mirrors with 6 m radius of curvature and reflectivities of 100% and 95.5% respectively, separated by 70 cm.

The electrical pulses were generated by means of a Blümleinlike RCL-network. The pulse width was in the order of 100 nsec FWHM. It did not vary significantly in the pressure range used in the experiments. The storage capacitor of 250 pF was charged up to 54 kV and the overvoltage-discharge was initiated with a spark gap. Preionization was achieved by connecting a wire, which was wrapped round the discharge tube, to one of the electrodes.

Some problems arose because of the high voltage used for providing the high electric field strength requiered. In gas fills with a total pressure of 1 bar, the discharge could not be initiated reproducibly with the available voltage of 60 kV. Therefor we restricted the experiments to gas fills in the range of 50-500 mbar total pressure.

At pressures of several 100 mbar, the rate of three-body collisions which contribute to the formation of RGH-excimers is negligibly small [6]. Though, high excimer densities are produced in a longitudinal discharge via reactions involving only two-body collisions, and can be maintained during a time long enough for laser oscillation to establish. Recently, laser operation in the XeF B--->X band transition at 352 nm could be achieved in a moderate-pressure longitudinal discharge[7,8].

The laser output as a function of gas pressure is summarized in Fig.1 for various gas mixtures. It is shown that the laser output is strongly dependent on the total gas pressure and optimizes at different pressures as the Xe content of the mixtures varies. We obtained maximum lasing output of 0.6 uJ at a total pressure of 280 mbar in a mixture of Xe: 1.5%, F_2: 3%, He: 95.5%. The output was measured with an energy meter (Laser Precision Rj 7100). The 352 nm laser pulse duration was 20 nsec FWHM. This corresponds to a peak power of 30 Watt.

In contrast to TE-devices, the axial discharge arrangement leads to a circular laser spot. Measured divergence was 1.5 mrad full angle which is about seven times the diffraction limit. The mirror alignement of the optical cavity proved to be rather critical. Lasing did not occur in the absence of one mirror.

Spectral properties of the XeF band emission were registered with a multichannel spectral analyser (Tracor Northern IDARSS/TN 1710). The band has a width of 10 nm in the case of spontaneous emission. During laser action, the band shrinks to a width of at most 2,5 nm, the resolution of our detection system beeing 1,7 nm.

As we used static gas fills in the laser tube, a significant decrease in laser output could be observed during the first discharges in a freshly prepared mixture. After a few minutes of continued discharges at a repetition frequency of about 1 Hz, the output was stable and

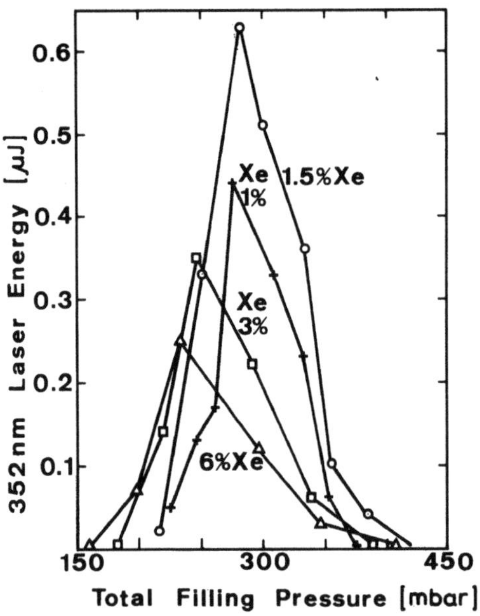

Fig.1 Laser energy output versus total gas pressure of the $He/F_2/Xe$ mixture. Parameter is the Xe content in the He/F_2 mixture of 97:3.

remained on the same level for more than half an hour. The output energy of 0,6 uJ was measured during this later stable phase. Thus, the laser output can be increased if the gas mixtures in the laser tube can continuously be exchanged.

Similar experiments with gas fills containing Kr or Ar instead of Xe did not lead to laser action in the KrF or ArF excimer band at 248 nm and 193 nm, respectively. Measuring the side light, however, it could be shown, that in the case of KrF 80% of the radiated energy is due to the KrF-band emission within a width of 5 nm FWHM (ref.9).Conversion of the stored electrical energy to UV-light below 280 nm was determined to have an efficiency up to 28%. In the case of ArF the efficiency was smaller by a factor of about 4 [6]. Fig. 2 shows the sidelight output of XeF, KrF, ArF for those mixtures with the best performance. Comparing the output of the spontaneous radiation obtained from discharges in different gas mixtures with the theoretically expected output, derived from a simple kinetical model [6], it

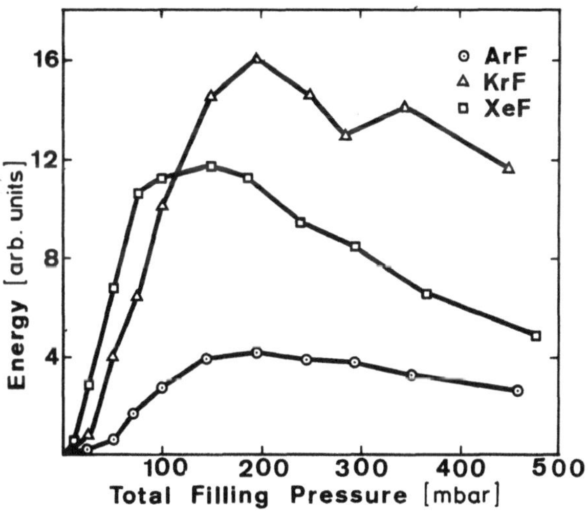

Fig.2 Comparison of the radiated energy by ArF, KrF, XeF for the best gas composition found. The mixtures are $Ar:F_2:He$ = 10:0.75:89.25, $Kr:F_2:He$ = 10:0.75:89.25 and $Xe:F_2:He$ = 5:2.25:92.75.

can be shown, that primary excitation of noble gas metastables by electron impact is more affected by a change of gas composition than the formation and quenching mechanismes of RGH-excimers. Therefore, the efficiency of RGH-production may considerably be increased, if the discharge performance can be improved by optimizing the discharge circuitry independently of the gas mixture used.

In conclusion, we have demonstrated, that the XeF-excimer laser can be operated in a longitudinal low-pressure discharge. A good beam quality is obtained. Moreover, longitudinal discharges seem to be an appropriate means to produce high KrF- and ArF-excimer densities and may therefore by used as efficient UV-flashlamps.

Acknowledgements

We would like to thank H. P. Weber for his stimulating interest in our work and E. Lenggenhager and H. J. Weder for technical assistance. This work was supported in part by the Swiss Commission for the Encouragement of Scientific Research.

References

1. Excimer Lasers, ed. C.K. Rhodes, Topics Appl. Phys. 30 (Springer Berlin, Heidelberg, New York, 1979)

2. J.J. Ewing, in: Laser Handbook, Vol. 3, ed. M.K. Stitch (North-Holland Publishing Company, Amsterdam, 1979)

3. R.T. Hawkins, H. Egger, J. Bokor, C.K. Rhodes Appl. Phys. Lett. 36, 391 (1980)

4. O.L. Bourne and P.E. Dyer Opt. Commun. 35, 242 (1980)

5. R. Girnus, Gasentladungen zur Anregung von Excimer-Lasern, Dissertation, Technische Hochschule Darmstadt,Darmstadt 1979

6. T. Gerber, XeF*-excimer laser at medium pressure excited by a longitudinal discharge, Dissertation, Universitätsdruckerei, Bern 1981

7. P.Burkhard, T. Gerber, W. Lüthy
Appl. Phys. Lett. 39, 19 (1981)

8. D. Cleschinsky, D. Dammasch, H.J. Eichler, J. Hammisch Opt. Commun. 39, 79 (1981)

9. T.Gerber, W. Lüthy and P. Burkhard
Opt. Commun. 35, 242 (1980)

POPULATION INVERSION OCCURING IN THE SOFT X-RAYS RANGE IN LITHIUM-LIKE ALUMINIUM IONS.

G. Jamelot, P. Jaeglé, A. Carillon, and H. Guennou, and A. Sureau

LSAI, Université Paris-Sud Bât. 350, 91405 ORSAY (France)
GRECO "ILM", Ecole Polytechnique, 91128 PALAISEAU (France)

ABSTRACT

The experimental study of radiative transfer in a line-shaped aluminium plasma produced by a high power laser flux (10^{12} W/cm^2) has shown an anomalous dependence of an unknown line intensity upon the plasma length. This line, at 105.7 Å, has been identified to the 3d-5f transitions of Al^{10+} ions. The plasma absorption determination gives the evidence of population inversion between 3d and 5f levels, which is corroborated by detailed calculations of the rate equations with a convenient plasma model.

INTRODUCTION

Laser produced plasmas are at present time generally considered as the most likely medium to achieve population inversion for amplification of soft X-rays. Such hot and dense plasmas contain highly ionized atoms, and beside classical inversion processes used in visible and UV range, such as collisional or optical pumping, varied recombination mechanisms may enhance populations of some peculiar levels, and so lead to inversion. Many theoritical inversion schemes have been considered in these plasmas, and several experiments have shown evidences of population inversion [1-10].

Owing to the lack of resonant cavities for wavelengths in the soft X-ray range, only a single-pass of the radiation through the plasma is possible. On the other hand, the plasma size is too small to support superradiance. So, it is not easy to clearly show up population inversions. One method [1-6] consists in measuring line intensities of a series of transitions, all of them having the same lower level. Population ratio between two upper levels may be calculated from intensity ratio of the two corresponding lines if the transition probabilities are accurately known and if the plasma is optically thin.

A more direct approach is given by the study of radiative transfer properties in the plasma at the proper wavelength associated with the transition under consideration. This technique requires a larger plasma volume than the former method. Two plasmas simultaneously produced may be used with this aim, the one being an absorbing sample while the other is used as an external radiation source [7,8]. Another mean consists in producing a line-shape homogeneous plasma by using a spherocylindrical focusing system, the plasma length being variable without changing the plasma parameters [9,10]. The plasma absorption coefficient $\alpha(\nu)$ follows from an intensity ratio by the expression :

$$\frac{I_L(\nu)}{I_l(\nu)} = \frac{1 - \exp\left[-\alpha(\nu)\,L\right]}{1 - \exp\left[-\alpha(\nu)\,l\right]} \tag{1}$$

where $I_L(\nu)$ and $I_l(\nu)$ are the intensities emitted at the frequency ν by two plasmas, the lengths of which being L and l. A population inversion results in a negative absorption coefficient, the gain being $G(\nu) = \alpha(\nu)$

In this paper, we present a spectroscopie study of a plasma produced by a 40 to 80 Joules, 20 nanosecond Nd laser pulse. The laser beam is focussed along a 8 x 0.1 mm. focal line onto an aluminium, solid, plane target. The soft X ray plasma emission is spatially resolved by the help of an astigmatic toroïdal mirror [11], and then analysed by using a grazing incidence spectrograph fitted with a 3600 grooves/mm holographic concave grating [12]. For the most part of the experiments, the detection of radiation is provided by Kodak SC-5 plates. Recently, we have used an OMA as a detector. In this case , .a scintillator an optical fibres transmit informations from vacuum spectrograph to OMA.

DEPENDENCE OF SOFT X-RAY EMISSION UPON THE PLASMA LENGTH

In a first way of plasma X rays investigation, the spectrum is recorded successively in axial and transverse directions. The advantage of this method consists in this, that the lengths covered by X-rays in plasma are very different from one case to the other one. As a consequence, the spectral features of interest are quite conspicuous. This is shown on figure 1 which presents the spectra observed in both directions for a 4 mm long plasma. As it can be seen, a major change in relative intensities of some lines, especially at 105.7 A, occurs when the observation axis is turn from transverse to axial direction. Now the question arises : is the observed intensity growth larger or smaller than the change of radiation path ? In other words,

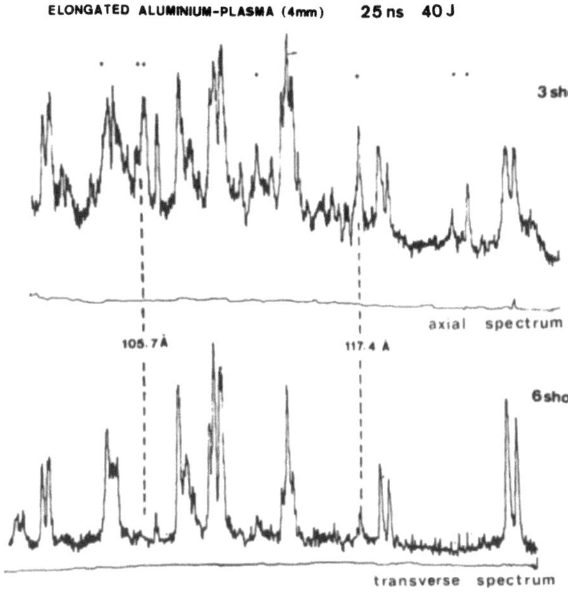

Figure 1. Comparison between axial and transverse spectra of a linear plasma produced from a massive aluminium plane target. Plasma length : 4 mm ; pulse duration : 25 ns ; shot energy : 40 J. An anomalous intensity increasing is clearly observed for a few-spectral lines, especially at 105.7 A (Al^{10+} ion).

is the feature exhibited by the spectra an actual evidence of gain, or just an effect of radiative transfer in a thin plasma ? A large uncertainty may affect the answer because errors are introduced by the plasma expansion. Indeed, the expansion of the plasma in mainly perpendicular to its axis ; this has been confirmed by the observation of the aluminium spot deposited on a protective film set between the focusing lens and the target. So, radiation path along the transverse direction is increased of an unknown amount. Moreover, the plasma is not homogeneous along this transverse direction, and equation 4 cannot be used to calculate the absorption coefficient.

The most part of this uncertainty can be removed with a second way of performing the investigation. The linear plasma axis is now kept as the direction of observation for all the measurements, and the plasma length is modified without change of illumination flux. Such a modification of the plasma length can be obtained either by sliding the target along its plane in order to a part of the laser beam escapes the target, or simplier by screening a part of the laser beam. Then, the lateral expansion is put aside from spectral intensity variations. An exemple of results obtained by this way is shown on figure 2. This figure exhibits recordings, between 100 and 114 A, of spectra emitted by aluminum plasmas of various

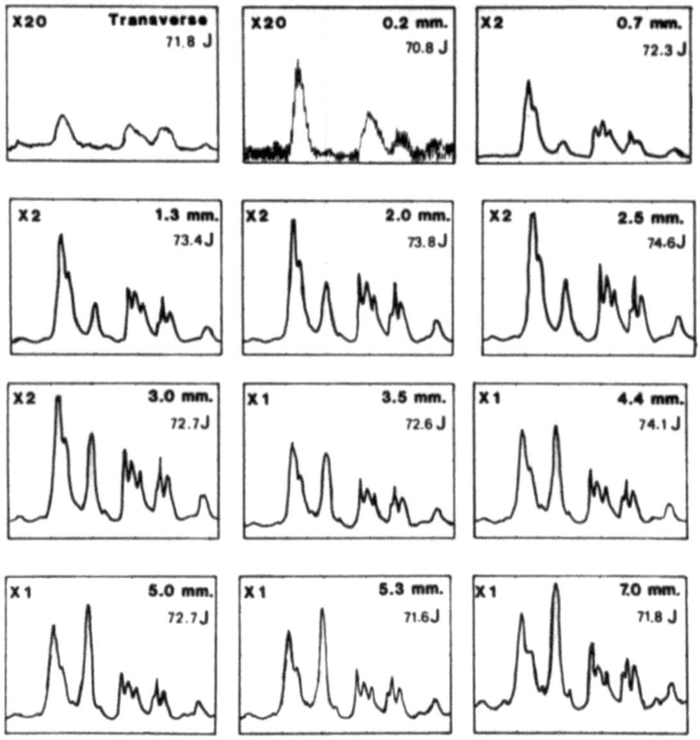

Figure 2 : OMA recordings of spectra emitted by linear plasmas of various lengths. Plasma length : between 0.2 and 7 mm ; pulse duration : 20 ns ; mean shot energy : 72.7 J ; wavelength range : 100 to 114 A. The considerable sensitivity of the line at 105.7 A to the plasma length is confirmed with this experiment.

lengths, comprised between 0.2 and 7 mm. The change in relative intensity of the 105.7 A line appears to be considerable, since this line is practically not emitted by the 0.2 mm long plasma, whereas it becomes the most intense line of the spectrum for the plasma lengths above 4 mm. These spectra have been recorded with OMA, the calibration of which being not well known at present time. Then, it is not yet possible to use these spectra for calculating the plasma absorption coefficient. However, the strong relative increasing of 105.7 A line constitutes a presumptive evidence of a highly populated upper level of the transition.

We have identified the unknown line at 105.7 A to 3d-5f transitions in Al^{10+} ion. Ab-initio Hartree-Fock calculations including spin-orbit interaction [13] provide energies, wave functions, and radiative dipolar transition probabilities concerning all $1s^2nl$ levels with $n < 7$. These calculations show that, for most of the transitions in Al^{10+} ions, the magnitude of transition probability is about 10^{10} sec^{-1}, with the exception of the 3d levels with a probability 100 times larger. Obviously, the fast depopulation of these levels may be a cause of population inversion with upper levels, like 5f are. Moreover, the 2p-nd transitions series of Al^{10+} ions has been studied in a laser produced plasma by Kononov et al [3]. It has been deduced from this work that population inversions occur between the 4d and 5d levels, on one hand, and the 3d one, on the other hand. Owing to the small energy interval between the levels of same main quantum number, the authors assumed the equality of populations of nf and nd levels, and they estimated the gain for the 3d-4f transitions (154.6 A) to be 0.05-0.2 cm^{-1}. However, they did not give direct experimental evidence of this gain.

NUMERICAL COMPUTATION OF LEVELS POPULATIONS IN Al^{10+} IONS

We have performed detailed computations to evaluate the population of energy levels in the Al^{10+} ion and to specify the range of plasma parameters suitable for producing inversions. The hypothesis included in the model are as follows :

- the plasma is in a steady state.
- the ratio of ground levels populations of Li-like to He-like ions is fixed by an adjustable parameter comprisable between LTE and coronal equilibrium values.
- the high quantum numbers levels (n=6 and n=7) are in thermal equilibrium with the He-like ground level.
- the populations of lower levels ($n \leqslant 5$) are governed by collisional and radiative deexcitation, and by collisional excitation. Other processes, such as excitation or de-excitation of optically forbidden transitions in electron-ion collisions, as well as charge exchange in ion-ion collisions, are neglected.

Under these assumptions, populations are gives by a system of coupled rate equations having the following form :

$$\frac{dN_i}{dt} = 0 = \sum_{j>i} N_j A_{ji} - N_i \sum_{j<i} A_{ij} + N_e \sum_j N_j \langle\sigma_{ji}.v\rangle - N_e N_i \sum_j \langle\sigma_{ij}.v\rangle$$

where N_i and N_j are the population densities of levels i and j, N_e the free electrons density, and A_{ij} the radiative transtions probabilities obtained with ab-initio calculations above mentionned. The collision rates $\langle\sigma_{ij}.v\rangle$ are calculated for the optical allowed transitions by using the Van Regemorter formula and our calculated oscillator strengths. About 200 transitions, involving 48 levels, are included in this computation.

An exemple of results obtained with this model is shown on figure 3, which presents the population ratios calculated for four pairs of levels. On the right part of the figure, the population ratios calculated with purely radiative cascades are presented. Results of complete calculations, taking into account both collisional and radiative transitions, are given on the left part,

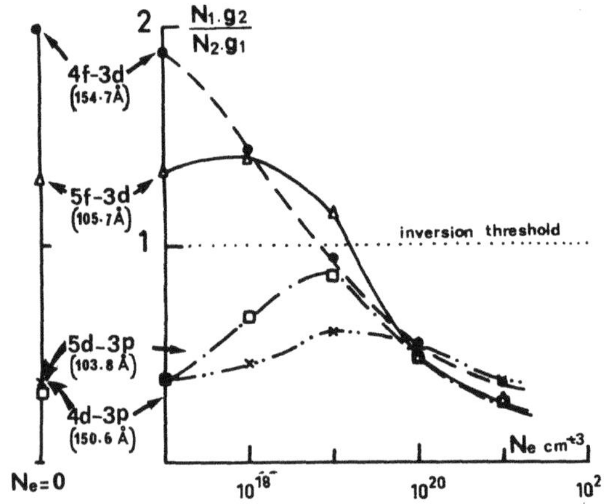

Figure 3 : Population ratios in Al^{10+} ions calculated for several levels as a function of plasma electron density, the temperature being 300 eV. The model predicts inversions between $4f_{7/2}$, $5f_{7/2}$ and $3d_{5/2}$ levels in a large density range.

versus the plasma density value. It can be seen that this model predicts indeed $4f_{7/2}$ and $5f_{7/2}$ populations to be inverted with respect to $3d_{5/2}$ population at electron densities less than about 10^{19} cm^{-3}. As an effect of collisional processes, the $4f_{7/2}$ inversion decreases with growing density. On the contrary, electron-ion collisions enhance the $5f_{7/2}$ inversion, as well as the 4d and 5d populations up to a density value of 10^{18} cm^{-3}. At higher densities, the $5f_{7/2}$ inversion is larger than the $4f_{7/2}$ one.

ABSORPTION MEASUREMENTS. EXPERIMENTAL EVIDENCE OF POPULATION INVERSION.

Experimentally, we find indeed that a population inversion occurs between the 5f and 3d levels. We used two couples of linear plasmas, with l=4mm, L=7mm, and l=5mm, L=9mm respectively. For this last case, a samll defocusing of the laser beam has been introduced in order to make the plasma density gradient longer in the direction of the spatial resolution, perpendiculary to the target plane. Laser energy was about 40 joules for this experiment. Several single shot spectra have been recorded on SC5 plates for each plasma length. After digitizing microdensitometry of each spectrum, averaged intensities have been calculated over a large wavelength range around the 3d-5f transition. Results are presented on figure 4 which shows the average plasma emission spectra and the absorption coefficient spectra, which are calculated at each wavelength with the help of equation 1. It can be seen that the 3d-5f transition gives rise to an inverted line on the absorption spectra, the minimum value being 0.5-2 cm^{-1} lower than the absorption coefficient value on the continuous spectrum near the line at 105.7 A. Then the proper share taken by the 3d-5f transitions in the plasma absorption is negative. That gives the experimental evidence of the population inversion of the 5f level with regard to the 3d one. However, the absorption scale is only approximately known owing to the large statistical error. Thus it is difficult to ascertain whether the final gain is effective or cancelled by continuous absorption due to plasma free-free transitions. It must be pointed out that inversion seems to be kept at larger densities than those predicted by the model, since it is observed at densities close to

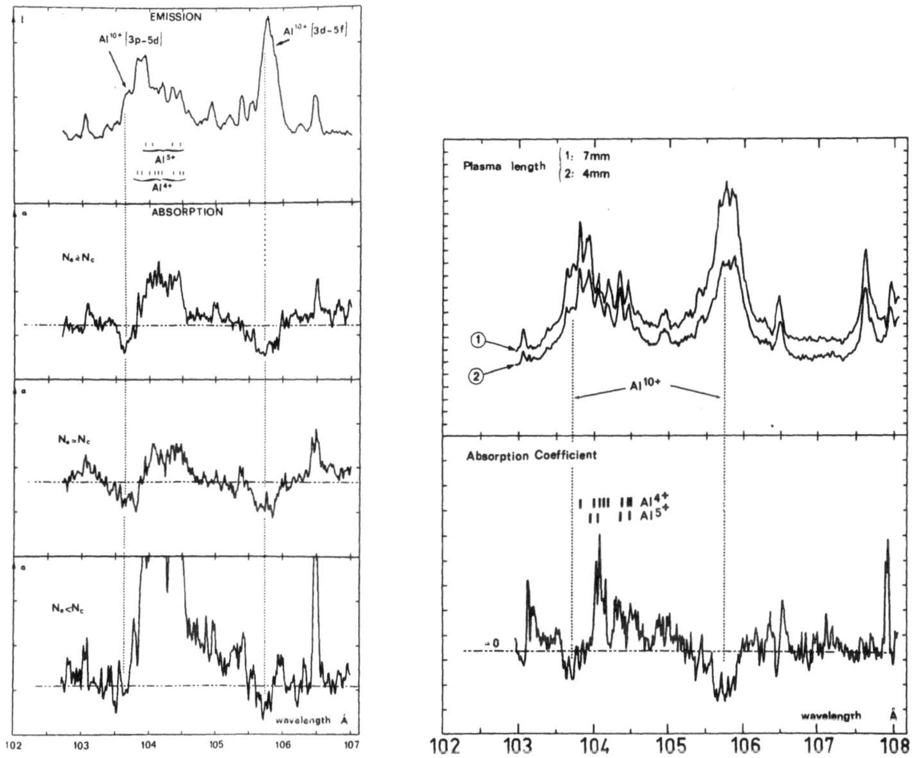

Figure 4 : Measurement of the plasma absorption coefficient spectrum in the range of 3d-5f transitions of Al^{10+} ions, at 105.7 Å. (a) Focussed laser beam. Plasma lengths: L=7 mm, l=4 mm. Emission spectra are represented for the two plasma lengths on the upper part of the figure. Population inversions are demonstrated by hollows in the absorption spectrum corresponding to emission lines. (b) Slightly defocussed laser beam. Plasma lengths L=9mm and l=5 mm. The absorption spectrum is calculated at three distances from the target, close to the zone of critical density.

the critical density, which is 10^{21} cm^{-3} for a Nd glass laser. Moreover, the inverted line on absorption spectra is slightly shifted to short wavelengths with regard to the emission line at 105.7 A. These facts require more information for being discussed.

We have also studied the plasma absorption in the wavelength range of the 3d-4f transition (154.6 Å). Results show that this intense line is optically thin, a very little population inversion being detected in some case. This is in accordance with our calculations, which predict the $4f_{7/2}$ population to lower than the $5f_{7/2}$ one at high densities.

In conclusion, our work has given a direct experimental evidence of population inversion of Li-like ions of aluminium. Now, precise measurements of plasma absorption in continuous spectrum are required to ascertain the possibility of producing an amplification of spontaneous emission.

ACKNOWLEDGEMENT

The authors are grateful to A. Bideau and C. Möller for their aid in recordings processing.

REFERENCES

1. F.E. Irons, N.J. Peacock, J. Phys. B7, 1109 (1974), 7, 2084 (1974)
2. R.J. Dewhurst, D. Jacoby, G.J. Pert, S.A. Ramsden, Phys. Rev. Letters 37, 1265 (1976)
3. E.Ya Kononov, K.N. Koshelev, Yu.A. Mevykin, Yu.V. Sidel'nikov, S.S. Vhurilov, Sov. J. of Quant. Electron. 6, 308 (1976)
4. R.H. Dixon, R.C. Elton, Phys. Rev. Letters 38, 1072 (1977)
5. V.A. Bhagavatula, B. Yaakobi, Optics Communications 24, 331 (1978)
6. M.H. Key, C.L.S. Lewis, M.J. Lamb, Optics Communications 28, 331 (1979)
7. P. Jaeglé, G. Jamelot, A. Carillon, A. Sureau, P. Dhez, Phys. Rev. Letters 33, 1070 (1974)
8. P. Jaeglé, G. Jamelot, A. Carillon, A. Sureau in "Laser Interaction and Related Plasma Phenomena", ed. by H. Hora and H.J. Schwarz, Plenum Publishing Corporation (NY) 4, 229 (1977)
9. G. Jamelot, P. Jaeglé, A. Carillon, C. Wehenkel, Le J. de Physique 40, C1-91 (1979)
10. D. Jacoby, G.J. Pert, S.A. Ramsden, L.D. Shorrock, G.J. Tallents, Optics Communications 37, 193 (1981)
11. A. Carillon, P. Jaeglé, G. Jamelot, C. Wehenkel, in "Laser Interaction and Related Plasma Phenomena" ed. by H.J. Schwarz, H. Hora, H.J. Lubin and B. Yaakobi, Plenum Press (NY) 5, 277 (1981)
12. P. Jaeglé, P. Dhez, F. Wuilleumier, Rev. Sci.Ins. 48, 978 (1977)
13. A. Sureau, H. Guennou, Atomic Spectroscopy Symposium, NBS, Gaithersburg (1975)

3-PHOTON EXCITATION OF XENON AND FREQUENCY TRIPLING BY NON-LINEAR EFFECT

Y.Salamero, A.Birot, H.Brunet, J.Galy, P.Millet, and J.Monso

Centre de Physique Atomique, Lab. Associé au CNRS n°277
Université Paul Sabatier, 118 route de Narbonne
31062 Toulouse Cedex, France

ABSTRACT

The results obtained in a xenon multiphoton excitation experiment using a pulsed dye laser are presented. At low pressures, the $^3P_1$ resonant state of xenon is created by 3-photon resonant absorption. At higher pressures (> 20 torr), the excitation spectrum presents 2 components situated on either side of the resonance line. Spectral analysis in this region showed that they were continually originating from the molecular states A $1u(^3\Sigma_u^+)$ and B $0_u^1(^1\Sigma_u^+)$ which are themselves related to the atomic $^3P_1$ state. In the zone of negative dispersion of the medium we demonstrated the existence of a phenomenon of frequency tripling of the excitation laser beam. The light obtained in the direction of incidence is intense, the frequency can be tuned, and the band width is very narrow.

The aim of this type of experiment is to selectively create various atomic states of inert gases to study the kinetic behaviour of the atomic and molecular excited species which emit a fluorescence spectrum in the far U.V. In the present paper we report the results obtained with xenon through excitation of the resonant $6s(^3P_1)$ state by 3-photon absorption.

At the same time, using the same experimental set-up we demonstrated a phenomenon of frequency tripling of the excitation laser beam when the wavelength corresponds to the region of negative dispersion of xenon with respect to the $^3P_1$ resonant state.

EXPERIMENTAL SET-UP

Figures 1a and 1b show the set-up used for both types of experiment. The excitation device and the technique used to analyse the VUV intensity were identical in both cases. The only difference was the direction of observation. Multiphoton excitation of the $^3P_1$ state of xenon was observed by recording the VUV fluorescence perpendicularly to the direction of the excitation laser beam (Fig. 1a).

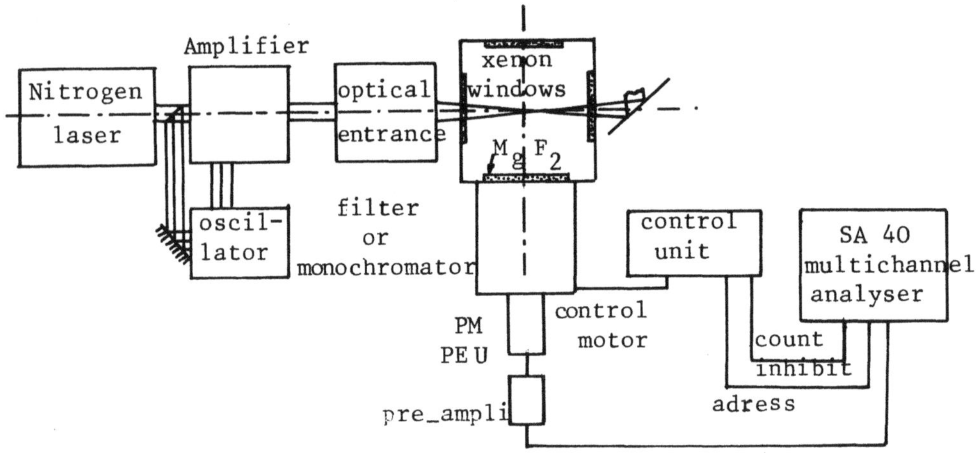

Fig. 1a Experimental set-up used for the observation of the multiphoton excitation of xenon.

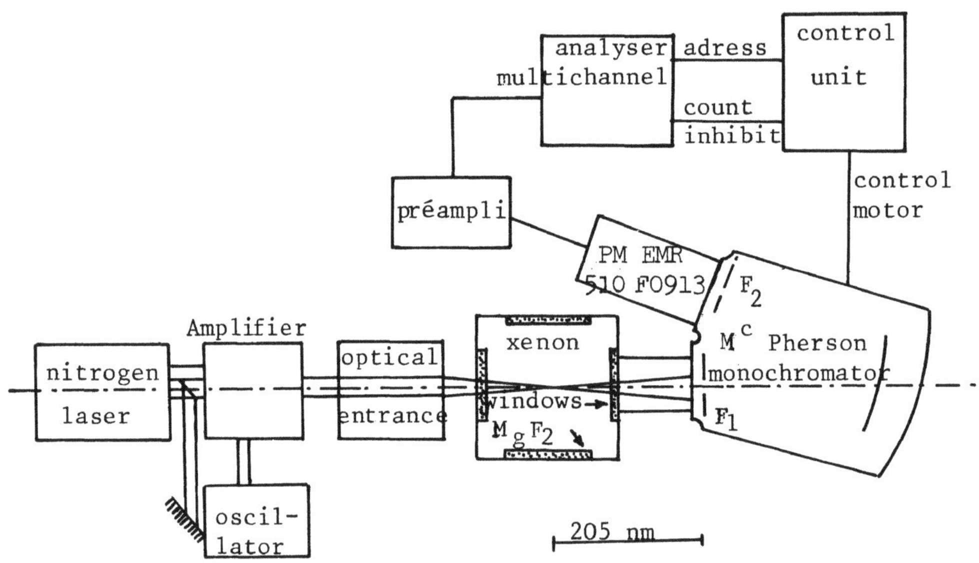

Fig. 1b Experimental set-up for frequency tripling.

The phenomenon of frequency tripling of the laser beam was observed in the same direction as the beam (Fig. 1b).

The laser beam was obtained from a SOPRA device composed of a tunable laser pumped by a nitrogen laser. The pumping laser supplied 5 ns impulses with a peak power of 250 kW and a firing rate of 50 Hz. These pulses excited an oscillator-amplifier system using Coumarine 440 dye. A grating and telescope were used for wavelength selection giving a band width of 10^{-2} nm at 440 nm. An intracavity Fabry-Perot interferometer further reduced the bandwith to 10^{-3} nm. The laser beam obtained had a peak power which could be varied between 4 to 20 kW with a pulse length of 5 ns.

A plano-convex lens (F=40 mm) was used to focus the beam in the centre of a stainless steel cell containing the high purity xenon at pressures between 10^{-3} and 500 torr. The focal spot was about 40 μm across. Luminescence was observed through MgF_2 windows by means of a solar blind P.M. (EMR 510 G or F).

At a right angle to the excitation beam axis, the phenomena was observed either directly to see the whole luminescence, or through a filter or monochromator used under vacuum.

The tripling frequency phenomenon was observed by a P.M. EMR 510 G 0913 *Extremely solar blind* set behind the monochromator Mc-Pherson type 218.

In all cases, the PM's were set to detect single photons; the spectra were automatically recorded on a multichannel selector counting the impulses.

MULTIPHOTON XENON EXCITATION

a) Excitation at low pressure

In the low pressure range (10^{-3} to 1 torr) the only observable emission is the resonance line emitted by the $^3P_1$ state (λ = = 146.9 nm). In order to avoid resorption of the line, the beam was focused very close to the observation window. Fluorescence was observed directly. The excitation spectrum of the $^3P_1$ state was obtained by varying the wavelength of the laser beam on increasing the pressure in the chamber containing the interferometer. Results obtained at a xenon pressure of 1.5×10^{-2} torr are presented in Fig.2. The peak of the spectrum, at 440.88 nm indeed corresponds to an energy which is one third that of the $^3P_1$ state (3-photon absorption) The width at mid-height of the excitation spectrum is 0.2 Å at this pressure.

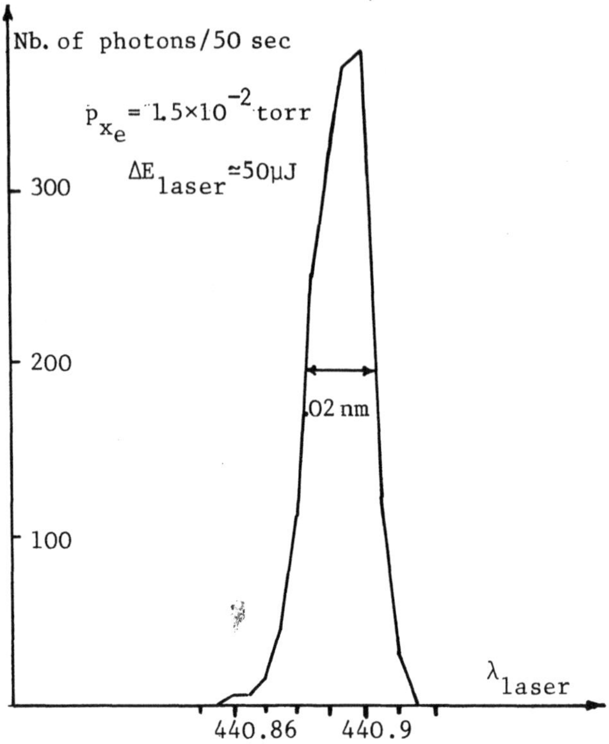

Fig.2 Excitation spectrum of the 1st $^3P_1$ resonant xenon state.

Measurement of the light intensity dependence against the energy of the laser beam confirms the multiphoton nature of $^3P_1$ state excitation. Fig. 3 shows the proportionality between the log of the counting rate against the log of the average energy of the beam. The slope always has a value of about 3 indicating that the intensity of the light emission is proportional to the cube of the energy E of the laser beam.

$$W_{g\to f} = K\ E^3$$

Furthermore it was checked that this emission was indeed centered at 147 nm by observing it through an interference filter.

b) Excitation at high pressure

As the pressure of xenon rises ($P > 10$ torr) the intensity of the fluorescence increases sharply. Fig.4 gives the emission spectrum of xenon. The excitation laser wavelength was that giving the

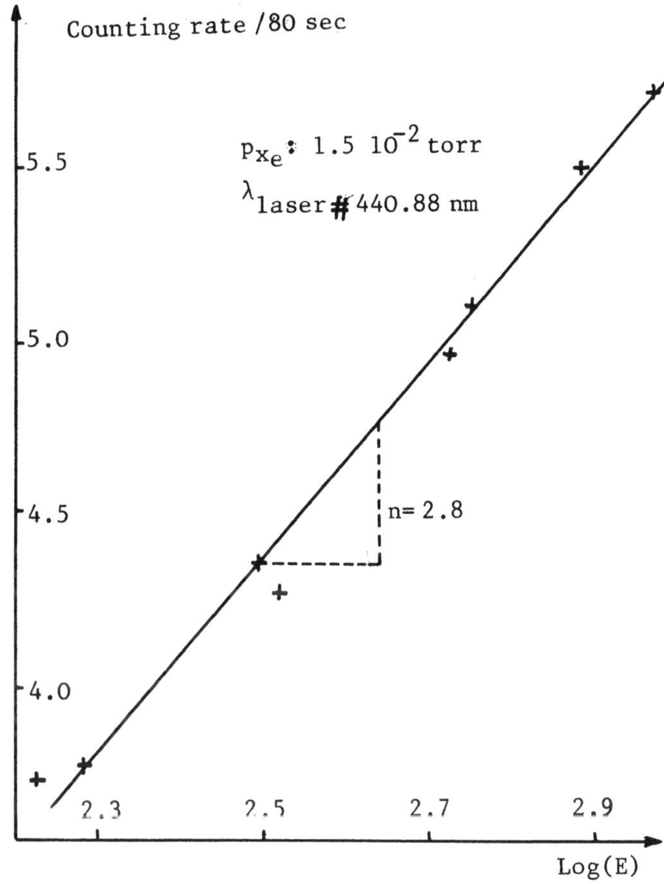

Fig.3 Fluorescence intensity against the energy of the laser beam.

highest luminescence intensity, it was determined from the excitation spectra (Fig.5a). Spectral analysis shows that the emissions are made up of the 1st and 2nd continua of xenon situated at 147 and 173 nm respectively. These emissions correspond to the radiative molecular transitions from the A $1_u(^3\Sigma_u^+)$ and B $0_u^+(^1\Sigma_u^+)$ states to the ground state X $0_g^+(^1\Sigma_g^+)$. Their behaviour with pressure is identical to that obtained using other means of excitation[6,2]. As the pressure increases, the intensity of the 2nd continuum rises at the expense of the first.

The first excitation spectra were plotted by either observing the first or the second continuum of xenon from behind a monochromator. For technical reasons, the laser beam was focused at about 5 mm from the observation window. The spectra show that over the

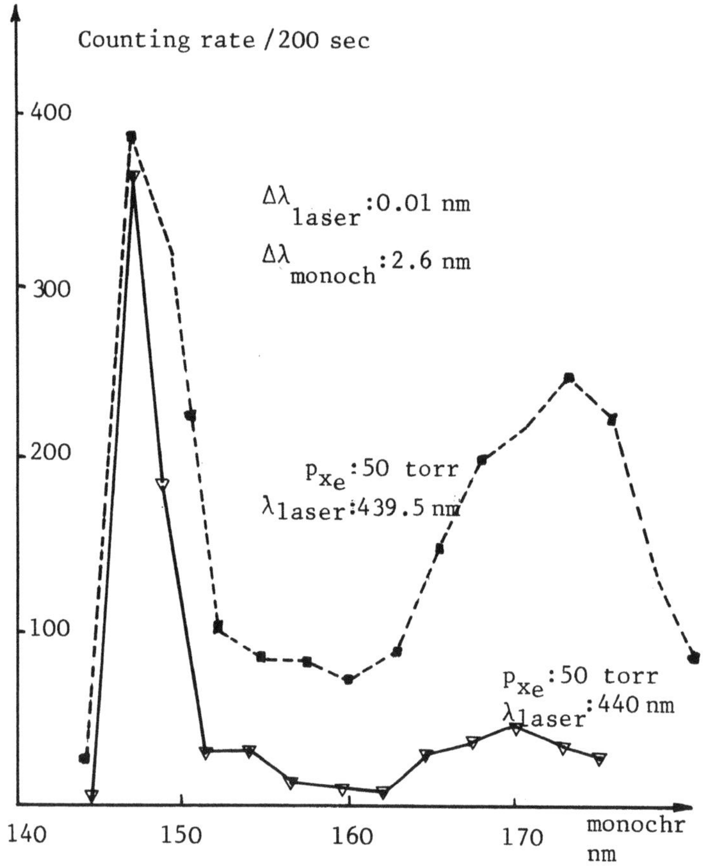

Fig.4 Emission spectrum of xenon excited by multiphoton absorption

pressure range where both continua exist, they behave in the same way with respect to the excitation wavelength. This indicates, considering the energy of the laser photons and the potential curves of the first molecular states of xenon, that the formation of the $^{1,3}\Sigma_u^+$ states must necessarily pass by the $^3P_1$ state. In these experimental conditions, no luminescence was observed for wavelengths over 440.88 nm (corresponding to 3-photon excitation of the $^3P_1$ state.

More recently, when observing overall luminescence with the

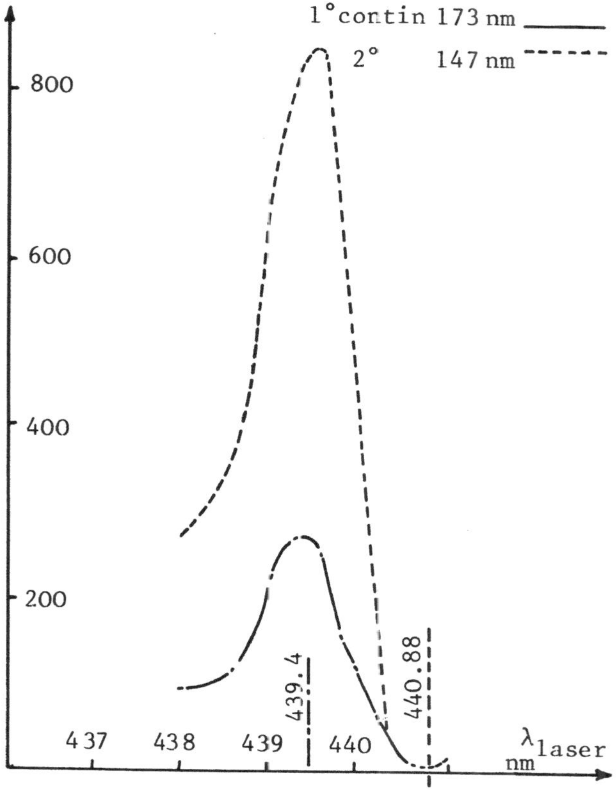

Fig.5a Laser excitation spectra correspondings to the 1st and 2nd continua of xenon (p_{Xe} = 50 torr).

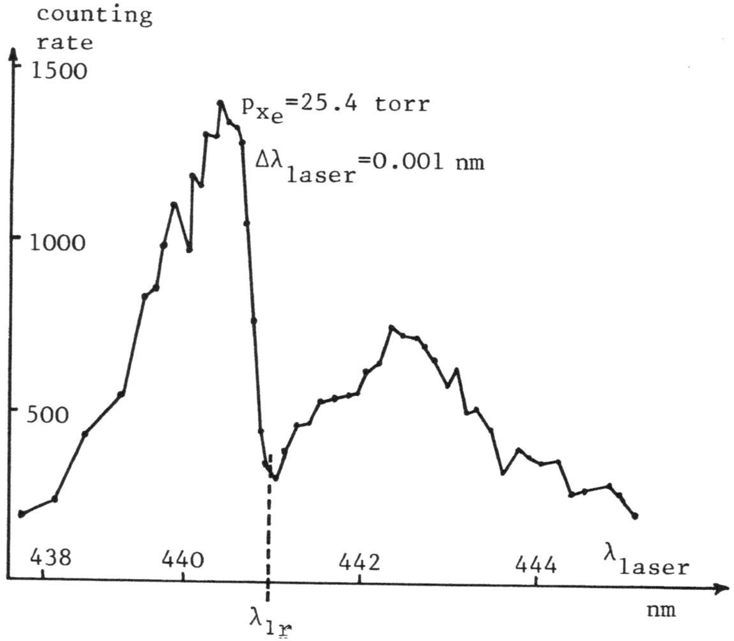

Fig.5b Excitation spectrum of xenon at p_{Xe}= 25.4 torr:the P.M. sees the whole light phenomenon directly.

excitation beam ($\Delta\lambda$ = 0.001 nm) focused as close as possible to the dome of a planoconvex window, an excitation spectrum was observed (Fig.5b) with a satellite on either side of the laser wavelength λ_{1r} = 440.88 nm.

The decrease in the intensity approaching λ_{1r} probably indicates the resorption of the resonance radiation at this pressure. A similar experiment was carried out in xenon by DUTUIT [4] who caused excitation on the gas near the $^3P_1$ state by absorption of 1 photon using synchrotron radiation. The excitation spectra obtained had a general form similar to ours buth with opposite dissymmetry i.e.the red satellite on the excitation spectra was more intense than the blue one.

It seems that the origin of these two components could be in a molecular type excitation. At pressures around those we used and at room temperature, the proportion of Xe_2 xenon molecules in the form of bound pairs is not negligible. A calculation made by CASTEX[3] did in fact show that the overall proportion of free and bound pairs represents about 0.5% of the total number of gas atoms. Analysis of the potential curves of the molecule show that considering the energy of the laser excitation photons, the blue satellite corresponds to a trasition from the ground state 0_g^+ towards the molecular dissociative state 1u correlated to the $^3P_1$ state. The red satellite produces a 3-photon transition from the 0_g^+ state to the bound molecular state 0_u^+ correlated to the resonant $^3P_1$ state.

To explain the difference in intensity of the two satellites we proposed the hypothesis that the red one was the result of a bound-bound transition (rotation-rotation) the same trapping phenomenon as that of a resonance line could occur but in much lesser proportions owing to the fact that the starting states are less numerous (population at the vibration-rotation level of the ground state). This phenomenon does not occur however for the blue satellite which is the result of a bound-free transition.

It must be added that there also exists a phenomenon of frequency tripling of the incident laser light which occurs for wavelengths shorter than λ_{1r} resonant. This was demonstrated in the present experiment. So a one-photon excitation of the 1_u state correlated to the $^3P_1$ state can be superimposed on the multiphoton excitation of the blue satellite. This additional excitation, which can only occur in the blue satellite is supported by our observations.

A temporal study of the luminescence decay is currently underway. It should allow evaluation of the apparent life-time of the $^3P_1$ state[5] and determination of the constants of the reaction lead-

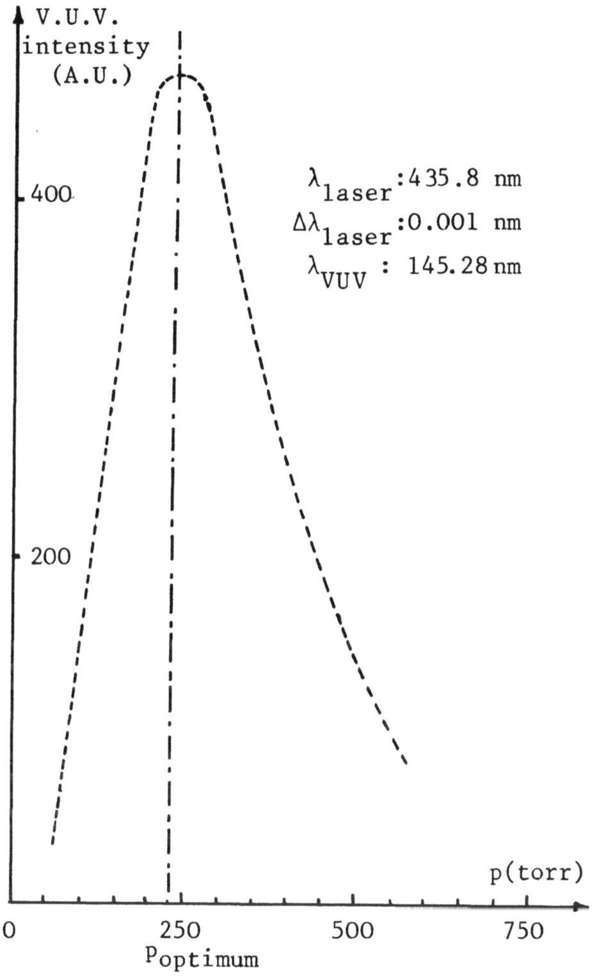

Fig.6 Variation of the VUV intensity obtained by frequency tripling against the xenon pressure.

ing to the formation of the $^{1,3}\Sigma_u^+$ excimers responsible for the xenon continua.

FREQUENCY TRIPLING IN XENON

When a power laser beam is focused in a chamber containing a rare gas, one can observe, in the direction of the excitation beam, coherent radiation with a wavelength one third that of the excitation laser. This phenomenon, dealt with in the theoretical study by BJORKLUND[1] and MAHON[8], occurs if the medium presents a phenomenon of negative dispersion i.e. if:

$$\Delta k = k_1 - 3k_3 = \frac{2\pi}{\lambda} |n_1 - n_3| < 0$$

where n_1 and n_3 are the indices of the medium at λ and 3λ respectively.

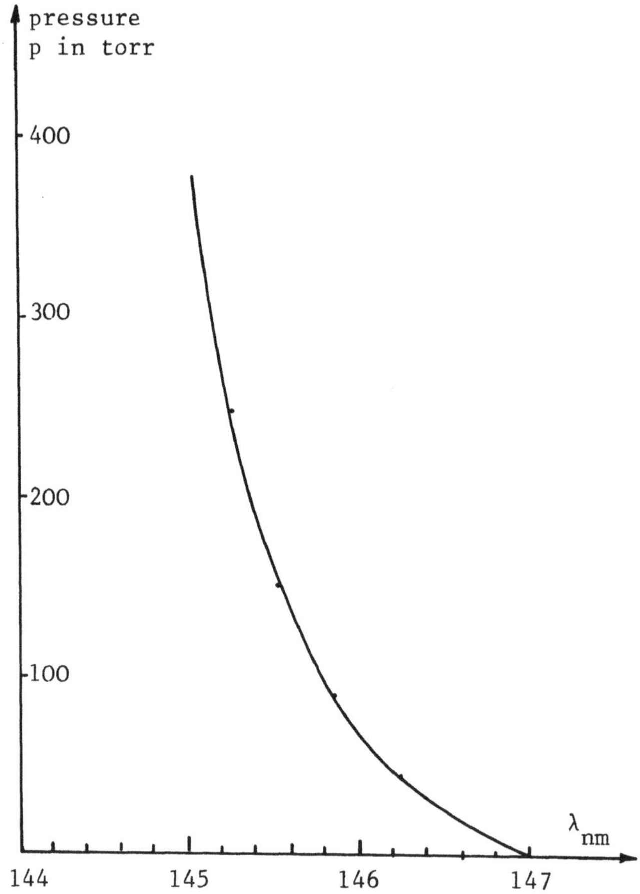

Fig.7 Variation of the optimal pressure with the VUV wavelength for the frequency tripling experiment.

We experimentally checked, as did MILLER et al.[7], that:

1. the phenomenon occurred at laser wavelengths shorter than 440.88 nm;
2. the phenomenon was tunable and

3. The spectral width of the VUV beam obtained was narrow as we obtained the monochromator band-width (FWHM = 1 Å).

The variation of the VUV intensity with the xenon pressure shows a maximum for a given excitation wavelength (Fig.6). This optimal pressure can be predicted using the theory developed by Mahon, with our experimental conditions, the agreement is good. After recording a whole series of curves we were able to determine the optimal pressure for the require VUV wavelength.

The measurements were carried out between 145 and 146.9 nm, limit which were imposed by the resonance line (Fig. 7).

Although the efficiency of the frequency tripling phenomenon is low (of the order of 10^{-6}) intense VUV light fluxes are obtained which can be applied to atomic and molecular spectroscopy experiments.

REFERENCES

1. G.S.Bjorklund, Journ.Quant.Elect.QE 11,287,1975
2. R.Brodmann & G.Zimmerer, J.Phys.B:Atom.Molec.Phys.,vol.10,n°17, 1977
3. M.C.Castex, Thèse d'Etat n°11038,Université P. et M.Curie,Paris 1975
4. O.Dutuit, Thèse de Doctorat ès-Sciences,Paris,1978
5. T.Holstein,Phys.Rev.83,6,1951
6. P.Millet,Thèse de Doctorat ès-Sciences,Toulouse,1979
7. J.C.Miller; R.N.Compton; M.C.Payne; W.W.Garret,Phys.Rev.Lett.45 114,1980
8. R.Mahon; McLlrath; V.M.Gerscough; D.W.Koopman, IEEE Journal of Quant.Elec.WE-5, 444,1979

A COMPACT RESISTIVELY STABILIZED EXCIMER LASER FOR INJECTION- AND MODE-LOCKING APPLICATIONS

Robert C. Sze

University of California, Los Alamos National Laboratory
P. O. Box 1663
Los Alamos, New Mexico 87545

ABSTRACT

A miniature, resistively-stabilized excimer laser has been developed. The laser has 86 separately stabilized discharges within a length of only 27.6 cm, which allows for a total cavity length of only 33 cm. Presently, with a 2.5-mm electrode separation, we have obtained over 1 mJ per pulse energy in XeCl and the laser has been operated over 70 Hz without gas flow. Pulse lengths as long as 40 ns full-width at half-maximum intensity and over 80 ns total lasing time has been observed. The beam divergence is measured at three times the diffraction limit. This laser is expected to be uniquely suited for injection-locking and mode-locking applications.

Recent work by Colin Webb's group at Oxford[1,2] on resistively stabilized long-pulse lasers shows great promise toward the construction of simple long-pulsed excimer lasers. However, the devices they report are characterized by very long cavity lengths, as well as poor efficiency ($\sim 3 \times 10^{-2}\%$). We report here the development of a resistively stablized laser that uses simple 0.5-W carbon resistors closely packed in an array that allows 87 separately stabilized discharges in a 27.6-cm long active length, resulting in a cavity length separation of only 33 cm.

A schematic of the resistively stabilized laser is shown in Fig. 1. The laser consists of a simple L-C inversion circuit with a segmented resistor array cathode. Preionization is accomplished

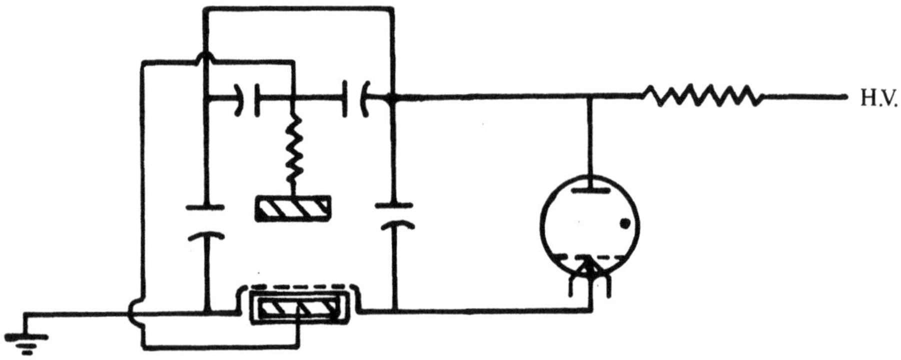

Figure 1

with a corona discharge using a square quartz tube located behind the screened anode and with a conductor inside the quartz tube connected to the opposite electrode. A photograph of the laser is given in Fig. 2. The two arms of the L-C inversion circuit are divided into two sections one on each side of the discharge. The switch is a EG&G 1102 thyratron located on one side of the discharge and connected to the L-C inversion arm on the other side of the discharge through a plate at the top of the laser device. Thus, it is expected that the second section of the L-C inversion circuit will deposit its energy at a somewhat later time. Each arm of the four sections of the L-C inversion circuits consists of twelve 500-pf barium titanate doorknob capacitors.

The cathode structure is constructed by soldering the resistor array to a solid bar. Epon 828 epoxy is then used to mold and hold the array. The bar is then segmented and the entire electrode structure is re-epoxied and milled to the dimension desired for the electrode separation. Figure 3 gives a photograph of the segmented cathode structure. The electrode is simply milled flat with no attention paid toward profiling. However, the discharge is observed to be uniform and, generally, encompassing the preionized area. Thus, the laser output gives a spot size of 4 mm × 2.5 mm.

Previous work[3] on high-repetition-rate mini-TEA lasers showed the need for a very fast circuit in order to insure that much of the stored energy is deposited within the stable discharge time to the laser medium. The stable discharge time is proportional to the electrode separation because the shorter the electrode separation

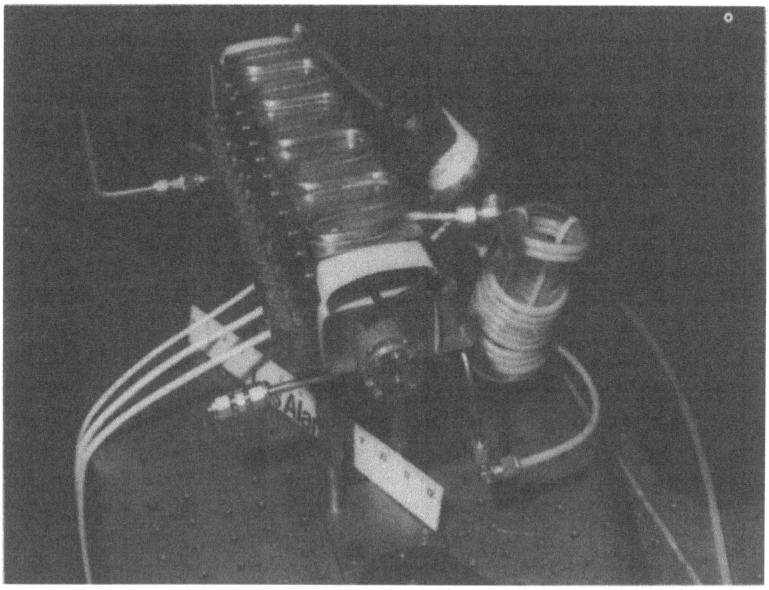

Figure 2

Figure 3

is, the less time it takes for arc streamers to propagate across the gap. In a Blumlein device in which great care was taken to make the discharge circuit inductance as small as possible, we were not able to obtain reliable continuous lasing at electrode separations of less than 4 mm. In the resistively stabilized laser, reliable lasing is observed with only a 2.5-mm electrode separation. Indeed, the inductance of the discharge circuit is substantially higher than the Blumlein mini-TEA, as we took no great care in constructing the simple L-C inversion circuit using doorknob capacitors.

Another impressive feature of the resistively stabilized laser is the capability of going to 70-Hz pulse repetition rate (the limit of our thyratron pulser) without gas flow. In the mini-TEA Blumlein laser,[3] the used gas must be cleared at every shot in order to avoid arcing. We do, however, observe some thermal blooming when the pulse repetition rate goes above 30 Hz. This thermal lensing should be easily correctable, as in moderate repetition-rate YAG lasers[4] by knowing the index change and making the proper optics corrections.

The operation characteristic of the miniature laser depends on the value of the stabilizing resistive array and the gas mixture. We have tried 4.7 Ω, 10 Ω, and 39 Ω resistive arrays for the cathode. For all values we observe the discharge to be stable. Impedence matching and the rate at which energy is deposited appears to be important factors determining the ultimate lasing energy and pulse width. The rate at which energy leaves the storage capacitors is partially determined by the value of the resistive array. Thus, for high values of resistance (39 Ω bank) the current flow is limited and substantial energy remains in the storage capacitors such that the voltage measured across the discharge gap plus the resistive array never goes negative (as in the case of the 4.7 Ω and 10 Ω resistor banks). This is shown in the voltage traces for the three resistor arrays given in Fig. 4. Thus, at optimum pressures for high energy laser operation the resulting laser pulse is short (approximately 8 ns and 1 mJ) for the low resistor arrays and long (approximately 50 ns full-width at half-maximum and 80 ns total lasing time at 0.2 mJ) for the 39 Ω resistor array. Figure 5 shows the long pulse operation for the 39 Ω resistor array giving the temporal traces of voltage, fluorescence and lasing. Long pulses can be obtained even for the small-valued resistor arrays but at nonoptimum gas mixtures. This generally means low HCl concentrations and the lasing (approximately 0.2 mJ) often has the appearance of two peaks, probably due to double pulsing resulting from the second ring in the voltage curve. All this points to the fact that the L-C inversion circuit is not the proper charging network to use. Since the discharge is stable even with the very low 4.7 Ω resistor array, we ought to be able to deposit substantially more energy per unit time and for much longer times.

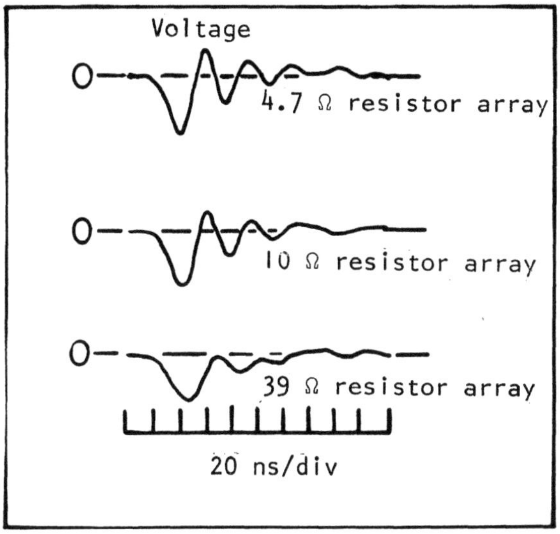

Figure 4

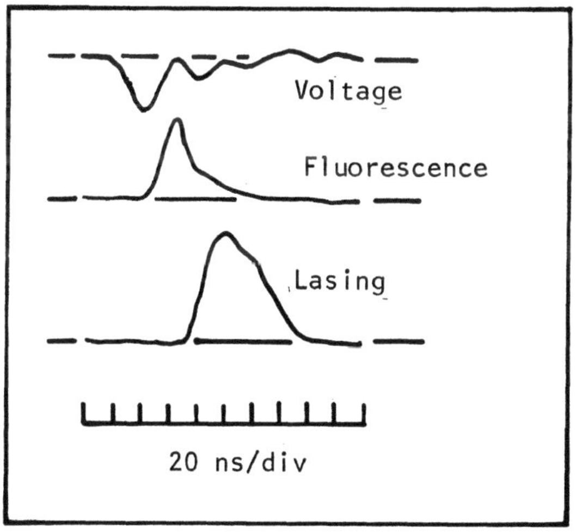

Figure 5

Preliminary results utilizing a discrete component charging line shows this to indeed be true and much improved performance has resulted.

The flat-flat optics give very impressive beam divergence. The uncorrected divergence out of the laser is measured to be approximately 0.5 mrad. Using a 2-m focal length collimating lens placed 28 cm in front of the output coupler, we measure a correctable divergence of 97 microradians or for the 2.5-mm lasing spot a beam divergence only three times the diffraction limit of a Gaussian beam.

REFERENCES

1. D. C. Hogan, A. J. Kearsley, C. E. Webb, J. Phys. D:Appl. Phys. 13, (1980), L225-8.
2. Kan-ich Fujii, A. J. Kearsley, A. J. Andrews, K. H. Errey, C. E. Webb, IEEE Jour. Quant. Elec. QE-17, 8, 1315-17 (1981).
3. R. C. Sze, E. Seegmiller, IEEE Jour. Quant. Elec. QE-17, 1, 81-91 (1981).
4. For example, the Quanta-Ray 10-Hz laser.

ATMOSPHERIC TURBULENCE SIMULATION CELL FOR OPTICAL PROPAGATION EXPERIMENT

Michel Billard, Guy Fertin, and Jean Claude Fontanella

Office National d'Etudes et de Recherches Aérospatiales

92320 Châtillon (France)

INTRODUCTION

Propagation of optical waves in the atmosphere is often perturbed by the presence of natural turbulence, which destroys the coherence properties of the beams.

Outdoor measurements of these effects on laser beams and experimentation of correction devices, like adaptive optics, are subject to two constraints:
- a complete characterization of the turbulence strenght (C_N) along the optical path generally requires a multistatic set-up;
- meteorological condition variability prevents to reproduce experiments.

In order to avoid these drawbacks, we decided to build a laboratory facility which simulates the natural turbulence effects on a laser beam. This facility has to be considered as the complement of the on-field experimentation.

TURBULENCE GENERATION

The index of refraction variations can be produced by three ways:

1 - heating of a liquid[1];
2 - mixing of two gas with different index of refraction[2];
3 - heating of a gas[3].

The first solution provides a rather good simulation of atmospheric turbulence, but liquids are opaque to infra-red radiation.

The second method seems uneasy to carry out on a few meters and the turbulence generated by this mean is difficult to characterize.

We have chosen a solution of the third kind which consists in mixing cold and hot air; the main drawback of this method is that the index of refraction gradients are weak so that a long path is needed.

To simulate the atmospheric turbulence, some conditions have to be fulfilled:

A - The turbulence has to follow the Kolmogorov law.

B - The effect of the laboratory generated turbulence on the beams has to be similar to the effect of the real atmosphere; the spread can be estimated from the theoretical relation[4,5]:

$$\frac{\omega_T^2}{\omega_D^2} = 0.909\,\frac{\omega_0^2 k^{13/6}}{N^2}\,C_N^2\,Z$$

where ω_T is the radius of the beam in the focal plane, ω_D is this radius without turbulence, ω_0 is the radius of the beam at the entrance of the cell, k is $2\pi/\lambda$, C_N the turbulence strenght, Z

Table 1 Comparison of the optical characteristics of the atmosphere with those of different simulation cells; ω_T/ω_D is the beam spread (see text), r_0 the Fried's diameter, $\sigma_I/\langle I\rangle$ the scintillation ratio.

Medium	Laser	λ/n (μm)	C_n^2 ($m^{-2/3}$)	Z (m)	ω_0 (mm)	ω_T (mm)	ω_D (mm)	ω_T/ω_D	r_0 (mm)	$\sigma_I/\langle I\rangle$
ATM	CO_2	10,6	10^{-13}	5000	250	160	34	4,7	75	2,0
Air cell	CO_2	10,6	10^{-10}	10	25	0,75	0,68	1,1	50	0,214
	CO_2	10,6	10^{-10}	30	25	3,6	2,0	1,76	26	0,59
	CO_2	10,6	3.10^{-10}	30	25	6,2	2,0	3,1	13	1,01
	DF	3,8	10^{-10}	10	25	0,82	0,24	3,4	14,5	0,39
	HeNe IR	3,3	10^{-10}	10	25	0,83	0,21	3,95	12	0,42
	HeNe	0,63	10^{-10}	10	25	0,95	0,04	24	1,7	1,1
	HeNe	0,63	10^{-10}	3	15	0,17	0,02	8,6	3,4	0,37
Water cell	HeNe	0,48	10^{-9}	1	5	0,12	0,015	7,7	1,2	0,5

the pathlength and N a quality factor of the beam.

The examination of the Table 1 shows that the atmospheric propagation of a CO_2 laser beam in strong turbulence conditions cannot be simulated in the laboratory with realistic turbulence strenght (10^{-5} $m^{-1/3}$) and pathlength (30 m). This simulation can only be achieved with a shorter wavelength (3.39 µm for example).

C - L_o (the outer scale of the turbulence) has to be much greater than ω_0 in order to simulate the *beam-wander* effect; this can be done by a simulation cell height of 20 cm and a beam of a few centimeters.

D - To obtain the similarity of temporal frequency spectrum, the ratio ω_0/v (v : wind speed) has to be reproduced; a wind speed of 1 m/s leads to realistic values.

E - The condition $\ell_0 \ll (\lambda Z)^{\frac{1}{2}}$ is generally obtained in the atmosphere (ℓ_0 : lower scale of turbulence); this is achieved in a simulation cell of a few meters length if ℓ_0 is about 1 mm.

DESIGN OF THE SIMULATION CELL

The general lay-out of the cell is presented on Fig.1.The hot air is produced by a commercial heater/blower whose heating power and flow can be independently adjusted. A thin grid is set fifteen centimeters after the entrance of the cell in order to homogenize the flow. The optical path is perpendicular to this flow; folding mirrors can be used to increase the pathlength up to thirty meters.

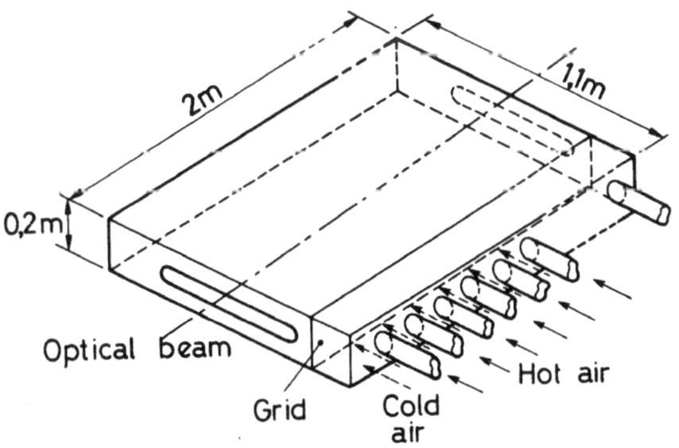

Fig.1 General lay out of the simulation cell.Hot air is injected by twelve circular nozzles, cold air is carried along into the cell by the hot flow.

CHARACTERIZATION OF THE SIMULATION CELL

1. Homogeneity of the wind speed and of the temperature fluctuations

The wind speed and the temperature fluctuations appear to be quite homogeneous (within 20%) in the volume located fifty centimeters beyond the entrance of the cell. The temperature fluctuations have been measured with a hot wire thermometer.

2. Temperature fluctuations spectrum (Fig.2)

For the wind speeds within the range 0.5 m/s to 2 m/s, the temperature spectrum exhibits the -5/3 slope characteristic of the inertial subrange, which proves that the turbulence obeys to the Kolmogorov law.

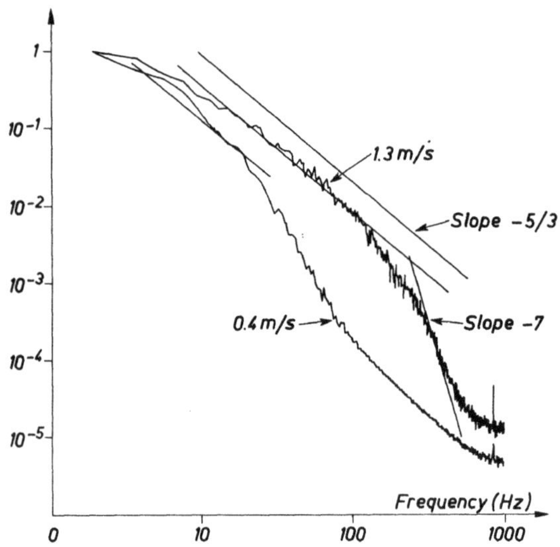

Fig.2 Temperature power spectrum (measured with a hot-wire thermometer).

3. Structure function of the temperature (Fig.3)

This function has been measured with microthermocouples; the high frequency cut-off of these probes (40 Hz) prohibits measurements for wind speeds larger than 1 m/s; the measurements show that the random mean square value of the temperature increases as the power 2/3 of the probes separation from about 1 mm and up to 10 cm. This is again a proof of the validity of the Kolmogorov law in our cell.

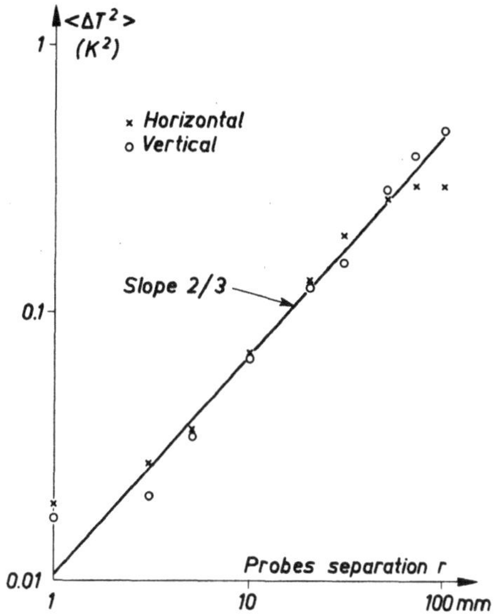

Fig.3 Temperature structure function (measured with microthermocouples).

4. C_N determination by scintillation

Scintillation has been measured using a collimated HeNe laser beam and a silicon photodiode. The C_N value deduced from the scintillation measurement is in good agreement (30%) with the C_N deduced from the temperature measurements for the wind speeds lower than 1 m/s. The disagreement observed below can be inferred to the reduced bandwidth of the microthermocouples.

The Fig.4 presents the C_N measurements as a function of the distance from the entrance of the cell. The higher value of C_N obtained is 10^{-5} $m^{-1/3}$ (with a heating power of 12 kW).

5. Spectrum of the log-irradiance of a point (Fig.5)

The scintillation measurements have been processed in order to obtain the log-irradiance spectrum of a point. This spectrum shows the -8/3 slope characteristic of the inertial subrange and, hence, of the applicability of the Kolmogorov theory to our cell.

The general behaviour and frequency cut-off of the spectra are similar to the atmospheric ones.

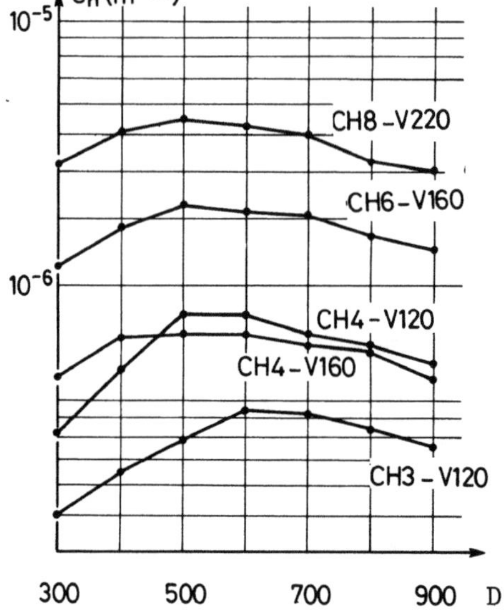

Fig.4 C_N as a function of the distance from the entrance of the cell. The different values of CH and V referred to different settings of the heating power and wind speed.

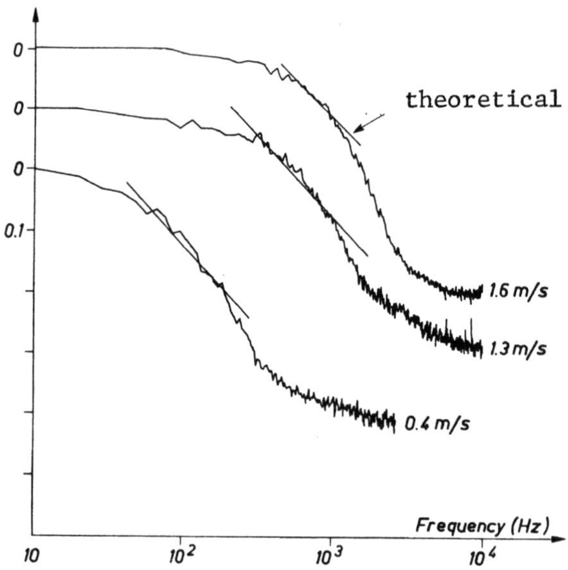

Fig.5 Power spectrum of the log.irradiance (for different speeds).

6. Spread of a laser beam

The long term spread of the HeNe laser spot has been evaluated using a T.V.camera, a T.V.line digitizer and averaging by a computer. The spread is in good agreement with theoretical predictions [4,5] (see Fig.6). This proves again the validity of the simulation and the fulfillment of condition B .

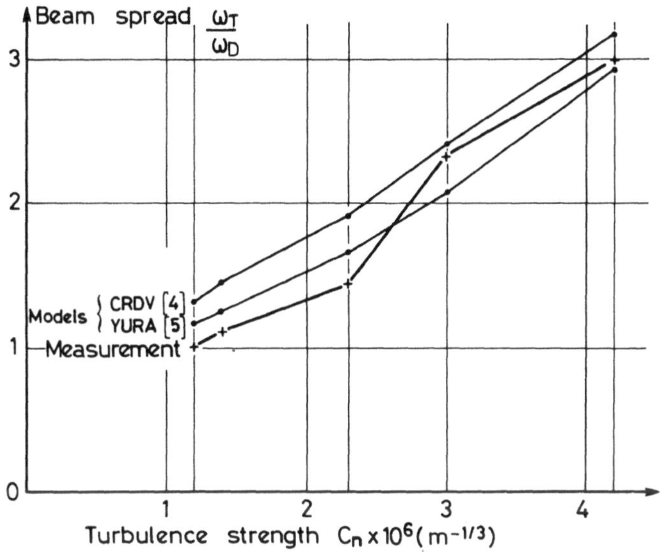

Fig.6 Experimental beam spread compared with theoretical predictions.

CONCLUSION

A facility for the simulation of the atmospheric turbulence has been built and characterized: C_N up to $10^{-5} m^{-1/3}$ can be generated, the maximum pathlength is 30 meters. The main features of atmospheric turbulence and their consequences on optical beam are reproduced.

Some points have still to be elucidated: precise value of ℓ_0, measurement of the structure function at high wind speeds.In parallel with this work, new measurements have been undertaken:fluctuations of the angle of arrival of the wavefront,coherence diameter,domain of isoplanetism.

REFERENCES

1. L.R.Bissonnette, Appl.Opt.,Vol.16,n°18 (1977)
2. R.S.Rohde & R.G.Buser, Appl.Opt.,Vol.18,n°5 (1979)
3. H.Gamo & A.K.Majumdar,Appl.Opt.,Vol.17,n°23 (1978)
4. L.R.Bissonnette,Private Communication
5. H.T.Yura, Appl.Opt.,Vol.10,n°12 (1971)

ANALYSIS OF WAVEGUIDE SYSTEMS FOR TRANSMITTING AND SHAPING INFRARED RADIATION

A. Cutolo, L. Palumbo, and S. Solimeno

Istituto Elettrotecnico, Facoltà di Ingegneria
Via Claudio 21, 80125 Napoli
Italy

One of the long-standing problems in many laser applications is the low loss transmission of radiation from source to target. In recent years many different kinds of waveguides have been extensively studied in the search for flexible systems capable of delivering the output of high-power CO_2 lasers [1,2]. Because their transverse dimension is many wavelengths large and they transmit efficiently only low-order modes, there is a limit to the divergence of the beams delivered. Correspondingly, alignment and matching of a laser beam with the guide input section are very critical. Moreover, a wide acceptance angle should allow the use of short focus optical systems with great benefit for system compactness.

The aim of the present communication is to discuss and compare the properties of a dielectric-coated metallic guide with those of an uncoated one.

With reference to the geometry illustrated in Figure 1, the dispersion relation for the propagation constant reads

$$k_x \tan(k_x b + (m-1)\pi/2) = \begin{cases} -\dfrac{n_1^2}{k_{x1}} \tan(k_{x1}\delta + \psi) & \text{TM modes} \\ k_{x1}/\tan(k_{x1}\delta + \psi) & \text{TE modes} \end{cases}, \qquad (1)$$

where m is an integer > 0, and

$$\psi = \begin{cases} \tan^{-1}(ik_{x1}n_2/k) & \text{TM modes} \\ \tan^{-1}(ik_{x1}/n_2 k) & \text{TE modes} \end{cases}, \qquad (2)$$

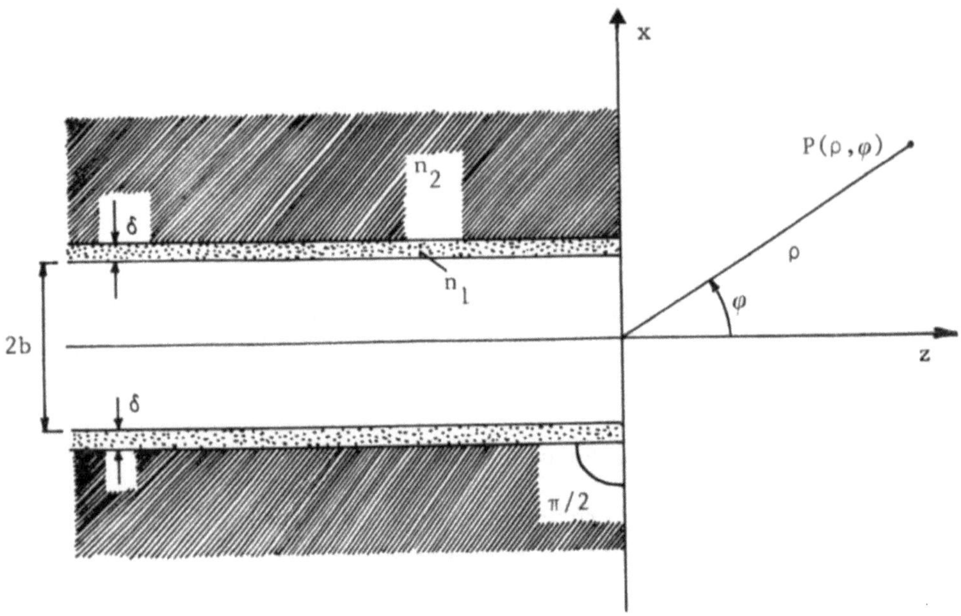

Fig. 1. The waveguide in its reference frame.

where n_1 and n_2 are the generally complex refractive index of the dielectric coating and of the metal substrate, respectively; δ is the dielectric thickness, and k_x and k_{x1} are the transverse propagation constants outside and inside the dielectric, respectively. Equation (1) rests on the field continuity condition at the interface air-dielectric and on the impedance boundary condition at the interface dielectric-metal. Equation (1) has been solved numerically and plots of k_x have been reported in Figure 2. For a more detailed analysis of Equation (1) see Reference 3.

The loss factor α_m^{TE} of the m-th mode has been calculated by solving Equation (1) and taking the imaginary part of the longitudinal propagation constant (see Figure 3).

The radiation pattern and the coupling of a cylindrical Gaussian beam into the uncoated metallic waveguide modes have been calculated by many authors [1] by assuming the field on the open end coincident with the mode itself. This assumption, corresponding to neglecting the reflection caused by the guide truncation, is no longer valid when there is a coating. In fact, a surface wave can be excited which, travelling backward, can detract power from the forward scattered wave. In this case the problem becomes particularly difficult and may be approximately solved by assimilating the waveguide walls as half-planes with two different surface impedances (see Figure 4) [3].

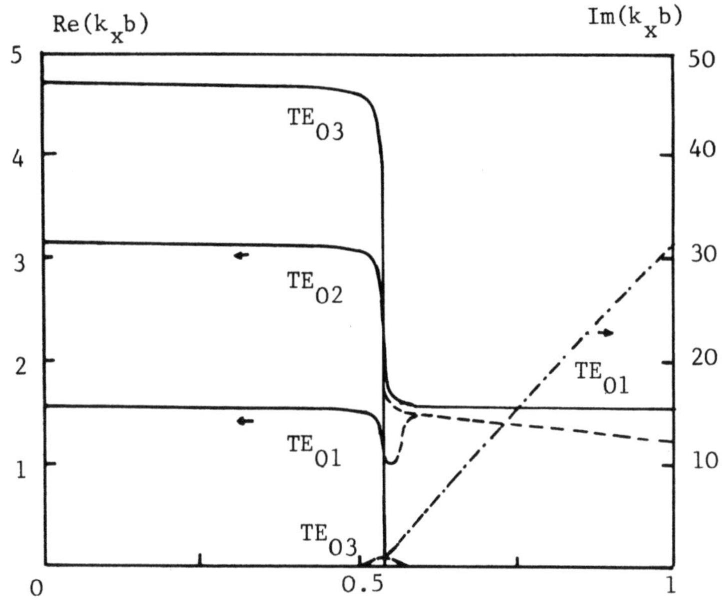

Fig. 2. Plots of Re(k_xb) and Im(k_xb) for TE_{01}, TE_{02}, TE_{03} modes for a waveguide with kb = 20, n_1 = 3 and n_2 = 7 + i60.

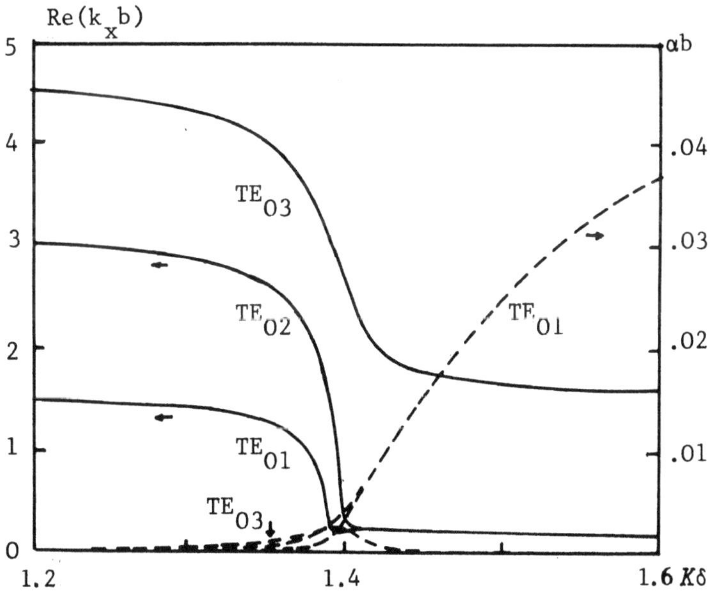

Fig. 3. Plots of Re(k_xb) (solid lines) and the loss factor α (dashed) lines) for the TE_{01}, TE_{02}, TE_{03} modes of a waveguide with kb = 150, n_1 = 3 and n_2 = 7 + i60.

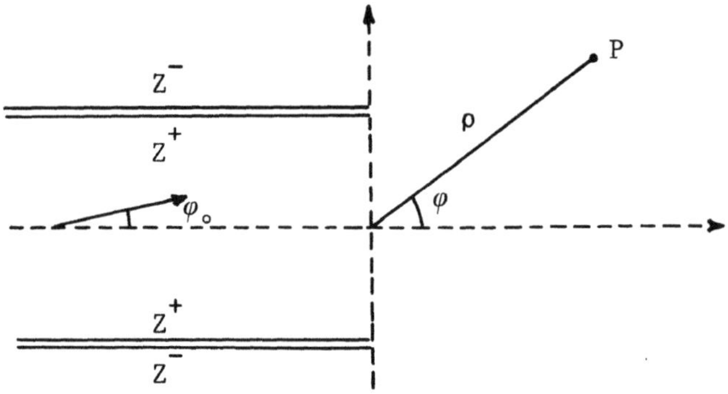

Fig. 4. The model of the waveguide with different surface impedances Z^+ and Z^-.

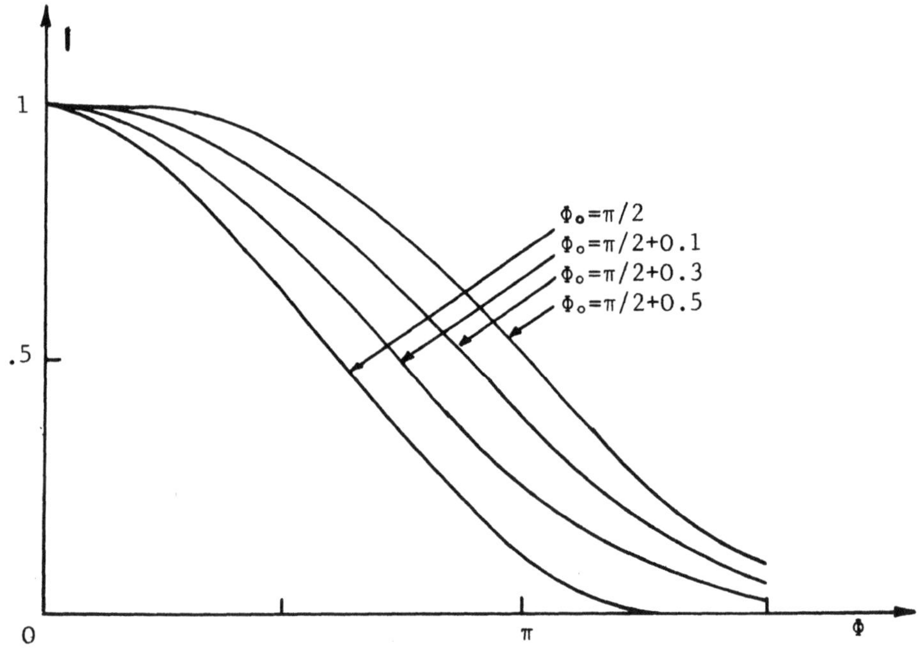

Fig. 5. The field radiated by a waveguide with dielectric loaded walls.

Taking advantage of the expression given by Maliuzhinets [4] for the field diffracted by a half-plane with two different surface impedances, neglecting the coupling between the edges and within the limit kb << 1, we get [3]

$$u_d(\rho,\phi) = -\frac{\exp(ik\rho+i\pi/4)}{(2\pi\rho/b)^{\frac{1}{2}}}\exp(i\Phi)\,\frac{\mathrm{sinc}(\Phi+\Phi_o)+\mathrm{sinc}(\Phi-\Phi_o)}{\mathrm{sinc}\Phi_o + i\,\frac{\Phi}{\Phi_o}\cos\Phi_o}, \tag{3}$$

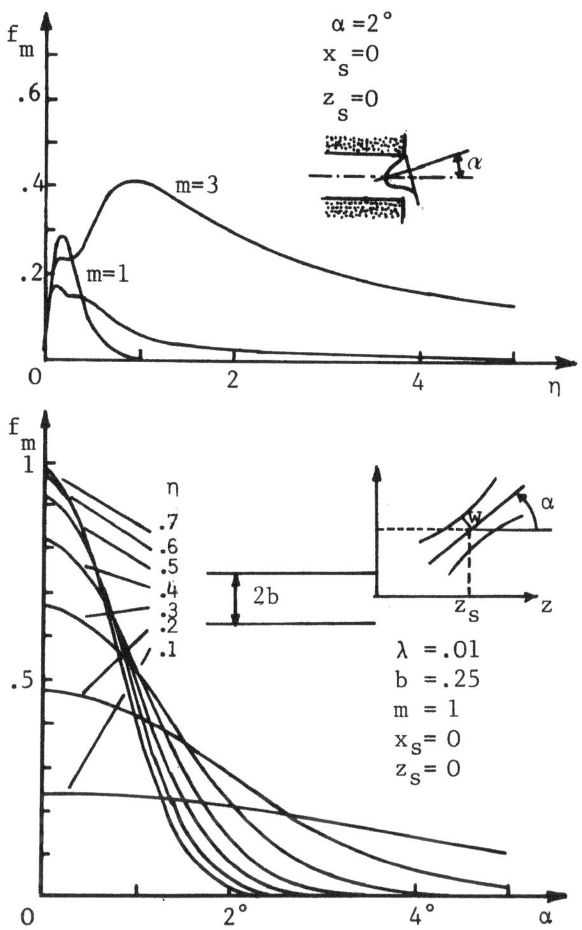

Fig. 6. Plots of the coupling efficiency f_m for an uncoated waveguide with kb = 150; η = w/b and all the dimensions are normalized to the wavelength.

where $\Phi_o = kb\sin\phi_o = k_x b$, $\Phi = kb\sin\phi$ and sinc(x) = sin(x)/x. Equation (3) has been plotted in Figure 5. The case $\Phi_o = \pi/2$ corresponds to an uncoated waveguide.

Now, we are able to calculate the coefficients of the modes excited by a cylindrical Gaussian beam into a waveguide. In fact, a cylindrical Gaussian beam can be thought of as generated by a current source $\underline{J}^s$ occupying a complex position (x_s, z_s). The waist center coordinates are given by $(\mathrm{Re}(x_s), \mathrm{Re}(z_s))$, while $\mathrm{Im}(x_s)$ and $\mathrm{Im}(z_s)$ are related to the waist dimension.

Relying on the reciprocity theorem, the amplitudes c_m of the m-th mode excited by a laser beam may be written as

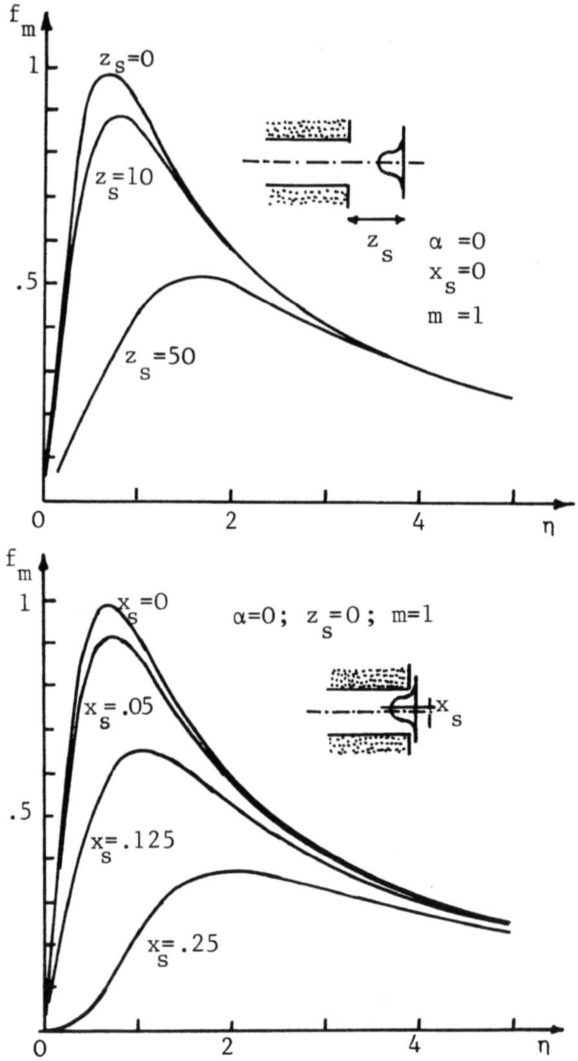

Fig. 6. Continued - see previous page for caption.

$$c_m = \frac{1}{2bE^s(w)} \underline{J}^s \cdot \underline{E}_m^r(x_s, z_s), \tag{4}$$

$E^s(w)$ being the field generated by the current source $\underline{J}^{(s)}$ at the waist of the incident Gaussian beam, and $\underline{E}_m^r$ the field radiated by the m-th mode and calculated at the complex point (x_s, z_s). Now, we define the coupling efficiency $F_m = |c_m|^2/P_i$ as the percentage of the incident power P_i coupled into the m-th mode. In Figure 6 we have plotted the coupling efficiency for an uncoated waveguide.

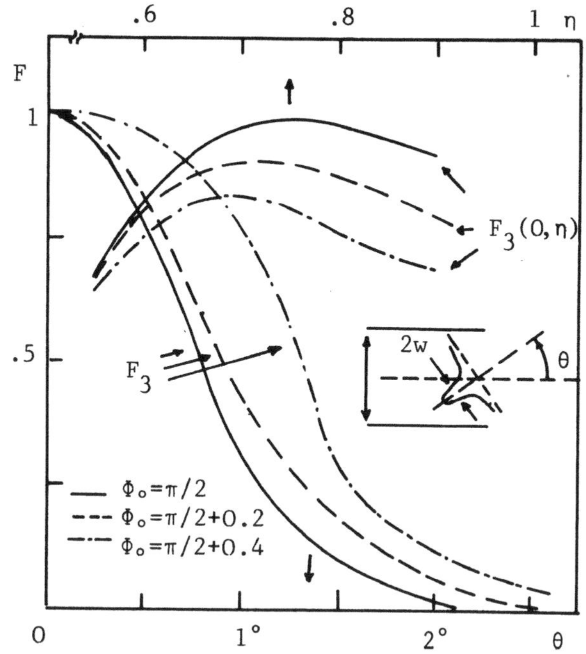

Fig. 7. Coupling efficiency $F(0,\eta)$ vs. $\eta = w/b$ and $\hat{F} = F(\Theta,\eta_{opt})/$ $/F(0,\eta_{opt})$ for different values of Φ_o. Θ is the angle between the waveguide axis and the laser axis. The waist center has been taken coincident with the guide input section center.

In Figure 7 the same quantity has been plotted for a waveguide with dielectric coated walls.

REFERENCES

1. E. Garmire, T. McMahon and M. Bass, IEEE J. Quant. Electron., QE-16:23 (1980).
2. L. Casperson, IEEE J. Quant. Electron., QE-15:491 (1979).
3. A. Cutolo and S. Solimeno, Optics Communications (in press).
4. G. D. Maliuzhinets, Soviet Phys. Dokl., 3:752 (1958).

RESONATORS FOR HIGH POWER LASERS

J. Dembowski, R. Hauck, H.P. Kortz, and H. Weber

Department of Physics
University of Kaiserslautern, Postfach 3049
D-6750-Kaiserslautern, W.-Germany

INTRODUCTION

High power fundamental mode laser systems require a careful design of the resonator. The simplified theory, neglecting gain saturation, mirror aperture, and mirror misalignment fails. To get reliable results, the Kirchhoff-Fresnel-integral equation has to be solved for limited mirrors, including mirror misalignment and gain saturation, which influences strongly the mode structure.

These effects will be discussed in two parts. The first part deals with the influence of the active medium and the mirror apertures on mode structure and farfield divergence. The second part gives a survey on the misalignment sensitivity of stable resonators.

GAIN SATURATION, MODE STRUCTURE, AND FARFIELD DIVERGENCE

Theory The active medium is characterized by the intensity dependent gain factor, which for homogeneous broadening reads:

$$G = \exp\left\{g_0(1 + J(r)/J_s)^{-1}\right\} \qquad G_0 = \exp\left\{g_0\right\} \tag{1}$$

with: g_0, G_0 = small signal gain, $J(r)$ = mode intensity, depending on the radial coordinate r, J_s = saturation intensity.

The intensity dependent gain G_0 has a strong influence on the mode structure, which is subject in this part of the paper.

For a homogeneous radial gain coefficient, the TEM 00 mode always dominates, since it has the lowest diffraction losses. But by the oscillating mode itself, the gain coefficient is saturated and a radial profile is produced. Consequently the oscillating modes are different from those of a homogeneous resonator and higher order modes predominate. To obtain the mode structure of the saturated resonator, the Kirchhoff-Fresnel integral equation has to be solved, taking into account the nonlinear gain coefficient[1]. For the numerical calculation, a symmetric, confocal resonator is considered, which is equivalent to the widely used semiconfocal system. The active medium is assumed as a thin slab at both mirrors. In this case, the integral equation for the field amplitude $\mathcal{E}_i$ on the mirror S_i reads:

$$\gamma \mathcal{E}_i = \hat{\imath}\pi F \int_0^1 \sqrt{RG}\, K(s_1,s_2)\, \mathcal{E}_j \, ds_j, \qquad i,j = 1,2 \quad (2)$$

with: γ= Eigenvalue, $|\gamma|$= 1 for stationary systems, $F = a^2/\lambda L$ = Fresnelnumber, K = kernel for circular symmetry, $s_j = r_j/a$ = normalized radial coordinate, a = radius of mirror aperture.

Results Eq. 2 was solved numerically by the well known Fox/Li method. In fig. 1 the radial intensity distributions on the mirrors for different apertures are shown. Two points are remarkable:

- with increasing s or Fresnelnumber $F = a^2/\lambda L = s^2/\pi$, resp., the higher order modes appear, without any additional assumption

- the modes differ strongly from those of a homogeneous resonator.

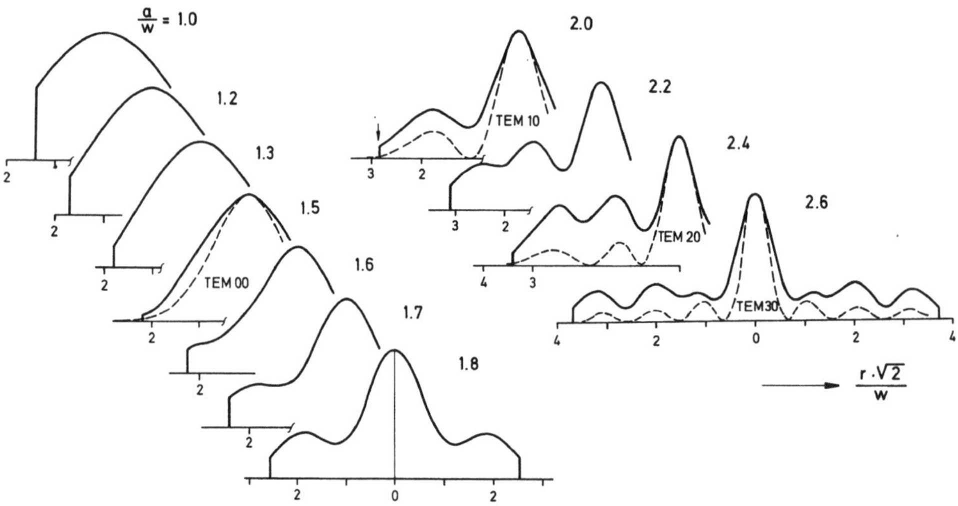

Fig. 1. Intensity distribution on the mirrors of a semiconfocal resonator. Parameter is the normalized pinhole radius s = a/w, w = Gaussian beam radius. Small gain factor G_0 = 2.7, reflectivities R_1 = 1, R_2 = 0.5.

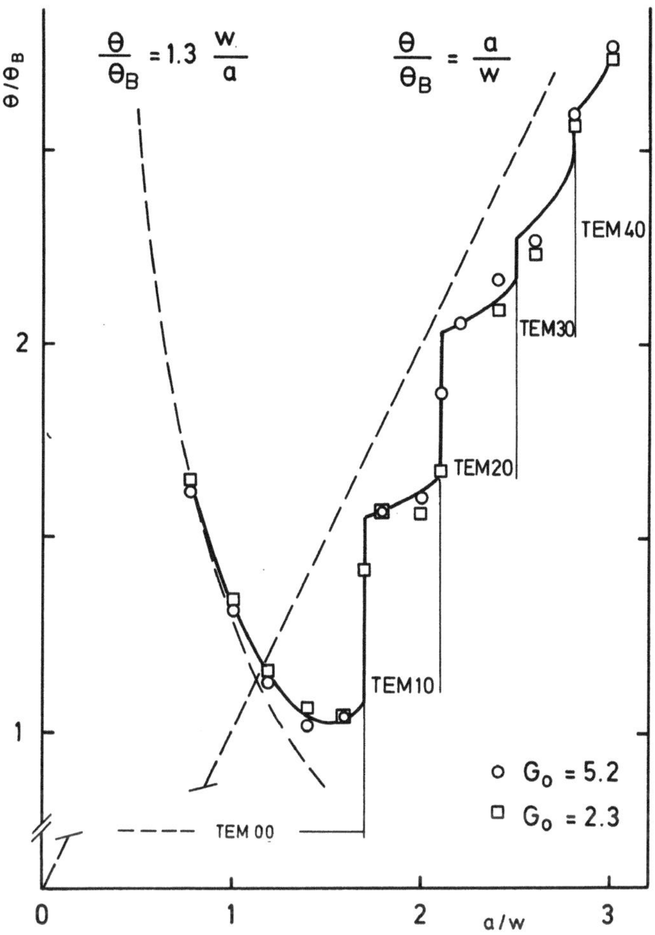

Fig. 2. Normalized farfield divergence Θ/Θ_B versus normalized pinhole radius a/w. Parameter is the small signal gain factor G_0. The dotted lines refer to eq. 3.

An important parameter is the farfield divergence Θ, which is obtained by Fouriertransform from the near field distribution. In fig. 2 the results for the semiconfocal resonator are compiled. The farfield divergence, normalized to the divergence Θ_B of the TEM 00 mode is plotted versus the normalized pinhole radius s. The dotted lines indicate approximation formulas, given by diffraction theory[2]

$$\Theta/\Theta_B = \begin{cases} 1.3/s & s \ll 1 \\ s & s \gg 1 \end{cases} \tag{3}$$

The main results are:

- an optimal value s_{opt} exists, which minimizes the beam divergence. $s_{opt} = 1 - 1.5$
- additional experiments have shown, that at s_{opt} the subsidiary diffraction peaks are minimized too.
- the appearance of higher modes is mainly determined by the Fresnelnumber.

Experiments were carried out with pulsed Nd-Glass and Nd-YAG lasers, operating in the spiking and active Q-switch mode. An example is given in fig. 3 and compared with the numerical calculation. The resulting valus s_{opt} are compiled in fig. 4.

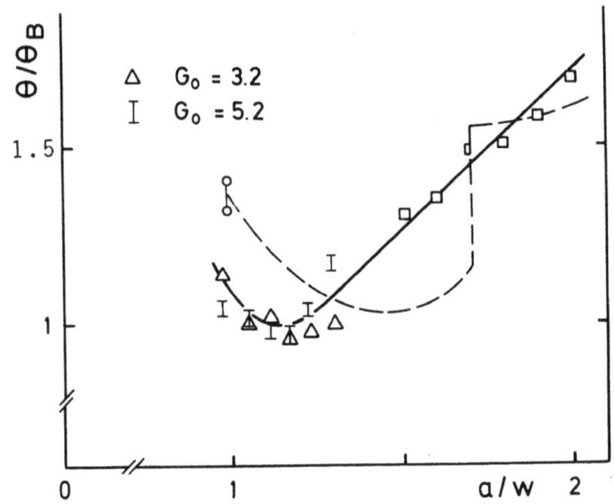

Fig. 3. Normalized farfield divergence Θ/Θ_B versus normalized pinhole radius a/w. --- theory, —— experiment, □: Nd-YAG (g_1=1, g_2=0.81); Δ,I: Nd-YAG (g_1=1, g_2=0,5).

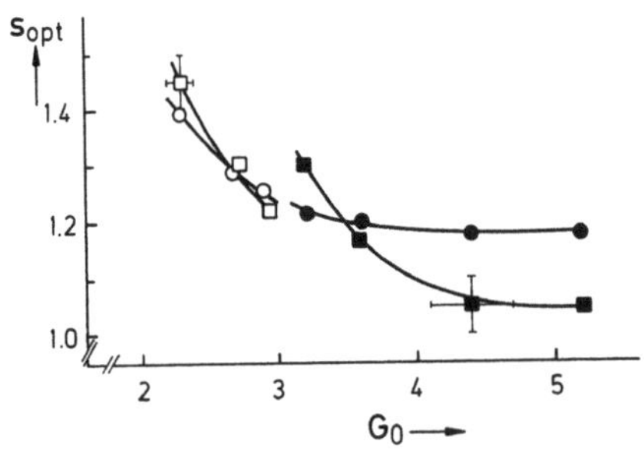

Fig. 4. Optimal values s_{opt} vs small signal gain G_0. □ : Q-switch Nd-glass, ■ : Q-switch Nd-YAG, o: spiking Nd-glass, ●: spiking Nd-YAG

From eq. 2, we also obtain the output power of the system. The efficiency η defined as:

$$\eta = \frac{\text{Output power calculated by eqs. 1,2 with } P = \int J \cdot 2\pi \cdot r dr}{\text{Output power for rectangular radial intensity distribution without diffraction losses}}$$

is plotted versus the normalized pinhole radius s in fig. 5. With larger s the diffraction losses decrease and the efficiency increases. In the region of s = 1,2...1,5, the TEM 00 mode changes into TEM 10 with higher diffraction losses and the efficiency decreases, but increases again with larger s.

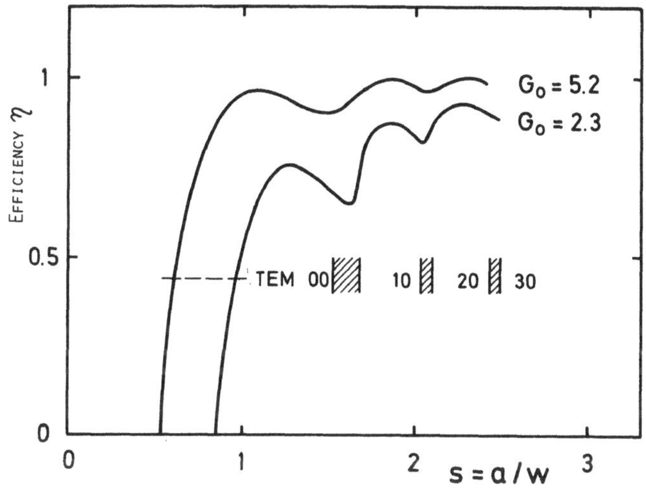

Fig. 5. Efficiency η as a function of normalized pinhole radius s = a/w. Parameter is the small signal gain G_o.

MISALIGNMENT SENSITIVITY

Theory If the mirror S_i of a spherical resonator is tilted by an angle α_i, the resonator axis is rotated by an angle Θ_i (see fig. 6), and the centers of the field intensity patterns are shifted relative to the aperture. Perturbation theory, using eq. (2) with G = 1 delivers [3,4] the loss factor per bounce:

$$V_i = V_o \cdot (1 - [\alpha_i \cdot D_i / f(s)]^2) = V_o (1 - [\alpha_i / \alpha_{oi}]^2) \qquad (4)$$

$$f(s) = [(\exp 2s^2 - 4)/8s^2]^{1/2} \qquad \alpha_{oi} = f(s)/D_i \qquad (5a,b)$$

V_o = loss factor (aligned resonator), D_i = misalignment sensitivity

α_{0i} is the characteristic tilt angle, which can be determined experimentally. An example is given in fig. 7. The sensitivity of the resonator is characterized by D_i, which is a function of the g_i-parameters:

$$D_i^2 = \frac{\lambda L}{\pi} \left(\frac{g_j}{g_i}\right)^{1/2} \frac{1 + g_1 g_2}{(1 - g_1 g_2)^{3/2}} \tag{6}$$

with: L = resonator length, $g_i = 1 - L/\rho_i$, ρ_i = radius of curvature of mirror S_i

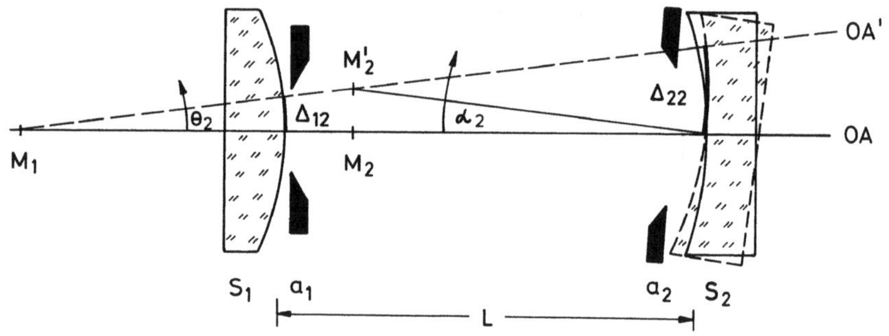

Fig. 6. The tilted, spherical resonator.

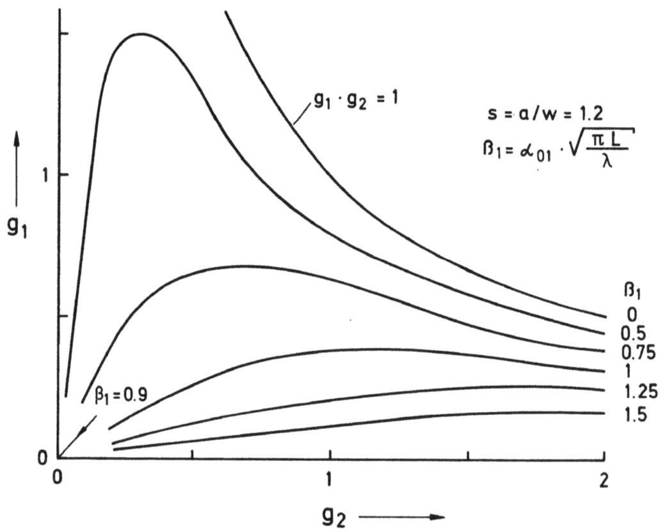

Fig. 7. Lines of constant angle β_1 in the g_i-diagram. β_1 is the normalized characteristic angle α_{01} according eq. 5b.

If both mirrors are misaligned, the losses proportional to D_i^2 are summed up. Therefore, the misalignment sensitivity of the complete system is defined as:

$$D = \left[D_1^2 + D_2^2\right]^{1/2} = \left[\frac{\lambda L}{\pi} \cdot \frac{1 + g_1g_2}{(1 - g_1g_2)^{3/2}} \quad \frac{|(g_1 + g_2)|}{(g_1g_2)^{1/2}}\right]^{1/2} \tag{7}$$

The relation between the resulting misalignment sensitivity D and the g_i-parameters is shown in fig. 8.

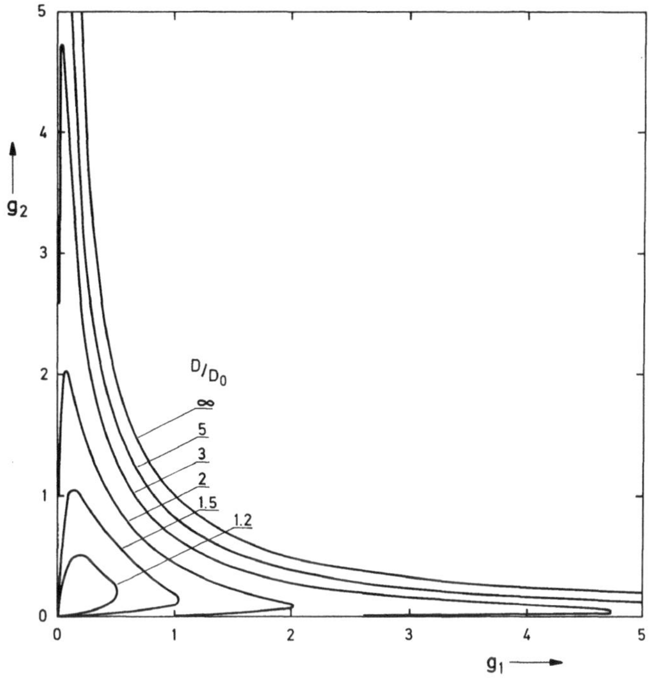

Fig. 8. Lines of constant misalignment sensitivity D/D_0 in the g_i-diagram. $D_0 = (2\pi L/\lambda)^{1/2}$.

Results For various resonators the characteristic values α_{0i} were determined experimentally and compared with perturbation theory according eq. (5b). The results are shown in fig. 9. To estimate the reliability of the perturbation theory, the losses of the tilted resonator were determined by solving eq. 2 numerically (diffraction theory) for some special resonators. Fig. 9 demonstrates, that the discrepancy between perturbation theory and exact theory is obvious. But anyhow, the simple formula of perturbation theory is sufficient to estimate the misalignment sensitivity of any stable resonator.

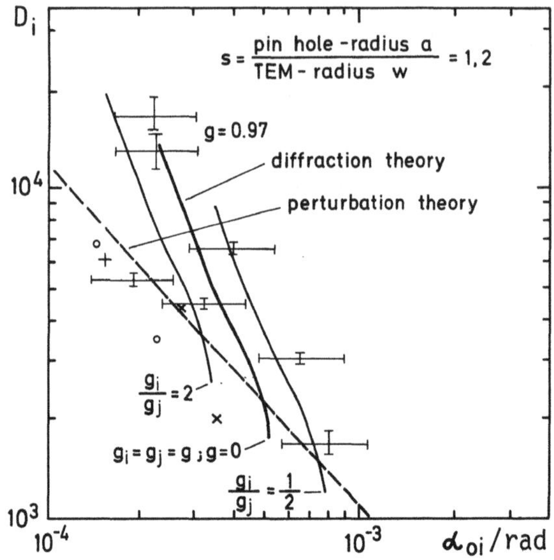

Fig. 9. Misalignment sensitivity D_i versus characteristic angle α_{oi} for s = a/w = 1.2
----- perturbation theory (eq. 5b) for s = 1.2
——— exact numerical results
⊢I⊣ experimental results (Nd-YAG, this experiment)
x,o,+ experimental results (Ar-Laser[5]).

This work was partially supported by the "Deutsche Forschungsgemeinschaft".

Literature

1. A.G. Fox and T. Li, Effect of Gain Saturation on the Oscillating Modes of Optical Masers, IEEE J. QE-2, 774 (1966)

2. J. Dembowski and H. Weber, Optimal Pinhole Radius for Fundamental Mode Operation, Opt. Comm. 42, 133 (1982)

3. N.K. Berger, N.A. Deryugin, Yu.N. Lukyanov and Yu.E. Studenikin, Open Misaligned Spherical Mirror Resonator, Opt. Spectr. (USSR) 43 (2), 176 (1977)

4. R. Hauck, H.P. Kortz and H. Weber, Misalignment Sensitivity of Optical Resonators, Appl. Optics 19, 598 (1980)

5. R.J. Freiberg and A.S. Halsted, Properties of Low Order Transverse Modes in Argon Ion Lasers, Appl. Optics 8, 355 (1969)

ABERRATIONS IN LASER BEAM EXPANDERS DUE TO TILT OF INPUT BEAM

R. W. Jones and Charles Cason

U. S. Army Missile Command
Redstone Arsenal, AL 35809

J. F. Perkins

M. K. Associates
2215 Mock Road
Huntsville, AL 35801

ABSTRACT

Tilt of the input beam to a beam expander, introduced for fast-steering purposes, leads to aberrations in the output beam. Properties of the aberrations in the output beam and their dependence on system parameters were investigated computationally by use of a recently-developed ray-tracing code. Methods for reducing the aberrations were also studied.

INTRODUCTION

In laser applications involving propagation of the beam over appreciable distances, it is often advantageous to expand the beam which emerges from the laser prior to propagating it to the target. Such a beam expander is similar to an optical telescope operated in reverse. A conceptual design of an overall laser beam-handling system is shown in Figure 1. The overall system includes, in addition to the beam expander, provisions for steering the beam. Commonly the major steering is done by a mirror or mirrors having the full size of the output beam, located downbeam of the expander. It is often almost essential that at least minor amounts of beam steering be done upbeam of the expander so that rapid-response steering can be done using a lightweight "fast-steering mirror (FSM)".

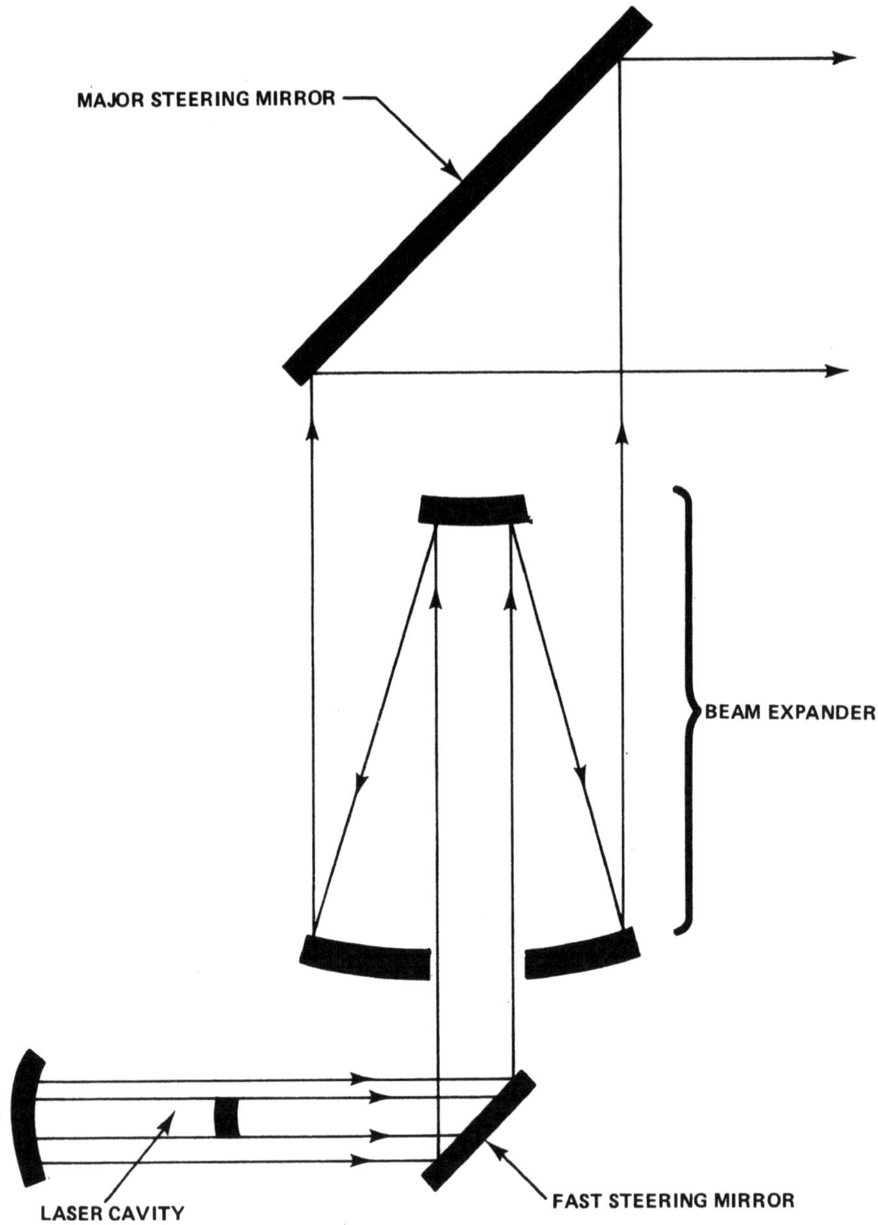

Fig. 1. Prototype beam handling system.

The tilt (introduced by the FSM or otherwise) of the input beam relative to the optical axis of the beam expander results in aberrations in the output beam. In the work reported here we have investigated the dependence of such aberrations on various system parameters, including: (1) transverse size of the beam, (2) tilt angle, and (3) axial position of the FSM. These investigations were carried out by use of a ray-tracing computer program which was recently developed (Cason, Jones, and Perkins, 1980; Perkins and Cason, 1980). A centered-alignment (i.e. Cassegrain) type of beam expander with expansion ratio of 2.23606 was considered. The prospects for reducing the resulting aberrations by introducing wave-front curvature into the input beam (e.g. by adaptive optics methods) were also investigated. It was found that the aberrations can be very greatly reduced.

METHOD

The basic beam-expander configuration was taken to be a confocal pair of paraboloidal mirrors whose optic axes are coincident. The nominal direction of the incident beam is parallel to the beam-expander axis in a direction such as to first strike the convex mirror, after passing through an opening in the concave mirror in a Cassegrainian arrangement). For simplicity the mirror separation has been taken to be 100.0 units; the absolute physical dimensions are not important since, except for a scale change, all beam expanders with the same relative values of dimensions will behave the same (such scaling was confirmed computationally, partially as a check on the overall computer program).

For most of the studies the input beam was assumed to essentially originate at an axial position coincident with the surface of the concave mirror (but variations from this position were later investigated).

The overall calculations proceed as follows. First a set of ray-tracing calculations are performed, using starting points which are equally spaced on an 11 by 11 square array. For each ray there is calculated the difference between the optical path length from its beginning point to its intersection with a final reference surface (normally a plane) and a reference length. This calculated array of optical path differences (OPD's) then serves as the basis for further calculations. The root-mean-square value of the optical path difference (RMSOPD) is then evaluated. Since an overall tilt of the beam relative to the final reference surface strongly affects the RMSOPD, one needs to be able to extract the effect of such a tilt. This can be done in either of two ways. Firstly, it can be done by simply repeating the calculations with various tilts of the final reference surface to achieve a minimal value of RMSOPD. Secondly, one can instead proceed as follows. A least-squares fit to the array of OPD's is made using three terms of a Taylor's series expansion, corresponding to piston and to tilt along two axes. The contribution corresponding to this least-squares fit is then subtracted from each OPD and the resulting REDUCED RMSOPD is evaluated. It has been numerically confirmed for a number of these calculations that the latter procedure leads to the same results as the former. The simpler procedure, involving REDUCED RMSOPD, was ordinarily used. For simplicity the optical-quality parameters RMSOPD, etc. are based on an unobscured beam. A more detailed evaluation of far-field intensity deterioration can be made by separate calculations which account for the obscuration. It was found that neglect of the obscuration does not greatly affect the estimate of optical-quality deterioration.

The nature and magnitude of the aberrations in the optical system can be investigated by performing a least-squares fit to a truncated Taylor's series expansion of the OPD's in the transverse coordinates, using as many as 10 terms. The various contributions to aberrations can be identified verbally as piston, tilts, defocus, spherical aberration, coma, etc. In the present studies it was found that the principal aberration (aside from piston and tilts) is simply defocus. This expansion term, of the form $(X^2 + Y^2)$, is the fourth term in the expansion, and the corresponding expansion coefficient obtained from the function fitting is labelled (e.g. in the following tables) as COF(4).

RESULTS

The variation of overall aberrations, as measured by the REDUCED RMSOPD with respect to beam size was first studied, using a fixed value of 10 milliradians for the input-beam tilt angle CXTLT. Results are shown in Table I. The parameters XFMAX and YFMAX represent the nominal halfwidths, along the X and Y coordinates, of the output beam. It turns out that the aberrations (as measured by REDUCED RMSOPD) scale very closely as the square of the aperture size, as can be seen from the right-hand column of the table.

The dependence of the aberrations on input-beam tilt angle CXTLT was investigated by carrying out calculations for several values of this tilt angle. It was found that both the magnitude of the overall aberrations as measured by REDUCED RMSOPD and the value of the defocus expansion coefficient COF(4) varied essentially as the square of the CXTLT. This dependence is directly illustrated for REDUCED RMSOPD in the last column of Table II, which displays results for various input beam tilt angles.

The starting position of the input beam as used in the calculations is denoted by ZCRD(1), with a value of zero representing the position of the concave mirror, and distance measured as positive toward the convex mirror, which is 100 units away. The value

Table I. Dependence of aberrations resulting from tilt of input beam on beam half-width, for input tilt angle of CXTLT = 0.010 radians. Beam expander has expansion ratio of 2.23606, and mirror separation of 100.

XFMAX= YFMAX	COF(4)	REDUCED RMSOPD	(REDUCED RMSOPD)/ (XFMAX**2)
6.66667	-0.34164918-7	0.75869445-6	0.17070608-7
8.33333	-0.34164849-7	0.11854580-5	0.17070609-7
10.0	-0.34164826-7	0.17070609-5	0.17070609-7
12.5	-0.34165034-7	0.26672828-5	0.17070610-7
16.66667	-0.34164966-7	0.47418388-5	0.17070613-7
20.0	-0.34164622-7	0.68282461-5	0.17070615-7
25.0	-0.34164894-7	0.10669138-4	0.17070621-7

Table II. Calculated values of REDUCED RMSOPD, the residual aberrations which remain after removal of the first three aberration contributions, i.e. piston and tilts, for various input-beam tilt angles. Nominal half-width of output beam is 25.0.

CXTLT	REDUCED RMSOPD	(REDUCED RMSOPD)/ (CXTLT**2)
0.001	0.10667818-6	0.10667818
0.002	0.42671432-6	0.10667858
0.005	0.26670344-5	0.10668137
0.010	0.10669138-4	0.10669138

of ZCRD(1) essentially represents the position of the FSM in an overall beam-handling system. The calculations reported above dealt with an axial position coincident with the concave mirror. It seemed appropriate to investigate possible effects of different axial positions of the FSM on the beam-expander aberrations. In Table III are given calculated values of COF(4) and REDUCED RMSOPD for ZCRD(1) values of -20 and +20. These results are directly comparable to values given in Table II for ZCRD(1) = 0. It is immediately evident that the aberrations are quite insensitive to the value of ZCRD(1).

The possibility of reducing the aberrations which result from input-beam tilt angle by adding an appropriate amount of wave-front curvature to the input beam was investigated numerically. The amount of wave-front curvature of the starting beam is defined in

Table III. Calculated values of COF(4) and of REDUCED RMSOPD for cases with tilted input beams which originate at axial positions ZCRD(1) which lie, respectively, outside and inside the region bounded by the primary and secondary mirror of the beam expander. Nominal half-width of output beam is 25.0.

CXTLT	ZCRD(1)	COF(4)	REDUCED RMSOPD
0.001	-20.0	-0.34163819-9	0.10667820-6
0.002	-20.0	-0.13665535-8	0.42671462-6
0.005	-20.0	-0.85410360-8	0.26670464-5
0.010	-20.0	-0.34164850-7	0.10669328-4
0.001	+20.0	-0.34163765-9	0.10667816-6
0.002	+20.0	-0.13665593-8	0.42671401-6
0.005	+20.0	-0.85410069-8	0.26670225-5
0.010	+20.0	-0.34164759-7	0.10668946-4

the computer program by the parameter CSPHR, the value of which can be assigned as desired. The expected value of CSPHR which would be required to cancel the aberrations can be shown to be given by

$$CSPHR = - 2 M^2 COF(4) \quad ,$$

where M is the expansion ratio, and COF(4) is the value of the "defocus" expansion coefficient which is calculated for an input beam with CSPHR = 0, i.e. without focus correction. Some results are shown in Table IV. The calculated values of REDUCED RMSOPD are to be compared to the corresponding values in Table II; it can be readily seen that the effects of aberrations are very greatly reduced. For these calculations we have also fairly carefully minimized the calculated value of RMSOPD with respect to the final reference surface angle ANGLE2(4) by repeated calculations. Since RMSOPD is directly calculated in a simple way from the array of OPD's it would be considered more reliable a priori than the REDUCED RMSOPD. It is clear that the deterioration of the overall optical quality of the output beam from a beam expander which results from tilting of the input beam relative to the optic axis can be very greatly reduced by introducing a suitable amount of wave-front curvature onto the beam which is input to the expander. Such a wave-front modification should be relatively simple to produce (especially as compared to a general multizonal adaptive optics scheme).

Table IV. Calculated values of RMSOPD and REDUCED RMSOPD for cases with input wave-front curvature (CSPHR parameter) chosen to reduce aberrations. Nominal half-width of output beam is 25.0. Values of RMSOPD are based on a reference-surface angle which approximately minimizes RMSOPD.

CXTLT	CSPHR	RMSOPD	REDUCED RMSOPD
0.001	0.3416-8	0.1408494-10	0.1405024-10
0.002	0.1366-7	0.18419451-9	0.18419422-9
0.005	0.8541-7	0.70341449-9	0.70340938-9
0.010	0.3416-6	0.64266967-8	0.64266947-8

It was also confirmed by direct calculations that introduction of appropriate phase-front curvature onto the input beam very greatly reduces the aberrations for cases with axial starting position (i.e. FSM positions) of -20 and +20.

CONCLUSIONS

The results obtained in these investigations indicate that the aberrations which result from tilt of the input beam to a laser beam expander: (1) scale as the square of the output aperture, (2) vary as the square of the input tilt angle, (3) consist very largely of wave-front curvature, i.e. defocus, (4) can be very greatly reduced in magnitude by introducing a suitable amount of wave-front curvature onto the beam which is input to the expander. The possibility of essentially eliminating tilt-induced aberrations by adding phase-front curvature seems to offer promise of making more pronounced use of beam steering ahead of the expander than would otherwise be practicable.

REFERENCES

Cason, C., Jones, R. W., and Perkins, J. F., 1980, Optical Quality of Tilted Spherical Mirror Unstable Resonators, J. de Physique C9: 385.

Perkins, J. F. and Cason, C., 1980, Spherical-Aberration-Free Unstable Resonator with Spherical Primary Mirror, J. de Physique C9: 393.

MIRRORS AND WINDOWS FOR HIGH-ENERGY LASERS:

THE WAVEFRONT DISTORTION PROBLEM

Claude A. Klein

Research Division
Raytheon Company
Lexington, Massachusetts 02173

ABSTRACT

High-energy laser (HEL) systems involve an optical train consisting of mirrors and windows, which may compromise the system's operation because of unavoidable wavefront distortions resulting from the absorption of some fraction of the incident laser-beam energy. The first part of the paper concerns actively cooled HEL reflectors; a generalized mirror/heat-exchanger model is used for evaluating both irradiance-mapping and coolant-pressure induced distortions. The theory of thermal lensing of HEL windows is reviewed in part 2; the purpose is to clarify the role of stress-induced birefringence effects and to apply the figure-of-merit concept to key laser-window material candidates.

INTRODUCTION

The development of high-energy laser (HEL) systems has stimulated a great deal of activity relating to the physics and the technology of optical components designed for handling high-power beams of infrared light. The reason for this is that typical HEL systems involve an optical train, which must be capable of transporting and directing the laser beam without seriously degrading the nominal performance of the system. Available experience indicates that catastrophic failure modes associated with thermally induced stresses or temperature excursions are not a major threat. The performance of the system, as measured in terms of achievable target irradiances, initially degrades on account of "thermal lensing," that is, the process of beam defocussing and beam distorting caused by thermally generated phase aberrations. Since HEL operation usually requires close-to-diffraction limited characteristics at the far-field focus, this lensing phenomenon must be carefully assessed in relation to specific features of the optical train. In this paper, I will summarize analytical investigations of relevance in evaluating how laser-driven mirror/window distortions may affect the performance of HEL systems; specifically, I will provide simple expressions for judging the relative merits of both mirror-substrate materials and window-material candidates with regard to enhancing the focal intensity.

In the absence of birefringence, the far-field intensity derives from the Huygens-Fresnel integral[1] and amounts to

$$I_{GF} = (k/R_o)^2 \left| \int_0^{D/2} E(\rho) \exp[i\delta\phi(\rho,t)]\rho d\rho \right|^2 \quad (1)$$

at the Gaussian focus, on assuming cylindrical geometry.* The phase-aberration function $\delta\phi$ maps the cumulative aberrations impressed on the beam while propagating through the optical system and relates to the optical path difference (OPD) simply by means of the propagation constant. For weak distortions, we may proceed by expanding $\exp(i\delta\phi)$ to second order, which leads to a highly compact expression:

$$I_{GF} = \kappa_o \frac{\pi(D/2)^2 P}{(\lambda R_o)^2} \left\{ 1 - k^2 \text{var}[\text{OPD}] \right\} \quad (2)$$

if the variance is defined in accord with $\text{var}[X] = \langle X^2 \rangle - \langle X \rangle^2$, the brackets symbolizing amplitude-weighted averages. Note that κ_o represents a beam-profile factor and accounts for the possibility that non-uniform beam intensities may cause some reduction in far-field irradiance, even in the absence of near-field phase aberrations. The degradation in focal intensity thus obeys a "Strehl-relation" type equation,[1] which implies that the optical train should remain nearly diffraction-limited as long as the square root of the OPD variance does not exceed $\lambda/14$. Throughout the rest of this paper, we will rely on the magnitude of var[OPD] for assessing the loss in focal irradiance caused by the emergence of phase aberrations in the near field.

THE MIRROR PROBLEM

Currently operational beam-relay/beam-steering mirrors consist of a thin, highly reflecting faceplate, which transfers to a coolant fluid the fractional laser energy that is deposited at the surface, and which is stiffly held to a stable backup structure that minimizes thermomechanical deformation modes. Thermally induced surface deformations resulting from spatial variations in beam intensity generate "irradiance-mapping" wavefront distortions and have been the subject

---

* For a listing of the symbols, refer to the Appendix.

of a number of investigations. Mirror "bowing" involves low-spatial frequency deformations, which are usually modelled as spherical aberrations[2]; here, we will assume that they can be compensated for by means of refocussing procedures and, thus, can be ignored in the context of assessing the performance of HEL mirrors. This, however, does does not apply to deformations of the faceplate resulting from the flow of pressurized coolant, unless they can be reduced to a negligible amount or essentially eliminated by figuring the mirror surface under a static pressure distribution that simulates the actual operating pressures of the heat exchanger. An evaluation of the variance of mirror-induced OPDs, therefore, must include the initial figuring error as well as high-spatial-frequency deformations caused by the incident beam and the pressurized coolant. At normal incidence, and since the mechanisms involved are uncorrelated, this means that

$$\sqrt{\mathrm{var}[\mathrm{OPD_M}]} = 2[(\tau_{F,M})^2 + (\Delta_{I,M})^2 + (\Delta_{P,M})^2]^{1/2} \tag{3}$$

if $\Delta_{I,M}$ and $\Delta_{P,M}$ refer to properly RMSsed mirror surface deformations associated with irradiance and pressure, respectively.

Beam-Induced Distortion

If a laser beam of intensity I impinges on a mirror surface of absorptance A_M, there is a constant flow of heat, $A_M I$, into the substrate. If, in addition, we postulate that the rear surface can be maintained at a constant temperature T(0), the steady-state temperature distribution through the faceplate becomes a linear function of the axial position[3]:

$$T(z)-T(0) = (A_M I/K)z. \tag{4}$$

In equilibrium, the temperature of the back surface, measured with respect to the temperature of the coolant, levels off at

$$T(0) = A_M I/h , \tag{5}$$

thus indicating that laser irradiance results in a temperature increase

$$\delta T(z) = A_M I(1/h + z/K). \tag{6}$$

Out-of-phase growth of the front surface caused by temperature increments $\delta T(z)$ derives from a single integration,

$$\delta \ell_I = \int_0^\ell \varepsilon_z(z)dz , \tag{7}$$

where $\varepsilon(z)$ is the differential strain for axial expansion and includes stress-related contributions in accord with Hooke's law.[4] Thermally induced stresses are difficult to evaluate in a situation as applies here, where the temperature is a function of both axial and radial coordinates; in the present case, however, thin-plate theory may provide good estimates and yields[5]

$$\varepsilon(z) = \alpha(1+\nu)\delta T(z) - \alpha\nu\delta T_{AV}(z) , \qquad (8)$$

where $\delta T_{AV}(z)$ refers to the z-dependent average temperature over the mirror area. Since a term proportional to δT_{AV} does not contribute to the variance, it follows from (6), (7), and (8) that

$$\sqrt{\mathrm{var}[\delta\ell_I]} = \alpha(1+\nu)[(\ell/h)+(\ell^2/2K)]A_M \sqrt{\mathrm{var}[I]} . \qquad (9)$$

In this simple model, the RMSsed surface deformation increases linearly with $\sqrt{\mathrm{var}[I]}$, the slope involving a combination of material parameters and design features that provides a measure of the performance of the reflector:

$$\Delta_{I,M} = (1/\mathrm{FoM}_M) \sqrt{\mathrm{var}[I]} . \qquad (10)$$

In this manner, we are introducing the concept of a "mirror figure of merit," which boils down to

$$\mathrm{FoM}_M = \frac{h/\ell}{A_M \cdot \xi} , \qquad (11)$$

with $\xi = \alpha(1+\nu)$, considering that, for realistic values of the heat-transfer coefficient ($h \leq 10\ \mathrm{Wcm^{-2}K^{-1}}$) and the faceplate thickness ($\ell \leq 1$ mm), the thermal conductivity is an essentially irrelevant parameter from the point of view of rating the performance of the mirror. In Table 1 are the figures of merit of high-performance HEL reflectors with multilayer dielectric coatings that are representative of the current state of the art; since Poisson's ratio ν does not affect the distortion coefficient ξ in a significant way, we conclude that the thermal expansion coefficient of the substrate represents the single most important material parameter if irradiance mapping is the only source of dynamic surface distortion.

Coolant-Induced Distortion

Distortions caused by the flow of coolant originate from bending and shear deformations induced by the pressure acting in the cooling channels but may also include a thermal component associated with frictional heating that scales with pressure. For mirror/heat-exchanger configurations as contemplated by Holt,[6] it appears that, averaged in the direction normal to the flow, the coolant-induced surface deflection at station $\underline{x}$ is

$$\delta\ell_P = P\ell[(19.7/Y)(1-|x/L_f|) + .0107\alpha|x/L_f|] , \qquad (12)$$

Table 1. Figure of Merit and Index of Goodness for HEL Mirror-Substrate Material Candidates.

Substrate Material	Figure of Merit[a]		Index of Goodness	
	GW/cm^3	Rank	Tpsi·K	Rank
Copper	9.6	6	1.07	5
Molybdenum	26.1	5	9.40	3
Tungsten	28.1	4	12.50	2
Carbon/Carbon	101.0	1	0.77	6
Silicon	50.7	2	6.15	4
Silicon Carbide	29.6	3	13.33	1

[a] Assumes $A_M = .1\%$, $h = 10\ Wcm^{-2}K^{-1}$, and $\ell = .5$ mm.

which reflects the observation that pressure-induced deflections peak at the coolant inlet ($x = 0$), whereas frictional heating peaks at the outlet ($x = \pm L_f$). In a first approximation, therefore, the RMS distortion amounts to

$$\Delta_{P,M} = (1/2\sqrt{3})P\ell[(19.7/Y) - .0107\alpha] , \tag{13}$$

which suggests that the effect of frictional heating should be minimal even with high-stiffness materials; for this reason, we conclude that Young's modulus is the only other material parameter to be considered if the situation requires that coolant-related deformations be minimized.

Guidelines for selecting the "best" mirror-material candidate thus emerge in a straightforward manner: The coefficient of thermal expansion, α, should be small enough to reduce beam-induced distortions, and the modulus of elasticity, Y, must be large enough to minimize coolant-induced deformations. In this light, it follows that a simple combination of these two material parameters can provide an appropriate "index of goodness,"

$$IoG = Y/\alpha, \tag{14}$$

which should be maximized in order to enhance the mirror performance. Table 1 shows how HEL mirror-substrate materials rate in terms of this index; silicon carbide and tungsten are the leading candidates while low-modulus composites such as carbon/carbon become just as undesirable as copper, which is handicapped by its high expansion coefficient.

THE WINDOW PROBLEM

Windows can distort an incident beam in a complex manner because laser-induced phaseshifts reflect changes in the optical pathlength arising from position-dependent variations in window thickness in addition to position-and polarization-dependent variations in refractive index.[7] This problem is now well understood in the sense that a theory of window thermal lensing has been developed and shown to be amenable to practical calculations for relatively simply configurations in terms of material characteristics as well as beam geometry. In its present form, the theory applies only if the following assumptions are verified: (a) The window is subjected to axially symmetric thermal loadings; (b) The elastic and photoelastic properties are isotropic in the plane of the window; and (c) The stress distribution obeys classical thin-plate theory.

Optical Distortion Coefficients

Under these conditions, and in the absence of mechanical loadings, unconstrained laser windows are subjected to stress distributions characterized by cylindrically symmetric radial and azimuthal components, σ_ρ and σ_θ, which relate to the temperature profile in a fairly simple manner.[4] Furthermore, the distortion of the beam can be described by means of two principal phaseshifts, $\delta\phi_\rho$ and $\delta\phi_\theta$, which are the phaseshifts experienced by normally incident light when polarized either in the radial or the azimuthal direction. Considering that σ_ρ and σ_θ differ everywhere except at the origin, it follows that, for windows made of stress-birefringent material, the two principal phaseshifts are also different. These phaseshifts are best expressed as follows[8]:

$$\delta\phi_\rho = \delta\phi_+ + \delta\phi_- \qquad (15a)$$
$$\delta\phi_\theta = \delta\phi_+ - \delta\phi_-, \qquad (15b)$$

where $\delta\phi_+$ and $\delta\phi_-$ represent two aberration functions, which can be specified in a compact manner,

$$\delta\phi_+ = (k\chi_+ L)\,\delta T \qquad (16a)$$
$$\delta\phi_- = (k\chi_- L)(\overline{\delta T} - \delta T), \qquad (16b)$$

on using the symbol $\overline{\delta T}$ to designate the mean temperature increase from the center of the window to the radial position of interest. On this basis, we define two optical distortion coefficients,

$$\chi_+ = (\partial n/\partial T)_{\sigma=0} + (n-1)\alpha(1+\nu) + (n^3\alpha Y/4)(q_\parallel + q_\perp) \qquad (17a)$$
$$\chi_- = (n^3\alpha Y/4)(q_\parallel - q_\perp), \qquad (17b)$$

which regroup all material properties required to assess the susceptibility to thermal lensing of stress-birefringent laser windows. The coefficient χ_+ reflects the temperature-induced isotropic change in index, the contribution due to surface bulging through constrained expansion, and the photo-elastic effect averaged over the two principal stress directions; the χ_- coefficient involves only the stress-induced birefringence. It follows immediately that the relative importance of birefringence-dependent effects will be determined by the ratio χ_-/χ_+. In the case of polycrystalline (fusion-cast) CaF_2, for example, a correct evaluation of the distortion coefficients at 3.39 μm yields[9] $\chi_+ = (0.2 \pm .3) \times 10^{-6}K^{-1}$ and $\chi_- = -0.5 \times 10^{-6}K^{-1}$, thus indicating that thermal lensing will be dominated by stress-birefringence effects, which explains why depolarization can become highly apparent with CaF_2 windows that are not [111] oriented. In the case of polycrystalline (hot-forged) KCl as well as polycrystalline (CVD) ZnSe, it is clear (see Table 2) that the small-birefringence condition is satisfied, which ensures polarization-independent lensing; note that the coefficient χ_+ of KCl takes on negative values, which is in accord with the observation that KCl windows behave in the manner of a negative lens.

Window Figure of Merit

If we choose to characterize the window-induced performance degradation on the basis of the reduction in light intensity at the Gaussian focus, we find that the applicable Strehl relation becomes[8]

$$i = 1-\{\mathrm{var}[\delta\phi_+] + \langle\delta\phi_-^2\rangle\}. \tag{18}$$

For convenience, we now relate the degradation in focal intensity to the variance of the temperature rise; in other words, we postulate that the amplitude-weighted RMSsed OPD caused by the distorted window is given by

$$\Delta_{I,W} = \chi L\sqrt{\mathrm{var}[\delta T]}. \tag{19}$$

This amounts to injecting the concept of an "effective optical distortion coefficient,"

$$\chi = |\chi_+|[1+G(\chi_-/\chi_+)^2]^{1/2}, \tag{20}$$

where G is a temperature-profile dependent factor, which is close to 0.8 for truncated Gaussian shapes.[8]

We now consider a pulsed-mode of HEL operation for which the time scales are such that lateral heat conduction and surface-cooling effects can be ignored. In this approximation, the window temperature images the incident beam fluence; in terms of the window absorptance, $A_W = 2\beta_S + \beta_V L$, the variance of the temperature increase is therefore

Table 2. Key Window-Material Characteristics at HF/DF Laser Wavelengths.

Item	Symbol	Units	CaF_2	KCl	ZnSe
Heat Capacity	(ρC_p)	$Jcm^{-3}K^{-1}$	2.71	1.35	1.79
Coating Absorptance	(β_S)	10^{-3}	0.2	0.2	0.75
Absorption Coefficient	(β_V)	$10^{-3}cm^{-1}$	0.1	0.5	1.5
Distortion Coefficient	(χ_+)	$10^{-6}K^{-1}$	0.2	-9.4	74
Distortion Coefficient	(χ_-)	$10^{-6}K^{-1}$	-0.5	-0.6	-3.5
Figure of Merit[a]	(FoM_W)	10^6Jcm^{-3}	$1.1x10^4$	160	8.1

[a] For 1-cm thick windows.

$$\mathrm{var}[\delta T] = \frac{(A_W/L)^2}{(\rho C_p)^2}\,\mathrm{var}[F] \,. \tag{21}$$

Returning now to Eq. (19), we may write

$$\Delta_{I,W} = (1/FoM_W)\,\sqrt{\mathrm{var}[F]} \,, \tag{22}$$

the "window figure of merit" FoM_W combining all window-related features that affect thermal lensing:

$$FoM_W = \frac{\rho C_p}{A_W \cdot \chi} \,. \tag{23}$$

The figures of merit and relevant properties of three window-material candidates for operation at HF/DF laser wavelengths are listed in Table 2; evidently, CaF_2 is by far the most promising candidate but it should be kept in mind that this figure-of-merit assessment disregards the impact of coherently additive phase aberrations on the overall system performance.[10]

Finally, if we make use of a beam-profile factor for relating the variance of the irradiance to the average power density over the aperture area, the Strehl relation (18) becomes

$$i = 1-k^2[(\kappa/FoM_W)F_{AV}]^2 \,. \tag{24}$$

On assuming beam-profile factors that are representative of realistic beam shapes ($\kappa \simeq 0.5$), the figures of merit in Table 2 indicate that, in principle, CaF_2 windows can pass fluences of the order of 300 kJ/cm^2, without excessive wavefront distortion, whereas ZnSe cannot tolerate more than 300 J/cm^2.

APPENDIX

A : optical absorptance (1)
D : aperture diameter (cm)
E : beam amplitude ($W^{1/2}cm^{-1}$)
F : beam fluence (Jcm^{-2})
h : heat-transfer coeff. ($Wcm^{-2}K^{-1}$)
i : Strehl ratio (1)
I : beam irradiance (Wcm^{-2})
k : propagation constant (cm^{-1})
K : thermal conductivity ($Wcm^{-1}K^{-1}$)
ℓ : faceplate thickness (cm)
L : window thickness (cm)
L_f: coolant pathlength (cm)
n : refractive index (1)
P : beam power (W)
P_f: coolant pressure (psi)
q : stress-optic coeff. (psi^{-1})
R_0: focal length (cm)
t : elapsed time (sec)
Y : Young's modulus (psi)
z : axial coordinate (cm)
α : expansion coeff. (K^{-1})
$\delta\phi$: phase aberration (1)
Δ : RMSsed deformation (cm)
κ : profile factor (1)
λ : laser wavelength (cm)
ν : Poisson's ratio (1)
ξ : mirror distortion coeff. (K^{-1})
ρ : radial coordinate (cm)
τ_f: figuring error (cm)
χ : window distortion coeff. (K^{-1})

REFERENCES

1. M. Born and E. Wolf, "Principles of Optics," Pergamon Press, New York (1975).
2. H. E. Bennett, Thermal distortion thresholds for optical trains handling high pulse powers, in: "Laser-Induced Damage in Optical Materials:1976," A. J. Glass and A. H. Guenther, eds., USGPO, Washington (1976).
3. H. S. Carslaw and J. C. Jaeger, "Conduction of Heat in Solids," Oxford University Press, Oxford (1959).
4. B. A. Boley and J. H. Weiner, "Theory of Thermal Stresses," John Wiley, New York (1960).
5. C. A. Klein, Mirror figure of merit and material index of goodness for high-power laser-beam reflectors, in: "Proceedings of the Los Alamos Conference on Optics:1981," D.H. Liebenberg, ed., SPIE, Bellingham/Washington (1981).
6. T. Holt, Trade study of parameters influencing high-spatial frequency distortion in water-cooled mirrors, in: "Proceedings of the High Power Laser Optical Components Meeting:1979," J. L. Stanford, ed., NWC, China Lake/California (1980).
7. M. Sparks, Optical distortion by heated windows in high-power laser systems, J. Appl. Phys. 42:5029 (1971).
8. C. A. Klein, Concept of an effective optical distortion parameter:application to KCl laser windows, Infrared Phys. 17:343 (1977).
9. C. A. Klein, Stress-induced birefringence, critical window orientation, and thermal lensing experiments, in: "Laser-Induced Damage in Optical Materials:1980," H.E. Bennett, A. J. Glass, A. H. Guenther, and B. E. Newnam, eds., USGPO, Washington (1981).
10. C. A. Klein, Thermally induced optical distortion in high-energy laser systems, Opt. Eng. 18:591 (1979).

AN ANALYSIS OF OPTICAL RESONATORS WITH TAPERED-REFLECTIVITY MIRRORS

Paolo Luchini and Salvatore Solimeno

Istituto Elettrotecnico, Facoltà di Ingegneria

Università di Napoli, Italy

INTRODUCTION

In recent years many attempts have been undertaken to devise approximate analytical descriptions of unstable optical resonators. The starting point for a large part of this work has been the analysis of plane resonators carried out by Weinstein[1,2].

According to Weinstein's method, a two-dimensional open resonator can be imagined as if it were cut out of an infinite metallic waveguide formed by two opposite infinite mirrors; when the Fresnel number is high, the resonator modes appear as superpositions of waveguide modes traveling back and forth between the open ends formed by the mirror edges. The calculation of the configuration of resonator modes and their resonance frequencies and damping factors acquires then the form of a consistency condition for the reflection at the opposite open ends.

Felsen and his collaborators[3,4,5,6] have extended the above method to cylindrical and spherical resonators by using the parabolic cylinder functions to express the propagation modes of these systems. They have extended the expressions for the reflection coefficients at the open ends found by Weinstein by using an optical approximation of the Green function of the cavity. The analysis of unstable optical resonators shows that the sharp discontinuity of the reflectivity at the mirror edges is the cause of a complicate dependence of the cavity loss and phase factors on the Fresnel number. A more physical analysis should be based on a continuous decrease of the reflectivity.

To this end we have extended the Weinstein method to cylindrical resonators with tapered-reflectivity mirrors, with particular attention to Gaussian tapering. In our analysis a W.K.B. approximation of the infinite-resonator modes is used, taking into account reflections from both the mirror edges and the modal caustics. The caustic reflection coefficients are calculated by standard complex-plane techniques[7], while for the edge reflection coefficients the expression given by Weinstein[2] is used. A method to improve the calculation of the reflection coefficients has been suggested in Ref. 8.

Aim of this paper is to present the results of this theory compared to those obtained by numerical solution of the Fox and Li equation. The satisfactory agreement between the results obtained for tapered-reflectivity resonators allows us to conclude that the Weinstein-Felsen one-mode theory works much better in the case of tapered-reflectivity resonators than in the case of abrupt mirror edges, since in the former case a single traveling mode is sufficient to describe the oscillation mode of the cavity.

Tapered-reflectivity resonators can moreover be used as a model of resonators with a tapered-gain active medium, as is often the case with real laser oscillators. The fact that a tapered-gain medium may prevent the excitation of higher-order modes and the irregular mode-loss behavior characteristic of unstable optical resonators could explain why this behavior is difficult to observe experimentally.

THE DIMENSIONLESS INTEGRAL EQUATION

By expressing the optical field at each mirror of a two-dimensional resonator as a functional of the field at the other mirror by means of the Fresnel diffraction integral, one obtains the classical Fox and Li integral equations[9]:

$$u_2(x') = \frac{e^{ikd}}{\sqrt{i\lambda d}} \int_{-W_1'/2}^{W_1'/2} u_1(\xi)\ \exp\ ik\left[\frac{(x'-\xi)^2}{2d} - \frac{x'^2}{4f_2} - \frac{\xi^2}{4f_1}\right] d\xi$$

$$u_1(x') = \frac{e^{ikd}}{\sqrt{i\lambda d}} \int_{-W_2'/2}^{W_2'/2} u_2(\xi)\ \exp\ ik\left[\frac{(x'-\xi)^2}{2d} - \frac{x'^2}{4f_1} - \frac{\xi^2}{4f_2}\right] d\xi \qquad (1)$$

$u_1(x')$ and $u_2(x')$ represent the scalar optical field, that is the component of the electromagnetic field along a given polarization, calculated at the mirrors 1 and 2 (a time dependence as $e^{-i\omega t}$ is assumed). The mirrors are assumed parabolic, coherently with the Fresnel approximation, with focal distance f_1 and f_2 and width W_1' and W_2'; d is the distance of the mirrors, λ the wavelength and k the wavenumber.

The system of equations (1) can be reduced to just one integral equation in two cases, when the two mirrors are equal (symmetric resonator), or when one of the mirrors can be assumed infinite. In the former case the two equations have identical kernels, and therefore the resonator modes must be either even modes, with $u_1(x') = u_2(x')$, or odd modes, with $u_1(x') = -u_2(x')$. In this case the equation determining the mode can be made dimensionless by posing:

$$x = (k/d)^{\frac{1}{2}} x', \qquad W = (k/d)^{\frac{1}{2}} W',$$

$$g = 1 - d/2f_1, \qquad kd = (n + 2p)\pi.$$

The resulting form of the integral equation is:

$$u(x) = \frac{e^{i2\pi p}}{\sqrt{2\pi i}} \int_{-W/2}^{W/2} u(\xi)\ \exp\ i\left[g\,\frac{x^2 + \xi^2}{2} - x\xi\right] d\xi \qquad (2)$$

The parameter g represents in this equation the mirror curvature, while W is the dimensionless mirror width. $W^2/4\pi$ is usually named the Fresnel number N_F of the cavity. When one mirror, say mirror 2, can be assumed infinite, as it often happens for the larger mirror in asymmetric unstable resonators, the two equations (1) can be combined, and the second integral, with infinite bounds, evaluated explicitly. The resulting equation can then be made dimensionless by posing:

$x= (k/2dg_2)^{\frac{1}{2}} x'$, $\quad W= (k/2dg_2)^{\frac{1}{2}} W'$,

$g_1= 1 - d/2f_1$, $\quad g_2= 1 - d/2f_2$, $\quad kd= (n + p)\pi$.

The dimensionless form of the integral equation thus obtained is identical to Eq.(2). The solution of Eq.(2) must determine the cavity modes and the corresponding eigenvalues of the parameter p, as a function of g and W. (p is determined only up to an arbitrary integer, because of its appearance as the factor $e^{i2\pi p}$. The choice can be made unique by imposing $-\frac{1}{2} < \mathrm{Re}(p) < \frac{1}{2}$.) Together with the longitudinal mode index n, p determines the mode frequency and loss factor. Notice however that the relationship of p to the wavenumber is different for symmetric and asymmetric resonators, as well as is different the meaning of g and the scale of the coordinate x.

When the mirrors present a tapered reflectivity, Eqs.(1) and (2) acquire a different form according to whether the incident or the reflected field is assumed as the unknown u(x). In particular the variable mirror reflectivity may appear inside or outside the integral on the right-hand side of Eq.(2). A symmetric form of the equation can be obtained by defining the unknown field u(x) as the geometrical mean of the incident and reflected field: $u(x)= (u_i(x)u_r(x))^{\frac{1}{2}}$. Eq.(2) is then written as:

$$u(x)= \frac{e^{i2\pi p}}{\sqrt{2\pi i}} \int_{-W/2}^{W/2} u(\xi)\ \sqrt{r(x)r(\xi)}\ \exp i\left[g\ \frac{(x^2 + \xi^2)}{2} - x\xi\right] d\xi$$

In the particular case of Gaussian tapering, $r(x)= e^{-\gamma x^2}$, on which we shall focus our attention in the following, the equation can be given a very compact form by defining an effective g, a complex number, as $g_{ef}= g + i\gamma$, so that the integral equation for tapered-reflectivity resonators becomes identical to Eq.(2), only with g_{ef} in the place of g. Up to this point we have assumed that we are considering either a symmetric resonator with equal tapering of both mirrors, or an asymmetric resonator where only the finite-size mirror has tapered reflectivity. An asymmetric resonator with Gaussian tapering of both mirrors can also be reduced to Eq.(2) by adsorbing the tapering constants in the parameters g_1 and g_2, as it can be shown by repeating the above reasoning on the system (1). Finally, it should be emphasized that the meaning of the dimensionless constant γ is different for symmetric and asymmetric resonators because of the different adimensionalization of the coordinate x.

THE ONE-MODE RESONANCE EQUATION

The analysis given by Weinstein[2] of plane two-dimensional resonators is based upon an expansion of a finite-resonator oscillation mode in a series of traveling modes of the infinite resonator, treated as an infinite metallic waveguide. These modes are considered to propagate up and down in the direction parallel to the mirror surfaces and appropriate reflection coefficients are found for the mode reflection and coupling at each mirror end; the consistency condition for reflection at the opposite open ends yields then a discrete infinite set of equations whose solution determines the oscillation mode and the eigenvalue of p. What makes this method feasible is that for high Fresnel numbers, i.e. for large W, only the fundamental propagating mode is found to contribute significantly to the oscillation mode, and the coupling with higher-order modes can be neglected. This circumstance gives very simple and accurate approximate analytical expressions for the eigenmodes and eigenvalues of the plane resonator.

In trying to extend this method to other than plane resonators two problems must be faced: the traveling modes of the infinite resonator have a more complicate expression and may present caustics, and the higher-order modes may no longer be negligible. Considering only cases where the fundamental traveling mode is still sufficient to represent the resonator oscillation mode, as is the case with tapered-reflectivity resonators, the problem of interaction between the caustics and the open ends remains to be solved. In general, the infinite-resonator modes, that is the modes of Eq.(2) when the integral is taken with infinite bounds, are known to be parabolic cylinder functions, solutions of the second-order equation:

$$\frac{d^2u}{dx^2} + (\lambda + \alpha^2 x^2)u = 0 \qquad (3)$$

where the costants α and λ are given functions of the mode index m, the parameter p and the quantity μ defined so that $g = \mathrm{Cos}(\mu)$, namely

$$\alpha = \mathrm{Sin}(\mu), \qquad \lambda = 4\pi \frac{\mathrm{Sin}(\mu)}{\mu}(p + m)$$

These quantities can also be given in terms of the parameter $M = e^{\mu}$, known in the literature as the cavity magnification. The solutions of Eq.(3) present inversion points (caustics) at the two, possibly complex, positions where $\lambda + \alpha^2 x^2 = 0$. Rather than using the exact solutions of Eq.(3), which should be obtained numerically or from tables, it is convenient to use a W.K.B. approximation of the solution,

$$u(x) \simeq s(x)^{-\frac{1}{2}} \exp\left[\pm i \int s(x)\, dx\right] \qquad (4)$$

where $s(x) = (\lambda + \alpha^2 x^2)^{\frac{1}{2}}$ is the local dimensionless spatial frequency of the mode. When this is done, however, partial reflection from the complex inversion points must be taken into account explicitly. As shown in Ref.7, when the determination of the square root is chosen so that the mode amplitude is exploding at infinity the mode is not reflected by a complex inversion point, while when the other determination is chosen, the appropriate reflection coefficient can be obtained by letting the field propagate, according to Eq.(4), along a path encompassing the inversion point. The situation is illustrated in Fig.1

Assuming now that the wave traveling from the center of the resonator towards each edge is partially reflected from the caustic, while the wave returning from the edge to the center is not, as is our case, one can calculate the total reflection coefficient at the center as the sum of the reflection coefficient due to the caustic and the reflection coefficient due to the mirror edge. The reflection coefficient due to the caustic is obtained calculating the integral in Eq.(4) along a path going from the origin round the inversion point and back to the origin, taking properly into account the factor 1/i coming from the multiplier $s(x)^{-\frac{1}{2}}$, and is:

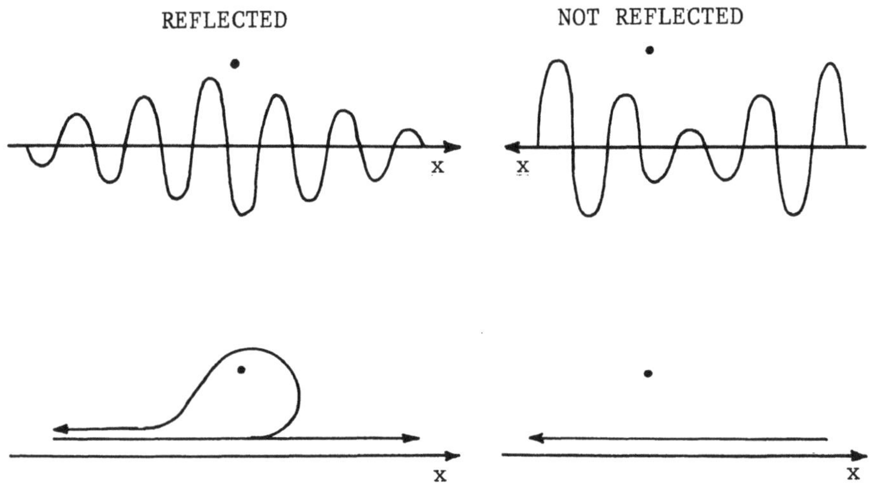

Fig. 1. Partial reflection from a complex inversion point.

$$R_{co} = \exp\left[-\frac{\pi}{2}\left(\frac{\lambda}{\alpha} + i\right)\right]$$

The reflection coefficient due to the mirror edge at the origin, R_{eo}, is obtained by multiplying the reflection coefficient at the edge R_e by the propagation factor P^2, where

$$P = \exp\left[i\int_0^{W/2} s(x)\, dx\right]$$

The resonance equation, determining the eigenvalues of p, is then obtained by imposing for the total reflection coefficient at the origin to be ±1, as required by symmetry:

$$R_{co} + P^2 R_e = \pm 1 \qquad (5)$$

This equation should in principle be written for each mode of the infinite resonator, and edge-coupling between different modes should be taken into account. However it is possible for plane, stable and tapered-reflectivity resonators of high Fresnel number to neglect higher-order modes; in this case Eq.(5) written for the m=0 mode is the complete resonance equation.

DISCUSSION OF THE RESONANCE EQUATION AND COMPARISON WITH NUMERICAL RESULTS

It is interesting to observe how Eq.(5) reduces to the known resonance equations in particular cases. For instance, for stable resonators the factor P^2 is rapidly vanishing with growing W, and already for moderate values of W Eq.(5) reduces to $R_{co} = \pm 1$, that is the reflection from the caustic is prevailing. This condition gives $\lambda/\alpha = i(2h + 1)$ for the h-th mode, which is the exact solution for an infinite stable resonator, in spite of the W.K.B. approximation. For plane resonators the caustic disappears and $R_{co} = 0$; P reduces to exp (i s W/2), with constant s, and Eq.(5) becomes the resonance condition found by Weinstein[2].

To test the validity of Eq.(5) when both reflection from the caustic and from the mirror edge are important, we compared the values of p given by Eq.(5) for a resonator with plane, tapered-reflectivity mirrors against a numerical solution of Eq.(2) for the same system. The results of this comparison are shown in Figs.2 and 3, for different values of W and the reflectivity tapering γ. The values of p corresponding to the resonance equation were also obtained by numerical inversion of Eq.(5), assuming for the reflection coefficient R_e the expression suggested by Weinstein ,

$$R_e = \exp\left[0.824\ i(1 + i)\ s(W/2)\right]$$

Of course the numerical inversion of Eq.(5) is a much less cumbersome task than the integration of Eq.(2), and can be handled by today's

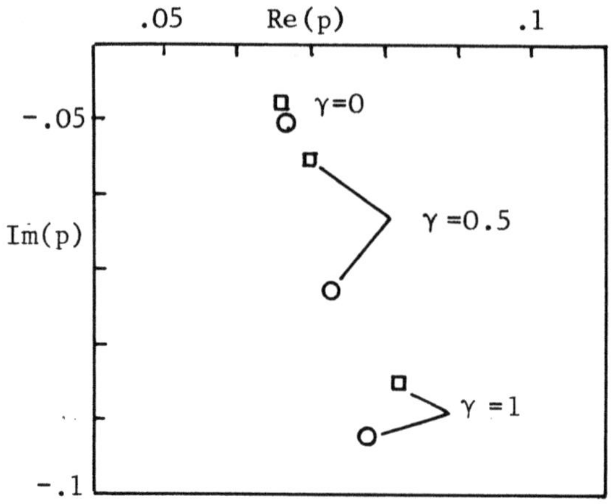

Fig. 2. Values of p for a plane tapered-reflectivity resonator of width W = 2. Points denoted by circles are obtained by numerical integration of Eq.(2); points denoted by squares are obtained from Eq.(5).

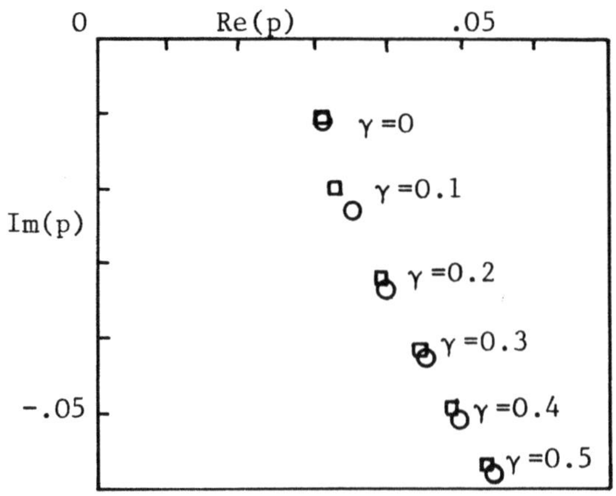

Fig. 3. Values of p for a plane tapered-reflectivity resonator of width W = 4. Same symbols as in Fig. 2 are used.

pocket calculators. The analysis of Figs.2-3 shows poor agreement between the two solutions for W = 2, and a rapid improvement for W = 4. For W = 6 and W = 8, which were also tried, the difference between the two solutions would hardly be distinguishable in a diagram. It should be noticed that the chosen values of γ range from a condition where edge reflection is dominant ($\gamma = 0$), to a condition where caustic reflection is prevailing ($Im(p) \simeq -Re(p)$).

As a result of this analysis it can be said that Eq.(5), a generalization of Weinstein's theory of plane resonators, yields very good approximate results for the resonance frequency and damping factor of plane, stable and tapered-reflectivity resonators of high Fresnel number. A one-mode model is known not to work for unstable resonators[3], however we expect that unstable resonators with tapered-reflectivity mirrors, having a lower coupling to higher-order modes than ordinary unstable resonators, should be fitted by this resonance equation. Finally we want to emphasize that a tapered-reflectivity resonator may serve as a model of a resonator with a non-homogeneous active medium, which is a much better approximation of a real laser oscillator.

REFERENCES

1. L. A. Weinstein, "The Theory of Diffraction and the Factorization Method", Golem Press, Boulder, Colorado (1969)
2. L. A. Weinstein, "Open Resonators and Open Waveguides", Golem Press, Boulder, Colorado (1969)
3. L. W. Chen, L. B. Felsen, Coupled-mode theory of unstable resonators, IEEE J. Q. E., 9,1102(1973)
4. S. H. Cho, L. B. Felsen, Ray-optical analysis of unstable resonators with sferical mirrors, J. Opt. Soc. Am., 69,1377(1979)
5. S. H. Cho, S. Y. Shin, L. B. Felsen, Ray-optical analysis of unstable resonators, J. Opt. Soc. Am., 63,1528(1973)
6. C. E. Santana, L. B. Felsen, Ray-optical calculation of edge diffraction in unstable resonators, IEEE Trans. M.T.T., 26,101(1978)
7. L. Landau, E. M. Lifschitz, "Mécanique Quantique", Editions Mir, Moscou (1967); sec. 47 and sec. 132.
8. P. Luchini, S. Solimeno, Indeterminate Padé approximations in the analysis of diffraction at the mirror edges of an optical resonator, Optics Letters, 7,259(1982)
9. A. G. Fox, T. Li, Resonant modes in a maser interferometer, Bell Syst. Tech. J., 40,453(1961)

EXPERIMENTAL AND THEORETICAL INVESTIGATIONS OF CO_2-GDL WITH A HIGH HYDROGEN CONTENT

S. S. Novikov, V. M. Doroshenko, and N. N. Kudryavtsev

Institute of Chemical Physics
Academy of Sciences of the USSR
Moscow 117977, USSR

INTRODUCTION

Investigations of CO_2 gas dynamic lasers (CO_2-GDL) on CO_2 + + N_2 + H_2 mixtures have been carried out in experimental [1-3] and theoretical [4] works. The admixtures of molecular hydrogen in CO_2-GDL active medium of about $\xi_{H_2} \leqslant 0.05$ mole fraction made it possible to attain a gain coefficient small-signal, K_o = 0.5 - 0.7 m^{-1}. At higher molecular hydrogen concentrations a sharp decrease in the gain coefficient to 0.1 - 0.2 m^{-1} at ξ_{H_2} = 0.2 occurs [1-4].

In this work, systematic studies and an improvement in the performance of CO_2-GDL with high hydrogen content (ξ_{H_2} = 0.4 ÷ 0.5) allow us to estimate the conditions under which the relatively high values of the gain coefficient K_o = 0.4 - 0.5 m^{-1} are realized. Use of these mixtures in CO_2-GDL is of interest on account of the availability of large quantities of hot hydrogen, a promising heat-carrier for nuclear power plants [5].

METHOD

Chemical Kinetics in CO_2 + N_2 + H_2 GDL Mixtures

A significant distinction between hydrogen-containing CO_2-GDL mixtures and conventional CO_2 + N_2 + He(H_2O) mixtures lies in the fact that chemical reactions proceed at gas temperatures in the stagnation region, $T_o \geqslant 1000$°K [2,3]. This considerably affects the population inversion generation in a subsequent gas expansion in the supersonic nozzle. Chemical reactions in CO_2 + N_2 + H_2 mixtures are determined by the interaction of carbon dioxide and hydrogen and described by the overall reaction [6]:

$$CO_2 + H_2 \rightleftarrows CO + H_2O - 12.8 \text{ kcal/mol.} \qquad (1)$$

The most important factors leading to the decrease of the gain coefficient in CO_2-GDL on $CO_2 + N_2 + H_2$ mixtures, occurring because of the reaction (1), are: formation of a significant concentration of water vapor, which considerably increases vibrational relaxation rates, and a decrease in the content of radiating molecules - carbon dioxide.

The calculation according to Reference 7 shows that typical stagnation gas parameters of CO_2-GDL using $CO_2 + N_2 + He(H_2O)$ mixtures will be unacceptable when operating on mixtures with molecular hydrogen, because of the above mentioned negative influence of the $CO_2 + H_2$ chemical reaction on population inversion generation. The rise in the hydrogen content results in a decrease of stagnation gas pressure and temperature corresponding to a sharp reduction of the gain coefficient owing to the chemical reaction (1). Therefore, it may be suggested that the optimization of the parameters of CO_2-GDL on high hydrogen content mixtures must be made by the "freezing" of chemical reactions, that is by decreasing gas pressure and temperature in the stagnation region, or by changing the mixture composition when the reaction (1) equilibrium shifts to the left. Both alternatives are used in the present investigation.

Method of CO_2 and CO Vibrational Temperature Measurements

For the study of population inversion of the CO_2 molecules laser levels (001) and (100) the method used enabled recovery of vibrational temperature of asymmetric T_3 and coupled - T_2 (symmetric and bending) CO_2 modes and hence the determination of the population of upper (001) and lower (100) laser levels by measuring spontaneous radiation intensity in the 4.3 μm CO_2 band and the gain coefficient at wavelength λ = 10.6 μm [8]. In the present work this method was modified, since we examine mixtures containing a considerable amount of carbon monoxide whose intensive 4.7 μm vibrational-rotational band in vibrational non-equilibrium conditions overlaps with the 4.3 μm CO_2 band [9]. Therefore, additional measurements of 4.7 μm CO and 4.3 μm CO_2 radiation are made in a spectral region near 4.7 μm. Simultaneous recording of radiation intensity in two spectral ranges corresponding to the central parts of 4.3 μm CO_2 and 4.7 μm CO bands and the gain coefficient K_o allows us to obtain a closed system of equations for the determination of three vibrational temperatures of carbon monoxide T_5 and carbon dioxide T_2 and T_3:

$$R_1 = a_1 \int_{\Delta\omega} c_1(\omega)[\bar{I}_{4.3}(\omega,T_3,T_2) + \bar{I}_{4.7}(\omega,T_5)]d\omega$$

$$R_2 = a_2 \int_{\Delta\omega} c_2(\omega)[\bar{I}_{4.3}(\omega,T_3,T_2) + \bar{I}_{4.7}(\omega,T_5)]d\omega \qquad (2)$$

$$K_o = K_o(T_3,T_2) ,$$

where R_1 and R_2 are the measured values of radiation intensity; a_1 and a_2 are constants for characterized sensitivity; $c_1(\omega)$ and $c_2(\omega)$ are spectral transmission of the first and second optical channels respectively; $\bar{I}_{4.3}(\omega,T_3,T_2)$ and $\bar{I}_{4.7}(\omega,T_5)$ are the spectral radiation intensity of 4.3 μm CO_2 and 4.7 μm CO bands averaged over rotational structure, calculated numerically according to Reference 9.

The random error of radiation intensity recording is defined from the results of a series of experiments made under similar conditions and is in the range 5 - 10%. The accuracy of intensities $\bar{I}_{4.3}$ and $\bar{I}_{4.7}$ calculations are 7 - 10% [8], and the effect of vibrational temperatures, T_3 and T_5, on integral intensity radiation in 4.3 μm $\partial \ell n I_{4.3}/\partial \ell n T_3 > 2 \div 3$ and 4.7 μm $\partial \ell n I_{4.7}/\partial \ell n T_5 > 2 \div 3$ bands respectively is the more considerable in comparison with the influence of other vibrationally non-equilibrated gas parameters. Taking into account the above errors in the recording of radiation intensity, calculations of the intensities $\bar{I}_{4.3}$ and $\bar{I}_{4.7}$ and uncertainties in the determinations of the parameters in a supersonic flow, the estimated accuracy of the recovery of vibrational temperatures T_5 and T_3 is 10 ÷ 15%.

The accuracy in determination of vibrational temperature T_2 depends considerably on the error of the gain coefficient measurements which in this work is 10 ÷ 20%. The estimation of a random error in the recovery of vibrational temperature T_2 obtained by solving Equations (2) with terms R_1 and R_2 changed by 10% and K_o by 20% is 15 ÷ 20%.

Calculations of the Gain Coefficient and Vibrational Temperatures

Together with the experimental measurements of the gain coefficient K_o and vibrational temperatures CO_2 - T_3 and T_2 and CO - - T_5, theoretical calculations for these parameters are carried out. The simultaneous numerical solution of a system of gas-dynamics and kinetics of vibrational relaxation and chemical reactions equations are conducted for the process of one-dimensional, isentropic gas expansion in the supersonic nozzle [10]. The choice of the vibrational relaxation rates in the mixture $CO_2 + N_2 + CO + H_2O$ are done in accordance with Reference 10. The following vibrational energy transfer processes in CO_2-GDL including molecular hydrogen are taken into consideration:

$$CO_2(001) + H_2 \overset{Q_{32}^{H_2}}{\rightleftarrows} CO_2(mno) + H_2 + \Delta E , \qquad (3)$$

where m = 0, n = 3 or m = n = 1, ΔE = 350 cm^{-1} at T > 400°K, m = 2 and n = 0; m = 1, n = 2 or m = 0, n = 4, ΔE = −300 cm^{-1} at T < 400°K [11],

$$CO_2(010) + H_2 \overset{Q_{20}^{H_2}}{\rightleftarrows} CO_2(000) + H_2 + 667\ cm^{-1} \tag{4}$$

$$N_2(1) + H_2 \overset{Q_{40}^{H_2}}{\rightleftarrows} N_2(0) + H_2 + 2330\ cm^{-1} \tag{5}$$

$$CO(1) + H_2 \overset{Q_{50}^{H_2}}{\rightleftarrows} CO(0) + H_2 + 2143\ cm^{-1} \tag{6}$$

The relaxation rates of these processes chosen according to recent investigations of various authors are presented for the temperature range T = 100 − 2000°K in the Table.

EXPERIMENTAL

Experimental investigations are carried out in a shock tube of 75 mm diameter having a supersonic nozzle at the end of the low pressure section. The gain coefficient and intensity of spontaneous IR radiation in two spectral bands are simultaneously recorded in the supersonic gas flow at the nozzle exit. Gas parameters and chemical reaction in the stagnation region are monitored by pressure

Table. Values of Vibrational Relaxation Rates (μs atm)$^{-1}$ for the Process (3)-(6).

$T^{-1/3}$ °K	T °K	$\ell g Q_{32}^{H_2}$	$\ell g Q_{20}^{H_2}$	$\ell g Q_{40}^{H_2}$	$\ell g Q_{50}^{H_2}$
0.08	1953	0.80	0.84	−0.20	−0.20
0.09	1372	0.75	1.00	−0.55	−0.20
0.10	1000	0.70	1.16	−1.00	−0.60
0.11	751	0.64	1.31	−1.40	−0.97
0.12	579	0.60	1.47	−1.80	−1.38
0.13	455	0.54	1.62	−2.25	−1.70
0.14	364	0.49	1.78	−2.65	−2.00
0.15	296	0.44	1.93	−3.05	−2.10
0.16	244	0.38	2.08	-	−2.20
0.17	203	0.30	2.24	-	−2.25
0.18	171	0.22	2.39	-	−2.30
0.19	146	0.13	2.55	-	−2.33
0.20	125	0.04	2.70	-	−2.38

and intensity of IR radiation (of H_2O molecules in 6.3 μm band) measurements. The parameters of a vibrational non-equilibrium gas are recorded at time interval t = 0.2 ÷ 0.3 mc, after the reflecting of a shock wave from the end of the tube. This corresponds to the undisturbed gas state in the stagnation region and stationary nozzle expansion. Purities of the gases used were: CO_2 - 99.6%, N_2 - 99.9%, CO - 99.7%, H_2 - 99.8%.

RESULTS

To decrease the water vapor content in $CO_2 + H_2$ mixture, carbon monoxide may be introduced, resulting in a shift of the reaction (1) equilibrium to the left. The measured results of the gain coefficient and vibrational temperatures in CO_2-GDL on 0.1 CO_2 + 0.4 CO + + 0.5 H_2 mixture with various gas parameters in the stagnation region are depicted in Figure 1a and 1b. We use the contoured supersonic nozzle with half angle, $\phi = 15^\circ$, throat height $h_* = 1.5$ mm, and area ratio $h/h_* = 23$. The measured gain coefficient in CO_2--GDL on this mixture has low values which do not exceed 0.1 m^{-1}, at temperature $T_o \simeq 1200^\circ K$ (Figure 1a). The rapid gain coefficient decrease at temperature $T_o \gtrsim 1100\text{-}1200^\circ K$ is due to water molecules in the mixture (curve 2).

The measured vibrational temperatures of CO_2 and CO confirm a significant increase in the vibrational energy of carbon monoxide and asymmetric mode of carbon dioxide relaxation in the presence of water vapor in the mixture forming in reaction (1).

The vibrational temperature of combined CO_2 mode T_2 coincides within experimental error with gas temperature in the gas flow. Results of the measurement of vibrational temperatures, CO_2 - T_3 and T_2, (Figure 1b) reveal that the decrease in the gain coefficient at $T_o \gtrsim 1200^\circ K$ is found by increasing the relaxation rates of the upper laser level when practically complete thermalization of the lower laser level occurs. The calculated vibrational temperatures (curves 3-5, Figure 1b) and the gain coefficient (curve 1, Figure 1a) are in good agreement with the experimental data. In this way, the results presented in Figure 1a and 1b show an insufficient efficiency of utilization of carbon monoxide admixtures in CO_2-GDL on mixtures with considerable hydrogen content. Another possibility in optimization of the CO_2-GDL on the performance of high hydrogen content mixtures is to reduce gas pressure and temperature in the stagnation region, in order to prevent chemical reactions.

The results of the gain coefficient measurements in 0.2 CO_2 + + 0.4 N_2 + 0.4 H_2 mixture at lower temperatures T_o and pressure P_o in the stagnation region are shown in Figure 1c. The maximum values of the gain coefficient K_o 0.4 - 0.45 m^{-1} are attained at temperature $T_o = 1200 \div 1400^\circ K$, and pressure $P_o \simeq 2$ atm. It should be noted

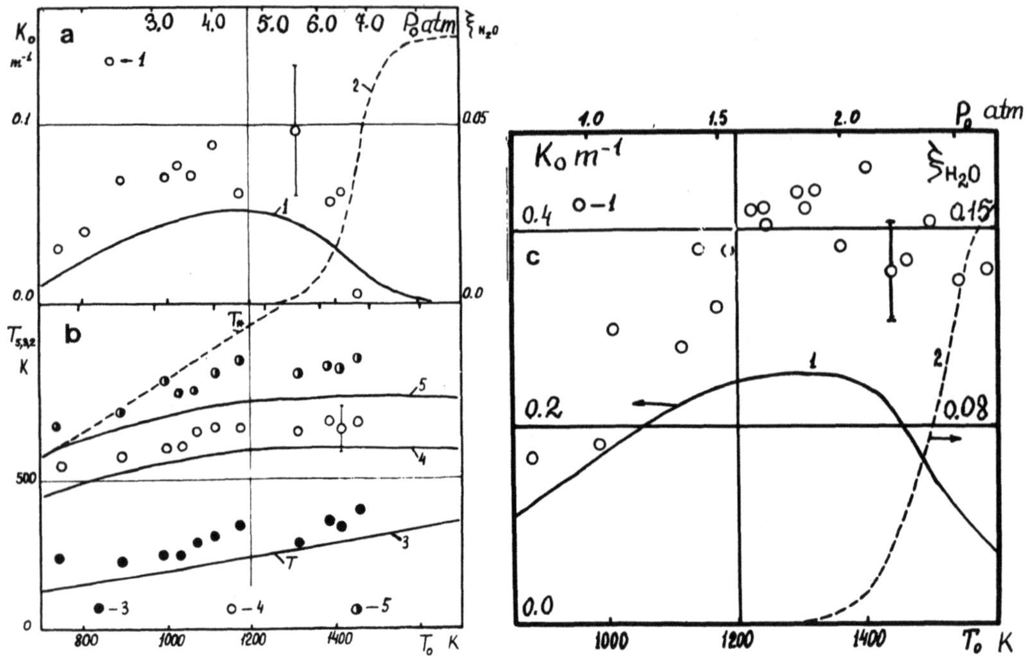

Fig. 1. The gain coefficient K_o (a) and vibrational temperatures T_5, T_3 and T_2 (b) in 0.1 CO_2 + 0.4 CO + 0.5 H_2 mixture and the gain coefficient K_o (c) in 0.2 CO_2 + 0.4 N_2 + 0.4 H_2 mixture versus pressure and gas temperature in the stagnation region. The following designation of experimental points and theoretical curves are used: 1 - K_0; 2 - ξ_{H_2O} (calculated according to Reference 7); 3 - T_2; 4 - T_3; 5 - T_5.

that the optimal values obtained for P_o and T_o are considerably lower than for the CO_2-GDL using CO_2 + N_2 + He(H_2O) mixtures: T_o = = 1500 - 1800°K, P_o = 10 ÷ 20 atm [10]. It should be stressed that in the optimal conditions for this CO_2-GDL all chemical reactions in the stagnation region are completely freezing and the influence of molecular hydrogen on population inversion generation is limited by the vibrational energy deactivation of CO_2, N_2 and CO molecules in processes (3) - (6).

The results of population inversion generation in this mixture expanding in a minimum-length, contoured nozzle [12] are represented in Figure 2a and 2b. The application of minimum length supersonic nozzles in the CO_2-GDL using non-reacting mixtures CO_2 + N_2 + He(H_2O) provide for a significant increase in the gain coefficient [13] up to $K_o = 1.5\ m^{-1}$. The maximum gain coefficient for CO_2 + N_2 + H_2 mixtures expanding in a minimum-length nozzle, K_o = 0.4 - 0.45 m^{-1} is achieved at temperature $T_o \simeq 1250$°K and pressure $P_o \simeq 2.5$ atm

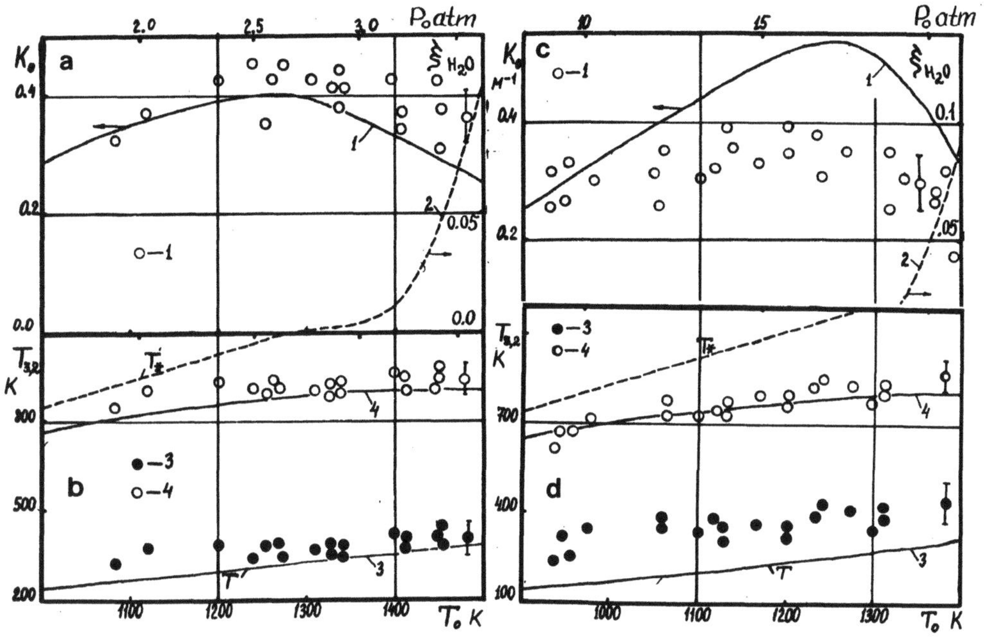

Fig. 2. The gain coefficient K_0 (a,c) and vibrational temperatures CO_2 - T_3 and T_2 (b,d) in 0.2 CO_2 + 0.4 N_2 + 0.4 H_2 mixture expanding in minimum-length contoured supersonic nozzle [12]. Designations correspond to those of Figure 1.

(Figure 2a); with accurate measurements this coincides with the values of K_0 which are obtained for a slit wedge nozzle (Figure 1c).

The vibrational temperature of the asymmetric mode rises from 800 to 1000°K when gas temperature in the stagnation region increases from 1100 to 1250°K, and it changes insignificantly with a further increase in T_0 up to 1500°K. The attained values of vibrational temperature T_3 are 100 - 200°K higher than in the mixture 0.1 CO_2 + + 0.4 CO + 0.5 H_2, where $T_3 \simeq 800$°K (Figure 1b).

The results represented in Figure 2a and 2b show that the temperature in stagnation region T_0 = 1200 ÷ 1300°K is optimal for performance of the CO_2-GDL with high hydrogen content 0.2 CO_2 + 0.4 N_2 + 0.4 H_2 mixtures. From the results of vibrational temperature measurements it follows that at concentration ξ_{H_2} = 0.4 the low, laser level is completely thermolized. The values of the vibrational temperature T_3 are $\simeq$ 900°K which practically coincides with the gas temperature in the nozzle throat. Chances of vibrational temperature increase, T_3, owing to a rise in the cooling rate at the nozzle throat region are nil, since the half-angle of contoured nozzle is maximum, preventing shock wave formation in the expanding flow.

The increase in gas temperature in the stagnation region T_o 1300°K, unlike a conventional non-reactive $CO_2 + N_2 + He(H_2O)$ mixture, does not cause a rise in the gain coefficient K_o, because of the considerable concentration of water vapor forming in reaction (1). In the practical implementation of CO_2-GDL on high hydrogen content mixtures an increase in gas pressure in the stagnation region is required. To compensate for the acceleration of vibrational relaxation of the upper laser level (nitrogen and asymmetric CO_2 mode), a nozzle with a reduced throat height is used. The minimum value of throat height, h_*, is restricted by the thickness of a boundary layer in the throat $\simeq 0.2 \div 0.3$ mm [10]. The increase in gas pressure in the stagnation region considerably affects the kinetics of chemical reactions and hence, may strongly influence population inversion generation, owing to variations in mixture composition.

Figure 2c and 2d gives the results of the investigations of $0.2\ CO_2 + 0.4\ N_2 + 0.4\ H_2$ mixtures at higher pressures in the stagnation region, $P_o \simeq 15 - 20$ atm, using a minimum-length contoured nozzle [12] with reduced throat height, $h_* = 0.4$ mm ($h/h_* = 50$). The maximum coefficient, $K_o \simeq 0.3 - 0.4\ m^{-1}$ is attained at temperature, $T_o = 1100 - 1300$°K, and pressure, $P_o \simeq 16$ atm (Figure 2c). The dependence of the gain coefficient K_o on temperature T_o in the region of $T_o = 900 - 1300$°K, is weak. The calculated values of K_o are somewhat higher by $0.1 - 0.15\ m^{-1}$ compared to the measured values and are characterized by a stronger dependence on temperature T_o.

The measurements, results and calculations of vibrational temperatures, $CO_2 - T_3$ and T_2, are shown in Figure 2d. With increasing temperature T_o, from 900°K to 1400°K the vibrational temperature, T_3, rises from 650°K to 750°K. Calculated and measured vibrational temperatures of asymmetric CO_2 modes coincide well. The measured vibrational temperatures of combined CO_2 mode are by $100 \div 150$°K higher than the calculated values, T_2, which are practically the same as the translational temperature of the gas in a supersonic flow.

CONCLUSION

Systematic experimental and theoretical studies have been undertaken for population inversion generation in the CO_2-GDL on high hydrogen content mixture. Values are obtained for the gain coefficient $K_o = 0.4 - 0.5\ m^{-1}$ in a mixture with mole fraction of hydrogen molecules $\xi_{H_2} = 0.4$.

Acknowledgments

The authors wish to thank Professors R. I. Soloukhin and W. A. Sheglov for valuable discussions and interest displayed in this study.

REFERENCES

1. A. V. Britan, S. A. Losev and O. P. Shatalov, Influence of Admixtures on Gain Coefficient in the CO_2- Gas Dynamic Lasers, Quantum Electronics (in Russian), 1:2620 (1974).
2. N. N. Kudryavtsev, S. S. Novikov and I. B. Svetlychyi, Effect of Molecular Hydrogen Additives on CO_2-Laser Radiation Gain Coefficient in Expanding Flow of Carbon Dioxide-Nitrogen Mixture, The Phys. of Comb. and Explosion (in Russian), 12:729 (1976).
3. I. Rom and I. Stricker, Effects of Chemical Reactions on the Performance of Gas Dynamic Lasers, Acta Astronautica, 1:1101 (1974).
4. V. N. Makarov and S. A. Losev, On the Influence of Admixtures on the Gain Coefficient for the Gas Expanding in Supersonic Nozzle, The Phys. of Comb. and Explosion (in Russian), 12: 804 (1976).
5. E. F. Ratnikov and S. D. Tetelbaum, "Gases as Heat Carriers for Nuclear Power Plants", Atomizdat, Moscow (1978).
6. V. N. Kondratiev, "Rate Constants of Reactions in Gas Phase", Nauka, Moscow (1970).
7. I. M. Burbacher, R. D. Kern and B. V. O'Gady, Reaction of Hydrogen and Carbon Dioxide behind Reflected Shock Waves, J. Phys. Chem., 80:1031 (1976).
8. N. N. Kudryavtsev, S. S. Novikov and I. B. Svetliching, On a Method of Measurements of Vibrational Temperatures in CO_2 Gas Dynamic Lasers, Quantum Electronics (in Russian), 6:690 (1979).
9. N. N. Kudryavtsev and S. S. Novikov, A Study of Infrared Radiation of Vibrationally Excited CO in the 4.7 μm Band and CO_2 in the 4.3 μm and 2.7 μm Bands, Rev. Phys. Appl., 16:49 (1981).
10. S. A. Losev, "Gasdynamic Lasers", Nauka, Moscow (1977).
11. G. D. Billing, Semiclassical Calculation of Energy Transfer in Polyatomic Molecules. VII Intra and Inter Molecular Energy Transfer in N_2 + CO_2, Chem. Phys., 67:35 (1982).
12. M. G. Ktalkherman, V. B. Malkov and N. A. Ruban, Experimental Research of Flow in the Gasdynamic Laser Nozzles, Mech. of Gas and Fluids (in Russian), 5:178 (1980).
13. D. M. Kuehn, Importance of Nozzle Geometry to High Pressure Gasdynamic Lasers, Appl. Phys. Lett., 21:112 (1972).

EFFECTS OF INJECTION HOLE IN CONCAVE MIRROR OF UNSTABLE RESONATOR WITH SATURABLE GAIN

J. F. Perkins

M K Associates
2215 Mock Road
Huntsville, AL 35801

Charles Cason and R. W. Jones

U.S. Army Missile Command
Redstone Arsenal, AL 35809

ABSTRACT

Operation of an unstable resonator can be controlled by injection of a weak signal through a hole in the concave mirror, but the presence of the hole results in some deterioration of output optical quality. The degree of deterioration as a function of resonator parameters (in the absence of an injected signal) has been investigated by wave-optics calculations as reported herein.

INTRODUCTION

In many applications, a high-energy laser is operated in an unstable oscillator configuration as a free-running oscillator. In some applications it becomes desirable to control the output beam in its frequency or mode content by injecting an externally generated low-energy laser control beam. Injection can be either by "walking" the beam in to the optic axis or by injecting it through a hole in a resonator mirror at the optic-axis position. Signal injection control has eliminated the gain spike and axial mode beating and demonstrated control of frequency and polarization (Cason et al., 1977; Buczek et al., 1973). The presence of a hole in the mirror will, however, perturb the resonator near-field phase (and intensity) distribution and therefore should produce some deterioration of output optical quality, to an extent which depends on the size of the injection hole.

The degree of deterioration of the optical quality of the output beam which is directly caused by the presence of injection holes of various sizes (with no input of a control beam) is a problem which is of interest but does not appear to have received any theoretical or computational attention in the literature. (We had earlier performed unpublished calculations of empty strip resonators with injection openings; results indicated pronounced beam-quality deterioration even for fairly small openings, but it was felt that these calculations were not at all realistic for three dimensional saturated-gain lasers). This problem was investigated in the work reported herein by numerical three dimensional wave-optics calculations for cases with a moderately high level of saturated gain. While the near-field distributions of intensity and phase have been considered, primary attention has been given to the overall degree of deterioration of the central intensity maximum at the far-field (or focal) position after vacuum propagation. The calculations deal specifically with steady-state operation, but the results are presumably relevant also to pulsed operation after a stable mode has formed.

Figure 1 illustrates the resonator system which is being analyzed. In the absence of the injection hole, the configuration is simply that of a positive-branch confocal unstable resonator with central alignment. An injection source and an opening in the concave mirror are shown in the top of Figure 1. The outer two squares in the bottom of Figure 1 indicate the nominal outer and inner boundary of the output beam, respectively. The dot in the center indicates the position of the optic axis. The injection hole in the convex mirror is assumed to be square.

WAVE-OPTICS COMPUTER CODE

The present calculations were carried out with a three dimensional wave-optics computer code, CAV3D, which was prepared largely for this work. It calculates the propagation between ideally aligned resonator mirrors with Fourier transforms, using the efficient FFT algorithm. Coordinate transformations are used such that the propagation in both the expanding-beam and the collimated-beam portions of the system can be treated with equations which are applicable to collimated-beam propagation (Perkins and Shatas, 1976; Sziklas and Siegman, 1975). Gain of the active medium is treated as if it were a thin sheet near the concave mirror; nonlinear gain-saturation effects are incorporated by using the equations of Rigrod (1965).

The output of the code includes separate character plots of the transverse distribution of near-field intensity and phase; vertical and horizontal directions along the printer paper correspond to X and Y coordinates transverse to the beam. Of greatest interest in the present work is the relative central far-field

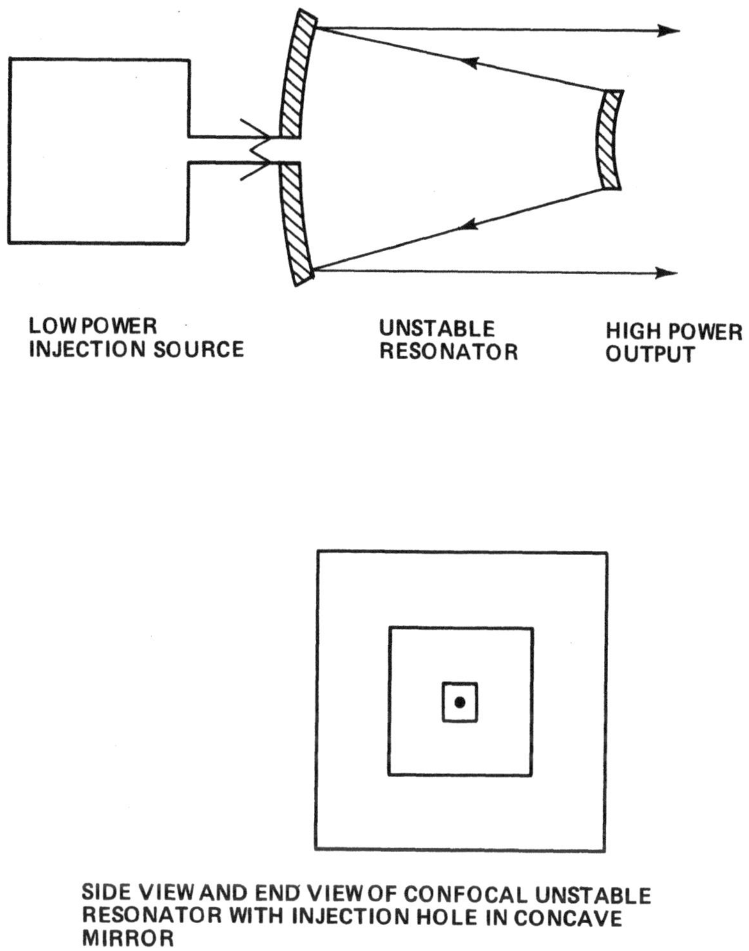

Fig. 1. Unstable resonator with hole in concave mirror at position of optic axis for purpose of injecting a control signal. Upper figure is side view; lower figure is end-on view.

intensity value, I/Io, i.e. the ratio of the central far-field (after vacuum propagation) intensity of the calculated beam to that of an idealized beam containing the same total power but with uniform intensity and constant phase at the laser output plane.

RESULTS

The oscillator configurations which were treated in this work are of positive-branch, confocal unstable resonator type, with magnification of 2.0 and equivalent Fresnel numbers of 2.4, 4.4, and 9.4. The integrated value of small signal gain over a round trip in the laser medium has been taken to have the value 5.0. Calculations were carried out with overall mesh sizes of 64 x 64 and 128 x 128. Concave mirror hole sizes of 0 x 0, 6 x 6, 8 x 8, 12 x 12, and 16 x 16 mesh points were included in the calculations.

Some computed results of intensity and phase at the position of the convex feedback mirror of resonators with an equivalent Fresnel number of 9.4 are presented in the form of character plots in Figs. 2-5. These calculations were carried out with an overall transverse grid size of 128 x 128. The plotted values of intensity are normalized such that the highest intensity value is unity, which is represented by the letter "A". Successive letters of the alphabet are used to indicate relative intensity values which are less than unity in increments of 0.04. A value smaller than 0.04 is indicated by a blank.

Figures 2 and 3 depict the intensity and phase, respectively, for a case with no hole in the mirror. The intensity is quite uniform in the center of the beam. The calculated value of I/Io for this case is 0.98.

Figures 4 and 5 depict intensity and phase for a case with a hole in the mirror of size corresponding to a 12 x 12 array of mesh points. The effect of the hole in the concave mirror is quite evident in the small intensity values in the beam center. The phase distribution is clearly quite nonuniform. The calculated value of I/Io was 0.60.

As an independent variable associated with the size of the mirror hole, we have chosen to use the Fresnel number of the hole. This is defined in a quite straightforward way by:

$$F_{HOLE} = b^2/(\lambda L) \ .$$

Here "b" represents the half-width of the hole, λ is the radiation wavelength, and L is the spacing between resonator mirrors. The calculated values of I/Io have been plotted as a function of this independent variable in Figure 6. The combination of number of total mesh points and equivalent Fresnel number of the resonator are indicated by the type of plot symbol, as explained in the figure legend.

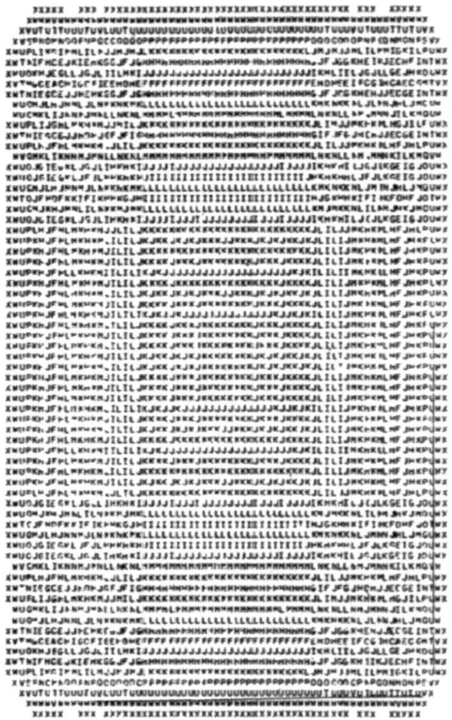

Fig.2 Intensity distribution at position of feedback mirror for a configuration with no hole in concave mirror.

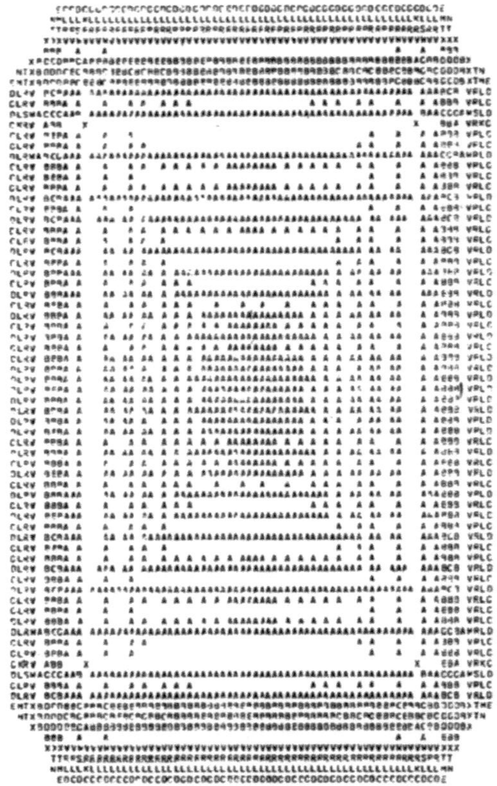

Fig.3 Phase distribution at position of feedback mirror for a configuration with no hole in concave mirror.

Fig.4 Intensity distribution at position of feedback mirror for a configuration with a 12 x 12 hole in concave mirror.

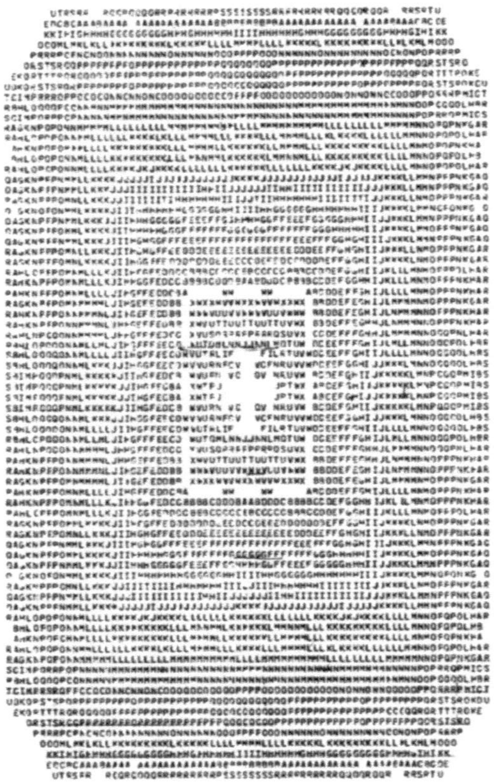

Fig.5 Phase distribution at position of feedback mirror for a configuration with a 12 x 12 hole in concave mirror.

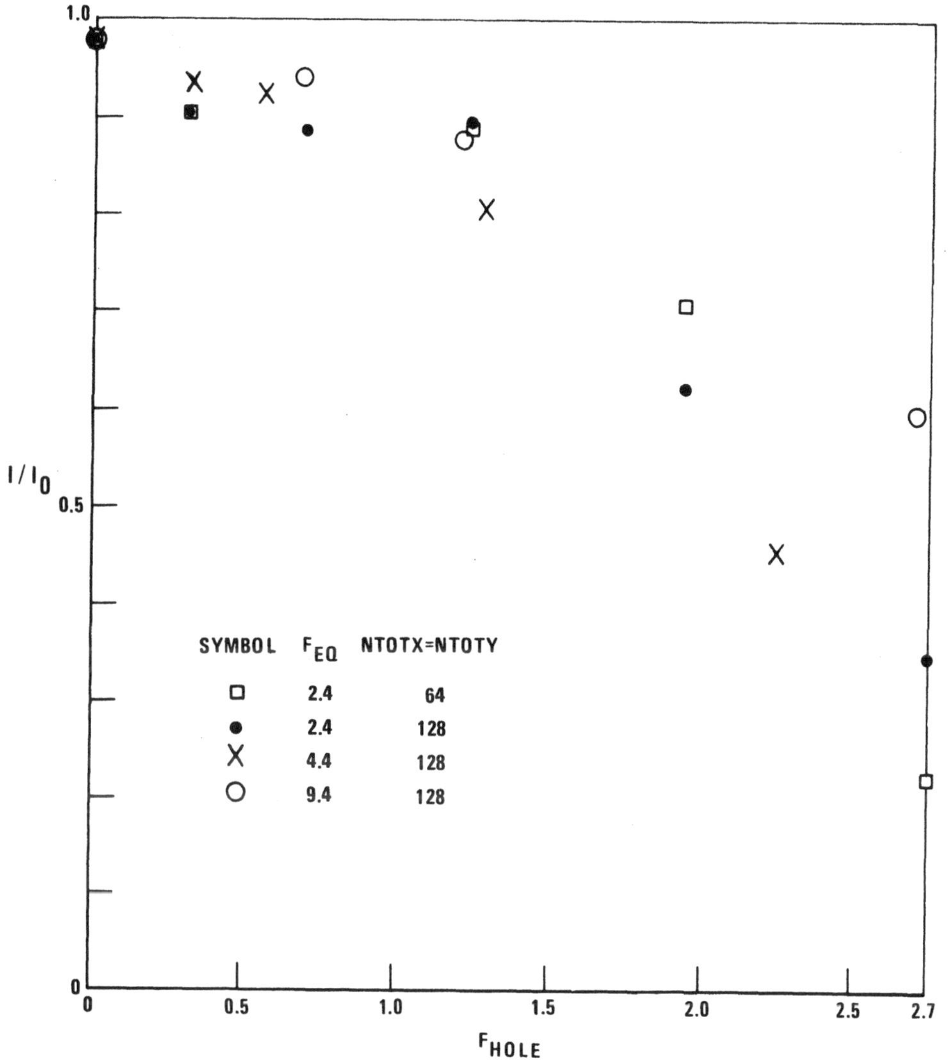

Fig. 6. Calculated values of relative central far-field intensity I/Io for various combinations of (1) Fresnel number associated with the hole, (2) equivalent Fresnel number of the resonator, and (3) number of mesh points in each of the two transverse dirextions for the overall calculations.

The most significant results of the present set of calculations are the calculated values of I/Io, which are plotted in Figure 6. Included in this figure are I/Io values for some calculations which had not converged; these are points to the right side of the figure, corresponding to a Fresnel number of the hole of more than 1.5. The principal features of these results are that: (1) although the beam quality begins to be deteriorated by the presence of a hole in the primary mirror which is fairly small, the deterioration remains moderate up to a hole size whose Fresnel number is roughly unity; (2) for hole sizes whose associated Fresnel number is substantially larger than unity, the beam-quality deterioration seems unacceptably pronounced.

The single most important conclusion of this study is that the beam-quality deterioration is not particularly serious for the smaller hole sizes which have been investigated.

REFERENCES

Buczek, C. J., Freiberg, R. J., and Skolnick, M. L., 1973, Laser Injection Locking, Proc. IEEE 61: 1411.

Cason, Charles, Werkheiser, A. H., Otto, W. F., and Jones, R. W., 1977, Gain Spiking and Mode-Beating Control by Signal Injection in CO_2 Lasers, J. Appl. Phys. 48: 2531.

Horwitz, P., 1973, Asymptotic Theory of Unstable Resonator Modes, J. Opt. Soc. Am. 63: 1528.

Perkins, J. F. and Shatas, R. A., 1976, Propagation in Unstable and Plane-Mirror Optical Resonators, Appl. Phys. 9: 343.

Perkins, J. F. and Cason, Charles, 1977, Effects of Small Misalignments in Empty Unstable Resonators, Appl. Phys. Lett. 31: 198.

Rigrod, W. M., 1965, Saturation Effects in High-Gain Lasers, J. Appl. Phys. 36: 2487.

Sziklas, E. A. and Siegman, A. E., 1975, Mode Calculations in Unstable Resonators with Flowing Saturable Gain. 2: Fast Fourier Transform Method, Appl. Opt. 14: 1874.

ANALYTICAL MODEL OF OPTICAL CAVITY FOR A GASDYNAMIC LASER

G. Rabczuk

Institute of Fluid Flow Machines
Polish Academy of Sciences, Gdansk, Poland

GDL MEDIUM MODEL

The corpuscular model of GDL operation[1] is described in the frames of the following approximations:

1) four-temperature model of the working medium in which we have $T(q)$- translational temperature (enthalpy) of gas molecules, T_p (q_p) - vibrational temperature (enthalpy) of N_2 and CO, $T_m(q_m)$ - vibrational temperature (enthalpy) of $CO_2(\nu_3)$ and $T_n(q_n)$ - vibrational temperature (enthalpy) of both $CO_2(\nu_1)$ and $CO_2(\nu_2)$,

2) one-dimensional model of the gas flow in the GDL channel with friction and heat exchange with walls taken into account,

3) kinetic model of interactions between the radiation field (represented by photons densities) and the optically active mixture (with pumping, relaxation and spatial variability of pressure, temperature and velocity taken into consideration,

4) the optical resonator model formed by two identical parallel flat mirrors with reflectivity $R(x)$ increasing downstream linearly.

The main equations used in the modelling of the GDL medium are as follows:

- *Momentum Equation*

$$\rho v\left(\frac{\partial v}{\partial x}\right) = -\left(\frac{\partial p}{\partial x}\right) - \left(\frac{\lambda_f}{2H}\right)\left(\frac{\rho v^2}{2}\right) \tag{1}$$

- *Energy Equation*

$$\frac{\partial}{\partial x}\left(q+\frac{v^2}{2}\right) = -\frac{\partial}{\partial x}\left(q_p(T_p)+q_m(T_m)+q_n(T_n)\right) - \frac{2\lambda_n}{\rho v H}(T-T_s) \qquad (2)$$

- *Equation of State*

$$p = \{(x-1)/x\}/\rho q \qquad (3)$$

- *Continuity Equation*

$$\rho v\ H\ L = G \qquad (4)$$

- *Relaxation Equations*

$$v\,\frac{\partial q_p(T_p)}{\partial x} = -\frac{q_p(T_p)-q_p(T_m)}{\tau_{pm}} - \frac{q_p(T_p)-q_p(T)}{\tau_{po}} \qquad (5)$$

$$v\,\frac{\partial q_m(T_m)}{\partial x} = -\frac{q_m(T_m)-q_m(T_p)}{\tau_{mp}} - \frac{q_m(T_m)-q_m(T_n)}{\tau_{mn}} - \frac{q_n(T_m)-q_m(T)}{\tau_{mo}} + $$

$$+ v\left(\frac{\partial q_m}{\partial x}\right)_{las} \qquad (6)$$

$$v\,\frac{\partial q_n(T_n)}{\partial x} = -\frac{q_n(T_n)-q_n(T_m)}{\tau_{nm}} - \frac{q_n(T_n)-q_n(T)}{\tau_{no}} + v\left(\frac{\partial q_n}{\partial x}\right)_{las} \qquad (7)$$

- *Lasing Equations*

$$v\left(\frac{\partial N_m}{\partial x}\right)_{las} = (\beta_{nm}N_n - \beta_{mn}N_m)(N_f^+ + N_f^-) \qquad (8)$$

$$v\left(\frac{\partial N_n}{\partial x}\right)_{las} = -v\left(\frac{\partial N_m}{\partial x}\right)_{las} \qquad (9)$$

$$C\,\frac{\partial N_f^+}{\partial z} = (\beta_{mn}N_m - \beta_{nm}N_n)N_f^+ \qquad (10)$$

$$C\,\frac{\partial N_f^-}{\partial z} = -(\beta_{mn}N_m - \beta_{nm}N_n)N_f^- \qquad (11)$$

$$N_f = N_f^+ + N_f^- \qquad (12)$$

$$N_f^+(x,0) = R(x)N_f^-(x,0) \qquad (13)$$

$$N_f^-(x,L) = R(x)N_f^+(x,L) \qquad (14)$$

The medium gain coefficient is given by

$$\alpha = \frac{1}{2}\frac{\partial \ln N_f}{\partial z} = \frac{1}{2C}(\beta_{mn}N_m - \beta_{nm}N_n) \tag{15}$$

In Eqs.(1-15) v,p,H,L indicate the gas mixture velocity , static pressure, height and width of the nozzle (L >> H), G is mass flow, x denotes isentropic exponent, τ_{ik} (i,k=m,n,p,o, o-ground state) stands for relaxation time corresponding to $s \to t$ change of the molecule vibrational state, N_f^+ , N_f^- are the densities of photons moving back and forth between the mirros (of lenght L_x).Coefficients of stimulated emission and absorption β_{mn} , β_{nm} are known functions of a gas mixture. Gas flow friction coefficient λ_f as well as the gas-wall heat exchange coefficiente λ_n are calculated based on the boundary layer theory.

The numerical solution of the system of equations(1-15)permits to determine the laser operation parameters but only in the case of the hypothetical resonator defined in assumption 4),as it was mentioned earlier.Also influence of the wave effects accompanying the lasing cannot be taken into account in the model. In that case it is not possible to properly calculate and design the laser optical system.

These difficulties of the corpuscular model of GDL can be bypassed with the aid of the wave theory of the resonator[2,3] by replacing. the conditions (13-14) describing the reflection of the photons on the mirrors of the hypothetical resonator with the wave equations describing the field distributions over the real resonator mirrors.

The following simplifying assumptions are adopted:

1) due to the fact that the GDL medium gain coefficient is relatively small and the field intensity inside the working region is a weak function of z (z-axis is perpendicular to the flow axis x) it can be assumed that

$$N_f(x,y,z) \cong \tfrac{1}{2}\,(N_f(x,y,0) + N_f(x,y,L)) \cong \tfrac{1}{2}\,(|U_1|^2 + |U_2|^2) \tag{16}$$

where U_1 and U_2 indicate radiation fields upon the mirror surfaces;

2) due to the influence of turbulence and diffusion perpendicular to the flow direction and the fact that population inversion is a weak function of y , the dependence of N_f on y in Eq.(16) can

be neglected, by finding the average value of $|U_1|^2$ and $|U_2|^2$, with respect to the y direction;

3) lasing occurs between rotational sublevels of the 00°1 and 10°0 vibrational energy levels of CO_2, characterized by the quantum number $J = J_{opt}$ corresponding, for given values of mixture state parameters, with the maximum gain value;
4) the influence of medium inhomogeneities is negligible (the gain coefficient is real valued).

GDL RESONATOR MODEL

An unstable resonator with oblong cylindrical mirrors forming side walls of the flow channel, Fig.1, is the simplest cavity configuration that seems to best satisfy requirements to be met by the gasdynamic laser optical system. Two problems arise in connection

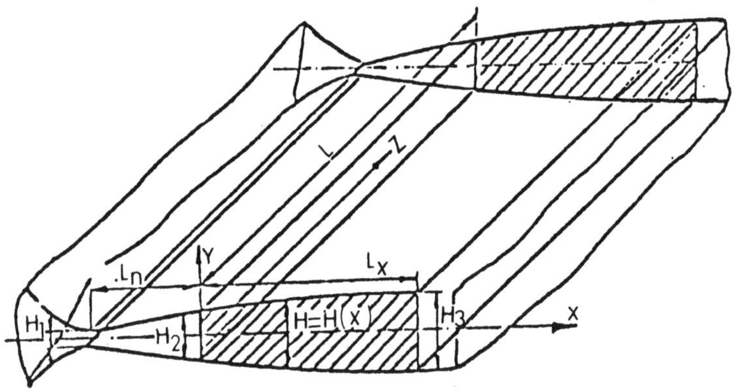

Fig.1 Scheme of the GDL channel.

with selection of such configuration, namely (1) in which point along the nozzle should the optical cavity begin, and (2) how far along the flow should it extend. The possibilities of solving the problem (1) are bounded by two extreme solutions. In the first case the working region i.e. the laser cavity may begin near the nozzle throat, covering the part of the channel where the small - signal gain has highest values. Here the cavity mirrors will have increasing height (H=var). The second case consists in locating the initial point of the cavity downstream of the nozzle where the channel (and mirror) height is already constant. As relaxation processes are predominant over pumping in this part of the flow, the gain is lower there. It follows from approximate optimizing calculations that cavity length (L_x) corresponds to Fresnel numbers (N_F) as large as 200-500.

The field distributions over the mirror surfaces are described by the system of integral equations:

$$\gamma U_1(x,y) = \int_0^{L_x} \int_{-H(x)/2}^{H(x)/2} k(x,y,x',y')G(x,x')U_2(x',y')R_{c2}(x',y')dy'dx' \quad (17)$$

$$U_2(x,y) = \int_0^{L_x} \int_{-H(x)/2}^{H(x)/2} k(x,y,x',y')G(x,x')U_1(x',y')R_{c1}(x',y')dy'dx' \quad (18)$$

where $0 \leq H(x) \leq H_{max}$ denotes the local height of the channel, $\alpha(s)$

$$G(x,x') \cong \exp\{\int_{p_1(x')}^{p_2(x)} \alpha(s)ds\} \qquad s = |p_1p_2|$$

is described by the formula (15),

$$k(x,y,x',y') = k_x(x,x')k_y(y,y') = \frac{i}{\lambda L} \exp\left\{\frac{\iota k\{(x-x')^2+(y-y')^2\}}{2L}\right\}$$

and

$$R_{ci}(i=1,2) = \{1-T_i\}\exp\left\{\iota k \frac{x^2}{R_{ix}} + \frac{y^2}{R_{ij}}\right\}$$

R_i are the mirror curvature radii measured in the directions x and y, respectively. Transmittivities T_i are considered as independent of the y coordinate.

Due to the fact that $L_x >> H$ (and $N_{Fx} > N_{Fz}$, $N_x = Lx^2/(4\lambda L)$ $N_z = H^2/(4\lambda L)$) and because of the assumption that gain coefficient can be regarded, in the first approximation, as a function of x alone the complex amplitude $U_i(x,y)$ can be written as

$$U_i(x,y) \cong v_i(x)W_i(x,y) \quad (19)$$

where $W_i(x,y)$ is a slow function of x.

The above condition permits one to transform the equations (17-18) into sets of one-dimensional equations: non-linear equations along the flow (along the x-axis) and linear ones across the flow (along the y-axis). The equations describing the field distributions along the y-axis do not contain the gain function being therefore solvable by means of the iterative procedure applied to a resonator with low value of the Fresnel number.

In a numerical code for finding the $v_i(x)$ distributions, the gain medium is treated in thin-gain-sheet approxiamtion. This approximation permits to introduce the medium corrections by multiplying the free-space complex amplitudes by the appropriate gain formula. The free-space amplitudes are found by applying a FFT propagation code.

NUMERICAL RESULTS

An algorithm for numerical solution of the resonator equations (17-18) combined with gasdynamic relations (1-15)has been developed. In the program the composition of the gas mixture, length, and height of the nozzle, its divergence angle and its shape are fixed. From the flow and relaxation equations the distributions of the velocity,

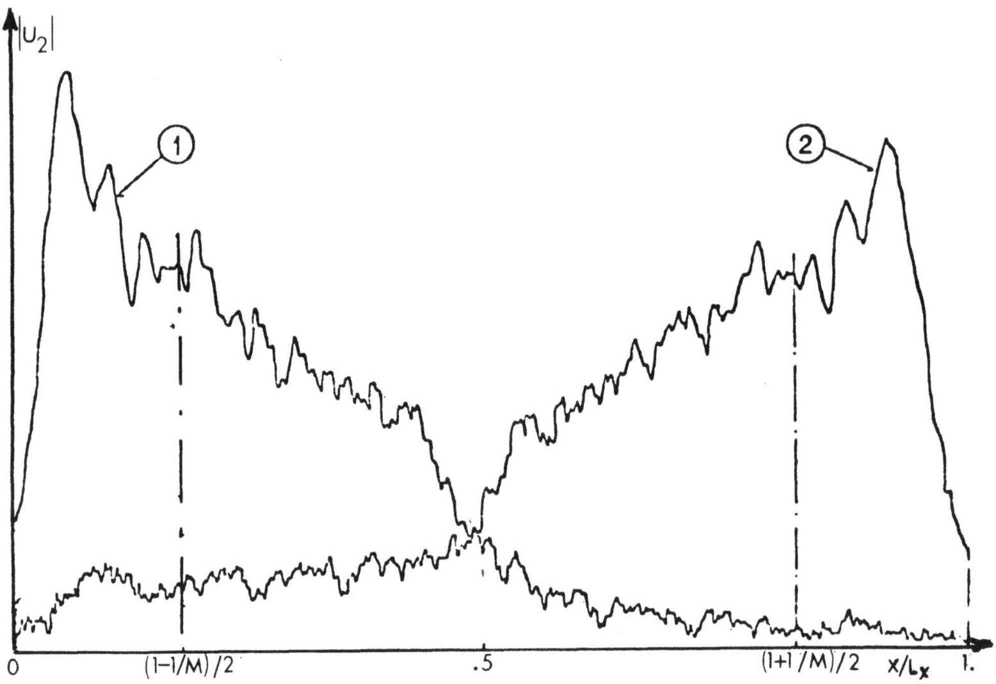

Fig.2 Field distributions appearing alternately at the GDL optical cavity mirrors for N_F = 222, M = 1.55.

the pressure, temperatures, population inversion and the small signal gain, along the nozzle are determined. With know location of the cavity inlet the first approximation of the medium gain distribution

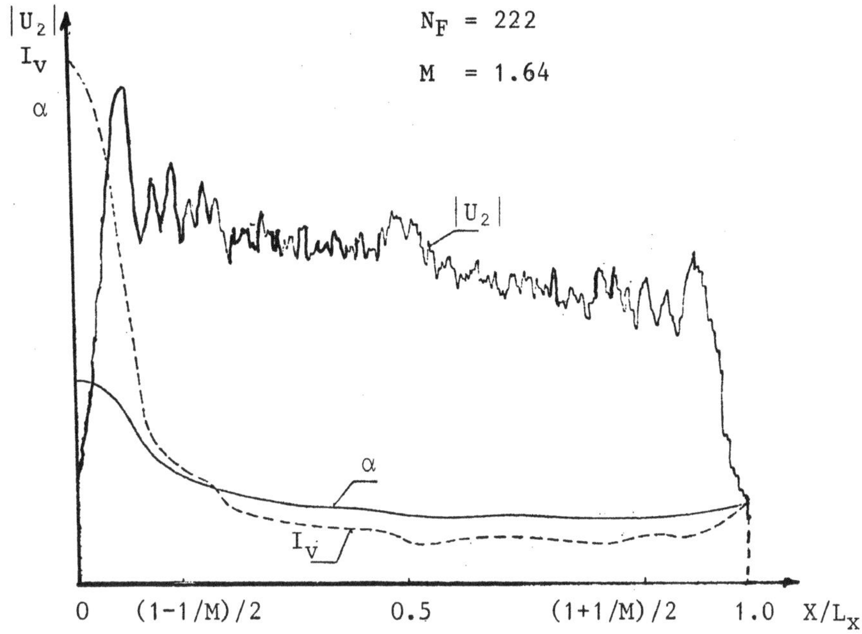

Fig.3 Distribution of field intensity $|U_2|$ (along the smaller mirror), gain α and population inversion I_v, N_F=222 & M=1.64

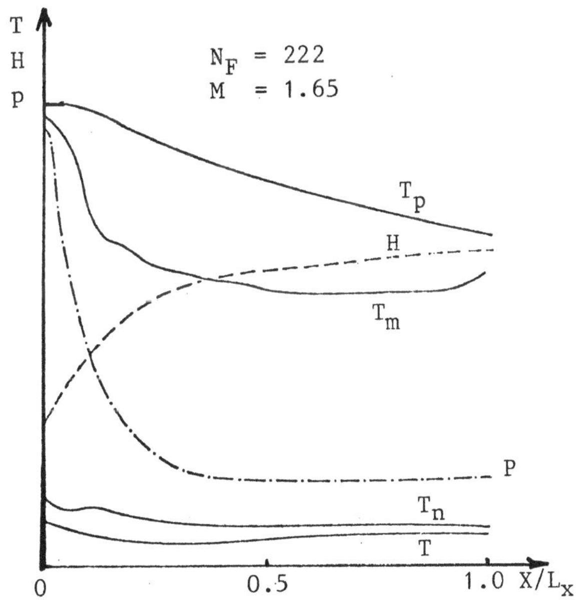

Fig.4 Distributions of gasdynamic parameters along the channel with increasing height.

along the nozzle is determined from equations (1-15) with the conditions (13-14). With $\alpha = \alpha(x)$ once determined the corresponding field distributions upon the mirror surfaces are found by solving Eqs.(17-18). With these distributions known, the radiation energy distribution within the whole resonator region can be determined. This distribution is then used to calculate a new more precise function $\alpha(x)$. The above procedure is repeated until both gasdynamic parameters, field amplitudes and phase attain their steady-state distributions.

The present calculations concern cavities of a CW-GDL driven by the benzene combustion in air (with the air fuel ratio $\alpha_s = 1.2$) at the stagnation pressure $p_o = 15$ atm. In these conditions T_o equals 1993 K, and the lasing medium is composed of $CO_2/N_2/H_2O/O_2/Ar$ in the following proportions: 13.5/75.3/6.7/3.5/1. This mixture is expanded through a nozzle having the throat height 0.1 mm and divergence angle 30 degr. An unstable, confocal, positive - branch cavity composed of two oblong cylindrical mirrors encloses a constant cross-section area channel (with 8 mm height), or a constant pressure (0.01 atm) with the height (variable) increasing from 4 to 8 mm. In the first case the cavity inlet is placed at a distance 4.1 cm from the nozzle throat, where the small-signal gain is equal to 0.7, and in the second case 0.85 cm from the nozzle throat, where the small-signal gain is 0.879.

In both cases the distance between mirrors equals 52.1 cm. Their length (7 cm) was chosen to give the Fresnel number equal to about 200. Examples of the results concerning the above described resonators are shown in Figs.2-4. In Table 1, the relation between the GDL output power P_g and the cavity magnification M is given.

Table 1

M	1.36	1.39	1.44	1.49	1.55	1.64	1.81
P_q (kJ/kg) (H=const)	*	23.36	*	20.4	*	18.245	15.45
P_q (kJ/kg) (H = var)	-	-	*	23.61	*	21.04	18.82

Numerical processes did not converge for $M \leqq 1.36$ (H = const) and $M \leqq 1.4$ (H variable). For cases marked * in Table 1 stable cavity operation could not be achieved. After some large number of iterations two unchanging distributions were obtained, alternating between subsequent passages. An example is shown in Fig.2. This numerical effect,occurring in the region of small M , does not have an adequate explanation till now. Figs.3-4 show the field distribution over the output (smaller) mirror surface, the gain, and some gas-dynamic parameters for a cavity with M = 1.64 and increasing height of mirrors. It follows from the calculations that the effect of channel form on the field distribution along the cavity mirrors is unnoticeable. There are some differences in distributions of gas-dynamic parameters and output power values (Table 1). For H variable pumping and relaxation result in increased value of the population inversion I_V (while the opposite is observed for H const), leading to higher laser output power values.

REFERENCES

1. M.Brunnè,J.Milewski,J.Stanco,A.Zielinski,AIIA Journal,Vol.14,n. 3,1976
2. A.G.Fox,T.Li,Bell Syst.Tech.J.,40,453,1964
3. E.A.Sziklas,A.E.Siegman,Appl.Opt.,Vol.14,1874,1975

PRODUCTION OF GRAM QUANTITIES OF HIGHLY ENRICHED CARBON-12 BY SELECTIVE INFRARED DISSOCIATION

G.I. Abdushelishvili, O.N. Avatkov, A.B. Bakhtadze, V.Yu. Baranov, V.S. Doljikov, S.A. Kazakov, A.G. Kudziev, V.S. Letokhov, S.V. Pigilski, G.I. Pismenni, E.A. Ryabov, G.I. Tkeshelashvili, and V.M. Vetsko

Stable Isotope Inst.,USSR,380086,Tbilissi 86
St.Kavtqradze 21

Selective infrared dissociation of molecules is one of the perspective industrial methods of isotope separation. This is due to the availability of rater powerful (≅ 1 kW) and economical ir sources high repetition rate (HRR) CO_2-lasers and also to the extensive experimental data on the selective ir dissociation of molecules[1].

In a number of cases experimental data have been obtained for the selectivity and ir dissociation yield which allow to suppose that the laser isotope separation will prove competitive as compared with such conventional methods as mass-diffusion, distillation, chemical isotopic exchange. The Table I illustrates the results of experiments on CF_3I molecules in a single pulse regime[2,3]. The experiments were carried out both on a pure gas and on the mixtures of CF_3I with scavengers. The obtained values of selectivity and dissociation yield allowed to conclude that the laser dissociation of CF_3I is a perspective process for producing the products with high carbon isotopic enrichment at a good degree of a desired isotope extraction from the natural isotopic mixture.

Realization of an isotope separation process is dependent whether the desired isotope is a greater part of natural isotopoc mixture or constitute the small fraction. In the first case is necessary that ir dissociation of nondesired isotope should be more effective than desired isotope dissociation. Original substance can be used both as a pure gas and as a mixture with scavengers. On the example of selective ir dissociation of CF_3I molecules consider the production of highly enriched carbon-12 which constitutes the greater part (98.9%) of natural isotopic mixture. CF_3I molecules are dissociated

Table I

GAS / Dissociation Characteristics	CF_3I(1 torr)	CF_3I(1 torr) + Scavenger
Selectivity	8*	7.5* Br_2 (5 torr)
		6.5* NO (15 torr)
$\alpha = \frac{\beta_{13}}{\beta_{12}}$	20**	21** NO (15 torr)
Dissociation Yield	0.18*	0.5* O_2 (8 torr)
		0.56* Br_2 (5 torr)
		0.52* NO (15 torr)
β_{13}	0.08**	0.38** NO (15 torr)

* Fluence 2 J cm^{-2} Gas temperature 300 K
** Fluence 2 J cm^{-2} Gas temperature 230 K

by laser radiation with production of CF_3 radicals which form C_2F_6 molecules (Fig.1). If the laser emission line is chosen so that the dissociation of $^{13}CF_3I$ is dominant compared with $^{12}CF_3I$ then as a result of radiation CF_3I molecules will be enriched in carbon-12 and C_2F_6-in carbon-13. In this case the advantage is that the absorption bands of C_2F_6 molecules do not coincide with CO_2 laser emission lines. Due to this fact C_2F_6 molecules are not dissociated by laser radiation. Therefore the production of highly enriched carbon-12 can be carried out in single cycle without gas change in the irradiated volume and without separation the CF_3I molecules from C_2F_6 ones. But for technology the continuous pumping of the original substance with a chosed gas channel is more advantageous.

The experimental installation used for the production of gram quantities of highly enriched carbon-12 consisted of high-repetition rate CO_2-laser (HRR laser) and laser separating cell (LSC).
HRR laser[4] had the following parameters: the average power ≅ 1 kW,

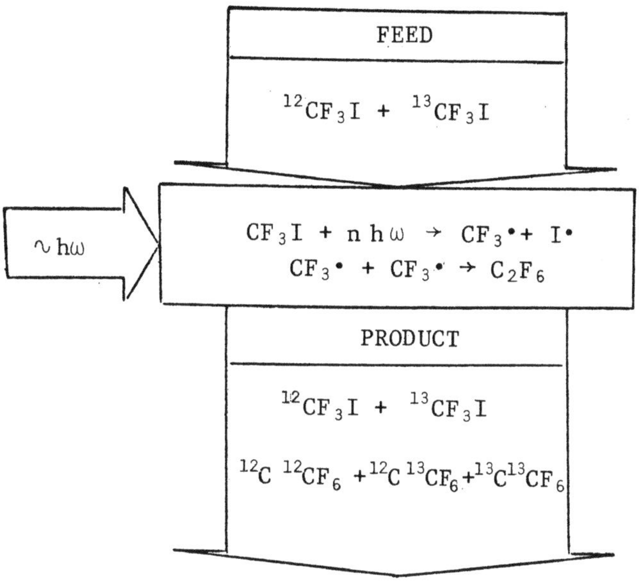

Fig.1

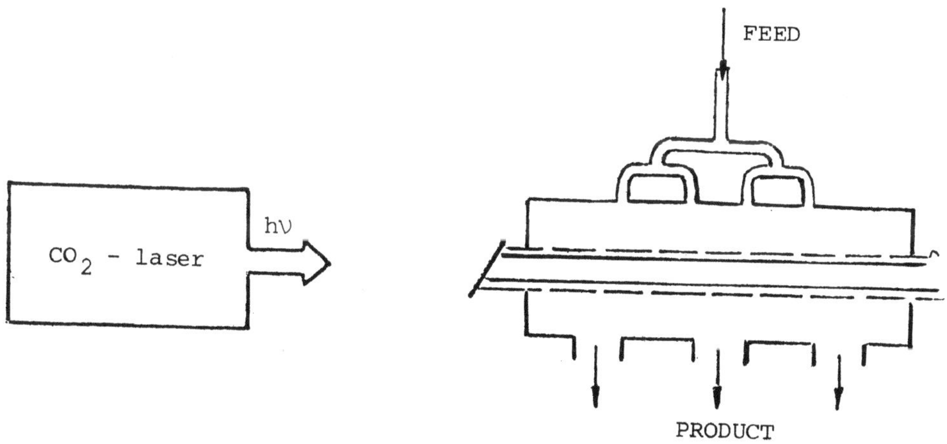

pulse energy - 6J
($CO_2:N_2:He$ = 1:1/4:5)
pulse duration - 1.2×10^{-7}s
fluence - 2 ÷ 3 J • cm^{-2}
pulse repetition rate - 100 Hz

length - 150 cm
irradiated volume - 400 cm^3
gas pressure - 0 ÷ 20 torr
linear gas flow speed-0 ÷ 3 m/s

Fig.2

the pulse energy 6J ($CO_2:N_2:He$ = 1:1/4:5) at pulse duration 1.2 10^{-7}s. The pulse repetition rate was determined by the gas flow speed in the LSC and its value in our experiments was up to 100 Hz.

In LSC a transverse gas flow has been used (Fig.2). The choice of transverse scheme is explained by the necessity of gas change in the irradiated volume between successive laser pulses under the condition of practically constant pressure in the whole irradiated volume. The complete change of the gas in the irradiated volume was mostly conditioned by the necessity of preventing the irradiation of the gas that has been heated by the previous pulse because of the resulting reduction of the selectivity. LSC parameters are the following: the cell length - 150 cm, the irradiated volume - 400 cm^3 , gas pressure - 0.1 ÷ 15 torr, gas flow speed - 0 ÷ 3 m/s. Before entering the LSC, laser beam had been collimated by a system of long-focus mirrors. Beam cross sectional-area in LSC was ≅ 2.5 cm^2.

The purpose of the first experiments on the LSC was to compare the results of the selective ir dissociation of CF_3I molecules in the regime of high pulse repetition rate with the results of the single pulse regime. The measurements showed that at the pulse repetition rate up to 100 Hz and the gas flow speed corresponding to the complete change of the gas at the irradiated volume between successive pulses, the absorbed energy and dissociation yield values practically coincide with the same values for the single pulse regime. But for the selectivity we observed drop of its value with increasing of pulse repetition rate f . The Fig.3 illustrates this fact. The gas flow speed in these experiments increased with f and corresponded to the change of one irradiated volume between the pulses. Only at f ≅ 10 Hz a value α^* = 17 is near to the value of α = 22 obtained in the single pulse regime. Further increase of the pulse repetition rate led to the selectivity drop.

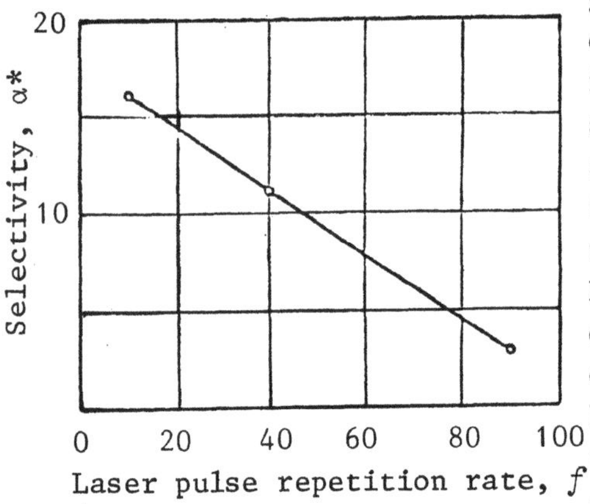

Fig.3

The selectivity drop with increasing of the pulse repetition rate is probably conditioned by gas heating. The Fig.4 illustrates dependence of selectivity on gas temperature. Probably at low pulse repetition rate ($f \cong 10$ Hz) during the time between successive pulses the gas is cooled because of the gas molecule diffusion on the LSC walls. At high pulse repetition rate the diffusion time is more than that between the successive laser pulses. The gas is not cooled that results in the selectivity drop.

The relation between the temperature dependence of selectivity and $\alpha^*(f)$ dependence is demonstrated by Fig.5. It is seen that the ratio of the selectivity in the regime of high pulse repetition rate α^* to the selectivity in the single pulse regime α is a approaching to 1 with increasing the gas flow speed under the condition of constant pulse repetition rate. The value $\alpha^*/\alpha = 1$ is obtained in that case when during the time between successive pulses the gas volume equal to two irradiated volumes leaves the LSC. The above experiments confirmed the feasibility of obtaining the high values of selectivity and CF_3I molecule dissociation yield in the regime of high pulse repetition rate. This allowed the process of obtaining gram quantities of highly enriched carbon-12 to be realized on the LSC.

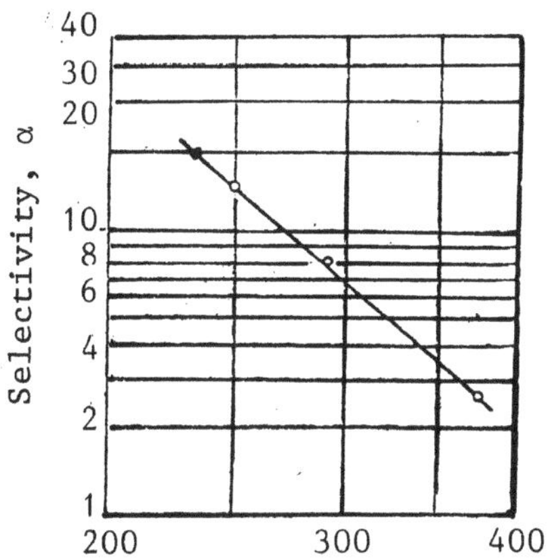

Fig.4

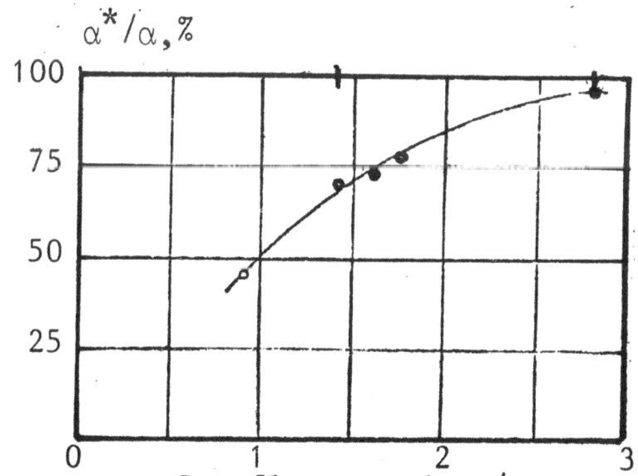

Fig.5

The process of obtaining the product enriched in carbon-12 was realized as follows (Fig.6). 400 g of natural CF_3I was used as an original substance. This gas was many times passed through LSC where it was irradiated. In each cycle CF_3I flow entered the LSC where the dissociation of CF_3I was done by laser radiation with production of C_2F_6 and I_2. Thus, the gas flow on the LSC output consisted of CF_3I, C_2F_6 and I_2 molecules. Because of special choice of CO_2 laser emission line the dissociation of $^{13}CF_3I$ molecules mainly took place ($\alpha^* = 6$ at $T = 300$ K). Therefore in the output gas flow in comparison with input gas flow the CF_3I molecules were enriched in carbon-12 and C_2F_6-in carbon-13. Then by a

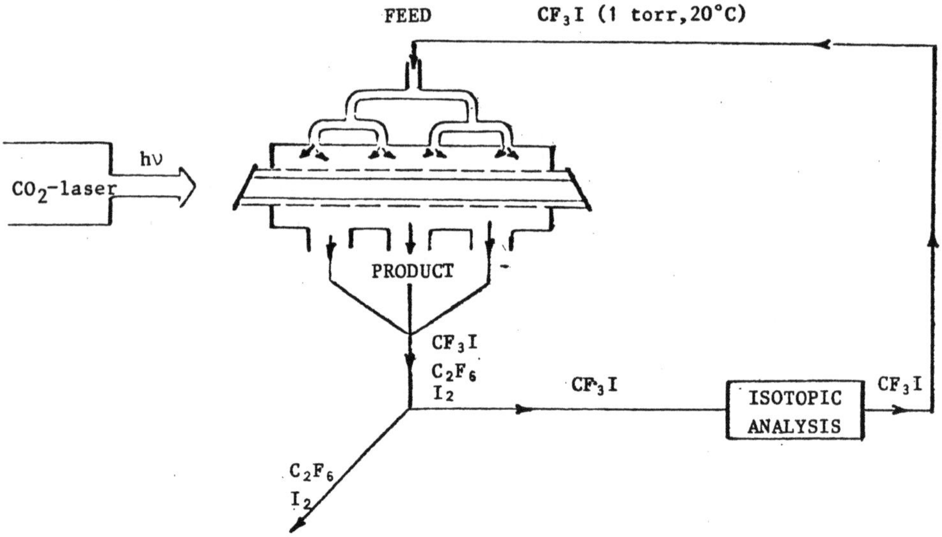

Fig.6

low temperature distillation CF_3I molecules were separated from C_2F_6 and I_2 molecules and were returned to the LSC input. The separation of the products after each cycle was done for mass-and ir -spectrometric analysis. In principle the gas could circulate passing many times through the irradiated zone till the concentration of carbon-12 in CF_3I reached a desired level. Finally 110 g of CF_3I enriched in carbon-12 to 99.99% has been obtained from 400 g of natural CF_3I where carbon-12 concentration is 98.9%.

The Fig.7 shows the dependence of ^{13}C concentration on the cycle number. The process consisted of 108 cycles. Finally the carbon -13 concentration having been reduced hundred times up to 0.01%. On the basis of the preliminary experiments on LSC the process parameters were chosen: the CF_3I pressure - 1 torr, the fluence - 1.7 and 3 $J \cdot cm^{-2}$, the gas temperature -300 K, the pulse repetition rate-100 Hz, the linear gas flow speed -3 m/s. During the experiment the regime was changed only once. The fluence was increased from 1.7 up to 3 $J \cdot cm^{-2}$ that led to the increase in $^{13}CF_3I$ dissociation yield from

0.12 to 0.3. The dependence of lg{^{13}C concentration} on the cycle number is linear (Fig.7). It confirms the conclusion that in the regime of high pulse repetition rate as in the case of single pulse regime, dissociation yield does not depend on concentration of resonance isotope. We have experimentally observed this in the range of ^{13}C concentration from 1.1% to 0.01 %.

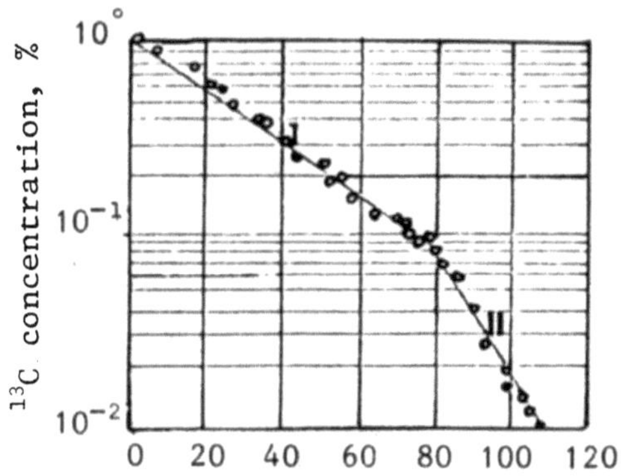

Fig.7

As it was mentioned above two irradiation regimes have been used. If from the beginning to the end, the process was conducted only in the first or only in the second regime it would take 148 cycles at the fluence 1.7 J • cm^{-2} and 62 cycles at 3 J • cm^{-2} . Expenditure of energy for the production of one CF_3I molecule at ^{12}C concentration 99.99%, estimated from the energy absorbed in the LSC is ≅ 20 eV/molecule. Experimental installation output is one gram of ^{12}C(99.99%) in hour, if the whole process is conducted at the fluence 3 J·cm^{-2} and gas circulates continuously through the irradiated zone till the desired ^{12}C concentration is obtained. It should be noted that the production of more pure ^{12}C makes no problem since C_2F_6 molecules at this fluence practically are not dissociated by laser radiation. The results obtained on the experimental installation allows to make a conclusion that production of highly enriched carbon-12 by laser method is quite feasible.

REFERENCES

1. V.N.Bagratashvili,V.S.Letokhov,A.A.Makarov,E.A.Ryabov:*Multiphoton Processes in Molecules at ir Laser Radiation*, VINITI Publ., Moscow, 1981
2. G.I.Abdushelishvili,O.N.Avatkov,V.I.Andrjushenko et al.Zn.Tekn. Fiz.Pisma Red.,5,14,849,1979
3. G.I.Abdushelishvili,O.N.Avatkov,A.B.Bakhtadze,V.M.Vetsko et al. Kvant.Electron.,8,3,534,1981
4. V.YuBaranov,E.P.Velikhov,S.A.Kazakov et al.,Kvant.Electron.,6,4 811,1979

APPLICATION OF MATHEMATICAL HEAT TRANSFER ANALYSIS TO HIGH POWER CO_2 LASER MATERIAL PROCESSING*

G. Alessandretti, P. Gay, and G. Manassero

Centro Ricerche FIAT
Orbassano (Italy)

INTRODUCTION

Industrial applications of CW CO_2 lasers in material processing include welding and cutting of various metals, surface heat treating and alloying.

In the case of welding and cutting the beam is focussed to a spot of about 1 mm diameter; typical power densities of the laser radiation at the interaction point are 10^6 W/cm^2, with interaction times of about 10^{-2} sec. A description of the relevant physical phenomena includes melting and vaporization processes, convection flows and "key holing" of the laser beam.

For heat treating and alloying, the laser spot has dimensions of the order of 1 cm: power densities of $10^3 \div 10^4$ W/cm^2 and interaction times of about 1 sec. Since the radiation is absorbed in a very thin layer, the subsequent processes can be conveniently described on the basis of heat conduction physics, by equating the laser beam to a heat flux entering the surface.

Analyses of different aspects of laser-material interactions, based on such hypotheses, are described in the present work. Attention is given to the practical use of mathematical computations and to a comparison between experimental and calculated results since many authors already described theoretical models[1,4] but only few experimental

* Work partially carried out under "Progetto Finalizzato Laser di Potenza" of Consiglio Nazionale delle Ricerche (C.N.R.)

applications have been reported[5].

Modelling can be shown to have two main aims: one regards the forecasting of working process parameters and the other the phenomenological understanding of laser processes. When persuing the first objective the model input parameters are the process characteristics, defined as type of work (cutting, welding, heat treatment, etc.), work piece geometry, type of material and the output data are the working process parameters such as laser power, beam shape and relative velocity between work piece and laser beam. In particular the input data considered are the depth and hardness of the heat treated zone for surface treatments and the thickness of the melted zone for alloying or surface melting.

When the model is used in order to understand laser processes then the inputs are the working process parameters and the obtained results, while the outputs are a number of information on the process such as phase transformation temperature, absorption coefficients (ratio between absorbed and incident energy) or the physical properties of the material. Presented in this work are applications of mathematical models of both types.

So far as the prediction of process parameters is concerned it is shown that for typical applications it is necessary to consider not only the finite source dimensions but also the combined effects of work piece translation and thermal diffusion. As an example of determination of material properties, the use of a numerical model is described to measure the absorption coefficients of metallic samples with different surface coatings.

Also discussed are various aspects concerning the change of physical parameters with temperature and phase transformation mechanisms.

NUMERICAL ANALYSIS FOR WORK PARAMETER PREDICTION

A first useful approach for the study of material-radiation interaction is offered by the one dimensional model. The treated material is assumed to occupy an infinite semispace with a constant incident energy flux at the surface, for a certain interaction time. The model uses a Fourier heat transfer equation with the following boundary conditions:

$$q_s = K \left. \frac{\partial T}{\partial X} \right|_{X=0} \tag{1}$$

where q_s is the entering heat flux, T is temperature and X the coordinate orthogonal to the surface. In this equation the main physical property of the material, the thermal conductivity K, appears. The second most important property is the thermal volumetric capacity $\rho \cdot C_p$ (ρ density, C_p specific heat). With the ratio between these parameters one can define the thermal diffusivity α connected to temperature propagation in the material. These parameters are considered constant to the first approximation.

The main results of the model are:

- prediction of the maximum surface temperature to be compared with the temperature of phase transformation, of melting or of vaporization;
- prediction of hardness depth, based on the thickness of material heated above the austenitic phase transformation temperature;
- prediction of heat penetration depth $D_c = 2\sqrt{\alpha \tau_s}$ in the interaction time τ_s; this is equal to the depth at which the temperature reaches 10% of the surface temperature.

Although the one dimensional model allows an understanding of the fundamental phenomena involved, comparison of the results with experimental data were judged unsatisfactory[6].

When the thickness of the workpiece or the linear dimensions of the laser spot are of the same order of magnitude as the depth of heat penetration then the one dimensional hypothesis is no longer adequate.

To show the effect of these factors a three dimensional model has been developed. This model considers a plate of finite thickness moving at constant velocity under a rectangular shaped incident laser beam of uniform power density. The initial temperature of the plate is constant and temperature field is calculated for the steady state case. The physical parameters are considered constant as temperature varies. The model is based on the image method and considers a reference system in translation. The calculations allow definition of temperature behaviour in time for each point and of heating and cooling speeds in the austenitic field, threreby estimating the depth and shape of the hardened zone. Some examples are here given to show a comparison between results obtained by calculation and by experiment for the case of flat medium steel samples of normalized AISI 1045, in the absence of surface melting. A 15 kW CW CO_2 laser, installed at Fiat Research Center, was used and working process parameters are reported in Table 1 (column 1).

Fig. 1 shows the calculated temperature distributions

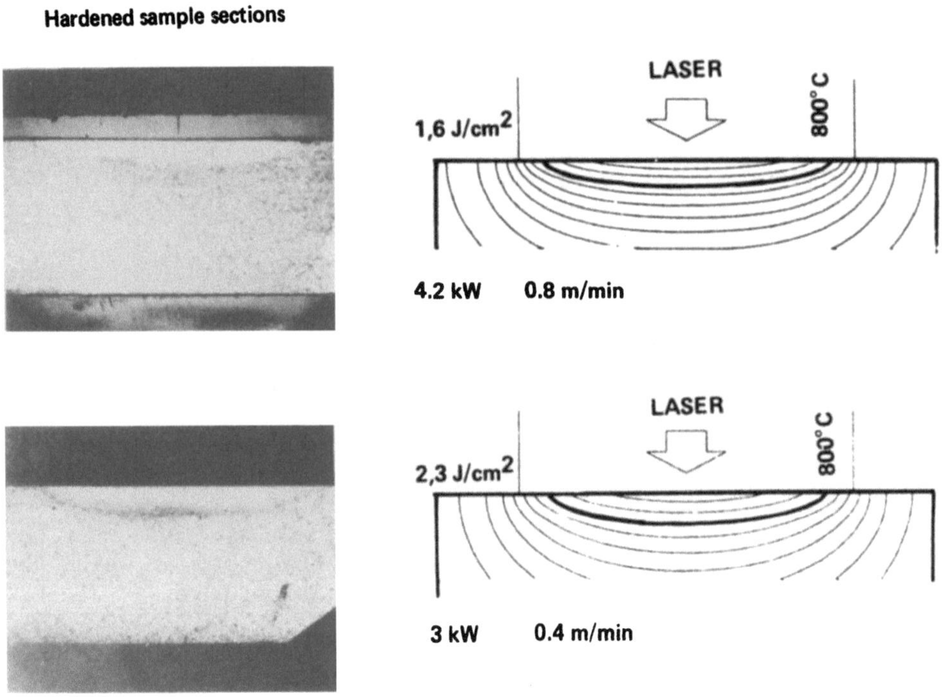

Fig. 1 - Section of treated samples and calculated isotherms

Table 1 - Experimental tests

Material	AISI 1045	AISI 1045
Coating	phosfate	graphite
Plate thickness (mm)	15	30
Beam dimension (mm)	16x16	10x10
Power (kW)	6÷10	3÷10
Velocity (m/min)	0.4÷1.5	0.2÷0.5
Melted depth (mm)	0	0.4÷1.8
Hardened depth (mm)	0÷1.7	1÷5.5

in the transverse sections corresponding to the maximum temperature. The 800°C isotherm, corresponding to the austenitic phase transformation temperature, is shown in the diagrams. The shapes delineated by these isotherms are seen to correspond to the hardened zones shown in the macrophotos.

Depths and widths are within 5-10% of the measured values, for different values of laser power.

The simplified one dimensional model, on the contrary, gives depth values 30 to 50% greater than those obtained experimentally.

APPLICATION OF NUMERICAL ANALYSIS TO THE DETERMINATION OF MATERIAL PROPERTIES

Absorption coefficients

An experimental program was carried out to determine the absorption coefficient at surface temperatures up to 800°C. Fig. 2 shows the experimental set up. A circular laser beam was directed into a cylindrical metal sample. The incident laser power was measured calorimetrically and the sample heating was evaluated using 2 thermocouples located at different depths.

Tests were performed on nodular cast iron and AISI 1045 steel samples. For the two materials different surface conditions were investigated (for example: with phosphate, graphite and titanium oxide coatings). The laser beam was focused on a circular spot (of 8 mm radius) with a nearly

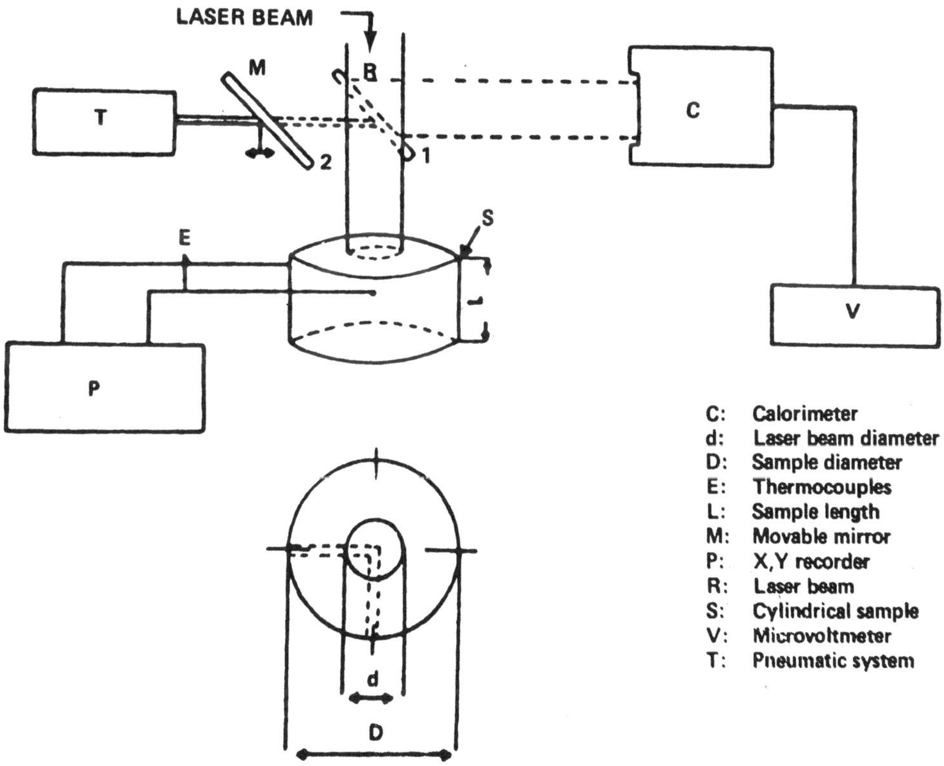

Fig. 2 - Schematic diagram of the experimental system used for absorption coefficient measurements

Table 2 - Summary of test data

Material	Surface treatment	Average Incident Power[W]	Absorption Coefficient [A] average	Maximum temperature[°C] at 1 mm below heated surface
Nodular Cast Iron	Phosphate	964	0.59±14%	486
"	Graphite	916	0.64±4 %	479
"	Sandblast	952	0.33±18%	265
"	Titanium Oxide	911	0.71±4%	564
AISI 1045	Phosphate	870	0.89±9%	436
"	Graphite	879	0.81±23%	396
"	Sandblast	932	0.46±18%	218

constant power density (about 500 W/cm^2) and the samples were heated for a few seconds. During the treatment, the temperatures at two points inside the piece were recorded with 0.5 mm diameter Chromel-Alumel thermocouples inserted at 1 mm and 4 mm from the surface on the axis of the laser spot.

Results were analysed with a model considering an homogeneous cylindrical body, heated on a circular area with axial symmetry. The top surface was supposed to exchange heat with surrounding environment, while the lateral and bottom surfaces were considered adiabatic. Thermal conductivity K(T) and heat capacity, as well as absorptance A were considered variable with temperature (a given function was furnished as a data). The temperature on a meridian plane was computed during heating and cooling.

The analysis was performed by comparing experimental and computed data of temperature versus time, having as variable parameters, absorption and thermal properties. For a preliminary study constant values were assumed. A more refined study was done by assuming thermal conductivity changing linearly (or changing linearly in steps) with temperature.

In Table 2 the absorption coefficient values determined for different coatings and materials are given. The absorption varies from a minimum of 33% to a maximum of 89% changing from sand blasting to phosphating.

The absorption coefficient also varies for different materials with the same surface preparation.

Material thermo-physical parameters

In the presence of surface melting it is necessary to consider the variation of the thermophysical properties with temperature. To evaluate the importance of such variations a one dimensional model was devoloped which takes into account this effect. The model considers a plate of finite thickness, and different physical properties for the solid and the liquid. Moreover, the latent heat of fusion is included in the calculations.

Laser treatments with surface melting were carried out on steel samples (the working parameters are shown in column 2 of Table 1).

The applicability of different models was investigated by studying the depth of the melted zone versus laser power. The diagram in Fig. 3 shows some results. Curve 1 is the forecast from the simple one dimensional model, curve 2 from the model using a moving source, while curve 3 represents the prediction of the one dimensional model with variation of the physical parameters. These results are much better than the previous ones but still not in agreement with experimental data.

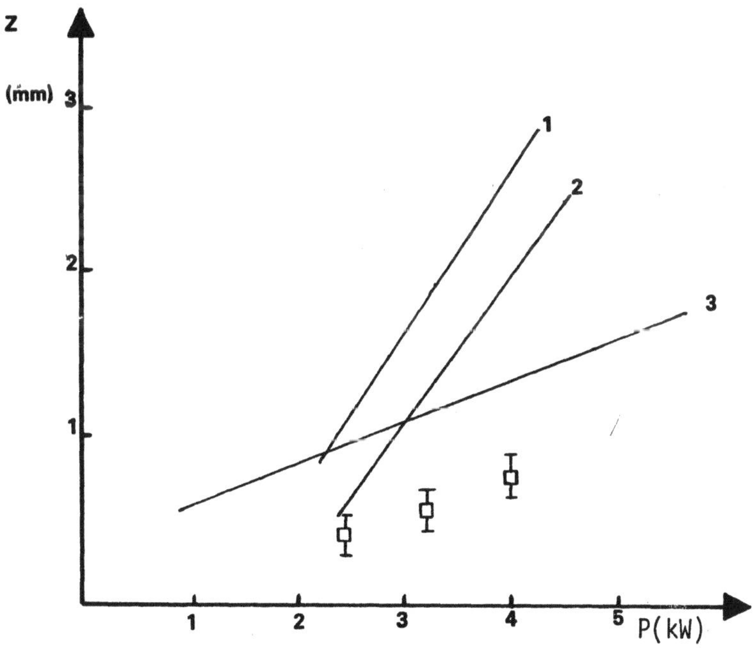

Fig. 3 - Change of melted depth as function of laser power
(1: simple one dimensional model
2: three dimensional model
3: one dimensional model with varying physical properties).

CONCLUSIONS

The field of mathematical heat transfer analysis to CO_2 high power applications is very wide.

Only a few examples are reported to emphasize its importance. Numerical analysis not only permits a better understanding of process mechanisms, but also reduces process definition time. This by an approximate determination of work parameters like laser power, beam dimensions and relative speed between laser and treated components.

A simple monodimensional model can be used when the ratio between depth of heat penetration, during the interaction time, and the specimen thickness or the beam dimension is less than about 0.5.

A three dimensional model which takes work piece movement into account is useful for the prediction of hardened zones in the absence of surface melting, using appropriate absorption coefficients and phase transformation temperature data.

Finally it was a fundamental requirement to measure the absorption coefficients and to know the variation of physical parameters due to temperature, especially in forecasting melt depth.

REFERENCES

1. A.V. La Rocca - First European Conference on Optical System and Applications, Brighton (U.K.) (1978).
2. Y. Arata, I. Miyamoto - Trans. J.W.S. 3, 163 (April 1972).
3. M. Sparks, J. Appl. Phys., 47, 837 (1976).
4. R. Harrach, J. Appl. Phys., 48, 2370 (1977).
5. W. Duley, D.J. Semple, J. Morency, M. Gravel - Opt. Las. Tech. 11, No. 6, 313 (1979).
6. P. Antona, A. Blarasin, M. Castagna, P. Gay - La Metallurgia Italiana, N. 1, 10 (1980).

A POWERFUL CW CO_2-OSCILLATOR-AMPLIFIER SYSTEM FOR RESEARCH AND MATERIAL PROCESSING

C. Carlhoff, S. Müller, H. Odenthal, J.H. Schäfer, and J. Uhlenbusch

Physikalisches Institut II, University of Düsseldorf
D-4000 Düsseldorf, FRG

INTRODUCTION

High power cw CO_2-lasers are frequently applied to material processing as welding, cutting and hardening and in more basic research for producing high pressure plasmas, so called optical discharges[1,2]. In all these investigations a good mode behaviour beside a sufficient laser power is the prior condition for a successful work. Thus our main intension was to develop a monomode high power CO_2-laser system. Experiences with the convectively cooled laser system COLA II fulfilling those conditions are summarized in the following. To increase the gain of the active medium a small test device was mounted for the local diagnostic of important laser discharge parameters as gas velocity and temperature. These data are obtained from resonance scattering experiments on Doppler-broadened (temperature) and shifted (velocity) spectral lines. Welding experiments with thin tin plates show that because of good beam quality a processing velocity of 50 cm/s can be reached.

THE LASER SYSTEM COLA II

A schematic view of the oscillator-amplifier laser system COLA II is plotted in fig. 1. The oscillator is a diffusion dominated low-power CO_2-laser tunable to different wave lengths with passive and active stabilization of the cavity. As it was carefully checked the system works in an OO mode in a power range of about 10 W. After amplifying the oscillator signal by a preamplifier (a convectively cooled CO_2-laser system, amplification factor of nearly 1.6) the laser beam is threaded into a 10 m long device consisting of 14 convectively cooled individual glass segments of 3.5 cm inner

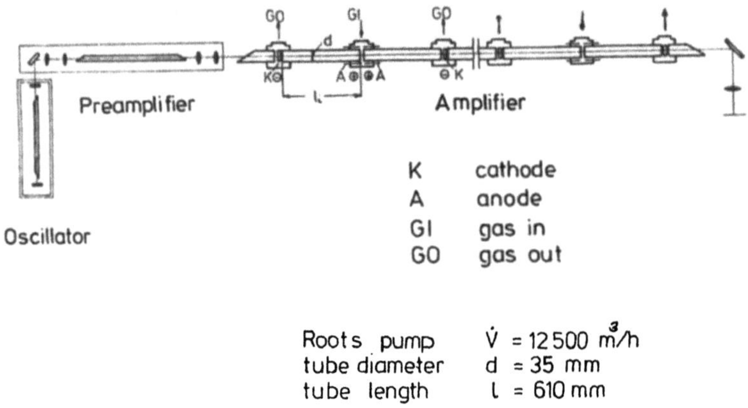

Fig. 1. Oscillator-amplifier system COLA II

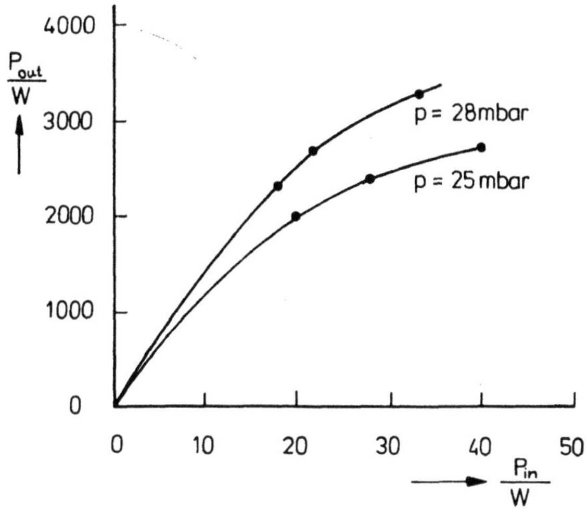

Fig. 2. Output power versus input power for gas pressures p = 25 mbar and p = 28 mbar

diameter and 60 cm length. The swirling flow in the glass tubes is produced by a Roots pump (12500 m^3/h) reaching an average velocity of 240 m/s. A current of 150 mA driven by a voltage of 22000 V excites the gas discharge of the active medium, a 5.7 : 3.3 : 1 He : N_2 : CO_2 mixture. The output power of the system reaches 3.5 kW corresponding to an estimated intensity of 10^7-10^8 W/cm^2 in the focal plane of a collimating system, see fig. 2.

OPTIMIZING THE LASER OUTPUT

Using a shortened and modified version of the main amplifier (2 segments of 2.4 cm inner diameter and a total length of 1 m) the most important properties of the discharge were studied by changing the discharge and flow parameters as pressure, gas composition, current, mass rate and conditions of the gas inlet (swirl superimposed by azimuthal configuration). As a main diagnostic technique local measurements of gas temperature and flow velocity were performed by

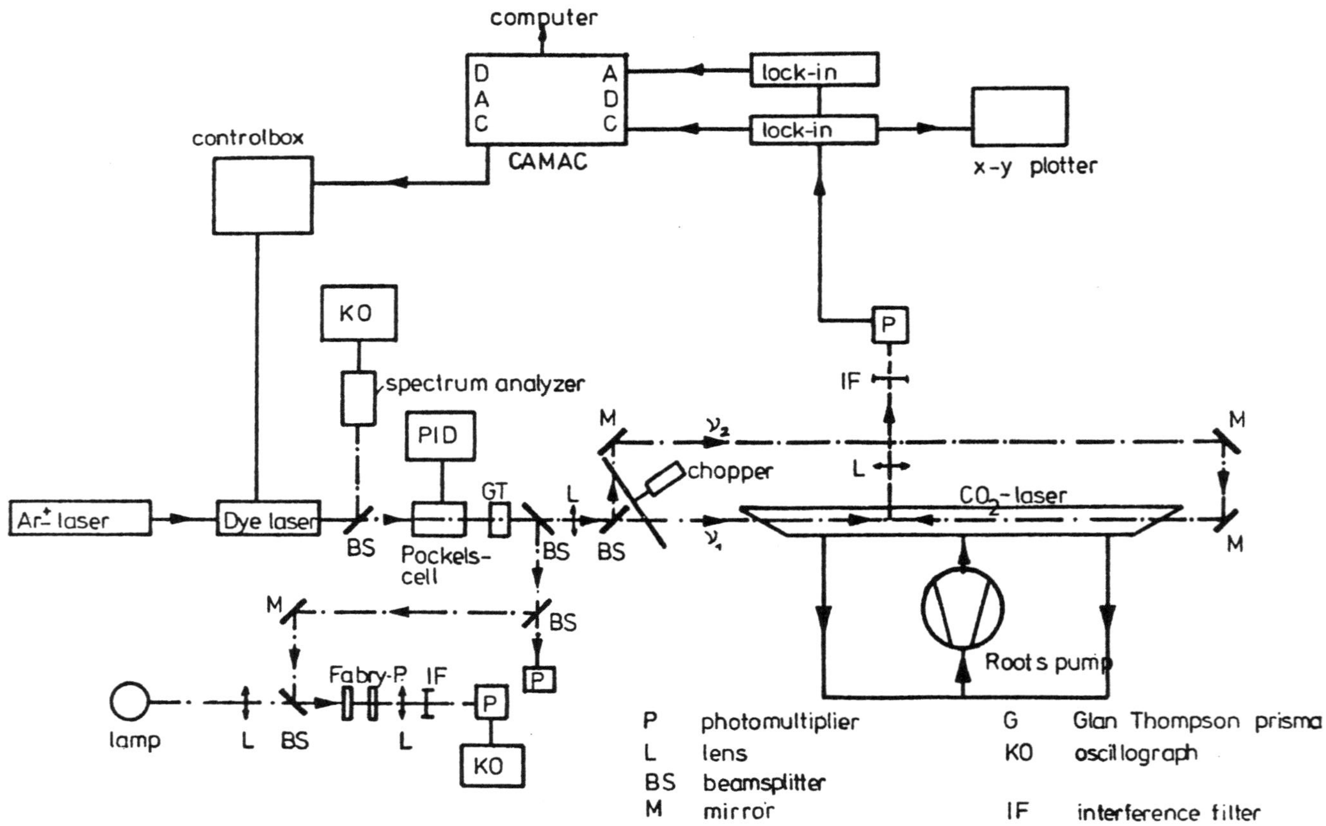

Fig. 3. Resonance scattering experiment

resonance scattering methods[3]. A schematic view of the experiment is shown in fig. 3. The output of a cw dye laser tuned on an atomic resonance is divided into two beams which propagate in opposite direction through the laser discharge. The fluorescence caused by the two beams is simultaneously detected by one multiplier, where a lock-in technique is applied. Two Doppler-shifted and Doppler-broadened line profiles are monitored. The shift of these profiles leads directly to the local velocity and the line width to the local gas temperature. At first, the experiments were done in a pure helium discharge using the λ = 587.6 nm line, because the excited helium levels are not sufficiently populated in the laser gas mixture. As a result fig. 4 shows the radial and axial distribution of the axial velocity component in the laser tube (volume flow of the Roots pump 1400 m^3/h). The appropriate local gas temperatures are drawn in fig. 5. These measurements correspond to a situation of optimal small signal gain of the laser. A rise of local gas temperature can be measured with increasing pressure, increasing current and decreasing mass rate. Flow velocities in pure helium range between 200 and 700 m/s, temperatures are about 300 to 600 K.

As the electron temperature in the laser discharge is too low in order to excite He-atoms an atomic species with low ionization energy must be injected into the laser discharge (e.g. sodium). Measurements applying sodium reproduce the experiments performed in helium.

The radial temperature and velocity profiles are extremely influenced by the gas inlet conditions. Due to the swirling flow a gas stream opposite to the main direction can exist, if the angular momentum of the vortex is enlarged. As a comparison of various inlet situations shows, optimum conditions are only established, if the radial distribution of axial flow velocity and gas temperature are nearly homogeneous across the diameter. With the optimized test device a maximum small signal gain of 1.2 %/cm and a maximum oscillator output power of 700 W could be reached with an active medium of 1 m length.

WELDING OF THIN TINPLATES

For welding experiments the monomode laser beam is focussed on the weld seam between two 0.2 mm thick tinplates. The sheets are carefully mutually aligned on a rotating drum. The welded joint has a good quality. Polished sections of the weld are presented in figs. 6 and 7. If the laser spot diameter is optimized, a welding speed of nearly 50 cm/s has been reached. The nearly linear relation between speed and laser power, see fig. 8, is in good agreement with the theoretical results[4].

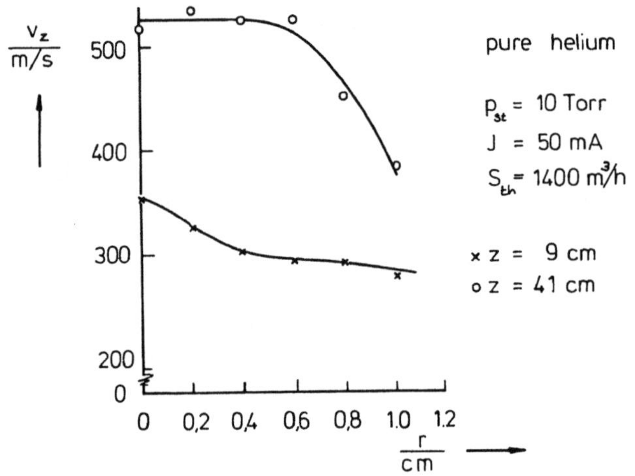

Fig. 4. Radial distribution of the axial velocity component for two different distances from the gas inlet

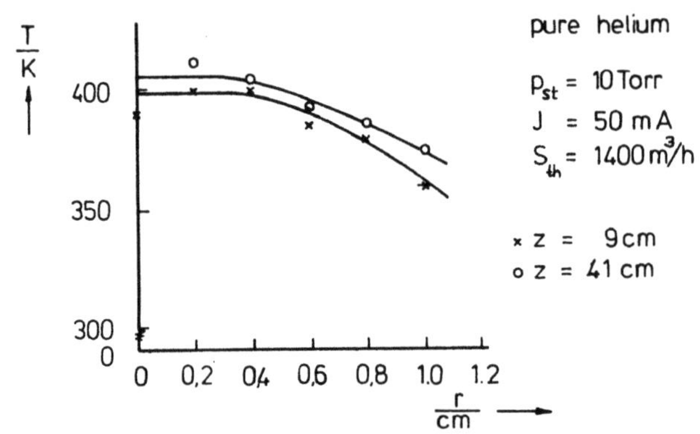

Fig. 5. Radial temperature distribution for two different distances from the gas inlet

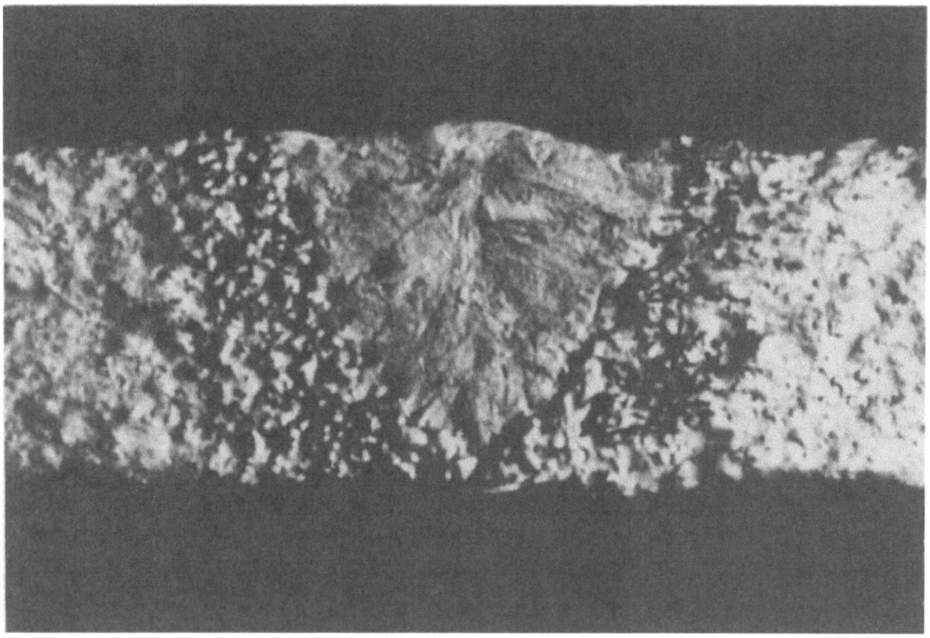

Fig. 6. Cross section of a bead-on-plate weld
$v = 17$ cm/s, $P_L = 1050$ W

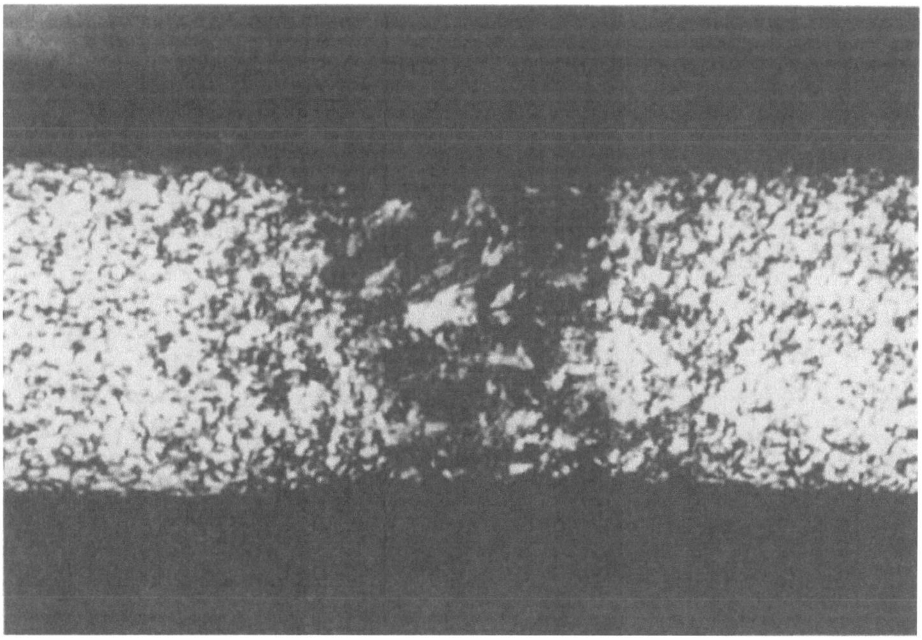

Fig. 7. Cross section of a bead-on-plate weld
$v = 30$ cm/s, $P_L = 1680$ W

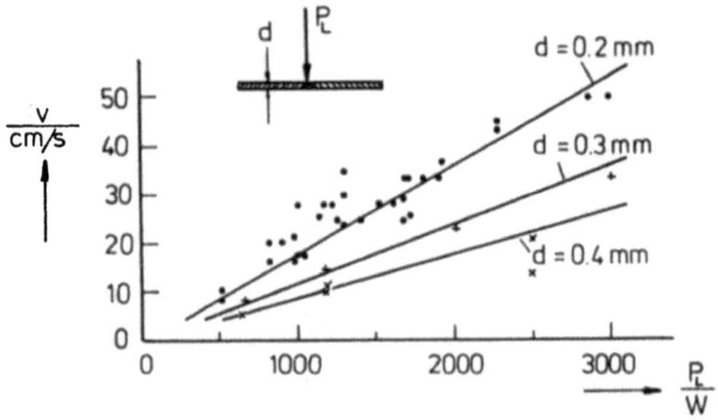

Fig. 8. Weld velocity versus laser power
experiment: • d = 0.2 mm, + d = 0.3 mm, x d = 0.4 mm
theory: ——

REFERENCES

1. C. Carlhoff, E. Krametz, J. H. Schäfer, K. Schildbach J. Uhlenbusch and D. Wroblewski, Continuous optical discharges at very high pressure, Physica 103 C:439 (1981)
2. S. Müller and J. Uhlenbusch, Theoretical model for a continuous optical discharge, Physica 112 C:259 (1982)
3. H. Odenthal and J. Uhlenbusch, Determination of temperature, velocity distribution and population densities of neutral helium by means of laser light tuned on atomic resonances, IEEE PS-8,4:431 (1980)
4. W. W. Duley, "CO_2-lasers - effects and application", Academic Press, New York (1976)

INERT-GAS-ASSISTED CUTTING OF LOW-CARBON STEEL SHEETS USING A HIGH-POWER LASER SYSTEM

*G. Dionoro,
**F. Memola Capece Minutolo, and V. Tagliaferri

*Istituto di Meccanica, Università di Cagliari
P.zza d'Armi, 07100 Cagliari, Italy
**Istituto di Tecnologie, Università di Napoli
P.le Tecchio, 80125 Napoli, Italy

Abstract

The high productivity that a laser system could have if employed in an integrated working system has induced the authors to confront the problem of cut surfaces obtained with laser technology. A cut of inferior quality would make a subsequent finishing operation necessary, which would weigh heavily on machining time.

To this end the influence of an inert gas on the morphology of the cut surface of a low-carbon steel sheet has been considered.

The absence of exothermic reaction due to the presence of the inert gas and the removal of heat due to the same gas makes the employment of a high-power laser (< 5kW CW) necessary.

It seems that the presence of line markings when an inert gas is used arises from the irregular absorption of the laser beam, owing to the surface condition of the material.

In this work the results of different kinds of tests on low-carbon steel samples cut with a 15 kW CW CO_2 laser system are presented. The cutting zone is covered with N_2 in different conditions of flow and pressure. The morphological characterization of cutting surfaces and the mechanical and structural characterization of the cut material are studied. The cut surface is studied by mechanical holographic techniques allowing for linear and tree-dimensional plotting of the cut surface. Metallographic and microhardness tests are carried out in order to examine the influence of the particular covering gas on the alteration of the structure and of the mechanical properties of the cut zone. Experimental tests show that there are no structural modifications in the thermally altered zone; the latter presents, on the other hand, a valuable increase in hardness. The morphology of the cut surfaces shows, for all values of working parameters, a double appearance: the entrance zone of laser beam presenting the typical line markings and the exit zone with a softer profile, like the surface resulting from the undisturbed solidification of a liquid metal.

INTRODUCTION

Users of a laser system in metal cutting cannot disregard the morphological and structural alterations that this new technology induces in the processed material. The intention is to obtain cut surfaces exempt from macroscopic irregularities that might compromise subsequent manufacturing processes.

The aim of this work is to evaluate the influence of an inert gas on the morphology of the cut surface of a low-carbon steel sheet obtained with a laser beam. The use of N_2 as a covering gas requires a high-power laser system, both for the absence of heat generations due to chemical reactions and for the powerful heat removal due to the gas flow. Therefore, within a research program on the technological use of a high-power laser system, we intend to study the influence of N_2, as a covering gas in laser cutting processes, taking into account the appearance and finishing of the cut surfaces and the mechanical properties of the worked material.

In this work the results of different kinds of tests on low-carbon steel samples cut with a 15 kW CO_2 laser system in N_2 inert ambient are presented.

EXPERIMENTAL

Thirteen samples of low-carbon steel 4 mm thick are examined. Samples are cut with several values of power and speed using N_2 as covering gas. The gas nozzle was coaxial with the laser beam. Table 1 illustrates the operative parameters (power, speed, gas pressure).

The morphological characterization of the cut surface is carried out through macrographic analysis and measurements on a Talysurf roughness tester. The roughness index R_a, the bearing surface S_p, the number of profile peaks HSC intersected by the mean line are indicated. In order to obtain a complete characterization of the cut surface, linear scanning surveys are carried out. The morphological characterization of the surface is completed both with the

Table 1. Cutting Parameters

Sample	Power(kW)	Speed(m/min)	Gas Pressure(MPa)
1	7	0.5	0.4
2	7	0.5	0.2
3	5	0.5	0.4
4	5	1.0	0.4
5	7	1.0	0.6
6	5	1.5	0.4
7	7	1.5	0.4
8	7	2.0	0.4
9	10	2.5	0.4
10	10	2.0	0.4
11	7	0.5	0.6
12	7	1.0	0.4
13	7	0.5	0.8

measurement of the size of the zone with line markings in comparison with the thickness of the sample and with the evaluation of the amount of dripped material in relation to gas pressure.

Subsequently, for every sample, the surfaces perpendicular to the cut surfaces are subjected to metallographic analysis.

Microhardness tests are carried out in different and characteristic zones of a surface perpendicular to the cut surface. The samples are subsequently etched and examined on a metallographic microscope with different magnifications.

RESULTS

Visual analysis of the cut zone shows two morphologically different zones: at the laser beam entrance zone it is possible to note the line markings typical both of other traditional fusion cutting processes and, in particular, of oxygen assisted laser cutting. In the exit zone, on the other hand, surface texture shows a less irregular periodicity and with a smaller frequency than the previous zone. The line markings that generally characterize the laser beam entrance zone (Figure 1) consist as usual of beads of melted and resolidified material whose width is practically constant along the whole thickness of the sample. They appear bulged with the concavity oriented in the direction of the feed. However, it is possible to note in some samples a better surface finish, almost like the finish of a fusion cutting carried out in the best conditions, i.e., characterized by a lack of line marking and a surface typical of the undisturbed soldification of a liquid metal. In Figure 2 is shown the way in which the percentage of material

Fig. 1. Cut surface.

thickness not showing line markings varies with feed speed for different values of laser beam power and of gas pressure. Three roughness measurements are carried out for every sample: the first on the laser beam entrance zone, the second on the central zone, the third on the exit zone (Figure 3). Sometimes it is not possible to measure the roughness index because it assumes such high values that the instrument cannot record it. The scanning direction is parallel to the feed of the sample. Three graphs of the bearing surface calculated at different levels in relation to the mean line and of the number of profile peaks intersected by the same level, obtained from the three above mentioned zones, are shown in Figure 4. In Figure 5a,b,c, the three-dimensional aspect of the cut surface of three different samples obtained with a linear scanning method is shown. The conditions of the adduction of the gas (flow and pressure) are different.

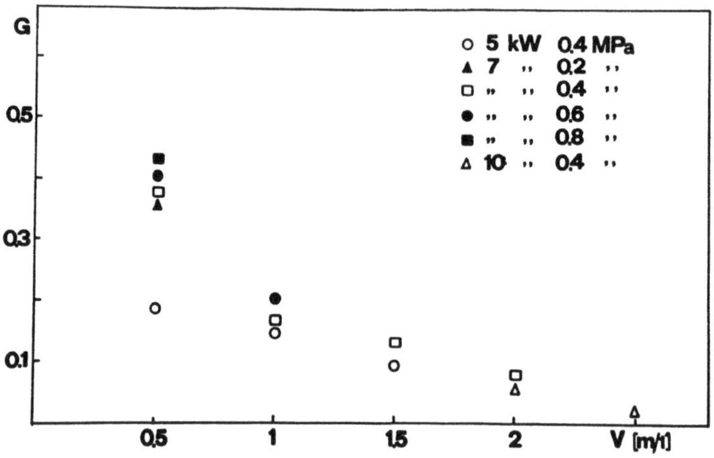

Fig. 2. Material thickness without line marking as a function of cutting parameters.

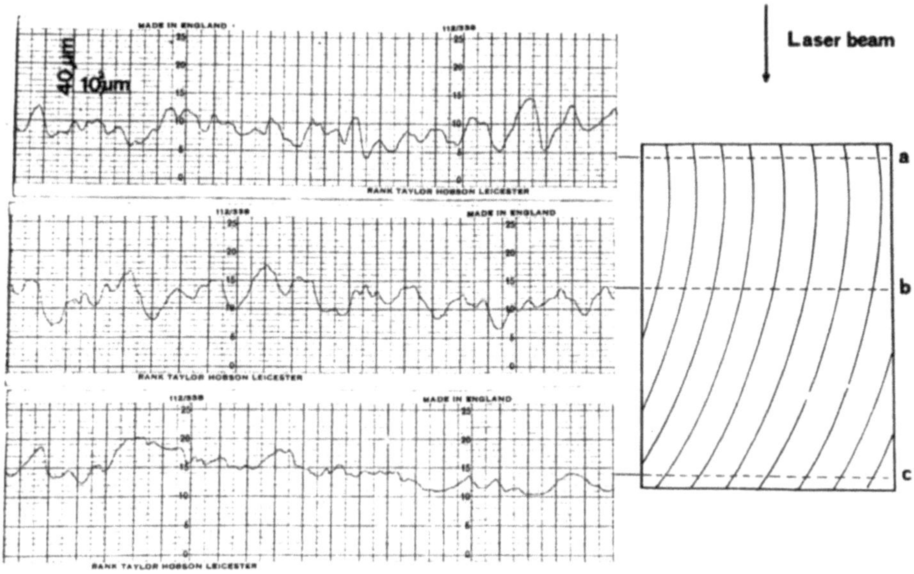

Fig. 3. Roughness measurements.

Metallographic tests are carried out on specimens taken from the cut surface. Material structure is examined with various magnifications in order to recognize possible structural alterations owing to the passage of the laser beam and to the action of the covering gas.

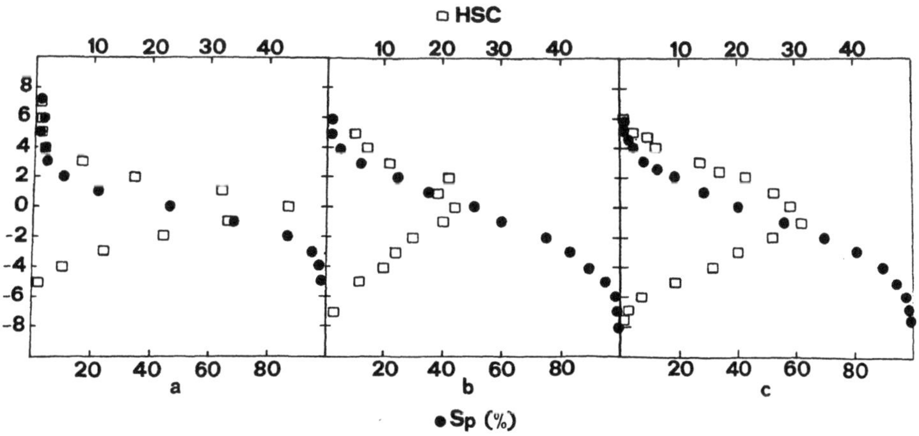

Fig. 4. Values of HSC and S_p at different levels; a) laser entrance zone; b) central zone; c) exit zone.

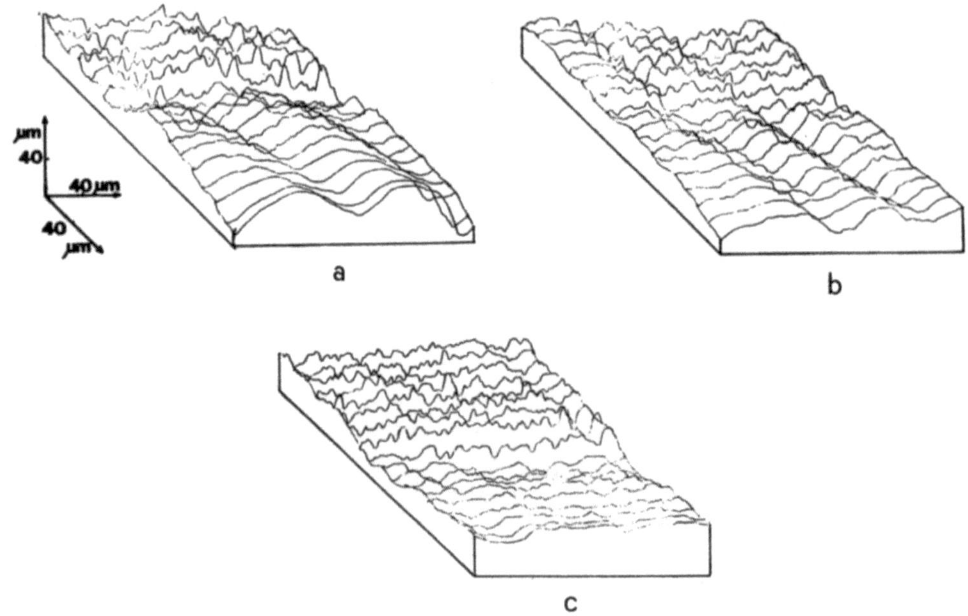

Fig. 5. Three-dimensional aspect of cut surface.

In Figure 6 it is possible to observe the structure of the base material and of a zone of the cut path.

Microhardness tests were carried out on sections perpendicular to the cut surface, at various heights and depths. Along the cut path thirty points 250 µm far from one another and 80 µm from the

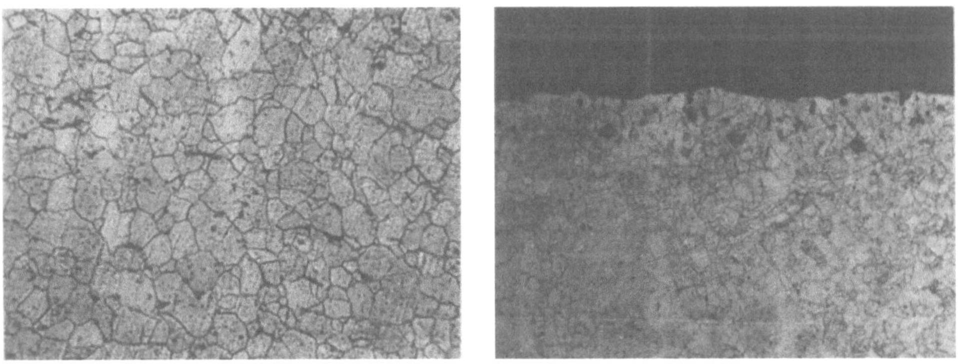

(a) (b)

Fig. 6. Micrographs of the analyzed structure: a) base material; b) cut path.

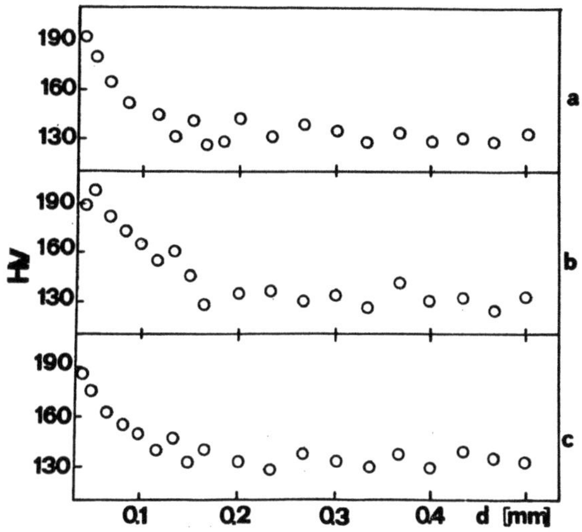

Fig. 7. Average values of microhardness towards the center of the sample: a) entrance zone; b) center zone; c) exit zone.

cut path were examined. In Figure 7 the average values of microhardness towards the center of the sample are shown.

DISCUSSION

Experimental tests carried out for the morphological characterization of the cut surfaces display two typical zones, as it is possible to observe from the holographs in Figure 5: the first zone shows the line marking previously discussed, the second shows a different testure, less irregular and with a smaller frequency than the other zone. It is probable that the presence of line markings, like the ones that may be found in other fusion cutting processes, is above all due to the periodical variation of heat transferred to the workpiece. The reason for this variability could be attributed to the screening effect of the laser beam by the metallic vapors. It is obvious that the more metal vaporizes the more heat is screened and therefore the less metal melts in a successive instant with a consequent lesser vapor formation. The absorption variation due to the formation of nitrides on the cut surface should be negligible.

Hardness values decrease from the thermally altered zone towards the center of the sample with a maximum value of approximately 200 HV near the thermally altered zone. It may be inferred that the thermal gradient assumes such high values as to produce structural alterations not observable with optical microscopy but such as to improve the material mechanical properties. The calculated hardness values are far from the values typical of nitrided surfaces, thereby justifying the hypothesis of an inert ambient around the cut zone.

The evaluation of the dripped material thickness d in mm in relation to the covering gas pressure is still more interesting.

In Figure 8 this quantity is shown as a function of covering gas pressure for a specific heat contribution of 840 kJ/m. If we examine the diagram it is possible to infer that the mechanical effect of the covering gas may not be negligible.

In Figure 9 it is noteworthy that by modifying power and speed, i.e., the specific heat contribution, we obtain a cut surface for which the dripped material is negligible.

CONCLUSIONS

The results obtained show how, in an inert gas flow laser beam assisted cutting, on a part of the cut material thickness, it is possible to obtain a superficial morphology with a double appearance: the zone with line marking typical of other fusion cutting processes is always present, but in this case it is accompanied by an exit zone whose appearance is like that of a fluid surface solidified in an undisturbed way.

Most of the cut surface shows an intolerable material dripping. However, it is noted that the extent of dripping may be reduced with an appropriate variation of the covering gas pressure. This improvement is obtained, nevertheless, preferably in the zone with a softer profile.

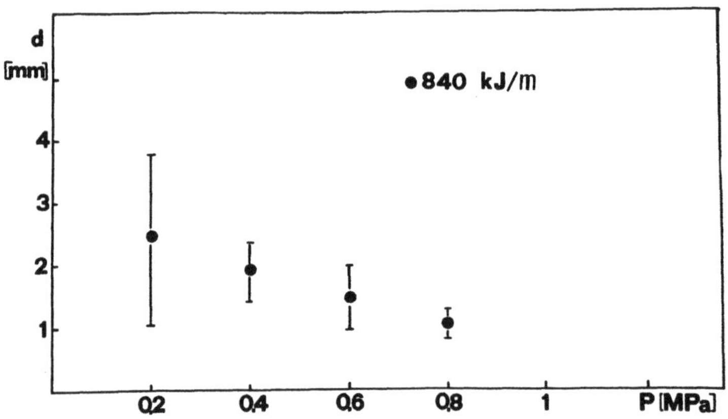

Fig. 8. Dripped material thickness as a function of gas pressure.

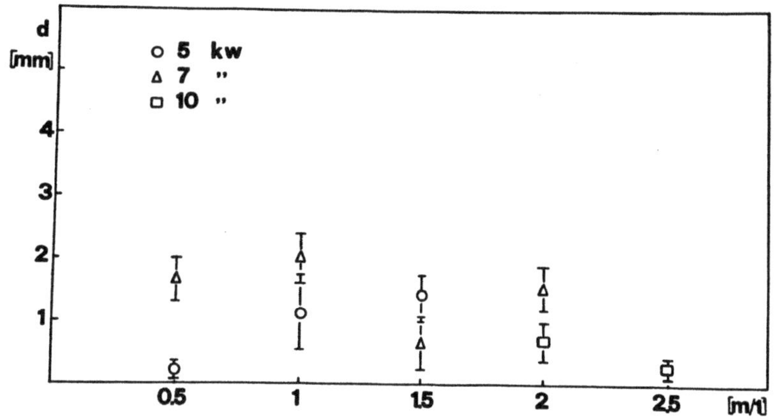

Fig. 9. Dripped material thickness as a function of cutting parameters.

Future research should aim at optimizing both laser working parameters and adduction parameters of covering gas.

Microhardness tests results and metallographic examination seem to confirm the hypothesis of an inert ambient around the cut zone. It is true that the covering technique adopted does not assure the complete absence of other gases, reactive with the metal. The achievement of this aim is problematic for a laser system because of the difficulty of source motion and achieving a completely inert ambient. It should be possible to confirm the hypotheses advanced in this work by operating under vacuum conditions, which would guarantee the absolute absence of the mechanical action of a covering gas. Nevertheless, producing a vacuum in the working zone would be more difficult than the obtaining of a complete inert ambient.

Therefore, the use of medium-high powers facilitates inert assisted laser cutting of low carbon steel. Remembering what was said about cut surface morphology, and the diagram of Figure 8, it may be understood that the amount of heat necessary to obtain a good surface finish is higher than that guaranteeing the simple separation of the edges.

The hardness increase found near to the cut zone should be ascribable to the thermal gradients generated.

The authors feel the need for more profound research on the cut surface in order to explain on the one hand the typical appearance of the surface texture on on the other hand to determine the structure resulting from the interaction between the covering gas, the laser beam and the material.

Acknowledgments

This work was sponsored by the Consiglio Nazionale delle Ricerche (C.N.R.) under the contract no. 80.1811.98, Progetto finalizzato "laser di potenza".

The authors wish to acknowledge the Istituto R.T.M. of Vico Canavese (Turin) for preparation of testing specimens.

REFERENCES

1. T. Miyazaki et al., "Cutting Steels and Steel Products with a 2.5 kW Multimode CO_2 Laser", Proc. of the first Int. Laser Processing Conf., 16-17 Nov., Anheim, California, U.S.A. (1981).
2. Y. Arata et al., "Improvement of Cut Quality in Laser-Gas-Cutting Stainless Steel", Proc. of the first Int. Laser Processing Conf., 16-17 Nov., Anheim, California, U.S.A. (1981).
3. A. Gallo et al., "Problemi di taglio dei metalli con raggio laser", La Meccanica Italiana, no. 148, February (1981).
4. E. Capone et al., "Rilevamento tridimensionale delle superfici", V Congresso Naz. AIMETA., Palermo, Italy, October (1980).

LASER PARAMETERS FOR SURFACE TREATMENTS OF METALS

Emilio Ramous and Luigi Giordano

Istituto di Chimica Industriale
Università di Padova
Via Marzolo, Italy

INTRODUCTION

The laser may be used in various ways as a heating device of high power density to produce surface treatments of metals and alloys. In localized transformation hardening the laser offers great advantages, such as the possibility of being precisely directed and focused, and therefore offering minimal thermal distortion of components and a very short operating time. However, in the laser hardening process the heating mode and the thermal cycle are very different from those of conventional hardening processes [1,2]. The fast heating and cooling rates can give complicated metallurgical responses and non-uniform microstructures [3]. For industrial applications of laser hardening a sufficient process reliability is required and can be evaluated in terms of reproducibility of microstructure and depth of hardened layer.

EXPERIMENTAL

Thermal Model

Generally, a surface layer of a steel workpiece may be hardened by a scanning high-power laser beam which heats the layer and produces a transformation upon subsequent rapid cooling. The thermal cycle induced in the affected layer will be dependent on the scanning velocity V, the beam intensity profile, the laser power level, and the absorption of laser energy by the surface.

The heat flow in the solid will be dominated by its thermal diffusivity D and specific heat. The temperature distribution can be calculated by conventional formulations of the thermal response of the treated material [4]. For example, a simple one-dimensional model may be used to predict the temperature in the central part of the cross-section of the heat-treated material:

$$\text{Heating:} \quad T(z,t) = \frac{\Sigma z W_o}{K} \sqrt{Kt} \cdot \text{ierfc}\left\{\frac{Z}{2\sqrt{Kt}}\right\} \tag{1}$$

$$\text{Cooling:} \quad T(z,t) = \frac{2W_o\sqrt{K}}{K}\left\{\sqrt{t}\ \text{ierfc}\ \frac{Z}{2\sqrt{Kt}} - \sqrt{t-te}\ \text{ierfc}\ \frac{Z}{2\sqrt{K(t-te)}}\right\}.$$

If a more exact thermal distribution is desired, the actual energy distribution in a three-dimensional model must be determined. For example, assuming a rectangular Gaussian distribution of the laser beam w(x,y),

$$T(x,y,z,t) = \frac{\sqrt{K}W}{4\pi Kb}\int_o^\infty \frac{1}{t(a^{*4}b^2+4Kt)} \exp\left[-\frac{(x+vt)^2}{a^{*4}b^2+4Kt} - \frac{Z}{4Kt}\right] \cdot \left[\text{ierf}\ \frac{y+b}{\sqrt{4Kt}} - \text{ierf}\ \frac{y-b}{\sqrt{4Kt}}\right] dt\ . \tag{2}$$

Equations (1) and (2) define the parameters influencing the temperature distribution in the affected layer. An example of the calculated temperature distribution is shown in Figure 1.

Approximately, the correspondence between the temperature profiles calculated with the two models is very good if v = 10 cm/sec or more, but fair if v = 1 cm/sec, with an error of about 10% using the one-dimensional model. These models allow prediction of the laser parameters in order to arrive at the thermal cycle for heat treatment on the basis of the metallurgical requirements discussed below.

Generally, the main requirements for lasers used in industrial applications include, besides a sufficient power level available at the workpiece surface:

(a) controlled intensity profile;
(b) sufficient reproducibility of power level, mode, pointing stability, etc.;
(c) reliability (e.g., long interval between service).

A quantitative evaluation, at least indicative of the necessary reproducibility, can be derived by considering the reproducibility required as "metallurgical results" and the influence of the laser parameters on the surface treatment.

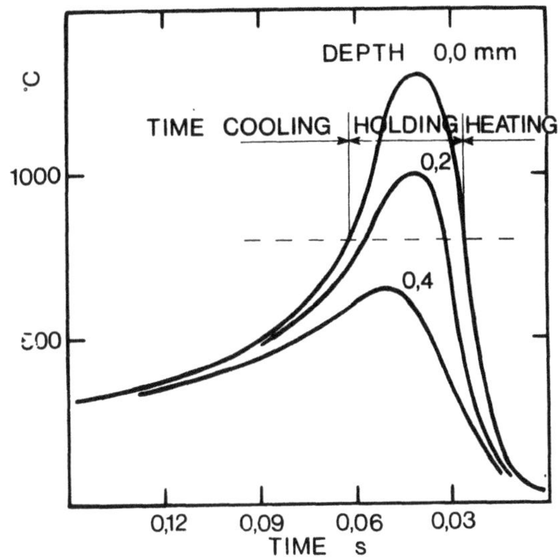

Fig. 1. Calculated temperature profile.

Metallurgical Conditions for Hardening

It is well-known that transformation hardening of steels requires a thermal cycle:

(a) heating above a minimum temperature value ($\alpha \rightarrow \gamma$ transformation) depending on steel composition;
(b) holding temperature a sufficiently long time to allow the dissolution of carbides and carbon diffusion;
(c) cooling at a rate above a critical value for martensitic transformation, as summarized in Figure 2.

The first condition can always be achieved by a correct choice of laser treatment parameters as power density (KW/cmq) and interaction time (τ) or coverage. Choice of values for these two parameters, however, must take into account the necessity of avoiding high superficial temperatures which might cause the formation of anomalous microstructures.

The achievement of the other two conditions depends on an appropriate choice of interaction time. Generally, there are no problems with the cooling velocity, which, with interaction times in tenths of seconds, is far higher than the critical hardening velocity on all the layer heated above the transformation temperature.

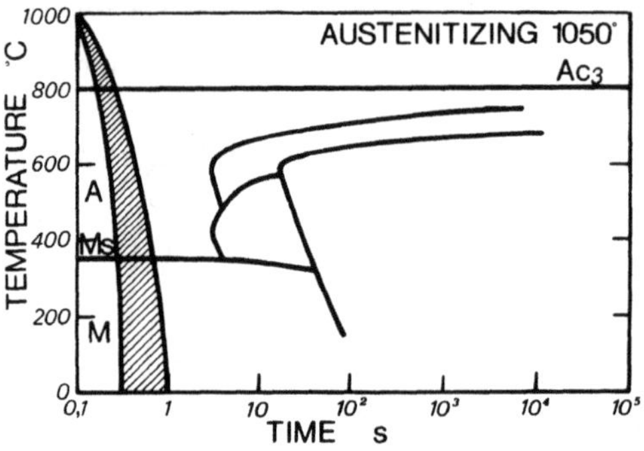

Fig. 2. Transformation diagram of UNI C 43 steel.

Correlation between Laser Parameters and Case Depth

The following considerations refer to the surface hardening of the steel UNI C 43 (with surface phosphatation). Experimental tests have been undertaken with the CO_2 AVCO laser from the Institute of RTM of Vico Canavese.

Similar considerations can be transferred to other steels concerning the differences of behavior on heating and above all on cooling. The thermic models mentioned afford, by calculation, some essential data for characterizing treatment and particularly:

- the distance from the surface at which the transformation temperature is achieved and
- the variation of cooling velocity in connection with the surface distance.

Practically, the first datum also gives the theoretical thickness of the hardened layer which should result.

Agreement between the experimental results and the calculation forecasts is good (see, for instance, Figure 3); this confirms the applicability of these thermic models. The depth of the hardened layer is surely the first control parameter of the surface treatment.

Reproducibility of the values of hardened thickness is important so that the surface treatment can have effective technological applications.

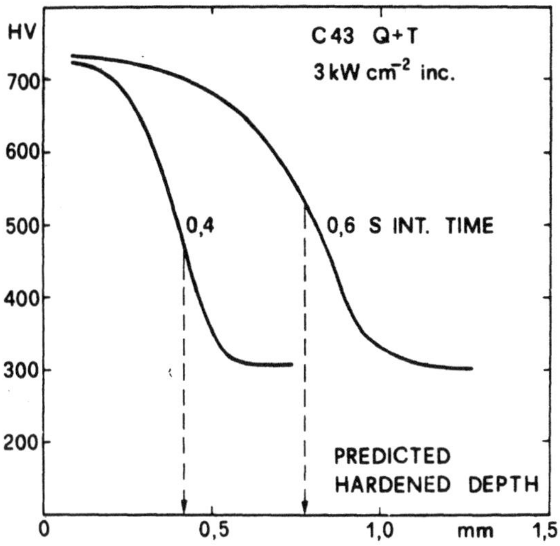

Fig. 3. Hardness profiles of laser treated samples.

The parameters which determine the temperature profile in the superficial area (and therefore the hardened layer) are the power density and the interaction time. This may be inferred from the relations which set up the thermic model of the laser treatment.

From the same equations it is also possible to draw (for the single kinds of steel), correlations between these treatment parameters and the thickness of the hardened layer.

The results have been graphically reported in the two Figures 4-5 for the interval of hardened thickness of greater applicative interest, from 0.3 to 1 mm. For indicative purposes only, the curves have been traced beyond conditions which cause fusion of the surface. Obviously, for a correct hardening treatment, the surface temperature must stay far below melting point.

Considering this necessity, the interaction times of practical interest are obtained with a velocity of 100 - 350 mm/sec (beam size 10 x 10) and the corresponding powers result near to about 2 KW/cmq. The correlation curves gradient, in that value's interval, also gives an indication of the influence of parameter treatment variation on the thickness of the hardened layer.

If, for instance, as is reasonable, for a hardened "nominal" thickness of 0.5 mm a maximum difference of ± 0.1 mm is accepted, it is found that the variations of the power density must be constrained within ± 0.10 - 0.15 KW.

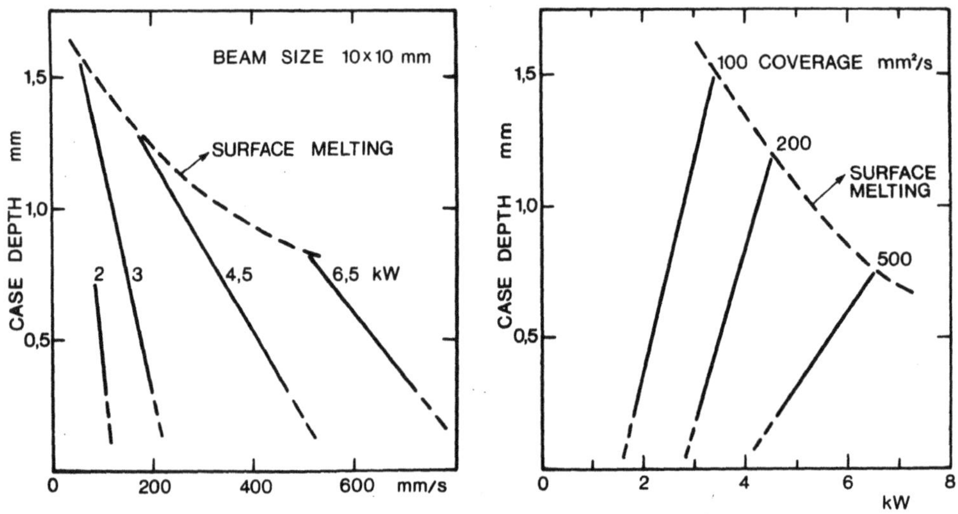

Figs. 4-5. Correlation between case depth and treatment parameters: power level and interaction time.

Clearly, this value has to be referred not only to the incident beam power, but also to the source-beam-surface system to be treated.

For example, similar variations of the hardened thickness occur if, even though the power of the incident beam remained constant, the reflectivity of the surface to be treated were to increase, with consequent decrease in the absorbed power.

These considerations also give information, at least of qualitative character, on the influence that an incidental lack of homogeneity of the incident beam can have on the distribution of the values of hardened thickness within the section of the treated tracks.

Therefore, some devices are proposed, and are under consideration, that should guarantee a homogeneous and constant distribution of the power inside the beam.

Microstructures of the Hardened Layers

For obtaining a satisfactory treatment of surface hardening, both thickness and microstructure of the hardened layer must be controlled. The final structure depends, in a complex way, both on the initial microstructure and on the thermic cycle undergone by the piece. We have already mentioned the need to limit the maximum temperature reached at the surface to avoid the presence of retained austenite, with a consequent decrease in hardness. This problem is more evident in the treatment of alloy steels.

From this "structural" necessity derives a limitation of the thickness of the hardenable layers; in fact, for increasing the thickness it is necessary to use higher power and times of interaction, with consequent increase in the superficial temperature. In practice, the obtainable hardened thickness does not exceed the mm, with superficial temperatures already at a level of 1400°C.

Other difficulties in laser hardening treatment derive from the short times of permanence at high temperature at which a homogeneous distribution of carbon for diffusion is not always achieved.

For this reason structures of homogeneous hardening are more easily obtained given a suitable initial structure of the steel. For example: Figure 6 shows structures obtained after laser hardening of UNI C 43 steel quenched and tempered (more homogeneous) and normalized (heterogeneous martensite).

Therefore, the choice of the treatment parameters should allow, for the same required thickness, higher interaction times with low power densities. Under these conditions the time of permanence at high temperature increases also; this allows the best structures to be obtained.

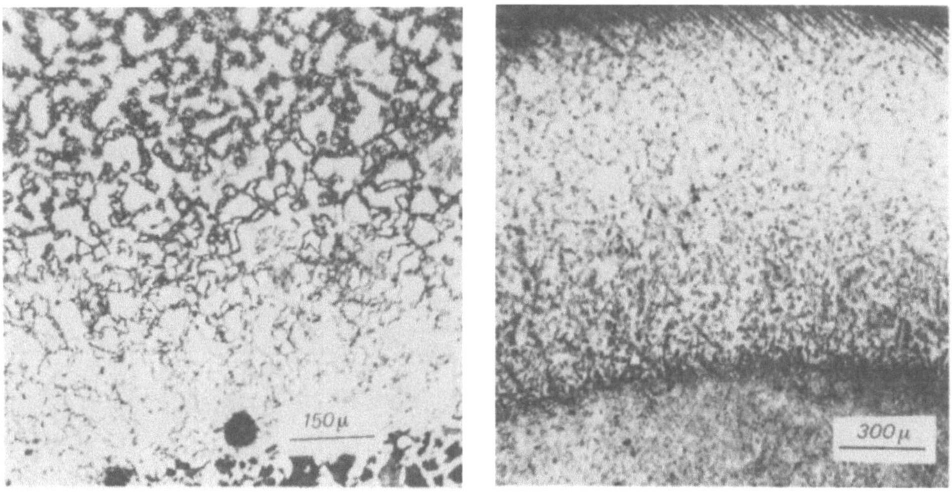

(a) (b)

Fig. 6. Case microstructures: a) C 43 quenched and tempered; b) C 43 normalized.

Moreover, from the curve of Figure 4 we see that increasing the interaction time increases the slope of the curves case-depth/ power density and a better reproducibility of power values is therefore required.

REFERENCES

1. B. H. Kear, E. M. Breinan and L. E. Greenwald, Metals Techn., 6:121 (1979).
2. W. M. Steen and C. Courtney, Metals Techn., 6:456 (1979).
3. R. Menin, E. Ramous and M. Magrini, Proc. 4th Int. Conf. on Rapidly Quenched Metals, Sendai, 193 (1981).
4. H. Marno, I. Miyamoto, T. Islude and A. Y. Arata, Proc. "First International Laser Processing Conference", Anaheim, USA (1981).

LASER CUTTING OF BULK STEEL (40 mm) DUE TO GUIDED FLOW OF RADIATION AND REACTIVE GAS IN THE WORKPIECE

Dieter Schuöcker

Institut für Nachrichtentechnik, TU Wien
Gusshausstraße 25
A-1040 Vienna, Austria

INTRODUCTION

The performance of a laser cutting machine for a specific material can be characterized by the cutting speed obtained at a given thickness. It is well-known that steel up to thicknesses of 10 mm can be cut very well with a laser power of roughly one kilowatt. In that case the product of cutting speed and thickness remains constant and describes the capacity of the machine.

Nevertheless, there are also very attractive applications for laser cutting of steel thicker than 10 mm. From the relationship mentioned above it could be argued that arbitrary thicknesses could be cut, if only the speed was correspondingly reduced. The general opinion to the contrary assumes an upper limit for the thickness of about 10 mm for steel. It is the purpose of this paper to report on experimental investigations and theoretical studies carried out to clarify the question of the maximum thickness obtainable with a given laser power.

EXPERIMENTAL

The experimental work has been done in the laser application laboratory of Voest-Alpine AG, Linz, Austria.

The laser cutting installation consists mainly of a carbon dioxide transverse flow laser that delivers 1200 W of infrared power; a cutting head containing the focusing optics and the nozzle for the reactive gas flow that can be moved automatically in the vertical direction and a coordinate table that moves the workpiece in the horizontal plane. All functions of the arrangement can be controlled automatically by a central processor.

With that set-up, initial cutting experiments have been carried out with a number of materials. Later on the experiments concentrated on steel. For thicknesses below 8 mm, the reciprocal relationship between thickness and speed has been verified; for greater thicknesses considerably lower speeds could be obtained than those predicted by the hyperbolic relationship. By optimizing all parameters of the cutting process, excellent cuts have been obtained for steel up to 20 mm thickness and lower quality cuts have been carried out in steel up to 40 mm thickness with a very low speed. Figure 1 shows the dependence of cutting speed on thickness, obtained by extrapolation of the hyperbolic relationship relevant below 8 mm and the actual values up to 40 mm.

Analysis of an Improved Model of Laser Cutting

A further clarification of the cutting process and its limitations may be obtained from a theoretical analysis.

Several models have been developed in the past [1,2]. All models describe laser cutting of a relatively thick workpiece and cannot be applied without modifications to bulk materials.

Thus, a more sophisticated model has to be used that takes account of bulk effects. In the model illustrated in Figure 2 the cut acts as a waveguide with three walls, whereas the two cut walls guide the laser beam and the nearly vertical cut front absorbs the beam at proper polarization parallel to the walls [3]. On the other hand, the reactive gas flow is also guided by the walls of the cut and absorbed by the vertical front of the cut.

Under the nearly vertical cut front, a thin layer of molten material covers the solid body of the workpiece. In that molten layer, a reaction with the cutting gas and absorption of laser radiation occur and heat up the liquid. The molten layer is also subject to several cooling mechanisms, such as heat conduction, evaporation, melting and convective cooling. Erosion is caused by ejection of molten material on the lower surface of the workpiece owing to friction of the melt with the cutting gas flow and to evaporation at the cut front.

The main input quantities of that model are the laser power P_L and the waveguide attenuation of the laser radiation α_L [4], the number of reactive gas particles impinging on the erosion front q_o, the fraction of reactive particles absorbed by the erosion front β and finally the reaction energy per particle ε_R. The main output quantities are the depth of the molten layer s, the densities of

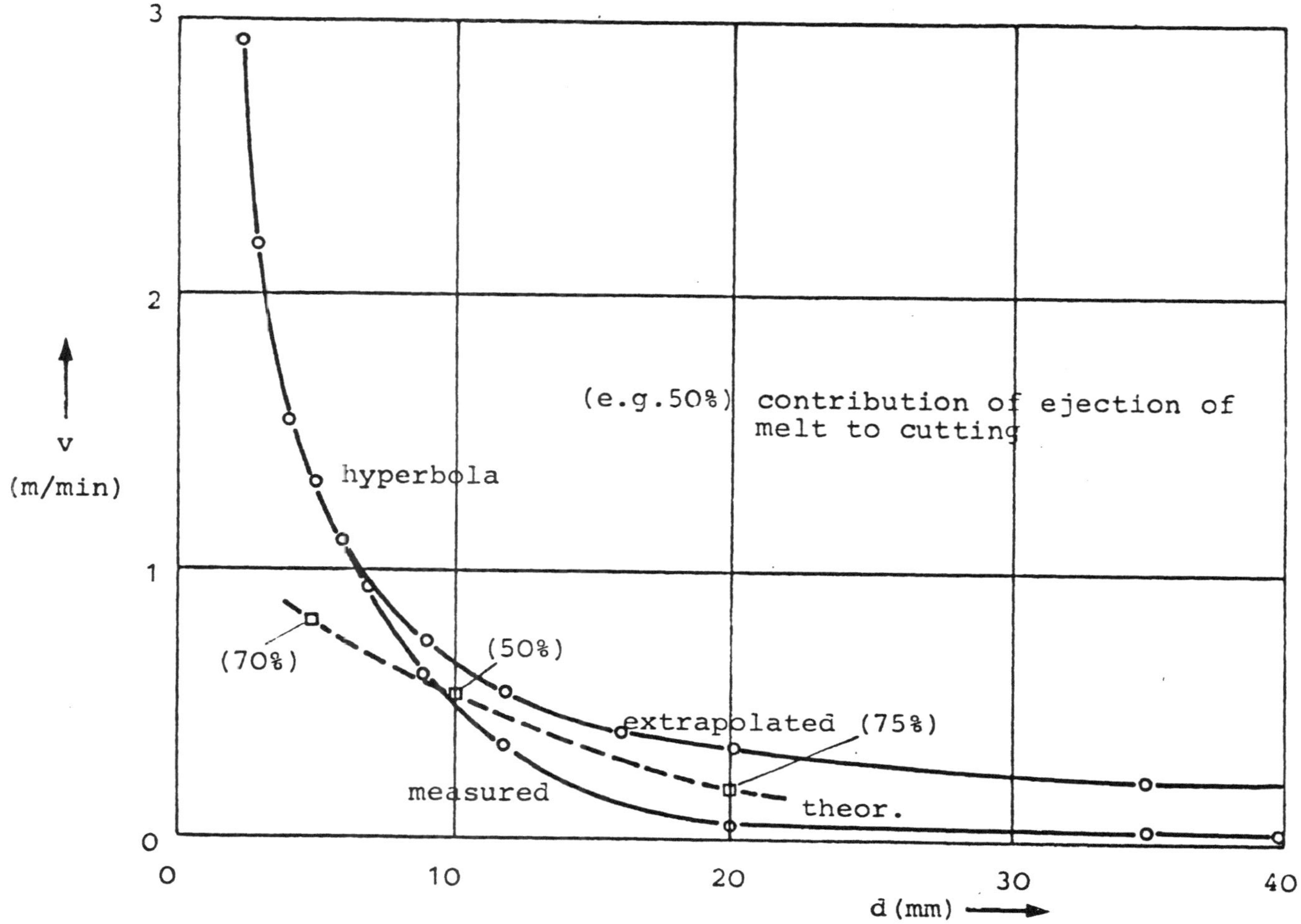

v
(m/min)
3
2
1
0
0
10
20
30
40
d (mm)
hyperbola
(e.g.50%) contribution of ejection of melt to cutting
(70%)
(50%)
extrapolated (75%)
measured
theor.

Fig. 1.

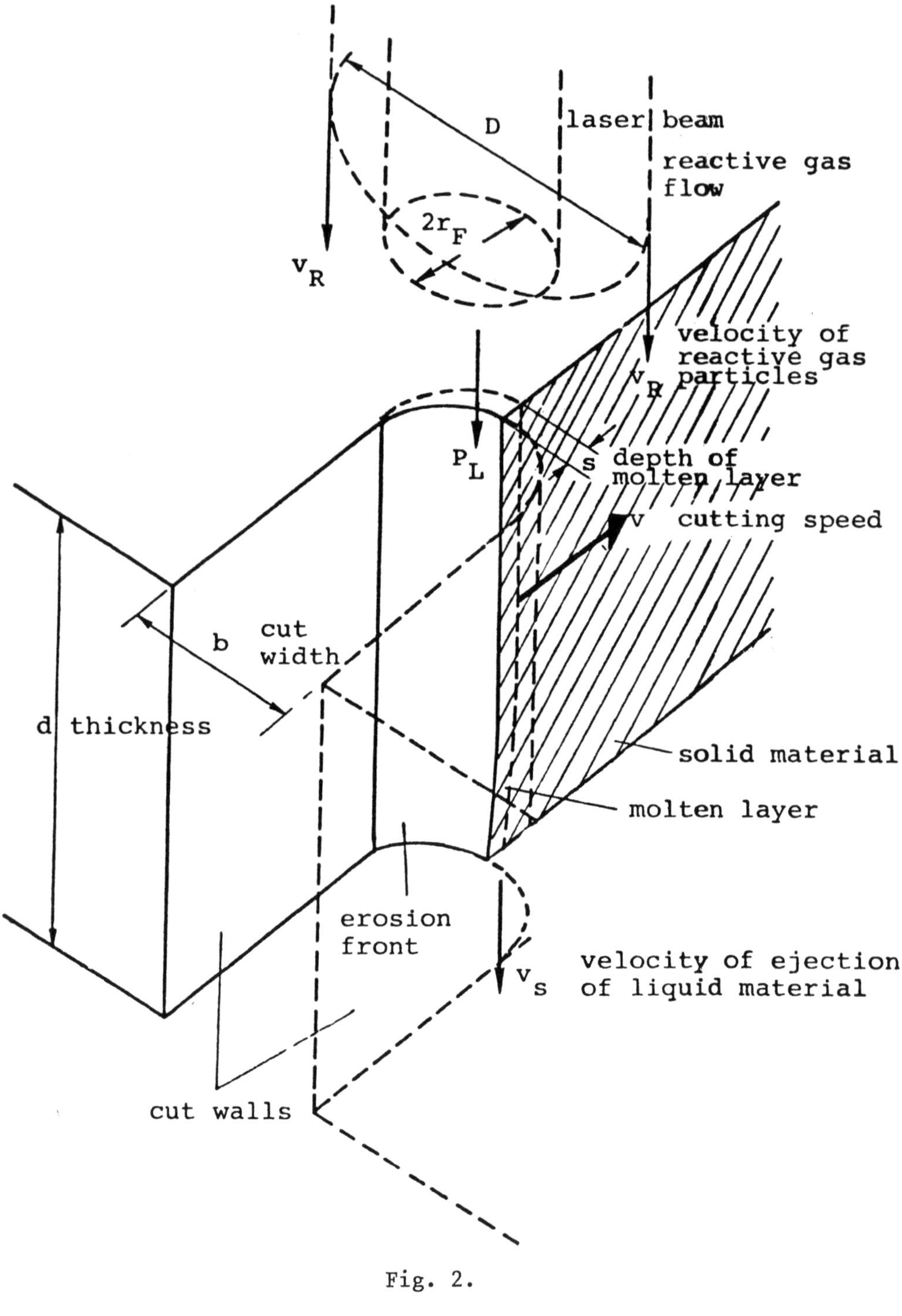
D
laser beam
reactive gas flow
$2r_F$
v_R
velocity of reactive gas particles
v_R
P_L
s depth of molten layer
v cutting speed
b
cut width
d thickness
solid material
molten layer
erosion front
velocity of ejection of liquid material
v_s
cut walls

Fig. 2.

reactive gas and pure metal particles in the molten layer n_R and n_s, the velocity of the particles ejected from the melt at the lower surface of the workpiece v_s, the temperature at the cut front T and the cutting velocity v.

For the determination of these six quantities, six equations are available, namely: two balances for the number of reactive gas and pure metal particles in the molten layer; the momentum balance for the direction perpendicular to the surface of the workpiece; the heat conduction equation; the energy balance of the molten layer and the mass balance of the molten layer.

The whole analysis is greatly simplified if the problem of reactive gas heating is split-off and treated in advance, since this reveals that a maximum energy can be supplied by reactive gas to the workpiece under properly adjusted conditions [5]:

$$\left(\frac{dE}{dt}\right)_m = \varepsilon_R n_o vbd, \tag{1}$$

where n_o is the density of particles in the solid material (for explanation of b,d see Figure 2). Further treatment may then be carried out using maximum energy supply by reactive gas.

That very simple expression is derived from the solution of the balance equations for the number of reactive gas and pure metal particles and may be easily understood, because it represents the maximum energy that is liberated if all pure metal particles that enter the molten region due to the cutting movement are burned.

The momentum balance of the molten layer in the vertical direction includes the gain by friction with the gas flow and the loss by ejection of melt at the lower surface of the workpiece. That balance yields an expression for the velocity of the ejected molten material (η_R dynamic viscosity of the reactive gas, v_R velocity of the reactive gas, δ_s density of the melt):

$$v_s = \sqrt{\frac{\eta_R}{\delta_s}\frac{d}{sb}\,v_R}. \tag{2}$$

The solution of the heat conduction equation for a moving point source with energy influx dE/dt and velocity v yields

$$T(z) = \frac{1}{2\pi Kd}\cdot\frac{dE}{dt}\cdot\exp\left(\frac{vz}{2\kappa}\right)\cdot K_o\left(\frac{vz}{2\kappa}\right), \tag{3}$$

where ρ = density of workpiece; c, specific heat; K, thermal conductivity; $\kappa = K/\ c$ thermal diffusity; z, distance to the point source in the direction of cutting and K_o, Besselfunction of the second kind and zero order [6]. Approximately, it may be assumed that at z = b/2 (at the erosion front) a rotational symmetry exists.

In that case, the energy lost from the cut front (temperature T) into the workpiece by heat conduction is found to be

$$\left(\frac{dE}{dt}\right)_K = 2\pi dKT \, \frac{\exp(-vb/4\kappa)}{K_o(vb/4\kappa)} \; . \qquad (4)$$

The depth of the molten layer may be determined from Equation (3) (T_s = melting point, $K_1 = -K_o'$):

$$s = \frac{2\kappa}{v}\left(1 - \frac{T_s}{T}\right) \frac{K_o(vb/4\kappa)}{K_1(vb/4\kappa)} \; . \qquad (5)$$

The molten layer gains energy by absorption of laser radiation and by reaction and loses energy by heat conduction as given by Equation (4), by evaporation from the erosion front, by melting solid material and by ejection of the melt at the lower surface of the workpiece. A numerical estimation shows that the last two terms can be neglected compared to the first and second term. If properly adjusted cutting gas conditions are assumed, the following energy balance can be found from Equations (1) and (4):

$$P_L(1-e^{-\alpha_L d}) + \varepsilon_R n_o vbd = 2\pi dKT \, \frac{\exp(-vb/4\kappa)}{K_o(vb/4\kappa)} + \varepsilon_v bd\left(\frac{133.3}{\sqrt{2\pi kTm_s}} \, 10^B T^C 10^{A/T}\right), \qquad (6)$$

where ε_v = evaporation energy per atom; m_s, atomic mass; k, Boltzmann constant and A,B,C evaporation constants. The mass balance of the molten layer includes mass entering the molten region due to the cutting movement and mass lost by evaporation and by ejection in liquid state [$(dE/dt)_v$ = energy loss at the erosion front due to evaporation as given in Equation (6) and n_s = average density of particles in the melt]:

$$bdn_o v = \frac{1}{\varepsilon_v}\left(\frac{dE}{dt}\right)_v + sbn_s v_s \, . \qquad (7)$$

With the approximation $n_s \sim n_o$ and using Equations (2), (5) and (6) the following equation results:

$$\frac{bv}{4\kappa} = \sqrt{\frac{\eta_R}{2\delta_s}} \, \frac{b}{4\kappa} \sqrt{\frac{v_R}{d}} \sqrt{1 - \frac{T_s}{T}} \, \frac{1}{\sqrt{vb/4\kappa}} \sqrt{\frac{K_o(vb/4\kappa)}{K_1(vb/4\kappa)}} + \frac{1}{n_o}\left(\frac{133.3}{\sqrt{2\pi kTm_s}} \, 10^B T^C 10^{A/T}\right) . (b/4x) . \qquad (8)$$

The temperature of the erosion front and the cutting velocity can now be determined by the simultaneous solution of the energy balance Equation (6) and the mass balance Equation (8). The numerical evaluation has been carried out with the data given in Table 1.

The evaluation yields the cutting speed v, the temperature of the erosion front T, the total power P_{tot} supplied to the workpiece by radiation and by gas, the relative contributions P_L and P_R of laser and gas to heating and the relative contributions P_K and P_v

Table 1. Data of Laser Radiation, Reactive Gas Flow (Oxygen) and of Workpiece (Steel).

Laser Radiation			Reactive Gas (Oxygen)		
P_L	α_L	$b^{1)}$	V_R	ε_R	η_R
W	m^{-1}	m	ms^{-1}	eV	Nsm^{-2}
1200	60	$5\cdot10^{-4}$	300	2.74	$2\cdot10^{-5}$

1) Approximately equal to focal spot diameter.

Workpiece (Iron)									
n_o	m_s	δ_s	K	κ	T_s	ε_v	A	B	C
m^{-3}	kg	kg/m^3	W/mK	m^2/s	K	eV	K	-	-
$7\cdot10^{28}$	$9.27\cdot10^{-26}$	$7.8\cdot10^3$	44	10^{-5}	1809	3.31	-19710	13.27	-1.27

of heat conduction and evaporation to cooling given in Table 2. Cutting speed and contribution of liquid ejection are shown in Figure 1.

CONCLUSIONS

The numerical results obtained with the improved model for reactive gas assisted laser cutting presented here show only a sligh downward deviation from a hyperbolic dependence of cutting speed on

Table 2. Results of the Numerical Evaluation.

			Heating		Cooling		
d	v	P_{tot}	P_L	P_R	P_K	P_v	T
cm	$m\ min^{-1}$	W	-	-	-	-	K
0.5	0.816	1460	21%	79%	67%	33%	3200
1	0.528	2030	26%	74%	80%	20%	3000
2	0.24	2190	38%	62%	96%	4%	2400

thickness for workpieces thicker than 2 cm. This explains why reactive gas assisted laser cutting of steel is possible for thicknesses as high as 4 cm as shown experimentally.

Acknowledgment

The author is indebted to Dr Jörg Sellner, Linz for steady support of this work sponsored by Finished Parts Division of Voest-Alpine AG, Linz, Austria.

REFERENCES

1. L. Bakowsky, G. Herziger and W. Peschko, Deutscher Verband für Schweißtechnik, DVS-Berichte, Band 63, p. 1 (1980).
2. T. Eneff and J. Ruge, Deutscher Verband für Schweißtechnik, DVS-Berichte, Band 63, p. 163 (1980).
3. F. O. Olsen, Deutscher Verband für Schweißtechnik, DVS-Berichte, Band 63, p. 197 (1980).
4. H. Opower, Laser und Optoelektronik, Band 2, p. 47 (1982).
5. D. Schuöcker and W. Abel, SPIE Proc. Nr. 455, Industrial Applications of High Power Lasers, p.
6. W. W. Duley, CO_2-Lasers, Effects and Applications, Academic Press, N.Y., p. 189 (1976).

OXYGEN ASSISTED LASER CUTTING OF LOW CARBON STEEL SHEETS

V. Sergi, V. Tagliaferri, and R. Teti

Istituto di Tecnologie
Università di Napoli
Piazzale Tecchio, 80125 Napoli, (Italy)

INTRODUCTION

Oxygen assisted laser cutting of low carbon steel sheets has been studied in order to establish the influence of this technological process on the properties of the material being cut. Hardness and microhardness tests, micrographic examinations and roughness measurements were carried out and their results have been comparatively analyzed. What this paper means to produce is a characterization both of the cut surface and of the adjacent material structure as thermally altered by the process. The possible implications of the presence of the oxygen jet are constantly born in mind and some observations aiming at the explanation of the typical drag lines formation are presented.

MATERIAL AND EXPERIMENTAL PROCEDURES

Low carbon steel sheet samples cut with a 500 W cw CO_2 laser (Valfivre) and oxygen jet coaxial with laser beam were examined. Steel composition is reported in tab.I and sample thickness was 4 mm. The laser beam was focalized, through a lens with a focal lenght f=127 mm, on the surface of the material being processed. Each sample was cut with different values of laser beam power and feed speed of the material, keeping constant both flow and pressure of the oxygen jet (tab.II). Both through the width and interrupted cuts were carried out for each power/speed combination. The feed motion was possessed by the sample and the laser beam impinged orthogonally on the sample surface. The diameter of the focal spot was 0.5 mm. For a power of 300 W, for example, a power density of 10^6 W/cm^2 was obtained. The temperature achieved at the sample

Table I. Material chemical composition.

C	Mn	P	S	Sn	Cu
0.05	0.28	0.009	0.018	0.012	0.05

surface was largely in excess to both ignition and melting temperatures for low carbon steel.

The following measurements were carried out:

- low magnification photographic examination of the cut path and cut surface
- microhardness tests along the cut path and in a direction orthogonal to it
- metallographic examination of the material in proximity of the cut path
- roughness measurements of the cut surface.

RESULTS AND DISCUSSION

Cut path and surface examination

The width of the cut path was measured for all the employed cutting parameters on the samples with interrupted cuts at the inlet surface of laser beam. Although width values had appeared little different for the various combinations of power and speed (their limiting values being 0.6 and 0.85 mm) in the graph of fig.1 it is possible to draw separate curves of width W vs. cutting speed V for the different powers. The curves seem to evidence a movement towards higher width values and a reduction of the distance between successive curves with increasing power and a decrease of slope with increasing speed. This result is in good agreement with the

Tab. II Cutting parameters and cut surface quality classification

Power (W)	Cutting speed (mm s^{-1})	Class
200	5/6/7/8	I/I/I/I
250	7/8/9/11/12/15	I/I/I/I/I/I
300	7/9/10/11	I/II/II/I
350	9/10/11/12/15	I/II/II/I/I
400	11/12/13/15	II/II/II/I
450	12/13/14/15/18/20	II/II/II/II/I/I

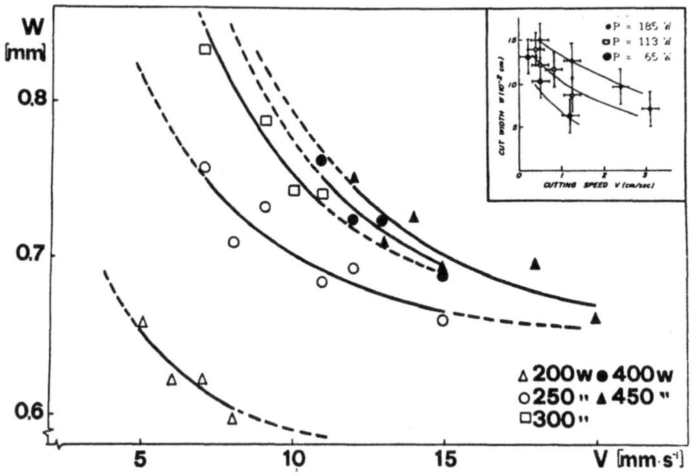

Fig. 1 Cut width W vs. cutting speed V for various powers. Right upper corner: experimental data and theoretical curves of W vs. V for various powers drawn from (1).

experimental data and theoretical curves of W vs. V presented by (1) for stainless steel thin sheets (0.05 mm thick). In our case the apparent constance of width values measured for different couples (P,V) is due to the overlapping of the projection of the experimental curves W-V in fig.1 on the W axis. This overlapping was caused by the different speed ranges necessarly employed for the various powers.

The appearence of the cut surface was the one typical of oxygen assisted laser cutting processes (fig. 2a), i.e. showing regularly spaced drag lines[2,3]. Seen from the above, the cut path (fig. 2b) evidences an edge made of circulars arcs and raised peaks. The number of peaks and valleys on the cut path edge, i.e. the number of drag lines on the cut surface, per unit time (or per unit lenght) increases with increasing speed (fig. 3a) and stays constant with power (fig. 3b). Cutting speed seems therefore to be the only parameter controlling the production of drag lines.

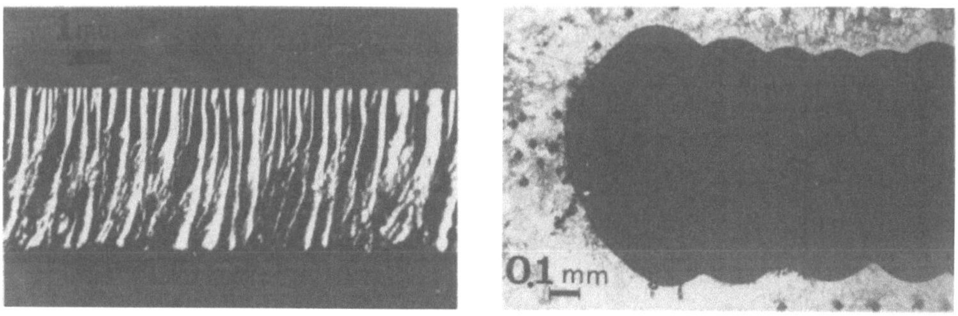

a) Cut surface appearence b) Contour of cut path

Fig. 2

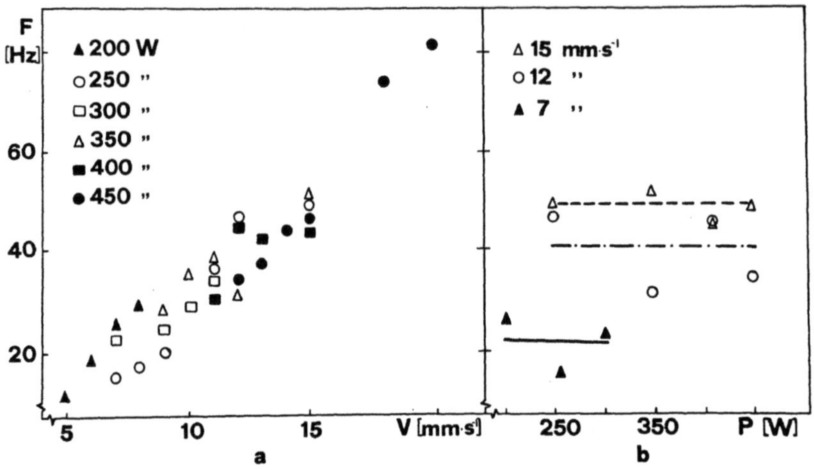

Fig. 3 Drag lines frequency (a) vs. cutting speed and (b) vs. power

The cut surface quality is significantly affected by the laser process parameters. Cut surface qualities of the examined samples were classified into two classes (tab. II) : only class II was considered acceptable. Fig. 4 shows the dependence of cut surface quality on cutting speed and laser power. It possible to notice that for low power values (P≦ 250 W) no speed value produced acceptable cuts. As power increases (P >250 W) a range of acceptable speeds becomes available: speeds beyond and below the limits of this range still produce unacceptable cuts. The range of acceptable speeds, circumscribed in fig. 4 by a thin continuous line rectangle, seems to spread and move towards higher speed with increasing power. This behaviour seems to be supported by analogous data drawn from the results presented by Arata et al.[2] for only two powers (200 W and 1 kW), and indicated in fig. 4 with a dotted line rectangle. Also for[2] at low power no acceptable cut was obtained whereas at high power the range of acceptable speed values was very wide with best results obtained beyond the critical speed value of 33 mm/s (see also following section).

Microhardness tests

Microhardness tests were first carried out, for all samples, parallel to the cut path at a constant distance of 80 μm from its typical wave edge configuration (fig. 2b). Microhardness values showed in all cases an evident periodicity about a mean value dependent on laser parameters. Fig.5 shows one of these characteristic graphs with reference to the sample with P=350 W and V=9 mm/s. The oscillating values of hardness witness an equally alternating, though limited, thermal alteration of the adjacent material structure. This kind

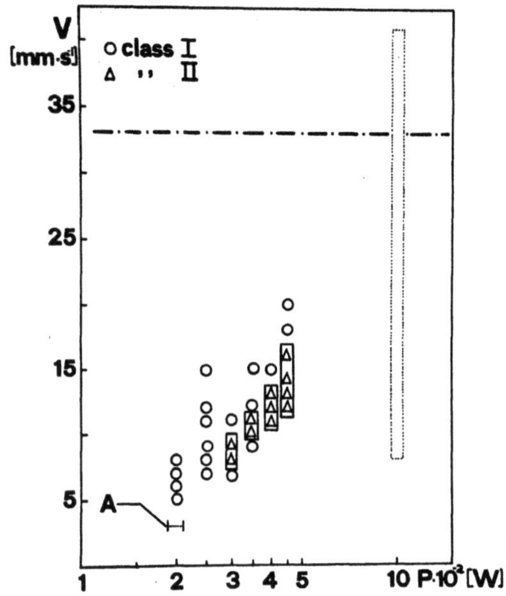

Fig. 4 Effect of laser cutting parameters on cut quality. The continuous line rectangles delimit the range of acceptable speeds for the examined samples. The dotted line rectangle indicates the same range drawn from the results of (3). The horizontal dash-and-dot line indicates the critical speed value according to (3).
A= no acceptable range exists according to (3).

of thermal influence seems to be well related to the simplified model of the mechanism of laser gas periodic cutting proposed by (2). Because of the relatively low maximum power (500 W) of the employed laser system, all adopted cutting speeds are lower than the speed of the oxidation reaction front, which according to (2) is 33 mm/s for low carbon steel. Under these condition the reaction front, started by

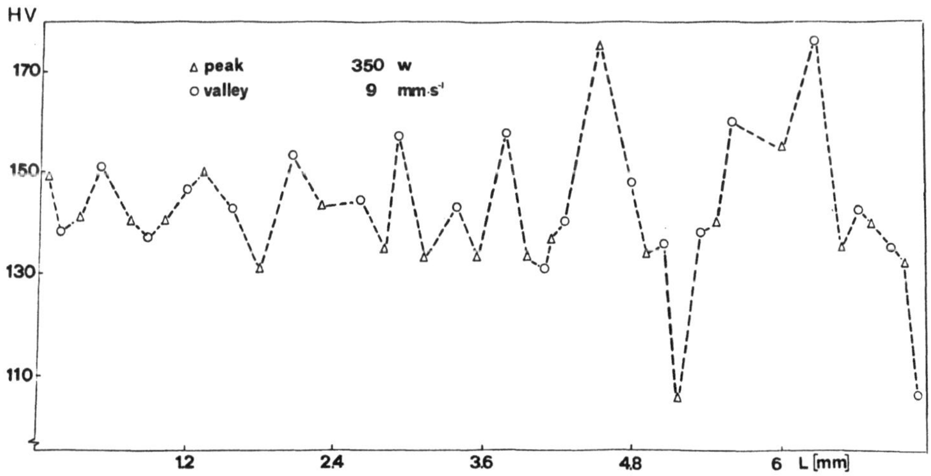

Fig. 5 Microhardness values parallel to the cut path

laser power density δ_i (fig. 6), procedes with its characteristic speed Vr, higher than the speed of the beam V, as long as the power density keeps higher than the value δ_S necessary for reaction sustentation : i.e. the reaction front moves from configuration I to configuration II. The next step in the process of material removal will be possible only when, as the beam progresses, laser power density $\delta_i > \delta_S$ is again available on the material to be processed: i.e. when the circle of center O' touches again the material. This results in a mechanism of material removal through periodically changing transfer of thermal energy. The successive enlargements and reductions of the oxidated zone, respectively determined by higher and lower laser absorption and exothermic energy release, provoke a thermal flux towards the inside of the material periodically varying and determining analogously periodical thermal alterations. The mean values of hardness measured parallel to the cut path seem to be slightly decreasing with feed speed and decisedly increasing with laser beam power (fig. 7, a and b). This data is in agreement with the fact that a higher power density cannot but induce higher themal alteration and a higher speed allows for less heat to be transfered to the adjacent material. The decrease of hardness with speed is rather small as speed variation is actually rather limited because of the already mentioned relatively low maximum power of the laser system.

Microhardness tests carried out starting from the cut path edge and moving orthogonally from it, showed that hardness remains generally constant (fig. 8, continuous line). This data seems to confirm the fact that in laser cutting the heat affected zone (HAZ) is practically negligible, as is typical of all laser processes[5,6]. A maximum in the hardness - distance graphs (fig. 8, dotted line) was observed only in a few cases where power was high and speed rather low. Such phenomenon seems therefore to be due to an excessive and useless storage of heat in the material adjacent to the cut path determining,

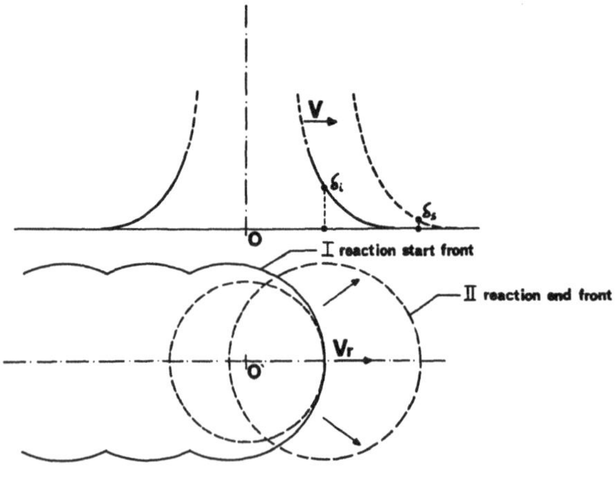

Fig. 6 Laser beam power density distribution.

δ_i = reaction start power density
δ_S = reaction end power density
Vr= oxidation reaction front speed.

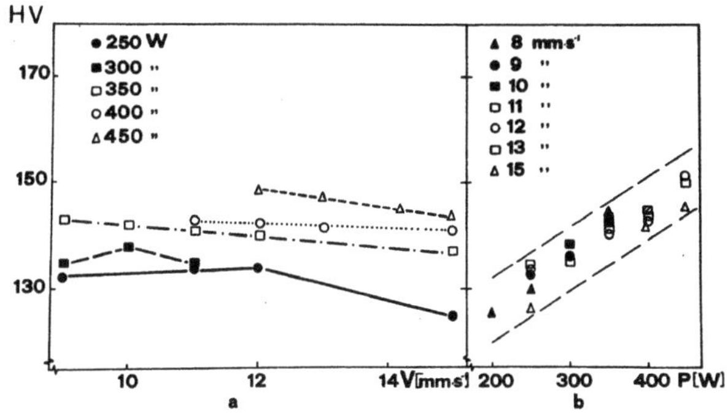

Fig. 7 Microhardness mean values parallel to the cut path (a) vs. cutting speed and (b) vs. power.

through a rapid dissipation, a more marked thermal alteration.

Metallographic examination

The typical microstructure of the base material is characterized by medium size ferritic grain. Both grain size estimations and metallographic examinations, carried out for all samples on zones progressively farther from the cut edge, do not evidence significant variations with laser parameters and with distance from the cut path. Grain size is in all cases practically constant (ASTM N.7) and structure remains essentially ferritic. A HAZ could not be visualized through micrographic examinations. These results are in good agreement with microhardness tests carried out orthogonally to the cut path confirming the limited thermal alteration induced into the material by the laser cutting process. On the other hand, microhardness oscillating values parallel to the cut path could not be justified resorting to simple optical microscopy as neither phase transformation could occur because of the very low carbon content and the absence of alloy elements, nor grain size variations actually occurred as experimentally observed.

Roughness measurements

Before analyzing roughness measurements results, it must be

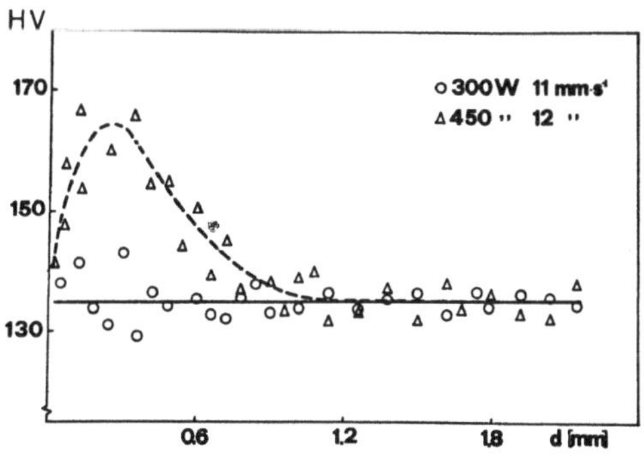

Fig. 8 Microhardness values vs. distance from tests carried out starting from the cut path edge and moving orthogonally from it.

observed that cut surface roughness was measured for all samples only on the zone of the cut surface actually allowing for such measurements. In fact it was not always possible to carry out roughness measurements also on the bottom part of the cut surface as the latter was in some cases particularly irregular.

From fig.9 it can be seen that the roughness of the cut surface upper portion decreased with increasing speed. Also the roughness of the middle and low portions varied in a similar way. On the other hand, roughness seems to stay almost constant with power, as shown in fig.10 with reference to speeds 12 mm/s and 15mm/s. On the basis of these results the cut surface roughness seems strictly related to a mechanism of drag line formation such as the one proposed by (2), according to which the portion of material interested by oxidation undergoes successive spreadings and reductions. As a matter of fact such mechanism predicts on one hand that drag lines formation is indipendent from power and on the other that, as speed increases, drag lines frequency increases and their height (amplitude) decreases. This goes on up to a cutting speed sufficiently high ($\geqq$ speed of oxidation reaction front) for frequency to become infinite, amplitude zero and cut surface smooth.

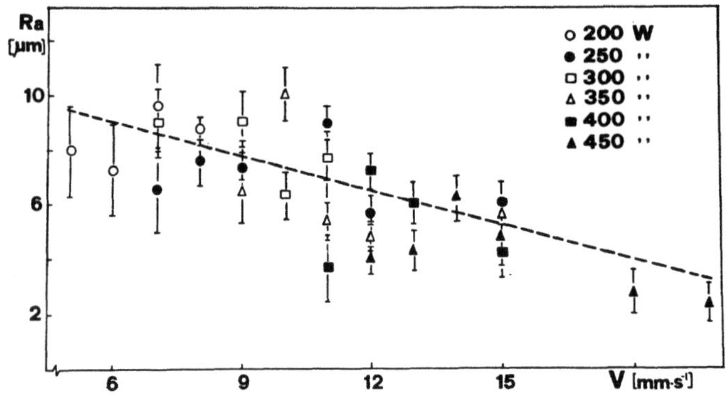

Fig.9 Roughness of the upper portion of cut surface vs. cutting speed.

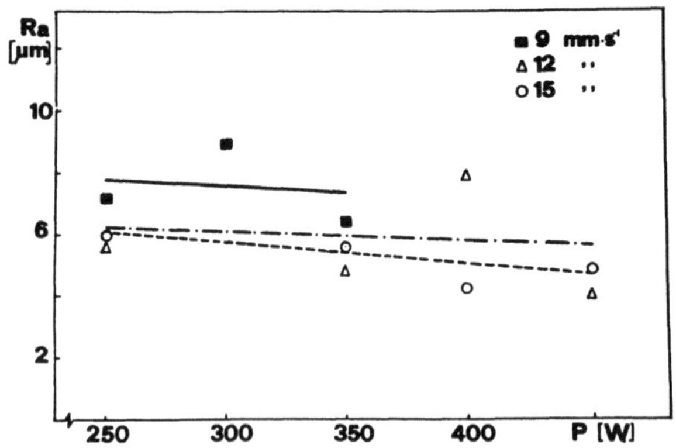

Fig.10 Roughness of the upper portion of cut surface vs. power.

CONCLUSIONS

From the above considerations the following conclusions can be drawn:

- Microhardness tests and metallographic examinations evidence that the HAZ in laser cutting is practically negligible.

- Cutting speed range yielding acceptable cut quality increases with power and moves towards higher speeds.

- Despite the simplifications (among which disregard of material surface condition influence on laser power absorption) introduced by (2) in order to devise a model of material removal mechanism through periodically changing transfer of thermal energy, this kind of model seems to be well related to the results of roughness measurements and the oscillating values of microhardness along and very close to the cut path.

- The parameter controlling cut surface quality seems to be the cutting speed. However, it must be considered that speed values are of course dependent on power; operation at cutting speeds sufficiently high to obtain smooth surface finishes is not feasible, for sheets of thickness 4 mm, with a 500 W cw max power CO_2 laser. It seems therefore necessary to resort to laser sources with maximum power so high as to allow for the cutting at speeds greater than the critical value set by the oxidation reaction front movement.

ACKNOWLEDGEMENT

This work was sponsored by the Consiglio Nazionale delle Ricerche (CNR) under Contract No. 80.1811.98-Progetto Finalizzato "Laser di potenza". The authors wish to acknoledge the Istituto R.T.M. of Vico Canavese (Turin) for preparation of testing specimens.

REFERENCES

1. J.N., Gonsalves, Ph. D. Thesis, York University, Toronto, 1973.
2. Y. Arata et al., Dynamic behaviour in laser gas cutting of mild steel, Trans. of JWRI, Vol.8, N.2,1979, pp. 15-26.
3. A. Gallo, F. Memola, V. Sergi, R. Teti and G. Dionoro, "Problemi di taglio dei metalli con raggio laser", La Meccanica Italiana, n. 148,febbraio 1981, pp.27-40.
4. W.W. Duley, "CO_2 Lasers-Effect and Applications", Academic Press, New York, 1976.
5. S.S. Charschan, "Lasers in Industry", Van Nostrand Reinhold Company, New York, 1972.

BROADENING OF CO_2-LASER BEAM DURING ITS PROPAGATION IN AEROSOL WITH ABSORBING SOLID PARTICLES

R.Kh.Almaev, L.P.Semenov, A.M.Skripkin, and P.N.Svirkunov

Institute of Experimental Meteorology
Obninsk, U.S.S.R.

INTRODUCTION

During the propagation of a laser radiation beam with the intensity below the break-down threshold in an aerosol medium with solid particles, in which evaporation is negligible, the main nonlinear effect defining the process of beam propagation is an interaction between radiation and thermal aureoles which are randomly distributed in space and formed in the vicinity of aerosol particles due to heat transfer[1,2]. As a result of this interaction, the radiation propagation is accompanied by the beam broadening and intensity redistribution in the cross-section.

This paper presents the calculation and measurement results of the CO_2-laser beam broadening during its propagation in the aerosol with solid particles surrounded by thermal aureoles.

THEORY

The radiation beam of radius a_0 is assumed to propagate along the z axis in the aerosol moving along the x axis with the mean velocity V_0. Variation of the transverse dimensions of electromagnetic radiation beam during its propagation in a randomly inhomogeneous medium can be usually determined with the help of the expression for the square of the effective beam radius

$$r_{ef}^2 = \langle R^2\rangle - \langle\vec{\rho}_c\rangle^2 = \frac{\iint d^2R \cdot R^2 \cdot \Gamma(z,\vec{R},0)}{\iint d^2R \cdot \Gamma(z,\vec{R},0)} - \langle\vec{\rho}_c\rangle^2 , \tag{1}$$

where $\vec{\rho}_c$ is the radius-vector of the beam gravity centre; $\Gamma(z,\vec{R},\vec{\rho}) = \langle E(z,\vec{R}+\frac{\vec{\rho}}{2}) \cdot E^*(z,\vec{R}-\frac{\vec{\rho}}{2})\rangle$ is the coherence function of the electrical wave field E ; $\vec{R} = \frac{\vec{\rho}_1+\vec{\rho}_2}{2}$; $\vec{\rho} = \vec{\rho}_1 - \vec{\rho}_2$; $\vec{\rho}_j$ is the radius-vector of the point in the beam cross-section(j =I,2). Angular brackets denote averaging over the assembly of the random field E realizations. From(I) it follows that to calculate the effective beam radius it is necessary to derive and solve the equation for the coherence function of electromagnetic wave propagating in a randomly inhomogeneous medium with thermal aureoles. This in its turn requires the knowledge of statistical characteristics of the dielectric constant field. The total variation of the medium dielectric constant due to thermal aureoles is the sum of individual local disturbances

$$\mathcal{E}_a = \sum_m \left(\frac{\partial\varepsilon}{\partial T}\right)_p \cdot T_m(\vec{z},\vec{R}_m,t) = \sum_m \Psi_a(\vec{z},\vec{R}_m,t) , \tag{2}$$

where $\vec{z}$ is the radius-vector of the space point; t is the time; $T_m(\vec{z},\vec{R}_m,t)$ is the local temperature disturbance near the m-th particle located in the point $\vec{R}_m$.

The air temperature variation in the vicinity of absorbing solid spherical particle can be found from the thermal conductivity equation. In the considered case the calculation of temperature field in the vicinity of the particle is simplified compared to case of a droplet aerosol,since the energy loss for particle evaporation is negligible. In the coordinate system moving together with the medium the following expression was derived for T_m :

$$T_m(\vec{z},\vec{R}_m,t) = \frac{\sigma_n(R_p)}{(2\sqrt{\pi\chi_T})^3 C_p\rho} \cdot \int_0^t dt' \frac{I(t-t',\vec{R}_m)}{t'^{\frac{3}{2}}} \cdot \left(1 - \frac{R_p}{|\vec{z}-\vec{R}_m|}\right) \times$$
$$\times \exp\left\{-\frac{(|\vec{z}-\vec{R}_m|-R_p)^2}{4\chi_T t'}\right\} , \quad |\vec{z}-\vec{R}_m| \ge R_p , \tag{3}$$

where I is the intensity of radiation falling on the particle; χ_T , C_p , ρ are the air temperature conductivity, specific heat capacity at constant pressure, and density, respectively; $\sigma_a(R_p)$ is the cross-section of radiation absorption by a particle of radius R_p. The value of temperature disturbance in the aureole substantially depends on the particle location in space, since radiation intensity is different in the beam cross-section and along the path (the latter is due to radiation extinction). Thus, the medium is not only locally-inhomogeneous, with spatial scales close to the aureole's radius R_a, but inhomogeneous on the average with scales of the order of a_o, α^{-1} (α is the coefficient of wave extinction by aerosol particles).

The equation for the coherence function Γ may be derived from the parabolic equation for the field E [3] in the approximation of δ -correlated along z -axis fluctuations of ε. It takes the form:

$$\frac{\partial \Gamma}{\partial z} - \frac{i}{2k}(\Delta_1 - \Delta_2)\Gamma + \alpha\Gamma(z, \vec{\rho}_1, \vec{\rho}_2, t) = \tag{4}$$
$$= ik\,[n_{ef}(\vec{\rho}_1, t) - n_{ef}(\vec{\rho}_2, t)]\,\Gamma + \frac{k^2 N_o}{16} G\Gamma \;,$$

where N_o is the aerosol particle concentration

$$n_{ef}(\vec{\rho}, z, t) = \frac{N_o}{2} \iint_0^{\infty} d\vec{z}'\, dR_p\, f(R_p) \Psi_a(\vec{\rho}, z, \vec{z}', R_p, t)\,,$$

$$G = -\int_0^{z} d\eta \iint_0^{\infty} d\vec{z}' dR_p \cdot f(R_p) \{ \Psi_a(\vec{\rho}_1, z, \vec{z}', R_p, t) \times$$
$$\times \Psi_a(\vec{\rho}_1, \eta, \vec{z}', R_p, t) + \Psi_a(\vec{\rho}_2, z, \vec{z}', R_p, t) \Psi_a(\vec{\rho}_2, \eta, \vec{z}', R_p, t) -$$
$$- 2\,\Psi_a(\vec{\rho}_1, z, \vec{z}', R_p, t) \cdot \Psi_a(\vec{\rho}_2, \eta, \vec{z}', R_p, t) \}\,.$$

When deriving Eq.(4), averaging over microvolumes $(\Delta V) \ll a_o^3, \alpha^{-3}$ containing a great quantity of particles was performed. Particle distribution in the volume was assumed to be given by the Poisson's law. It should be noted that Eq.(4) can be written in the form of the transfer equation for Fourier transform of the function Γ over ρ, if the transverse scale of the field coherence is small in comparison with a_o^2.

Solution of the nonlinear equation(4) followed by

calculation of the beam broadening was made (a) by the perturbation method and (b) in a paraxial approximation using the averaged geometrical and angular characteristics of the beam.

In the paraxial approximation, the solution for the Gaussian beam takes the following form:

$$\Gamma(z,\vec{R},\vec{\rho},t)=\frac{P_0}{\pi<R^2>}\cdot\exp[-\alpha z-\frac{R^2}{<R^2>}+i\frac{<\vec{R}\vec{\omega}>}{<R^2>}\vec{R}\vec{\rho}-$$
$$-\frac{K^2\rho^2}{4}(<\omega^2>-\frac{<\vec{R}\vec{\omega}>}{<R^2>})] \quad , \tag{5}$$

where $\vec{\omega}=\frac{d\vec{R}}{dz}$, $<\vec{R}\vec{\omega}>=P^{-1}\int d\vec{R}\vec{R}\vec{\omega}\cdot I(\vec{\omega},\vec{R},z)$ defines the focal distance of the refracted beam $F=k<R^2>/<\vec{R}\vec{\omega}>$; $<\omega^2>=P^{-1}\int d\vec{R}\cdot d\vec{\omega}\,\omega^2\cdot I(\vec{\omega},\vec{R},z)$ is the total angular width; $P=\int d\vec{R}\cdot d\vec{\omega} I(\vec{\omega},\vec{R},z)$, $I(\vec{\omega},\vec{R},z)=\int d\vec{\rho}\, e^{i\vec{\omega}\vec{\rho}}\Gamma(z,\vec{R},\vec{\rho})$
Using(5) and the expression $n_{ef}=B+AR^2/2$,we can derive from(4) a set of equations for the averaged characteristics:

$$\frac{d<R^2>}{dz}=2<\vec{R}\vec{\omega}> \quad , \tag{6}$$

$$\frac{d<\vec{R}\vec{\omega}>}{dz}=A(z)<R^2>+<\omega^2> , \tag{7}$$

$$\frac{d<\omega^2>}{dz}=2A(z)<\vec{R}\vec{\omega}>+\left(\frac{d\omega^2}{dz}\right)_S \tag{8}$$

Here $\left(\frac{d\omega^2}{dz}\right)_S=N_0\int d\vec{\omega}'\bar{\sigma}_S(\vec{\omega},\vec{\omega}')(\vec{\omega}-\vec{\omega}')^2$ is the value characterizing the angular broadening due to aureole scattering; $\bar{\sigma}_S$ is the averaged scattering cross-section of an aerosol particle with a thermal aureole. An addition to the square of the effective beam radius due to the interaction between radiation and thermal aureoles was calculated by two methods which gave the identical results. Within the time range $t<a_0/V_0$ (nonstationary problems),

$$\Delta \tau_{ef}^2=\tau_{ef}^2-a_0^2=\tau_m^2+\tau_a^2 \quad , \tag{9}$$

$$r_a^2 = \frac{D^2 \cdot I_o^2(0)}{192\pi \chi_T^2 \alpha^2} \cdot \frac{\bar{\sigma}_a^2}{\bar{\sigma}_e^2} \cdot \ln\left[\frac{(t+t_o)^2}{4tt_o}\right](2\tau^2 - 2\tau + 1 - e^{-2\tau}) , \qquad (10)$$

$$r_m^2 = \frac{D \cdot I_o(0)}{2\alpha} \cdot \frac{\bar{\sigma}_a}{\bar{\sigma}_e} \cdot (t - t_o)(\tau - 1 + e^{-\tau}) , \qquad (11)$$

where $D = 1/C_p \cdot \rho \times (\partial \varepsilon / \partial T)_p$, $\bar{\sigma}_{a,e} = \int^{\infty} dR_p \cdot \sigma_{a,e}(R_p) f(R_p)$, σ_a, σ_e is the cross-section of radiation extinction by an individual particle; $f(R_p)$ is the function of aerosol particle distribution to the radii R_p ; $I_o(0)$ is the radiation intensity on the beam axis; $\tau = \alpha z$ is the optical depth of the aerosol layer with the length z ; r_m is the beam broadening due to radiation refraction on the mean profile ε_a ; r_a is the beam broadening due to aureole scattering. The time t_o limits the aureole minimum sizes for which the relation(I) is true.

In another limit, when $t > \frac{a_o}{V_o}$ (stationary problem), a contribution of the value r_a^2 to the beam broadening, as a rule, is negligible, and the expression for r_m^2 differs from(II) by substitution of $\sqrt{2}a_o/V_o$ for $t - t_o$.

From (IO) and (II) it follows that the value Δr_{ef}^2 increases proportionally with α at $\tau \ll$ I. As α increases at $\tau > 1$, Δr_{ef}^2 tends to saturation which results from radiation extinction and, therefore, decrease of the disturbances ε at the end of the path.

EXPERIMENTAL DATA

Experimental study of the laser beam broadening in an aerosol medium with absorbing solid particles was carried out in an aerosol chamber 90 cm in length. A variation of the beam width depending on microphysical properties of the medium and its optical depth τ was studied. A continuous CO_2-laser radiation was collimated to a beam 0,6 cm in diameter with an intensity I-2·10^3 W/cm^2. Al_2O_3 particles often occurring in nature were used as aerosol particles. Microphysical properties of aerosol particles were controlled with the help of a microscope by the sample analysis on slides. Optical depths were measured at the 0,63 μm and IO,6μm wavelengths. Transverse dimensions of the CO_2-laser beam were registered at the chamber exit.

The figure presents dependence of the square of the effective CO_2-laser beam broadening on the optical depth τ_λ of the aerosol medium at the wavelength λ = 0.63 μm and the fixed path length for the aerosol with large($R_p \sim 5 \div 10$ μm ,curves 2,2′) and small

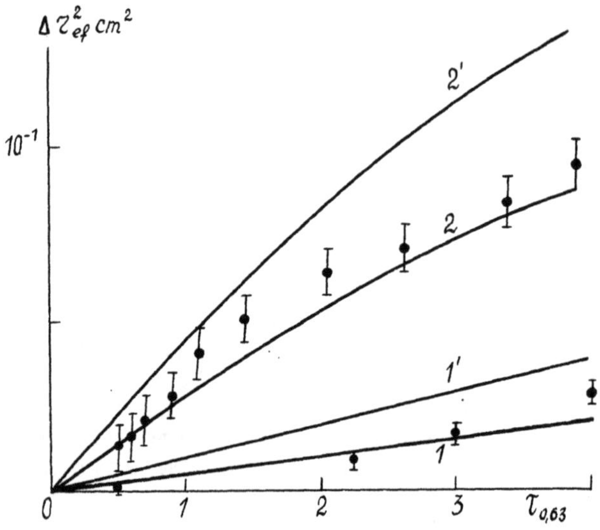

Figure. Dependence of the radiation beam broadening on the optical depth at a fixed path length.

($R_p \sim 0{,}1 \div 1\ \mu m$, curves I, I$'$) particles of Al_2O_3. Experimental data are indicated by dots. Solid curves correspond to calculations with formulae (9)-(II) at radiation intensities $I = 10^3\ W/cm^2$ (curves I$'$, 2$'$), $I = 2\cdot 10^3\ W/cm^2$ (curve I) and $I = 1{,}5\cdot 10^3\ W/cm^2$ (curve 2).

In calculations it was assumed that particles had a spherical form and their optical constants were temperature independent. The absorption and extinction efficiency factors were computerized with the help of the Mie-theory. Measurements were made for the stationary case $(t \geqslant a_0/V_0)$.

The estimates show that in this situation the contribution of r_a to the effective beam broadening is small, and the value Δr_{ef}^2 is mainly determined by refraction on the average profile ε_a .

One can see from the figure that there occurs a monotonous increase of the refraction broadening with increasing τ_λ in the aerosol with small particles which optical depths τ at $\lambda = 10{,}6\ \mu m$ are very small ($\tau = 3\cdot 10^{-3}\,\tau_\lambda \ll 1$). In the aerosol with large particles the tendency to saturation of the value Δr_{ef}^2 is observed at great values of τ_λ, though the maximum value τ ($\tau = 0{,}2\,\tau_\lambda$ in this case), does not exceed unity. This behaviour of the experimental values Δr_{ef}^2 is in good agreement with that of Δr_{ef}^2 in accordance with formula(9).

It should be noted that in spite of a substantial difference in optical depths τ of the aerosol with large and small particles(τ_{max} of large particles is 70 times as great as that of small ones), the broadening values differ only by a factor of three.

REFERENCES

1. R.Kh.Almaev, A.M.Skripkin. Thermal refraction of laser radiation in the aerosol medium containing solid particles. Proceedings of the IEM, issue 26(99), 92 (1978).
2. P.N.Svirkunov. On thermal self-broadening of laser beams propagating in aerosol media. The All-Union Conference on the optical radiation propagation in the dispersion medium. Obninsk, 224 (1978).
3. R.Kh.Almaev, L.P.Semenov. Broadening of radiation beam propagating in a randomly-inhomogeneous cleared cloudy medium. Izvestiya Vuzov, Radiophysica, vol.21, 408 (1978).

THE ROLE OF CO_2-LASER BEAM DIVERGENCE IN CLOUD CLEARING

R.Kh.Almaev, L.P.Semenov, A.G.Slesarev, and O.A.Volkovitsky

Institute of Experimental Meteorology
Obninsk, U.S.S.R

As a CO_2-laser beam interacts with a cloud medium, the initial beam divergence is an important parameter that determines characteristics of the cleared cloud. Its influence on the characteristics of the cleared zone manifests itself mainly through the thermal action function q which allows one to calculate the distribution of mean values of temperature T, water content W, dielectric constant ε and transmittance of the medium Π_λ as well as fluctuations of these parameters occurring due to clearing.[1-4]

Based on an asymptotic analysis and numerical solution of the equation for the thermal action function of a diverging beam, we propose in this paper an approximated formula for q that adequately describes an exact solution over a wide range of parameter variations. The effect of initial divergence on characteristics of a cleared medium is analyzed, using as example transmittance and temperature fluctuations. Calculated values of transmittance of the medium cleared by a diverging beam are compared with experimental data.

We consider the problem of steady-state clearing of a droplet aerosol medium by a CO_2-laser beam with the characteristic initial size $2 \cdot a_0$ and initial divergence

$$\vec{\theta}_0 = \left\{ \frac{x}{F_x (1 + z/F_x)} , \frac{y}{F_y (1 + z/F_y)} \right\} , \qquad (1)$$

where x and y are coordinates transverse to the direction of radiation propagation z ; F_j is the initial radius of wave front curvature in the plane(j, z); and $j = x, y$. Mean wind velocity $\vec{V}_0$ is assumed to be directed along the x axis.

A complete set of equations describing steady-state clearing of a cloud medium[3] (after solving the equations for T and W in approximation to water content, geometrical optics and small induced fluctuations) can be reduced to the following equations for the thermal action function q and eikonal S :

$$\frac{\partial q}{\partial z} + \theta_x \frac{\partial q}{\partial x} + \int_{-\infty}^{x} \frac{\partial}{\partial y}\left(\theta_y \cdot \frac{\partial q}{\partial x'}\right) dx' + \alpha_e^{\circ}\left(1 - e^{-q}\right) = 0, \quad (2)$$

$$2\frac{\partial S}{\partial z} + (\nabla_{\perp} S)^2 = \left(\frac{\partial \varepsilon}{\partial T}\right)_p \cdot \frac{A \cdot \alpha_e^{\circ}}{C_e \mu} \cdot \left(1 - e^{-q}\right), \quad (3)$$

where $q(\vec{\tau}) = \frac{\mu}{V_0} \int_{-\infty}^{x} I(x', y, z)\, dx'$, $\vec{\tau} = (x, y, z)$, $\alpha_e^{\circ} = C_e W_0$; $\mu = \frac{C_a \beta_T}{L}$; $C_a = \frac{3 B_a}{4 \rho_w}$, $C_e = \frac{3 B_e}{4 \rho_w}$,

B_a, B_e are the proportionality coefficients between the factors of droplet absorption and extinction and its radius in water content approximation; ρ_w is the the water density; W_0 is the water content of the undisturbed medium; L is the specific evaporation heat of water; β_T is the factor characterizing the efficiency of the use of energy absorbed by a droplet; $\theta_x = \frac{\partial S}{\partial x}$ and $\theta_y = \frac{\partial S}{\partial y}$ are the divergence angles of beam in the medium; $A = \frac{(1-\beta_T) C_a}{\rho C_p}$; ρC_p is the heat capacity per air unit volume; and I is the intensity of CO_2-laser radiation. The initial value $q_0 = q|_{z=0}$ is assumed to be specified.

To separate the effects caused by the initial beam divergence $\bar{\theta}_0$ we solve the problem for those pathes where the nonlinear divergence $\bar{\theta}_n$ is small compared to the initial divergence $\bar{\theta}_0$. In this case the equation for q takes the form

$$\frac{dq(\tau, x(x_0, \tau), y(y_0, \tau))}{d\tau} + \frac{\beta_y}{1 + \beta_y \tau} q + 1 - e^{-q} = 0. \quad (4)$$

For convenience, Eq.(4) is written on the trajectory $x = x_0(1 + \beta_x \tau)$, $y = y_0(1 + \beta_y \tau)$, where $\beta_j = (\alpha_e^{\circ} F_j)^{-1}$ is the parameter characterizing the divergence of radiation beam in the cleared medium; $\tau = \alpha_e^{\circ} \cdot z$ is the optical depth of the undisturbed medium at the wavelength of CO_2-laser radiation; and $X_0 = X|_{\tau=0}$, $y_0 = y|_{\tau=0}$.

Asymptotic analysis of Eq.(4) enables one to write the following expressions for the thermal action function:

$$q(\tau,\vec{\rho}) = \left[q_0(\tilde{\vec{\rho}}) - \tau\left(1+\frac{\beta_y\tau}{2}\right)\right]\cdot(1+\beta_y\tau)^{-1} , \qquad (5)$$

valid in the clearance zone ($q > 2$) and

$$q(\tau,\vec{\rho}) = q_0(\tilde{\vec{\rho}})e^{-\tau}\cdot(1+\beta_y\tau)^{-1} , \qquad (6)$$

valid at small q ($q \ll 1$),where $\vec{\rho}=(x,y)$, $\tilde{\vec{\rho}}=(\tilde{x},\tilde{y})$, $\tilde{x}=\frac{x}{1+\beta_x\tau}$ and $\tilde{y}=\frac{y}{1+\beta_y\tau}$. Using asymptotic expressions(5),(6) and the results of numerical analysis of Eq.(4) at intermediate values of q , we can propose the approximated formula for the thermal action function

$$q(\tau,\vec{\rho}(\vec{\rho}_0,\tau)) = \frac{1}{1+\beta_y\tau}\cdot \ln\left[1+\left(e^{q_0(\vec{\rho}_0)}-1\right)e^{-\tau\left(1+\frac{\beta_y\tau}{2}\right)}\right] + 1{,}5\,\frac{(1+\beta_y\tilde{\tau})}{(1+\beta_y\tau)}\cdot\exp\left\{-\left[\,|\tau-\tilde{\tau}|+\left(\frac{\tilde{\tau}}{\tau}\right)^5+\left(\frac{2}{\tau}\right)^2\right]\right\} , \qquad (7)$$

where $\tilde{\tau} = -\frac{(1+\beta_y)}{\beta_y} + \sqrt{\left(\frac{1+\beta_y}{\beta_y}\right)^2 + \frac{2(q_0-1)}{\beta_y}}$.

The extent to which the approximated formula(7) approaches the exact solution of Eq.(4) is illustrated in Fig.I, which presents calculation results for q variation along the ray trajectory. Solid lines show the results of calculation by Eq.(7) and crosses - the results of numerical solution of Eq.(4). As seen from the figure,formula(7) is in good agreement with the exact solution. It also should be noted that to describe cloud medium clearing by a diverging radiation beam at small values of τ and β one can use the approach developed by the authors earlier[2].

Using the obtained expressions for q ,we can analyze the influence of the initial divergence of CO_2-laser beam on characteristics of the cleared cloud medium such as its transmittance to visible radiation and induced fluctuations of temperature(dielectric constant) which are of interest from the viewpoint of the problems of laser radiation propagation in the atmosphere. For simplicity, we assume that $\beta_x = \beta_y = \beta$.

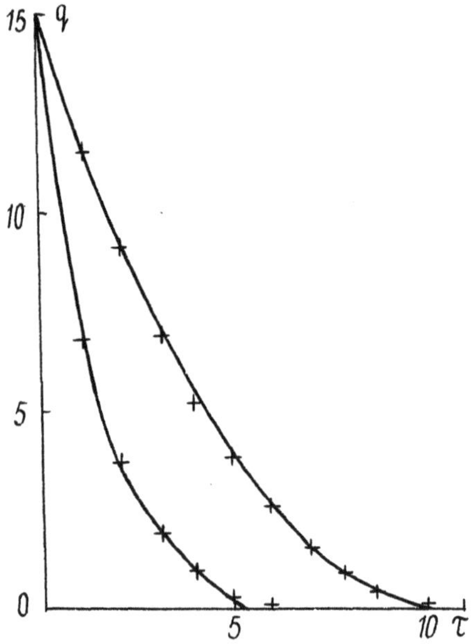

Fig.I. Variation in the thermal action function along the ray trajectory depending on the optical depth for parameters $q_o(X_o, y_o) = 15$, $\beta_y = 0.2$ (I), $\beta_y = 1$ (2).

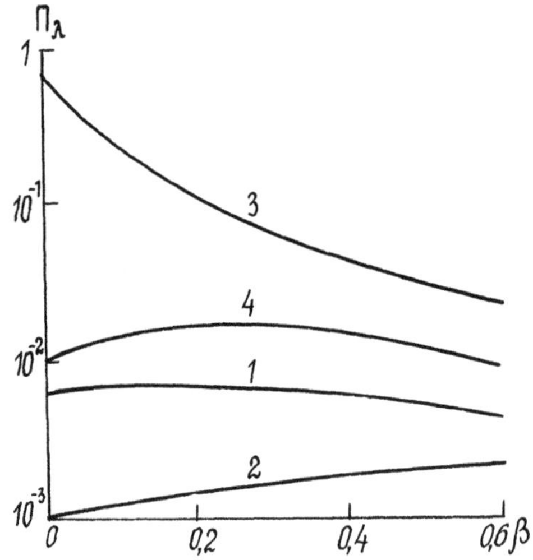

Fig.2. Variation in the transmittance of the medium depending on the radiation divergence parameter.

Transmittance function Π_λ which characterizes the transmittance of cloud medium at the wavelength λ are given by

$$\Pi_\lambda = \frac{I_\lambda(z)}{I_{\lambda 0}} = e^{-\tau_\lambda} \quad , \quad \tau_\lambda = \int_0^\ell \alpha_\lambda(z')\,dz' \, , \qquad (8)$$

where τ_λ and α_λ are the optical depth and the extinction coefficient of the cleared aerosol medium for probing radiation. Integration in (8) is performed along the trajectory of probing beam propagation.

Under the assumptions adopted (water content approximation for CO_2-laser radiation and large droplet approximation for probing radiation), the value α_λ in the cleared medium can be expressed in terms of q as follows: $\alpha_\lambda = \gamma_o \cdot C_e W_o e^{-\frac{2q}{3}}$, (9)

where $\gamma_o = \frac{\alpha^o_{e\lambda}}{\alpha^o_e}$ (α^o_e and $\alpha^o_{e\lambda}$ are the coefficients of extinction of CO_2-laser and probing radiations by undisturbed aerosol)[5]. For gamma droplet size distribution, parameter γ_o is equal to

$$\gamma_o = \frac{2\sqrt{(\tilde{\mu}+1)(\tilde{\mu}+2)}}{B_e \cdot R_{02} \cdot (\tilde{\mu}+3)} ,$$

where $\tilde{\mu}$ is the parameter of gamma distribution and R_{02} is the mean square radius of droplets in the undisturbed medium.

Calculations of the transmittance of droplet aerosol medium cleared by a diverging CO_2-laser beam were performed for a narrow, collimated beam of probing radiation which propagated parallel to the z -axis in various areas of the action zone. Characteristic dependences of the cleared medium transmittance Π_λ on the divergence parameter β are shown in Fig.2. The calculations were performed for the initial profile of CO_2-laser beam intensity

$$I_o(\vec{\rho}) = I_o(0)e^{-\eta^2} \cdot \mathcal{X}(1-|\xi|) , \tag{10}$$

for which the initial thermal action function takes the form

$$q_o(\vec{\rho}) = q_o(0)e^{-\eta^2}(1+\xi) \; ; (\xi \in [-1,1]) . \tag{11}$$

where $\xi = \frac{x}{a_o}$, $\eta = \frac{y}{a_o}$, and $\mathcal{X}(X)$ = I if $X \geq 0$ and $\mathcal{X}(X) = 0$ if $X < 0$. In these calculations γ_o = 0.6; τ = 5; and $q_o(0)$ = 5. The curves in the figure correspond to the following trajectories of probing: I – ξ = -0.5; η = 0; 2 – ξ = -0.5; η = I; 3 – ξ = 0.5; η = 0; 4 – ξ = 0.5; η = I. It is seen that the dependence of transmittance on the beam divergence is different for various trajectories. For example, on the trajectory ξ = 0.5, η = 0 there occurs a rapid decrease in transmittance of the medium, as the beam divergence increases. On other trajectories(curves I,2,4) transmittance initially increases until a maximum is reached and then decreases.

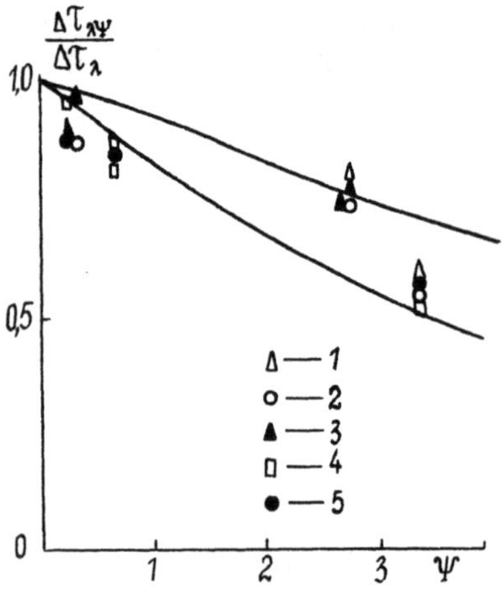

Fig.3. Dependence of the extent of medium clearing on the radiation beam divergence. I - 5 - τ_λ°=I.8; 2.4; 2.8; 3.2 and 4.0, respectively.

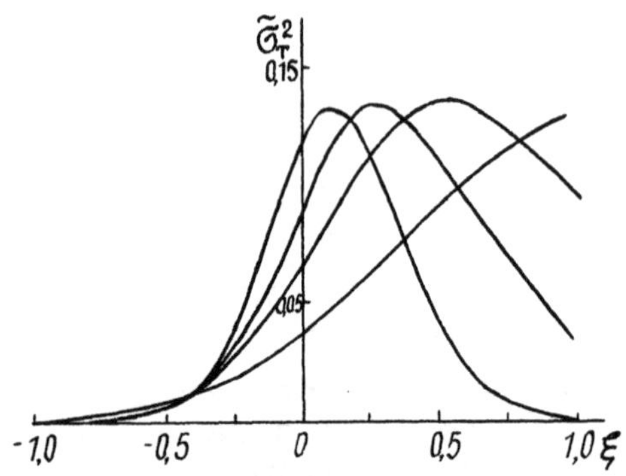

Fig.4. Distribution of the normalized dispersion of temperature fluctuations $\tilde{\sigma}_T^2$ in the cross-section $y = 0$, $\tau = 5$.

Experimental study of the influence of CO_2-laser beam divergence on the extent of cloud medium clearing was conducted in the aerosol chamber installed at the Institute of Experimental Meteorology that adequately simulates natural clouds and fogs by the droplet spectrum. In these experiments the values of the thermal action function $q_o(0)$ varied about $q_o(0) = 5.6$ ($q_o(0) = 5.6 \pm 0.7$). The optical depth of undisturbed aerosol τ_λ° ranged from I.8 to 4, and the geometric parameter of divergence $\psi = \beta\tau = \frac{z}{F}$ assumed the values 0.25, 0.62, 2.7 and 3.3.

Variation in the optical depth of the cleared droplet aerosol medium $\Delta\tau_\lambda = \tau_\lambda^\circ - \tau_\lambda$ was measured in the clearance zone on the axis of CO_2-laser beam using the probing He-Ne laser radiation.

The results of the experimental study of the influence of CO_2-laser beam divergence on the clearing of moving droplet aerosol are given in Fig.3 as the dependence of $\frac{\Delta\tau_{\lambda\psi}}{\Delta\tau_\lambda}$ on ψ . Here $\Delta\tau_{\lambda\psi}$ is the variation in the optical depth of the medium τ_λ ,as it is cleared by the radiation beam with the initial divergence $\psi \neq 0$, and $\Delta\tau_\lambda$ is the calculated variation in τ_λ ,as the medium is cleared by the nondiverging beam($\psi = 0$). Solid lines show the calculation results for the value $\frac{\Delta\tau_{\lambda\Psi}}{\Delta\tau_\lambda}$ at τ°_λ = I.8, $q_o(0)$= 6.3(upper curve) and at τ°_λ = 4, $q_o(0)$ = 4.9(bottom curve). As seen from the figure, there is a good agreement between the experimental and theoretical data.

In conclusion we estimate variations in the distribution of induced temperature fluctuations of the cleared medium associated with the divergence of CO_2-laser beam. Dispersion of temperature fluctuations σ_T^2 was calculated using the relation from [3] for the cloud medium clearing by a CO_2-laser beam with the initial distribution q of the form(II) at $q_o(0)$ = 5. Fig.4 shows the results of σ_T^2 calculation in the cross-section y = 0, τ = 5 at various values of the divergence parameter β . Curves I,2,3 and 4 in the figure correspond to β = 0; 0.05; 0.I and 0.2. It is seen that as the beam divergence increases, the maximum of fluctuations shifts to the downwind boundary of the beam, and the shift increases more rapidly with increasing β than the beam width.

REFERENCES

I. A.F.Nerushev, L.P.Semenov. Light beam propagation through a vaporizing liquid droplet medium in the presence of wind refraction. - Kvantovaya Electronika,vol.3, I226 (I976).

2. O.A.Volkovitsky. Experimental study of CO_2-laser radiation effects on a cloud droplet medium. Meteorologiya i Gidrologiya,No.9,I2 (I977).

3. R.Kh.Almaev,S.D.Pinchuk, L.P.Semenov,O.A.Volkovitsky. Propagation of CO_2-laser radiation in a turbulent cloud medium. Journale de Physique, colloque C9, supplement au No.II,Tome 4I, Novembre, C9-I07 (I980).

4. R.Kh.Almaev,A.G.Slesarev. On the role of laser radiation beam divergence during the clearing of aerosol droplet medium. Proceedings of the Institute of Experimental Meteorology,issue 26 (99), 22 (I98I).

5. O.A.Volkovitsky. On the variation of water aerosol transmittance at the wavelength of probing radiation in the zone of its clearing by a CO_2-laser beam. Radiotechnika i Electronika,No.7, I379 (I977).

ON THE INTERACTION OF HIGH-POWER LASER BEAM WITH METAL TARGETS

M.Cantello
R T M Vico Canavese, Italy

V.Donati, L.Garifo, R.Menin, and F.Pandarese
C I S E, Milano, Italy

A.V.La Rocca
F I A T, Torino, Italy

M.Onorato and P.Savorelli
Politecnico di Torino, Italy

INTRODUCTION

The interaction of a CO_2 cw high power laser beam with a target material in air has been subject of intensive studies during the past few years[1-17]. The laser-metal interaction problem is of great importance in laser metalworking because the presence of vapors above the surface of the workpiece and the formation of plasma in the vapor and in the surrounding air may influence strongly the thermal coupling between the radiation and the workpiece.

When the intensity of the laser beam is higher than 10^4 W/cm^2 as required e.g. in welding, interaction is dominated by laser sustained absorption waves generated in metal vapors and in the air volume above the material. More specifically at laser intensities higher than 10^7 W/cm^2 laser supported detonation waves can be generated, while for intensities in the range $3 \times 10^4 \div 10^7$ W/cm^2 laser sustained combustion waves can be observed.

An experimental research program is carried out on these phenomena by the present authors, using a CO_2 laser capable of up to 15 kW cw power, under experimental conditions typical of actual metal working. In Ref.15 some fluid-dynamical aspects of laser-metal interaction are discussed and results on the development and motion of absorption waves obtained by high speed photographic records are shown. In the same paper physical properties of the waves, examined by time integrated absorption measurements and spectroscopic analysis are reported. In Ref.16 measurements of radiation diffused by the vapor and the target are shown.

In the present paper new results on the development and motion of absorption waves obtained by high speed photographic technique are reported. These refer to the case in which no penetration in the metal target is observed.

Preliminary results of time resolved absorption measurements are also shown.

EXPERIMENTS AND RESULTS

Experiments have been performed using an AVCO-HPL cw CO_2 laser capable of up to 15 kW power.

For a better knowledge of the characteristics of this laser previous inspection of the beam power fluctuations was performed. The experimental apparatus is shown in Fig.1. The analysed beam impinges on a scattering plate. The detector is located at a suitable angle and distance in order not to exceed maximum rated power

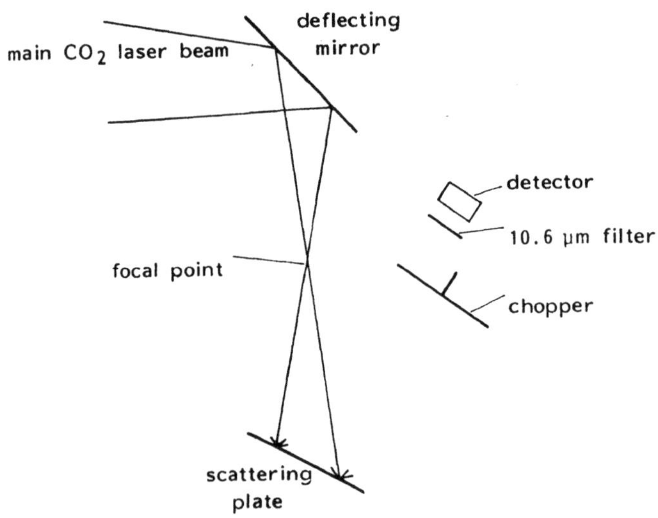

Fig.1 Schematic diagram of the experimental apparatus for beam power fluctuation measurements.

at the detector and for it to see the laser spot on the plate as a point source. The detector is a pyroelectric diode, Molectron P1-72 with a suitable amplifier, with overall band width 10 ÷ 2000 Hz. In front of the detector is a 10.6 μm filter and a chopper whose purpose is to measure the cw power level because, as well known, pyroelectric detectors are only sensitive to intensity variations.

In Fig.2 the output waveform from the detector amplifier is shown for the laser operated at 10 kW.

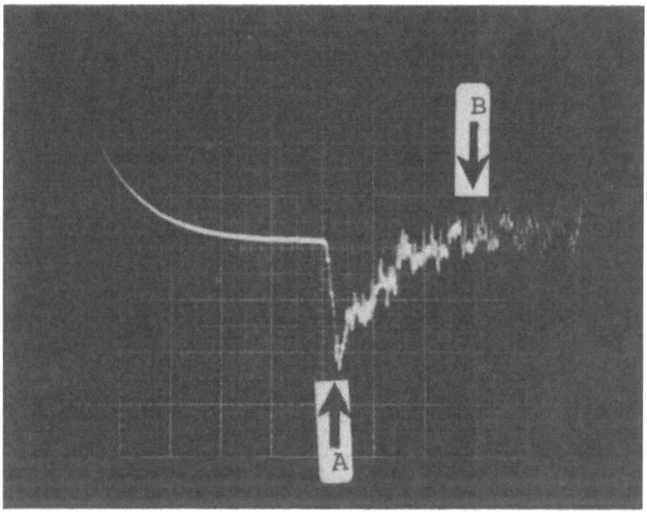

Fig.2 Beam power fluctuations.Laser operated at 10 kW.

The peak labelled A is proportional to instantaneous power, fluctuations are shown as B . The ratio between fluctuations and total power is about 35%. The spectrum of fluctuations is reported in Fig.3. The dominant frequency component is 300 Hz. At lower laser averaged power, 1-2 kW, depth of modulation of 100% have been detected.

The high speed photographic results have been obtained using a rotating prism camera at 4000 frames per second.The photographic sequence, here shown, refers to laser interaction with an aluminum target. The laser beam averaged intensity is 10^5 W/cm^2, with the 90% of the power concentrated in a focal spot having a diameter of about 1.5 mm. The translational workpiece speed is 4 cm/s. In these conditions no penetration below the target surface is observed. As it will be seen, this influences strongly the absorption wave life time.

Fig.4 is a sequence of pictures, extracted from the film, showing developing absorption waves travelling along the direction of the impinging laser beam. The leading and the trailing edge of the wave

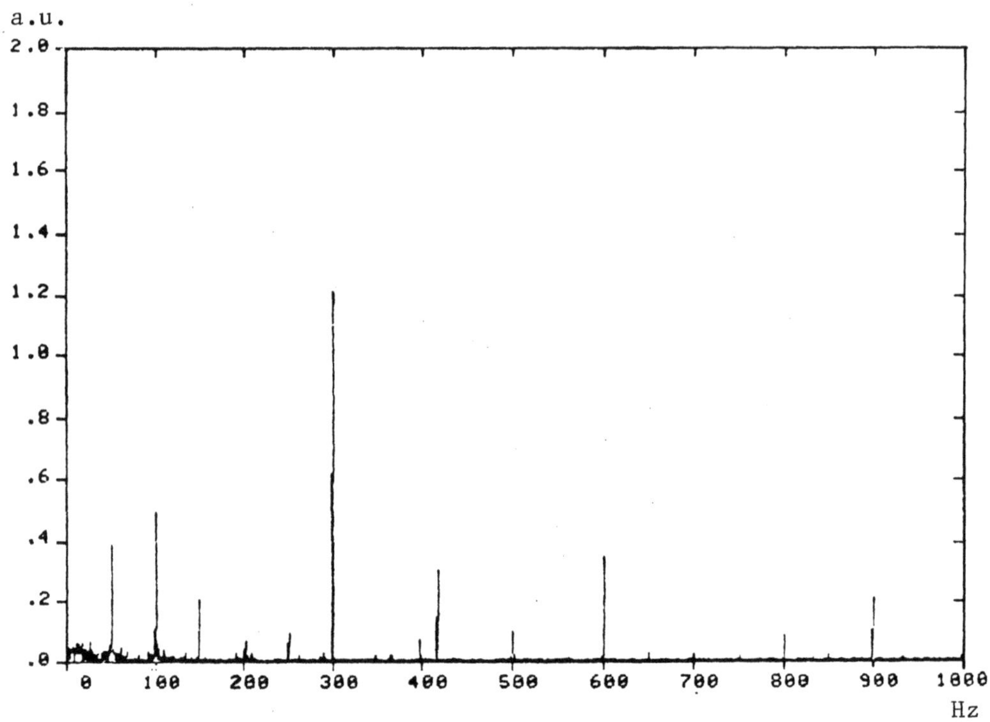

Fig.3 Spectrum of laser beam power fluctuations. Laser operated at 10 kW.

can be identified with the upper and the lower boundaries of the more luminous ball within the less luminous cloud. In Fig.4a this *fireball* is standing at a distance of few centimeters from the target and clearly uncouples the laser radiation from the target, as can be inferred from the absence of a bright spot in the region where the laser should impinge on the target. In Figg.4b, 4c and 4d it may be observed the extinction of this absorption wave and the generation of a new one from the target.This new fireball increases, in size, leaves the target surface (next pictures) and stands at a certain distance from the workpiece; then the process starts over again.

From the analysis of this repetitive phenomenon it can be deduced that the averaged life time of the absorption waves is about 400 ms.

Fig.4a

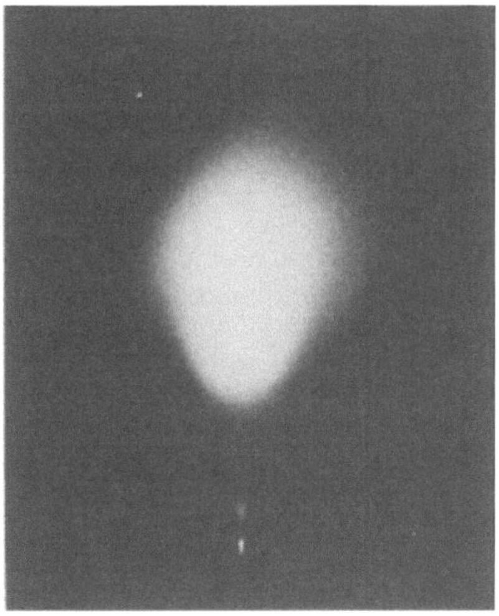

Fig.4b

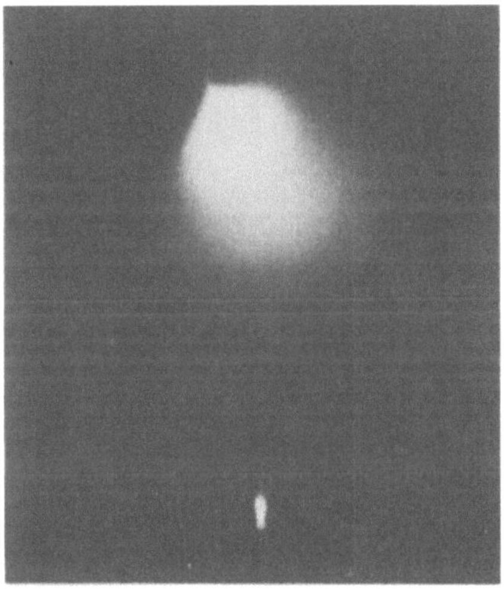

Fig.4c

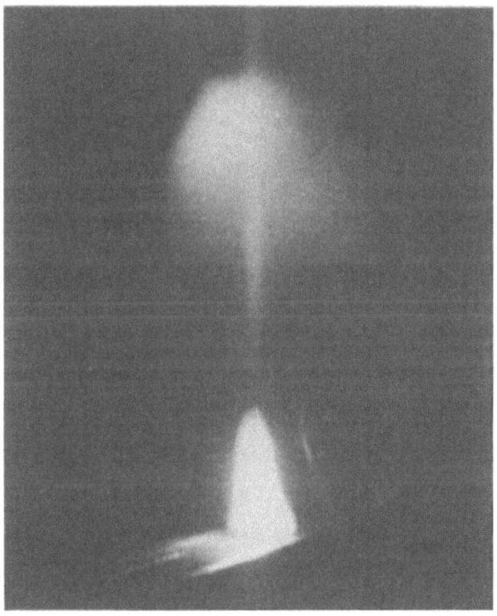

Fig.4d

Fig.4e

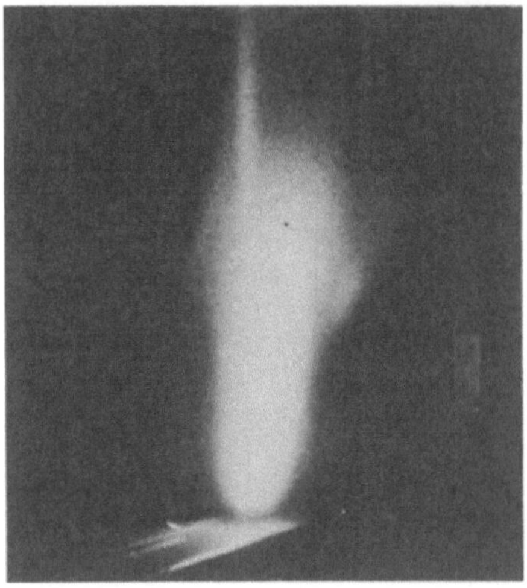

Fig.4f

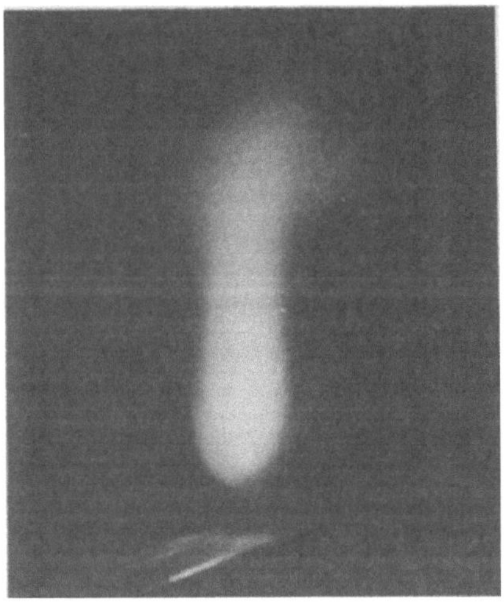

Fig.4g

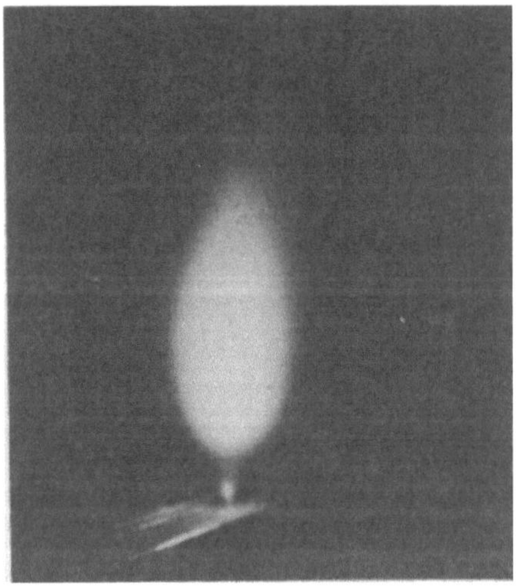

Fig.4h

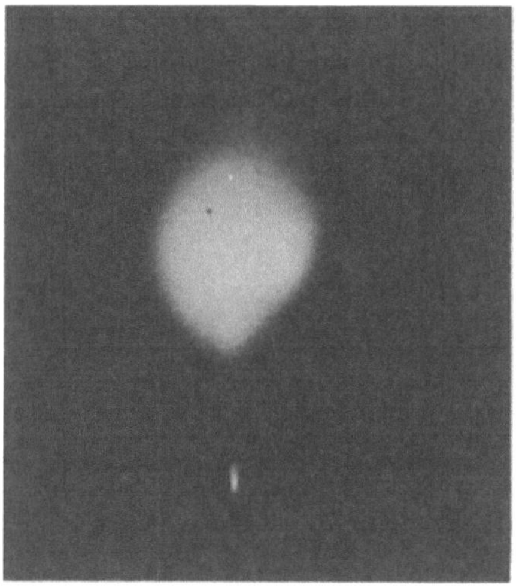

Fig.4i

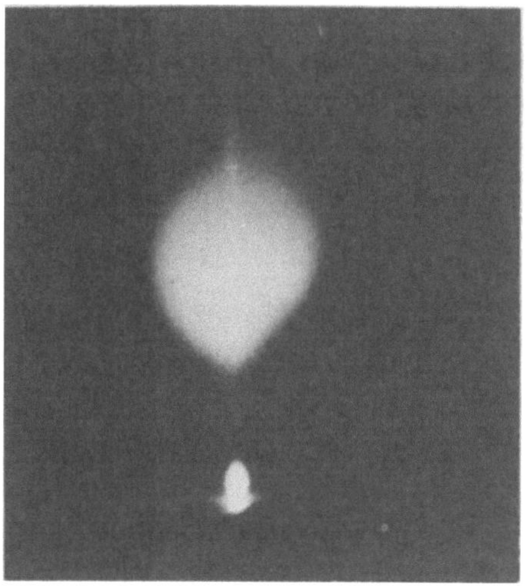

Fig.4j

This life time is 20 times higher than that reported in Ref.15, for aluminum at about the same power of the incident laser beam, but at a target translational speed of only 1 cm/s. In this case a penetration of few millimeters into the metal surface was obtained.

Probably disturbances originating from the penetration hole, due to thermal phenomena and vapor emission, influence the stability conditions of the absorption wave, that becomes extinguished in a shorter time.

In order to understand some physical properties of these absorption waves space and time integrated absorption measurements were carried out and results are reported in Refs.15 and 16. Both for aluminum and steel metals time averaged absorption coefficients of about 0.3 cm^{-1} were found at the wave standing distance from the target. For a better understanding of this time-dependent phenomenon, time resolved absorption measurements have been recently performed and preliminary results are shown here.

The measurements have been carried out according to the scheme reported in Fig.5. The beam of 2 W cw CO_2 laser is used as probe crossing the interaction region at various distances H above the

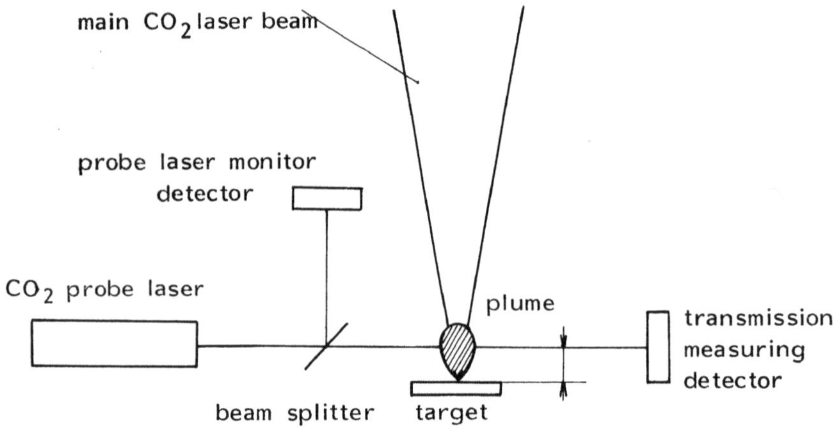

Fig.5 Schematic diagram of the experimental time resolved apparatus for absorption measurements.

surface of the target. A detector (of the type of that in Fig.1) measures the transmitted radiation. The probe beam is monitored by means of a beamsplitter and a detector. The amplified output signals are sent to a memory oscilloscope for monitoring and to a fast, from 0 to 2000 Hz, U.V. chart recorder (not shown in Fig.5).

A sample of the time behaviour of probe laser beam transmitted through the *plume*, at a distance of 10 mm from the target along the optical axis of the main laser beam, is shown in Fig.6, upper trace. The lower trace represents the unwanted modulation of the probe laser beam. Indicated by the arrow is one of the many time intervals of zero transmitted probe beam radiation.

It is clear that the observed reduction of radiation transmission by time integrated measurements[15], takes place mainly in many steps of no transmission, each lasting a few milliseconds, followed by intervals of total transmission.

Changing the height H the overall time of zero transmission changes, decreasing for both lower and higher values of H .

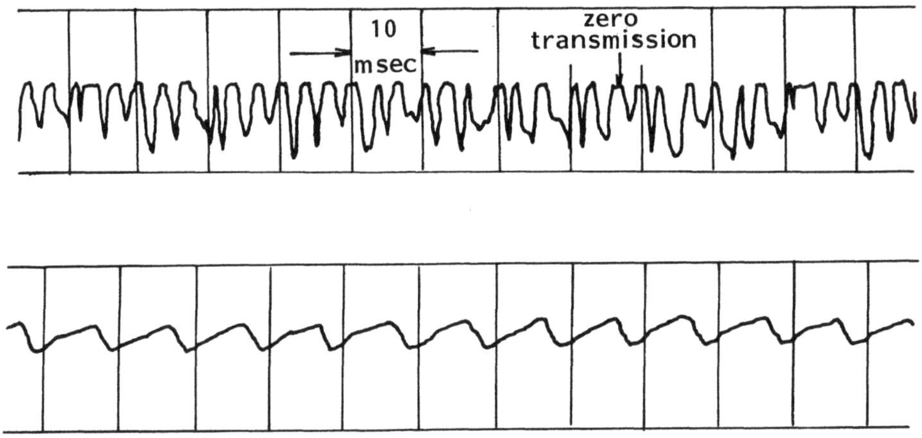

Fig.6 Upper trace: sample of the time behaviour of probe laser beam transmitted through the plume
Lower trace: undesired ripple of the probe beam.

The sample here reported refers to AISI 304 stainless steel target at a translational speed of 4 cm/s. The laser beam intensity is 5×10^5 W/cm^2.

Penetration of few millimeters into the metal surface is observed.

CONCLUSIONS

Fluctuations of about 35% of the total power have been detected operating the AVCO-HPL laser at 10 kW and a dominant frequency component of 300 Hz have been observed.

The influence of these fluctuations on the interaction phenomena is not known and no sufficient attention on this effect has so far been paid in the available literature. Certainly these fluctuations act as a disturbance on the stability conditions of the absorption waves and influence their development and life time.

Absorption waves obtained in conditions of no laser beam penetration into the metal surface have been observed. In this case the wave life time is at least one order of magnitude greater that in the case when penètration is present. This is probably due to the absence of disturbances originating from the penetration hole.

Preliminary results from time resolved absorption measurements have been shown. Time intervals of zero transmission are followed by time intervals of quasi total transmission. Time interval amplitudes depend on the distance from the target.

For a bètter understanding of the phenomena, time resolved absorption measurements coupled and time correlated with high speed photography and monitoring of the high power incident laser beam intensity fluctuations are planned.

REFERENCES

1. C.J.Knight,AIAA J.,Vol.17,n.5 (1979)
2. M.Germano; M.S.Oggiano, VI Congresso Nazionale AIDAA,Roma(1981)
3. P.D.Thomas,AIAA J.,Vol.13,n.10 (1975)
4. E.L.Klosterman; S.R.Byron,AFWL-TR-74-003,Math.Soc.NW,Inc.,Seattle,Wash(1973)
5. A.N.Pirri,AIAA J.,Vol.15,n.1 (1977)
6. E.L.Klosterman,MSNW-75-123-2,Math.Sc.NW Rept.,Seattle,Wash(1975)
7. M.C.Fowler;D.C.Smith,J.AppL.Phys.,Vol.46,n.1 (1975)
8. Y.P.Raizer,Soviet Phys.JEPT,Vol.31,n.6 (1970)
9. R.G.Root,3th Int.Symp.on Gas Flow & Chem.Lasers,Marseille(1980)
10. E.L.Klosterman,J.Appl.Phys.,Vol.45,n.11 (1074)
11. D.C.Smith,Optical Engineering,vol.20,n.6 (1981)
12. G.Herziger,4th GCL Int.Syp.on Gas Flow & Chem.Lasers,Stresa(1982)
13. J.A.Wodroffe,4th GCL Int.Symp.on Gas Flow & Chemical Laser,Stresa(1982)
14. J.F.Ready,Proc.of the IEEE,Vol.70,n.6 (1982)
15. M.Cantello;V.Donati;L.Garifo;A.V.La Rocca;R.Menin;M.Onorato,13th Int.Symp.on Shock Tubes & Waves, Niagara Falls (1981)

16. V.Donati; L.Garifo; A.V.La Rocca; R.Menin; M.Onorato, VI Congr. Nazionale AIMETA,Genova(1982)

17. G.C.Lim; W.M.Steen, Optics and Laser Technology (1982)

EXPERIMENTAL INVESTIGATION OF THE THERMOMECHANICAL EFFECTS INDUCED BY A CO_2 LASER ON VARIOUS MATERIALS IN A RAREFIED ATMOSPHERE*

Claude Créput, Daniel Dufresne, Gilles Avakian, and François Puech

Institut de Mécanique des Fluides de Marseille
1, Rue Honnorat, 13003 Marseille, France

INTRODUCTION

When a solid target is irradiated by a high-power laser beam (up to 10^9 W/cm^2), various phenomena may occur in front of the target during the interaction. These phenomena principally depend on the value of the pressure in the surrounding medium; the behavior of the hydrodynamical flow in front of the target is now fully known when the interaction occurs at atmospheric pressure (1 bar), where the flow is governed by L.S.D. mechanisms. However, the structure of the flow in the range of ambient pressures between 400 mbar and 10^{-4} mbar (low vacuum) has not been completely determined. Several hydrodynamic regimes take place in the different ranges of surrounding pressure, from laser supported waves (L.S.A.W.) at medium densities to molecular plumes in low vacuum. To our present knowledge, exhaustive experimental studies of the hydrodynamic structure of the laser-induced flow in front of various targets located in vacuum have not been done, and only a few results about flow velocity and pressure measurements have been published [1-4], and these results are not in agreement. In this paper we report an experimental study of the thermomechanical effects induced by a CO_2 laser beam on solid targets in vacuum. Two materials with opposite physical properties have been irradiated:

- pure aluminium with low ionization potential, low latent heat of vaporization, and high reflectance;

- carbon, with high ionization potential, high latent heat of vaporization, and low reflectance.

* This work is supported by French D.R.E.T.

This study has involved radial and axial flow velocity measurements by means of a fast framing camera, and strioscopy visualizations of the shock waves inside the flow. The different regimes governing the interaction have thus been determined over the range of ambient pressure explored (400 mbar - 10^{-4} mbar).

EXPERIMENTAL

Laser

The experiments have been carried out with a T.E.A. CO_2 laser, which can deliver from 300 J to 2000 J in 1 to 10 μs duration pulses. A laser pulse is composed of a high power peak of 250 ns FWHM followed by a high energy tail of 8 μs duration. The output beam is focused onto the targets by a spherical mirror of 13 m focal length which provides a focal spot of about 2 cm^2, with a maximal intensity of up to 1.5 10^9 W/cm^2 at the center of the spot.

Flow Velosity Measurements and Strioscopy Visualizations

In order to determine the radial as well as axial velocity of luminous fronts, a fast framing camera IMACON 790, able to provide from 500,000 to 20 million frames per second, was used. In all the experiments, the framing rate of 5 million frames per second has been used to record the phenomena for a sufficiently long duration.

The inner hydrodynamic structure of the flow and the velocity of the different shock front have been investigated by means of fast strioscopy visualizations. These visualizations were done by the fast framing camera and the luminous probe was an ionic argon laser, with 4 W maximum power.

Some experiments involving double wavelenght Mach-Zehnder interferometric measurements have also been done, using either the fast framing camera or photomultipliers to record the evolution of the density in the flow.

Experimental Conditions

Typical laser pulses used in the experiments were 1000 J pulses, with a peak flux of $7 \cdot 10^8$ W/cm^2 and a flux of 1.4 10^8 W/cm^2 in the tail. Some experiments have been carried out with pulses of 300 J, corresponding to $3 \cdot 10^8$ W/cm^2 in the spike and $4 \cdot 10^7$ W/cm^2 in the tail.

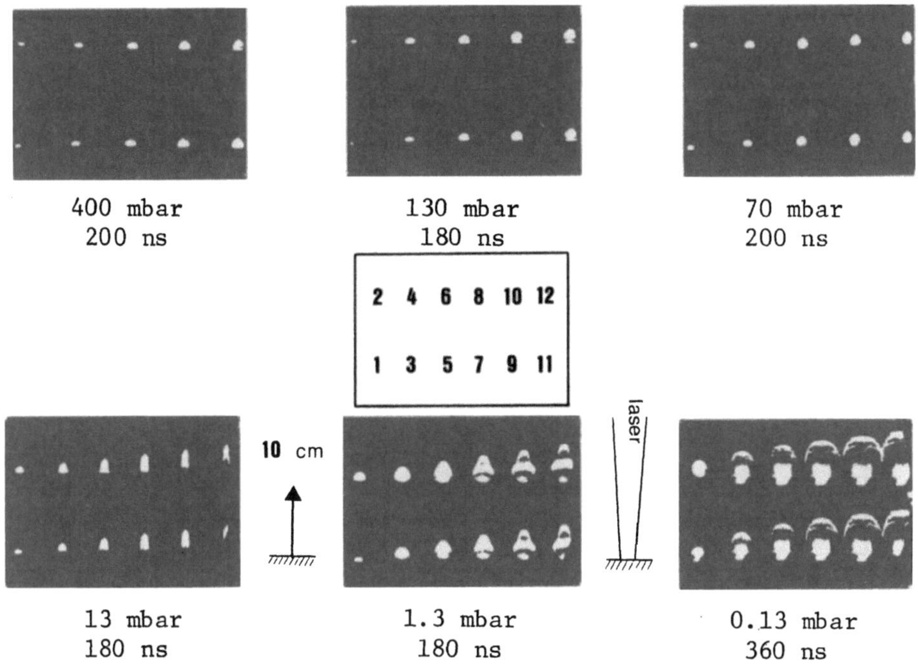

Fig. 1. Visualizations of the flow above a carbon target at different air pressures. Framing rate: $5 \cdot 10^6$ frames per second.

RESULTS

Flow Visualization and Stereoscopic Measurements

Characteristic visualizations of the luminous front above a carbon target at ambient pressures between 400 mb and 0.13 mb are presented in Figure 1.

These pictures have been recorded at a framing rate of $5 \cdot 10^6$ pictures per second, with an exposure time of 40 ns. The laser pulses used for these shots were complete (peak + tail, and 8 µs duration). The peak power was $2 \cdot 10^8$ W/cm^2 and the tail power $4 \cdot 10^7$ W/cm^2. These visualizations show that plasma expansion depends on ambient air pressure: as this pressure decreases, the axial velocity of the plasma increases, while the radial development becomes more and more important. Under 1 mbar this radial expansion is pre-eminent, as shown by the framing sequence taken at 0.13 mbar. These visualizations permit us to plot the axial expansion versus ambient pressure with a given laser flux, i.e., at a fixed instant of the laser pulse. This result is given in Figure 2, both for carbon (a) and for aluminium (b), whose behavior has been found to

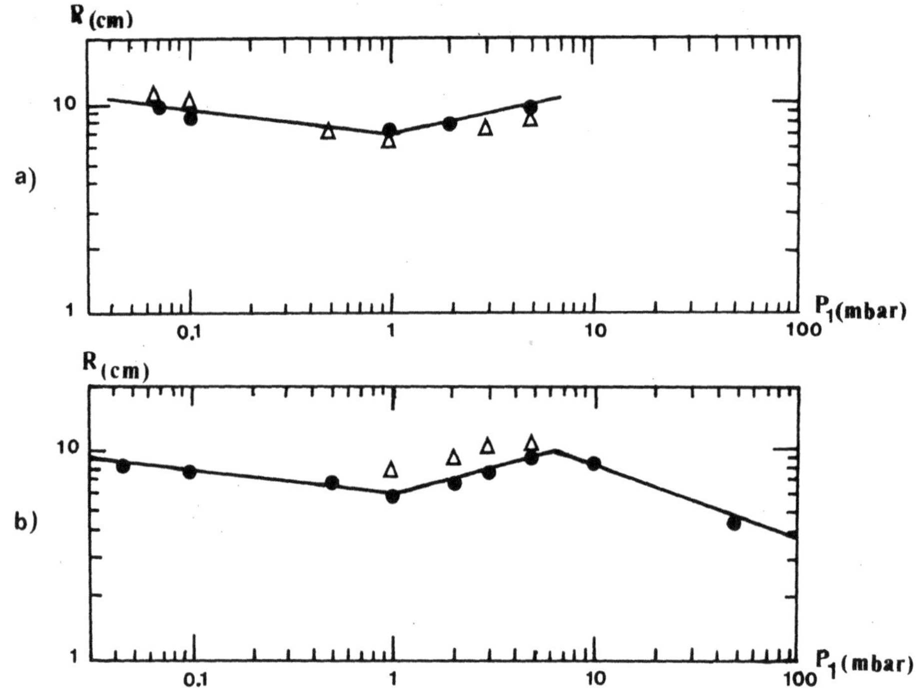

Fig. 2. Axial expansion of the flow versus ambient pressure for a given laser flux of $4 \cdot 10^7$ W/cm^2: a) carbon; b) aluminium.

be the same as that of the carbon. We note that the axial expansion of the flow both for aluminium and carbon is governed by a $\rho^{-1/3}$ law between 400 mbar and 10 mbar, a $\rho^{1/4}$ between 10 mbar and 1 mbar and a $\rho^{-1/5}$ law between 1 mbar and 10^{-2} mbar.

On the other hand, we have determined the temporal evolution of the distance covered by the luminous front during the first 3 microseconds of the laser pulse. As shown in Figure 3a this distance is clearly proportional to the time, independently of the ambient pressure between 400 mbar and 13 mbar. For the same range of ambient pressure, the axial velocity of the front has been determined as a function of the parameter I_0/ρ, ratio of the incident laser flux to the surrounding density. This has been done in order to compare the experimental results to the L.S.D. theory predictions in the range 400 mbar - 10 mbar. The result of this study is shown in Figure 3b, and it appears that the flow velocity measurement agrees to L.S.D. calculations within 15%.

All these results allow us to distinguish three hydrodynamic regimes in the flow:

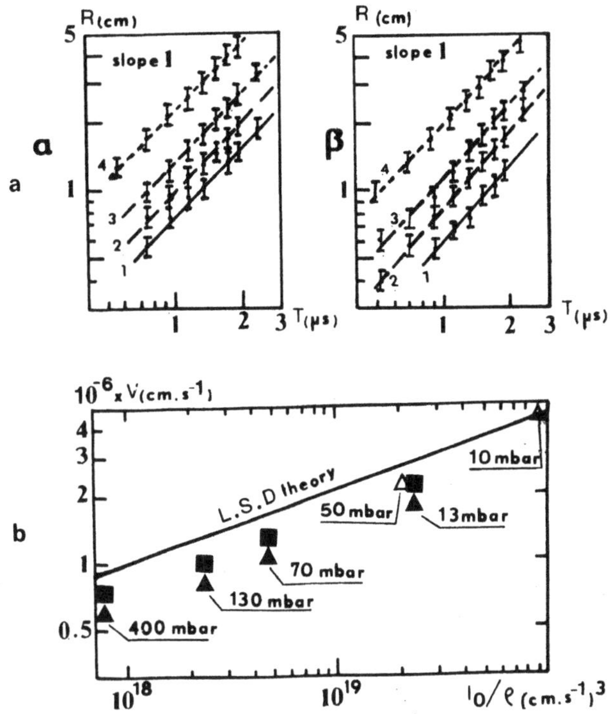

Fig. 3. a) Axial expansion as a function of time for (α) carbon and (β) aluminium at 1 - 400 mbar, 2 - 130 mbar, 3 - 70 mbar and 4 - 13 mbar; b) Velocity of the plasma front as a function of I_o/ρ parameter.

- from 400 to 10 mbar, the flow seems to be governed by laser-supported absorption waves theory;

- from 10 mbar to 1 mbar, the flow may be described by two types of waves: in a first step, the laser spike induces the formation of a spherical blast wave; then, an absorption wave initiated in the irradiated part of the blast wave is supported by the energy of the laser pulse tail;

- from 1 mbar to 0.01 mbar, a spherical blast wave regime is established, and this wave remains transparent to the incident infrared radiation.

In order to specify the intermediate regime (between 10 mbar and 1 mbar) with more accuracy, strioscopy visualizations of the flow have been done in this range both on aluminium and carbon targets. These visualizations are presented in Figure 4, for ambient pressures of 5 mbar, 3 mbar, 2 mbar and 1 mbar. The 5 mbar sequence shows clearly the existence of an expanding axial front with fast

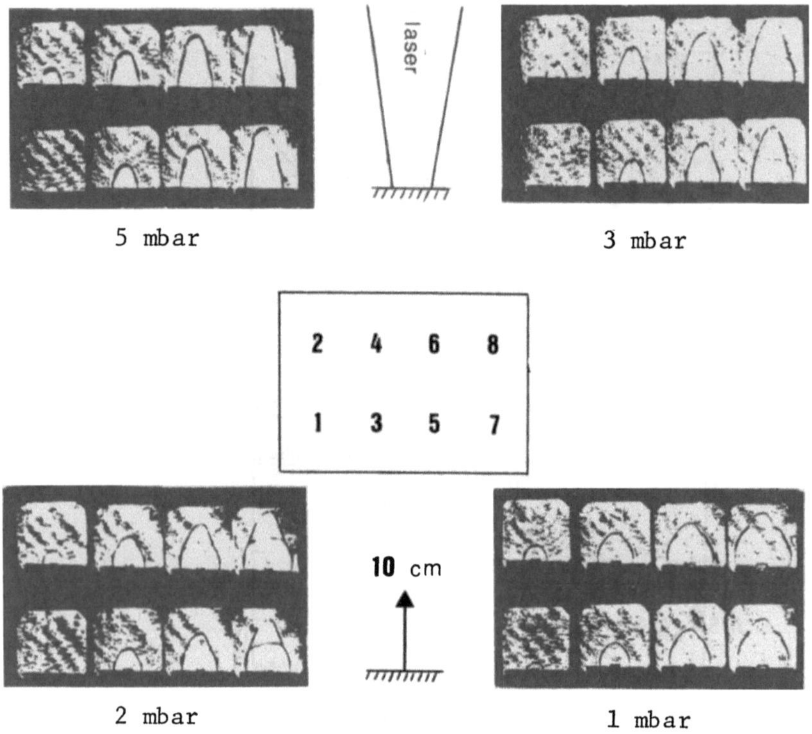

Fig. 4. Strioscopy visualizations of the flow above a carbon target. Framing rate: $5 \cdot 10^6$ frames per second.

velocity. When air density decreases (3 mbar sequence), a spherical wave appears near the target as soon as the sample is irradiated, and expands radially with relatively high velocity. An absorption front then grows on this spherical wave during the tail of the laser pulse; the velocity of this front becomes smaller when the ambient pressure decreases to 2 mbar, while the radial expansion and intensity of the spherical wave become greater. Finally, at 1 mbar, the surrounding density is no longer able to induce an absorption front over the spherical wave, and only the blast wave regime can hold.

The characterization of the flow has been completed using laser pulses with single peak of intensity, in order to point out the role of the tail in the establishment of laser-supported absorption waves. Figure 5 presents the comparison between the behavior of the flow with and without the laser pulse tail for a carbon target at 1.3 mbar. These pictures show that the temporal evolution of the plasma follows the laser pulse:

- for complete laser pulse, the axial development of the plasma is important during the whole visualization;

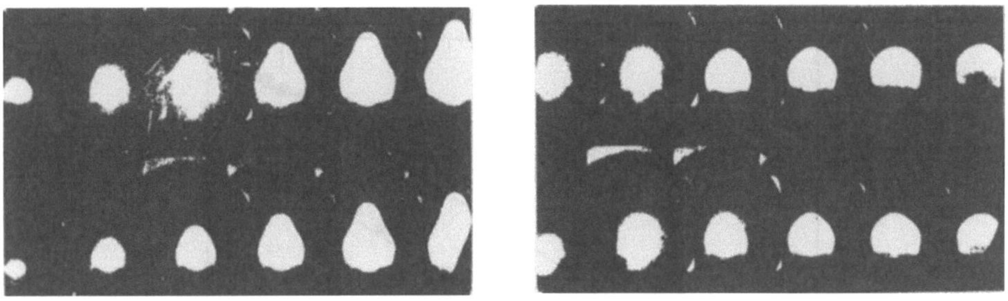

pulse 1 pulse 2

1.3 mbar

Fig. 5. Visualizations of the flow above a carbon taget at 1.3 mbar ambient pressure (framing rate: $5 \cdot 10^6$ frames per second); a) Complete laser pulse; b) laser pulse with spike only.

- for single peak laser pulse, the plasma expansion stops after 1 μs.

These results confirm the existence of two types of waves co-existing during the laser irradiation in the range 10 - 1 mbar.

Experimental results have been compared to theoretical predictions from a variable energy blast-wave model of Dabora [5] and have been found in good agreement for surrounding densities between 1 mbar and 0.01 mbar, where the spherical wave regime seems to be pre-eminent. The comparison is presented in Figure 6 for both aluminium and carbon samples.

Double Wavelength Interferometric Density Measurements

Measurements of neutral and electronic densities have been carried out by means of double wavelength interferometric apparatus (temporal study with photomultipliers recording and radial study by fast camera recording). Figure 7 shows the evolution of the maximal values of these densities on the range 130 - 0.13 mbar. The atomic density varies from about 10^{19} cm^{-3} at 0.13 mbar to $7 \cdot 10^{19}$ cm^{-3} at 130 mbar, while the electronic density grows from $7 \cdot 10^{17}$ cm^{-3} at 0.13 mbar to $2 \cdot 10^{18}$ cm^{-3} at 130 mbar.

CONCLUSION

The experimental study presented in this paper has provided better knowledge of the interaction between great energy laser beams and solid targets in the range of ambient pressure 400 mbar - 0.01

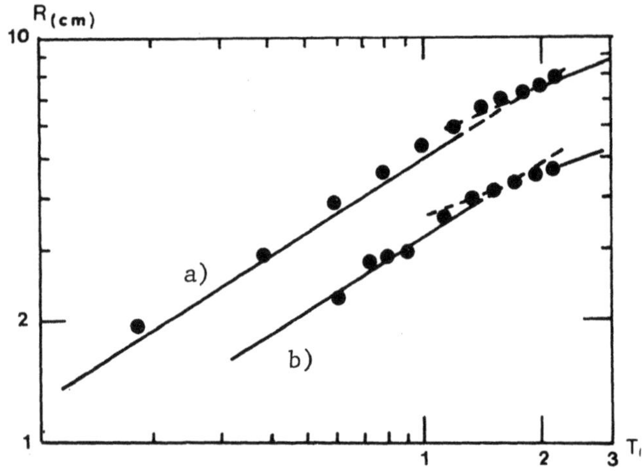

Fig. 6. Comparison between experimental results and the prediction of variable energy blast-wave model at 0.13 mbar. a) Carbon; b) aluminium.

mbar. The different hydrodynamic regimes governing the interaction have been characterized and the thresholds of surrounding density have been indicated both for aluminium and carbon targets. The behavior of the flow appears to be fully independent of the target nature in the range 400 mbar - 10 mbar, and only dependent on the ambient medium density and nature, as shown by complementary visualisations done with helium instead of air as the surrounding medium. The regime of LSAW no longer holds under 100 mbar of helium, owing to the high value of the ionization potential of this gas. Physical

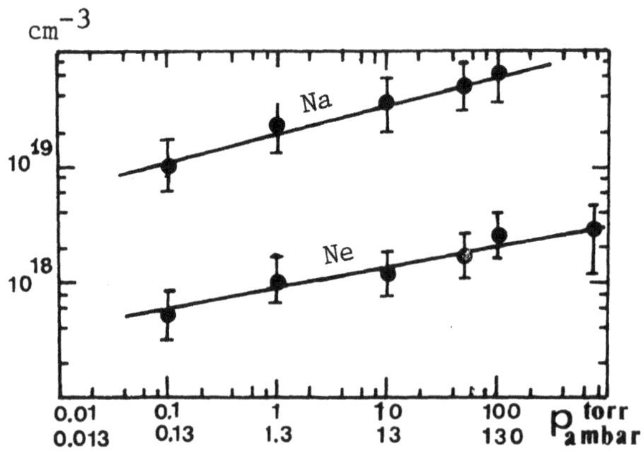

Fig. 7. Electronic and neutral densities as a function of ambient pressure.

properties of the target begin to play a role under 10 mbar of air. Under 1 mbar, the laser spike induces a spherical blast-wave. New space-resolved interferometric measurements would allow a better knowledge of densities and temperature of the fronts.

REFERENCES

1. J. L. Bobin et al., J.A.P., Vol. 39, No. 9, August (1968).
2. D. C. Emmony et al., Brit. J. Appl. Phys., (J. Phys. D.), Ser. 2, Vol. 2 (1969).
3. D. Dufresne and C. Créput et al., AIAA., 14th Fluid and Plasma Dynamics Conference, Palo Alto (1981).
4. M. Hugenschmidt et al., ISL report R118/78 (1978), (unpublished).
5. E. K. Dabora, AIAA Journal, Vol. 10, No. 10, pp. 1384-1396 (1972).

LASER-METAL INTERACTION: THE GASDYNAMIC OF THE EMITTED VAPOURS

Massimo Germano

Politecnico di Torino
C.so Duca Abruzzi,24
10129 Torino (Italy)

Maria Sandra Oggiano

Centro Studi Dinamica Fluidi
C.so Duca Abruzzi,24
10129 Torino (Italy)

INTRODUCTION

The surface vaporization into a surrounding atmosphere due to the interaction of a laser beam with a solid target has been the subject of intensive study both theoretical and experimental in the past years [1,4]. The behaviour of the vapour clouds due to the vaporization is strongly dependent on the laser flux and the ambient pressure as well as on the material properties. In the case of CO_2 lasers, and at intensities $< 10^5$ Watt/cm^2 the laser surface interaction is mainly an ablative process, in which after the initial attainment of the vaporization temperature the vapour emitted acts like a pushing piston. As a consequence a shock wave propagates in the ambient atmosphere that is separated from the vapour by a contact surface. At intensities $> 10^5$ Watt/cm^2 or more, depending on the surrounding ambient pressure, the laser surface interaction is controlled by the formation of a laser supported absorption wave. This formation is caused by the laser heating of the gas, occurs first in the vapour and then in the shocked air, and has the final effect of reducing the incident laser intensity on the target.

Thomas[3] has performed complete calculations that cover the three typical phases of the event, the initial attainment of the vaporization temperature, the gasdynamic evolution of a shock wave in the ambient atmosphere due to the pushing vapour and the formation of an absorption wave due to radiative heating that finally shields the target from further emission.

Knight[5] has developed a theoretical model particularly fitted to the study of the second phase. He assumes that the flow is one-dimensional and joins the vapour conditions at the surface with the ambient conditions via jump relations at the surface Knudsen layer, at the contact surface between vapour and gas and at the leading shock wave. By assuming a uniform laser emission it is possible to solve analytically the flow for given laser flux and ambient conditions. Two different flow structures are possible, due to the fact that nonequilibrium processes within the Knudsen layer cannot drive the flow to velocities greater than sonic at the surface. The first one, Fig.1, consists in three uniform regions, the unperturbed ambient gas, the shock heated gas, the emitted vapour, whose Mach number is in this case less than unity. The second one, Fig.2, corresponds to stronger shock waves and consists in four regions, the

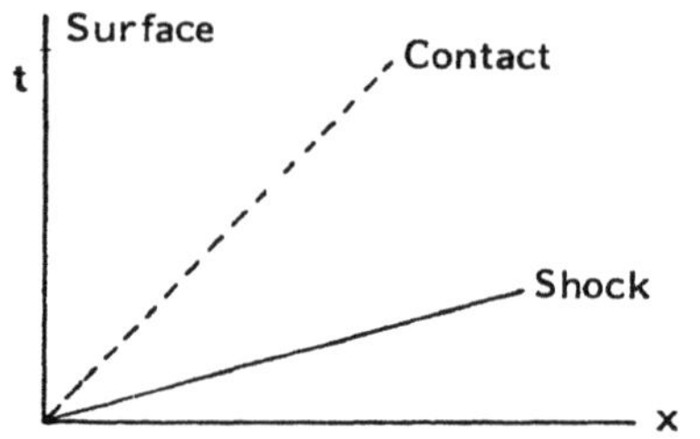

Fig.1 Knight Model. Subsonic flow structure.

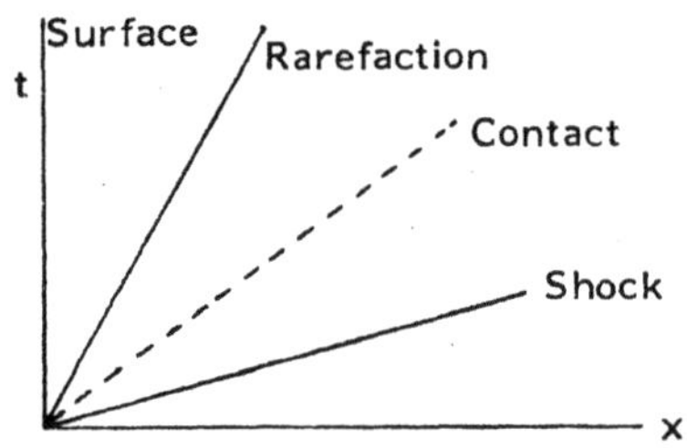

Fig.2 Knight Model. Supersonic flow structure.

three regions illustrated in the first case plus a rarefaction fan region that matches the vapour conditions behind the contact surface, whose Mach number is now greater than one, with the vapour conditions at the surface where the Mach number $M = 1$. This last flow structure is characteristic of higher laser fluxes and lower ambient pressures, and is peculiar of laser interaction with targets at high altitudes or in vacuum.

In our paper we have extended the Knight calculations to a general transient flow due to a variation of the incident laser flux with time. We have adopted his surface emission model, that practically consists in a state relation connecting vapour temperature, pressure and Mach number at the surface, and we have developed a numerical procedure for the study of the vapour and the shocked gas previously applied by the authors to the evolution of strong radiating blast waves[6,7].

EQUATIONS AND NUMERICAL CALCULATIONS

Let us consider a surface invested by a laser flux of intensity ϕ . The phase change from liquid to vapour is achieved in a very thin Knudsen layer, and the Knight model essentially provides the jump relations between the surface side quantities, index s , and the vapour side conditions, index v , (see Fig.3). We have

$$\alpha = \frac{T_v}{T_s} = \left\{ \sqrt{1 + \pi\left(\frac{\gamma_v - 1}{\gamma_v + 1}\,\frac{m_v}{2}\right)^2} - \sqrt{\pi}\,\frac{\gamma_v - 1}{\gamma_v + 1}\,\frac{m_v}{2} \right\}^2 \qquad (1)$$

$$\beta = \frac{\rho_v}{\rho_s} = \sqrt{\frac{T_s}{T_v}}\left((m_v^2 + \tfrac{1}{2})\, e^{m_v^2}\, \mathrm{erfc}(m_v) - \frac{m_v}{\sqrt{\pi}}\right) + \frac{1}{2}\,\frac{T_s}{T_v}\left(1 - \sqrt{\pi}\, m_v\, e^{m_v^2}\, \mathrm{erfc}(m_v)\right) \qquad (2)$$

where it has been assumed that the vapour follows the perfect gas law on both sides, s and v ,

$$p = \rho\,\frac{R}{\mathcal{M}_v}\,T \qquad (3)$$

with T absolute temperature, ρ density , p pressure, $\mathcal{M}_v$ the molecular mass, γ_v the ratio of specific heats, and R the gas constant. In (1) and (2) $m_v = u_v/\sqrt{2RT_v}$ where u_v is the velocity of the emitted vapour, and care must be given to the fact that nonequilibrium processes in the Knudsen layer cannot drive the flow to supersonic speed, so that the Mach number $M_v = u_v/\sqrt{\gamma_v R T_v} = m_v\sqrt{2/\gamma_v}$ is restricted to $M_v \leq 1$. The surface conditions are the saturated ones, so that the Clausius-Clapeyron relationship gives

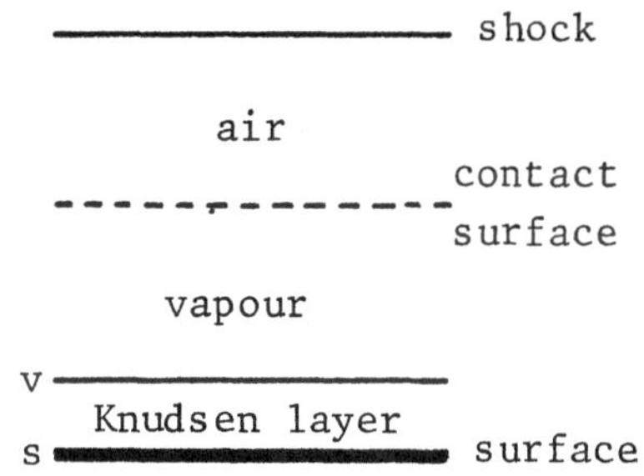

Fig.3

$$p_s = C\,\exp(-B/T_s) \qquad (4)$$

where B and C are constant for a given material, and we can derive from these relations a boundary condition for the vapour emitted in the form

$$T_v = T_v(M_v, p_v) = \alpha B / \ln(\alpha\beta C / p_v) \qquad M_v \leq 1 \qquad (5)$$

Another condition relates the incident laser flux ϕ to the flux of vapour emitted. If L is the latent heat of vaporization we have

$$L \rho_V u_V = \phi - \phi_c \tag{6}$$

where ϕ_c is the conductive heat loss at the surface. On the basis of the relations (5) and (6) Knight relates the conditions at the surface with the conditions at the contact surface and then at the shock, describing a unique steady flow state. This is possible if ϕ does not change in time, otherwise relations (5) and (6) remain valid but the flow must be calculated starting from the unsteady one-dimensional equations in the spatial coordinate x and the time t

$$\begin{aligned} &\frac{\partial \rho}{\partial t} + u \frac{\partial \rho}{\partial x} + \rho \frac{\partial u}{\partial x} = 0 \\ &\rho \frac{\partial u}{\partial t} + \rho u \frac{\partial u}{\partial x} + \frac{\partial p}{\partial x} = 0 \\ &\frac{\partial s}{\partial t} + u \frac{\partial s}{\partial x} = \frac{q}{T} \end{aligned} \tag{7}$$

where u, p, ρ, T, s are the velocity, the pressure, the density, the temperature, the entropy per unit mass and q is the heat added per unit time and unit mass as a consequence of an eventual absorption of laser radiation. The distribution of velocity, pressure and density at the time $t + \Delta t$ can be computed using a two level (predictor-corrector) finite difference scheme due to Mac Cormack[8], and three special points are explicitly treated in a different way and with particular care: the shock, the contact between vapour and gas and the surface. Referring to Fig.4, let us suppose that at the generic instant t the shock and the contact positions are known. We consider the characteristics reaching the shock, the contact surface

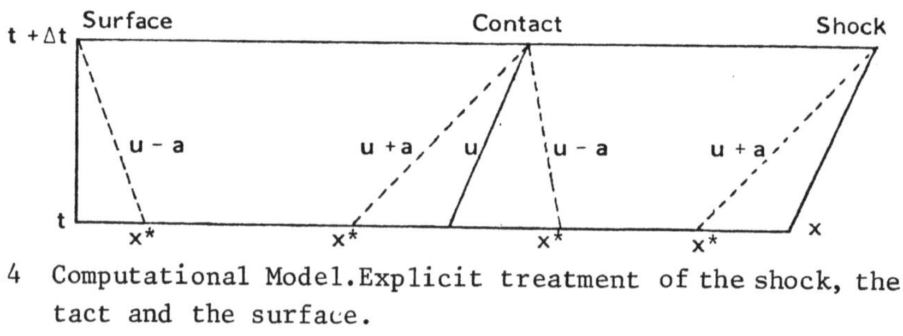

Fig.4 Computational Model. Explicit treatment of the shock, the contact and the surface.

and the wall at the time $t+\Delta t$, and the various points x^* can be found by geometrical relations, while the values of u,p,ρ,T in these points are obtained by linear interpolations at the time t. Along the characteristics $u \pm a$ the following differential equations hold

$$\frac{Dp}{Dt} \pm \rho a \frac{Du}{Dt} = (\gamma-1)\rho q \tag{8}$$

where a stands for the sound velocity, $a^2=(\partial p/\partial \rho)_s$, and γ is the specific heat ratio and we can calculate the new positions of the shock wave and of the contact surface by coupling the relations (8) with the jump conditions respectively at the shock and at the contact surface. The values at the wall are obtained by couplingthe relations (8) with the Knight relations (5) and (6) for the vapour condition at the surface,but it must be noticed that when $u=a$, $(M=1)$, the characteristic does not reach the wall, so that in this case $u=a$ is the real condition that must be satisfied at the wall and that substituted the relations (8).

To the end of having regions of fixed length the physical coordinate x is mapped in a transformed space, in which the shock position is normalized to one at any time, and stretching coefficients are used for a better computation at the shock and at the wall.

RESULTS

As examples of our calculations we report three particular cases. In all cases ϕ_c, the conductive heat loss at the surface,has been posed equal to zero, but as a matter of fact this condition is inessential to calculations, provided that we always intend, in the following that ϕ indicates the net incident radiation flux carried away by vapour leaving the surface. In all cases we have considered an aluminum target, and the ambient atmosphere is air.

In the first case we have a rapid variation of the laser flux, Fig.5, from 10^3 to 10^4 W/cm^2 at an ambient pressure of 10^{-3}atm and at a temperature of 300°K. The steady situation changes from an initial subsonic flow structure to a final supersonic flow structure, (see Fig.1,2), and in Fig.6, 7 we

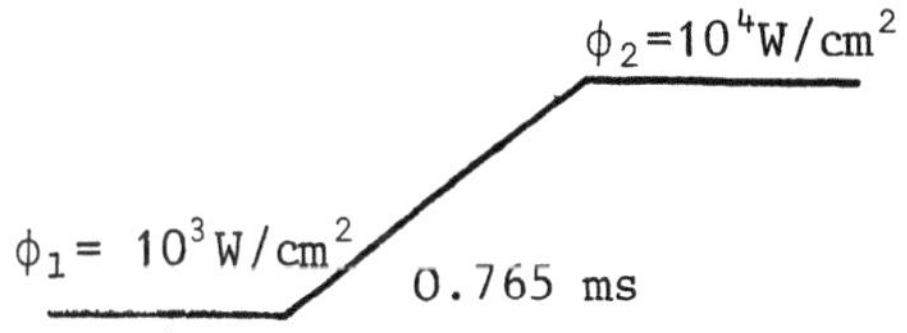

Fig.5

have represented the profiles of the pressure and the temperature in the vapour and in the gas. The distance target-shock has been normalized to one, and the letter C indicates the relative position of the contact surface, and times are reported on the different profiles. The dotted lines represent the steady final flow as calculated by the method of Knight, and it is interesting to note the overshoot in the temperature and in the position of the contact surface due to the rapidity of the event.

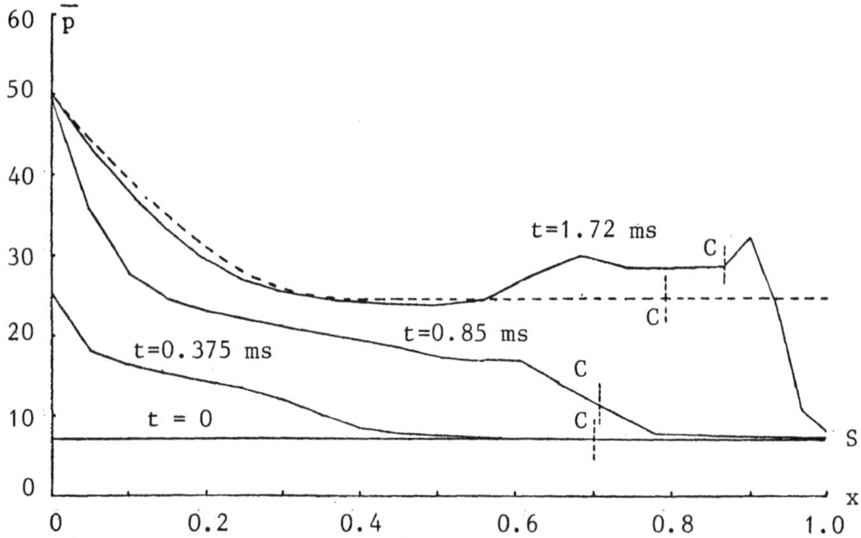

Fig.6 Case 1. Time variation of the dimensionless pressure $\overline{p} = p/p_o$. $p_o = 10^{-3}$atm. C contact. S shock.

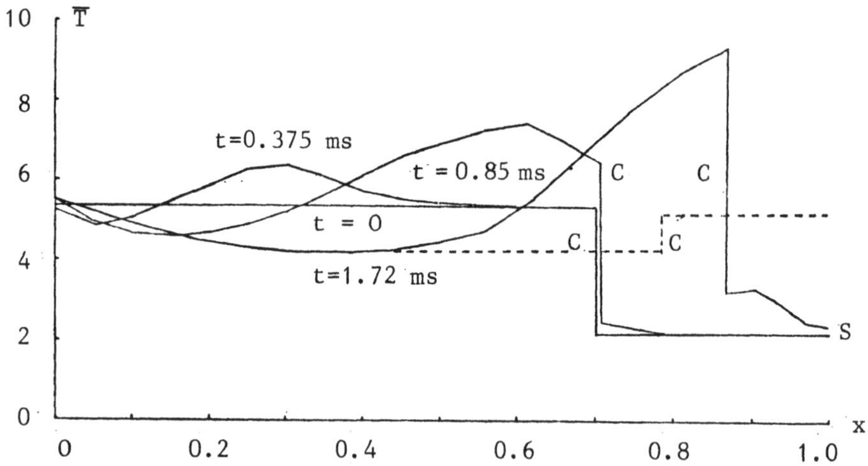

Fig.7 Case 1. Time variation of the dimensionless temperature $\overline{T} = T/T_o$. $T_o = 300°K$ C contact. S shock.

In the second case we have considered the reverse situation, Fig.8, in which the laser flux decreases very rapidly. In Fig.9,10 are represented the profiles of the pressure and the temperature at various times. It was out of scope in this preliminary work devoted to the testing of a numerical method applied to the gasdynamic of the emitted vapours, to study in details the radiative absorption of the laser light. It has been found by Thomas[3] that the absorption wave is triggered by nonequilibrium processes in the vapour, as avalanche ionization , and then develops very rapidly with the final result that no laser radiation ultimately reaches the target.

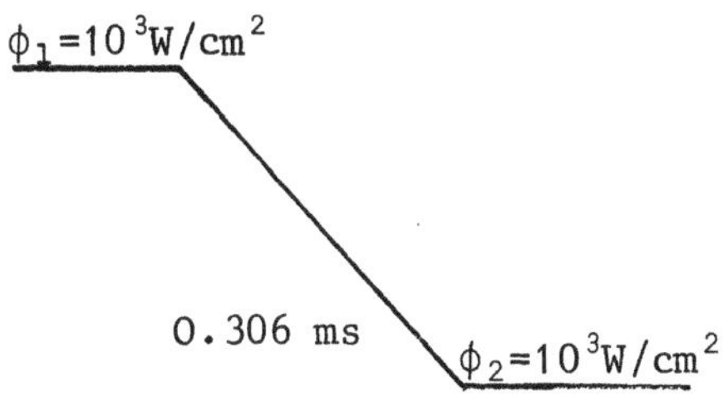

Fig.8

As a rough approximation we have computed a third case where has been assumed that the initial laser radiation is uniformly and completely absorbed in the vapour. The subsequent evolution of the vapour and the shocked air is reported in Fig.11,12 where the profiles of the pressure and of the velocity are represented at different times. In this case the ambient air is at a pressure of 1atm,

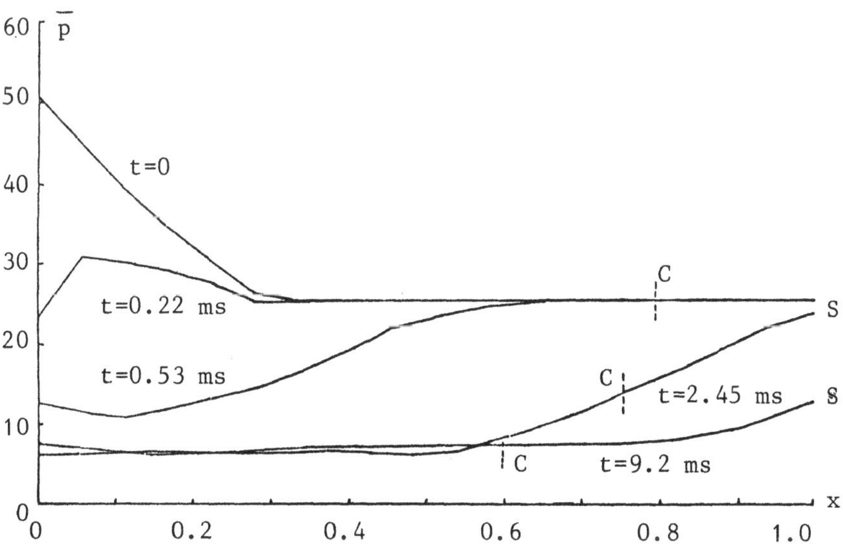

Fig.9 Case 2. Time variation of the dimensionless pressure $\overline{p} = p/p_o$. $p_o = 10^{-3}$ atm. C contact. S shock.

the initial flux is 10^5Watt/cm^2, and we can see from Fig.11 the development of a strong pressure wave that reflects at the target and

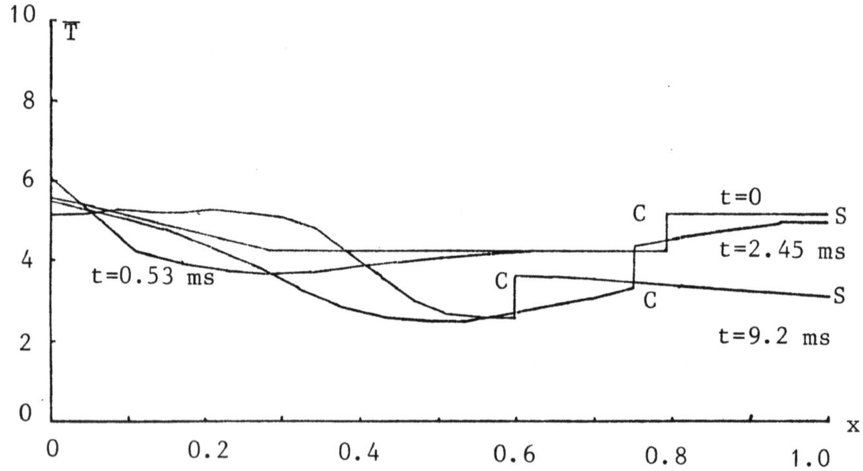

Fig.10 Case 2. Time variation of the dimensionless temperature $\overline{T}=T/T_o$. T_o = 300°K. C contact. S shock.

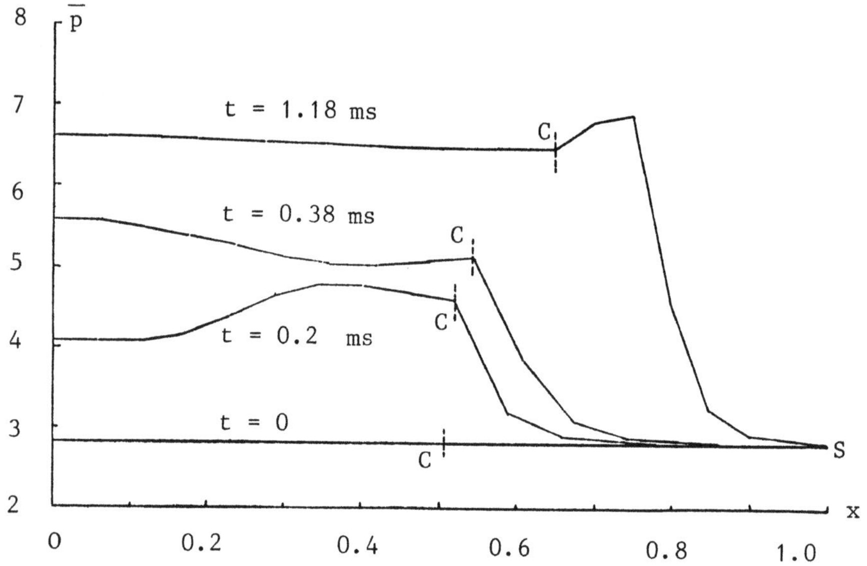

Fig.11 Case 3. Time variation of the dimensionless pressure $\overline{p}=p/p_o$. p_o = 1 atm. C contact. S shock.

reaches at later times a nearly uniform value. Very interesting are the profiles of the velocity, developing from the initial constant value to a linear ramp, and giving support to the hypothesis that in the long term a self similar solution applies to this case.

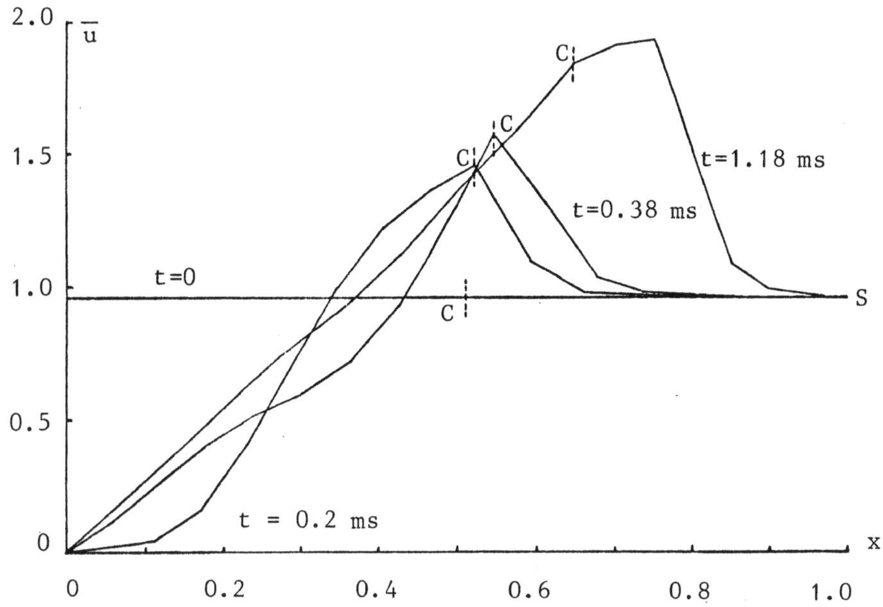

Fig.12 Case 3. Time variations of the dimensionless velocity $\bar{u}=u/u_o$. $u_o = \sqrt{RT_o/\mathcal{M}_g}$ = 290 ms. C contact. S shock.

CONCLUSIONS

The extension of the Knight model to a general transient flow has been performed by adopting a numerical procedure in which the wall, the contact surface between the vapour and the shocked gas , and the shock have been explicitly considered. Some typical cases have been evaluated, at low pressures and low fluxes without absorption of radiation and at higher pressures and fluxes with absorption of radiation. This absorption has been simulated as an instantaneous and uniform absorption in the vapour, and in a subsequent work a complete radiative model will be developed and coherently applied to our program.

REFERENCES

1. Y.P.Raizer,*Subsonic Propagation of a Light Spark and Threshold Condition for the Maintenance of a Plasma by Radiation*,Soviet Physics JEPT,31:6(1970)
2. M.C.Fowler & D.C.Smith,*Ignition and Maintenance ofSubsonic Plasma Waves in Atmospheric Pressure Air by CW CO_2 Laser Radiation*

and Their Effects on Laser Beam Propagation,J.Appl.Phys. 46:1 (1975)
3. P.D.Thomas,*Laser Absorption Wave Formation*, AIAA J.,13:10(1975)
4. A.N.Pirri,*Analytic Solutions for Laser-Supported CombustionWave Ignition above Surface*, AIAA J.,15:1(1977
5. C.J.Knight,*Theoretical Modeling of Rapid Surface Vaporization With Back Pressure*, AIAA J.,17:5 (1979)
6. M.Germano, M.S.Oggiano, M.Onorato, M.Pandolfi,*Calcolo numerico di onde esplosive*,"Omaggio a Carlo Ferrari",Levrotto & Bella Torino (1974)
7. M.Germano, M.S.Oggiano, M.Onorato:*Effect of Bremsstrahlung Radiation upon Blast Wave*, Proc. X Int.Shock Tube Symp., Kyoto (1975)
8. R.W.Mac Cormak, AIAA Paper n.69, AIAA 7th Aero.Sci.Meeting(1969)

ADVANCED LIGHTWEIGHT NUCLEAR POWER SOURCES FOR LASER BEAM POWER APPLICATIONS

A. Hertzberg

Aerospace and Energetics Research Program
University of Washington
Seattle, WA 98195

INTRODUCTION

In a period of less than 20 years the power levels of continuous gas lasers have advanced from a few watts to 1 megawatt, with several types scalable to even higher powers. These advances have brought within reach one of the persistent dreams of engineers: the wireless transmission of power. While lasers are impractical for terrestrial power transmission because of the variability of weather, laser power transmission may play a revolutionary role in space. Applications range from the beaming of power from a central power satellite to other satellites to bold concepts involving the remote operation of vehicles. In particular, energy for rocket or aircraft propulsion supplied by a power plant orbiting in space is considered by the author to be one of the more important potential applications of the lasers discussed at this meeting.

A barrier which has prevented studies of potential beam power transmission applications has been the absence of energy sources for space power systems. Several years ago, serious consideration was given to large orbiting solar photovoltaic power stations which would beam microwave power to earth. These studies failed to establish the economic feasibility of space solar energy as a useful source of megawatt power, and for all practical purposes the studies have been discontinued.

At the University of Washington, technologies which might offer a significant improvement over the photovoltaics have been under study for several years. Among these, solar thermal systems involving concentrators and dynamic energy converters were found to be competitive with photovoltaics, but required thermal peak cycle temperatures where the materials technology was largely unproven. Even when working at these high peak temperatures where heat rejection temperatures are also relatively high, the mass of the conventional header and heat pipe radiator proved to be a large part of the system mass. While the lower heat rejection temperatures provided more efficient energy conversion, the radiator size and mass for fixed temperature radiators increases as $1/T^4_{rej}$.

In an attempt to reduce the mass burden of heat rejection in space power systems a space radiator has been proposed at the University of Washington which replaces the solid radiating surface of a conventional radiator by a recirculating free stream of liquid droplets. By taking advantage of the large surface to volume ratio of small droplets, the radiating element of a liquid droplet radiator (LDR) can be made many times lighter than the lightest conventional radiators (heat pipes). This creates the opportunity for using relatively low rejection temperatures to achieve high conversion efficiencies in space power systems.

The application of the liquid droplet radiator to space-based solar thermal systems does appear attractive and serves to make solar thermal systems competitive with photovoltaic systems. However, it has become apparent that the features of compact storage and deployability which are inherent in the LDR appear particularly effective when combined with high energy density sources such as nuclear reactors. This combination offers the possibility of launching multimegawatt power modules in a single shuttle package.

A design study of a large (~10 MW) nuclear power system using the LDR was undertaken by the author and students in the space design sequence of the Department of Aeronautics and Astronautics in 1982. The objective of this study was to determine whether the LDR can be combined with a high power nuclear source in such a manner as to be launched into low earth orbit by a single shuttle. The power package would then be boosted by conventional rockets to a higher stable orbit (above 400 miles) where it would be deployed and activated. This exploratory design used conservative choices of materials and technology with the exception of the as yet unproven LDR. The results indicate that it is indeed within the capability of the shuttle to launch into orbit nuclear power modules producing 10 MW of continuous electrical power. The author feels that the existence of this potential makes it proper to reexamine beam power applications which have been proposed in the past.

In the following sections of this paper we will review the proposed applications of laser power beaming for space applications, review the fundamentals of the liquid droplet radiator, and summarize the results of the design study for a nuclear multimegawatt power station incorporating the LDR.

SPACE BASED LASER PROPULSION

Rocket Propulsion

The use of lasers to heat a gas in a propulsive system has previously been suggested. Advantages are obvious in that the energy content of the propulsive fluid is no longer limited by chemical or nuclear heating. The transmission station, if placed in space, would carry the entire burden of the massive energy generation equipment. The propulsive fluid would be laser-heated to extremely high temperatures. Considerations have been made of the advantages of such systems for orbital transfer from low earth orbit to synchronous orbit or the initiation of deep space exploration missions. It is in this flight regime that high specific impulse is of paramount importance and chemical propulsion is inefficient.

Aircraft Propulsion

The concept of powering an aircraft from space, while equally bold, appears nevertheless to be possible using the current laser technology.[1,2] A synchronous power satellite of the solar or nuclear type placed in orbit at some convenient point would be able to transmit power to vehicles flying at high altitudes over a hemisphere (Fig. 1). An aircraft would only be required to carry fuel for takeoff, emergency reserve, and landing maneuvers. At cruising

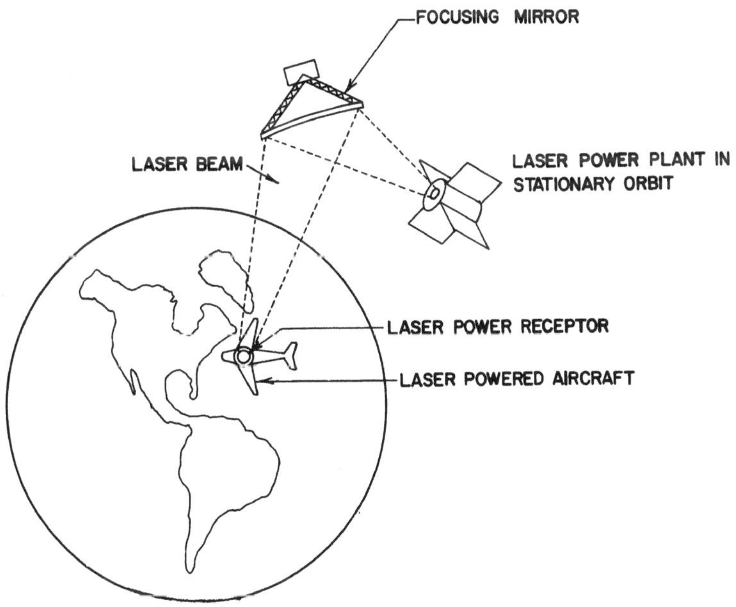

Fig. 1. Laser powered aircraft.

altitude, commensurate with present SST technology, the aircraft has nearly all the atmosphere beneath it. The remaining atmosphere offers little impediment to the laser power beam.

Current aircraft are powered by compact gas turbines driving fan propulsion devices. The propulsion thrust from the fan supplies nearly all of the required thrust for the vehicle. Even with these advanced propulsion systems, the cost of fuel for aircraft is rising to forbidding levels and concern has been expressed about the increasing quantities of exhaust products in the upper atmosphere. The fuel weight of existing high performance aircraft exceeds the payload at transcontinental ranges. By eliminating the fuel necessary to maintain cruise power and most of the climb to cruising altitude, the aircraft could accommodate a larger payload, and cruise indefinitely. Aircraft of special design and new flight strategies would have to be developed to optimize these advantages.

A design study for such a laser-powered aircraft was carried out by Hertzberg and Sun at the 1978 GCL Symposium which explored a laser powered commercial aircraft transportation system. The system employed a solar powered electrical laser system and a fleet of conventional aircraft each equipped with laser driven turbofans. The laser power satellite is deployed in a low earth, sun synchronous orbit and converts solar radiation into infrared laser radiation which is beamed to relay satellites. The relay satellites are deployed in elliptical orbits and redirect the beam towards either a cruising airplane or another relay. At the airplane, laser energy is converted into thermal energy used to drive the turbofan engines. The aircraft engines are modified, kerosene-burning turbofans with the laser powered heat exchanger placed ahead of the combustor so that either kerosene or laser radiation can be used as an energy source.

The results indicated only marginal economic advantages over a conventional kerosene-powered aircraft transportation system except at a kerosene cost above $.40/liter. A major cost item proved to be the complex optics of the relay system. Utilizing nuclear powered satellites, the relay system might be avoided, and the author plans to review this work in view of the potential of nuclear power complexes which employ the liquid droplet radiator.

LIQUID DROPLET RADIATOR

A key element in achieving a lightweight power system for space lasers is the liquid droplet radiator.[3,4] This means of heat rejection, which has been under study for the past two years at the University of Washington, offers the advantages of low mass, small storage volume, low vulnerability to damage from micrometeoroids, and adaptability to a wide range of heat rejection requirements (wide power and temperature range). These advantages arise from using a sheet of heated liquid droplets to radiate energy into space rather than relying on solid radiating surfaces which conduct a coolant.

The fundamental idea behind the LDR is to make use of the large surface to volume ratio of small droplets. The mass per area, $\rho D/6$, of a 100 μm diameter droplet of silicone oil (appropriate for the LDR because of its low vapor pressure) is 0.017 kg/m^2, that is, over 300 times smaller than the mass per area of the lightest solid surface radiators (5-10 kg/m^2). While this factor is somewhat reduced for a sheet of droplets where there is some vignetting, and although the devices for generating and collecting droplets and for pressurizing the liquid may outweigh the radiating droplet sheet itself, the LDR is still many times lighter and certainly far more compact before deployment than solid surface radiators.

The LDR consists of a droplet generator to produce multiple streams of heated liquid which comprise the radiating sheet, a collector into which the droplets are aimed and which repressurizes the liquid, a primary pump, and a heat exchanger which transfers waste energy from the power cycle to the LDR liquid. Figure 2 shows an LDR configuration which uses two radiating sheets and eliminates the need for a long return pipe. Such a pipe would have to be heavily shielded to minimize the possibility of single point failure due to puncture by micrometeoroids. By contrast, it has been shown[4] that the loss of fluid over durations of decades due to micrometeoroids passing through an LDR droplet cloud is a negligible fraction of the mass of the cloud itself.

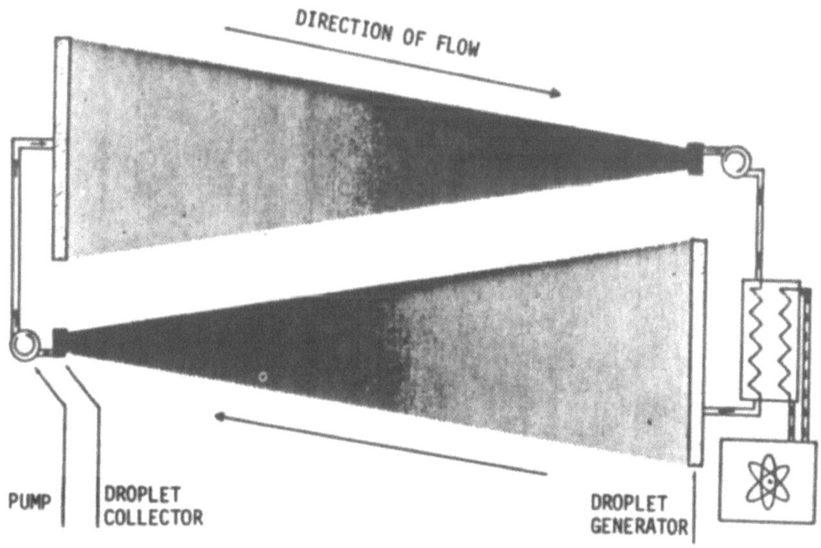

Fig. 2. Two-pass droplet radiator configuration.

The droplet generator is simply a pressurized plenum having orifices through which the liquid emerges as jets. These break up into droplets by surface tension instability. This type of device has been used to create very accurately directed streams of ink droplets <100 μm in diameter for ink jet printing,[5] and the technology has been carried further in the fusion target development program at Lawrence Livermore Laboratories. The LDR generator may be constructed as a long thin tube fitted with orifice plates and need not have a large volume.

A conceptual droplet collection device is shown in Fig. 3. The focused stream of droplets strikes a rotating antisplash target surface and the liquid migrates by centrifugal accelleration to a rotating collection drum where the rotation produces a slight pressure. The liquid is collected from the drum by an array of scoops. Although the collector is probably the least understood component of the LDR, a preliminary analysis has shown that it may comprise only a small fraction of the LDR system mass.

The choice of LDR fluids is constrained by the requirement of low vapor pressure to keep replenishment stocks to a minimum and to minimize spacecraft contamination. We have found that for operating lifetimes of ∿10 years, the vapor pressure at the peak droplet temperature should be $<10^{-7}$ Torr if evaporation losses are not to exceed the in-flight liquid mass. To avoid problems of collecting and reheating frozen droplets, the LDR medium should exist as a liquid at low pressure over a temperature range of at least 50°K. Figure 4 shows the temperature ranges of several candidate LDR liquids. The lower temperature limit is the freezing (or high viscosity) point, and the upper limit corresponds to an evaporation loss of 0.03 kg/m^2-yr. These liquids cover a range 250-1000°K, suitable for virtually any heat rejection requirement in space.

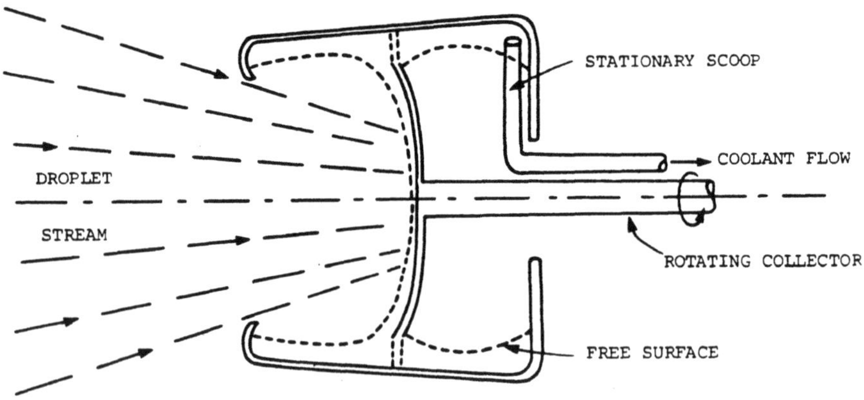

Fig. 3. Rotating droplet collector for the LDR.

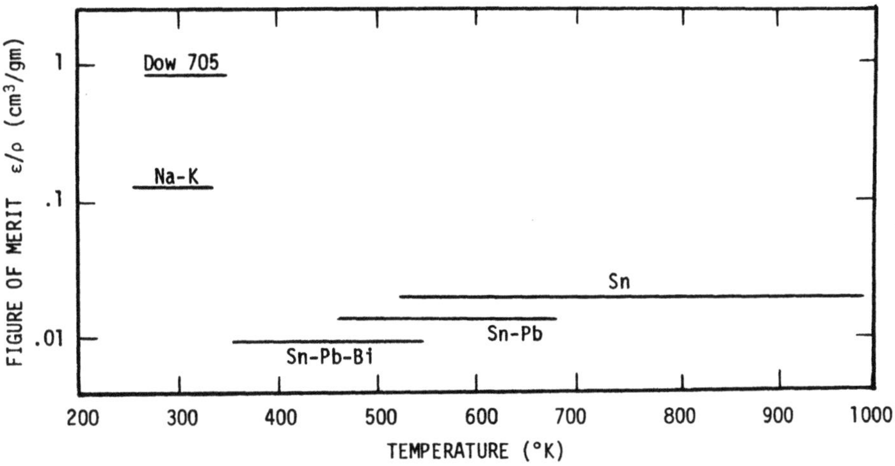

Fig. 4. Temperature ranges for LDR fluids. Figure of merit is proportional to power/mass. Dow 705 is a silicone oil.

While there are many technical issues yet to be resolved and hardware components to be developed for the LDR, this radiator does offer exceptional promise for space applications. The LDR must project at least as large a radiating area as a solid surface radiator, but the mass burden is much smaller. Thus, larger areas and consequently lower rejection temperatures may be used to improve power cycle performance and reliability.

MULTIMEGAWATT ELECTRIC NUCLEAR POWER MODULES

At the University of Washington design studies of advanced space power systems employing the LDR have been carried out.[6] Both solar thermal and nuclear systems were considered, and the nuclear-LDR combination appeared particularly attractive since each can be transported very compactly. This section describes an 8.5 MW_e nuclear-LDR design which can be transported to LEO in one shuttle launch. The basic components of this system to be discussed below consist of: high density nuclear reactor core, shadow shield, energy conversion unit, liquid droplet radiator, and deployment structure.

Reactor and Shadow Shield

The best known approach to a space nuclear power system is the SPAR reactor of Los Alamos National Laboratories. Similar approaches are currently being utilized by the USSR in several satellites. Although designs using a SPAR reactor and thermionic or thermoelectric conversion have been conceived in which a 100 kW_e system could be carried in a single shuttle launch, scaling to higher powers is limited by the extraction of heat from the reactor by heat pipes. The reactor approaches lifting capacity of the shuttle at little more than 10 MW thermal, leaving little mass or volume for the remaining components.

The limitations imposed by conductive thermal energy transport suggested reactor concepts utilizing forced convection. A reactor concept being studied by Dr. Powell and his colleagues at Brookhaven National Laboratories termed the fixed bed reactor (FBR) was examined as an example of a convectively cooled reactor. The FBR uses a cylindrical core moderator encircled by a relatively thin porous fuel bed which is in turn surrounded by a neutron reflector and pressure vessel, as shown in Fig. 5. Because of the effectiveness of convective heat transfer in this type of reactor, the FBR shows significant mass advantages over SPAR above 2 MW thermal, and was thus selected as the power source for the 8.5 MW_e design. A conventional composite shadow shield consisting of lithium hydride-tungsten was chosen to protect payload and crew. Despite the relatively small diameter of the FBR compared to SPAR, the mass for a 4π shield is too heavy to include with the remaining system in a single shuttle launch.

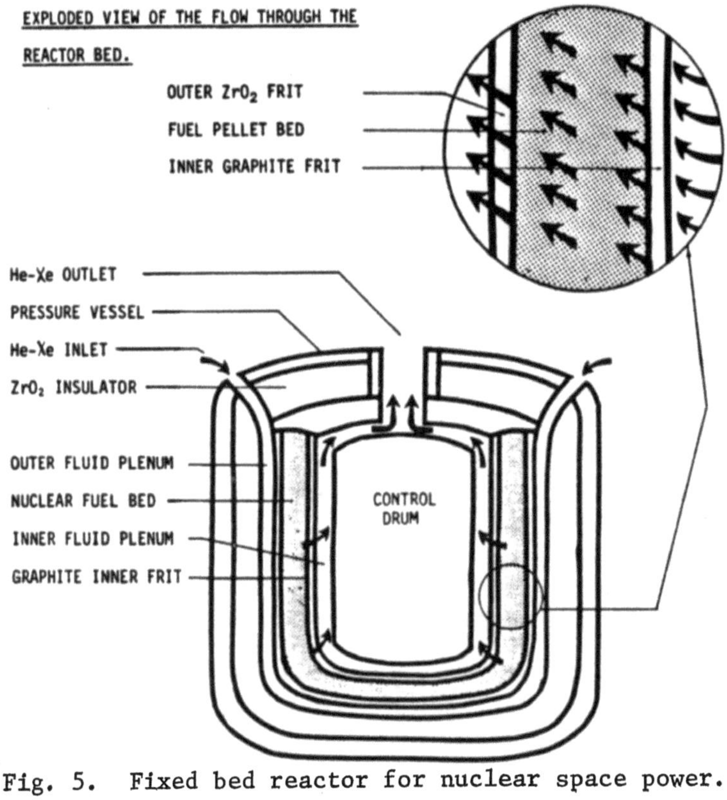

Fig. 5. Fixed bed reactor for nuclear space power.

Energy Conversion Subsystem--Brayton Cycle

Dynamic energy conversion systems were considered for the power module because of their advantages of high energy density and efficiency over thermionic or thermoelectric systems. After a preliminary examination of both the Rankine cycle and the Brayton cycle, the Brayton cycle was selected due to its ability to directly use the helium-xenon FBR coolant without an intermediate heat exchanger. In addition, the anticipated operating temperature ranges for the reactor and liquid droplet radiator are more suitable for the Brayton cycle than for the Rankine cycle.

In space power designs using Brayton cycles, regeneration is normally used due to the high heat rejection temperatures required to minimize the mass of a conventional heat pipe radiator. At the relatively low rejection temperatures possible using the LDR, however, the thermodynamic advantages of the regenerator were not found sufficient to warrant its inclusion. This appears to be a particularly important finding since the mass and volume of the regenerator of a space-based unit tend to dominate the unit. In addition, the regenerator tends to be most vulnerable to single point failure. Thus, a slightly lower cycle efficiency (23.4% vs 28.1%) was accepted to avoid this severe reliability penalty. In addition low temperature heat rejection also lent itself well to housekeeping functions related to the energy conversion system, such as alternator cooling.

Liquid Droplet Radiator and Deployment Structure

The power system utilized the LDR for heat rejection with Dow 705 silicone oil chosen as the radiating fluid. The effective radiator temperature for the LDR is 335°K and the emissivity of the droplet sheet was 0.87. With the radiator located such as to minimize the solar and earth radiation fluxes, the specific power is 0.75 kW/m^2, and the area necessary to radiate 37 MW (the maximum thermal output of the reactor) is ≃ 60,000 m^2. A solid surface radiator of such large area obviously could not be stored within the shuttle bay and it is an advantage of the LDR that the fluid can be stored in a small cannister and distributed through the system after the supporting structure is deployed.

The radiating mass of droplets must be located in a rigid frame with the basic radiator configuration as shown in Fig. 6. A two-pass system was selected with the droplets passing outward from a supporting beam traversing 300 m of space to a collector and then returning via another emitter for recycling through the power plant. The primary function of the deployment structure is to separate the respective droplet sheet generator and collector pairs and it must remain stable under thermal and mechanical loads.

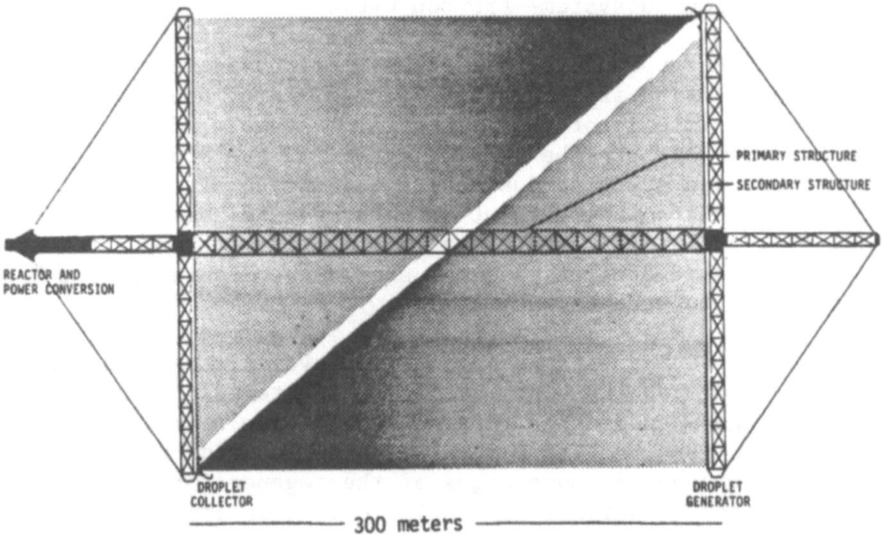

Fig. 6. Deployment structure for 8.5 MW FBR-LDR power module.

The structural system selected was a development of the collapsible truss concept conceived by the Astro Research Corporation and used in successful space flight missions such as the Pioneer missions to Jupiter. These structures are particularly lightweight, rigid and, most importantly, easily deployable. The stowed length of the truss is about 1/50 of the deployed length. The Astromast has a deep triangular truss cross section; individual members of the Astromast are articulated where the members need the longerons and battens. The joints rotate in the direction of deployment and when fully extended are self-locking and rigid under compression and bending. The materials used for this Astromast are graphite epoxy for the members and Kevlar for the wires. Both of these materials have low densities and graphite epoxy has a high modulus of elasticity and compressive strength, while Kevlar has little compressive strength and has an extraordinarily high tensile strength.

The Astromast beams are connected to form a structural frame supporting the LDR components. The frame consists of a main Astromast, four secondary Astromasts and four boom type Astromasts with supporting Kevlar cables. The frame forms a large rectangular area for the droplet sheet and the secondary masts support the LDR generators and collectors. Between the opposite stationary mast and booms, Kevlar cables are used to increase the rigidity and torsional stiffness of the entire structure. The length of the main mast is 300 m, each secondary mast 100 m, and each boom 20 m. This structure is easily deployed and can be stored very compactly. Prior to deployment the dimensions of the package (rectangular box) are 6.7 m x 2.6 m x 2.6 m. With the other components of the power module, the entire system can easily be placed in the shuttle bay of dimensions 18.3 m x 4.5 m x 4.6 m, and the system mass of 20,000 kg is also within the payload capacity of the shuttle. Table 1 gives a summary of the power system.

Table 1. Summary of 8.5 MW space power system

Component	Mass	Characteristics
Reactor	2200 kg	Fixed bed type; 37 MW thermal; T = 1100°K
Shadow Shield	500 kg	LiH-W composite; partial (not 4π)
Power Cycle	8550 kg	He-Xe Brayton; T_{max}=1100°K, T_{min}=350°K; Non-regenerative; η = 23.4%
Radiator	5300 kg	Silicone oil LDR; T_{max}=344°K, T_{min}=280°K; Emissivity = 0.87; Area = 60000 m^2
Structure	3500 kg	Deployable Astromast booms; Stored dimensions = 6.7 m x 2.6 m x 2.6 m
System	20050 kg	Within shuttle mass and volume capability

CONCLUSIONS

While this design may only be considered exploratory, it allows ample margin for error in mass and volume, and points the way to power capabilities deployable by the shuttle in space which had heretofore not been conceivable. The author is not aware of any other power system either solar or nuclear that can offer this promise. The availability of 10 MW power modules indicates that a few shuttle launches may be used to develop power complexes in space exceeding 50 MW and a high degree of reliability can be assured by additional launches containing redundant power modules. In addition, the utilization of power modules would allow a flexible output power. The technology demands are relatively modest.

Such power complexes would most certainly have a positive impact on the applications of laser beam power in space. For example, in the earlier study of a commercial aircraft transportation system utilizing lasers powered by solar energy, much of the cost owed to the large mass of the power plant, the high cost of space construction, and the complex, relatively inefficient relay system. Using an FBR-LDR power module as the direct energy source, it is possible to conceive of a far more flexible laser-powered aircraft system building on the work of Hertzberg and Sun, which may be cost effective with the present cost of kerosene. The specific combination of the liquid droplet radiator and an FBR nuclear power plant to provide energy for a dynamic energy conversion cycle appears to be a particularly attractive area of study and offers an important future potential for laser beam power transmission.

ACKNOWLEDGMENTS

The author commends the students who worked with him in the space design sequence at the University of Washington during the winter quarter of 1982 to develop the preliminary sizing of the space power system. The current design has been the specific work of Gary Beals, Julie Chin, John Day, Cary Gibson, Kurt Grevstad, and David Treiber. The author is deeply grateful for the unstinting work of Mr. Gregory Lee Larson, who served as a teaching assistant and contributed much to the development of this concept.

The work reported here is largely taken from the data base developed under NASA Grant NAG 3-16 and AFRPL Contract FO-4611-81-K-0040.

REFERENCES

1. A. Hertzberg, K.C. Sun, and W.S. Jones,"A Laser-Powered Flight Transportation System," AIAA preprint 78-1484, AIAA Aircraft Systems and Technology Conference, August 21-23, 1978.

2. A. Hertzberg, K.C. Sun, and W.S. Jones, "Laser Aircraft," Astronautics and Aeronautics, 17:41-49 (1979).

3. A.T. Mattick and A. Hertzberg, "Liquid Droplet Radiators for Heat Rejection in Space," Journal of Energy, 5:387-393 (1981).

4. A.T. Mattick and A. Hertzberg, "The Liquid Droplet Radiator--An Ultralightweight Heat Rejection System for Efficient Energy Conversion in Space," Acta Astronautica, 9:165-172 (1982).

5. L. Kuhn and R.A. Meyers, "Ink Jet Printing," Scientific American, 240:162-178 (April 1979).

6. G. Beals, J. Chin, J. Day, C. Gibson, K. Grevstad, G. Larson, D. Treiber, and A. Hertzberg, "Preliminary Design Investigation into a Nuclear-Electric Orbital Transfer Vehicle," (to be published).

HYDRODYNAMIC PERTURBATIONS IN

RADIATION-INDUCED VAPORIZATION OF METALS

A.I.Korotchenko, N.I.Popov, A.A.Samokhin,
and B.M.Zhirjakov

P.N.Lebedev Physical Institute
Academy of Sciences of the USSR
Moscow 117924

Radiation-induced vaporization of metals usually involves melt formation which can affect a laser metal interaction in several aspects. It is well known that laser drilling velocity is determined by the melt expulsion [1,2]. Hydrodynamic perturbations affect also a surface temperature T_0 and vaporization pressure behaviour which is very sensitive to the surface temperature variations.

We investigate the pressure history during vaporization of Bi and Pb by submillisecond Nd-laser pulses in air. The metal targets radius, thickness and the radiation spot radius are respectively r_0= 2 mm, h = 1.5 mm and r = = 0.8 mm. A piezoelectric transducer (r_1 = 2 mm, h_1 = 1 mm) is used for monitoring pressure behaviour.

Fig. 1 shows the pressure behaviour for Bi at different values of the radiation intensity I . If the intensity is sufficiently high the pressure $p(t)$ attains a stationary level after a shot transient interval so that $p(t)$ curves duplicate the laser pulses (Fig. 1, a-c). At lower intensities the transient interval increases and the shape of $p(t)$ curves changes appreciably (Fig. 1, e,f).

A plasma plume appears above the target surface at $I \geqslant 1.3$ MW/cm^2. It affects the pressure behaviour only slightly in the case of Bi. For Pb targets the plasma effects are more pronounced [3]. At the threshold region the plasma plume is unstable and it strongly affects the pressure behaviour (Fig. 2).

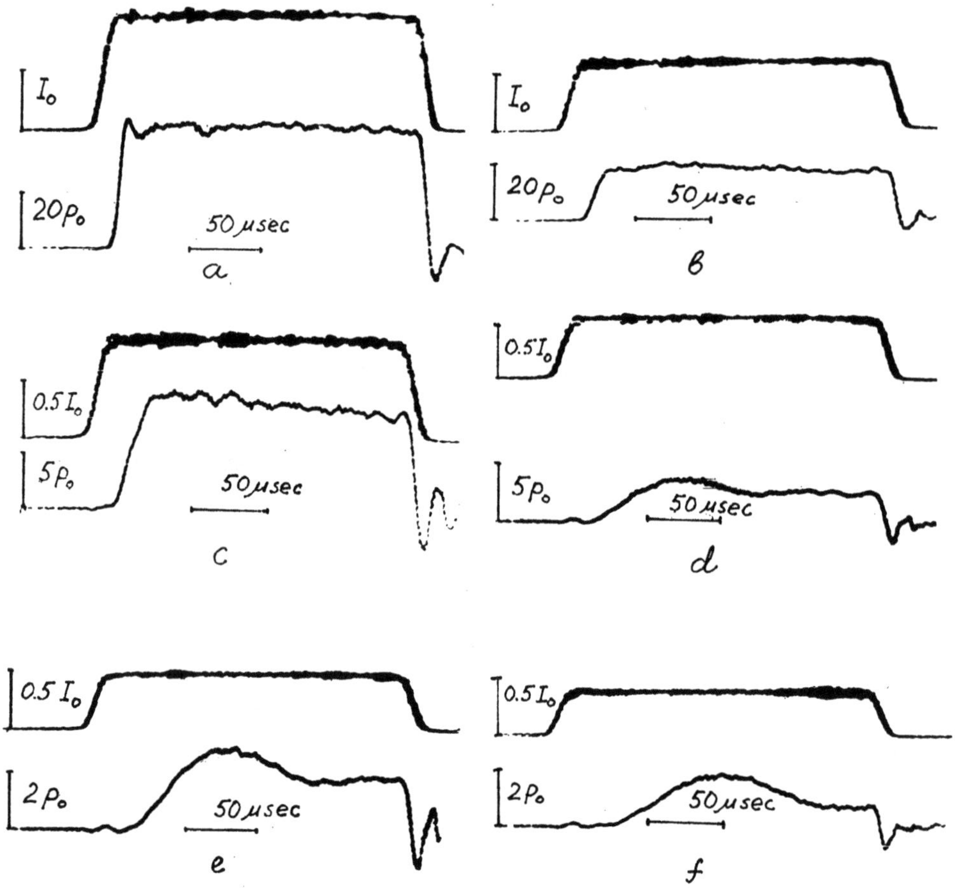

Fig.1 The laser pulses (upper curves) and transducer signals for Bi. $I_o = 1\ MW/cm^2$, $p_o = 1$ bar.

Fig.2 The pressure behaviour on Pb targets at the plasma formation threshold. $I_o = 1\ MW/cm^2$, $p_o = 1$ bar.

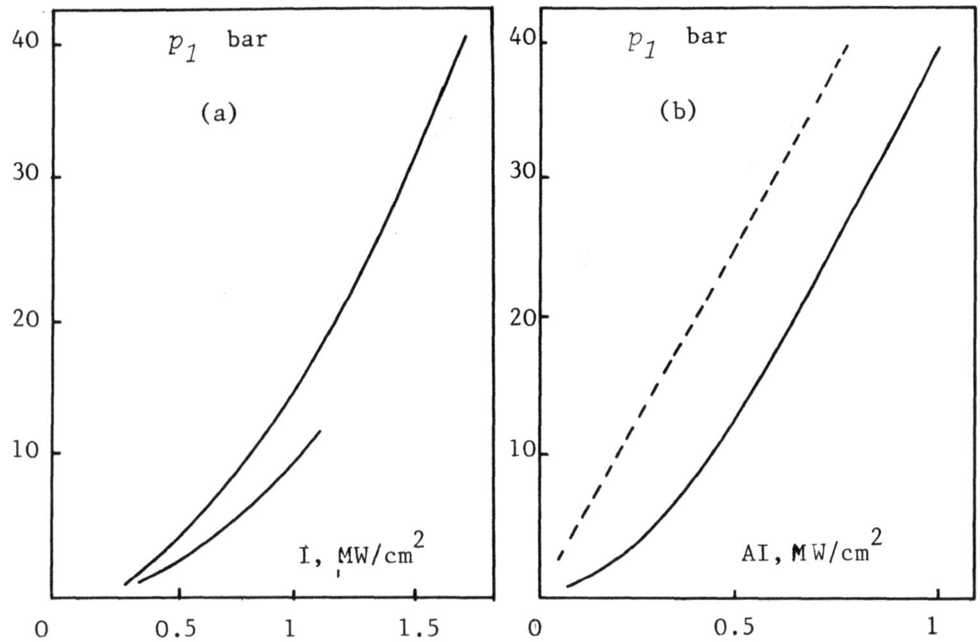

Fig. 3 Experimental (a) and theoretical (b) dependence of p on I and AI. The dotted curve gives $p(AI)$ dependence without hydrodynamic perturbations.

Below the plasma threshold pressure curves $p(t)$ are qualitatively the same for Pb and Bi though in the latter case the pressure magnitude p is somewhat higher. The pressure p depends nonlinearly on the radiation intensity I (Fig. 3a).

The nonlinear dependence of $p(I)$ disagrees with a heat-conduction model of the vaporization process where hydrodynamic perturbations are not taken into account. In this model the ratio p/I is approximately constant if one dimensional description is valid. It should be mentioned that in our case the heat diffusion length $\sqrt{\chi t}$ is small compared to τ. However, the one dimensional energy balance equation of the vaporization process is essentially modified because of the melt expulsion from the irradiated spot.

Energy losses due to radial melt flow can be approximately allowed for in the same way as in [1,2]. For the steady-state energy balance in a medium with density ρ, specific heat c, latent heat of vaporization (melting) L_v (L_m) and absorptivity A this gives

$$AI = \rho V(L_v + L_m + cT_0) + \rho V_1\left(L_m + \frac{c}{z_m}\int_0^{z_m} T\,dz\right),$$
$$V_1\tau = 2z_m V_\tau\,, \quad V_\tau = 2\sqrt{p/\rho}\,, \tag{1}$$
$$p = \rho V\sqrt{\pi kT/m}$$

where $V+V_1$ is the effective front velocity due to evaporation and melt expulsion, $p(p_s)$ is the recoil (saturation) pressure. The depth of melt layer z_m is determined by the condition that $T(z_m)$ is equal to the melting temperature T_m. We approximate the temperature distribution in the melt layer by the expression

$$T = T_0 \exp\left[-\left(V+\frac{V_1}{a}\right)\frac{z}{\chi}\right] \tag{2}$$

where the coefficient $a = 2$ is introduced to take into account the velocity distribution in z direction. From (1), (2) it follows

$$z_m = \frac{a\chi}{aV+V_1}\ln\frac{T_0}{T_m}\,, \tag{3}$$

$$V_1^2 + aVV_1 + 2aV_\tau\frac{\chi}{\tau}\ln\frac{T_0}{T_m} = 0$$

Fig. 3b shows calculated $p(AI)$ dependence for Bi. The following values are used in the calculations $\rho L_v = 14\rho L_m = 7\cdot10^3\,J/cm^3$, $\rho c = 1.45\ J/cm^3{\cdot}K$, $\chi = 0.08\ cm^2/s$, $\rho = 9.7\ g/cm^3$, $T_B = 5.76 T_m = 1560$ C, $V = 2.1$ cm/s at $p = p_B = 1$ bar and $p_s = p_B \exp[12.5(1 - T_B/T)]$. For Pb the corresponding parameters have nearly the same values and $p(AI)$ dependence is displased slightly to the higher intensities.

The calculated $p(AI)$ curve coincides within 15% with the experimental $p(I)$ dependence if $A = 0.56$. Such an agreement can be considered as satisfactory because the model used is very simple.

The results of our investigations show that the observed difference in the steady-state $p(I)$ dependence from the predictions of the heat-conduction model can be attributed to the radial melt flow effect on the energy balance in the irradiation zone. At low intensities the time-dependent $p(t)$ behaviour is more complicated which is probably connected with the increasing role of hydrodynamic perturbations in this region. Further experiments are needed to study the transient regime and to clarify the ambient pressure effect on the vaporization process.

It should be mentioned also that evaporating liquid surface can be unstable with respect to ripple-like perturbations. Hydrodynamic effects give an important contribution to the evaporation front behaviour. In particular, for the case of surface radiation absorption the plane evaporation front, which is stable for solids, becomes unstable for liquids because of the differential pressure recoil.

In the linear approximation the small-scale ripple perturbations do not contribute to the average recoil pressure p which can be affected only in nonlinear stage. The evaporating surface stability problem is rather difficult even in the linear approximation because it involves gas dynamics and kinetic aspects of the evaporation process. Evaporation regimes are studied by many authors while the instability effects are less investigated both theoretically and experimentally (see, e.g., references cited in [4]).

The photographs of Pb and Bi targets taken during the laser pulses show that the evaporating surface is disturbed by small-scale (tens of microns) perturbations. A considerable intensity reduction of the back reflected laser light is also observed at the biginning of the pressure rise. Details of the investigation will be published elsewhere.

REFERENCES

1. V.A.Batanov, V.B.Fedorov. Zh. Exp. Teor. Fiz. Pis'ma, 17, 348, 1973
2. M. von Allmen. J. Appl. Phys., 47, 5460, 1976
3. B.M.Zhirjakov, N.I.Popov, A.A.Samokhin. Zh. Exp. Teor. Fiz., 75, 494, 1978
4. A.I.Korotchenko, A.A.Samokhin, A.P.Gus'kov. Appl. Phys., A27, 121, 1982

LASER-INDUCED FLOWS IN AEROSOL

V. Ya. Korovin, S. D. Pinchuk, L. P. Semenov,
P. N. Svirkunov, and O. A. Volkovitsky

Institute of Experimental Meteorology
Obninsk, USSR

As an intensive laser radiation propagates through the turbid atmosphere, various movements can occur in the area of beam propagation, such as flow of liquid inside a drop, motion of individual aerosol particles, and induced convective flow in the medium. This paper presents the results of experimental and theoretical studies on the movements initiated during the interaction of a continuous CO_2-laser beam with drop and solid aerosols.

In the experiments, an electrodischarge slow-pumped CO_2-laser with a radiation power up to 1 kW was used. The regularities of the motion of individual aerosol particles in the radiation field were studied on freely falling water drops of micron size and on solid refractory particles.

Aerosol particles in the area of laser beam propagation move under the influence of the sum of forces differing according to their physical nature.

For water drops, one of these forces is associated with thermocapillary convection, caused by inhomogeneous heating of a drop during radiation absorption. Variations in the surface tension coefficient caused by temperature differences are responsible for the occurrence of tangential stresses and internal flow. The flow speed may be estimated on the basis of the energy balance relationship between the work of surface tension forces per unit time and the rate of kinetic energy dissipation due to water viscosity. The flow speed is assumed to be given by the expression corresponding to the Hill vortex, and Peclet number for the internal heat problem $Pe = R_d \upsilon_c / \kappa > 1$, where R_d is the drop radius, υ_c is the characteristic flow speed at the surface, and κ is the temperature conductivity of water. Under these assumptions, the speed of thermocapillary convection inside a drop is equal to

$$\upsilon_c = \left[0.3\,\frac{IK_\alpha}{\lambda\eta}\cdot\left|\frac{d\sigma}{dT}\right|\right]^{2/3}\left(\frac{\kappa R_d}{2\pi}\right)^{1/3}. \tag{1}$$

Here I is the intensity of laser radiation; K_α is the effectiveness factor of radiation absorption by a drop; T is the temperature of drop surface; σ, λ, and η are the coefficients of surface tension, heat conduction and water viscosity, respectively.

Circulation of liquid inside a drop results in velocity field disturbances in the ambient air. In this case, the drop is affected by a force due to air viscosity. This force, subsequently referred to as the thermocapillary force, gives to the drop a momentum opposite in direction to the laser beam [1]. For $I = 500\ W/cm^2$ and $R_d = 35\ \mu m$, the velocity of drop motion registered in the experiment was $\upsilon_d \simeq 10$ cm/sec. This agrees as to order of magnitude with the theoretical estimate obtained from Equation (1) and the relation $\upsilon_d = 2/3\ \upsilon_c$ valid for small Pe numbers in the environment.

The thermocapillary force $\vec{F}_t$ competes with the oppositely directed light-reactive force $\vec{F}_\ell$ which occurs due to vapor flow recoil from the nonuniformly heated surface of a drop evaporating in the radiation field. Based on a qualitative analysis of the processes of heat and mass exchange in a drop and in the environment, as well as on the results of high-speed microphotography of the behavior of drops with radius $R_d \gtrsim 10\ \mu m$ (Figure 1), the semi-empirical expressions for $\vec{F}_t$ and $\vec{F}_\ell$ were obtained in Reference 2 at the intensities of laser radiation corresponding to the diffusive-convective regime of drop evaporation ($5\cdot10^2\ W/cm^2 < I < 10^4\ W/cm^2$). The relation between I and R_d was found, as the equality of two forces studied is attained:

$$q^{5/4}(\ell^q-1) = 0.86\left[\left(\frac{R_d}{R^*}\right)^3 K_\alpha(R^*)\right]^{1/4}, \tag{2}$$

where $q = 0.55\ IR_dK_\alpha(R_d)/I^*R^*$, $I^* = 10^3\ W/cm^2$ and $R^* = 10\ \mu m$. Formula (2) gives in the plane of parameters (I,R_d) an equation of the curve (Figure 2) which demarcates the predominance regions of forces $\vec{F}_\ell$ and $\vec{F}_t$; in the area above the curve, $|\vec{F}_\ell| > |\vec{F}_t|$

At radiation intensities close to the threshold of drop fragmentation (for $R_d = 10\ \mu m$ the threshold intensity $I \sim 10^4\ W/cm^2$) the value $|\vec{F}_\ell|$ is several orders of magnitude higher than $|\vec{F}_t|$. Analysis of the experimental data [2] on the motion of drops of radius 3 to 25 μm enabled us to obtain an approximated expression for the light-reactive force at drop surface temperatures near to water boiling temperature:

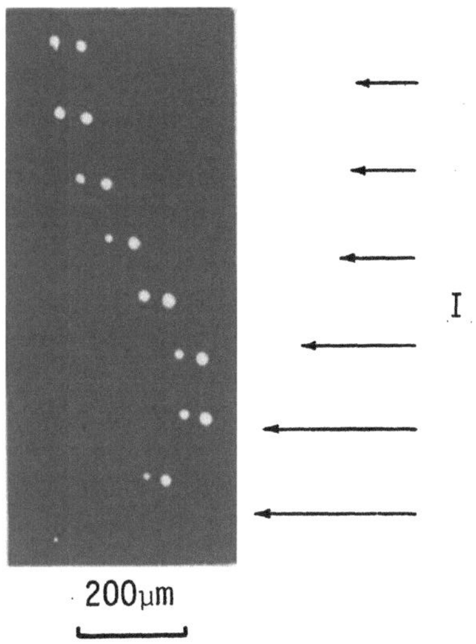

Fig. 1. Water drop motion trajectory under the influence of thermocapillary and light-reactive forces (the radiation propagates from right to left).

$$F_{\ell} = \frac{\pi}{3} \upsilon^{*} \frac{I K_{\alpha} R_d^2}{L + \varepsilon} \beta. \tag{3}$$

Here $\upsilon^{*} = 250$ cm/sec and $\varepsilon = 0.35 \cdot 10^{10}$ cm^2/sec^2 are the dimensional constants, L is the specific heat of water evaporation, $\beta = \beta(Q_i, Q_s)$, Q_i and Q_s are the radiation energy densities absorbed per unit time on the irradiated and shady sides of a drop, respectively. The dependence $\beta(R_d)$ can easily be found from the results of calculations of the optical field inside a drop performed on the bases of the Mie theory. From Equation (3) it follows that for $I \approx 10^4$ W/cm^2 the motion velocity of drops of the above size may be as great as ~ 1 m/sec under the action of the light-reactive force. The direction of drop motion with respect to the radiation source is determined by the sign β.

The action of gradient force manifests itself at high radiation intensities. In this case, water drops evaporate rapidly and, therefore, the effect of the gradient force may be adequately traced if we consider the behavior of solid refractory particles in the radiation field (Figure 3). Glowing particles fluoresce and, acquiring a velocity component transverse in direction to the radiation propagation, are ejected beyond the beam [3]. For $I \sim 5 \cdot 10^5$ W/cm^2 the maximum value of the transverse component of motion velocity for a particle of AL_2O_3 4 μm in diameter is ~ 2 m/sec.

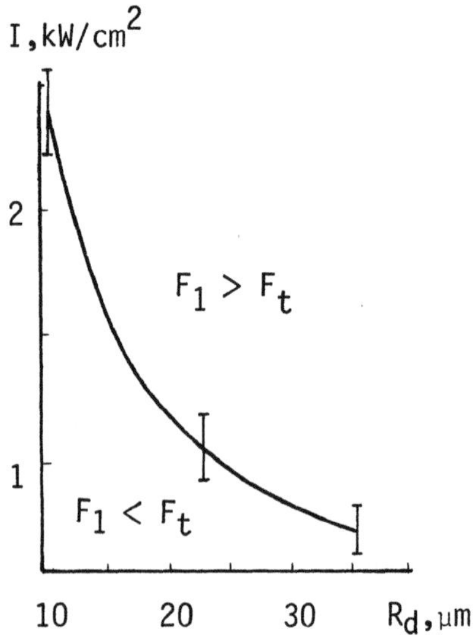

Fig. 2. Relationship between the radiation intensity and water drop radius determining the equality condition for thermocapillary and light-reactive forces.

In the area of laser beam propagation, a local thermal disturbance occurs owing to heat rejection from the heated aerosol particles, which causes the aerosol medium to move in a vertical direction under the action of Archimedes' buoyant force. The speed of convective flow induced by radiation and the character of its motion depend on the power density of heat sources and laser beam orientation.

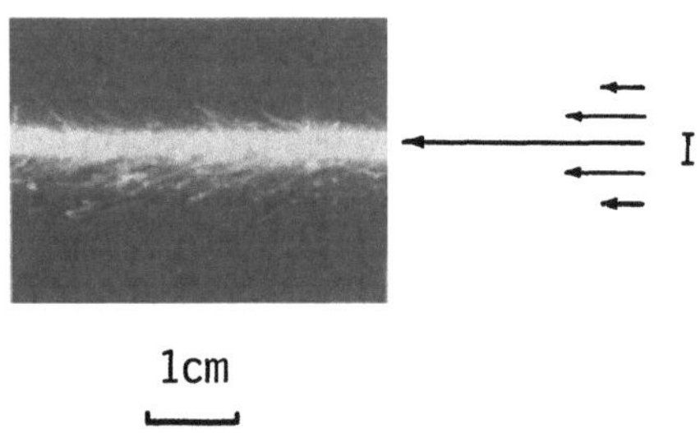

Fig. 3. Photograph of the motion of solid particles in the area of laser beam propagation with an intensity 10^6 W/cm^2 (radiation propagates from right to left).

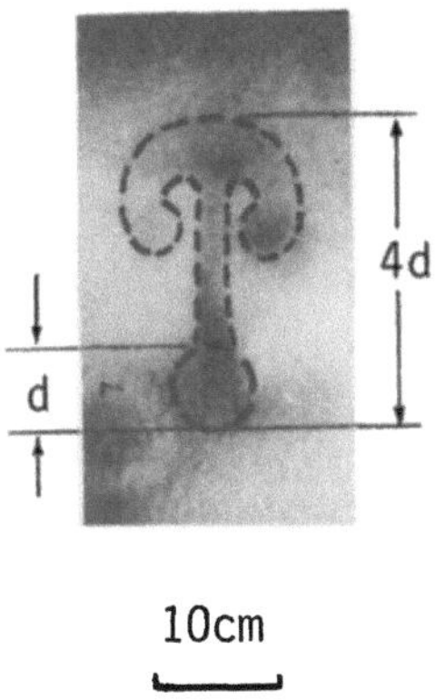

Fig. 4. The convective flow induced by radiation with an intensity 50 W/cm^2 during the clearing of a water aerosol at a concentration of drops of about 10^2 cm^{-3}.

Experiments to study convection [4-6] were carried out in a model cloud medium generated in an aerosol chamber [7]. Figure 4 shows the photograph of convection during the aerosol clearing taken from the monitor screen.

In the case of horizontal orientation of the laser beam, the dynamics of flow formation (buoyancy of the heated zone) at distances of several beam diameters d may be characterized by a quadratic function of the distance covered by the zone versus time: $z \sim t^2$. A stationary state of convection is maintained by a continuous drawing of the environment in the beam area. Measurement results for the speed of stationary convective flow (Figure 5) are adequately given by a formula similar to that for the movement of a thermal: $\upsilon \sim (\Delta T)^{\frac{1}{2}}$, where ΔT is the overheat of the zone. If the overheat is small, the characteristic time of convective heat transport beyond the beam $t_\upsilon = d/\upsilon$ is larger than that for conduction $t_\chi = d^2/4\chi$ (χ is the coefficient of air temperature conductivity), and conditions of weak convection are realized. In this case, asymmetry of the vertical temperature profile in the area of radiation propagation is insignificant: the ratio of medium temperature increments above the beam (δT_α) to those below it (δT_β) is less than 2. When $t_\upsilon <$ $< t_\chi$, conditions of developed convection are realized with $\delta T_\alpha/\delta T_\beta >$ > 2. As a parameter determining the transfer from the former conditions of convection to the latter, one may use the critical Rayleigh number for the Benard problem with free boundary conditions $R\alpha^* =$ $= \gamma g \Delta T d^3/\chi\nu = 658$, where γ and ν are the coefficients of thermal expansion and air viscosity, respectively, and g is the acceleration of gravity.

In conditions of developed convection, the flow induced by laser radiation can be either laminar or turbulent. It was found

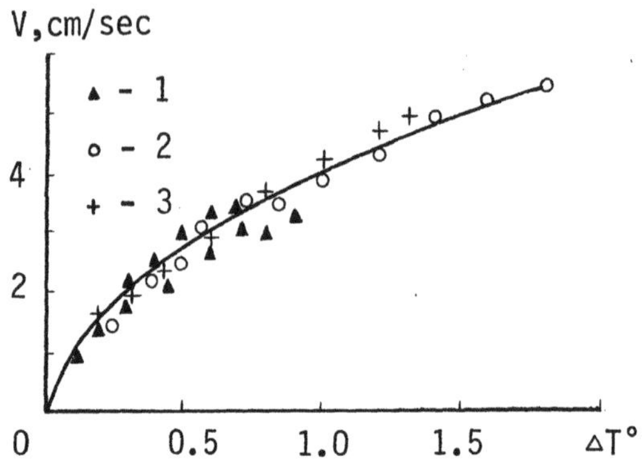

Fig. 5. Dependence of the speed of convective flow induced by laser radiation on the overheat of the zone: solid line - calculation; 1,2,3 - experimental data under various conditions:
1 - d = 1.6 cm, I = 70 W/cm^2
2 - d = 2.2 cm, I = 250 W/cm^2
3 - d = 5.0 cm, I = 50 W/cm^2.

experimentally that, at Reynolds number $Re^* = \upsilon D/\nu \sim 60$, where D is the characteristic size of the rising heated area, and Grashof number $Gr^* = \gamma g \Delta T D^3/\nu^2 \sim 4 \cdot 10^4$, the convective flow loses its stability. As a result, the pulsation component of speed occurs together with the mean one. At $Re > Re^*$ the flow induced by laser radiation is turbulent and the minimum scale of vortexes may be estimated with formula: $\ell_o = 2d(Re^*/Re)^{\frac{3}{4}}$. The degree of flow turbulence near the beam does not exceed 50%.

The exact solution for the movement velocity of a convective jet initiated by laser radiation can be obtained only in the specific case of the vertical orientation of an axisymmetric beam [8]. The initial set of equations incorporates the free convection equations in an approximation of Boussinesque and the equations which describe the process of aerosol clearing in an approximation of liquid-water content [9]. It is assumed that heat is transferred mainly by convection, that movement is laminar, the role of viscosity insignificant, and clearing almost complete. In this case, the overheat of the medium in the area of radiation propagation is

$$\Delta T = \frac{1-\beta_T}{\beta_T} \frac{L}{\rho C_p} W_o, \tag{4}$$

where β_T is the radiation energy absorbed by drops, which is expended in evaporation [9]; ρ and C_p are respectively the density and the specific heat capacity of the air; and W_o is the initial liquid-water content of the aerosol.

The expressions for the vertical w and radial u velocity components of convective jet stationary movement take the form

$$w = (2\gamma g\Delta T^* z)^{\frac{1}{2}} \cos\left[\frac{\pi}{2}\left(\frac{\zeta}{R}\right)^2\right],$$

$$u = -\left(\frac{2\gamma g\Delta T^*}{z}\right)^{\frac{1}{2}} \frac{R^2}{\pi\zeta} \sin\left[\frac{\pi}{2}\left(\frac{\zeta}{R}\right)^2\right] \tag{5}$$

at $\zeta < R$, and

$$w = 0,$$

$$u = -\left(\frac{2\gamma g\Delta T^*}{z}\right)^{\frac{1}{2}} \frac{R^2}{\pi\zeta} \tag{6}$$

at $\zeta > R$. Here ζ is the radial coordinate and R is the beam radius. It should be emphasized that Equations (5) and (6) are valid asymptotically for $z \gg R$, with the dependence $w(z) \sim z^{\frac{1}{2}}$ taking place only in the case of complete aerosol clearing. In the other extreme case of a nonevaporating (for example, solid) aerosol, it is possible to obtain the following qualitative relationships, based on considerations of dimension: $\Delta T(z) \sim z^{1/3}$ and $w(z) \sim z^{2/3}$.

When analyzing the dynamics of convection development, we have restricted ourselves to the discussion of the vertical velocity component of the jet $\bar{w}$ averaged over the beam cross-section:

$$\bar{w} = \gamma g\Delta T^* t\,\theta(z - \gamma g\Delta T^* t^2) + (\gamma g\Delta T^* z)^{\frac{1}{2}}\theta(\gamma g\Delta T^* t^2 - z), \tag{7}$$

where $\theta(x)$ is the Heaviside function. As seen from (7), the behavior of $\bar{w}$ differs substantially over the areas, demarcated by the boundary $z^* = \gamma g\Delta T^* t^2$. In the first of these ($z < z^*$) movement is steady, and $\bar{w}$ is dependent on z according to Equation (5). In the second area $z > z^*$, movement is uniformly accelerated, indicating that the heated air column rises by itself, without involving the environment.

In conclusion it should be noted that convection induced by intensive laser radiation may prevent fog dissipation (or hole-boring). On the other hand, this effect can be useful in the removal of aerosol pollutants from limited volumes.

REFERENCES

1. E. V. Ivanov, M. P. Kolomeev, N. K. Kraskovsky, P. N. Svirkunov and L. P. Semenov, Motion of drops in the field of laser radiation, Proc. IEM., Issue 11 (54):19 (1975).
2. E. V. Ivanov, V. Ya. Korovin and Yu. S. Sedunov, Motion of optically dense drops of liquid in the field of laser radiation, Kvantovaja Electronika, Vol. 4:1873 (1977).

3. S. V. Zakharchenko, S. D. Pinchuk and A. M. Skripkin, On the nonlinear propagation of laser radiation in a solid aerosol, Kvantovaya Electronika, Vol. 5:934 (1978).
4. N. K. Didenko, N. K. Kraskovsky and L. P. Semenov, Experimental study of the thermal characteristics of cloud medium in the zone exposed to CO_2-laser radiation, Proc. IEM., Issue 5(43): 92 (1974).
5. O. A. Volkovitsky, N. K. Didenko and S. D. Pinchuk, Experimental study of the convection initiated in fog by a CO_2-laser beam, IEM., Issue 18 (71):94 (1978).
6. L. G. Akulshina and S. D. Pinchuk, On the fog clearing at free convection, Trans. All-Union Conf., Prop. Opt. Rad., Moscow, 239 (1978).
7. O. A. Volkovitsky, Complex of experimental installations for geophysical research, Meteorologiya i Gidrologiya, No. 6:46 (1965).
8. P. N. Svirkunov, Convection induced by thermal action of vertical laser beam on cloud medium, Isv. AN SSSR., Physica Atmosphery i Okeana, Vol. 15:907 (1979).
9. A. V. Korotin, D. E. Svetogorov, Yu. S. Sedunov and L. P. Semenov, Formation of clearance zones in clouds and fogs, Doklady AN SSSR., Vol. 220:829 (1975).

COHERENT PUMP DYNAMICS AND PULSE EVOLUTION IN THREE-LEVEL SUPERFLUORESCENCE AND CONTROL OF LIGHT BY LIGHT

F. P. Mattar

Polytechnic Institute of New York, Brooklyn, NY 11201; and Massachusetts Institute of Technology, Cambridge, MA 02139*

and

C. M. Bowden

Research Directorate, US Army Missile Laboratory, US Army Missile Command, Redstone Arsenal, AL 35898

ABSTRACT

Results of a model calculation are presented for the dynamical evaluation of superfluorescence from a coherently-pumped three-level system with the full propagational, transverse and diffraction effects taken into account. It is demonstrated by two examples that the injected coherent pump pulse initial temporal and radial width variations can have significant deterministic effects on the superfluorescence pulse delay time and intensity optimization. Thus, by specifying certain initial properties of the injected pump pulse, the superfluorescent pulse can be shaped and altered. These results predict some of the conditions under which an injected light pulse of a given frequency can be used to shape and control a second light pulse of a different frequency via a nonlinear medium, thus demonstrating the phenomenon of light control by light.

* This work partially supported by the US Army Research Office; Battelle Columbus Laboratories; and the Office of Naval Research.

INTRODUCTION

Superfluorescence[1] (SF), is the dynamical radiation process which evolves from a collection of atoms or molecules prepared initially in the fully inverted state, and which subsequently undergoes collective, spontaneous relaxation[2]. Since Dicke's early work[2], much theoretical and experimental effort has been devoted to this subject[3].

With the exception of the more recent work of Bowden and Sung[4], all theoretical treatments have dealt exclusively with the relaxation process from a prepared state of complete inversion in a two-level manifold of atomic energy levels, and thus do not consider the dynamical effects of the pumping process. Yet, all reported experimental work[5-10] has utilized optical pumping on a minimum manifold of three atomic or molecular energy levels by laser pulse injection into the nonlinear medium, which subsequently superfluorsces.

It was pointed out by Bowden and Sung[4] that for a system otherwise satisfying the conditions for SF emission, unless the characteristic SF time[1], τ_R, is much greater than the pump pulse temporal duration, τ_W, i.e., $\tau_R >> \tau_W$, the pump dynamics can have a dramatic effect[4]. This is a condition which has not been realized over the full range of reported data.

In this paper, we present several calculational results and associated analysis for the effects of coherent pump dynamics, propagation, transverse and diffraction effects on SF emission from an optically-pumped three-level system. The full, nonlinear, copropagational aspects of the injected pump pulse, together with the SF which evolves are explicitly treated in the calculation. Not only do our results relate strongly to previous calculations and experimental results in SF, but we demonstrate a new manifestation of light control by light. We show how characteristics of the SF can be controlled and shaped by specifying certain initial characteristics of the injection pulse in the regime where $\tau_R/\tau_W < 1$.

The model upon which the calculation is based is comprised of a collection of identical three-level atoms, each having the energy level scheme shown in Figure 1, such that the $1 \leftrightarrow 3$ transition is induced by a coherent electromagnetic field injection pulse of frequency ω_0 tuned to the transition. The properties of this pumping pulse are specified initially. The transition $3 \leftrightarrow 2$ evolves by spontaneous emission at a much lower frequency ω. It is assumed that the energy level spacing is such that $\varepsilon_3 > \varepsilon_2 >> \varepsilon_1$ so that the fields at frequencies ω_0 and ω can be treated by separate wave equations. The energy levels $2 \leftrightarrow 1$ are not coupled radiatively due to parity considerations, and spontaneous relaxation from $3 \leftrightarrow 2$ is simulated, in the semi-classical approximation, by the choice of a small, uniform initial transverse polarization[11] characterized by the parameter $\phi_0 \sim 10^{-3}$. This calculation, therefore, does not deal with the statistics of the quantum initiation process[12], which results in temporal fluctuations in the results. Consequently, the results reported here are regarded as expectation values from the ensemble whose average spontaneously-induced transverse polarization is characterized by ϕ_0. Our results, at least qualitatively, do not depend upon nominal variations of this parameter. The initial condition is chosen consistent with the particular choice of ϕ_0, with nearly all the population in the ground state, and the initial values of the other atomic variables chosen consistently[4,13] according to the initial equilibrium properties of the system[11]. The full statistical treatment of the quantum initiation process[14] with resulting temporal fluctuations[15] will be presented in future developments.

We use the electric dipole and rotating wave approximations and couple the atomic dipole moments to classical field amplitudes which are determined from Maxwell's equations, which include transverse contributions. The equations of motion which describe the field matter interaction for this system comprising N atoms, with longitudinal and radial spacial distribution, are presented elsewhere[16,17]. The matter equations of motion which describe the atomic system are coupled to Maxwell's equations through the off-diagonal matrix elements associated with each transition field. The details of the calculation are presented in detail in our previous work[17]. Transverse effects are included in the calculation and diffraction is explicitly taken into account by the radial boundary condition that $r = r_{max}$ corresponds to completely absorbing walls. This computation is quite similar, in principle, to that reported for transverse effects for SF in a two-level atomic medium[18] (i.e., without coherent pump dynamics).

Calculational methods applied to this model[16,17] and discussed elsewhere[19,20] were used to compute the effects on SF pulse evolution for various conditions for the injection signal such that $\tau_R << \tau_p$, thus demonstrating control of the SF signal by control of the input signal initial conditions[4]. Some examples follow.

In Figures 2 and 3 are shown results of the numerical calculations for the radially integrated intensity profiles vs. retarded time, $\tau = t - z/c$, for the co-propagating SF and injected pulses at equal penetration depth in the nonlinear medium for a range of values for the injected pulse initial characteristics. These profiles correspond to what would be observed with a wide aperture,

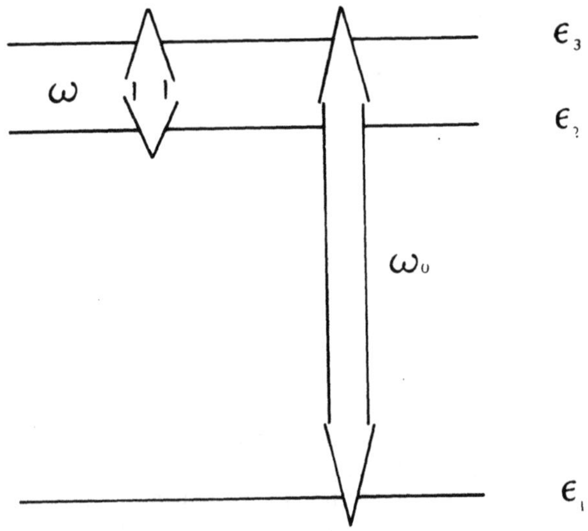

Figure 1. Model three-level system and electric field tunings under consideration. For the results reported here, the injected pulse is tuned to the 1 ↔ 3 transition.

fast, energy detector. The pumping pulses are labeled by capital letters and the corresponding SF pulses are labeled by the corresponding lower case letters. Each set of curves represents a different set of values for the initial characteristics of the injected pulse as indicated, with all other initial parameters remaining invariant among the different sets of curves for each figure.

The initial radial width, r_p , of the Gaussian injected pulse was varied and the effect upon the SF pulse evolution is shown in Figure 2. There is clearly indicated an optimum value for r_p for which the SF peak intensity is a maximum and the SF temporal width τ_s is a minimum. From the figure it is seen that optimization occurs for a value for the conventional Fresnel number F_s, for the transition, $F_s \approx 1$, where

$$F_s = \frac{2\pi r_p^2}{\lambda_s z} \tag{1}$$

and λ_s is the wavelength for the SF transition and z is the penetration of the excitation into the medium corresponding to Figure 2.

The quantity which determines the contribution from transverse effects in Maxwell's equation[17] is the "gain length" Fresnel number $\mathcal{F}_s$,

$$\mathcal{F}_s = \frac{2\pi r_p^2}{\lambda_s g^{-1}} \qquad (2)$$

where g is the on-axis effective gain for the SF transition. It is seen that (1) and (2) are related by the total gain, gz ,

$$\mathcal{F}_s / F_s = gz \quad . \qquad (3)$$

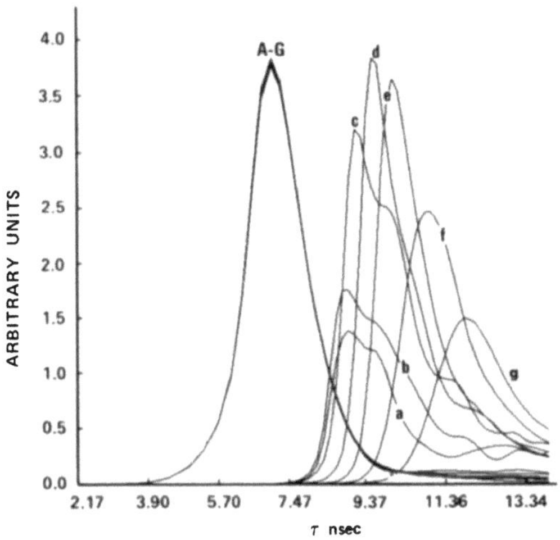

Figure 2. Radially integrated normalized intensity profiles for the SF and injected pulses at $z = 5.3$ cm penetration depth for seven different values for the injected pulse initial radial width at half-maximum r_p. The SF pulses are indicated by lower case letters, whereas the corresponding injected pump pulses are labeled by the corresponding upper case letters. The injection pulses are initially radially and temporally Gaussian with initial on-axis area $\theta_p = 2\pi$ with temporal widths $\tau_w = 4$ nsec. The level spacings are such that $(\varepsilon_3 - \varepsilon_1)/(\varepsilon_3 - \varepsilon_2) = 126.6$. The relaxation and dephasing times are taken as identical for all transitions, and are given as $T_1 = 80$ nsec and $T_2 = 70$ nsec, respectively. The SF effective gain $g_s = 758.3\ cm^{-1}$ and the pump transition effective gain $g_p = 14.6\ cm^{-1}$. The initial radial widths at half-maximum for the injected pulses are: a) $r_p = 0.57$ cm; b) $r_p = 0.43$ cm; c) $r_p = 0.24$ cm; d) $r_p = 0.18$ cm; e) $r_p = 0.15$ cm; f) $r_p = 0.11$ cm; g) $r_p = 0.09$ cm. The corresponding geometrical Fresnel numbers are: a) $F = 8.46$; b) $F = 4.79$; c) $F = 1.47$; d) $F = 0.85$; e) $F = 0.57$; f) $F = 0.35$; g) $F = 0.21$.

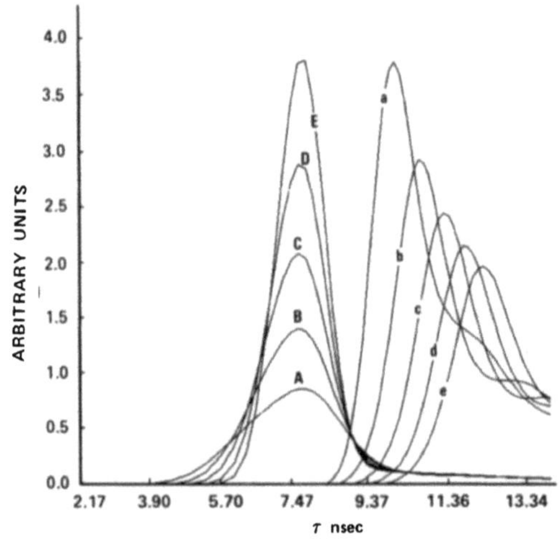

Figure 3. Radially integrated normalized intensity profiles for the SF and injected pulses at z = 5.3 cm penetration depth for five different values for the initial temporal width of the injected pulse. The initial on-axis area of the injected pulse $\theta_p = \pi$ and the pump transition and SF effective gains are $g_p = 17.5\ cm^{-1}$ and $g_s = 641.7\ cm^{-1}$, respectively. All other parameters are the same as those for Figure 2. The injected pulse initial temporal widths at half-maximum are: A) τ_W = 4 nsec; B) τ_W = 3.3 nsec; C) τ_W = 2.9 nsec; D) τ_W = 2.5 nsec; E) τ_W = 2.2 nsec.

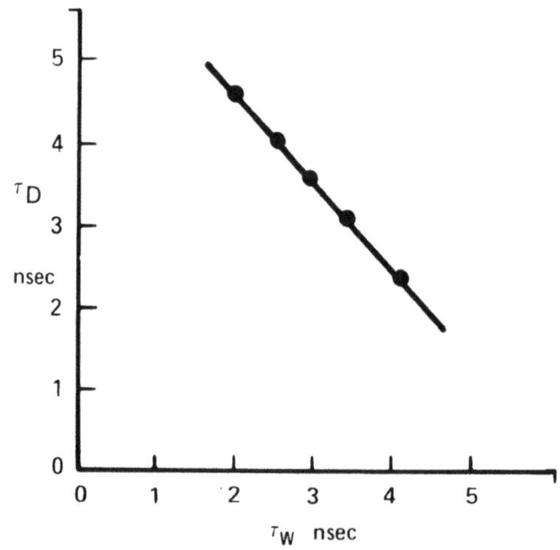

Figure 4. Delay time τ_D of the SF peak intensity from the corresponding pump pulse peak intensity vs. the pump pulse initial full temporal width at half-maximum τ_W according to Figure 3.

From Figure 2, the SF intensity optimizes for $F_s \approx 1$ and (3) becomes,

$$\mathcal{F}_s = g\, z_{max} \tag{4}$$

for the gain-length Fresnel number. Since $F_s \sim 1/z$, the implication is that Eq. (4) gives the penetration depth, z_{max}, at which the SF peak intensity reaches a maximum in terms of the ratio $\mathcal{F}_s/g$. Since this takes both transverse and diffraction explicitly into account as well as propagation, this is indeed a profound statement.

Further insight into the implication of (4) can be obtained by considering a one-spacial dimension analogy. If the linear field loss is taken to be entirely due to diffraction, then the one-dimensional linear loss κ_s corresponding to the two-dimensional case specified by $\mathcal{F}_s$, is given by

$$\kappa_s = \lambda_s/2\pi r_0^2 \quad . \tag{5}$$

Then, from (2)

$$\mathcal{F}_s = g/\kappa \tag{6}$$

is the effective gain-to-loss ratio. From the condition (4),

$$z_{max} = (\kappa^{-1}) \quad , \tag{7}$$

i.e., z_{max} is the penetration depth at which the SF peak intensity is a maximum and corresponds to one effective diffraction length, as defined by (5).

Figure 3 shows the effect upon the SF pulse of variation in the initial temporal width at half-maximum intensity for the pumping pulse. As the initial temporal width of the injected pulse τ_W becomes smaller, the SF delay time τ_D (the temporal separation between the pump intensity peak and the SF intensity peak), increases, whereas the peak SF intensity decreases and the SF temporal width τ_s remains very closely fixed.

It is clear from these results that there exists an approximate linear relationship between the time delay τ_D, between the peak SF intensity and the corresponding pump pulse intensity, and the initial temporal width τ_W of the pump pulse.

This linear relationship is shown in Figure 4, where the time delay τ_D is plotted vs. the corresponding pump pulse initial temporal width, from Figure 3. These results generate the following formula[4] for τ_D as a function of τ_W in one parameter ϕ_0:

$$\tau_D = 0.375\,\tau_R \left[\log\,(4\pi/\phi_o)\right]^2 - 4\tau_R\gamma_\perp(\gamma_R/4\gamma_\perp - 1)\tau_w \quad (8)$$

where[21]

$$\tau_R = \frac{2\,T_2}{g_s z} \quad (9)$$

and $\phi_o \approx 10^{-8}$ represents the average quantum initiation[14].

CONCLUSIONS

We have shown here two ways of shaping the SF pulse by controlling corresponding properties of the injection pulse in coherent optical pumping on a three-level system, where propagation, transverse effects and diffraction are precisely taken into account. We have thus demonstrated calculationally several[14] dynamical nonlinear effects of coherent optical pumping on SF pulse generation as well as the nonlinear control of light with light via a nonlinear medium for the regime where $\tau_w/\tau_R > 1$.

The initial process for SF pulse generation was treated in this calculation completely classically and non-statistically. Therefore, these results must be taken as expectation values, or ensemble averages. These results for a three-level system can thus be viewed as a logical extension of similar calculational results for a two-level system contained in reference 18. The statistical treatment with quantum initiation will be treated in future publications[14,15].

ACKNOWLEDGEMENT

The programming assistance of Y. Claude and M. Cormier is gratefully acknowledged.

REFERENCES

1. R. Bonifacio and L. A. Lugiato, Phys. Rev. A11, 1507 (1975); A12, 587 (1975).

2. R. H. Dicke, Phys. Rev. 93, 99 (1954).

3. See papers and references contained in Cooperative Effects in Matter and Radiation, edited by C. M. Bowden, D. W. Howgate and H. R. Robl, (Plenum, New York, 1977).

4. (a) C. M. Bowden and C. C. Sung, Phys. Rev. A18, 1558 (1978); (b) Phys. Rev. A20, 2033 (1979).

5. N. Skribanowitz, I. P. Herman, J. C. MacGillivray and M. S. Feld, Phys. Rev. Lett. 30, 309 (1973).
6. H. M. Gibbs, Q. H. F. Vrehen and H. M. J. Hickspoors, Phys. Rev. Lett. 39, 547 (1977).
7. Q. H. F. Vrehen, in Cooperative Effects in Matter and Radiation, edited by C. M. Bowden, D. W. Howgate and H. R. Robl, (Plenum, New York, 1977), p. 79.
8. M. Gross, C. Fabre, P. Pillet and S. Haroche, Phys. Rev. Lett. 36, 1035 (1976).
9. A. Flusberg, F. Mossberg and S. R. Hartmann, in Cooperative Effects in Matter and Radiation, edited by C. M. Bowden, D. W. Howgate and H. R. Robl, (Plenum, New York, 1977), p. 37.
10. A. T. Rosenberger and T. A. DeTemple, Phys. Rev. A24, 868 (1981).
11. J. R. R. Leite, R. S. Sheffield, M. Ducloy, R. D. Sharma and M. S. Feld, Phys. Rev. A14, 1151 (1976).
12. D. Polder, M. F. H. Schuurmans, Q. H. F. Vrehen, Phys. Rev. A19, 1192 (1979); F. Haake, H. King, G. Schroeder, J. Haus, R. Glauber, Phys. Rev. A20, 2047 (1979).
13. F. T. Hioe and J. H. Eberly, Phys. Rev. Lett. 47, 838 (1981).
14. C. M. Bowden and C. C. Sung, "Initiation of Superfluorescence in Coherently-pumped Three-level Systems", to be published.
15. F. P. Mattar and C. M. Bowden, to be published.
16. F. P. Mattar and C. M. Bowden, in Topics in Current Physics: Photon Dissociation of Polyatomic Molecules, edited by C. D. Cantrell (Springer-Verlag, 1982), to appear.
17. F. P. Mattar and C. M. Bowden, "Coherent Pump Dynamics, Propagation, Transverse and Diffraction Effects in Three-level Superfluorescence and Control of Light by Light"- Phys. Rev. A, to be published.
18. F. P. Mattar, H. M. Gibbs, S. L. McCall and M. S. Feld, Phys. Rev. Lett. 46, 1123 (1981).
19. F. P. Mattar, in Optical Bistability, edited by C. M. Bowden, M. Ciftan and H. R. Robl (Plenum, New York, 1981), p. 503; in Proceedings, Tenth Simulation and Modeling Conference, Pittsburgh, 1978, edited by W. Vogt and M. Mickle, Publ. Inst. Soc. Am. (1979), Pittsburgh, PA.
20. F. P. Mattar and M. C. Newstein, in Cooperative Effects in Matter and Radiation, edited by C. M. Bowden, D.W. Howgate and H. R. Robl (Plenum, New York, 1977), p. 139.
21. R. Friedberg and S. R. Hartmann, Phys. Rev. A13, 495 (1976).

EVALUATION OF LASER BEAM PROFILE UNIFORMITY BY MEANS OF A COMPUTERIZED THERMOGRAPHIC METHOD

R. Monti and C. Mamone Capria

Umberto Nobile Institute of Aerodynamics
University of Naples
Naples, Italy

Abstract

A method is proposed for the measurement of laser beam power profiles by means of a thermographic system which has been designed, realized and tested. The calibrations performed show that the method has good space resolution and measurement accuracy. Preliminary tests are presented which simulate the radiation flux of a laser beam and show the accuracy of the entire system capable of an almost on-line measurement of power distribution in a laser beam.

NOTATION

a_ℓ	plate absorptivity
c	heat capacity
G	thermograph output
h	convective heat transfer coefficient
k	thermal conductivity
L	scale for q distribution
n	coordinate orthogonal to plate surface
q	radiative heat flux
s	plate thickness
t	time
$T(x,y,n,t)$	plate temperature
T	ambient temperature
x,y	coordinates in the thin plate plane
α	thermal diffusivity
β	dimensionless parameter (11)
γ	dimensionless parameter (11)
ε	emissivity coefficient

θ	temperature defined in (10)
θ_c	temperature defined in (10)
ρ	plate density
τ	characteristic time defined in (8)

The overbar indicates mean values.

Subscripts

a	ambient	c	during cooling phase
ℓ	of the laser beam	n	normal to the plate surface
0	surface n = 0	s	surface n = s
t	along plate surface plane		
1	the first picture	2	the second picture

INTRODUCTION

Knowledge of laser beam power profiles is of importance for many laser applications and may give an insight into a number of processes occurring in the laser cavity. In laser applications, phenomena like atmospheric absorption or thermal blooming are of interest and could be evaluated by measuring beam power profiles at different stations along the beam axis. Therefore, the need arises for a reliable and accurate experimental method for the evaluation of both qualitative and quantitative laser beam power profiles.

In the present paper a method is proposed for the measurement of laser beam power density profiles, based on the elaboration of thermographic pictures of a thin plate on which the laser beam impinges. The temperature increase of the plate in a given time interval, when properly corrected, is proportional to the local radiative heat flux.

Numerical tests have supplied the necessary information on the most appropriate choice for the thermophysical properties of the plate material; for the plate thickness and for the best experimental procedure. Based on these indications an experimental method has been devised, designed, realized and tested. The experimental equipment consists of a thermographic system capable of computing the local radiative heat fluxes impinging on a thin plate.

A calibration of the hardware and of the software is performed by means of a known temperature source located at different distances from a flat sheet of known thickness and material properties [1]. The calibration gives an excellent agreement with the numerical results expected and shows both the feasibility and the accuracy of the proposed method. Finally, preliminary tests are presented which simulate the radiation flux from a laser beam.

THEORETICAL MODEL

A solid thin flat plate of thickness s is exposed to a radiative heat flux on one of its surfaces; the temperature distribution inside the plate $T(x,y,n,t)$ is governed by the conductive heat transfer equation which, by separating the contributions of the fluxes normal to the plate (direction n) and along the plate surface (direction x,y) (see Figure 1), reads:

$$\rho c \frac{\partial T}{\partial t} = k_n \frac{\partial^2 T}{\partial t} + k \, \nabla_t^2 T, \tag{1}$$

where ρc is the volumetric heat capacity, k and k_n are the two heat conductivities (in the plane x,y and along the normal n) and

$$\nabla_t^2 T = \frac{\partial^2 T}{\partial x^2} + \frac{\partial^2 T}{\partial y^2} .$$

Let us assume that the plate material properties are uniform; in a small temperature range ($10 \div 20^\circ$C) we may also assume that the thermophysical properties ($\rho c, k_n, k$) remain constant so that, by integrating Equation (1), we get:

$$\rho c \int_o^s \frac{\partial T}{\partial t} \, dn = k_n \left(\frac{\partial T}{\partial t}\right)_s - k_n \left(\frac{\partial T}{\partial t}\right)_o + k \int_o^s \nabla_t^2 T \, dn. \tag{2}$$

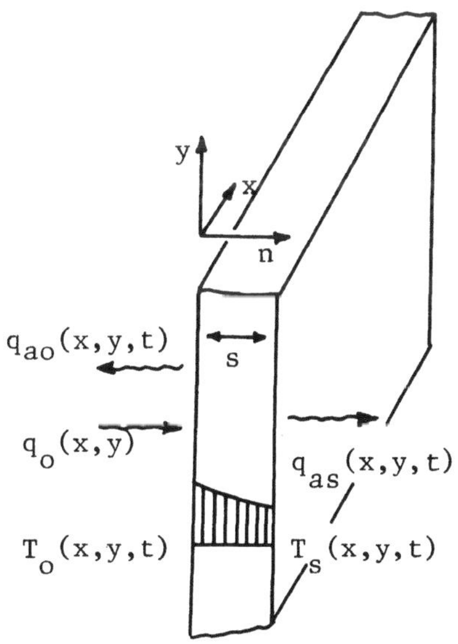

Fig. 1. Thin plate coordinates, heat fluxes and temperature distribution.

Boundary conditions on the two surfaces relate heat fluxes to the temperature gradients normal to the surfaces ($n = 0$, $n = s$)

$$n = 0 \qquad q_o(x,y,t) = -k_n\left(\frac{\partial T}{\partial n}\right)_o$$

$$n = s \qquad q_s(x,y,t) = k_n\left(\frac{\partial T}{\partial n}\right)_s . \tag{3}$$

The total net heat flux ($q_o + q_s$) exchanged between the ambient and the plate can be expressed by:

$$q_o + q_s = \rho c\int_o^s \frac{\partial T}{\partial t}\, dn - k\int_o^s \nabla_t^2 T\, dn. \tag{4}$$

The numerical results of Section 2 show that, if the temperature time derivatives are taken at times $> s^2/2\alpha_n$, one gets [1]:

$$\frac{1}{s}\int_o^s \frac{\partial T}{\partial t}\, dn = \frac{\partial \overline{T}}{\partial t} \cong \frac{\partial T_s}{\partial t} ,$$

which allows the first term in Equation (4) to be measured by a thermograph looking at the thin plate back surface.

The total fluxes at the plate surfaces are formed by the laser beam radiation q_ℓ (to be measured) and by the radiative and convective exchanges with the ambient (q_a)

$$q_o = a_\ell q_\ell + q_{ao}; \qquad q_s = q_{as}$$

$$q_{ao} = \varepsilon_o\sigma(T_o^4 - T_a^4) + h_{co}(T_o - T_a) \simeq h_o(T_o - T_a)$$

$$q_{as} = \varepsilon_s\sigma(T_s^4 - T_a^4) + h_{cs}(T_s - T_a) \simeq h_s(T_s - T_a),$$

where $a_\ell q_\ell$ is the amount of laser beam power absorbed at the plate, ε are the emissivities of the surface and h_c the convective heat transfer coefficient, h_o and h_s are the two total heat transfer coefficients.

In order to evaluate q_ℓ from Equation (4) by measurement of the plate temperature time variation it is necessary to find an experimental procedure and data elaboration able to take into account

(a) the presence of q_{ao}; $q_{as} \neq 0$

(b) a non negligible value of $k\int_o^s \nabla_t^2 T\, dn$ (which would be difficult to measure).

Suppose that after a heating phase the laser is shut-off (or better, its beam is diverted out of the calorimetric plate surface) at a time $t = t_{co}$; the plate will now cool off towards the equilibrium ambient temperature at a rate which depends on the heat fluxes q_{ao}, q_{as}.

Let us write the integrated heat balance equations during the heating and the cooling off phase (subscript c):

$$\rho c s \frac{\partial T_s}{\partial t} = a_\ell q_\ell - 2h\left(\frac{T_s+T_o}{2} - T_a\right) + ks\nabla_t^2 T_{os} \tag{5}$$

$$\rho c s \frac{\partial T_c}{\partial t} = -2h_c(T_c - T_a) + ks\nabla_t^2 T_c, \tag{6}$$

where $T_{os} \simeq (T_o+T_s)/2$. We assume that T_a = const and that the total heat transfer coefficients are the same during the heating and cooling stage ($h = h_c$ = const).

Subtraction of Equation (6) from (5) yields:

$$a_\ell q_\ell = \rho c s\left(\frac{\partial T_s}{\partial t} - \frac{\partial T_c}{\partial t}\right) + 2h(T_{os} - T_c) - ks\nabla^2(T_{os} - T_c). \tag{7}$$

If a proper time t_c is selected during the cooling off and $T_{os} \simeq T_c$ for all points, the second and third terms of Equation (7) may be neglected with respect to the first term. In general, however, when $T_{os} \neq T_c$ we need to evaluate the unknown value of h. Substituting the expression h from Equation (6) into (5), we get:

$$a_\ell q_\ell = \rho c s\left(\frac{\partial T_s}{\partial t} - \frac{\partial T}{\partial t}\right) + \left(-\rho c s\left(\frac{\partial T_s}{\partial t}\right)_c + k_s\nabla^2 T_c\right)\frac{T_{os} - T_c}{T_c - T_a} + \\ -ks\nabla^2(T_{os} - T_c). \tag{8}$$

In Equation (8) all the surface temperatures (at n = s) can be measured by means of a thermograph; the only quantity which cannot be measured is T_{os} and in order to obtain a working formula it is necessary to write a relation between the temperature T_o and T_s; it can be shown that, with good approximation, this reads:

$$T_o \simeq T_s + \frac{a_\ell q_\ell s}{2k_n}. \tag{9}$$

Let us define the following temperature differences which can be measured by the thermographic system:

$$\theta_c = T_c - T_a; \quad \theta = T_s - T_c \tag{10}$$

and let us introduce the following dimensionless parameters:

$$\gamma = \frac{\theta}{\theta_c}; \quad \beta = \frac{-\partial T_c/\partial t}{\partial T_s/\partial t}. \tag{11}$$

Substituting Equation (9) in (8) leads to the following expression for $a_\ell q_\ell$:

$$a_\ell q_\ell(x,y) = \rho c s \frac{\frac{\partial T_s}{\partial t}\left[1+\beta(1+\gamma)\right] - \alpha\left[\nabla^2\theta - \gamma\nabla^2\theta_c\right]}{1 - \frac{s^2}{4\alpha_n}\frac{\partial T_s}{\partial t}\frac{\beta}{\theta_c} + \frac{s^2}{4}\frac{\alpha}{\alpha_n}\left(\frac{\nabla^2 q_\ell}{q_\ell} - \frac{\nabla^2\theta_c}{\theta_c}\right)}, \quad (12)$$

where $\alpha = k/\rho c$; $\alpha_n = k_n/\rho c$ are the plate thermal diffusivities in the plane x,y and along the normal n.

Equation (12) takes into account the effects of the heat convected and radiated to the ambient (β terms) and the heat conducted in the x,y plane (∇^2 terms).

If the cooling phase is suitably chosen we can make the value of θ (and γ) very small. In practise the value of $\theta(x,y)$ is not zero at every point and the value of q must be corrected accordingly. The corrective terms at the denominator go to zero for an ideal calorimeter (very small value of s^2/α_n); the third term, however, has relevance at the laser beam edges where $\nabla^2 q_\ell/q_\ell$ is very large. Experimentally, where this term is important, we may follow an iterative process for its computation.

Equation (12) may be evaluated experimentally by measuring the surface temperature distribution $T_s(x,y,t)$ and $T_c(x,y,t)$ at two instants of time (t, $t+\Delta t$ and t_c, $t_c + \Delta t_c$ respectively). The finite difference approximations for the temperatures and the parameters appearing in Equation (12) are ($\Delta t = \Delta t_c$):

$$\frac{\partial T_s}{\partial t} = \frac{T_{s2}-T_{s1}}{\Delta t}; \quad \beta = \frac{T_{c1}-T_{c2}}{T_{s2}-T_{s1}}; \quad \theta = \frac{T_{s1}+T_{s2}-T_{c1}-T_{c2}}{2}$$

$$\theta_c = \frac{T_{c1}+T_{c2}}{2} - T_a.$$

The computerized thermographic system realized allows us to evaluate Equation (12) on a microcomputer.

Experimentally, it is convenient to choose the times (t_1,t_2, t_{c1},t_{c2}), so that the value of $\theta(x,y)$ is as small as possible and consequently so also are the "corrective" terms in Equation (12).

EXPERIMENTAL

1. Numerical Experiments

A computer code has been prepared in order to check the accuracy of expression (12) and to optimize the experimental conditions, i.e.,

(a) plate characteristics ρ, c, k_n, k_t

(b) procedure: t_1, Δt, t_{co}, t_{cl}, Δt_c.

The finite element method described in Reference 1 has been adopted and an axisymmetric plane geometry has been analyzed.

A laser beam with a diameter of 9.5 cm and with a uniform power profile of 1500 W/m has been considered. Boundary conditions take into account the radiative and the convective fluxes to the ambient. The paste-board thermographic plate characteristics are:

$$s = 8.5 \cdot 10^{-2}\ \text{cm} \qquad \rho c = 0.41\ \text{cal/cm}^3$$

$$k_n = k = 4 \cdot 10^{-2}\ \frac{\text{cal}}{\text{cm s °K}} \qquad \alpha_n = 9.7 \cdot 10^{-4}\ \text{cm}^2/\text{s}.$$

Figures 2 and 3 show the numerical results of the temperature and the temperature time derivative distributions during the thermographic plate heating (16 seconds) and during the laser shut-off. In Figure 2 the temperature T_s (i.e., of the plate surface which is not exposed to the laser beam) is plotted versus the non-dimensional distance from the axis of symmetry (r/R) (where R is the laser beam radius) at different times; the region near the beam edge experiences large values of the in-plane conductive heat transfer. In Figure 3 the time derivatives at both surfaces are plotted; only after 5 seconds the two surface temperature time derivatives become equal ($t > s^2/2\alpha_n \simeq 4s$).

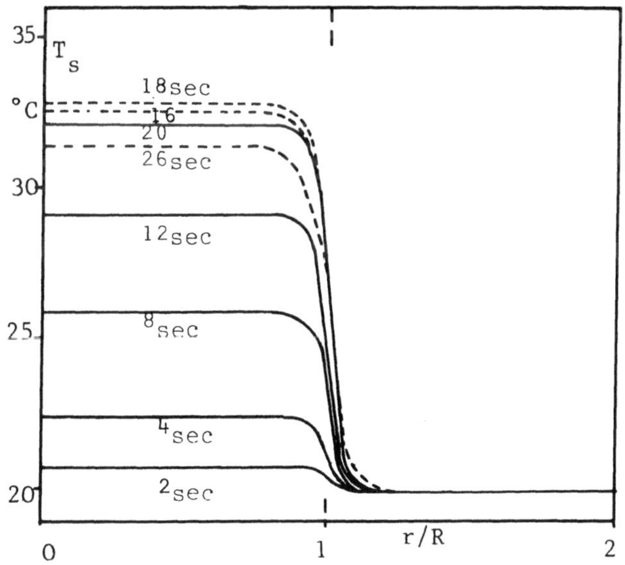

Fig. 2. Temperature of the non-exposed plate surface versus the non-dimensionalized distance from the symmetry axis at different times in the heating and cooling phases.

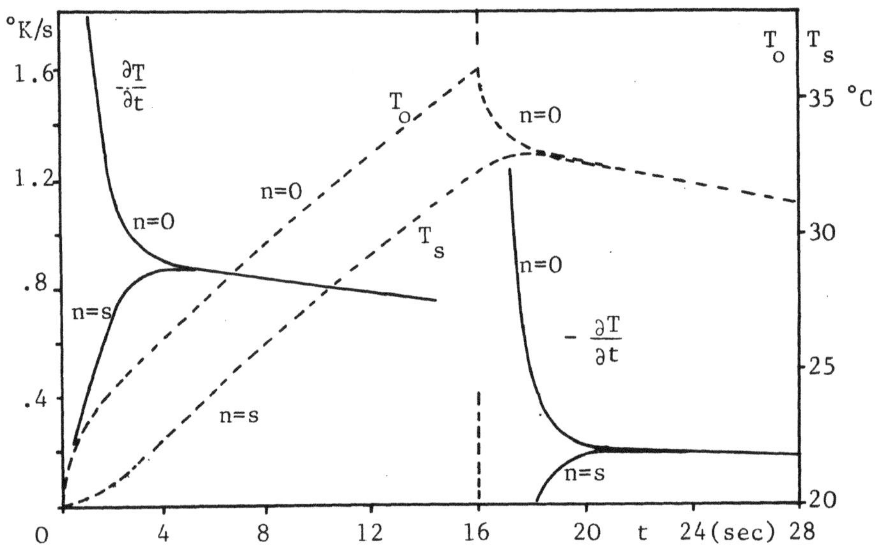

Fig. 3. Temperature time derivatives (solid curves) and temperature of exposed and unexposed plate surfaces versus time during heating ($t < t_{co} = 16$ sec.) and cooling ($t > t_{co}$) ($r/R = 0.2$).

Figure 4 shows the difference between the computed value of the power (q_N) and the incident value of q by means of Equation (12) at different stations (r/R); it can be seen that average errors are of the order of 1% (the method appears very insensitive with respect to the times at which the measurements are made). In Figure 4 is shown also the effect of the corrective terms over the main leading term $\rho c s\, \partial T_s/\partial t$. The corrections are important for both the exposed ($r/R < 1$) and the unexposed areas of the plate. For the points belonging to the area exposed to the laser beam the main correction is given by $\beta \partial T_s/\partial t$. The other terms give a remarkable contribution for the points near the spot edge on both sides. These terms which involve (∇^2) are necessary for obtaining a sharp beam spot edge power profile.

It must be pointed out that accuracy of the order of 1% for the points $r/R < 0.75$ is also due to the finite element integration model and to the way the surface time derivative is approximated ($\Delta t = \Delta t_c = 2.5$ s). If both the element sizes and the time steps are decreased, Equation (12) will give more accurate results.

2. Experimental Apparatus and Procedure

The experimental device consists in a Thermograph (AGA 680) which looks at the back of the exposed paste-board surface. An

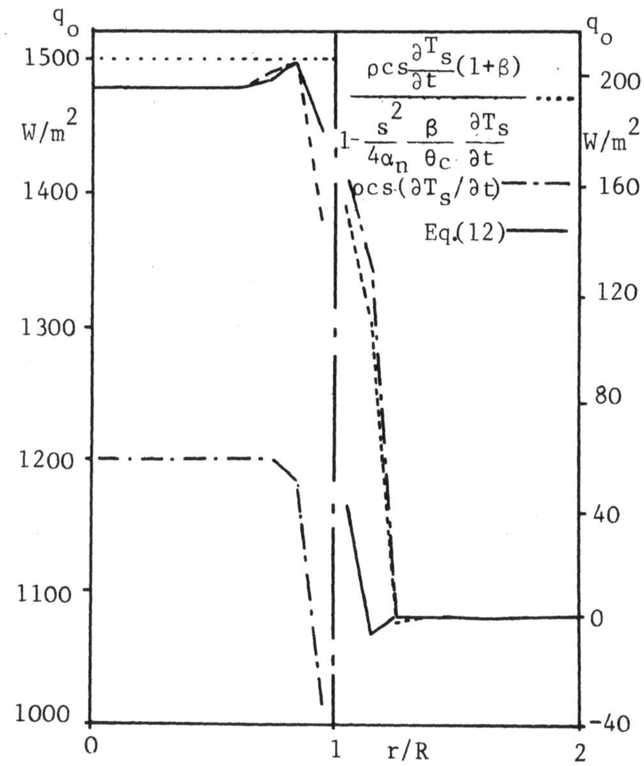

Fig. 4. Effects of the corrective terms in the evaluation of expression (12).

A/D converter (Digimen - Avioradio) interfaces the TG camera to a microcomputer (Apple II) and/or to a digital recorder (Penny & Giles) [1].

The operational procedure makes use of DIGIMEN RAMs to store two matrices $G(I,J,t_1)$, $G(I,J,t_2)$ of the Thermograph output taken at a predefined time interval $\Delta t = t_2 - t_1$ during the heating phase. Two other matrices $G(I,J,t_{c1})$, $G(I,J,t_{c2})$ are taken during the cooling phase with the same Δt or another $\Delta t_c = t_{c2} - t_{c1}$, after the two previous matrices have been transferred to the micro-computer.

The digitalized thermographic picture matrices $G(I,J)$ can be converted in temperature matrices $T(I,J)$ by means of a calibration curve built into the computer. The term $\nabla^2 q_\ell / q_\ell$ is evaluated, when necessary, by an iterative procedure.

As discussed in the previous section it is convenient that the time t_{co} (instant of the laser shut-off) and the time t_{c1} be tailored with the times t_1 and t_2 in order to obtain a value of $\theta(I,J) =$

$= T_{s1} + T_{s2} - (T_{c1} + T_{c2})/2$ as small as possible at all the points; if this is possible then the corrective terms at the numerator of Equation (12) (which are to be computed by finite difference approximations) are small and the evaluation of $q_\ell(x,y)$ can be more accurate.

RESULTS

A number of experimental tests have been performed by using in all tests a black ($a_\ell \simeq 1$) paste-board thin slab having the characteristics indicated in Section 2 (used in numerical experiments). To simulate a uniform and sharp edged beam, a stereopticon has been used. In order to realize a sharp transition, a metallic plate has been interposed very near the paste-board plate ($\sim$ 5 mm) to minimize diffraction effects. The metallic plate cuts off the circular beam spot along the diameter (see Figure 5).

In Figure 6 the power density profile along the radius normal to the cut-off diameter is as shown by the microcomputer output. Note the good spatial resolution of the beam cut-off. In the same Figure 6 the basic term $\rho cs\ \partial T/\partial t$ is also plotted to show the improvement due to the corrective terms. Calorimetric measurements of the total beam power confirm that the average power is in the range of 2200 W/m (as indicated in Figure 6).

CONCLUSIONS

The method proposed is capable of measuring laser beam power profiles with a sufficiently high accuracy even in the presence of

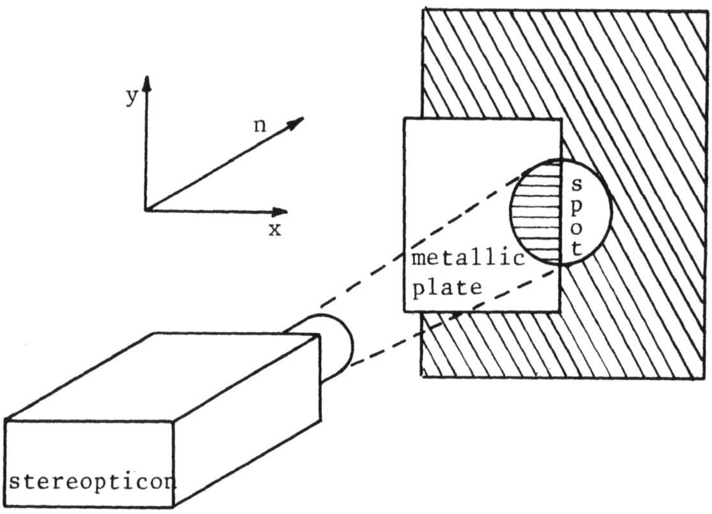

Fig. 5. Experimental arrangement for the laser simulation.

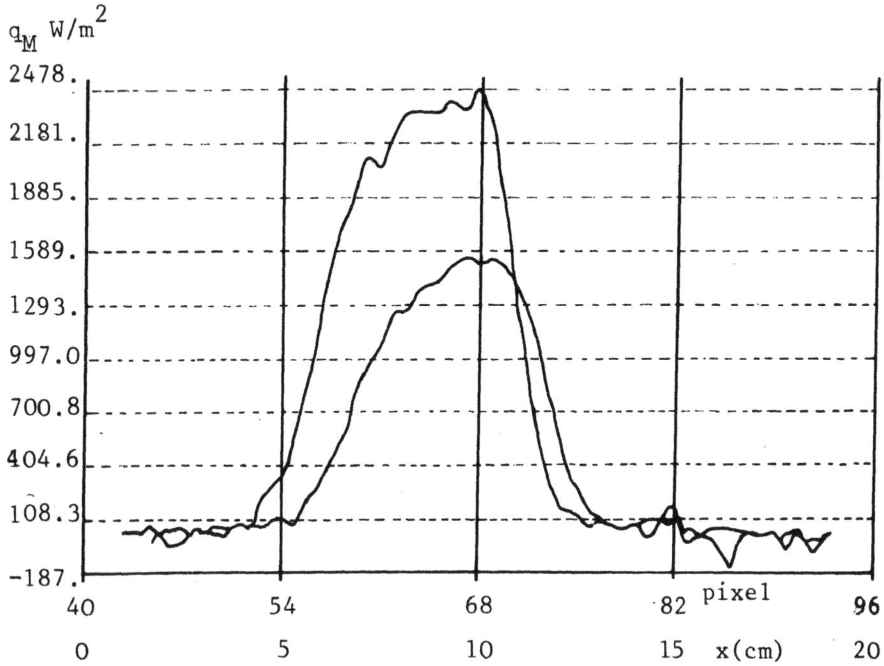

Fig. 6. Experimental measurement of the beam power profile (q_M) (upper curve, computer output) and improvement due to the corrective terms in Equation (12).

very large power density gradients; this has been shown both numerically and by calibration tests.

The preliminary tests on simulated laser beams have shown the operational feasibility of the hardware and of the software; three to five minutes are necessary for obtaining the power density profile plot (Figure 6) with an estimated accuracy of 1%. When measuring an actual laser beam profile, it seems convenient to work with a beam of relatively large cross-section. This 'expanded' beam can be found upstream of the optical lenses (or mirrors) or can be expanded by suitable mirrors downstream.

There are two reasons for performing the thermographic measurements at large beam diameters:

(a) the spatial resolution increases with the ratio of D/s, where D is the laser beam characteristic dimension and s is the plate thickness;
(b) the reduced power density allows measurements to be taken at larger time intervals, i.e., at conditions for which the assumptions of the analysis hold.

Further experiments with laser beams will be carried out and laser parameters such as efficiency, power, uniformity will be measured with the present method.

REFERENCES

1. R. Monti, G. Mannara and A. Russo Sorge, "Experimental evaluation of the radiative heat flux from rockets plumes on satellite structures", XXXIII Congress IAF., Paris, France, September 26 - October 2 (1982).
2. R. Monti and G. Mannara, "Determinazione sperimentale dei coefficienti di trasmissione del calore con metodi termografici", Aerotecnica, Missili e Spazio', Vol. 60, No. 2, Giugno (1981).
3. R. Monti, "Estimation of the soil composition by IR observation of the earth by satellites", XXIX Congress IAF., Dubrovnik, Yugoslavia, October (1978).

HYDRODYNAMICAL MODELS OF AEROSOL INDUCED BREAKDOWN

P. Vigliano, M. Autric, J.P. Caressa, V. Chhim, and G. Inglésakis

Institut de Mécanique des Fluides, Université Aix - Marseille II, 1 rue Honnorat, 13003 Marseille, France

INTRODUCTION

Numerous previously published experimental and theoretical investigations on the aerosol-induced breakdown have been conducted for 1.06 μm (Nd), 2.8 μm (HF), 3.8 μm (DF) and 10.6 μm (CO_2) wavelengths [1÷11]. When the laser intensity and fluence are high enough to cause breakdown on the aerosol particles naturally suspended in the air, the plasma created around each particle absorbs the incident laser radiation according to the inverse bremsstrahlung absorption process and subsequently expands axially and radially. This phenomenon leads to a laser beam blockage.It is of interest for propagation of high energy pulsed laser through the atmosphere to deterimine the development of the air plasma and the attenuation of the transmission.

A computed numerical simulation of laser beam propagation in these conditions has been worked out [12] In this simulation, we use an individual hydrodynamical development of plasma for each breakdown site. Because there is a lot of sites we must optimize the calculating time and model accuracy. Consequently, a comparative study of three hydrodynamical models of a laser supported shock wave is presented here :

- Laser supported detonation wave described by Raizer [13].
- Results of Edwards and Fleck numerical model [14].
- Boni and Su theory [15].

Experimental results are presented and compared with these different models.

I. RAIZER THEORY

When the laser intensity is sufficient ($\phi > 10^7$ W/cm^2) the temperature elevation behind the shock front is high enough to cause total absorption of the laser beam. This wave regime is called Laser Supported Detonation (L.S.D) wave and has been studied at I.M.F.M.

Following the classical Raizer theory, we use the conservation equations for the system in the stationary monodimensional plane case. Using also classical thermodynamic relationships between pressure, temperature and density on the ideal gas case, and the shock conditions, one gets the analytical relation :

$$v = (2\ (\gamma^2-1)\ \frac{\phi}{\rho})^{1/3}$$

where v is velocity, ρ density and ϕ laser intensity.

The interest of this model is to obtain an analytical solution which can be easily introduced in a numerical simulation. However, stationary, monodimensional and plane hypothesis are not very well adapted to aerosol induced breakdown. At this point, a semi-empirical correction has been introduced by Raizer which takes into account the radial expansion and gets the radial velocity equal to v/2.

II. EDWARDS AND FLECK MODEL

This model has been recently described by using optical considerations and three temperature nonequilibrium thermodynamic models of air plasma and two dimensional Lagrangian hydrodynamics. This assumption takes into account a mixture of ideal gases in equilibrium which is constituted by many monatomic gases, diatomic and triatomic molecules. The authors obtain the bidimensional expansion of plasma for the entire regime of laser supported wave development. This model yields the ratio 0.73 for radial velocity to axial velocity. This investigationed model has proven very useful but it is now too complex for our simulation.

III. BONI AND SU MODEL

According to the Bach and Lee theory[16], which has produced accurate results for the entire regime of blast wave propagation, Boni and Su have developped a model describing the maintenance of the laser driven shock waves. We have used the unsteady spherical wave model which is the most appropriate. The system of equations is numerically integrated by the fourth order Runge-Kutta formula.

Figure 1 illustrates any developments of a shock front for different laser pulse shapes as indicated.

The principal limitations of this model are :

- a power law density assumption
- only total energy conservation
- unidimensional symmetry

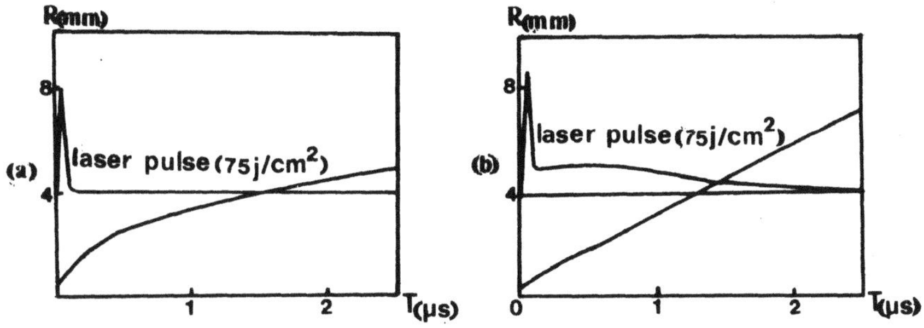

Fig. 1. Shock wave development versus time for Boni and Su model with different pulse shapes ; a: spike / b : spike + tail

In the case of a pulse shape of figure 1b, figure 2 shows temporal evolutions of shock wave velocity. The curves are obtained from different average laser intensities.

The figure 3 shows different hydrodynamical regimes which correspond to different stages of energy addition

IV. COMPARISON OF THE THREE MODELS

Figure 4 represents the velocities versus intensity with the different models. The three curves converge to a power law ($\phi^{1/3}$) of the velocity versus laser intensity for a high enough radiation field ($\phi > 3.10^7$ W/cm^2). Therefore, a reasonable agreement between Edwards-Fleck and Boni-Su models can be observed for axial developments. Concerning the radial development, discordance between the two models appear because of the monodimensional assumption of Boni and Su model.

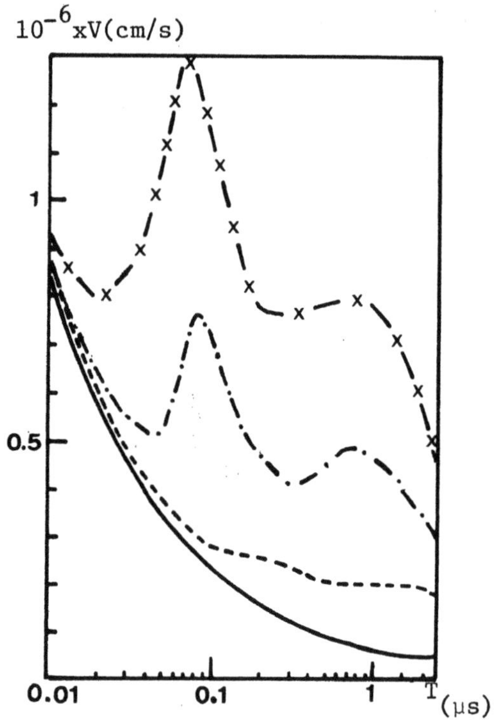

Fig. 2. Shock wave velocity versus time for different average intensities with Boni and Su model:
—x— 10^9 W/cm^2 / — . — 5.10^8 W/cm^2
------- 5.10^7 W/cm^2 / ——— 0

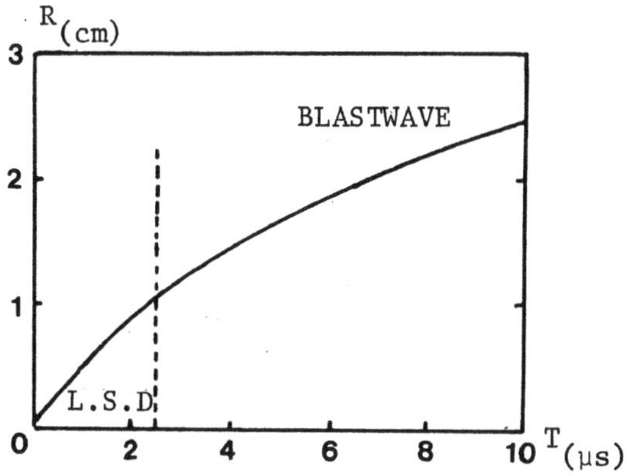

Fig. 3. Different regimes of shock wave propagation for Boni and Su model. The dotted line represents the end of the laser pulse (ϕ_{peak} = 10^9 W/cm^2).

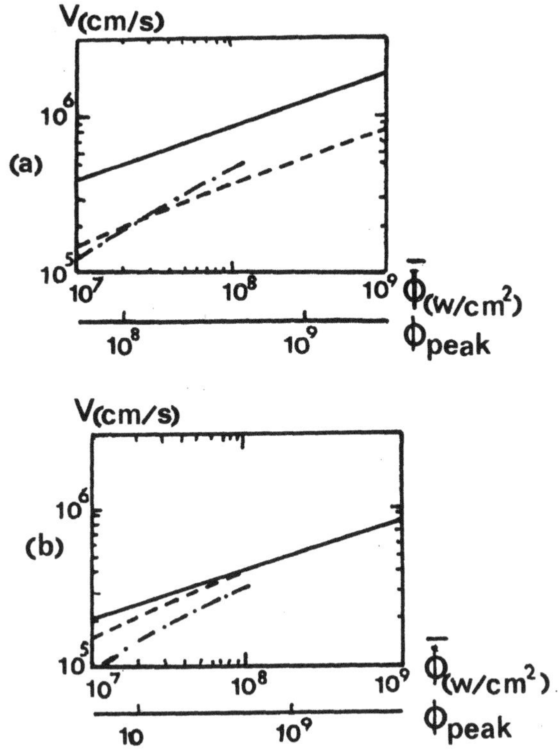

Fig. 4. Comparison of the three different models :
——— Raizer / —·— Edward-Fleck /--- Boni-Su
Axial velocity (a) and radial velocity (b) as a function of the laser intensity.

V. EXPERIMENTAL INVESTIGATIONS

At the same time, experimental investigations have been carried out. The following experiments have been performed with a TEA CO_2 laser which is capable of producing up to 200 J with a pulse duration of 2.5 μsec.

A scheme of the experimental set-up used for air breakdown experiments is shown in fig. 5.

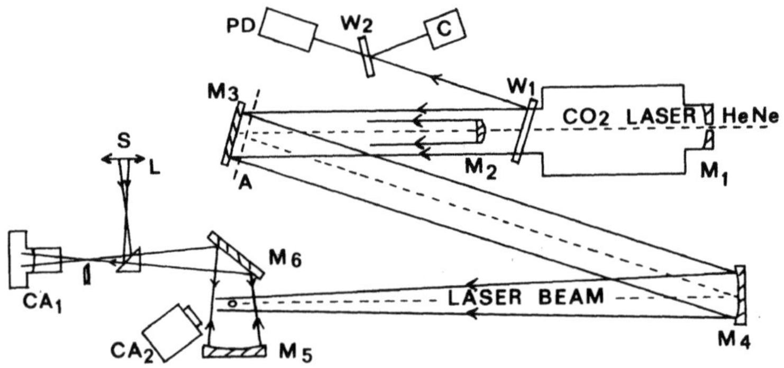

Fig. 5. Experimental set-up.

A Attenuators ; C Calorimeter ; CA_1 CA_2 cameras ; L lens ; M_1 to M_6 plane or concave mirrors ; PD ir detector ; S light source ; W_1,W_2, NaCl windows.

Fig. 6 (a) shows an streak photograph taken at an average laser intensity in the focal region of about 10^8 W/cm^2 (ϕ_{peak} $\simeq$ 5.10^8 W/cm^2). The photograph records the luminosity of the plasma and shows the fast axial development of the luminous front towards the laser source and in the same direction of the laser beam.

The fig. 6 b) presents strioscopic photographs taken at different times after the onset of the breakdown showing the axial and radial development of the shock waves surrounding the aerosol (the first image has been taken 4 μsec after the initiation of the breakdown).

These photographs show the nearly-spherical shock waves with develop until they coalesce to form a cylindrical channel.

In the fig. 7, we present the experimental results of the plasma development compared with calculated results obtained from Raizer, Edwards-Fleck and Boni-Su model.

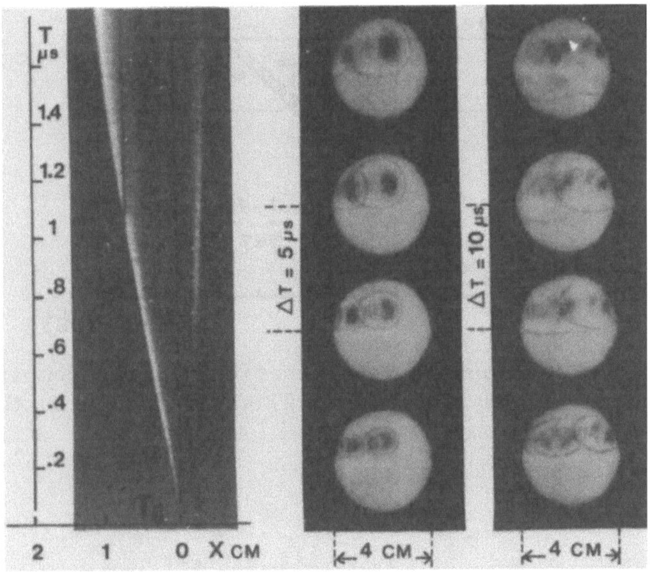

a) b)

Fig. 6. Air breakdown. The laser beam is incident from the left.
a) streak photograph of the plasma. T_c time of breakdown $\bar{\phi} = 10^8$ W/cm^2.
b) Strioscopic photographs of the shock waves.
$\bar{\phi} = 7.10^7$ W/cm^2 ($\phi_{peak} \simeq 3.5\ 10^8$ W/cm^2) $\Delta t = 5$ μsec.

VI. CONCLUSION

The model of Raizer is the most simple but not very well adapted to aerosol-induced breakdown.

Edwards-Fleck model is certainly the most accurate but also the most complexe. It is not envisageable to introduce it in our simulation.

The Boni-Su model is a good compromise between these two situations. For concluding, we present fig. 8 a compilation of results which have been obtained by different authors and which are compared with Boni and Su theory. One can see a good agreement between theoretical and experimental results.

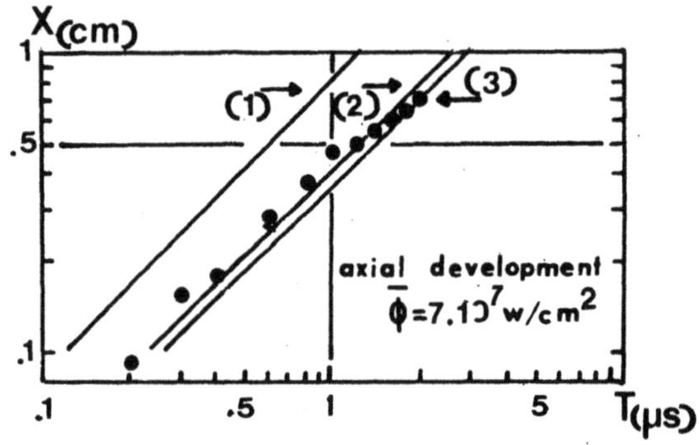

Fig. 7. Plasma development as a function of time compared with Raizer (1) Edwards-Fleck (2) and Boni-Su (3) theories. $\bar{\phi} = 7.10^7$ W/cm^2.

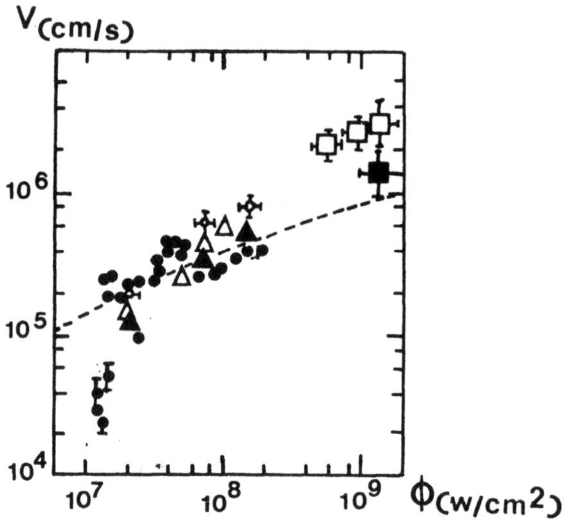

Fig. 8. Plasma expansion velocities observed in different aerosol-induced breakdown experiments as a function of the laser intensity

O axial velocity ▲ radial velocity (Lencioni, et al CO_2)
□ axial velocity ■ radial velocity (Amimoto, et al DF)
● mean velocity (Reilly, et al CO_2)
Δ axial velocity (I.M.F.M. CO_2)
-- theory (Boni - Su).

ACKNOWLEDGMENTS

We would like to thank J.P. FRAGASSI for his assistance throughout the course of the experiments.

This work is supported by the "Direction des Recherches, Etudes et Techniques".

REFERENCES

1. G.H. Canavan, W.A. Proctor, P.E. Nielsen, S.D. Rockwood, I.E.E.E. J. Quantum Electron, QE8, p.564 (1972).
2. N. Kroll, K.M. Watson, Phys. Rev. A5, p.1883-1905 (1972).
3. L.C. Marquet, R.J. Hull and D.E. Lencioni, I.E.E.E. Journal of Quantum Electronics, Vol.QE 8, p.564 (1972).
4. D.E. Lencioni, Applied Physics Letters, Vol.23, pp.12-14 (1973).
5. D.C. Smith and R.T. Brown, Journal of Applied Physics, Vol.46, p.1146-1154 (1975).
6. D.E. Lencioni, AGARD Conference Proceedings, n° 183, p.32.1-32.12 (1975).
7. D.E. Lencioni and L.C. Pettingill, Journal of Applied Physics, Vol.48, p.1848-1851 (1977).
8. D.C. Smith, Journal of Applied Physics, Vol.48, p.2217-2225 (1977).
9. J.P. Reilly, P. Singh and G. Weyl, AIAA Paper 77-697, AIAA 10th Fluid and Plasma Dynamics Conference, Albuquerque (1977).
10. M. Autric, J.P. Caressa, Ph. Bournot, D. Dufresne and M.Sarazin, AIAA Journal, Vol.19, n°11, pp.1415-1421 (1981).
11. S.T. Amimoto, J.S. Whittier, F.G. Ronkowski, P.R. Valenzuela, G.N. Harper, R. Hofland Jr., E.L. Segundo, G.L. Trusty, T.Cosden, D. Lesli, AIAA Paper n° 82.0894, AIAA/ASME 3rd Joint Thermophysics, Fluids, Plasma and Heat Transfer Conference, St.Louis (1982).
12. M. Autric, J.P. Caressa, P. Vigliano, Ph. Bournot, D. Dufresne, G. Inglésakis, Journal de Physique, C9, n°11, T.41, p.129-135 (1980).
13. Yu. P. Raizer, Soviet Physics J.E.T.P., Vol.21, n°5, p.1009-1017 (1965).
14. A.L. Edwards and J.A. Fleck Jr., Journal of Applied Physics, Vol.50, p.4308-4313 (1979).
15. A.A. Boni and F.Y. Su, Acta Astronautica, Vol.1, p.761-780, (1974).
16. G.G. Bach and J.H.S. Lee, AIAA Journal 8, p.271-275 (1970).

Index

MIX
Papier aus verantwortungsvollen Quellen
Paper from responsible sources
FSC® C105338

If you have any concerns about our products,
you can contact us on
ProductSafety@springernature.com

In case Publisher is established outside the EU,
the EU authorized representative is:
Springer Nature Customer Service Center GmbH
Europaplatz 3, 69115 Heidelberg, Germany

Printed by Libri Plureos GmbH
in Hamburg, Germany